# Long Term Socio-Ecological Research

# Human-Environment Interactions

## VOLUME 2

**Series Editor:**

**Professor Emilio F. Moran**, Michigan State University (Geography)

**Editorial Board:**

**Barbara Entwisle**, Univ. of North Carolina (Sociology)
**David Foster**, Harvard University (Ecology)
**Helmut Haberl**, Alpen-Adria Universitaet Klagenfurt, Wien, Graz
        (Socio-ecological System Science)
**Elinor Ostrom**[†], Indiana University (Political Science)
**Billie Lee Turner II**, Arizona State University (Geography)
**Peter H. Verburg**, University of Amsterdam (Environmental Sciences, Modeling)

For further volumes:
http://www.springer.com/series/8599

Simron Jit Singh • Helmut Haberl
Marian Chertow • Michael Mirtl
Martin Schmid
Editors

# Long Term Socio-Ecological Research

Studies in Society-Nature Interactions Across Spatial and Temporal Scales

Springer

*Editors*
Simron Jit Singh
Institute of Social Ecology Vienna (SEC)
Alpen-Adria Universitaet Klagenfurt,
  Wien, Graz
Vienna, Austria

Helmut Haberl
Institute of Social Ecology Vienna (SEC)
Alpen-Adria Universitaet Klagenfurt,
  Wien, Graz
Vienna, Austria

Marian Chertow
Center for Industrial Ecology
Yale School of Forestry
  and Environmental Studies
Yale University
New Haven, CT, USA

Michael Mirtl
Ecosystem Research and Monitoring
Environment Agency Austria
Vienna, Austria

Martin Schmid
Institute of Social Ecology Vienna (SEC)
Alpen-Adria Universitaet Klagenfurt,
  Wien, Graz
Vienna, Austria

The translation/the editing of foreign language was prepared with financial support from the Austrian Science Fund (FWF)

FWF
Der Wissenschaftsfonds.

ALTER-Net (A Long-Term Ecosystem and Biodiversity Research Network) provided financial supported for the editorial work

ALTER-Net

ISBN 978-94-007-1176-1            ISBN 978-94-007-1177-8 (eBook)
DOI 10.1007/978-94-007-1177-8
Springer Dordrecht Heidelberg New York London

Library of Congress Control Number: 2012953254

© Springer Science+Business Media Dordrecht 2013
This work is subject to copyright. All rights are reserved by the Publisher, whether the whole or part of the material is concerned, specifically the rights of translation, reprinting, reuse of illustrations, recitation, broadcasting, reproduction on microfilms or in any other physical way, and transmission or information storage and retrieval, electronic adaptation, computer software, or by similar or dissimilar methodology now known or hereafter developed. Exempted from this legal reservation are brief excerpts in connection with reviews or scholarly analysis or material supplied specifically for the purpose of being entered and executed on a computer system, for exclusive use by the purchaser of the work. Duplication of this publication or parts thereof is permitted only under the provisions of the Copyright Law of the Publisher's location, in its current version, and permission for use must always be obtained from Springer. Permissions for use may be obtained through RightsLink at the Copyright Clearance Center. Violations are liable to prosecution under the respective Copyright Law.
The use of general descriptive names, registered names, trademarks, service marks, etc. in this publication does not imply, even in the absence of a specific statement, that such names are exempt from the relevant protective laws and regulations and therefore free for general use.
While the advice and information in this book are believed to be true and accurate at the date of publication, neither the authors nor the editors nor the publisher can accept any legal responsibility for any errors or omissions that may be made. The publisher makes no warranty, express or implied, with respect to the material contained herein.

Printed on acid-free paper

Springer is part of Springer Science+Business Media (www.springer.com)

# Foreword

**People and their changing environment:
how to deal with complexity**

The broad field of ecology expanded during the twentieth century as a sub-discipline of biology, in order to combine the fundamental curiosity of scientists who wished to uncover the relationships between organisms and their environment with a growing societal awareness of the fact that we are now changing these relationships, on every single square metre of this planet. Nearly all of this change is to the detriment of the functioning of plants, animals and the communities they live in. As such, ecology can be seen as a success story: environmental legislation, first in the US during the 1960s and later also in Europe, began to be informed by ecological research. Now, ecologists form a large and mature community, drawing students to most universities world-wide. However, the environment keeps changing, and environmental policies very frequently fail to take into account even the simplest concepts of ecology. For example, it seems as though few, if any, nations had established an official assessment of their own natural capital and ecosystem services before Norway recently did so.

Most dramatically, we find ourselves helplessly witnessing the loss of species at an accelerating rate, thereby eradicating the fundamental "software" that might provide essential functions ("services") from our changed environment. In addition, the level of pollutants and other disturbing compounds in the environment is increasing in most places, with improved conditions only where the impacts were seen as "too lethal" (such as in European acidified lakes during the 1980s or for chlorinated hydrocarbons in North America during the 1960s). Finally, we still do not really know where the changes in our environment are affecting people in the most direct way, and which impacts might last longer than others.

Hence, while ecology often portrays itself as being helpful to society and policy makers, most often the link between published scientific findings and societal problems is not made. Instead, many ecologists express their concern to media and policy makers with a single and undifferentiated message: stop changing our environment, cut greenhouse gas emissions, ban the destruction of the deep sea marine ecosystem, enlarge all protected areas, etc. Nearly all public debate in response to these calls merely succeeds in generating feelings of guilt among some portions of society and opposition in other sectors, while often producing little or no policy action and only temporary reductions in the scale of environmental degradation.

One key reason for this failure is that the root cause analysis of the problem is often incomplete. Frequently, any change of the so-called "natural state" is portrayed as negative by ecologists. But even hunting and gathering of food from ecosystems inevitably has an impact on species and communities. Agriculture, in the sense of either cultivating plant species on cleared land, or herding animals in open landscapes, is more intensive, covering a broad range from low impacts to the much higher ones of agro-industrial complexes. If society is to benefit from enhanced scientific knowledge about such impacts in a useful way, then systems must be analysed from a more comprehensive, interdisciplinary and human perspective – e.g., the perspective of Long-Term Socio-Ecological Research (LTSER). LTSER benefits from the conceptual advances in social ecology, which derive from the full range of interdisciplinary approaches that have developed, and are developing, to address the complexity of systems of nature and society over long periods of time.

From this viewpoint, the aspect of benefits, or "usefulness" (which is often relegated to managers or "applied research") of scientific efforts should be distinguished from pragmatism and advocacy. Aiming to directly address public concerns in the human-environment relationship does not imply asking less profound questions than those in other fields of science. Aiming to arrive at an objective analysis of human land use and the associated changes in the composition of species, as well as their population and community dynamics, demands substantial efforts in terms of conceptual development, multi-scale gathering of data and complex interpretation. Just as putting the "S" for "Society" into "Long-Term Ecological Research" means adding an important layer to an already complex set of studies, it also means that new types of topics enter the scene, such as socioeconomics, security, equity and gender issues. In this sense, while it might be more pragmatic to document a physico-chemical change (for example, the acidification of lakes or oceans) and the associated loss of biological function, extrapolate both into the future, and then complain loudly about society's lack of willingness to "do something", a more challenging in-depth analysis would include the study of the way in which the problem is perceived together with society's willingness to act, as part of the same investigation.

This book performs a remarkable "tour de table" of modern LTSER and related studies. Why the long-term? Clearly, from a human perspective, our agricultural life support system has been attuned to a geological period of particular stability over many millennia. Anthropogenic environmental change must be seen against these rather special conditions which have caused the evolution of highly specific ways of relating to the environment (at least on northern temperate latitudes). To adjust to the dynamics now introduced into the physical and biological environment requires an understanding of systemic behaviour on a range of time-scales, at a minimum of several decades. Gathering knowledge about the longer term situation, and observing systems over periods that extend beyond the scope of a single PhD thesis or research grant is therefore essential to the analysis of social ecology.

The book also reveals that there is not a single unified theory for LTSER. In some studies, the actual analysis of social dynamics goes much deeper than it does in others. We may view this rather as an asset than as a limitation. If anything, this demonstrates that there is plenty of scope for further research developments and creativity, using the work assembled here as an inspiration rather than a straitjacket.

A key aspect of developing the field of LTSER is cooperation – among disciplines of course, but also among like-minded teams in different locations. In times of limited financial resources, international cooperation in particular may provide ways to enhance the value of the various contributions. The international community presently benefits from several platforms for such cooperation, two of which are directly associated with much of the work presented in this book. At the European level, the Network of Excellence ALTER-Net, funded by the European Unions 6th Framework Programme for the Environment (2004–2009) continues to provide crucial support for the development of the LTSER concept, including the training of a large number of next-generation scientists, many of whom are now familiar with concepts of social ecology. In the United States, the LTER network is becoming more interdisciplinary, adding expertise in demography, economics, geography, political science and sociology. At the global scale, the International Council of Science (ICSU) now builds on the achievements of its Earth System Science Partnership (ESSP) by developing a new Programme on Ecosystem Change and Society (PECS) to create global linkages between scientists addressing the human-environment relationship. We have no doubt that this book will provide substantial inspiration for anyone participating in these programmes – indeed, we hope that the programmes themselves will be enhanced by the material presented here.

Wolfgang Cramer<br>
Stephen R. Carpenter

# Foreword

In the pages of this book you will find a collaborative effort uniting many disciplines to understand humanity's long relationship with nature. It is a scientific enterprise in the broadest sense, including experts in social as well as natural fields. We can be hopeful that this effort marks a major turning point in consciousness and applied intelligence.

Ecology stands at the very centre of this book, a science that has grown in scope and importance since it was first named in 1866 by Ernst Haeckel, the leading German disciple of Charles Darwin. Haeckel derived the name from the Greek word *oikos*, or household, so that ecology was meant to be the study of Nature's household, or the natural economy, including the interactions of plants and animals, their relations to the soils and atmosphere. In this book, however, ecology moves decisively beyond the purely natural to encompass human society as well. "Long-Term Socio-Ecological Research" aims to achieve a comprehensive understanding of how humans have lived within and changed ecosystems over time. Why has this new, enlarged ecology become so necessary in our time? Because the changes going on across the earth are so cataclysmic and yet so poorly understood that we ignore them at our peril. Because they require a deep historical understanding of where we have been to know where we are going.

Over the past 500 years, good science has somehow advanced against the most powerful opposition, winning more battles than it has lost. It has driven not one or two but multiple revolutions, and at this moment the interdisciplinary study of ecology may be driving us toward still another intellectual revolution. The outcome will be not merely a better understanding of the interrelationships between society and nature but also a better understanding of where our limits lie.

In their concluding commentary on the book *Limits to Growth*, published in 1972, the executive committee of the Club of Rome wrote: "The concept of a society in a steady state of economic and ecological equilibrium may appear easy to grasp, although the reality is so distant from our experience as to require a Copernican revolution of the mind." That concept of society in a steady state of equilibrium seems implicit in the very notion of LTSER; if so, it will require an intellectual revolution before it is achieved.

The call for a new Copernican revolution appears more than once in recent writing: for example, in a paper that H. J. Schnellnhuber of the Potsdam Institute for Climate Impact Research in Germany published in Nature in December, 1999. Schnellnhuber argues that just as "optical amplification techniques brought about the great Copernican revolution, which finally put the Earth in its correct astrophysical context," so "sophisticated information-compressing techniques including simulation modeling are now ushering in a second 'Copernican' revolution." We are learning to see, for the first time, that the planet is "one single, complex, dissipative, dynamic entity, far from thermodynamic equilibrium—the 'Earth system.'"

So what was the Copernican revolution about, and what might a new Copernican revolution look like? Just 50 years after Columbus's first voyage to the New World, the Polish astronomer Nicholas Copernicus published his last and greatest work, *On the Revolutions of the Heavenly Sphere.* Before Copernicus, the earth had been the fixed centre of the universe, just as Europe had considered itself the fixed centre of human history. A 100 years later, astronomers had finally accepted that the earth was only one of several planets in motion around the sun, and that the universe was far more grand and infinite in its dimensions than anyone had realised. But it was far from easy to make that shift in consciousness, and Copernican ideas would bring fierce controversy in religion, philosophy, economics and politics that would not end for centuries to come. We are still struggling with their implications today.

Can we be sure that another, post-Copernican revolution is in the making? Do we have enough information to judge? The idea of a comprehensive perspective of "socio-ecology" does seem to be emerging, a science to which ecologists, geologists, climatologists, historians, geographers and others are contributing. It promises to provide a new understanding of the natural world and of our place in it. Whether this awareness adds up to a revolutionary change in understanding, to a new human way of thinking that accepts the ecosphere's limits and conserves its systems, we will not know for a long time to come. But such a revolution is possible, and we might even say inevitable. We are being driven by material changes that render old ideas outdated and even dangerous to our survival.

Donald Worster

# Acknowledgments

With most creative and intellectual processes, achievements can rarely be credited to the efforts of single authors or editors alone. This volume is no exception. The current work is an outcome of unwavering support from a number of institutions, research networks and individuals who all deserve our heartfelt thanks.

The genesis of this effort can largely be attributed to the ALTER-Net process – A Long-Term Ecosystem and Biodiversity Research network of excellence mobilised and established with generous funding from the European Commission's 6th Framework Programme (2004–2009). Led mainly by natural scientists, integrating the social dimensions in the European Long-Term Ecological Research (LTER) was no easy task. In the years that followed, intense dialogues between the natural and social scientists involved eventually led to promising outcomes thereby enhancing the utility of LTER for society, denoted by the expansion of "ecological" in LTER to "socio-ecological" in LTSER. We would like to thank the ALTER-Net project and network co-ordinators, Terry Parr and Allan Watt, as well as the commission's project evaluator, Martin Sharman, for their support in providing this impetus.

A crucial role in spearheading these early discussions, besides some of the editors, was played by Verena Winiwarter, Sander van der Leeuw, Angheluta Vadineanu and Eeva Furmann, who paved the way for a European LTSER agenda, while conceptual rigour was provided by Marina Fischer-Kowalski and Anette Reenberg. Alongside this, the constant feedback received from the 30 members of the LTER-Europe Expert Panel on LTSER, mainly composed of LTSER Platform managers and primary investigators, cannot be underestimated. They actively tested the LTSER approach across Europe and fed their experiences back into further refining the LTSER concept.

However, it was not only developments in Europe that inspired this work in the first place. The goal of this volume at the outset was to crystallize the state-of-the-art in LTSER research and this would not have been achieved without the support of a number of pioneering colleagues in the US. We particularly want to thank Charles Redman, Nancy Grimm, Morgan Grove and Carole Crumley for their encouragement in the project and/or contributions to this book. The current volume has also greatly benefitted from the knowledge generated in a number of European research projects over the past years. Indeed, several contributions to this book draw heavily on research undertaken in such projects. We particularly want to thank the Austrian Science Fund (FWF) for their financial support for the projects 'Global HANPP' (P16692), 'Analysing Global HANPP' (P20812-G11) and 'GLOMETRA' (P21012 – G11), and to the European Commission's 7th Framework Programme for the project 'VOLANTE' (265104). We are grateful for continuing support by the Austrian Federal Ministry of Science and Research (BMWF) within the proVISION programme as well as by the Austrian Academy of Science (ÖAW) in supporting the implementation of the first explicit LTSER Platform worldwide. Across the Atlantic, we sincerely wish to acknowledge the financial support of the US National Science Foundation ULTRA-Ex Program and the American Recovery and Reinvestment Act.

The editors are extremely grateful to two donors for their financial support in the editorial process that has enhanced the quality of this book enormously. The English language check was made possible by support from the Austrian Science Fund's (FWF) Translation and foreign language editing of stand-alone publications programme. The ALTER-Net New Initiative Fund supported the initial editorial and exploratory phase of this book project. We extend our heartfelt thanks to Ursula Lindenberg for her competent language editing skills which she delivered meticulously and on time. The editors are especially grateful to Irene Pallua and Georg Schendl for their painstaking and careful proof-reading of the manuscript ensuring compliance with Springer guidelines. This is by no means a trivial task. Without the enduring and dedicated efforts of these colleagues the book would have fared poorly. Last but not least, we would like to thank Jennifer Baka from the Yale School of Forestry and Environmental Studies, for her careful reading of the chapters and critical feedback, and for her help with the figures in the introduction to this book.

# Contents

**1 Introduction** ........................................................................................... 1
Simron Jit Singh, Helmut Haberl, Marian Chertow, Michael Mirtl, and Martin Schmid

**Part I LTSER Concepts, Methods and Linkages**

**2 Socioeconomic Metabolism and the Human Appropriation of Net Primary Production: What Promise Do They Hold for LTSER?** ................................................................. 29
Helmut Haberl, Karl-Heinz Erb, Veronika Gaube, Simone Gingrich, and Simron Jit Singh

**3 Using Integrated Models to Analyse Socio-ecological System Dynamics in Long-Term Socio-ecological Research – Austrian Experiences** .......................................... 53
Veronika Gaube and Helmut Haberl

**4 Modelling Transport as a Key Constraint to Urbanisation in Pre-industrial Societies** ............................................................. 77
Marina Fischer-Kowalski, Fridolin Krausmann, and Barbara Smetschka

**5 The Environmental History of the Danube River Basin as an Issue of Long-Term Socio-ecological Research** ......................... 103
Verena Winiwarter, Martin Schmid, Severin Hohensinner, and Gertrud Haidvogl

**6 Critical Scales for Long-Term Socio-ecological Biodiversity Research** ........................................................................... 123
Thomas Dirnböck, Peter Bezák, Stefan Dullinger, Helmut Haberl, Hermann Lotze-Campen, Michael Mirtl, Johannes Peterseil, Stephan Redpath, Simron Jit Singh, Justin Travis, and Sander M.J. Wijdeven

| | | |
|---|---|---|
| 7 | **Human Biohistory** | 139 |
| | Stephen Boyden | |
| 8 | **Geographic Approaches to LTSER: Principal Themes and Concepts with a Case Study of Andes-Amazon Watersheds** | 163 |
| | Karl S. Zimmerer | |
| 9 | **The Contribution of Anthropology to Concepts Guiding LTSER Research** | 189 |
| | Ted L. Gragson | |

**Part II   LTSER Applications Across Ecosystems, Time and Space**

| | | |
|---|---|---|
| 10 | **Viewing the Urban Socio-ecological System Through a Sustainability Lens: Lessons and Prospects from the Central Arizona–Phoenix LTER Programme** | 217 |
| | Nancy B. Grimm, Charles L. Redman, Christopher G. Boone, Daniel L. Childers, Sharon L. Harlan, and B.L. Turner II | |
| 11 | **A City and Its Hinterland: Vienna's Energy Metabolism 1800–2006** | 247 |
| | Fridolin Krausmann | |
| 12 | **Sustaining Agricultural Systems in the Old and New Worlds: A Long-Term Socio-Ecological Comparison** | 269 |
| | Geoff Cunfer and Fridolin Krausmann | |
| 13 | **How Material and Energy Flows Change Socio-natural Arrangements: The Transformation of Agriculture in the Eisenwurzen Region, 1860–2000** | 297 |
| | Simone Gingrich, Martin Schmid, Markus Gradwohl, and Fridolin Krausmann | |
| 14 | **The Intimacy of Human-Nature Interactions on Islands** | 315 |
| | Marian Chertow, Ezekiel Fugate, and Weslynne Ashton | |
| 15 | **Global Socio-metabolic Transitions** | 339 |
| | Fridolin Krausmann and Marina Fischer-Kowalski | |

**Part III   LTSER Formations and the Transdisciplinary Challenge**

| | | |
|---|---|---|
| 16 | **Building an Urban LTSER: The Case of the Baltimore Ecosystem Study and the D.C./B.C. ULTRA-Ex Project** | 369 |
| | J. Morgan Grove, Steward T.A. Pickett, Ali Whitmer, and Mary L. Cadenasso | |

| | | |
|---|---|---|
| 17 | **Development of LTSER Platforms in LTER-Europe: Challenges and Experiences in Implementing Place-Based Long-Term Socio-ecological Research in Selected Regions** ............... Michael Mirtl, Daniel E. Orenstein, Martin Wildenberg, Johannes Peterseil, and Mark Frenzel | 409 |
| 18 | **Developing Socio-ecological Research in Finland: Challenges and Progress Towards a Thriving LTSER Network**................................................................. Eeva Furman and Taru Peltola | 443 |
| 19 | **The Eisenwurzen LTSER Platform (Austria) – Implementation and Services**................................................. Johannes Peterseil, Angelika Neuner, Andrea Stocker-Kiss, Veronika Gaube, and Michael Mirtl | 461 |
| 20 | **Fostering Research into Coupled Long-Term Dynamics of Climate, Land Use, Ecosystems and Ecosystem Services in the Central French Alps**...................................... Sandra Lavorel, Thomas Spiegelberger, Isabelle Mauz, Sylvain Bigot, Céline Granjou, Laurent Dobremez, Baptiste Nettier, Wilfried Thuiller, Jean-Jacques Brun, and Philippe Cozic | 485 |
| 21 | **Long-Term Socio-ecological Research in Mountain Regions: Perspectives from the Tyrolean Alps**..................................................... Ulrike Tappeiner, Axel Borsdorf, and Michael Bahn | 505 |
| 22 | **Integrated Monitoring and Sustainability Assessment in the Tyrolean Alps: Experiences in Transdisciplinarity**................... Willi Haas, Simron Jit Singh, Brigitta Erschbamer, Karl Reiter, and Ariane Walz | 527 |
| 23 | **Conclusions**....................................................................................... Marian Chertow, Simron Jit Singh, Helmut Haberl, Michael Mirtl, and Martin Schmid | 555 |

**About the Contributors** ............................................................................. 563

**Index**............................................................................................................ 579

# Acronyms

| | |
|---|---|
| ALTER-Net | A Long-Term Biodiversity, Ecosystem and Awareness Research Network |
| BEC | Baltimore Ecosystem Study |
| CBD | Convention on Biodiversity |
| CBNA | Alpine National Botanical Conservatory |
| CEM | Commission on Ecosystem Management |
| CZO | Critical Zone Observatory |
| DEHI | Danube Environmental History Initiative |
| DIVERSITAS | An international Programme on Biodiversity science |
| DPSIR | Driver-Pressure-Impact-State-Response |
| EBONE | European Biodiversity Observation Network |
| EEA | European Environment Agency |
| e-MORIS | Electronic-Monitoring and Research Information System |
| EnvEurope | Environmental quality and pressures assessment across Europe |
| ERA | European Research Area |
| ESEE | European Society for Ecological Economics |
| ESI | Ecosystem Service Initiative |
| ESSP | Earth System Science Partnership |
| EVALUWET | European Valuation and Assessment Tools Supporting Wetland Ecosystem |
| EXPEER | Experimentation in Ecosystem Research |
| ExtremAqua | Influences of Extreme Weather Conditions on Aquatic Ecosystems |
| FCM | Fuzzy Cognitive Maps |
| FP | Framework Programme |
| GEA | Global Energy Assessment |
| GISP | Global Invasive Species Programme |
| GLEON | Global Lake Ecological Observatory Network |
| GLORIA | Global Observation Research Initiative in Alpine Environments |
| GLP | Global Land Project |
| GMBA | Global Mountain Biodiversity Assessment |
| HEF | Human Ecosystem Framework |

| | |
|---|---|
| IBP | International Biological Programme |
| ICP | International Cooperative Programme |
| ICPDR | International Commission for the Protection of the Danube River |
| ICSU | International Council for Science |
| IGBP | International Geosphere-Biosphere Programme |
| ILTER | International Long-Term Ecological Research Network |
| INSPIRE | Infrastructure for Spatial Information in the European Community |
| IPBES | International Platform on Biodiversity and Ecosystem Service |
| IPCB | International Press Centre for Biodiversity |
| IPCC | Intergovernmental Panel on Climate Change |
| ISEE | International Society for Ecological Economics |
| ISSE | Integrative Science for Society and the Environment |
| IUCN | International Union for Conservation of Nature |
| JPI | Joint Programming Initiative |
| LECA | Laboratory of Alpine Ecology |
| LIFE+ | The Financial Instrument for the Environment |
| LTER | Long-Term Ecosystem Research or Long-Term Ecological Research |
| LTER-Europe | Long-Term Ecosystem Research Network Europe |
| LTER Site | Long-Term Ecosystem Research Site |
| LTSER | Long-Term Socio-Ecological Research |
| LTSER Platform | Long-Term Socio-Ecological Research Platform |
| MAB | Man and the Biosphere Programme |
| MEA | Millennium Ecosystem Assessment |
| MoU | Memorandum of Understanding |
| MSP | Math-Science Partnership |
| NCEAS | US National Center for Ecological Analysis and Synthesis |
| OOI | Oceans Observatory Institute |
| NAS | National Academy of Sciences |
| Natura2000 | An ecological network of protected areas within the European Union |
| NCA | National Climate Assessment |
| NEHN | Nordic Environmental History Network |
| NEON | National Ecological Observatory Network |
| NESS | Nordic Environmental Social Science |
| NoE | Network of Excellence |
| NSF | National Science Foundation |
| PAME | Participatory Assessment, Monitoring and Evaluation |
| PECS | Programme on Ecosystem Change and Society |
| PPD | Press-Pulse Dynamics Framework |
| PTA | Participatory Technology Assessments |
| PVA | Population Viability Analysis |
| RCN | Research Coordination Network |
| SCOPE | Scientific Committee on Problems of the Environment |
| SEBI | Streamlining European Biodiversity Indicator |

| | |
|---|---|
| SEIS | Shared Environmental Information System |
| SERD | Simulation of Ecological Compatibility of Regional Development |
| SMCE | Social Multi-Criteria Evaluation |
| TEEB | The Economics of Ecosystems and Biodiversity |
| TERENO | Terrestrial Environmental Observatories |
| TFRN | Task Force on Reactive Nitrogen |
| ULTRA-Ex | Urban Long-Term Research Areas Exploratory Projects |
| UNECE | United Nations Economic Commission for Europe |
| UNEP | United National Environment Programme |
| UNESCO | United Nations Educational, Scientific and Cultural Organisation |
| URGE | Urban Rural Gradient Ecology project |
| UTC | Urban Tree Canopy |
| WFD | EU Water Framework Directive |
| WSSD | World Summit on Sustainable Development |

# List of Boxes

| | | |
|---|---|---|
| Box 1.1 | The International Long-Term Ecological Research (ILTER) Network | 7 |
| Box 7.1 | Human Health Needs | 158 |
| Box 7.2 | The Health Needs of Ecosystems | 158 |
| Box 20.1 | Structure and Governance of the Central French Alps LTSER Platform | 488 |
| Box 22.1 | Challenges of Transdisciplinary Research | 549 |

# List of Figures

| | | |
|---|---|---|
| Fig. 1.1 | Intellectual genealogies of LTSER............................................... | 3 |
| Fig. 1.2 | Case studies in this volume across time and space ..................... | 15 |
| Fig. 2.1 | Basic approaches to analyse socioeconomic metabolism. (a) Systemic approaches account for all physical flows (materials, energy, substances) required for reproduction and functioning of socioeconomic stocks. (b) Life-cycle analysis accounts for resource requirements or emissions from one unit of product or service throughout its entire life cycle ('cradle to grave')................................................................... | 32 |
| Fig. 2.2 | Scheme of economy-wide (national-level) material and energy flow (MEFA) accounts................................................ | 34 |
| Fig. 2.3 | Definition of HANPP – see text for explanation ........................... | 38 |
| Fig. 2.4 | Stocks and flows of carbon in Austria 1830–2000. (a) Socioeconomic carbon flows per year. (b) Carbon stocks in biota and soils in billion tonnes of carbon. (c) Net carbon exchange between atmosphere and biota/soils. (d) Net carbon emissions considering the terrestrial carbon sink............................ | 42 |
| Fig. 2.5 | Preliminary causal loop model of the land use model for Austria 1830–2000 – see text for explanation ........................... | 44 |
| Fig. 3.1 | Generic heuristic model of socio-ecological systems at the interface of natural and cultural spheres of causation [Reprinted from Haberl et al. (2004). With permission from Elsevier]................................................................................ | 54 |
| Fig. 3.2 | Sustainability triangle of farmsteads simulated by SERD ............... | 61 |
| Fig. 3.3 | Decision tree for the 'farm' agent type. *Round light grey boxes*: status of a farm calculated by an automatically induced yearly evaluation of the relation between income and workload. Depending on the status, each farm decides | |

xxiii

|  |  |  |
|---|---|---|
|  | with certain probabilities to take one of the defined actions or to stay passive (that is mostly the case). *Angled boxes*: Options for decisions available in each state. A decision for one of the options changes income, work load, land use, substance flows, etc | 63 |
| Fig. 3.4 | Interface of the Reichraming model simulating the development of different parameters (socioeconomic and ecological) under different framework conditions (changeable as sliders – *grey boxes* at the *bottom*) | 67 |
| Fig. 4.1 | Territory and transport in hunting and gathering societies | 81 |
| Fig. 4.2 | Territory and transport in agrarian societies, village level | 84 |
| Fig. 4.3 | Territory and transport in agrarian civilisations, by spatial scale | 85 |
| Fig. 4.4 | The overall structure of the model | 86 |
| Fig. 4.5 | Detailed model structure of the rural subsystem | 89 |
| Fig. 4.6 | Urban subsystem details | 92 |
| Fig. 4.7 | Per capita material use (*DMC*), and per capita transport effort (*ML, MM*) for hunters & gatherers and for agrarian societies by size of urban centre to be sustained (standard productivity assumptions) | 96 |
| Fig. 4.8 | The share of urban (in contrast to rural, agricultural) population in the total system, and the share of the urban population required for transportation, in relation to size of urban centre (standard productivity assumptions) | 96 |
| Fig. 4.9 | The relation between agricultural productivity and human transport effort, by size of urban centre. Results from the standard, a low and a high productivity scenario. In the low productivity scenario we assumed a 40% lower area and 15% lower labour productivity as compared to the standard scenario (see Table 4.2). In the high productivity scenario we assumed corresponding increases in area and labour productivity | 97 |
| Fig. 5.1 | Location of the Danube sections Machland and Struden, Lower and Upper Austria (Map modified from Hohensinner et al. 2011) | 109 |
| Fig. 5.2 | (a–f) Historical development of the Danube in the Machland floodplain 1715–2006: (a) and (b) prior to channelisation in 1715 and 1812, respectively, (c) at the beginning of the channelisation programme in 1829 (*red circle*: first major river engineering measure). (d) excavation of the cut-off channel 1832, (e) at the end of the channelisation programme in 1859, (f) after channelisation and hydropower plant construction in 2006 (Hohensinner 2008) | 110 |

List of Figures

| Fig. 5.3 | The flood in Regensburg in 1784 (Angerer 2008) (© Museen der Stadt Regensburg – Historisches Museum) | 111 |
| Fig. 5.4 | Arrangement for catching the Beluga sturgeon (*Huso huso*) at the Iron Gate. | 115 |
| Fig. 7.1 | Biohistorical pyramid | 141 |
| Fig. 7.2 | Biohistorical conceptual framework | 142 |
| Fig. 7.3 | Biosensitivity triangle | 153 |
| Fig. 7.4 | The transition framework | 157 |
| Fig. 8.1 | Visualisation of the principal intellectual spaces of Nature-Society Geography | 165 |
| Fig. 9.1 | Model of "circumstances remote in time and of a general order" from de Toqueville's analysis of the French Revolution – factors not in parenthesis are features identified by de Toqueville; those in parenthesis are psychological assumptions or assertions (After Smelser 1976) | 201 |
| Fig. 9.2 | Coweeta LTER project area scales and research activities. (a) SE U.S. showing relation of project areas. (b) Coweeta LTER Project area in relation to state boundaries and intensive research sites. (c) Macon County NC with research watersheds by development type. Insets show watersheds classified by land ownership persistence (*LOPI*) – darker values indicate greater persistence. (d) Example of land required for architecture, agriculture, firewood, and mast harvest in proximity to a group of Cherokee villages based on "best" assumptions of per capita requirements (After Bolstad and Gragson 2008) | 204 |
| Fig. 10.1 | Conceptual framework for an urban socio-ecological system (*SES*) that can be used to visualize human-environment interactions at multiple scales. These interactions operate continuously in this multiscalar space, as shown by the *faint, gray, elliptical arrows* (Modified for the urban SES based the framework presented in Collins et al. 2011) | 221 |
| Fig. 10.2 | Map of Arizona, USA, showing extent of the Sun Corridor Megapolitan (*blue shading*) and the CAP study area within it (*red shading*). *Gray lines* are county boundaries | 222 |
| Fig. 10.3 | Analysis of land transitions in central Arizona between 1970 and 2000 (After Keys et al. (2007) based on a graphic designed by B. Trapido-Lurie; used with permission from Taylor & Francis) | 226 |

Fig. 10.4  Spatial distribution of heat intensity in Phoenix, AZ, in July 2005. (a) Hours in a 4-day period that temperature exceeded 110 °F (43 °C), (b) demographic characteristics of the population in low-, medium- and high-exposure areas, and (c) a graphic representation of the heat exposure "riskscape" for the region (Map in (a) and data in (b) used with permission from Ruddell et al. 2010) ............................................................... 229

Fig. 10.5  Maps of the urbanised central Arizona, USA region, showing the spatial distribution of plant phenological variables, start of growth, rate of greenup, end of growth, and rate of senescence. The *thick black lines* are the area freeways, which approximately enclose the urbanised/suburbanised portion of the region, comprising ~24 municipalities of the Phoenix metropolitan area. The large area that is differentiated from the surrounding desert to the north and south of the east-west freeway extending to west from the metropolitan area is an agricultural region that has yet to become urbanised. Seasonal parameters for the initial (spring) growth period were extracted from 2004 to 2005 normalised difference vegetation index (NDVI) data that were filtered with a Savitsky-Golay technique. NDVI data provide an index of greenness from remote imagery that can be correlated with vegetation biomass; changes in NDVI with time reflect growth (*increasing greenness*) or senescence (*decreasing greenness*). Dates are displayed as day of year (year 2005 days are shown in *parentheses*). Rates are calculated as tangent of slope between 20 and 80% levels of NDVI (Data and figures from Buyantuyev 2009) ........................................ 230

Fig. 10.6  Paired aerial photographs showing the change in land cover and use in Indian Bend Wash, Scottsdale and Tempe, Arizona between 1949 (*top*) and 2003 (*bottom*). The 1949 image shows a landscape dominated by farm fields, with country roads (part of the characteristic grid pattern of metropolitan Phoenix) evident as *light gray lines*. These same roads can be seen in the 2003 image, but farm fields have been replaced by housing developments and commercial (*right side of image*) and institutional (*upper left part of image*) land uses. The ephemeral stream (1949) and designed lake chain (2003) can be seen bisecting the images from *top to bottom*. Note the wide, shrub and tree-covered channel in the *upper image*; although by 2003 it was replaced by parks, lakes, and streams, the relatively wide channel still contains flash floods (see text for further description) ................................ 232

Fig. 10.7  Model showing major compartments of the biogeochemical cycles (Reprinted from Kaye et al. 2006, with permission from Elsevier) ........................................................................................... 234

| | | |
|---|---|---|
| Fig. 10.8 | Lead concentration (µg/kg) measured in 2005 in the surface soil (1–10 cm) across the CAP LTER study area. *Brown lines* show major freeways; the urbanised region is encircled by these roads (Reproduced with permission from Zhuo 2010) | 235 |
| Fig. 10.9 | Spatial distributions of (a) (*top*) criteria air pollutants and (b) percentage of the population that is in the Hispanic ethnic group, in the central Arizona–Phoenix region (Reproduced with permission from Grineski et al. 2007) | 236 |
| Fig. 11.1 | Population development, Vienna 1800–2000. System boundary: 1800–1890 territory of districts 1–9; from 1890 on the respective administrative boundaries, see text | 252 |
| Fig. 11.2 | Energy consumption (DEC) in the city of Vienna, 1800–2000: DEC in PJ/year (a) and share of energy carriers in DEC (b) | 254 |
| Fig. 11.3 | Vienna's electricity supply by source 1926–2006 | 256 |
| Fig. 11.4 | Population and energy consumption 1800–2000. (a) Development of population, domestic energy consumption (DEC) and DEC per capita and year (indexed 1800=1); (b) DEC per capita and year by main energy carriers in GJ | 256 |
| Fig. 11.5 | Actual and virtual forest area required to supply Vienna with wood and coal | 261 |
| Fig. 12.1 | Austro-Hungarian immigrant farms, including the Thir farm, situated within Finley Township. Small locator maps show the location of Kansas within the United States and of Decatur County and Finley Township within the state of Kansas | 272 |
| Fig. 12.2 | Theyern land management; (a) Small meadows and orchards clustered closely around residential house lots, while cropland surrounded the village. On the outskirts of the community, woodlands prevailed on poor soils not suitable for cropping; (b) The cropland portion of the agro-ecosystem rotated annually through a three-field sequence. Family farms consisted of scattered plots distributed across all parts of the village, as illustrated here for the Gill family, one of the larger holdings (ca. 13 ha farmland) | 275 |
| Fig. 12.3 | A conceptual model of agriculture as a coupled socioeconomic and natural system (See text for explanation) | 278 |
| Fig. 12.4 | People and space, Theyern, 1829 and Finley Township and Thir farm, 1895–1940; (a) population density; (b) average farm size; (c) land availability | 284 |
| Fig. 12.5 | Annual farm productivity, Theyern 1829 and Finley Township and Thir farm, 1895–1940; (a) grain yield; (b) area productivity; (c) labour productivity; (d) marketable crop production | 286 |

| | | |
|---|---|---|
| Fig. 12.6 | Livestock and nutrient management, Theyern, 1829 and Finley Township and Thir farm, 1895–1940; (a) livestock density; (b) nitrogen return .......................................... | 289 |
| Fig. 13.1 | Geographic position of the two case study regions ......................... | 300 |
| Fig. 13.2 | Conceptual scheme of the applied system boundaries and the energy flows calculated in this study ................................. | 302 |
| Fig. 13.3 | (a) Land use in Sankt Florian, 1864–2000; (b) Land use in Grünburg, 1864–2000. "Additional forest area" in Grünburg region was modelled based on the differences between cadastral forest records available only in 1995 and the statistical data used in all other points in time ................... | 308 |
| Fig. 14.1 | Line of sight along the dynamic metabolic interface between natural and human systems where quantitative measurements of socio-ecological variables are made .................. | 320 |
| Fig. 14.2 | Oahu's dependence on various types of imported goods by percentage (Eckelman and Chertow 2009b) ............................ | 328 |
| Fig. 14.3 | Number of manufacturing enterprises by sector and successional stage in Barceloneta, PR, 1950–2005 (Based on Ashton 2009) ................................................................ | 332 |
| Fig. 15.1 | The development of coal use (a), pig-iron production (b) and the railway network (c) in selected countries from 1750/1830 to 1910 and coal use in the United Kingdom (UK) as virtual forest area (d) (Datasources: Authors' calculations based on Mitchell 2003; Maddison 2008; Schandl and Krausmann 2007). To convert coal use into virtual forest area (d), it was assumed that a quantity of fuelwood with the equivalent energy content to the coal used can be provided through sustainable forest management (i.e. through the use of annual growth and not standing timber mass). The forest area required to produce this volume of fuelwood is presented as a virtual forest area. Accordingly, by 1900, coal use in the United Kingdom represented a forest area five times the size of the entire country ......................................... | 346 |
| Fig. 15.2 | The establishment of new energy sources in the United Kingdom (1750–2000) (a) and worldwide (1850–2005). (b) In this diagram, the share of total primary energy supply represented by the three fractions biomass, coal and oil/natural gas (including other energy forms) is depicted. The biomass fraction includes all biomass used as food for humans and livestock and biomass used for all other purposes, together with fuelwood .................... | 351 |
| Fig. 15.3 | Motor vehicle stocks (a) and electricity generation (b) in the 20th century (Data sources: Authors' calculations based on Mitchell 2003; Maddison 2008 ) ..................................... | 353 |

List of Figures

| Fig. 15.4 | The development of global energetic and material societal metabolism. (a) Per capita energy consumption in the UK and Austria, (b) Global per capita energy and material consumption, (c) Global primary energy consumption by technology, and (d) Global material consumption | 357 |
| Fig. 15.5 | $CO_2$ emissions resulting from combustion of fossil energy sources and cement production in selected countries. Data given in tonnes of carbon (C) per capita and year. (a) Industrialised countries (b) Southern hemisphere countries | 361 |
| Fig. 16.1 | In *Pasteur's Quadrant*, Stokes categorises four different types of research. Most research associated with BES would be located in Pasteur's quadrant: Use-inspired basic research | 375 |
| Fig. 16.2 | The human ecosystem concept, bounded by the *bold line*, showing its expansion from the bioecological concept of the ecosystem as proposed originally by Tansley (1935) in the *dashed line*. The expansion incorporates a social complex, which consists of the social components and a built complex, which includes land modifications, buildings, infrastructure, and other artefacts. Both the biotic and the physical environmental complexes of urban systems are expected to differ from those in non-urban ecosystems (Figure copyright BES LTER and used by permission (Pickett and Grove 2009) | 378 |
| Fig. 16.3 | The human ecosystem framework. This conceptual framework identifies the components of the resource and human social systems required by inhabited ecosystems. The resource system is comprised of both biophysical and social resources. The human social system includes social institutions, cycles, and the factors that generate social order. This is a framework from which models and testable hypotheses suitable for a particular situation can be developed. It is used to organise thinking and research and is a valuable integrating tool for the BES (Re-drawn from Machlis et al. 1997) | 380 |
| Fig. 16.4 | Press–pulse dynamics framework (PPD). The PPD framework provides a basis for long-term, integrated, socio–ecological research. The *right-hand side* represents the domain of traditional ecological research; the *left-hand side* represents traditional social research associated with environmental change; the two are linked by pulse and press events influenced or caused by human behaviour and by ecosystem services, top and bottom, respectively (Collins et al. 2011). Individual items shown in the diagram are illustrative and not exhaustive | 381 |

Fig. 16.5  Framework for complexity of socio-ecological systems. The three dimensions of complexity are spatial heterogeneity, organisational connectivity and temporal contingencies. Components of the framework are arrayed along each axis increasing in complexity. For example, a more complex understanding of spatial heterogeneity is achieved as quantification moves from patch richness, frequency and configuration to patch change and the shift in the patch mosaic. Complexity in organisational connectivity increases from within unit process to the interaction of units and the regulation of that interaction to functional patch dynamics. Finally, historical contingencies increase in complexity from contemporary direct effect through lags and legacies to slowly emerging indirect effects. The *arrows* on the *left* of each illustration of contingency represent time. While not shown in the figure, connectivity can be assessed within and between levels of organisation (Cadenasso et al. 2006) .................................................................. 386

Fig. 16.6  Examples of socio-ecological data types organized by scale and intensity of analysis. Data types marked in *green* are data that LTSER sites must typically acquire, document, and archive. Data types marked in *red* are typically collected by LTSER sites (Zimmerman et al. 2009) ..................................... 388

Fig. 16.7  Example of non-census data sets with spatial reference to Baltimore City, 1800–2000 (Figure developed for Baltimore LTER Figure copyright BES LTER and used by permission from Boone) ............................................. 389

Fig. 16.8  BES has instrumented a set of nested and reference watersheds that vary in current, historical, and future land use and condition (Figure copyright BES LTER and used by permission from O'Neil-Dunne) ............................... 392

Fig. 16.9  LTSER Platforms are similar to a table with four legs essential to the integrity of the whole: long-term monitoring, experimentation, comparative analyses, and modelling (Figure adapted from Carpenter 1988) ........................................... 393

Fig. 16.10  The LTSER data temple, with specific BES research themes included (Figure copyright BES LTER and used by permission from Grove) ............................................. 395

Fig. 16.11  Linkages between decision making, science, and monitoring & assessment (Figure copyright BES LTER and used by permission from Grove) ........................................... 397

Fig. 16.12  An abstracted cycle of interaction between research and management. The cycle begins with the separate disciplines of ecology, economics and social sciences interacting with a management or policy concern. In the past, ecology

List of Figures xxxi

has neglected the urban realm as a subject of study, leaving other disciplines to interpret how ecological understanding would apply to an urban setting. A management or policy action (Action$_z$) results. Management monitors the results of the action to determine whether the motivating concern was satisfied. Contemporary urban ecology, which integrates with economics and social sciences, is now available to conduct research that recognises the meshing of natural processes with management and policy actions. Combining this broad, human ecosystem and landscape perspective with the concerns of managers can generate a partnership to enhance the evaluation of management actions. New or alternative management actions can result (Actions$_{z+1}$) (Pickett et al. 2007) .................. 398

Fig. 16.13 An example of the management-research interaction in Baltimore City watersheds. Traditional ecological information indicated that riparian zones are nitrate sinks. The management concern was to decrease nitrate loading into the Chesapeake Bay. In an effort to achieve that goal, an action of planting trees in riparian zones was proposed. Management monitoring indicated that progress toward decreasing Bay nitrate loadings was slow. Results from BES research suggested that stream channel incision in urban areas has resulted in riparian zones functioning as nitrate sources rather than sinks. In partnership with managers and policy makers in Baltimore City and the Maryland Department of Natural Resources, a re-evaluation of strategies to mitigate nitrate loading was conducted. This led to a decision to increase tree canopy throughout the entire Chesapeake Bay watershed. Baltimore City adopted an Urban Tree Canopy goal, recognising both the storm water mitigation and other ecological services such canopy would provide (Pickett et al. 2007) ............................................................ 400

Fig. 17.1 The functional components of LTSER Platforms .................. 416
Fig. 17.2 Infrastructural elements of LTSER Platforms across spatial scales within a LTSER Platform region ............................ 417
Fig. 17.3 Hierarchy of research projects in LTSER Platforms .................. 419
Fig. 17.4 Schematic fuzzy cognitive map derived from two interviews in the Eisenwurzen LTSER Platform – Austria ............................ 424
Fig. 17.5 Simplified model of the Critical Ecosystem Services of the Eisenwurzen LTSER Platform and their interaction *left*: *green* = positive, *red* = negative) and scenario of their future importance (*right*: light *blue* = historical situation, dark *blue* = current situation) (Austrian contribution to the ILTER Ecosystem Service Initiative) .................................. 425

| | | |
|---|---|---|
| Fig. 17.6 | Interactions of key elements and factors in the socio-ecological system across sectors (environment in *greens* and *blue*, economy and society in *white* and *grey*) and scales in the LTSER Eisenwurzen (Austrian contribution to the ILTER Ecosystem Service Initiative) | 425 |
| Fig. 17.7 | Socio-ecological profile of the LTSER Eisenwurzen Platform according to the ISSE/PPD framework (Collins et al. 2007, 2011): The conceptual elements, described by Grove et al. (Chap. 16 in this volume) are parameterised based on comprehensive analyses combining disciplinary scientific expertise and primary stakeholders perception (Fuzzy Cognitive Mapping) | 426 |
| Fig.17.8 | Thresholds (T) and their interactions (I) across sectors (environment, economy, society) and scales in the Eisenwurzen LTSER region (Austrian contribution to the ILTER Ecosystem Service Initiative according to Kinzig et al. (2006)) | 427 |
| Fig. 17.9 | Geopolitical coverage of LTER-Europe (as of 2010; Mirtl et al. 2010) | 432 |
| Fig. 17.10 | Location of 31 European LTSER Platforms in 2010 (including five preliminary Platforms). The map reflects the 48 socio-ecological systems of Europe (Metzger et al. 2010). Environmental zones are colour-coded. The brightness of each colour varies according to the economic density, varying between $< 0.1$ Mio €/km$^2$ (lightest) and $> 0.1$ Mio €/km$^2$ (darkest). The Platform labels are the unique LTER-Europe site codes. According to these site codes, details for each Platform can be found in Table 17.1 | 433 |
| Fig. 17.11 | Conflicting priorities in LTSER Platform implementation. *Left side*: Cases requiring complex approaches in creating the framework for socio-ecological research. *Right side*: Less demand for matrix functionalities and supporting services due to simpler settings | 435 |
| Fig. 19.1 | Overview of the process and development of the Eisenwurzen LTSER Platform | 468 |
| Fig. 19.2 | The area of the Eisenwurzen LTSER Platform, including infrastructure elements | 470 |
| Fig. 19.3 | Interlinkage between the elements of the Eisenwurzen LTSER Platform | 471 |
| Fig. 19.4 | Data management concept for Eisenwurzen LTSER Platform | 478 |

List of Figures

| | | |
|---|---|---|
| Fig. 20.1 | Location map for the Central French Alps LTSER Platform and meteorological stations. The Platform includes areas with strict protection status (*dark grey*): the Vercors High Plateaux Natural Reserve and some of the core area of the Ecrins National Park, as well as inhabited areas managed by agriculture and forestry (*light grey*): the Vercors Natural Regional Park and part of the peripheral area of the Ecrins National Park. Meteorological monitoring stations set up by the LTSER Platform are located using different symbols depending on the equipment in place | 490 |
| Fig. 20.2 | Conceptual presentation of the Central French Alps LTSER research questions | 492 |
| Fig. 20.3 | Recent climatic trends over the Vercors High Plateaux. Interannual variability of air temperature (at 850 hPa level; in °C) and water precipitation (in kg/m$^2$) calculated for the Vercors site on the 1948–2010 period (anomalies were calculated from NCAR-NCEP reanalysis; time-series anomalies are smoothed using a moving average over 12 months) | 492 |
| Fig. 20.4 | Potential ecosystem service supply and actual provision of agronomic services to farmers of Villar d'Arène (Hautes Alpes). (a) Green biomass (tons/ha) (b) Potential agronomic value (unitless) calculated as a combination of different functions (green biomass, digestibility and phenology), and actual benefits: (c) the number of days of livestock units/ha and (d) hay production (tons/ha). Roads and tracks are added on maps as they are important elements of analysis (Modified from Lamarque et al. 2011) | 499 |
| Fig. 21.1 | LTER Sites in the Tyrolean Alps LTSER Platform. *Numbers* refer to Table 21.1 | 508 |
| Fig. 21.2 | Climate, land cover and indicators of development in the Tyrolean Alps LTSER Platform (For details see Tappeiner et al. 2008b) | 511 |
| Fig. 21.3 | Environmental and socioeconomic factors affecting ecosystems in the Tyrolean Alps | 513 |
| Fig. 22.1 | Integrated monitoring of natural and social spheres as a foundation for negotiating development options (Adapted from Fischer-Kowalski et al. 2008) | 534 |
| Fig. 22.2 | (a) Development of number of inhabitants, buildings and beds in relation to the index year 1970 (equals 1) up to 2005. (b) Number of overnight stays for summer and winter tourism and the total of both of these in 1,000 stays from 1977 to 2006 | 538 |

| | | |
|---|---|---|
| Fig. 22.3 | Settlement area in 1973 and 2003. The *red spots* in the aerial photos are buildings and the *blue line* is the boundary of the settlement area | 538 |
| Fig. 22.4 | (a) Comparing a photomontage from 1974 projecting the year 2000 (*left*), with (b) an actual photo taken in 2007 (*right*) | 539 |
| Fig. 22.5 | Comparison of the occurrence of species of two alpine plant communities (snow bed, alpine grassland with Carex curvula) in 1970s and 2006 | 541 |
| Fig. 22.6 | Land cover change visible in aerial photos for 1972 and 2003 | 542 |
| Fig. 22.7 | Map of stakeholders, grouping persons or groups according to their relation to nature and their predominant scale of action | 544 |
| Fig. 22.8 | Causal model linking influential external and internal variables – 13 in all – with system parameters and ultimately with the outcome dimension of sustainability | 546 |

# List of Tables

| | | |
|---|---|---|
| Table 3.1 | Elements of the stakeholder process conducted in Reichraming over a period of 3 years | 65 |
| Table 3.2 | Storylines of the three scenarios trend, globalisation and local policy described by the stakeholders and calculated by the integrated socio-ecological model | 66 |
| Table 3.3 | Results of the three scenarios in comparison with the initial value (all values are in total for the whole region) | 68 |
| Table 4.1 | Basic model assumptions of the human food calculator | 87 |
| Table 4.2 | Basic model assumptions and characteristic parameters of the biomass production calculator (standard and low productivity scenario), compared with empirical data on nineteenth century Austrian land use systems (three case studies) | 88 |
| Table 4.3 | Basic model assumptions on the relation of food production and agricultural biomass flows (biomass flows per unit of food output) and assumptions on transport stages of different biomass categories | 88 |
| Table 4.4 | Basic model assumptions to calculate transport indicators | 91 |
| Table 4.5 | Modelling results for material flows and transport in agrarian societies | 94 |
| Table 6.1 | Definition of scale mismatches between biodiversity, biodiversity management, and biodiversity-relevant policy (upper two rows). Definition of scale mismatches of research, monitoring, and evaluation carried out within each of these parts (third row) | 127 |
| Table 8.1 | Core themes of this study and the levels of correspondence to principal areas (Human-environment interactions and nature-society relations) of the geographic sub-field (see also Fig. 8.1) | 166 |

| | | |
|---|---|---|
| Table 8.2 | Case study-based illustrations of concepts at the intersection of geography and LTSER | 175 |
| Table 8.3 | Examples of temporal scaling of human-environment interactions and water-resource/agricultural land use in Bolivia Table (19th and 20th centuries) | 176 |
| Table 8.4 | Principal characteristics of a spatial network of potential LTSER sites (Central Bolivia) | 178 |
| Table 10.1 | LTER ecological core areas, proposed LTER social science core areas, and CAP LTER integrated project areas | 224 |
| Table 11.1 | Phases of the urban energy transition: Average annual growth rates of population and energy consumption (DEC) and the share of biomass and coal in total energy consumption | 264 |
| Table 12.1 | Population, land use, livestock and crop production in Finley Township, 1895–1940 | 280 |
| Table 12.2 | Population, land use, livestock and crop production on the Thir farm, 1895–1940 | 280 |
| Table 12.3 | Socio-ecological characteristics, Finley Township, 1895–1940 | 281 |
| Table 12.4 | Socio-ecological characteristics, Thir farm, 1895–1940 | 282 |
| Table 12.5 | Population, land use, livestock and crop production in Theyern municipality, 1829 | 282 |
| Table 12.6 | Socio-ecological characteristics, Theyern municipality, 1829 | 283 |
| Table 13.1 | Data sources | 302 |
| Table 13.2 | Structural parameters characterising the case studies: Number of population, farms, and livestock; land availability and use; average yields and productivity | 304 |
| Table 13.3 | Biomass flows in the agricultural sector of the studied land use systems: Domestic extraction (DE), Imports (Im), Direct Input (DI), Exports (Ex) and Domestic Consumption (DC) of biomass; per capita (GJ/cap) and per agricultural area (GJ/$ha_{agr}$) | 305 |
| Table 14.1 | Characteristics of the selected islands | 319 |
| Table 14.2 | Puerto Rico's top five export Commodities, USD million | 326 |
| Table 14.3 | Singapore's top five import commodities, USD million | 326 |
| Table 14.4 | Singapore's top five export commodities, USD million | 327 |
| Table 15.1 | Sociometabolic profile of selected countries in 2000 | 358 |
| Table 17.1 | Overview of European LTSER Platforms, status as of 2010. The labels of platforms in Fig. 17.10 refer to the column "Site_Code" in this table | 434 |

# List of Tables

| | | |
|---|---|---|
| Table 18.1 | Interdisciplinary thematic research funding programmes of the Academy of Finland.................................. | 447 |
| Table 19.1 | Biophysical characteristics of the Eisenwurzen region............... | 464 |
| Table 19.2 | Land-cover characteristics of the Eisenwurzen region................ | 464 |
| Table 19.3 | Inhabitants and population density in the Eisenwurzen region......................................................... | 465 |
| Table 19.4 | Exemplary list of research projects within the Eisenwurzen LTSER Platform allocated to the three research areas................ | 474 |
| Table 20.1 | Partners and their disciplines involved in the Central French Alps LTSER Platform Aps............................................ | 488 |
| Table 21.1 | Location, site characteristics and research focus of current research sites at the Tyrolean Alps LTSER Platform.................. | 509 |
| Table 22.1 | Variants of "transdisciplinary" ("td") research and main features........................................................... | 530 |
| Table 22.2 | Critical indicators investigated in the biosphere reserve Gurgler Kamm ................................................. | 537 |
| Table 22.3 | Alpine winter sporting days for village and mountains of Obergurgl for a recent decade and the decade starting in 2041.................................................................. | 542 |
| Table 22.4 | Sustainability indicators for describing and assessing the scenarios.......................................... | 547 |
| Table 22.5 | Initial sustainability assessment of the four scenarios for Gurgl 2020 ......................... | 548 |

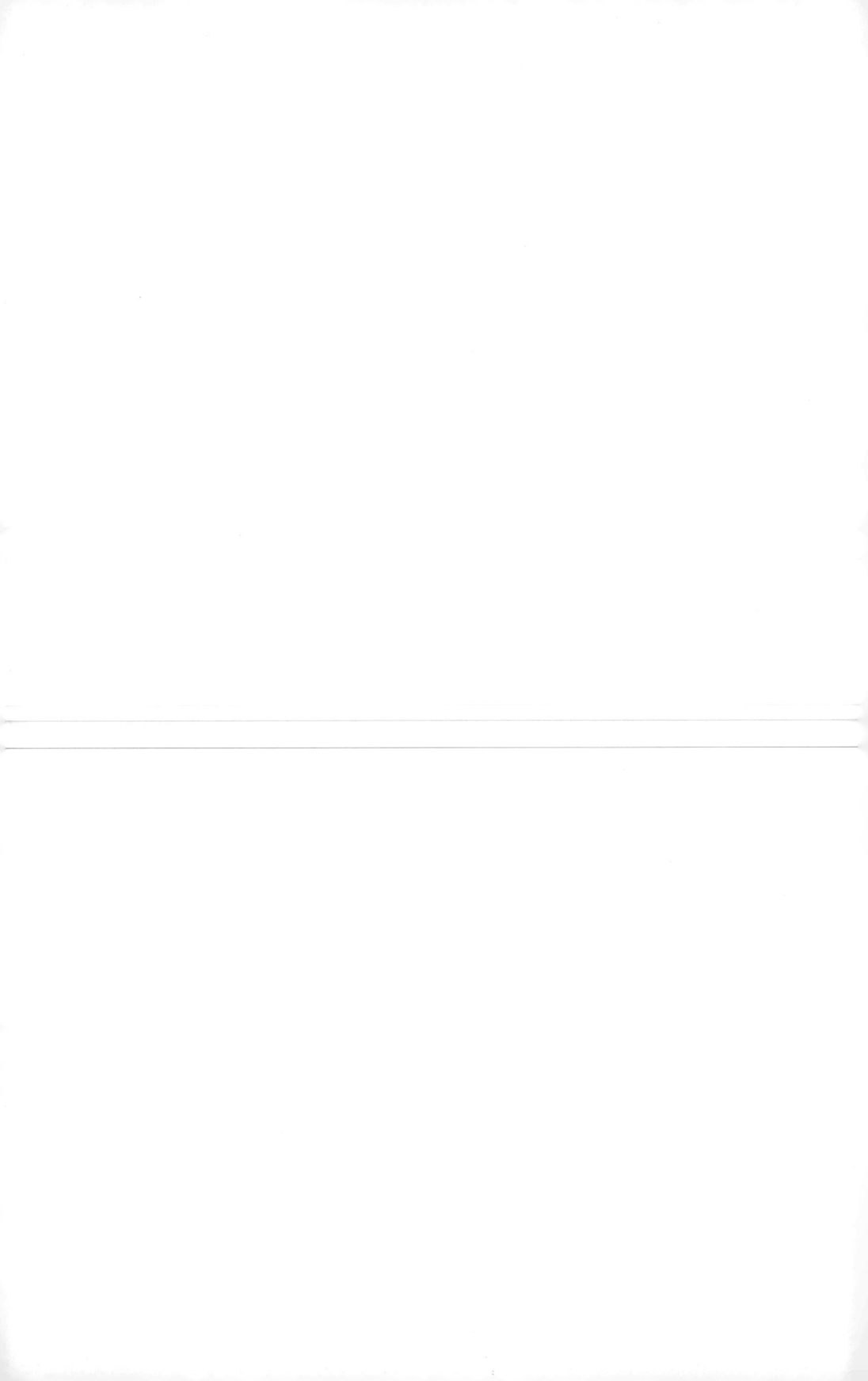

# Chapter 1
# Introduction

Simron Jit Singh, Helmut Haberl, Marian Chertow, Michael Mirtl, and Martin Schmid

## 1.1 Long-Term Socio-Ecological Research (LTSER): An Emerging Field of Research

Over the last half century, exceptional changes in the natural environment attributed to human activities have placed renewed importance on the study of society-nature interactions. Contemporary problems such as climate change, loss of biodiversity and valuable ecosystems, and resource depletion have been greatly exacerbated by the unsustainable ways in which humans interact with their environment. Indeed, the magnitude of the problems we now face is an outcome of a much longer process, accelerated by industrialisation since the nineteenth century. There is evidence that ecosystems are increasingly challenged by coping with human demands (Millennium Ecosystem Assessment 2005) and that costs and benefits of the use of nature's

---

S.J. Singh, Ph.D. (✉) • H. Haberl, Ph.D.
Institute of Social Ecology Vienna (SEC), Alpen-Adria Universitaet Klagenfurt,
Wien, Graz, Schottenfeldgasse 29/5, Vienna 1070, Austria
e-mail: simron.singh@aau.at; helmut.haberl@aau.at

M. Chertow, Ph.D.
Center for Industrial Ecology, Yale School of Forestry and Environmental Studies,
Yale University, 195 Prospect Street, New Haven, CT 06511, USA
e-mail: marian.chertow@yale.edu

M. Mirtl, Ph.D.
Department of Ecosystem Research and Monitoring,
Environment Agency Austria, Spittelauer Lände 5, Vienna 1090, Austria
e-mail: michael.mirtl@umweltbundesamt.at

M. Schmid, Ph.D.
Center for Environmental History, Institute of Social Ecology Vienna (SEC),
Alpen-Adria Universitaet Klagenfurt, Vienna, Graz,
Schottenfeldgasse 29/5, Vienna 1070, Austria
e-mail: martin.schmid@aau.at

bounty are unequally distributed socially and geographically, inducing great potential for social conflict (Hornborg et al. 2007; Martinez-Alier et al. 2010). In this sense, the present problems are not only "ecological" but also "socio-ecological" since the effect of how societies interact with their environment has a bearing not only on ecosystems but also upon social systems and human wellbeing.

In response to these concerns, also articulated in the Brundtland report (World Commission on Environment and Development 1987), many new interdisciplinary research fields have arisen including industrial ecology (Graedel and Allenby 1995; Erkman 1997), ecological economics (Martinez-Alier 1987), and most recently "sustainability science" (Kates et al. 2001; Parris and Kates 2003; Clark et al. 2004). Their research agendas share an aim to assess more deeply the delicate relation between society and nature, including the sustainable and equitable access to natural resources and ecosystem services for current and future generations. Concerned primarily with questions of socio-ecological sustainability and global environmental change, the emerging interdisciplinary field of Long-Term Socio-Ecological Research (LTSER) aims to observe, analyse, understand and model changes in coupled socio-ecological (or human-environment) systems over decadal, sometimes even centennial, periods of time. By including long-term monitoring, historical research, forecasting and scenario building, empirical and conceptual research as well as participatory approaches, LTSER aims to provide a knowledge base that helps to reorient socioeconomic trajectories towards more sustainable pathways.

Interest in the study of the environmental relations of human society is not new and goes back to prominent writings of classical thinkers such as Hippocrates, Aristotle, Cicero, Vitruvius and Pliny. The impulse to explain something as fundamental as why people are different across the world led over time to a number of theoretical propositions on human-environment relations. The earliest among these and also the one with the longest influence that survived well into the twentieth century is *environmental determinism*. According to this theory, the varying natural environments were seen to be the main cause for sociocultural differences across the world. Classical Greek, Roman and Arab theories attributed the superiority of their respective civilizations to a perfectly balanced climate and/or geopolitical location. In contrast, populations in the humid tropics were described as lethargic, less courageous and intelligent and destined to be ruled by others. This approach found its way well into the nineteenth and twentieth century, wherein geographers such as Friedrich Ratzel, William Morris Davis, Ellsworth Huntington and Griffith Taylor were chiefly concerned with documenting the influence of the environment in determining the formation of human societies and culture (Moran 2000).

Throughout the twentieth century, the study of human-environment relations has received elaborate treatment in both the social and natural sciences (Fig. 1.1). Geographers, in reacting to environmental determinism, soon concerned themselves with documenting the impact of human activities on the landscape. The ideas of Carl Sauer (1925) at Berkeley in the first half of the twentieth century were influential. Sauer proposed landscape as a unit of analysis, one that carries both physical and cultural associations (Grossman 1977). Around the same time, the Chicago School of hazards research under Gilbert White was another important

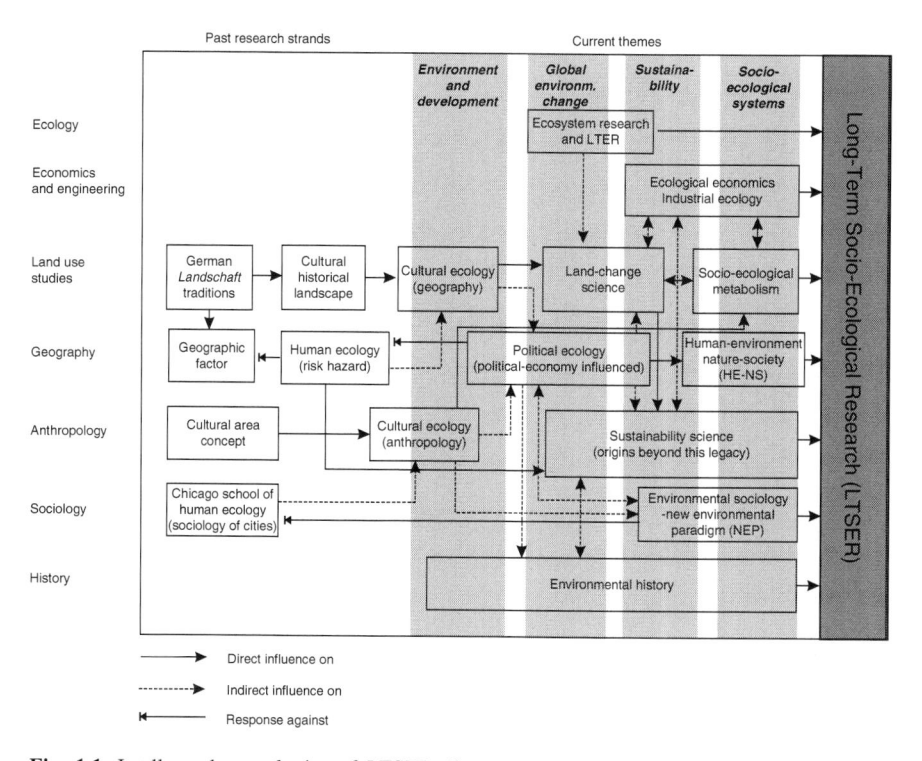

# 1 Introduction

**Fig. 1.1** Intellectual genealogies of LTSER (Source: Adapted and redrawn after Turner and Robbins 2008)

influence in nature-society geography that focussed on governmental planning and public awareness (Zimmerer 2010). The symposium "Man's role in changing the face of the Earth" was a landmark in the study of human-environment relations for revealing the impact of socioeconomic activities on the global environment (Thomas 1956). Research undertaken since the 1960s aimed at studying processes of environmental change caused by socioeconomic activities (subsumed under the term *cultural ecology*) today comprises the largest, most active and popular grouping within American geography (Zimmerer 2004). These studies lie at the interface of human dimensions of global change.

For most of the twentieth century, (environmental) anthropologists have primarily been concerned with examining the role of culture and nature in the formation of human societies. Also rejecting the notion of environmental determinism, anthropologists Franz Boas, Alfred Kroeber and C. D. Forde found no correlation between the environment, the economy and the society (Boas 1896, 1911; Kroeber 1917; Forde 1934). After World War II, under the influence of Leslie White and Julian Steward, a renewed interest in the role of the environment in shaping human societies arose. For Leslie White, cultural evolution was determined by the levels of energy use extracted from nature and its conversion efficiency (White 1949) – a notion that had already been put forward almost half a century earlier by Nobel laureate

Wilhelm Ostwald, then causing heated controversy with sociologists (Ostwald 1909); see Martinez-Alier (1987) for an in-depth discussion of that dispute. Steward, on the other hand, argued that a society's social and cultural template is a response to the adaptation process with the environment for the purpose of survival and procurement of food. His method of 'cultural ecology' sought to analyse the technologies, behaviour patterns and the social organisation of societies that have a direct bearing on utilising nature optimally for subsistence (Steward 1955). Since the late 1960s, inspired by Odum's *Fundamentals of Ecology* (Odum 1959), a number of anthropologists directed their interest in studying small-scale production (subsistence) systems, tracing the flows of materials and energy between society and the environment (Rappaport 1968; Ellen 1982; Netting 1981, 1993). Only recently have anthropologists begun to investigate contemporary social and environmental problems (Bodley 2008). Taking up pressing environmental issues of the day, several anthropologists have challenged the nature – culture dualism that informs most conservation and sustainable development policy (Croll and Parkin 1992; Descola and Palsson 1996; Fairhead and Leach 1996). In its current form, *political ecology* attributes environmental degradation and human impoverishment in developing countries to the politics of natural resource management and capitalism in the North (Greenberg and Park 1994; Robbins 2004). While most of these studies relate to small-scale systems, an increasing number of scholars now direct their attention to the study of human-environment interactions at varying spatial and temporal scales that are promising for LTSER (Crumley 1993; Hornborg and Crumley 2006; Moran et al. 1994; Gragson, Chap. 9 in this volume).

The first group of researchers to claim the notion of 'human ecology' was the so-called 'Chicago School,' a group of sociologists around R. E. Park and E. W. Burgess working on social relations in cities. Some of their early statements suggest considerable relevance even for current human-environment studies. For example, it was stated that "man has, by means of invention and technical devices of the most diverse sorts, enormously increased his capacity for reacting upon and remaking, not only his habitat but his world" (Park 1936, p. 12), and human ecology was defined as the study of "spatial and sustenance relations in which human beings are organised" (McKenzie 1926, p. 141). Despite such programmatic enunciations, the Chicago School, which gained considerable prominence in US sociology from the 1920s to the 1940s, was mostly focused on the analysis of interrelations among humans in urban regions, and paid little, if any, attention to society-nature interaction in the modern sense (Vaillancourt 1995). This strand of human ecology used ecological metaphors such as competition, succession, climax, and others for the analysis of human societies (Teherani-Krönner 1992; Young 1974), an approach that lost credibility in the late 1930s and 1940s, following devastating critiques of the uncritical application of biological analogies and theoretical vagueness (Beus 1993). When sociological interest in environmental issues reappeared in the early 1970s, scholars were eager to distance themselves from the Chicago School and labelled their enterprise "Environmental Sociology", the nomenclature still used today. While a large part of environmental sociology is focused on environmental awareness, the sociology of environmental movements and ecological modernisation, there are also authors who call for a "new environmental paradigm" (Catton and Dunlap 1978; Dunlap and Catton 1994) that would go beyond the narrow sociological 'human exemptionalism paradigm' and underpin broader, perhaps inter- and transdisciplinary research into human-environment interactions and sustainability.

Environmental history is yet another (sub-) discipline within the humanities where human–environment interactions have received considerable attention. Environmental history as a self-confident and self-conscious academic field emerged in the wake of modern environmentalism somewhere in the decades between the 1960s and 1990s. Similar to anthropology and geography, the question of nature's role, importance or even agency in the course of history had its origins in the intellectual debates among historians at least from the 1700s onwards. After several intellectual 'turns' (linguistic, cultural, etc.) that influenced the humanities in general during the late twentieth century, the majority of historians have lost sight of nature in the different ways they view the world. Environmental historians, in contrast to most of their fellow historians, assume that *'human history has always and will always unfold within a larger biological and physical context, and that [that] context evolves in its own right'* (McNeill 2003, p. 6). Compared to other branches of historical studies (social, economic or political history to name but a few), environmental history as a sub-discipline of history still forms a small community. Nevertheless environmental history, in particular the strand of 'material' environmental history (for that notion and the two other strands, 'cultural/intellectual' and 'political' environmental history, cp. McNeill 2003) has contributed, in line with LTSER, to a better understanding of how nature, society and culture have been intertwined for millennia. Some of those environmental historians who craft environmental history as a genuinely interdisciplinary field have also been involved in and contributed to the emergence of LTSER (cp. Haberl et al. 2006).

Efforts on the other side of the 'great divide' are of equal relevance to further research on society–nature interactions having a bearing on LTSER. Despite some early calls for an "ecology of man" (Adams 1935; Darling 1956; Sears 1953), biological ecology hardly contributed to the inquiry of human-environment interactions prior to the environmental debate of the 1970s (Young 1974). Moreover, when the first influential texts of biological ecologists on human ecology appeared (e.g. Ehrlich and Ehrlich 1970; Ehrlich et al. 1973), they focused on the role of humans as agents of disturbance in ecosystems. This resulted in an insufficiently complex concept of human agency and did not facilitate interdisciplinary approaches toward the analysis of society-nature interaction. That ecologists were inclined to study "natural" ecosystems rather than "human-dominated" ones (Likens 1997) may have contributed to this bias. Probably the most important factor was that many biological ecologists at that time did not recognise the need to develop a more complex approach toward "humans as components of ecosystems" (McDonnell and Pickett 1997), which recognises that socioeconomic systems are qualitatively different from natural systems and hence need to be analysed in joint efforts with scientists from disciplines such as sociology or economics. Moreover, neo-Malthusian concepts played an important role in bio-ecological approaches toward human ecology of that time, which further limited cooperation with social scientists.

A prominent framework developed in the 1970s and still used in studies of society-nature interaction is the so-called IPAT equation (Holdren and Ehrlich 1974). It conceptualises environmental impact (I) as the product of population (P), affluence (A) and technology (T). Although this equation neglects feedbacks between these factors, its simplicity makes it still useful to guide analyses of trajectories of society-nature interaction (Chertow 2001; Dietz and Rosa 1994; Dietz et al. 2007; Fischer-Kowalski and Amann 2001; Haberl and Krausmann 2001).

Other initiatives contributed strongly to shaping current interdisciplinary scientific endeavours such as LTSER. One was the creation of the section "human adaptability" by the International Biological Programme (IBP) in the late 1960s (see Moran 1993), another the establishment of the *Man and the Biosphere* (MAB) programme by UNESCO in the early 1970s. The "Hong Kong Human Ecology Programme" chaired by the Australian ecologist Stephen Boyden resulted in conceptual models of society-nature interaction (Boyden 1992, 1993, 2001) that have strongly influenced current socio-ecological approaches (Fischer-Kowalski and Weisz 1999; Fischer-Kowalski and Haberl 2007) and LTSER (Haberl et al. 2006). In particular, these approaches led to modern methods and concepts for analysing socioeconomic metabolism that are broadly used in Ecological Economics and Industrial Ecology (see below and Haberl et al., Chap. 2 in this volume).

The current volume is a first effort to bring together prominent scholarly traditions explicitly concerned with Long-Term Socio-Ecological Research within the context of sustainability. This volume covers contributions from Europe and North America, regions where LTSER first began to be discussed and implemented in its current form. Contributing scholars were selected from a variety of disciplinary and interdisciplinary fields of research, such as social and human ecology, industrial ecology, sociology, environmental history, human geography and anthropology. It has been our sincere effort to reveal the inter-and transdisciplinary facet of LTSER in response to understanding and seeking solutions to contemporary societal and environmental challenges. To encourage the use of this volume for university teaching, editors and authors have taken care to ensure its presentation and readability, along with an attractive subsidy of more than 50% by the publishers to enhance the affordability of the book.

## 1.2 From LTER to LTSER

Long-term socio-ecological research (LTSER) is an extension of Long-Term Ecological Research (LTER), a strand of research that has gained prominence since the early 1980s among scholars in the United States and subsequently in Europe, concerned with questions of global environmental change. Pioneered largely by the natural sciences, LTER aims to better understand, analyse, and monitor ecosystem changes with respect to patterns and processes over extended periods of time. It is acknowledged that collecting evidence on the impacts of global changes in climate and other important environmental variables of ecosystems requires a long-term approach, not only because many variables are changing slowly, but also because the spatial and temporal variability of some of these variables (e.g. seasonal temperatures, rainfall) makes it difficult to distinguish the 'signal' of global environmental change from the 'background noise'. The classical LTER approach is thus often based on the consistent monitoring of a large number of variables that characterise patterns and processes in ecosystems over long periods of time in spatially explicit areas, often with little or no direct human influence.[1]

For LTER to be effective, it was imperative to look beyond the usual project duration of 3–5 years, and to organise into a network of sites using comparable methods and approaches for a meaningful interpretation of ongoing processes of global environmental change. The first national LTER network was established in the United States in the 1980s with support from the US National Science Foundation (NSF). Over the next 20 years, some 1,100 scientists were part of this network working on 26 LTER sites (NSF 2002, 2011b). The US example became a trigger for the creation of more national and regional networks, culminating in the founding of a global network of research sites (International Long-term Ecological Research – ILTER) encompassing highly varied ecosystem types and climatic zones worldwide. Established in 1992, ILTER presently comprises 43 national networks organised into regional networks such as *LTER-North America*, *LTER-Europe*, and *LTER-East Asia Pacific*, to create a unique long-term data system (Box 1.1).

---

**Box 1.1** The International Long-Term Ecological Research (ILTER) Network

**Scope and Future Trends**

The International Long-term Ecological Research (ILTER) Network is a global international research network with a unique capability for long-term site-based ecosystem research. Its mission is to improve understanding of ecosystem research around the globe and to inform solutions to current and future environmental problems. It does this through the efforts of many thousands of scientists and information managers within a global community of member networks and by working in partnership with other major programmes and organizations such as the Global Land Project, the Global Biodiversity Observation Network, the Global Biodiversity Information Facility and UNEP's Global Adaptation Network for Climate Change.

ILTER's objectives are: (i) to coordinate long-term ecological research networks at local, regional and global scales; (ii) to improve the comparability of long-term ecological data; (iii) to deliver scientific information to scientists,

(continued)

---

[1] While North American LTER sites are all over 30 km², European LTER Sites are much smaller and range between 0.01 and 10 km².

**Box 1.1** (continued)

policymakers and the public to meet the needs of decision makers at multiple levels; and (iv) to educate the next generation of long-term scientists and resource managers. Established in 1992, ILTER is now an expanding network with over 600 sites in 43 member networks around the globe. ILTER still has significant gaps in its coverage that limit its capability to address issues in some areas. Therefore, a key part of ILTER's current strategy is to build new capacity through training, the development of strategic partnerships and the addition of new sites and member networks. It is doing this through a targeted accumulation of new sites and networks that add value and help answer critical policy and science-driven questions. Of particular importance is the need to improve the coverage of sites in the Southern Hemisphere, particularly in Africa and South America.

In the beginning, ILTER was predominantly an ecological research network, but over the last decade it has increasingly recognised the need to include the human dimension in its work on environmental change and is now working extensively with social scientists, economists and science communicators to do so. In some countries, urban LTER Sites have been developed, while in other, larger LTER Sites, incorporating social and economic processes and the active engagement of stakeholders in research is now common. As a result of this trend, ILTER's Sites and networks are well placed to address many contemporary environmental issues, particularly climate change, sustainable development, biodiversity, the sustainable use of resources, ecosystem management (including water resource management), environmental hazards and disasters.

All of these issues have local, national, regional and global implications. ILTER's new emphasis on LTSER combined with its established multi-scale structure, based on sites, national networks and regional networks all co-ordinated within a global framework, put it in a powerful position to inform choices, solutions, and decisions across all of these scales.

More on ILTER, including information on sites, research activities, information management and training, can be found on the ILTER website at: www.ilter.edu.net.

Terry Parr, Chair of ILTER (2007–2012)

While LTER does provide us with relevant signals and explanations of change in ecosystems, there is increasing evidence that the traditional LTER approach is limited when it comes to guiding action to conserve ecosystems especially from large scale, often human-induced perturbations resulting in loss of ecosystem services, function and biodiversity. Towards this end, a shift in disciplinary approaches to support a healthy fertilisation of concepts from natural and social sciences is required. This was recognised rather early by American LTER scholars who met in 1998 in Madison, Wisconsin, to seek ways of integrating relevant social science concepts into LTER (Redman et al. 2004). The NSF in its 20-year review of LTER called for collaboration between LTER scholars and social scientists to contribute towards an environmental policy based on a better understanding of the reciprocal effects of human activities and ecosystems (NSF 2002).

Hereafter, an increasing number of scholars began to emphasise "coupled socio-ecological systems" as their unit of analysis for understanding processes of global environmental change as opposed to studying ecological and social systems in isolation. It was argued that if classical LTER is to contribute to finding solutions to socio-ecological problems (that is, sustainability problems), it must go beyond its focus of monitoring ecosystem processes and patterns to include an understanding of socioeconomic activities that actively use and change ecosystems (Redman et al. 2004; Haberl et al. 2006; Singh et al. 2010; Mirtl 2010; Collins et al. 2011). An influential article in *Science* (Liu et al. 2007) pointed to two problems underlying the difficulty in understanding socio-ecological systems as, first, the traditional separation of ecological science and social science and second, the need for incorporation of a complex adaptive systems view especially since work in this area has been more theoretical than empirical. Under the banner of "coupled human and natural systems," the US National Science Foundation has provided substantial funding for cross-cutting socio-ecological projects since 2001 for which an integrated, quantitative, systems-level method of inquiry is essential (NSF 2011a).

The diffusion of the US LTER concept to Europe in 2003 also entailed a practical problem.[2] Europe, with its high population density and long history of intensive land use meant that very few areas were available that could be classified as natural or relatively unaffected by human activities. Thus, it was logical to place the human utilisation of nature as an important aspect of the European LTER approach. As such, in 2003, Long-Term Socio-Ecological Research (LTSER) was promulgated as an integral part of European LTER, where the "socio" component became synonymous with the human dimensions of environmental change and sustainability research (Mirtl et al. 2009). LTSER requires an understanding of society-nature interactions at multiple scales and of the cumulative effects of each to illuminate emergent properties that can shed light on broader approaches to global environmental change. In undertaking LTSER studies, scientists are confronted with a complex interplay of various ecosystem and societal dynamics. Important insights are gained when these dynamics are captured and analysed over long periods of time and their trajectories mapped. Although this entails considerable challenges and efforts (conceptually and methodologically) in dealing with this complexity, it has important benefits.

---

[2] Given European fragmentation and heterogeneity, it is not easy to specify when exactly the LTER concept was introduced in Europe. Individual researchers and countries were in contact with the emerging LTER in the USA from early on. In the late 1990s, Hungary and the Czech Republic, for example, already had their LTER Sites, networks and a 'Central European regional group'. However, since the first European LTER meeting took place in Copenhagen in 2003, this can be said to be the year when LTER was introduced on a European-wide scale (Mirtl et al. 2009).

The LTSER concept, being an integral part of European LTER, was also an opportunity for Europe to take up the challenge of creating a scientific basis for the sustainable management of ecosystems and of sustainable development in general. In 2007 a trans-European network – LTER-Europe – was founded as part of a project funded under the European Commission's Sixth Framework Programme, ALTER-Net.[3] The in-situ network LTER-Europe currently covers 21 member countries, 400 LTER Sites and 31 so-called LTSER Platforms (Mirtl et al., Chap. 17 in this volume). With a size of 100–10,000 km² or more, LTSER Platforms are extensive landscapes characterised by manifold interactions between society and nature, ranging from strict conservation areas to intensively used ones (Mirtl et al. 2010). LTSER Platforms were first started in Austria, Finland, Hungary, Germany, Romania and Spain, providing experiences that could be fed back into conceptual work within ALTER-Net (Mirtl and Krauze 2007).

By 2010, progress and bottlenecks in setting up LTSER Platforms across Europe were reported (Mirtl 2010). It became clear that LTSER Platforms should cover not only major biogeographic regions, but rather "socio-ecological" regions with varying economic conditions and population densities. This was seen as a precondition for the representativeness of LTSER Platforms in terms of socio-ecological systems at a European scale. Working towards this target, a stratification of European Socio-Ecological Regions (SER) was performed (Metzger et al. 2010) and the coverage by LTSER Platforms tested. With regard to their large sizes, it was found that most LTSER Platforms are comprised of more than one socio-ecological regions. Therefore, of the 48 SERs identified in Europe, 43 of them occur in at least one Platform.

In recent years, efforts to strengthen and bring together the LTSER community in the United States and Europe have become more visible. The US-EU LTER conference held in July 2003 in Motz, France, was a milestone in this process, stimulated largely by preparatory efforts for the LTER-Europe initiative. At this conference, "Multifunctional Research Platforms" (MFRPs) were presented as conceptual predecessors of LTSER Platforms and strong overlaps – specifically with larger and urban US sites – were identified (LTER Europe 2011a).

In February 2005, a joint workshop of pioneering scholars from North America and Europe from various academic backgrounds met at a workshop held at the Institute of Social Ecology in Vienna (Austria) to discuss the integration of social science concepts into LTER. Results were published in a paper (Haberl et al. 2006) that coined the term 'Long-Term Socio-Ecological Research' as well as the acronym

---

[3] ALTER-Net – "A Long-Term Ecosystem and Biodiversity Research Network" (http://www.alter-net.info/) was launched in 2004 for a period of five years to create a network (of excellence) comprising prominent European institutions located in 17 countries engaged in long-term ecosystem research. An important goal of this project was to establish synergy in terms of infrastructure and data sharing on biodiversity and ecosystem change at a European level. The network continues, along with a secretariat and a regular summer school. An important product of this project was the formal foundation of the LTER-Europe network (http://www.lter-europe.net/) in 2007 with a strong LTSER Expert Panel.

LTSER (discussed in more detail in the next section). Building on experiences in implementing LTSER Platforms and traditionally strong links between the US-LTER and Israeli scientists, the first LTER Middle East conference was held in 2009 in Aqaba (Jordan), facilitated by experts from Europe and the US (LTER Europe 2011c).

More recently, bringing together active scholars from natural and social sciences alike, the LTSER session in the 2010 Global Land Project Open Science conference (Phoenix, Arizona) was another step in this direction. The discussions at this conference have shown the need for concepts that can improve integration of research from different scientific disciplines, such as the development of conceptual models. They have also highlighted the importance of concepts such as socioeconomic metabolism and ecosystem services for future LTSER (Shibata and Bourgeron 2011; Singh and Haberl 2011), as also discussed in depth in various chapters in this volume. In June 2011, under the auspices of ALTER-Net and LTER-Europe, a LTSER workshop was held in Helsinki (Finland), inviting scholars and practitioners from Europe and North America to discuss the state-of-the-art in LTSER science and practice. Addressing conceptual, methodological, legal and socio-political challenges related to LTSER goals, the conference was another milestone in strengthening the European LTSER community and in seeking synergies between LTSER science and practice towards sustainability (LTER Europe 2011b).

## 1.3 Introducing the Social Dimension in LTER: Conceptual Frameworks

Incorporating the human dimension into LTER has been a difficult task and is a work in progress. In the 10 years following the NSF (2002) report, few scholars have proposed integrative frameworks towards this end. Inspired by the 1998 Madison meeting and several workshops thereafter, Charles Redman and colleagues (2004) in the United States were among the early ones to propose an integrative conceptual framework for including the human dimension into LTER. Drawing on research and concepts from sociology, anthropology and the Resilience Alliance in general and from earlier writings of Machlis et al. (1997) and Burch and DeLuca (1984) in particular, they propose "socio-ecological systems" (SES) as the unit of analysis in LTER studies. In this view, SES is characterised by "(1) a coherent system of biophysical and social factors that regularly interact in a resilient, sustained manner; (2) a system that is defined at several spatial, temporal, and organizational scales, which may be hierarchically linked; (3) a set of critical resources (natural, socioeconomic, and cultural) whose flow and use is regulated by a combination of ecological and social systems; and (4) a perpetually dynamic, complex system with continuous adaptation" (Redman et al. 2004, p. 163).

As an integrative framework for interdisciplinary research on SES, the authors suggest a focus on "interactive" activities that lie at the interface of social and ecological elements such as land-use decisions affecting changes in land cover, land surface and biodiversity, production and consumption systems and on the networks of waste management and disposal. These complex system interactions are in turn affected by "patterns and processes" that also fit within traditional social and ecological sciences. In the case of the former, these would be to describe and monitor changes in demography, economy, technology, culture, institutions and information. Some classic ecological variables concern the patterns and control of primary production, frequency of site disturbances, accumulation and movements of organic, inorganic and sediment materials, and the spatial and temporal distribution of populations in the trophic structure. Finally, the conceptual model must also collect background information on "external framework conditions" both at the biogeophysical level (such as climate change or large-scale land changes) and political and economic levels (such as global market prices, changes in governance and political systems) that affect SES as a whole.

Redman and colleagues propose four meta-questions to investigate long-term dynamics of socio-ecological systems in the context of LTER. These are: "(1) How have past ecological systems and social patterns conditioned current options through legacies and boundary conditions? (2) How do current characteristics of ecological systems in the region under study influence the emerging social patterns and processes? (3) How do current social patterns and processes influence the use and management of ecological resources? (4) How have these interactions changed over time, and what does this mean for future possible states of the SES?" (Redman et al. 2004, p. 166).

The joint workshop of scholars from both North America and Europe held in Vienna in February 2005, as mentioned previously, was a follow-up to these ideas. In this workshop (the outcome of which was a multi-authored paper), LTSER was described as an extension of LTER concerned with the analysis of long-term changes in socio-ecological systems, defined as "complex, integrated systems that emerge through the continuous interaction of human societies with ecosystems" (Haberl et al. 2006, p. 2). The authors identified several challenges resulting from the interdisciplinary nature of the LTSER approach, such as:

- the need to conceptualise the interactive process between society (human systems integrated by communication) and ecosystems (biophysical systems integrated by material and energy flows);
- the challenges resulting from scaling issues, e.g. the problem that natural and administrative boundaries often do not match;
- the difficulty of distinguishing site-specific and general dynamics; and
- the need to integrate not only explanatory and predictive modelling, but also monitoring, reconstructions and empirical approaches with 'soft knowledge' from the humanities.

The paper discusses several conceptual requirements of LTSER, including the need for sound research design grounded in epistemologically sound concepts of society-nature interaction and the need to analyse socio-ecological transitions, i.e. fundamental changes in the relation between natural and social systems (Fischer-Kowalski and Haberl 2007; Haberl et al. 2011). The authors also argued that participatory approaches are needed to involve stakeholders in the research process. They identified four themes of LTSER:

- Socio-ecological metabolism – the energy and material flows in socio-ecological systems;
- Land use and landscapes – the analysis of integrated land systems that are simultaneously shaped by natural and socioeconomic drivers;
- Governance and decision-making – the institutional aspects of human use of natural resources such as energy, materials, water or land;
- Communication, knowledge and transdisciplinarity – the role of self-reflexivity in LTSER projects that allows them to be placed properly within their respective contexts.

A subsequent conceptual update in LTSER was published by Simron J. Singh and colleagues (2010) in a Springer volume edited by German scholars synthesising the state-of-the-art on long-term ecosystem research in Europe (Müller et al. 2010). Re-emphasising the role of LTSER in seeking solutions to societal problems, the chapter highlights relevant theoretical and conceptual approaches to society-nature interactions as found in environmental sociology, ecological anthropology, ecological economics, participative modelling and decision support science. The discussion of the structure and dynamics of socio-ecological systems addresses the "long-term" dimension of LTSER, where the authors compare the natural science approach of "adaptive cycles" (Holling 1986) with that of the more interdisciplinary "socio-metabolic transitions" notion (Fischer-Kowalski and Haberl 2007). The authors propose "socio-economic metabolism" (Fischer-Kowalski and Weisz 1999) along with its operational tool, the "MEFA framework – Material and Energy Flow Accounting" (Haberl et al. 2004) as promising for the LTSER community. The paper ends with a case study of the Eisenwurzen LTSER Platform in Austria to illustrate an innovative effort where agent-based modelling was integrated with a biophysical stocks and flows module for participatory decision making at the regional level (see also Peterseil et al., Chap. 19 in this volume and Gaube and Haberl, Chap. 3 in this volume).

More recently, Scott Collins and colleagues (2011) in the United States have proposed a new integrated conceptual framework for long-term socio-ecological research. Building on the European DPSIR (Drivers – Pressures – State – Impact – Response) model proposed by the European Environmental Agency (EEA 1998), the authors include an explicit focus on ecosystem services (MEA 2005). Termed the "Press-Pulse Dynamics" (PPD) framework by the authors, this links the social domain (characterised by socioeconomic activities) with the biophysical one (characterised by ecosystem structure and functions) under the premise that the dynamics within the biophysical domain are driven either by pulse events (such as floods, fire, and storms) or by press events that are sustained and chronic (such as climate change, sea-level rise and nutrient loading). In both instances, these events might be natural or human induced, or a combination of both. Over time, presses, pulses and press-pulse interactions alter the relationship between the biotic structure and ecosystem functioning, which in turn affects essential services humans obtain from ecosystems. The authors argue that the PPD framework provides a roadmap for an interdisciplinary approach to long-term socio-ecological research to help build a knowledge base for addressing current and future environmental challenges (see also Grove et al., Chap. 16 in this volume; Grimm et al., Chap. 10 in this volume; Mirtl et al., Chap. 17 in this volume).

## 1.4 The Contributions in This Volume

The current volume is a rich and varied collection of 23 contributions (co-) authored by nearly 70 scholars engaged in some way with the field of LTSER. To aid reading as well as to present the chapters within a meaningful structure, the book is organised in three thematic parts. The first is intended to provide an overview of concepts, methods and disciplinary linkages that we consider relevant for LTSER. The second part reports on a number of case studies that illustrate LTSER applications across ecosystems, time and space. The final part is an assemblage of valuable experiences and challenges confronted by colleagues in setting up LTSER research projects and Platforms, both in the United States and Europe, with an apparent bias towards the latter. A clear categorisation of chapters is often not possible. The placement of chapters into sections was based on the key orientation of their contents and on the main message intended by the authors. Eventually, the volume must be seen as a single piece of work that aims to crystallise the state-of-the-art in LTSER across time and space – as the title of this volume suggests (Fig. 1.2).

Part I begins with the contribution by Haberl and colleagues (*Chap. 2*) in which they review methods to analyse the 'metabolism' of socioeconomic systems consistently across space and time. Current sustainability problems such as climate change or biodiversity loss are seen to be closely related with socioeconomic use of natural resources, such as materials, energy or land. Pioneered several decades ago, material and energy flow analysis has developed an increasingly standardised tool kit that currently provides basic data for scientific fields such as social ecology, industrial ecology and ecological economics, as well as resource-use policies in various domains. The chapter extends this approach to the broader concept of socio-ecological metabolism, which additionally considers changes in ecological material flows related to socioeconomic metabolism – thereby establishing a link to global biogeochemical cycle research. It discusses a prominent indicator, the human appropriation of net primary production, or HANPP, which has gained importance in that field in recent years. The chapter applies the socio-ecological metabolism approach to the case of the Austrian trajectory from 1830 to 2000, demonstrating the power of a consistent stock-flow approach in analysing Austria's long-term carbon metabolism. It concludes by showing how the complex interactions between social, economic, institutional and ecological components in system transitions can be analysed using integrated socio-ecological models.

*Chapter 3* by Gaube and Haberl takes up the issue of integrated socio-ecological modelling, demonstrating how models can be developed and used in participatory

# 1 Introduction

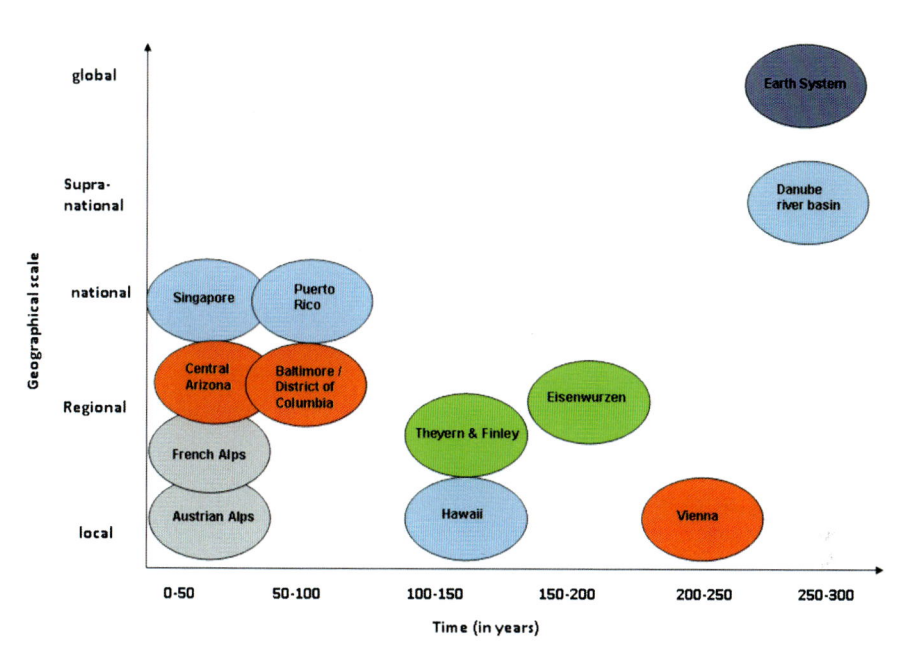

**Fig. 1.2** Case studies in this volume across time and space

processes with intensive stakeholder involvement. They discuss insights gained from combining an agent-based decision model with land-use maps and an integrated socio-ecological stock-flow component. The model was developed in a rural Austrian village based on intensive discussions with farmers, the local administration and many other experts and stakeholders. The model is capable of simulating decisions of local actors such as farmers or the municipal administration under different assumptions on social, economic or political framework conditions (e.g. agricultural subsidies and product prices) as well as on local decisions (e.g. cooperation, direct marketing). The outcomes can be analysed in terms of land-use change maps as well as changes in carbon and nitrogen flows and greenhouse gas emissions. One of the main conclusions in this chapter is that participatory model development, if done well, can support learning and empowerment of local actors and simultaneously generate relevant and stimulating scientific insights on the functioning of the interactions between society and ecosystems – thereby supporting LTSER.

*Chapter 4* by Fischer-Kowalski and colleagues presents a model capable of analysing constraints to historical urbanisation processes resulting from transport limitations. The model is useful for analysing how the spatial configurations of resources and their appropriation through different socioeconomic strategies (hunting and gathering, agriculture) may have interacted in enabling and constraining spatial patterns of human activity in prehistoric and historic times. Combining data on socioeconomic metabolism in different socio-ecological regimes with information on transport infrastructures and technologies available to those societies shows how energetic needs and limitations interacted in determining settlement patterns, e.g. villages and regional centres, as well as centre-hinterland relations. This research theme, although currently not in the mainstream of LTSER, is highly important for a proper understanding of society-ecosystem interactions across space and time, and would therefore deserve more attention in future LTSER programmes.

Winiwarter and colleagues in *Chap. 5* take the Danube, Europe's second longest river and the world's most international river basin, as their case to demonstrate how environmental historians can contribute to LTSER. Going back to early modern and even medieval times, they tell a story of fundamental changes of the river Danube. The chapter uses and introduces the concept of "socio-natural sites" to show how these fundamental changes are together a consequence of past human practices and biophysical arrangements. The authors emphasise the decisive role of energy in this process of transformation and with their background in history, they argue that a long-term perspective, covering at least centuries, is urgently needed to improve understanding of what it might mean to manage complex socio-natural sites in a post-fossil fuel age.

Dirnböck and colleagues in *Chap. 6* explore the question of which spatial and temporal scales are most relevant for assessing and monitoring the sustainability of socio-ecological systems. Their findings are based on an earlier meta-analysis of 18 biodiversity case studies. They identify a severe mismatch between three distinct scales of observation and action that are currently major obstacles in nature conservation and in reducing biodiversity loss. These are (a) scales on which ecological processes are observed and analysed, (b) the scales on which these processes are managed, and (c) the scale on which environmental policies are implemented. To address this, the authors suggest 'landscape' as an appropriate unit of analysis for LTSER, allowing for interdisciplinary research since the landscape's structure and processes is an outcome of the interplay of nature and society. The chapter reviews a set of methods useful for addressing scale mismatches within socio-ecological systems, the most promising being modelling.

Summarising the work done since 1965, Stephen Boyden in *Chap. 7* introduces the "biohistorical paradigm" to demonstrate how this can inform LTSER and how it can aid in transitioning to a sustainable society. Tracing the history of humans in biological and historical perspective, Boyden suggests that the emergence of human culture with a key ability to invent, learn, communicate and store a symbolic spoken language allowed the exchange and storage of useful information about the environment. This ability, while well adapted for sustainable use of nature during most of human history, has often resulted under conditions of modern civilisation in "cultural maladaptations", damaging both humans and other forms of life. According to Boyden, the understanding of biohistory is an essential prerequisite for the future wellbeing of humankind, in what he terms a "biosensitive society".

Karl Zimmerer in *Chap. 8* makes a compelling case for how geography can enrich LTSER, drawing upon human-environment and nature-society (HE-NS) geography. The themes of HE-NS geography lie at the interface of human dimensions of global change, cutting across six areas as follows: (a) Coupled Human-Environment Interactions (b) Sustainability Science, Social-Ecological Adaptive

Capacity, and Vulnerability (c) Land-Use and Land-Cover Change/Land Change Science (d) Environmental Governance and Political Ecology (e) Environmental Landscape History and Ideas, and (f) Environmental Scientific Concepts in Models, Management, and Policy. Zimmerer concludes that "geography and LTSER share a significant degree of general similarity that promises ample and potentially vital opportunities for future crossover, collaboration and shared directions."

In Chap. 9, Ted Gragson highlights the importance of anthropology as a crucial link between the social and biological sciences. Despite the increasing importance of interdisciplinary research in dealing with contemporary environmental challenges, Gragson argues that there is still the demand for place-based, long-term, cross-scale and comparative research that anthropology can provide. Anthropology has long been a problem-oriented discipline interested in policy relevant issues such as poverty and underdevelopment. According to Gragson, anthropology "challenges us to understand what it means to be human through the study of culture in place as well as by comparison to other cultures." Going beyond the savage, anthropology's usefulness to contemporary LTSER is seen in its extensive research on institutions and governance of common pool resources, as well as in the sub-discipline of cognitive anthropology that provides insights into how the organisation of knowledge and the perception of environment are linked in complex ways. The chapter concludes with two case studies from ongoing U.S. research in the Coweeta LTER Project in Southern Appalachia that relies on anthropology.

Part II begins with the contribution of Nancy Grimm and colleagues in the United States where urban LTER sites are focal areas for the analysis of socio-ecological (or human-environment) interactions and sustainability. *Chap. 10* reviews research on urban sustainability carried out in the Central Arizona Phoenix (CAP) LTER programme in recent years. This work is based on the premise that cities can be conceptualised as socio-ecological systems (SES) as a way to improve understanding of their sustainability problems. Drawing on a huge array of empirical work, modelling and data analysis, as well as GIS work, the chapter convincingly demonstrates the power of the Press-Pulse Dynamics (PPD) framework (Collins et al. 2011) to reveal the manifold linkages between environmental conditions – in this case, the arid environment features prominently – and human dynamics in creating both biophysical as well as socioeconomic outcomes. Phoenix, Arizona, is certainly a challenging example owing to its rapid development and harsh environment – and is therefore highly relevant for sustainability analysis and hence for LTSER.

In another example of an urban system, Fridolin Krausmann in *Chap. 11* provides a socio-ecological history of Vienna from the year 1800 to 2006 to better characterise city-hinterland relations. In the context of sustainability, insights into city-hinterland relationships are useful to find ways to "minimise inefficient patterns of resource supply and use as well as negative environmental impacts of urban consumption in distant regions, where they are not visible to the urban consumer." Drawing upon the concept of social metabolism and material and energy flow analysis (see also Chap. 2), Krausmann argues that urbanisation is intrinsically linked to the emergence of fossil fuel-based energy systems. In the case of Vienna, the transition from biomass to fossil fuels characterised the shifting pattern of relationships between cities and the hinterlands that supply urban areas with energy and food resources. However, growth in urban resource use did not cause an equal growth in the spatial imprint of urban consumption. The results point out that the size and spatial location of the resource-supplying hinterland is the combined result of various dynamic processes, including transport technology and agricultural productivity.

Geoff Cunfer and Fridolin Krausmann in *Chap. 12* compare using a long-term perspective of environmental history and the dynamics of agricultural change between the Old World (represented by the village of Theyern in Austria) and the New World (represented by Finley Township in Kansas, U.S.). The narrative – woven into the historical context of a migration wave from Europe to the United States during the nineteenth century – traces the destiny of the Thir family that moved out of Austria to make Kansas their new home in 1884. Recognising that agriculture is a coupled human-environment system, the chapter draws on the concept of social metabolism and derived indicators of material and energy flow accounting (see also Chap. 2) to ask how the farming system that immigrants left behind compares with what they found (and created) on the Great Plains frontier. Illustrated by a wealth of data and indicators of material and energy flows as well as land and labour productivity, this comparative research provides a socio-ecological interpretation of agricultural history that is often overlooked or downplayed by economists. In effect, the authors reveal the significance of environmental history in understanding long-term dynamics of coupled human-environment systems – from a biomass-based (tightly linking soil, plants, animals and people into a single complex and highly evolved system) to a fossil fuel-driven regime that is highly fragmented, vulnerable and unsustainable.

In *Chap. 13*, Simone Gingrich and colleagues use a sociometabolic approach to reconstruct material and energy flows through agricultural systems from the late nineteenth to the turn of the twenty-first century. By comparing two different locales in today's LTSER Platform Eisenwurzen, they show that the fundamental transformation of agricultural systems, in particular after World War II, had effects on local scales. The second aim is to develop a conceptual framework for LTSER that allows cooperation between natural scientists and humanities (including history, some strands of anthropology and other disciplines interested in the cultural dimensions of society-nature interactions) to be strengthened. For that purpose, the authors adapt the concept of "socio-natural sites" with its focus on human practices and biophysical arrangements to critically discuss their own empirical results and to raise questions for future collaborative research. The chapter is a first step towards a LTSER that consequently combines a birds-eye or systems perspective with a close-up view of (historical) actors, their motives and perceptions on changes in agricultural systems and landscapes.

Marian Chertow and colleagues in *Chap. 14* provide cross-cutting reflections on human-nature interactions based on the examination of four islands: Singapore, Puerto Rico, Hawaii and O'ahu Islands (the last two in the state of Hawaii, USA). Isolation, vulnerability to disruption, and constraints on the availability of natural resources add urgency to island sustainability questions with limited solution sets.

Over the course of the twentieth century, each of these islands became heavily dependent on imports such as water, food, and/or fuel to sustain basic human needs and modern economic functions. Within the last decade, each has consciously sought to restructure its socio-ecological configuration by using more locally available resources in one or more of its metabolic linkages. This pattern has the potential to reconnect island economies with their natural systems while simultaneously enhancing relationships and increasing resilience.

While LTSER is often seen as a primarily place-based endeavour, development of its full potential requires the ability to draw general conclusions relevant for larger areas, from nation states to the globe. *Chap. 15*, by Fridolin Krausmann and Marina Fischer-Kowalski, demonstrates how the analysis of resource use over long periods of time can improve understanding of global sustainability problems. Based on the notion of three socio-metabolic regimes: hunter-gatherers, agrarian, and then industrial society, this chapter synthesises a vast array of empirical research on socioeconomic material and energy flows to discuss sustainability challenges stemming from the currently ongoing agrarian-industrial transition (see also Fischer-Kowalski and Haberl 2007). The authors convincingly show that the current "industrial metabolism" prevailing in the world's industrialised centres cannot be replicated to the non-industrialised world – thereby raising the question of how a transition to more sustainable, yet equitable patterns of resource use could be achieved globally. It provides a powerful example of how the synthesis of insights from case studies, be they regional (as in most of LTSER), national or even supranational, can provide a fresh perspective to old problems and shed light on options for, but also constraints of, widely debated policy proposals.

Part III begins with the contribution of Morgan Grove and colleagues (*Chap. 16*) in which they share their experience building a LTSER with the Baltimore Ecosystem Study, one of two US-designated urban LTER sites. They use an architectural metaphor to describe how to build site context, how to build a durable structure, and how to build process and maintenance into the project. They discuss several research tools that help to shift from an orientation of ecology *in* cities to one of ecology *of* cities that is much more focused on the biophysical and human components of the broader system and how these interact. The authors describe the major research tools to facilitate this shift to an "Ecology of Cities" including press-pulse dynamics and patch dynamics. Finally, the chapter describes the four legs of the Carpenter table and how these have been applied in Baltimore, specifically, long-term monitoring, modelling, and conducting experiments and comparative analyses.

Michael Mirtl and colleagues in *Chap. 17* report on the conceptualisation, design and implementation of Long-Term Socio-Ecological Research (LTSER) in the 31 LTSER Platforms in Europe. LTSER Platforms are hot spot areas for interdisciplinary research emphasising the integration of the four pillars of LTER-Europe´s science strategy (systems approach, process-oriented, long-term and site-based). By applying acknowledged conceptual models like ISSE (see Chap. 16) to LTSER Platforms, "regional socio-ecological profiles" can be generated. Such profiles distil the multiple social and ecological variables and their complex interactions operating within the Platform and identify key topics of study. The experiences gathered over 6 years of practical work in the Austrian Eisenwurzen LTSER Platform (Chap. 19), the Central French Alps LTSER Platform (Chap. 20), the Austrian Tyrolean Alps LTSER Platform (Chap. 21) and in the Finnish LTER network (Chap. 18) are used to assess weaknesses and strengths of two LTSER platform implementation strategies (evolutionary vs. strategically managed) and to derive recommendations for the future. The chapter represents the close of the first substantive loop of LTSER research that began in 2003 from conceptualisation to implementation and, through the introspective analysis here, a reconsideration of the central concepts.

Eeva Furman and Taru Peltola in *Chap. 18* outline internal and external factors that catalysed the adoption of LTSER in Finland. The authors analyse the initiation phase, the very first steps taken by the Platforms and the challenges faced during this period. The strategic decisions to undertake a demanding bottom-up process and to adopt a strong interdisciplinary approach to the development of the FinLTSER network are seen to be rewarding. The authors feel that though problem-oriented environmental research is highly acknowledged by many funding agencies, there exists an epistemological and methodological friction when moving from research with a short-term mono-disciplinary focus to a long-term one with emphasis on interdisciplinarity.

Johannes Peterseil and colleagues in *Chap. 19* synthesise their 7 years of experience with the Austrian Eisenwurzen LTSER Platform, one of the flagship projects of the European LTSER Platform concept. The authors describe the challenges of the implementation process and its current management structure. The Eisenwurzen LTSER Platform provides a framework for transdisciplinary socio-ecological research in the region, based on a Memorandum of Understanding (MoU) among 32 parties. The authors stress the importance of the time factor in the successful operation of a LTSER Platform: expectations held by the stakeholders are often too high and unrealistic, given the lengths of the cycle from identifying an appropriate research question, writing the proposal, obtaining funding and undertaking research and analysis to finally translating the outcomes for management and policy. Such a process may take from 5 to 10 years or more, but the longer stakeholders have to wait for research outcomes and anticipated benefits, the higher the risk that their interest and financial support will be lost and the greater the effort needed to keep them motivated.

Sandra Lavorel and colleagues in *Chap. 20* report on the Central French Alps LTSER Platform with a focus on coupled dynamics of alpine ecosystems. The French LTSER Platform has fostered three important advances: (a) Long-term data consolidation and sharing; (b) establishing interdisciplinary research projects; and (c) engaging stakeholders in transdisciplinary projects, including climate change adaptation in the French Alps. In line with other experiences in establishing LTSER Platforms, the authors highlight the importance of time and patience in such a process, but find it to be rewarding. In only a short time since the creation of the French LTSER Platform, individuals and partner institutions concluded that the platform structure provided them with an important framework to formalise already ongoing collaborations across different disciplines. More recently, interdisciplinary research projects in which social sciences are truly incorporated into the design of the research framework are evolving. The issue of climate change in particular has emerged as a common theme fostering transdisciplinary collaboration.

In *Chap. 21*, Ulrike Tappeiner and colleagues unravel the history, future perspectives and challenges of setting up LTSER research in the Austrian (Tyrolean) Alps, the second Austrian LTSER Platform. Sharing some aspects of the French experience (Chap. 20), the authors emphasise policy-oriented research and monitoring of sensitive mountain habitats highly vulnerable to changes in land use and climate, which are occurring at increasingly rapid rates in the Alps. Long-term ecological research in the Tyrolean Alps has a long tradition going back more than a century, with pioneering research on plant geography, alpine botany and chemical ecology. By comparison, socio-ecological and socioeconomic studies are fewer and more recent, beginning in the 1970s with the launch of UNESCO's Man and [the] Biosphere Programme in the region. The authors note some of the challenges faced in implementing LTSER: delivering sufficient funding, defining a coherent LTSER research agenda, establishing an appropriate management structure for the platform, forming interdisciplinary research teams, and installing transdisciplinary research processes across scientists, stakeholders and the local population.

The final chapter, by Willi Haas and colleagues, explores in depth the transdisciplinary challenge within LTSER. Drawing on several years of research experience in the Gurgler Kamm Biosphere Reserve, part of the Tyrolean LTSER Platform in Austria, the authors present outcomes from integrated monitoring and sustainability assessment. They suggest that in the context of evaluating sustainability, monitoring efforts should be concentrated not only on natural processes, but also on those elements of the social sphere that have a direct causal effect on the ecosystem. In particular, the emphasis should be on understanding the dynamics of society-nature interactions, the results of which will have to be assessed by the relevant stakeholders from time to time, including management in the light of policy goals and targets. Thus, the goal of the research project outlined in this chapter was not only scientific, but has high societal relevance insofar as the investigation would have to take into account past and current socioeconomic trends as well as future development options compatible with regional sustainability. Although this is easier said than done, the chapter provides valuable insights into the challenges of such a research process, both in setting up an interdisciplinary research team and in engaging stakeholders.

We hope that these varied and rich contributions – spelled out in more detail in the following pages – will allow us to take stock of the emerging field of LTSER both in the United States and in Europe and will also inspire future collaborative research elsewhere. The environmental and sustainability concerns of our times are indeed pressing and there is an urgent need to act with enhanced clarity and proficiency. We hope that this volume will give the reader a greater sense of where we are and what still needs to be done to engage in and make meaning from long-term, place-based and cross-disciplinary engagements with socio-ecological systems.

# References

Adams, C. C. (1935). The relation of general ecology to human ecology. *Ecology, 16*, 316–335.
Beus, C. E. (1993). Sociology, human ecology, and ecology. *Advances in Human Ecology, 2*, 93–132.
Boas, F. (1896). The limitations of the comparative method of anthropology. *Science, 4*, 901–908.
Boas, F. (Ed.). (1911). *The mind of primitive man. A course of lectures delivered before the Lowell Institute, Boston, Mass, and the National University of Mexico, 1910–1911*. New York: Macmillan.
Bodley, J. H. (Ed.). (2008). *Anthropology and contemporary human problems* (5th ed.). Plymouth: AltMira Press.
Boyden, S. (Ed.). (1992). *Biohistory: The interplay between human society and the biosphere. Past and present*. Paris: United Nations Educational, Scientific and Cultural Organization (UNESCO).
Boyden, S. (1993). The human component of ecosystems. In M. J. McDonnell & S. T. A. Pickett (Eds.), *Humans as components of ecosystems, the ecology of subtle human effects and populated areas* (pp. 72–78). New York: Springer.
Boyden, S. (2001). Nature, society, history and global change. *Innovation – The European Journal of Social Sciences, 14*, 103–116.
Burch, W. R., Jr., & DeLuca, D. R. (Eds.). (1984). *Measuring the social impact of natural resource policies*. Albuquerque: University of New Mexico Press.
Catton, W. R., Jr., & Dunlap, R. E. (1978). Environmental sociology: A new paradigm. *The American Sociologist, 13*, 41–49.
Chertow, M. (2001). The IPAT equation and its variants: Changing views of technology and environmental impact. *Journal of Industrial Ecology, 4*(4), 13–29.
Clark, W. C., Crutzen, P. J., & Schellnhuber, H.-J. (2004). Science for global sustainability. In H.-J. Schellnhuber, P. J. Crutzen, W. C. Clark, M. Claussen, & H. Held (Eds.), *Earth system analysis for sustainability* (Report of the 91st Dahlem workshop, pp. 1–28). Cambridge, MA/London: MIT Press.
Collins, S. L., Carpenter, S. R., Swinton, S. M., Orenstein, D. E., Childers, D. L., Gragson, T. L., Grimm, N. B., Grove, J. M., Harlan, J. P., Kaye, J. P., Knapp, A. K., Kofinas, G. P., Magnuson, J. J., McDowell, W. H., Melack, J. M., Ogden, L. A., Robertson, G. P., Smith, M. D., & Whitmer, A. C. (2011). An integrated conceptual framework for long-term social-ecological research. *Frontiers in Ecology and the Environment, 9*, 351–357.
Croll, E., & Parkin, D. (1992). Anthropology, the environment and development. In E. Croll & D. Parkin (Eds.), *Bush base: Forest farm*. London: Routledge.
Crumley, C. L. (Ed.). (1993). *Historical ecology: Cultural knowledge and changing landscapes*. Santa Fe: School of American Research Press.
Darling, F. F. (1956). The ecology of man. *American Scholar, 25*, 38–46.
Descola, P., & Palsson, G. (Eds.). (1996). *Nature and society: Anthropological perspectives*. London: Routledge.
Dietz, T., & Rosa, E. A. (1994). Rethinking the environmental impacts of population, affluence and technology. *Human Ecology Review, 1*, 277–300.
Dietz, T., Rosa, E. A., & York, R. (2007). Driving the human ecological footprint. *Frontiers in Ecology and the Environment, 5*, 13–18.
Dunlap, R. E., & Catton, W. R., Jr. (1994). Struggling with human exemptionalism: The rise, decline and revitalizaton of environmental sociology. *The American Sociologist, 25*, 5–29.
EEA. (1998). *Europe´s environment, the dobrís assessment*. Luxembourg: European Environment Agency, Office for Official Publications of the European Communities (on CD).
Ehrlich, P. R., & Ehrlich, A. H. (Eds.). (1970). *Population, resources, environment: Issues in human ecology*. San Francisco: Freeman.
Ehrlich, P. R., Ehrlich, A. H., & Holdren, J. H. (Eds.). (1973). *Human ecology, problems and solutions*. San Francisco: Freeman.
Ellen, R. (Ed.). (1982). *Environment, subsistence and system. The ecology of small-scale social formations*. Cambridge: Cambridge University Press.

Erkman, S. (1997). Industrial ecology: An historical view. *Journal of Cleaner Production, 5*, 1–10.
Fairhead, J., & Leach, M. (Eds.). (1996). *Misreading the African landscape: Society and ecology in a forest-savanna mosaic*. Cambridge: Cambridge University Press.
Fischer-Kowalski, M., & Amann, C. (2001). Beyond IPAT and Kuznets curves: Globalization as a vital factor in analysing the environmental impact of socio-economic metabolism. *Population and Environment, 23*, 7–47.
Fischer-Kowalski, M., & Haberl, H. (Eds.). (2007). *Socioecological transitions and global change: Trajectories of social metabolism and land use*. Cheltenham/Northhampton: Edward Elgar.
Fischer-Kowalski, M., & Weisz, H. (1999). Society as a hybrid between material and symbolic realms. Toward a theoretical framework of society-nature interaction. *Advances in Human Ecology, 8*, 215–251.
Forde, C. D. (Ed.). (1934). *Habitat, economy, society*. London: Methuen.
Graedel, T. E., & Allenby, B. (Eds.). (1995). *Industrial ecology*. Upper Saddle River: Prentice-Hall.
Greenberg, J. B., & Park, T. K. (1994). Political ecology. *Journal of Political Ecology, 1*, 1–12.
Grossman, L. (1977). Man-environment relations in anthropology and geography. *Annals of the Association of American Geographers, 67*, 126–144.
Haberl, H., & Krausmann, F. (2001). Changes in population, affluence and environmental pressures during industrialization. The case of Austria 1830–1995. *Population and Environment, 23*, 49–69.
Haberl, H., Fischer-Kowalski, M., Krausmann, F., Weisz, H., & Winiwarter, V. (2004). Progress towards sustainability? What the conceptual framework of material and energy flow accounting (MEFA) can offer. *Land Use Policy, 21*, 199–213.
Haberl, H., Winiwarter, V., Andersson, K., Ayres, R. U., Boone, C. G., Castillio, A., Cunfer, G., Fischer-Kowalski, M., Freudenburg, W. R., Furman, E., Kaufmann, R., Krausmann, F., Langthaler, E., Lotze-Campen, H., Mirtl, M., Redman, C. A., Reenberg, A., Wardell, A.D., Warr, B., & Zechmeister, H. (2006). From LTER to LTSER: Conceptualizing the socio-economic dimension of long-term socio-ecological research. *Ecology and Society, 11*, 13. (Online), http://www.ecologyandsociety.org/vol11/iss2/art13/
Haberl, H., Fischer-Kowalski, M., Krausmann, F., Martinez-Alier, J., & Winiwarter, V. (2011). A socio-metabolic transition towards sustainability? Challenges for another great transformation. *Sustainable Development, 19*, 1–14.
Holdren, J. P., & Ehrlich, P. R. (1974). Human population and the global environment. Population growth, rising per capita material consumption, and disruptive technologies have made civilization a global ecological force. *American Scientist, 62*, 282–292.
Holling, C. S. (1986). The resilience of terrestrial ecosystems: Local surprise and global change. In W. C. Clark & R. E. Munn (Eds.), *Sustainable development of the biosphere* (pp. 292–320). Cambridge: Cambridge University Press.
Hornborg, A., & Crumley, C. (Eds.). (2006). *The world system and the Earth system: Global socioenvironmental change and sustainability since the Neolithic*. London: Walnut.
Hornborg, A., McNeill, J. R., & Martinez-Alier, J. (Eds.). (2007). *Rethinking environmental history: World system history and global environmental change*. Lanham: Alta Mira Press.
Kates, R. W., Clark, W. C., Corell, R., Hall, J. M., Jaeger, C. C., Lowe, I., McCarthy, J. J., Schellnhuber, H. J., Bolin, B., Dickson, N. M., Faucheux, S., Gallopin, G. C., Grübler, A., Huntley, B., Jäger, J., Jodha, N. S., Kasperson, R. E., Mabogunje, A., Matson, P. A., Mooney, P., Mooney, H. A., Moore, B., III, O'Riordan, T., & Svedin, U. (2001). Environment and development: Sustainability science. *Science, 292*, 641–642.
Kroeber, A. L. (1917). The superorganic. *American Anthropologist, 19*, 163–213.
Likens, G. E. (1997). Preface. In M. J. McDonnell & S. T. A. Pickett (Eds.), *Humans as components of ecosystems, the ecology of subtle human effects and populated areas* (p. xi). New York: Springer.
Liu, J. G., Dietz, T., Carpenter, S. R., Alberti, M., Folke, C., Moran, E., Pell, A. N., Deadman, P., Kratz, T., Lubchenco, J., Ostrom, E., Ouyang, Z., Provencher, W., Redman, C. L., Schneider, S. H., & Taylor, W. W. (2007). Complexity of coupled human and natural systems. *Science, 317*, 1513–1516.

LTER Europe. (2011a). *EU-US-LTER conference in Motz/France*. http://www.lter-europe.net/events/european-american-workshop-ltser. Accessed 19 Dec 2011.

LTER Europe. (2011b). *Long term socio-ecological research: What do we know from science and practice?* http://www.lter-europe.net/events/lter-events-links/long-term-socio-ecological-research-what-do-we-know-from-science-and-practice. Accessed 19 Dec 2011.

LTER Europe. (2011c). *Middle East LTER startup meeting in Aqaba/Jordan*. http://www.lter-europe.net/events/middle-east-lter-start-up. Accessed 19 Dec 2011.

Machlis, G. E., Force, J. E., & Burch, W. R., Jr. (1997). The human ecosystem, part I, the human ecosystem as an organizing concept in ecosystem management. *Society and Natural Resources, 10*, 347–367.

Martinez-Alier, J. (Ed.). (1987). *Ecological economics. Energy, environment and society*. Oxford: Blackwell.

Martinez-Alier, J., Kallis, G., Veuthey, S., Walter, M., & Temper, L. (2010). Social metabolism, ecological distribution conflicts, and valuation languages. *Ecological Economics, 70*, 153–158.

McDonnell, M. J., & Pickett, S. T. A. (Eds.). (1997). *Humans as components of ecosystems, the ecology of subtle human effects and populated areas*. New York: Springer.

McKenzie, R. D. (1926). The scope of human ecology. In American Sociological Society (Ed.), *20th Annual meeting, 1925, papers and proceedings* (Vol. 20, pp. 141–154). Washington, DC: American Sociological Society.

McNeill, J. R. (2003). Observations on the nature and culture of environmental history. *History and Theory, 42*, 5–43.

Metzger, M. J., Bunce, R. G. H., van Eupen, M., & Mirtl, M. (2010). An assessment of long term ecosystem research activities across European socio-ecological gradients. *Journal of Environmental Management, 91*, 1357–1365.

Millennium Ecosystem Assessment (Ed.). (2005). *Ecosystems and human well-being, volume 1: Current state and trends*. Washington/Covelo/London: Island Press.

Mirtl, M. (2010). Introducing the next generation of ecosystem research in Europe: LTER-Europe's multi-functional and multi-scale approach. In F. Müller, C. Baessler, H. Schubert, & S. Klotz (Eds.), *Long-term ecological research: Between theory and application* (pp. 75–94). Dordrecht: Springer.

Mirtl, M., & Krauze, K. (2007). Developing a new strategy for environmental research, monitoring and management: The European long-term ecological research network's (LTER-Europe) role and perspectives. In T. Chmielewski (Ed.), *Nature conservation management: From idea to practical results* (pp. 36–52). Lublin/Lodz/Helsinki/Aarhus: ALTERnet.

Mirtl, M., Boamrane, M., Braat, L., Furman, E., Krauze, K., Frenzel, M., Gaube, V., Groner, E., Hester, A., Klotz, S., Los, W., Mautz, I., Peterseil, J., Richter, A., Schenz, H., Schleidt, K., Schmid, M., Sier, A. R. J., Stadler, J., Uhel, R., Wildenberg, M., & Zacharias, S. (2009). *LTER-Europe design and implementation report – Enabling "Next generation ecological science": Report on the design and implementation phase of LTER-Europe under ALTER-Net & Management Plan 2009/2010*. Vienna: Umweltbundesamt (Environment Agency Austria). http://www.lter-europe.net. Accessed June 2011.

Mirtl, M., Bahn, M., Battin, T., Borsdorf, A., Englisch, M., Gaube, V., Grabherr, G., Gratzer, G., Kreiner, D., Haberl, H., Richter, A., Schindler, S., Tappeiner, U., Winiwarter, V., & Zink, R. (2010). *LTER-Austria white paper – "Next generation LTER" in Austria, on the status and orientation of process oriented ecosystem research, biodiversity and conservation research and socio-ecological research in Austria*. Vienna: LTER Austria. http://www.lter-austria.at. Accessed June 2011

Moran, E. F. (Ed.). (1993). *The ecosystem approach in anthropology, from concept to practice*. Ann Arbor: University of Michigan Press.

Moran, E. F. (Ed.). (2000). *Human adaptability: An introduction to ecological anthropology*. Colorado: Westview Press.

Moran, E. F., Brondizio, E. S., Mausel, P., & Wu, Y. (1994). Integrating Amazonian vegetation, land-use and satellite data. *BioScience, 44*, 458–476.

Müller, F., Baessler, C., Schubert, H., & Klotz, S. (Eds.). (2010). *Long-term ecological research. Between theory and application*. Dordrecht: Springer.

Netting, R. M. (Ed.). (1981). *Balancing on an Alp. Ecological change and continuity in a Swiss mountain community*. London/New York/New Rochelle/Melbourne/Sydney: Cambridge University Press.

Netting, R. M. (Ed.). (1993). *Smallholders, householders. Farm families and the ecology of intensive, sustainable agriculture*. Stanford: Stanford University Press.

NSF. (2002). *Long-term ecological research twenty year review*. National Science Foundation (NSF). http://intranet.lternet.edu/archives/documents/reports/20_yr_review

NSF. (2011a). *Dynamics of coupled natural and human systems (CNH)*. National Science Foundation (NSF). http://www.nsf.gov/funding/pgm_summ.jsp?pims_id=13681. Accessed 25 Nov 2011.

NSF. (2011b). *The US long term ecological research network*. National Science Foundation (NSF). http://www.lternet.edu/overview/. Accessed 25 Nov 2011.

Odum, E. P. (Ed.). (1959). *Fundamentals of ecology*. Philadelphia: Saunders.

Ostwald, W. (Ed.). (1909). *Energetische Grundlagen der Kulturwissenschaften*. Leipzig: Dr. Werner Klinkhardt Verlag.

Park, R. E. (1936). Human ecology. *The American Journal of Sociology, 42*, 1–15.

Parris, T. M., & Kates, R. W. (2003). Characterizing and measuring sustainable development. *Annual Review of Environment and Resources, 28*, 559–586.

Rappaport, R. A. (Ed.). (1968). *Pigs for the ancestors*. New Haven: Yale University Press.

Redman, C. L., Grove, J. M., & Kuby, L. H. (2004). Integrating social science into the long-term ecological research (LTER) network: Social dimensions of ecological change and ecological dimensions of social change. *Ecosystems, 7*, 161–171.

Robbins, P. (Ed.). (2004). *Political ecology. A critical introduction*. Oxford: Blackwell Publishing.

Sauer, C. O. (1925). The morphology of landscape. *University of California Publications in Geography, 2*, 19–53.

Sears, P. B. (1953). Human ecology, a problem in synthesis. *Science, 120*, 959–963.

Shibata, H., & Bourgeron, P. (2011). Challenge of international long-term ecological research network (ILTER) for socio-ecological land sciences. *GLP News, 7*, 13–14.

Singh, S. J., & Haberl, H. (2011). Long-term socio-ecological research (LTSER) across temporal and spatial scales. *GLP News, 7*, 15–16.

Singh, S. J., Haberl, H., Gaube, V., Grünbühel, C. M., Lisievici, P., Lutz, J., Matthews, R., Mirtl, M., Vadineanu, A., & Wildenberg, M. (2010). Conceptualising long-term socio-ecological research (LTSER): Integrating the social dimension. In F. Müller, C. Baessler, H. Schubert, & S. Klotz (Eds.), *Long-term ecological research, between theory and application* (pp. 377–398). Dordrecht/Heidelberg/London/New York: Springer.

Steward, J. H. (Ed.). (1955). *Theory of culture change. The methodology of multilinear evolution*. Urbana: University of Illinois Press.

Teherani-Krönner, P. (1992). Von der Humanökologie der Chicagoer Schule zur Kulturökologie. In B. Glaeser & P. Teherani-Krönner (Eds.), *Humanökologie und Kulturökologie, Grundlagen, Ansätze, Praxis* (pp. 15–43). Opladen: Westdeutscher Verlag.

Thomas, W. L., Jr. (Ed.). (1956). *Man's role in changing the face of the Earth*. Chicago: The Chicago University Press.

Turner, B. L., & Robbins, P. (2008). Land-change science and political ecology: Similarities, differences, and implications for sustainability science. *Annual Review of Environment and Resources, 33*, 295–316.

Vaillancourt, J.-G. (1995). Sociology of the environment, from human ecology to ecosociology. In M. D. Mehta & É. Oúellet (Eds.), *Environmental sociology, theory and practice* (pp. 3–32). North York: Captús Press.

White, L. A. (1949). Energy and the evolution of culture. In L. A. White (Ed.), *The science of culture: A study of man and civilization* (pp. 363–393). New York: Grove.

World Commission on Environment and Development (Ed.). (1987). *Our common future, the Brundtland-report*. Oxford: Oxford University Press.
Young, G. L. (1974). Human ecology as an interdisciplinary concept: A critical inquiry. In A. Macfayden (Ed.), *Advances in ecological research* (pp. 1–105). London/New York: Academic.
Zimmerer, K. S. (2004). Cultural ecology: Placing households in human-environment studies – The cases of tropical forest transitions and agrobiodiversity change. *Progress in Human Geography, 28*, 795–806.
Zimmerer, K. S. (2010). Retrospective on nature – Society geography: Tracing trajectories (1911–2010) and reflecting on translations. *Annals of the Association of American Geographers, 100*, 1076–1094.

# Part I
# LTSER Concepts, Methods and Linkages

# Chapter 2
# Socioeconomic Metabolism and the Human Appropriation of Net Primary Production: What Promise Do They Hold for LTSER?

Helmut Haberl, Karl-Heinz Erb, Veronika Gaube,
Simone Gingrich, and Simron Jit Singh

**Abstract** This chapter reviews approaches to analysing the 'metabolism' of socioeconomic systems consistently across space and time. Socioeconomic metabolism refers to the material, substance or energy throughput of socioeconomic systems, i.e. all the biophysical resources required for production, consumption, trade and transportation. We also introduce the broader concept of socio-ecological metabolism, which additionally considers human-induced changes in material, substance or energy flows in ecosystems. An indicator related to this broader approach is the human appropriation of net primary production (HANPP). We discuss how these approaches can be used to analyse society-nature interaction at different spatial and temporal scales, thereby representing one indispensible part of the methodological tool box of LTSER. These approaches are complimentary to other methods from the social sciences and humanities, as well as to genuinely transdisciplinary approaches. Using Austria's sociometabolic transition from agrarian to industrial society from 1830 to 2000 as an example, we demonstrate the necessity of including a comprehensive stock-flow framework in order to use the full potential of the socio-ecological metabolism approach in LTSER studies. We demonstrate how this approach can be implemented in integrated socio-ecological models that can improve understanding of changes in society-nature interrelations through time, another highly important objective of LTSER.

**Keywords** Human appropriation of net primary production • Socioeconomic metabolism • Material flow analysis • Carbon stock • Carbon flow

---

H. Haberl, Ph.D. (✉) • K-H. Erb , Ph.D. • V. Gaube, Ph.D.
S. Gingrich, Ph.D. • S.J. Singh, Ph.D.
Institute of Social Ecology Vienna (SEC), Alpen-Adria Universitaet Klagenfurt,
Wien, Graz, Schottenfeldgasse 29/5, Vienna 1070, Austria
e-mail: helmut.haberl@aau.at; karlheinz.erb@aau.at; veronika.gaube@aau.at;
simone.gingrich@aau.at; simron.singh@aau.at

## 2.1 Introduction

One of the central aims of Long-Term Socio-Ecological Research (LTSER) is to provide scientific insights within the field of global environmental change to support transitions towards sustainability (Fischer-Kowalski and Rotmans 2009; see the Introduction Chap. 1 in this volume). Changes in stocks and flows of carbon, water, nitrogen and many other compounds are crucial aspects of global environmental change. Climate change, for example, is driven by the accumulation of gases in the atmosphere that alter the energy balance of the global climate system. Changes in the concentration of such greenhouse gases (abbreviated as GHG, the most important of these being $CO_2$, $CH_4$ and $N_2O$) in the atmosphere resulting from human activities are very likely responsible for most of the observed growth in global mean temperature since the mid-twentieth century (IPCC 2007). Likewise, emissions of toxic substances into water bodies or the atmosphere influence ecosystems, including agro-ecosystems and forestry systems, humans as well as buildings and other valuable artefacts at regional or even global scales (Akimoto 2003).

Human-induced changes in global biogeochemical cycles also contribute to biodiversity loss, both directly and indirectly. Nitrogen enrichment has been shown to reduce species diversity in many environments (Vitousek et al. 1997). There is evidence that avian species richness is positively related to biomass stocks in ecosystems (Hatanaka et al. 2011), implying that a reduction of biomass stocks (e.g. through deforestation) would contribute to species loss. Empirical studies suggest that species richness is lower in ecosystems where human activities reduce biomass availability (Haberl et al. 2004b, 2005). There is empirical evidence that land use is the most prominent driver of biodiversity loss, followed by climate change (Sala et al. 2000).

Mitigating climate change and reducing biodiversity loss are two cornerstones of global sustainability policies, at least since the Conventions on Biological Diversity and the United Nations Framework Convention on Climate Change, both adopted at the 1992 United Nations Conference on Environment and Development (UNCED) in Rio de Janeiro. Understanding the social and economic processes that contribute to changes in global stocks and flows of materials, energy and chemical elements allow us to gain insights into the drivers of these two highly important global environmental sustainability concerns. In other words, the concept of tracing changes in socio-ecological stocks and flows of materials and energy across time and space is a central approach of LTSER. Several chapters of this volume are built upon this approach (see Fischer-Kowalski et al., Chap. 4 in this volume; Krausmann, Chap. 11 in this volume; Gingrich et al., Chap. 13 in this volume; Krausmann and Fischer-Kowalski, Chap. 15 in this volume).

In this contribution, we outline concepts and methods suitable for analysing biophysical stocks and flows, e.g. material and energy flow accounting (often referred to as MEFA, e.g. Haberl et al. 2004a), the human appropriation of net primary production (HANPP) and related approaches for analysing socioeconomic and, more broadly, socio-ecological metabolism useful for LTSER studies. Based on a case study of Austria (1830–2000), we discuss how these methods can be used to analyse and improve understanding of socio-ecological transitions (Fischer-Kowalski and Haberl 2007; see Krausmann, Chap. 11 in this volume; Gingrich et al., Chap. 13 in this volume; Krausmann and Fischer-Kowalski, Chap. 15 in this volume).

The sociometabolic approaches discussed in this chapter are complimentary to concepts and methods used and discussed in other chapters in this volume: from environmental history, biohistory, geography and social sciences, to transdisciplinary methods as well as approaches based on modelling. Links between data-based, empirical approaches to understand material and energy flows and system-dynamic modelling methods are explicitly discussed in Sect. 2.4 of this chapter.

## 2.2 Socioeconomic Metabolism: Material and Energy Flow Analysis (MEFA)

All human activities depend on inputs of materials and energy from the natural environment. At the very least, food is needed to keep humans alive, healthy and able to perform work. But many activities require much more than this 'basic' or 'endosomatic' metabolism (Boyden 1992; Fischer-Kowalski and Haberl 1997; Giampietro et al. 2001). Economic activities such as production, consumption, trade, transportation, and even services need buildings, infrastructures or machinery that in turn require inputs of raw materials or manufactured goods as well as inputs of energy, be it in the form of human or animal labour or as technical energy carriers such as electricity, fuels or heat. Therefore, by thermodynamic necessity, economic processes result in outputs not only in the form of products, but also as wastes and emissions (Hall et al. 2001).

The study of biophysical flows associated with socioeconomic processes has a long-standing tradition in social and human ecology, ecological anthropology, ecological economics, industrial ecology and many other interdisciplinary fields of inquiry focused on processes of society-nature interaction. As several excellent reviews are available (e.g. Fischer-Kowalski 1998; Fischer-Kowalski and Hüttler 1998; Martinez-Alier 1987), we will not attempt a full review here; instead we focus on the use of these concepts within LTSER.

The concept of socioeconomic metabolism (Ayres and Kneese 1969; Ayres and Simonis 1994; Boulding 1972; Fischer-Kowalski and Haberl 1997; Martinez-Alier 1987) has been developed as an approach to study the extraction of materials or energy from the environment, their conversion in production and consumption processes, and the resulting outputs to the environment. Accordingly, the unit of analysis is the socioeconomic system (or some of its components), treated as a systemic entity, in analogy to an organism or a sophisticated machine that requires material and energy inputs from the natural environment in order to carry out certain defined functions and that results in outputs as wastes and emissions. At a very basic level, one can distinguish between two sociometabolic approaches: one that aims at forging a comprehensive account of all biophysical flows needed to build up, sustain

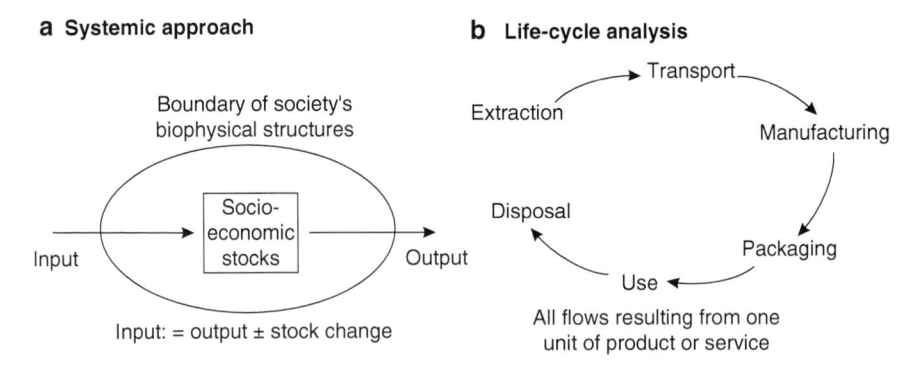

**Fig. 2.1** Basic approaches to analyse socioeconomic metabolism. (**a**) Systemic approaches account for all physical flows (materials, energy, substances) required for reproduction and functioning of socioeconomic stocks. (**b**) Life-cycle analysis accounts for resource requirements or emissions from one unit of product or service throughout its entire life cycle ('cradle to grave')

and operate a defined set of socioeconomic stocks for a given reference system identified by scale (global, national, regional) or by function (household, economic sector or commercial enterprise); the other is the life-cycle analysis (LCA) approach that aims to account for resource requirements as well as wastes and emissions resulting from a single unit of product or service (Fig. 2.1).

In both cases, the essential question is how to define the system boundaries. Systemic approaches such as economy-wide material and energy flow analysis (see below) usually focus on three compartments of 'society's biophysical structures' (Fischer-Kowalski and Weisz 1999): humans, livestock (all animals kept and used by humans) and artefacts (all non-living structures constructed, maintained and used by humans, i.e. infrastructures, buildings, machinery and other durables). Human labour is the main determinant in the choice of compartments. In other words, all that is created and maintained by human labour is considered as part of society's biophysical structures or stocks.[1] Only those biophysical flows are accounted for that serve to build up, maintain or use these structures (Fischer-Kowalski 1998). While systemic metabolism approaches are used to account for and analyse metabolic flows of societies across time and space, LCA is so far mainly used for a quite different purpose, namely to optimise chains of production. Accordingly, its system boundaries are different. In LCA, the 'functional unit' may be a service such as 'movement of one person from A to B' or a defined amount and quality of a product such as 'one kilogram of fresh tomatoes' (Rebitzer et al. 2004). Although LCA might become relevant to LTSER in the future, it has not been widely used in LTSER so far, to our knowledge, and will not be further discussed here.

---

[1] Agricultural fields are excluded from the definition of society's biophysical stocks even though they are produced and maintained by human labour, for accounting reasons, among others. For a detailed discussion on conceptual and methodological considerations, see Fischer-Kowalski and Weisz (1999) and Eurostat (2007).

Systemic approaches to analyse socioeconomic metabolism are able to trace the flows of materials (material flow analysis or MFA), individual substances (substance flow analysis or SFA) or energy (energy flow analysis or EFA) through biophysical structures of society (humans, livestock, artefacts). MFA attempts to establish comprehensive accounts of the material throughput of a defined societal subsystem, spatially and functionally, e.g. a national economy, a city or village, a household or an economic sector. In this context, the notion of 'materials' refers to broad aggregates such as construction materials, industrial minerals and ores, biomass, fossil energy carriers or traded manufactured goods (Krausmann et al. 2009a; Weisz et al. 2006). National-level (or economy-wide) MFA takes into account all those materials used by national economies. Economy-wide MFA has meanwhile become fairly standardised and is implemented as part of national environmental statistics and book-keeping (Eurostat 2007; OECD 2008).

In contrast to the flows of broad aggregates of materials discerned in MFA, substance flow analyses (SFA) account for the flows of defined substances or even chemical elements, e.g. nitrogen (N), carbon (C) or metals such as iron (Fe), Zinc (Zn), Copper (Cu) and others (e.g., Billen et al. 2009; Erb et al. 2008; Graedel and Cao 2010; Wang et al. 2007). MFA and SFA are seen as complimentary approaches to analyse socioeconomic use of resources: While MFA provides a comprehensive picture of total resource use (with concerns over depletion of natural resources and disruption of habitats during extraction), SFA can be more easily connected to specific scarcities or environmental problems, e.g. climate change in the context of carbon flows or alteration of global biogeochemical cycles in the case of nitrogen.

In physical terms, energy is the ability to perform work. Energy is less 'tangible' and more abstract than materials (measured in kilograms, kg) or substances (measured in kg of the relevant substance, e.g. kg C in carbon flow accounts or kg N in a study on nitrogen flows). Nevertheless, scholars have long been interested in human use of energy (for an excellent review see Martinez-Alier 1987). Data on socioeconomic use of technical energy (i.e., energy flowing through artefacts) in national economies (i.e. on the country level) are readily available in conventional energy statistics and energy balances (e.g. IEA 2010). Such statistics provide indicators such as Total Primary Energy Supply (TPES) or Final Energy Use (i.e. the energy used in industry, services and households for all purposes except for the production of other energy carriers).

However, these statistics by definition exclude human consumption of food as well as feed consumption of livestock, that is, the most important energy flows of agrarian societies. Energy flow accounting (EFA) methods that fully consider biomass flows have therefore been proposed based on the same system boundaries as MFA (Haberl 2001a, b). These methods have been used to reconstruct long time-series of socioeconomic energy use on several scales, from local to national and global (Haberl 2006; Haberl et al. 2006a; Haberl and Krausmann 2007; Krausmann and Haberl 2002, 2007). EFA is therefore useful to analyse the transition from the agrarian to the industrial sociometabolic regime (Fischer-Kowalski and Haberl 2007, see below). As the changes in energy systems connected to this transition are related to many current sustainability problems, e.g. climate change, depletion of resources or biodiversity loss, they are also relevant for LTSER.

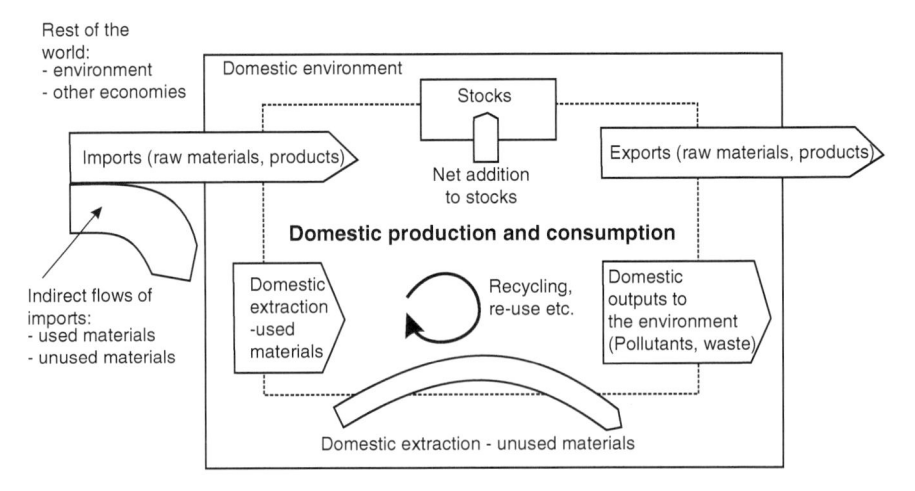

**Fig. 2.2** Scheme of economy-wide (national-level) material and energy flow (MEFA) accounts (Source: Adapted from OECD 2008)

No matter whether one is interested in materials, substances or energy, it is important to note that the simple representation of socioeconomic metabolism in Fig. 2.1a becomes quite complex when applied to concrete cases. There are many reasons for this. First, at every lower scale than the global, inputs can be generated by extracting materials or energy from natural systems on one's own territory or by importing raw materials or even manufactured goods from elsewhere. The same holds for outputs which may be wastes and emissions or goods exported to other territories. Second, one needs definitions of which flows to include or exclude in the accounting system. For example, economy-wide MFA in principle includes all materials, but not air and water – except for the water contained in products.[2] MEFA accounts distinguish between those biophysical flows that directly enter the economy and those that are physically moved at an early stage of extraction and production only but are not economically useful, e.g. agricultural residues left in the field or overburden in mining (Fig. 2.2). Important indicators include the 'direct material input' (i.e. imports plus domestic extraction of used materials) and 'domestic material consumption' (i.e. direct material input minus exports) (Eurostat 2007; OECD 2008). The same indicators can be calculated for energy (Haberl 2001a). Third, the complexity of the accounts and the difficulties of avoiding double-counting increase quickly if one tries to disaggregate material or energy flow, for example to economic

---

[2] Water and air together comprise 85–90% of all total material input. In order not to drown other "economically valued" materials in water and air, the latter are excluded from MFA. Another reason for their exclusion is the low environmental impact of their use, a supposition which is now beginning to be questioned in the context of discussions on ecosystem services (see http://www.teebweb.org/).

sectors. Physical input-output tables have been used as a method towards that end (Hoekstra and van den Bergh 2006; Suh 2005; Weisz and Duchin 2006). Indeed, full material balances that explicitly link inputs to outputs in the manner described in Fig. 2.2 are rare. Moreover, the quantification of socioeconomic material stocks and also of the stock changes is unfortunately still in its infancy (but see Matthews et al. 2000; Kovanda et al. 2007).

Complimentary to EFA that can be used to assess the quantity and quality of energy 'metabolised' by society, analysts have long been interested in the 'energy return on investment' (abbreviated as EROI) of different energy sources used by humans (Cleveland et al. 1984; Hall et al. 1986; Odum 1971; Pimentel et al. 1973; Rappaport 1971). The EROI is defined as the ratio between the amount of energy invested by society into a process and the amount of energy gained from it:

$$\text{EROI} = \frac{\text{Energy gained [J]}}{\text{Energy invested [J]}}$$

Obviously, an energy resource can only deliver a surplus of energy (a positive amount of 'net energy') if society has to invest less than it gains, i.e. if EROI is larger than 1. Under certain circumstances, societies may decide to use energy resources even at EROI < 1, but they can do so only if possessing other energy sources with EROI > 1 to be able to provide for these 'energetic subsidies'. For example, it has been observed that the EROI of many food products used in industrialised societies is far below 1 (Pimentel et al. 1990), but these societies can afford to subsidise these products because they have fossil fuels that have a much larger EROI at their disposal. By contrast, agrarian societies vitally depend on the EROI of agriculture being substantially larger than 1, as biomass is their most important source of net energy. Empirical analyses conducted in many places and on different spatial levels have consistently produced empirical support for this hypothesis (e.g., Pimentel et al. 1990; Krausmann 2004; Sieferle et al. 2006). Indeed, such changes in socioeconomic energy systems played a crucial role in facilitating the sociometabolic transition from agrarian to industrial society (Fischer-Kowalski and Haberl 1997, 2007; Krausmann and Haberl 2002; Haberl et al. 2011).

The above-reviewed methods to account for socioeconomic metabolism are important for sustainability science and LTSER for the following reasons (Haberl et al. 2004a): First, they provide a consistent accounting framework to assess biophysical flows associated with human activities at many levels of societal organisation, from the individual to households, towns and cities to national and supranational levels. Second, this accounting framework can be consistently applied to trace changes across historical social formations – material, substance and energy flows can be assessed for hunter-gatherers and agrarian societies as well as industrial societies, and the analyses of the changes in these biophysical flows have proven to be immensely useful in understanding differences in sustainability challenges across time and space (e.g., Dearing et al. 2007; Fischer-Kowalski and Haberl 1997, 2007; Haberl et al. 2011). Third, they provide a crucial framework to consistently link socioeconomic drivers such as decisions of actors, policies, institutions, prices or technology, to name but a few, with biophysical flows that have an obvious ecological significance, be it due to their toxicity (e.g. emissions of $NO_x$, $SO_2$, lead or dioxin), their ability to impact upon biological processes such as plant growth (e.g. reactive nitrogen) or their function as greenhouse gases (e.g., $CO_2$, $CH_4$ or $N_2O$).

The cumulative insights from these studies have resulted in the development of the theories of 'sociometabolic regimes' and 'sociometabolic transitions'. In the former, systematic interrelations between resource use profile, demographic trends, settlement patterns, governance structures and related environmental impacts are observed for a given mode of production. The transition from one sociometabolic regime to another, on the other hand, implies both a fundamental shift in terms of its resource use potential and environmental impacts as well as the qualitative attributes of the social system and its environmental impacts (Fischer-Kowalski and Haberl 2007; Fischer-Kowalski and Rotmans 2009; Fischer-Kowalski 2011). From an LTSER point of view, these concepts are not only useful in understanding the historical and ongoing transitions that affect global environmental change, but may also inform and aid a transition to a more sustainable future.

Accounting for socioeconomic metabolism is not sufficient, however, if one aims to understand the impact of human activities on stocks and flows of materials and energy in the biosphere. Many human activities are deliberately altering biophysical properties of ecosystems in order to increase useful output and in doing so are inducing changes in stocks and flows of materials and energy in ecosystems. The sum of these activities is denoted as land use – and is increasingly being recognised as a pervasive driver of global environmental change (Foley et al. 2005). Agriculture and forestry, but also the use of land for infrastructure and for deposition or waste absorption, almost always results in changes in stocks and flows of materials and energy in ecosystems. For example, converting natural ecosystems to cropland or managed grasslands affects not only the species composition of the ecosystem, but also water and nutrient flows, stocks and flows of carbon, water flows and retention capacity, etc. (Haberl et al. 2001; Hoekstra and Chapagain 2008; Vitousek et al. 1997). Many of these changes are associated with changes in land cover, e.g. conversions of forested land to agricultural fields, and can thus be monitored from space by remote sensing, but many other changes do not relate to such apparent alterations and are thus much more difficult to quantify, map or assess, despite their far reaching consequences for socio-ecological systems (Erb et al. 2009a; Lambin et al. 2001; Verburg et al. 2010). These processes can be analysed by using approaches to account for socio-ecological metabolism, e.g. the 'human appropriation of net primary production' (abbreviated as HANPP). Such approaches will be discussed in the following section.

## 2.3 Socio-ecological Metabolism: HANPP and Related Approaches

Socio-ecological metabolism (Haberl et al. 2006a) is an extension of the socioeconomic metabolism approach that aims to account for changes in both socioeconomic and ecological systems resulting from human activities. One such change particularly relevant in the LTSER context is that of biological productivity – that is, the annual net biomass production of green plants through photosynthesis (gross primary production minus plant respiration, i.e. net assimilation).

Net Primary Production (NPP) is a key parameter of ecosystem functioning (Lieth and Whittaker 1975; Lindeman 1942; Whittaker and Likens 1973). NPP determines the amount of trophic energy available for all heterotroph organisms (animals, fungi, microorganisms) in ecosystems. Many important processes such as nutrient cycling, build-up of organic material in soils or in above ground biomass stocks, vitally depend on NPP. NPP is connected to the resilience of ecosystems and to their capacity to provide services, such as biomass supply through agriculture and forestry, but also the buffering capacity or the absorption capacity for wastes and emissions (Millennium Ecosystem Assessment 2005). Human alterations of the availability of NPP in ecosystems are therefore ecologically relevant almost by definition (Gaston 2000; Kay et al. 1999; Vitousek et al. 1986; Wright 1983, 1990).

One of the indicators to measure human impact on biological productivity is the 'human appropriation of net primary production', abbreviated as HANPP. HANPP provides a framework to account for changes in biomass flows in ecosystems resulting from land use (Vitousek et al. 1986; Haberl et al. 2007a). There are two equivalent definitions of HANPP:

$$\text{HANPP} := \text{NPP}_0 - \text{NPP}_t \tag{1}$$

$$\text{HANPP} := \Delta \text{NPP}_{LC} + \text{NPP}_h \tag{2}$$

Definition (1) represents an ecological perspective: It defines HANPP as the change in biomass availability in ecosystems resulting from land use, i.e. as the difference between the NPP of potential natural vegetation ($\text{NPP}_0$) – the NPP of the vegetation assumed to exist in the absence of human interventions under current climate conditions – and the fraction of the actual vegetation NPP (abbreviated $\text{NPP}_{act}$) remaining in ecosystems after harvest. Harvest is denoted as $\text{NPP}_h$ and the amount of NPP remaining in the ecosystem as $\text{NPP}_t$ (see Fig. 2.3). Definition (2) is equivalent, but defines HANPP from a socioeconomic perspective: Land use changes the NPP of the vegetation by supplanting potential vegetation with actual vegetation – the difference between $\text{NPP}_0$ and $\text{NPP}_{act}$ is denoted as $\Delta \text{NPP}_{LC}$. In addition, harvest ($\text{NPP}_h$) removes NPP from the ecosystem, thereby reducing the amount of biomass remaining available in the ecosystem for all heterotrophic food chains or for biomass accumulation.

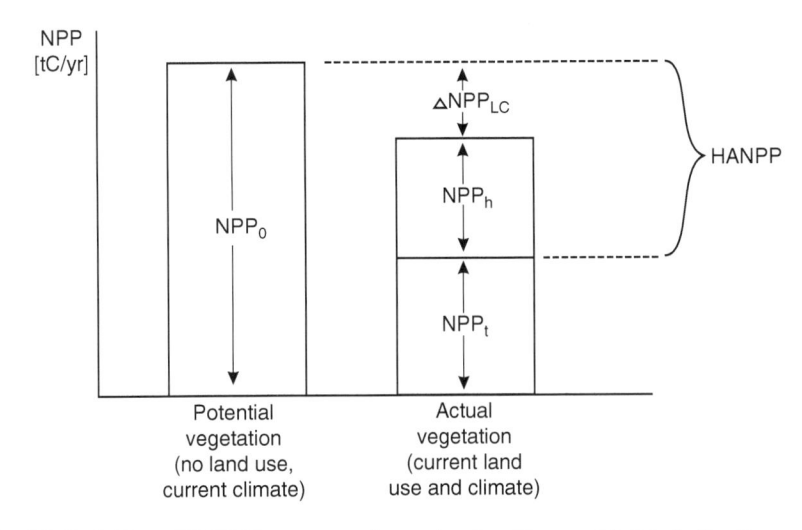

**Fig. 2.3** Definition of HANPP – see text for explanation (Source: Modified after Krausmann et al. 2009b)

HANPP is related to sustainability issues such as food supply from ecosystems to society,[3] the conversion of valuable ecosystems (e.g., forests) to infrastructure, cropland or grazing land (FAO 2004; Millennium Ecosystem Assessment 2005; Lambin and Geist 2006), with detrimental consequences for biodiversity (Haberl et al. 2004b, 2005). HANPP is connected to changes in global water flows (Gerten et al. 2005), carbon flows (DeFries et al. 1999; McGuire et al. 2001) and – as biomass contains nitrogen (N), and N fertiliser is an important factor for agricultural productivity – also N flows. HANPP is therefore directly related to global, human-induced alterations of biogeochemical cycles (Steffen et al. 2004).

Current global HANPP levels are at approximately one quarter of $NPP_0$ (referring to the year 2000; Haberl et al. 2007a), underpinning the notion that human activities have begun to overwhelm the great forces of nature, thereby driving the earth system into a new geological era, the 'anthropocene' (Crutzen and Steffen 2003; Steffen et al. 2007). Recent research suggests that HANPP could become a potent indicator of human pressures on biodiversity (Haberl et al. 2004b, 2005). Despite a broad acknowledgement of a strong interrelation between the NPP flow in ecosystems and biodiversity, however, there are discussions on the mathematical form of this interrelation (Waide et al. 1999; Haberl et al. 2009b). Empirical findings so far indicate

---

[3] For example, converting natural ecosystems to cropland increases HANPP. Increasing yields per unit area and year or reducing losses in the production chain allows the HANPP per unit of final product to be reduced and therefore HANPP to be 'decoupled' from supply of food or other land-dependent products.

that high HANPP levels do not correlate with high biodiversity levels, giving indirect evidence that HANPP can be used as an indicator for socioeconomic pressures on biodiversity (Haberl et al. 2007b).

Recent research has demonstrated that HANPP can be assessed with reasonable effort and precision at many spatial and temporal levels. Global maps of terrestrial HANPP in the year 2000 at a resolution of approximately 10 km are readily available (Haberl et al. 2007a).[4] Several long-term (decadal to centennial) country-level HANPP studies have meanwhile been conducted (e.g. Krausmann 2001; Kastner 2009; Musel 2009; Schwarzlmüller 2009, see e.g. Erb et al. 2009b).[5] Such studies have proven to be valuable in improving understanding of ecological implications of sociometabolic transitions (Fischer-Kowalski and Haberl 2007, see next section), an issue of paramount importance for LTSER. Of course, HANPP is no panacea. For example, while HANPP is a suitable indicator of overall land-use intensity, it is not well-suited to capturing cropland intensification: Intensification drives up $NPP_{act}$ and $NPP_h$ in parallel. In effect, even large increases in yields do not show up as an increase in HANPP. Additional indicators such as nutrient balances and EROI are required to make such effects visible (Erb et al. 2009b).

The HANPP framework can be extended in at least two directions that are relevant in LTSER. First, in addition to the HANPP on a defined territory, one can also calculate the HANPP caused by, or 'embodied in', the products consumed by a population. This is captured by the concept of 'embodied HANPP', abbreviated as eHANPP (Erb et al. 2009c; Haberl et al. 2009a). The eHANPP concept is related to approaches such as 'virtual water' (Allan 1998) and the 'water footprint' (Hoekstra and Chapagain 2007; Gerbens-Leenes et al. 2009). Embodied HANPP is the HANPP resulting from the consumption of all products used by a population. National eHANPP can be calculated by adding to the HANPP on a country's own territory the HANPP resulting from imports and subtracting the HANPP resulting from exports (Erb et al. 2009c; Haberl et al. 2009a).

Let us consider the example of Australia: This highly industrialised, but sparsely populated country (two inhabitants per square kilometre) has an HANPP on its national territory of 708 million tonnes of dry-matter biomass (Mt/year), but the eHANPP related to the consumption of its population is only 177 Mt/year. In other words, net biomass trade results in a 'net export' of three-quarters of Australia's HANPP. By contrast, Japan, with its high population density (330 inhabitants per square kilometre) has an HANPP on its own territory of 113 Mt/year, but the eHANPP related to its consumption is more than five times higher and amounts to 581 Mt/year – hence, Japan obviously could not generate sufficient supplies on its

---

[4] Gridded HANPP data can be freely downloaded at http://www.uni-klu.ac.at/socec/inhalt/1191.htm

[5] HANPP studies are not restricted to terrestrial ecosystems, but can also be used to analyse trends, trajectories and the magnitude of human impacts on e.g. marine ecosystems (Pauly et al. 2005; Swartz et al. 2010). The utility of these approaches in LTSER has so far not been explored.

own territories, at least at its current consumption levels. Such data allow us to analyse the 'teleconnections' between cities and their hinterlands or between exporting and importing countries (Haberl et al. 2009a).

The eHANPP concept thus allows us to explicitly analyse the impacts of consumption on terrestrial systems in terms of their trophic energy flows. Second, one can also link the flows accounted for in HANPP assessments to stocks of biomass and carbon in biota and soils (Erb 2004; Erb et al. 2008; Haberl et al. 2001; Gingrich et al. 2007; see next section). This is particularly relevant as it allows us to establish full carbon balances thus providing a comprehensive picture of the carbon stocks and flows in a defined country or region that considers not only C-flows resulting from socioeconomic metabolism but also those resulting from land-use change.

Approaches that are conceptually related to HANPP can be developed for other relevant resources as well. For example, one can calculate the 'human appropriation of freshwater' (Postel et al. 1996; Weiß et al. 2009) and human-induced changes in river runoff (Vörösmarty et al. 1997). Another related concept is the mapping of the relation between human-induced and natural metal flows (Rauch and Pacyna 2009; Rauch 2010).

## 2.4 Austria 1830–2000: Towards a System-Dynamic Model of Carbon Stocks and Flows

In this section we summarise recent research on the stocks and flows of carbon in Austria from 1830 to 2000 and propose how this transition might be analysed using a system-dynamic model. The analysis of changes in carbon stocks and flows in Austria through almost two centuries provides an example of how the above-discussed methods and approaches can help to integrate empirical, data-driven and analytic, system-dynamic approaches for LTSER. Integrating system-dynamic modelling with data generation and interpretation allows us to test hypotheses on the relative importance of drivers and on interrelations between important factors and is therefore an important approach in LTSER (van der Leeuw 2004).

We focus in this example on carbon, not only because these data are available from previous research (Erb 2004; Erb et al. 2007, 2008; Gingrich et al. 2007) but also because changes in stocks and flows of carbon have an immediate bearing upon many contemporary sustainability challenges. Carbon is an essential chemical element indispensable not only for all living organisms (about half of dry-matter biomass is carbon), but also a major constituent of many materials, most prominently fossil fuels. Its concentration in the atmosphere is a major determinant of the earth's climate system because $CO_2$ absorbs infrared radiation and can thereby alter the earth's radiation balance. Carbon is therefore highly important for socioeconomic and ecological systems alike.

During the period 1830–2000, Austria underwent an almost complete sociometabolic transition from an agrarian to an industrial society. Population grew by a factor of 2.3 from 3.6 to 8.1 million. The agrarian share of the population (i.e. farmers and their families) dropped from 75 to 5%. The contribution of agriculture to GDP even declined to 1.4% in the year 2000, while total GDP rose by a factor of 28 and per-capita GDP by a factor of 12 (Krausmann and Haberl 2007).

At the beginning of the period in question, Austria was still a predominantly agrarian country.[6] In 1830, biomass accounted for 99% of the socioeconomic energy input for food, feed and fibre but also for mechanical work, light and heat. Hydropower was used by water mills that were important for processes such as grain milling or metal works, but the amount of energy gained through this process was below 1% of total socioeconomic energy input. Similarly, some coal was already used at that time, but the amount was almost negligible compared to biomass.

The first phase of Austria's industrialisation until World War I was largely powered by coal. At that time, the abundant coal reserves of Bohemia and southern Poland were 'domestic' resources: Although the coal had to be 'imported' to the current Austrian territory, the coal in fact came from other parts of the same country; that is, the Austro-Hungarian monarchy. This changed abruptly with World War I, after which most coal had to be imported from what were now independent countries. This in effect resulted in a restructuring of the Austrian industry, with less emphasis on heavy industry after the war and lower levels of coal use. After World War II, Austria's rapid economic growth was mostly powered by oil products, later by natural gas and by a considerable hydropower programme that led to the utilisation of about three-quarters of the economically usable potential, continuing into the present day (for detail see Krausmann and Haberl 2002, 2007).

All of this resulted in major changes in Austria's socioeconomic carbon flows. In 1830, almost all of the carbon metabolised by Austrian society came from biomass harvested on Austria's own territory through either agriculture or forestry (coal was insignificant at that point in time, contributing less than 1% to Austria's total energetic metabolism). By contrast, in the year 2000, fossil fuels played a major role, although the carbon contained in biomass was still by no means negligible (Fig. 2.4a). Almost all the carbon metabolised by the Austrian economy flowed to the atmosphere, mostly as $CO_2$, but at the same time plant growth also removed $CO_2$ from the atmosphere through photosynthesis. Carbon from biomass is often exempted on these grounds from greenhouse gas accounts, but the general assumption that the release of carbon in biomass to the atmosphere would be 'carbon neutral' because it is balanced by plant growth has long been recognised as being imprecise (Schlamadinger et al. 1997). In fact, this assumption may even result in major flaws, in particular in cases where large stock changes are triggered, such as with the conversion of pristine forests to used forests or to agricultural fields – a recent recognition that mandates revision of GHG accounting rules (Searchinger et al. 2009).

---

[6] Note that before 1918 the current territory of Austria was part of the much larger Austro-Hungarian monarchy. For this period, we were obliged to use data that refer to a territory that is similar, but not exactly identical to Austria's current territory. These data were used to extrapolate to Austria in its current boundaries, in order to generate a consistent time series (see Krausmann and Haberl 2007 for detail).

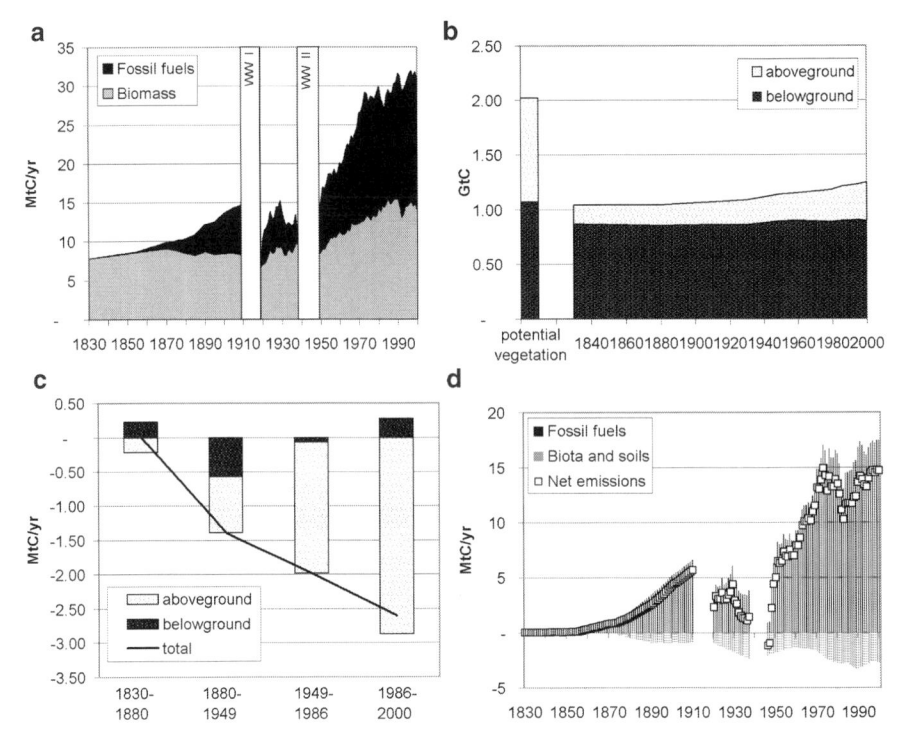

**Fig. 2.4** Stocks and flows of carbon in Austria 1830–2000. (**a**) Socioeconomic carbon flows per year. (**b**) Carbon stocks in biota and soils in billion tonnes of carbon. (**c**) Net carbon exchange between atmosphere and biota/soils. (**d**) Net carbon emissions considering the terrestrial carbon sink (Source: Redrawn after Erb et al. 2008; Gingrich et al. 2007)

Correct treatment of this critical issue requires a clear distinction between stocks and flows (Körner 2009). Most of the carbon absorbed by green plants during photosynthesis is metabolised either by plants or by heterotrophic organisms and therefore released back to the atmosphere. Compared to these yearly flows, net changes in stocks – either in the soil, e.g. as soil organic carbon (SOC), or aboveground in the carbon content of standing biomass stocks ('standing crop') – are comparably small. Estimating the net release ('source') or net absorption ('sink') of carbon therefore requires the assessment of carbon stocks in biota and soils at different points in time. If the stock is growing, one can assume that biota and soils have acted as a carbon sink, while in the opposite case they have acted as a source, i.e. emitted carbon to the atmosphere.

In Austria, carbon stocks in biota and soils are substantially lower than they would be in the absence of human use of the land (Fig. 2.4b). The reason is that most of Austria's area would be forested if not used by humans, whereas a considerable proportion of these natural forests have been replaced by agro-ecosystems (cropland, grasslands) by humans mainly from the Middle Ages onwards (in addition, natural forests have been almost entirely replaced by managed forests). As shown in

Fig. 2.4b, carbon stocks in biota and soils have been steadily increasing since 1880, thereby resulting in a considerable net uptake of carbon: While biota and soils were almost balanced from 1830 to 1880, the net carbon uptake increased to approximately 2.9 million tC per year in the period 1996–2000 (Fig. 2.4c). The reason for this phenomenon – which is typical for many industrial economies and is known as 'forest transition' (Mather and Fairbairn 1990; Kauppi et al. 2006; Meyfroidt et al. 2010) – is that cropland and grassland areas are shrinking and forests are growing both in terms of area and in terms of stocking density, i.e. in carbon stocks per unit area (Erb et al. 2008; Gingrich et al. 2007). In Austria, forest area grew by more than one-fifth in the last 170 years. In that period, infrastructure areas grew by a factor of four, while cropland area was reduced by one-third and pastures and meadows by one-fifth (Krausmann 2001).

That biota and soils in Austria absorb more carbon than they release can, at least in a first-order approach, be interpreted as justification for assuming that carbon releases through biomass combustion to the atmosphere were indeed 'carbon neutral'. In fact, Austrian ecosystems not only produced all that biomass, they even sequestered carbon at the same time. However, there are some caveats. First, this view neglects the possibility that Austria's biomass consumption causes carbon releases elsewhere. The analysis of this issue is still in its infancy, and considering such flows might well influence the overall balance (Gavrilova et al. 2010; Kastner et al. 2011). Second, it would also be necessary to consider the counterfactual. For example, if wood harvest in Austrian forests were to be reduced, the forests would sequester considerably more carbon. This effect is substantial and might well cancel out any emission reduction if additional wood were to be harvested in order to burn it instead of fossil fuels (Haberl et al. 2003).

The most important caveat, however, is the following one: The reduction in farmland was made possible by massive technological change in agriculture that helped to increase yields and conversion efficiencies in the livestock sector (e.g. feed to meat ratios) by large margins. These changes were massive enough to allow for surges in agricultural yields and a 70% increase in primary biomass harvests on the Austrian territory from 1830 to 2000, without increasing HANPP. In fact, an empirical analysis suggests that aboveground HANPP fell by some 15–20% over this period. This was so largely because, due to agricultural intensification, $NPP_{act}$ increased more than $NPP_h$, and because the fraction of $NPP_h$ that could be used as commercial product increased by large margins as well (Krausmann 2001). These technological improvements were only possible due to large-scale inputs of fossil fuels in agriculture, both directly (e.g. tractors) and indirectly (e.g. artificial N fertiliser). These changes have resulted in a massive reduction of the EROI of agriculture from around 6:1 in 1830 to approximately 1:1 in the year 2000 (Krausmann 2004). Ironically, the very same input of fossil fuels that resulted in the massive increases in total GHG emissions also helped to turn Austria's biota and soils into a carbon sink. It is therefore fully justified to speak of a 'fossil-fuel powered carbon sink' (Erb et al. 2007).

Similar trajectories of forest cover and carbon stocks are described for many countries (e.g., Kauppi et al. 2006; Kuemmerle et al. 2011), which suggests that such complex interrelations and feedback loops between land intensification, forest

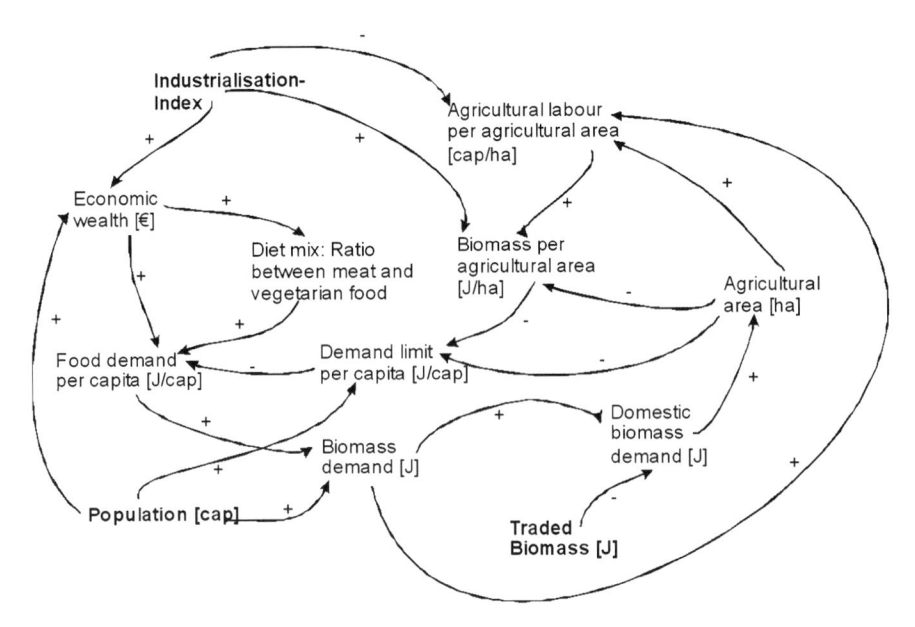

**Fig. 2.5** Preliminary causal loop model of the land use model for Austria 1830–2000 – see text for explanation (Source: authors' own figure)

growth, and the overall socioeconomic energy system are ubiquitous. Our understanding of the spatial and temporal interrelation of these feedback loops, however, is still limited, as many parameter and causal chains show time lags and are subject to trajectories that operate at other spatial scales, e.g. mega-trends such as the globalisation of production and consumption.

The development of algorithmic system-dynamic models has a high potential to advance our current understanding of the complex mechanisms underlying land-use change during socio-ecological transitions from agrarian to industrial society. System-dynamic models have been found to be useful heuristic tools that allow advances in the causal understanding of complex system change: they entail a well-considered reductionism, pragmatism and a clarity of definitions and assumptions at the same time (van der Leeuw 2004). Simple algorithmic formulation of the causal relationships and feedback loops between the highly interlinked factors can be implemented in readily available system-dynamic modelling software.[7]

System-dynamic modelling requires a definition of a so-called causal diagram which serves as the basis of technical implementation of mathematically described interrelations between system components. This might already deliver crucial insights, because causal diagrams depict all key elements of the system under study and require the explicit definition of the relationships between them (Garcia 2006). An example of such a diagram is displayed in Fig. 2.5. Once the variables

---

[7] Free software is readily available, for example Vensim, http://www.vensim.com/

of the system are defined the hypothetical relationships can be represented as arrows between them (Fig. 2.5), indicating directions of causal interdependencies. Each arrow is marked with a plus (+) or minus (−) sign that indicates if a change in the influencing variable will produce a change of the same direction in the target variable or if the effect will be the opposite. Such causal simulation models are capable of reconstructing the trajectory of human-driven land-use change (Lambin et al. 2000; Verburg et al. 2000).

In the model scheme displayed in Fig. 2.5, the four major socioeconomic factors that influence patterns and dynamics of land use are: (1) population, including changes over time, (2) changes in food consumption, (3) technological change, especially in agriculture, and (4) changes in international trade. Biomass harvest is directly influenced by national biomass demand and supply, moderated by trade. Biomass demand is a function of population and the consumption pattern of the population – e.g. diet behaviour. Biomass supply depends not only on natural conditions, but also on the dynamic interplay of labour, capital, livestock and land. External factors and dynamics as input variables used in the model could be (1) Industrialisation-Index indicating the technological change, (2) population numbers and (3) the traded biomass (all of these are highlighted in bold letters in Fig. 2.5), but different notions would also be valid (e.g. population numbers as an endogenous variable). Such causal models can be tested against historical statistics on land-use change, socioeconomic metabolism and land-cover change and are suitable heuristic tools for advancing our understanding of long-term socio-ecological changes (Haberl et al. 2006b; Turner et al. 2007).

## 2.5 Outlook

The study of global environmental change requires a long-term scientific perspective of society-nature interactions. Careful conceptual and methodological considerations are crucial in outlining a scientific agenda for this emerging field of LTSER. In this contribution, we have tried to show the analytical power of socioeconomic and socio-ecological metabolism approaches for understanding local, regional and global environmental changes. These approaches provide tools to assess and monitor socio-ecological interactions and provide insights into the cumulative effects of human activities and their sustainability challenges from a cross-scalar perspective.

Gauging from historical examples of various social formations and modes of production, it becomes evident that the study of the systemic interrelations between biophysical and socio-cultural attributes is key (Fischer-Kowalski and Haberl 2007): Insights into these dynamics are of high relevance for a sustainability science agenda within LTSER, not only in terms of mapping biophysical flows, but also in understanding feedback loops between these and other social, cultural, economic and political variables (Haberl et al. 2006b). This research gap will hopefully be a major focus of future LTSER. The sociometabolic approaches discussed in this chapter thus have to be seen as complimentary to other approaches, e.g. those from social sciences and humanities, and as representing an important part of the methodological toolbox of LTSER. In our view, the research discussed in this chapter shows that LTSER is maturing, developing and extending methods and has the potential to synthesise such methods and approaches in analysing and interpreting long-term changes.

Long-term socio-ecological research requires interdisciplinary efforts, dealing as it does with a plethora of paradigms and methods that require us to bring together not only the social and natural sciences, but also civil society and policy makers as major stakeholders to be considered. This chapter offers promising perspectives in dealing with some of the conceptual and methodological challenges in LTSER. Thus far, however, research is still biased towards understanding the biophysical aspects of society-nature interactions. Notwithstanding, there is an urgent need for more social science input, integrative and transdisciplinary research, as well as the establishment of effective communication pathways between scientists and other stakeholders, including the political system, to be able to influence policy and human behaviour effectively with respect to the choices we make. Global sustainability depends upon moving beyond purely ecological considerations only and towards a system that presupposes the equitable distribution of resources, both in quantitative and qualitative terms, for the current as well as for future generations.

**Acknowledgments** This chapter has profited from research funded by the Austrian Science Fund (FWF), project P20812-G11, by the Austrian Ministry of Science within the research programme proVISION, and from the FP7 Project Volante. It contributes to the Global Land Project (http://www.globallandproject.org) and to long-term socio-ecological research (LTSER) initiatives within LTER Europe (http://www.lter-europe.ceh.ac.uk/).

## References

Akimoto, H. (2003). Global air quality and pollution. *Science, 302*, 1716–1719.
Allan, J. A. (1998). Virtual water: A strategic resource. Global solutions to regional deficits. *Ground Water, 36*, 545–546.
Ayres, R. U., & Kneese, A. V. (1969). Production, consumption and externalities. *The American Economic Review, 59*, 282–297.
Ayres, R. U., & Simonis, U. E. (Eds.). (1994). *Industrial metabolism: Restructuring for sustainable development*. Tokyo/New York/Paris: United Nations University Press.
Billen, G., Barles, S., Garnier, J., Rouillard, J., & Benoit, P. (2009). The food-print of Paris: Long-term reconstruction of the nitrogen flows imported into the city from its rural hinterland. *Regional Environmental Change, 9*, 13–24.
Boulding, K. E. (1972). The economics of the coming spaceship Earth. In H. E. Daly (Ed.), *Steady state economics* (pp. 121–132). San Francisco: W.H. Freeman.
Boyden, S. V. (Ed.). (1992). *Biohistory: The interplay between human society and the biosphere – Past and present*. Paris/Casterton Hall/Park Ridge: UNESCO and Parthenon Publishing Group.
Cleveland, C. J., Costanza, R., Hall, C. A. S., Kaufmann, R. K., & Stern, D. I. (1984). Energy and the U.S. economy: A biophysical perspective. *Science, 225*, 890–897.
Crutzen, P. J., & Steffen, W. (2003). How long have we been in the anthropocene era? *Climatic Change, 61*, 251–257.

Dearing, J. A., Graumlich, L. J., Grove, R., Grübler, A., Haberl, H., Hole, F., Pfister, C., & van der Leeuw, S. E. (2007). Integrating socio-environment interactions over centennial timescales: Needs and issues. In R. Costanza, L. J. Graumlich, & W. Steffen (Eds.), *Sustainability or collapse? An integrated history and future of people on Earth* (pp. 243–274). Cambridge, MA/ London: The MIT Press.

DeFries, R. S., Field, C. B., Fung, I., Collatz, G. J., & Bounoua, L. (1999). Combining satellite data and biogeochemical models to estimate global effects of human-induced land cover change on carbon emissions and primary productivity. *Global Biogeochemical Cycles, 13*, 803–815.

Erb, K.-H. (2004). Land-use related changes in aboveground carbon stocks of Austria's terrestrial ecosystems. *Ecosystems, 7*, 563–572.

Erb, K.-H., Haberl, H., & Krausmann, F. (2007). The fossil-fuel powered carbon sink. Carbon flows and Austria's energetic metabolism in a long-term perspective. In M. Fischer-Kowalski & H. Haberl (Eds.), *Socioecological transitions and global change: Trajectories of social metabolism and land use* (pp. 60–82). Cheltenham/Northampton: Edward Elgar.

Erb, K.-H., Gingrich, S., Krausmann, F., & Haberl, H. (2008). Industrialization, fossil fuels and the transformation of land use: An integrated analysis of carbon flows in Austria 1830–2000. *Journal of Industrial Ecology, 12*, 686–703.

Erb, K.-H., Krausmann, F., Gaube, V., Gingrich, S., Bondeau, A., Fischer-Kowalski, M., & Haberl, H. (2009a). Analyzing the global human appropriation of net primary production – Processes, trajectories, implications. An introduction. *Ecological Economics, 69*, 250–259.

Erb, K.-H., Krausmann, F., & Haberl, H. (Eds.) (2009b). *Analyzing the global human appropriation of net primary production – Processes, trajectories, implications* (Special section of Ecological Economics 69(2)). Amsterdam: Elsevier.

Erb, K.-H., Krausmann, F., Lucht, W., & Haberl, H. (2009c). Embodied HANPP: Mapping the spatial disconnect between global biomass production and consumption. *Ecological Economics, 69*, 328–334.

Eurostat (Ed.). (2007). *Economy-wide material flow accounting. A compilation guide.* Luxembourg: European Statistical Office.

FAO. (2004). *FAOSTAT 2004, FAO statistical databases: Agriculture, fisheries, forestry, nutrition.* Rome: FAO.

Fischer-Kowalski, M. (1998). Society's metabolism. The intellectual history of material flow analysis, part I: 1860–1970. *Journal of Industrial Ecology, 2*, 61–78.

Fischer-Kowalski, M. (2011). Analyzing sustainability transitions as a shift between socio-metabolic regimes. *Environmental Innovation and Societal Transitions, 1*, 152–159.

Fischer-Kowalski, M., & Haberl, H. (1997). Tons, joules and money: Modes of production and their sustainability problems. *Society and Natural Resources, 10*, 61–85.

Fischer-Kowalski, M., & Haberl, H. (Eds.). (2007). *Socioecological transitions and global change: Trajectories of social metabolism and land use.* Cheltenham/Northhampton: Edward Elgar.

Fischer-Kowalski, M., & Hüttler, W. (1998). Society's Metabolism. The intellectual history of material flow analysis, part II: 1970–1998. *Journal of Industrial Ecology, 2*, 107–137.

Fischer-Kowalski, M., & Rotmans, J. (2009). Conceptualizing, observing and influencing social-ecological transitions. *Ecology and Society, 14*, 3. (Online) URL: http://www.ecologyandsociety.org/vol4/iss2/art3

Fischer-Kowalski, M., & Weisz, H. (1999). Society as a hybrid between material and symbolic realms. Toward a theoretical framework of society-nature interaction. *Advances in Human Ecology, 8*, 215–251.

Foley, J. A., DeFries, R., Asner, G. P., Barford, C., Bonan, G., Carpenter, S. R., Chapin, F. S., Coe, M. T., Daily, G. C., Gibbs, H. K., Helkowski, J. H., Holloway, T., Howard, E. A., Kucharik, C. J., Monfreda, C., Patz, J. A., Prentice, I. C., Ramankutty, N., & Snyder, P. K. (2005). Global consequences of land use. *Science, 309*, 570–574.

Garcia, J. M. (Ed.). (2006). *Theory and practical exercises of system dynamics.* Barcelona: Juan Martin Garcia.

Gaston, K. J. (2000). Global patterns in biodiversity. *Nature, 405*, 220–227.

Gavrilova, O., Jonas, M., Erb, K.-H., & Haberl, H. (2010). International trade and Austria's livestock system: Direct and hidden carbon emission flows associated with production and consumption of products. *Ecological Economics, 69*, 920–929.

Gerbens-Leenes, W., Hoekstra, A. Y., & van der Meer, T. H. (2009). The water footprint of bioenergy. *Proceedings of the National Academy of Sciences, 106*, 10219–10223.

Gerten, D., Hoff, H., Bondeau, A., Lucht, W., Smith, P., & Zaehle, S. (2005). Contemporary "green" water flows: Simulations with a dynamic global vegetation and water balance model. *Physics and Chemistry of the Earth, Parts A/B/C, 30*, 334–338.

Giampietro, M., Mayumi, K., & Bukkens, S. G. F. (2001). Multiple-scale integrated assessment of societal metabolism: An analytical tool to study development and sustainability. *Environment, Development and Sustainability, 3*, 275–307.

Gingrich, S., Erb, K.-H., Krausmann, F., Gaube, V., & Haberl, H. (2007). Long-term dynamics of terrestrial carbon stocks in Austria. A comprehensive assessment of the time period from 1830 to 2000. *Regional Environmental Change, 7*, 37–47.

Graedel, T. E., & Cao, J. (2010). Metal spectra as indicators of development. *Proceedings of the National Academy of Sciences of the United States of America, 107*, 20905–20910.

Haberl, H. (2001a). The energetic metabolism of societies, part I: Accounting concepts. *Journal of Industrial Ecology, 5*, 11–33.

Haberl, H. (2001b). The energetic metabolism of societies, part II: Empirical examples. *Journal of Industrial Ecology, 5*, 71–88.

Haberl, H. (2006). The global socioeconomic energetic metabolism as a sustainability problem. *Energy – The International Journal, 31*, 87–99.

Haberl, H., & Krausmann, F. (2007). The local base of the historical agrarian-industrial transition, and the interaction between scales. In M. Fischer-Kowalski & H. Haberl (Eds.), *Socioecological transitions and global change: Trajectories of social metabolism and land use* (pp. 116–138). Cheltenham/Northampton: Edward Elgar.

Haberl, H., Erb, K.-H., Krausmann, F., Loibl, W., Schulz, N. B., & Weisz, H. (2001). Changes in ecosystem processes induced by land use: Human appropriation of net primary production and its influence on standing crop in Austria. *Global Biogeochemical Cycles, 15*, 929–942.

Haberl, H., Erb, K.-H., Krausmann, F., Adensam, H., & Schulz, N. B. (2003). Land-use change and socioeconomic metabolism in Austria. Part II: Land-use scenarios for 2020. *Land Use Policy, 20*, 21–39.

Haberl, H., Fischer-Kowalski, M., Krausmann, F., Weisz, H., & Winiwarter, V. (2004a). Progress Towards sustainability? What the conceptual framework of material and energy flow accounting (MEFA) can offer. *Land Use Policy, 21*, 199–213.

Haberl, H., Schulz, N. B., Plutzar, C., Erb, K.-H., Krausmann, F., Loibl, W., Moser, D., Sauberer, N., Weisz, H., Zechmeister, H. G., & Zulka, P. (2004b). Human appropriation of net primary production and species diversity in agricultural landscapes. *Agriculture, Ecosystems and Environment, 102*, 213–218.

Haberl, H., Plutzar, C., Erb, K.-H., Gaube, V., Pollheimer, M., & Schulz, N. B. (2005). Human appropriation of net primary production as determinant of avifauna diversity in Austria. *Agriculture, Ecosystems and Environment, 110*, 119–131.

Haberl, H., Weisz, H., Amann, C., Bondeau, A., Eisenmenger, N., Erb, K.-H., Fischer-Kowalski, M., & Krausmann, F. (2006a). The energetic metabolism of the European Union and the United States: Decadal energy input time-series with an emphasis on biomass. *Journal of Industrial Ecology, 10*, 151–171.

Haberl, H., Winiwarter, V., Andersson, K., Ayres, R. U., Boone, C. G., Castillio, A., Cunfer, G., Fischer-Kowalski, M., Freudenburg, W. R., Furman, E., Kaufmann, R., Krausmann, F., Langthaler, E., Lotze-Campen, H., Mirtl, M., Redman, C. A., Reenberg, A., Wardell, A. D., Warr, B., & Zechmeister, H. (2006b). From LTER to LTSER: Conceptualizing the socio-economic dimension of long-term socio-ecological research. *Ecology and Society, 11*, 13. (Online), http://www.ecologyandsociety.org/vol11/iss2/art13/

Haberl, H., Erb, K.-H., Krausmann, F., Gaube, V., Bondeau, A., Plutzar, C., Gingrich, S., Lucht, W., & Fischer-Kowalski, M. (2007a). Quantifying and mapping the human appropriation of net primary production in earth's terrestrial ecosystems. *Proceedings of the National Academy of Sciences of the United States of America, 104*, 12942–12947.

Haberl, H., Erb, K.-H., Plutzar, C., Fischer-Kowalski, M., Krausmann, F., Hak, T., Moldan, B., & Dahl, A. L. (2007b). Human appropriation of net primary production (HANPP) as indicator for pressures on biodiversity. In *Sustainability indicators. A scientific assessment* (pp. 271–288). Washington, DC, Covelo/London: SCOPE, Island Press.

Haberl, H., Erb, K.-H., Krausmann, F., Berecz, S., Ludwiczek, N., Musel, A., Schaffartzik, A., & Martinez-Alier, J. (2009a). Using embodied HANPP to analyze teleconnections in the global land system: Conceptual considerations. *Geografisk Tidsskrift – Danish Journal of Geography, 109*, 119–130.

Haberl, H., Gaube, V., Díaz-Delgado, R., Krauze, K., Neuner, A., Peterseil, J., Plutzar, C., Singh, S. J., & Vadineanu, A. (2009b). Towards an integrated model of socioeconomic biodiversity drivers, pressures and impacts. A feasibility study based on three European long-term socio-ecological research platforms. *Ecological Economics, 68*, 1797–1812.

Haberl, H., Fischer-Kowalski, M., Krausmann, F., Martinez-Alier, J., & Winiwarter, V. (2011). A socio-metabolic transition towards sustainability? Challenges for another great transformation. *Sustainable Development, 19*, 1–14.

Hall, C. A. S., Cleveland, C. J., & Kaufmann, R. K. (Eds.). (1986). *Energy and resource quality. The ecology of the economic process*. New York: Wiley-Interscience.

Hall, C. A. S., Lindenberger, D., Kümmel, R., Kroeger, T., & Eichhorn, W. (2001). The need to reintegrate the natural sciences into economics. *BioScience, 51*, 663–673.

Hatanaka, N., Wright, W., Loyin, R. H., & MacNally, R. (2011). 'Ecologically complex carbon' – Linking biodiversity values, carbon storage and habitat structure in some austral temperate forests. *Global Ecology and Biogeography, 20*, 260–271.

Hoekstra, A. Y., & Chapagain, A. K. (2007). Water footprints of nations: Water use by people as a function of their consumption pattern. *Water and Resource Management, 21*, 35–48.

Hoekstra, A. Y., & Chapagain, A. K. (Eds.). (2008). *Globalization of water. Sharing the planet's freshwater resources*. Malden: Blackwell Publishing.

Hoekstra, R., & van den Bergh, J. C. J. M. (2006). Constructing physical input-output tables for environmental modeling and accounting: Framework and illustrations. *Ecological Economics, 59*(3), 375–393.

IEA. (2010, December 19). "World energy statistics", IEA World energy statistics and balances (database). doi: 10.1787/data-00510-en. OECD Library (Klagenfurt University). International Energy Agency (IEA).

IPCC. (2007). *Climate change 2007. Synthesis report. Contribution of Working Groups I, II and III to the fourth assessment report of the Intergovernmental Panel on Climate Change*. Cambridge/New York: Cambridge University Press.

Kastner, T. (2009). Trajectories in human domination of ecosystems: Human appropriation of net primary production in the Philippines during the 20th century. *Ecological Economics, 69*, 260–269.

Kastner, T., Kastner, M., & Nonhebel, S. (2011). Tracing distant environmental impacts of agricultural products from a consumer perspective. *Ecological Economics, 70*, 1032–1040.

Kauppi, P. E., Ausubel, J. H., Fang, J., Mather, A. S., Sedjo, R. A., & Waggoner, P. E. (2006). Returning forests analyzed with the forest identity. *Proceedings of the National Academy of Sciences of the United States of America, 103*, 17574–17579.

Kay, J. J., Regier, H. A., Boyle, M., & Francis, G. (1999). An ecosystem approach for sustainability: Addressing the challenge of complexity. *Futures, 31*, 721–742.

Körner, C. (2009). Biologische Kohlenstoffsenken: Umsatz und Kapital nicht verwechseln! *Gaia, 18*, 288–293.

Kovanda, J., Havranek, M., & Hak, T. (2007). Calculation of the "Net additions to stock" indicator for the Czech Republic using a direct method. *Journal of Industrial Ecology, 11*, 140–154.

Krausmann, F. (2001). Land use and industrial modernization: An empirical analysis of human influence on the functioning of ecosystems in Austria 1830–1995. *Land Use Policy, 18*, 17–26.

Krausmann, F. (2004). Milk, manure and muscular power. Livestock and the industrialization of agriculture. *Human Ecology, 32*, 735–773.

Krausmann, F., & Haberl, H. (2002). The process of industrialization from the perspective of energetic metabolism. Socioeconomic energy flows in Austria 1830–1995. *Ecological Economics, 41*, 177–201.

Krausmann, F., & Haberl, H. (2007). Land-use change and socio-economic metabolism. A macro view of Austria 1830–2000. In M. Fischer-Kowalski & H. Haberl (Eds.), *Socioecological transitions and global change: Trajectories of social metabolism and land use* (pp. 31–59). Cheltenham/Northampton: Edward Elgar.

Krausmann, F., Gingrich, S., Eisenmenger, N., Erb, K.-H., Haberl, H., & Fischer-Kowalski, M. (2009a). Growth in global materials use, GDP and population during the 20th century. *Ecological Economics, 68*, 2696–2705.

Krausmann, F., Haberl, H., Erb, K.-H., Wiesinger, M., Gaube, V., & Gingrich, S. (2009b). What determines spatial patterns of the global human appropriation of net primary production? *Journal of Land Use Science, 4*, 15–34.

Kuemmerle, T., Olofsson, P., Chaskovskyy, O., Baumann, M., Ostapowicz, K., Woodcock, C. E., Houghton, R. A., Hostert, P., Keeton, W. S., & Radeloff, V. C. (2011). Post-Soviet farmland abandonment, forest recovery, and carbon sequestration in western Ukraine. *Global Change Biology, 17*, 1335–1349.

Lambin, E. F., & Geist, H. J. (Eds.). (2006). *Land-use and land-cover change. Local processes and global impacts*. Berlin: Springer.

Lambin, E. F., Rounsevell, M. D. A., & Geist, H. J. (2000). Are agricultural land-use models able to predict changes in land-use intensity? *Agriculture, Ecosystems and Environment, 82*, 321–331.

Lambin, E. F., Turner, B. L. I., Geist, H. J., Agbola, S. B., Angelsen, A., Bruce, J. W., Coomes, O. T., Dirzo, R., Fischer, G., Folke, C., George, P. S., Homewood, K., Imbernon, J., Leemans, R., Li, X., Moran, E. F., Mortimore, M., Ramakrishnan, P. S., Richards, J. F., Skanes, H., Steffen, W., Stone, G. D., Svedin, U., Veldkamp, T. A., Vogel, C., & Xu, J. (2001). The causes of land-use and land-cover change: Moving beyond the myths. *Global Environmental Change, 11*, 261–269.

Lieth, H., & Whittaker, R. H. (Eds.). (1975). *Primary productivity of the biosphere*. Berlin/Heidelberg/New York: Springer.

Lindeman, R. L. (1942). The trophic-dynamic aspect of ecology. *Ecology, 23*, 399–417.

Martinez-Alier, J. (Ed.). (1987). *Ecological economics. Energy, environment and society*. Oxford: Blackwell.

Mather, A. S., & Fairbairn, J. (1990). From floods to reforestation: The forest transition in Switzerland. *The American Historical Review, 95*, 693–714.

Matthews, E., Amann, C., Fischer-Kowalski, M., Bringezu, S., Hüttler, W., Kleijn, R., Moriguchi, Y., Ottke, C., Rodenburg, E., Rogich, D., Schandl, H., Schütz, H., van der Voet, E., & Weisz, H. (Eds.). (2000). *The weight of nations: Material outflows from industrial economies*. Washington, DC: World Resources Institute.

McGuire, A. D., Sitch, S., Clein, J. S., Dargaville, R., Esser, G., Foley, J. A., Heimann, M., Joos, F., Kaplan, J., Kicklighter, D. W., Meier, R. A., Melillo, J. M., Moore, B., III, Prentice, I. C., Ramankutty, N., Reichenau, T., Schloss, A., Tian, H., Williams, L. J., & Wittenberg, U. (2001). Carbon balance of the terrestrial biosphere in the twentieth century: Analyses of $CO_2$, climate and land-use effects with four process-based ecosystem models. *Global Biogeochemical Cycles, 15*, 183–206.

Meyfroidt, P., Rudel, T. K., & Lambin, E. F. (2010). Forest transitions, trade, and the global displacement of land use. *Proceedings of the National Academy of Sciences of the United States of America, 107*, 20917–20922.

Millennium Ecosystem Assessment (Ed.). (2005). *Ecosystems and human well-being – Our human planet. Summary for decision makers*. Washington, DC: Island Press.

Musel, A. (2009). Human appropriation of net primary production in the United Kingdom, 1800–2000. Changes in society's impact on ecological energy flows during the agrarian-industrial transition. *Ecological Economics, 69*, 270–281.

Odum, H. T. (Ed.). (1971). *Environment, power, and society*. New York: Wiley-Interscience.

OECD. (2008). *Measuring material flows and resource productivity. Synthesis report*. Paris: Organisation for Economic Co-operation and Development (OECD).

Pauly, D., Watson, R., & Alder, J. (2005). Global trends in world fisheries: Impacts on marine ecosystems and food security. *Philosophical Transactions: Biological Sciences, 360*, 5–12.

Pimentel, D., Hurd, L. E., Bellotti, A. C., Forster, M. J., Oka, I. N., Sholes, O. D., & Whitman, R. J. (1973). Food production and the energy crisis. *Science, 182*, 443–449.

Pimentel, D., Dazhong, W., & Giampietro, M. (1990). Technological changes in energy use in U.S. agricultural production. In S. R. Gliessman (Ed.), *Agroecology, researching the ecological basis for sustainable agriculture* (pp. 305–321). New York: Springer.

Postel, S. L., Daily, G. C., & Ehrlich, P. R. (1996). Human appropriation of renewable fresh water. *Science, 271*, 785–788.

Rappaport, R. A. (1971). The flow of energy in an agricultural society. *Scientific American, 225*, 117–133.

Rauch, J. N. (2010). Global spatial indexing of the human impact on Al, Cu, Fe, and Zn mobilization. *Environmental Science and Technology, 44*, 5728–5734.

Rauch, J. N., & Pacyna, J. M. (2009). Earth's global Ag, Al, Cr, Cu, Fe, Ni, Pb, and Zn cycles. *Global Biogeochemical Cycles, 23*, GB2001. doi:10.1029/2008GB003376.

Rebitzer, G., Ekvall, T., Frischknecht, R., Hunkeler, D., Norris, G., Rydberg, T., Schmidt, W.-P., Suh, S., Weidema, B. P., & Pennington, D. W. (2004). Life cycle assessment: Part 1: Framework, goal and scope definition, inventory analysis, and applications. *Environment International, 30*, 701–720.

Sala, O. E., Chapin, F. S., III, Armesto, J. J., Berlow, E., Bloomfield, J., Dirzo, R., Huber-Sannwald, E., Huennecke, L. F., Jackson, R. B., Kinzig, A., Leemans, R., Lodge, D. M., Mooney, H. A., Oesterheld, M., Poff, N. L., Sykes, M. T., Walker, B., Walker, M., & Wall, D. H. (2000). Global biodiversity scenarios for the year 2100. *Science, 287*, 1770–1774.

Schlamadinger, B., Apps, M. J., Bohlin, F., Gustavsson, L., Jungmeier, G., Marland, K., Pingoud, K., & Savolainen, I. (1997). Towards a standard methodology for greenhouse gas balances of bioenergy systems and comparison with fossil energy systems. *Biomass and Bioenergy, 13*, 359–375.

Schwarzlmüller, E. (2009). Human appropriation of aboveground net primary production in Spain, 1955–2003: An empirical analysis of the industrialization of land use. *Ecological Economics, 69*, 282–291.

Searchinger, T. D., Hamburg, S. P., Melillo, J., Chameides, W., Havlik, P., Kammen, D. M., Likens, G. E., Lubowski, R. N., Obersteiner, M., Oppenheimer, M., Philip Robertson, G., Schlesinger, W. H., & vid Tilman, G. (2009). Fixing a critical climate accounting error. *Science, 326*, 527–528.

Sieferle, R. P., Krausmann, F., Schandl, H., & Winiwarter, V. (2006). *Das Ende der Fläche. Zum gesellschaftlichen Stoffwechsel der Industrialisierung*. Köln: Böhlau.

Steffen, W., Sanderson, A., Tyson, P. D., Jäger, J., Matson, P. A., Moore, B., III, Oldfield, F., Richardson, K., Schellnhuber, H. J., Turner, B. L., II, & Wasson, R. J. (Eds.). (2004). *Global change and the Earth system. A planet under pressure*. Berlin: Springer.

Steffen, W., Crutzen, P. J., & McNeill, J. R. (2007). The Anthropocene: Are humans now overwhelming the great forces of nature. *Ambio, 36*, 614–621.

Suh, S. (2005). Theory of materials and energy flow analysis in ecology and economics. *Ecological Modelling, 189*, 251–269.

Swartz, W., Sala, E., Tracey, S., Watson, R., & Pauly, D. (2010). The spatial expansion and ecological footprint of fisheries (1950 to present). *PLoS ONE, 5*, e15154.

Turner, B. L. I., Lambin, E. F., & Reenberg, A. (2007). The emergence of land change science for global environmental change and sustainability. *Proceedings of the National Academy of Sciences of the United States of America, 104*, 20666–20671.

van der Leeuw, S. E. (2004). Why model? *Cybernetics and Systems, 35*, 117–128.

Verburg, P. H., Chen, Y., & Veldkamp, T. (2000). Spatial explorations of land use change and grain production in China. *Agriculture, Ecosystems and Environment, 82*, 333–354.
Verburg, P. H., Neumann, K., & Nol, L. (2010). Challenges in using land use and land cover data for global change studies. *Global Change Biology, 17*, 974–989.
Vitousek, P. M., Ehrlich, P. R., Ehrlich, A. H., & Matson, P. A. (1986). Human appropriation of the products of photosynthesis. *BioScience, 36*, 363–373.
Vitousek, P. M., Aber, J. D., Howarth, R. W., Likens, G. E., Matson, P. A., Schindler, D. W., Schlesinger, W. H., & Tilman, D. G. (1997). *Human alteration of the global nitrogen cycle: Causes and consequences* (Issues in ecology, 1). Washington, DC: Ecological Society of America.
Vörösmarty, C. J., Sharma, K. P., Fekete, V. M., Copeland, A. H., Holden, J., Marble, J., & Lough, J. A. (1997). The storage and aging of continental runoff in large reservoir systems of the world. *Ambio, 26*, 210–219.
Waide, R. B., Willig, M. R., Steiner, C. F., Mittelbach, G., Gough, L., Dodson, S. I., Juday, G. P., & Parmenter, R. (1999). The relationship between productivity and species richness. *Annual Review of Ecology and Systematics, 30*, 257–300.
Wang, T., Müller, D. B., & Graedel, T. E. (2007). Forging the anthropogenic iron cycle. *Environmental Science and Technology, 41*, 5120–5129.
Weiß, M., Schaldach, R., Alcamo, J., & Flörke, M. (2009). Quantifying the human appropriation of fresh water by African agriculture. *Ecology and Society, 14*, 25. (Online) URL: http://www.ecologyandsociety.org/vol14/iss2/art25/
Weisz, H., & Duchin, F. (2006). Physical and monetary input-output analysis: What makes the difference? *Ecological Economics, 57*, 534–541.
Weisz, H., Krausmann, F., Amann, C., Eisenmenger, N., Erb, K.-H., Hubacek, K., & Fischer-Kowalski, M. (2006). The physical economy of the European Union: Cross-country comparison and determinants of material consumption. *Ecological Economics, 58*, 676–698.
Whittaker, R. H., & Likens, G. E. (1973). Primary production: The biosphere and man. *Human Ecology, 1*, 357–369.
Wright, D. H. (1983). Species-energy theorie: An extension of the species-area theory. *Oikos, 41*, 495–506.
Wright, D. H. (1990). Human impacts on the energy flow through natural ecosystems, and implications for species endangerment. *Ambio, 19*, 189–194.

# Chapter 3
# Using Integrated Models to Analyse Socio-ecological System Dynamics in Long-Term Socio-ecological Research – Austrian Experiences

**Veronika Gaube and Helmut Haberl**

**Abstract** Society-nature interaction is an inherently complex process the analysis of which requires inter- and transdisciplinary efforts. Integrated socio-ecological modelling is an approach to synthesize concepts and insights from various scientific disciplines into a coherent picture and thereby better understand the interrelations between various drivers behind the trajectories of socio-ecological systems. We here discuss insights gained in developing the integrated model SERD (Simulation of Ecological Compatibility of Regional Development) for the municipality of Reichraming in the centre of the Austrian LTSER platform Eisenwurzen. The model includes an agent-based actor module coupled with a spatially explicit land use module and a biophysical stock-flow module capable of simulating socio-ecological material flows (C and N). The model was developed, implemented and used in a transdisciplinary research process together with relevant stakeholders. We conclude that the development of such models is highly attractive for LTSER due to their ability to integrate contributions from various scientific disciplines and stakeholders and support learning on interactions in complex socio-ecological systems. Integrated socio-ecological models can therefore also support the study of sustainability-related issues in land-change science.

**Keywords** Agent-based modelling • Austria • Decision-making • Eisenwurzen • Integrated socio-ecological analyses • Participatory process

V. Gaube, Ph.D. (✉) • H. Haberl, Ph.D.
Institute of Social Ecology Vienna (SEC), Alpen-Adria Universitaet Klagenfurt,
Wien, Graz, Schottenfeldgasse 29/5, Vienna 1070, Austria
e-mail: veronika.gaube@aau.at; helmut.haberl@aau.at

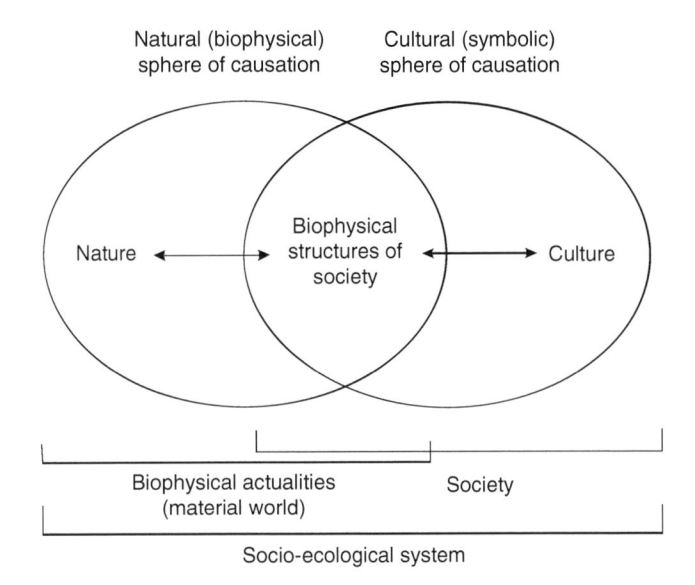

**Fig. 3.1** Generic heuristic model of socio-ecological systems at the interface of natural and cultural spheres of causation [Reprinted from Haberl et al. (2004). With permission from Elsevier]

## 3.1 Introduction

The notion that scientists need to better understand processes of society-nature interaction in order to be able to deliver the kind of knowledge required to support decision-making on sustainability issues is widely accepted (e.g. Kates et al. 2001; Ostrom 2009; Reid et al. 2010). Attempts to define sustainability have emphasised different aspects of this broad concept, mostly depending on scientific disciplines, professional backgrounds and personal interests of the researchers involved (see, for example, Brandt 1997; Clark and Munn 1986; Haberl et al. 2011; Holling 1986; Pearce et al. 1990; WCED 1987). However, any definition of sustainability that relates to society-nature interaction implies a need to observe societies and natural systems as well as their interaction over time and space, thereby seeking to answer questions such as 'which changes in natural systems are caused by socioeconomic activities?', 'what are the drivers of these changes?', 'how will these changes affect natural and socioeconomic systems' and 'how can we influence these trajectories in a more sustainable direction?' (Haberl et al. 2004). Such questions have inspired moves to develop the concept of Long-Term Socio-Ecological Research (LTSER) as outlined in recent publications (Redman et al. 2004; Haberl et al. 2006; Singh et al. 2010) and broadly exemplified in this volume.

Sustainability science hence focuses on socio-ecological (or human-environment) systems (Fischer-Kowalski and Haberl 2007; Folke and Gunderson 2006; GLP 2005; Holling 2001; Young et al. 2006a). Socio-ecological systems emerge through the interaction of societies with their natural environment (Fischer-Kowalski and Weisz 1999, see Fig. 3.1 below). Socio-ecological systems are inherently complex assemblages that are spatially heterogeneous and change over time according to an intricate interplay of a large number of biophysical as well as socioeconomic factors, including human decision-making. Attempts to better understand the spatio-temporal variability of socio-ecological systems need to be based on interdisciplinary approaches that allow us to take biophysical, social and economic factors as well as their interactions into account. Addressing this complexity is a prerequisite for developing a sufficiently inclusive understanding of the system dynamics at hand, and therefore also for deriving insights that are useful in supporting decision-making processes and in assessing why particular interventions may or may not be successful (Matthews 2006). Building and applying models is an approach that has proven useful in order to tackle this complexity (van der Leeuw 2004) and to support interdisciplinary collaboration and communication (Gaube et al. 2009a; Newig et al. 2008a; Verburg et al. 2004).

Land use is a prominent example of such complex and dynamic socio-ecological interaction processes. The recognition of the importance of adopting integrated approaches in studying land use has recently resulted in the notion of 'integrated land-system science' (GLP 2005; Turner et al. 2007). Land use directly affects biota, soils, water and the atmosphere and is therefore environmentally highly relevant (Meyer and Turner 1994). Land use is currently changing around the globe, often rapidly, and these changes are a pervasive driver of global environmental change (Foley et al. 2005), contributing to environmental problems such as biodiversity loss, greenhouse-gas emissions, degradation of ecosystems and loss of ecosystem service delivery and so on (Lambin and Geist 2006; Millennium Ecosystem Assessment 2005b). At the same time, the need to adequately nourish a growing world population, reduce hunger and malnutrition, provide renewable energy as well as raw materials and accommodate growing demands for living space and infrastructure add up to a formidable challenge, perhaps one of the toughest issues currently facing humanity (Erb et al. 2009; Godfray et al. 2010).

Dynamics and patterns of land use are simultaneously influenced by natural factors such as vegetation, land forms, climate or soil, as well as by socioeconomic characteristics such as family structures, diets, economic incentives and preferences, the structure of the economy in terms of global price developments of agricultural products, property rights, subsidies, markets and many others (Lambin et al. 2001; Reenberg and Lund 2004; Wrbka et al. 2004). Land-use change models can help in gaining a more detailed understanding of the interactions between these factors by analysing causes and consequences of land-use change (Matthews and Selman 2006; Verburg et al. 2004). Moreover, such models can also support the analysis of decision-making processes influencing land-use change and they can help to structure participatory processes and even decision-making in administration, policy-making or other settings (Newig et al. 2008b; Gaube et al. 2009a). Land-use change models used for these purposes address the question of how a socio-ecological system has evolved into its current state and how it might change in the future. In other words, how are interactions between the social and the natural system changing, what implications do these changes have for the state of the socio-ecological system, and how might the trajectory be influenced (Agarwal et al. 2002)?

Decisions of relevant actors shape land use. Decisions are made at many spatial and temporal scales, and the question of which actors are most relevant is strongly scale-dependent (Dirnböck et al. 2008; see Dirnböck et al., Chap. 6 in this volume). Moreover, feedbacks between spatial scales are abundant, and these feedbacks act at a different pace on various relevant scales (Gibson et al. 2000; Gunderson and Holling 2002). Agent-based models are a powerful approach to analyse these complex, context- and scale-dependent decision-making processes that influence land-use change (Berger 2001; Gaube et al. 2009a; Parker et al. 2003). Agent-based models are currently used as a tool for understanding the dynamics of socio-ecological systems in which the decisions of actors influence biophysical dynamics, such as socioeconomic metabolism and land use, and vice versa (e.g. McConnell 2001; Janssen 2004; Manson and Evans 2007). They are attractive in that respect because they are capable of simulating the aggregate outcomes resulting from the decisions of many individual actors.

In order to be able to focus research processes on the concrete needs of key actors or stakeholders, approaches are needed that allow mutually beneficial interaction processes between scientists and stakeholders. In many cases, it is necessary to entertain participative processes throughout the entire research process, from problem definition to the planning of concrete interventions or measures. Participation of this kind is key to enabling social actors or social systems to learn from or to be stimulated by the research process (Hare and Pahl-Wostl 2002; Newig et al. 2008a; Pahl-Wostl 2002a).

In this chapter we discuss how integrated models that comprise agent-based, land-use and biophysical stock-flow modules can be applied in LTSER. We argue that integrated socio-ecological models can contribute to local and regional sustainability studies by supporting transdisciplinary research and by structuring participatory processes that involve local stakeholders. We show that agent-based modelling helps in gaining a detailed and structured knowledge of socio-ecological systems and both facilitates and depends on the integration of relevant local stakeholders. We will discuss a case study examining how social and political interventions affect patterns of land use as well as socio-economic conditions in the Reichraming municipality, located in the LTSER region of Eisenwurzen in Austria.

## 3.2 Sustainability Science and Integrated Socio-ecological Modelling

The concept of socio-ecological systems outlined in Fig. 3.1 provides a heuristic basis for linking different disciplinary approaches in analysing biophysical, symbolic and social systems as well as their interactions. Socio-ecological systems can be depicted as two overlapping spheres, one representing a 'natural' or 'biophysical' sphere of causation governed by natural laws and a second one representing a 'cultural' or 'symbolic' sphere of causation reproduced by symbolic communication. The overlap between the two spheres constitutes the 'biophysical structures of society' that are part of both the cultural and the natural sphere of causation. According to this model, society continuously reproduces its symbolic as well as its biophysical structures by interacting with its biophysical environment (Fischer-Kowalski and Weisz 1999; Haberl et al. 2004, 2006; Singh et al. 2010; Weisz et al. 2001).

One approach to empirically describe the biophysical components of the interaction process between socioeconomic and natural systems is 'socioeconomic metabolism' (Ayres and Simonis 1994; Fischer-Kowalski 1997; Fischer-Kowalski et al. 1997; Martinez-Alier 1987; Matthews et al. 2000). This approach is focused on the analysis of material and energy flows between the biophysical structures of society and other components of the biophysical sphere of causation (see Haberl et al., Chap. 2 in this volume). The central idea of the metabolism approach is to conceptualise society as a physical input-output system that draws material and energy from its environment, maintains internal physical processes and dissipates wastes, emissions and low-quality energy to the environment (Georgescu-Roegen 1971; Daly 1973).

The analysis of material and energy flows related to economic activities is, however, not sufficient to capture society-nature interactions fully. Another important aspect of society-nature interaction is land use (Meyer and Turner 1994; Foley et al. 2005). Land use has been described as a form of 'colonisation of natural processes' (Fischer-Kowalski and Haberl 1997; Haberl et al. 2001; Weisz et al. 2001). The notion of colonisation emphasises that humans deliberately alter ecosystems and organisms in order to obtain goods (e.g. agricultural or forestry products) or services (e.g. flood prevention) that society wants. The effects of colonisation on ecosystems can be analysed empirically by comparing currently prevailing ecosystem patterns and processes with those that would be expected without human intervention, for example by using indicators such as the 'human appropriation of net primary production' or HANPP (Vitousek et al. 1986; Wright 1990; see Haberl et al., Chap. 2 in this volume).

The approaches of metabolism and colonisation were the conceptual basis for several empirical studies on land use, material and energy flows, their interactions and long-term trajectories (e.g., Fischer-Kowalski and Haberl 2007; Sieferle et al. 2006, see Haberl et al., Chap. 2 in this volume; Krausmann, Chap. 11 in this volume; Gingrich et al., Chap. 13 in this volume; Krausmann and Fischer-Kowalski, Chap. 15 in this volume).

These studies focused on very different scales and periods, from the local, regional and national to the global level, and from decadal to centennial historical studies as well as analyses of current conditions and future scenarios. These concepts provide a sound theoretical and empirical basis for integrating approaches from natural sciences, social sciences and the humanities in analysing long-term changes in society-nature interaction in LTSER (Haberl et al. 2006; Singh et al. 2010).

## 3.2.1 *LTSER Platforms and Models*

In particular, research in LTSER platforms needs models to synthesise contributions from different scientific disciplines in a coherent picture, to analyse and interpret patterns in space and dynamics over time, to reconstruct historical states based on incomplete data and to support stakeholders in their decision-making relevant for future sustainability (Gaube et al. 2009a; Haberl et al. 2006, 2009; Krausmann 2004; Newig et al. 2008a; van der Leeuw 2004). The goal for LTSER platforms should be to develop specific LTSER models that are sensitive to the characteristics of their respective sites or regions and research teams. Four general themes are important in this context (Haberl et al. 2006): (1) socio-ecological metabolism, (2) land use and landscapes, (3) governance and (4) communication. All four issues require interdisciplinary approaches, as they are crosscutting aspects of the interactions between social and natural systems. These themes should guide the development of models while also recognising the widely different objectives and history of LTSER platforms around the world, thereby providing flexibility to researchers. Developing and applying integrated socio-ecological models can help to combine the expertise of scientists from various disciplines with needs and insights of local stakeholders in tackling these questions and should thus be a priority of LTSER.

Existing models, even those applied to address environmental or sustainability issues, are often based on theories and concepts drawn from only one scientific discipline. Many models focus either on biophysical (e.g. ecological or climatic) or on economic aspects. Even models addressing broader questions often have at their core either an ecological or an economic model, perhaps extended by modules capable of establishing relations to other aspects of socio-ecological systems (Parker et al. 2002; Milne et al. 2009). Another approach is based on the coupling of existing models from various disciplines in order to be able to consider social, economic and ecological factors in land systems (Rounsevell et al. 2006; Schaldach and Priess 2008; Verburg et al. 2008). However, a shortcoming of such a model-coupling approach is that these models were often developed for different purposes and have different levels of complexity and basic procedures (Milne et al. 2009).

We here focus on a different approach, namely the construction of new, integrated socio-ecological models more or less from scratch that comprise social, economic and ecological factors in their basic design. In such models, socioeconomic and ecological components have a similar level of complexity, following the premise that these models should focus on society-nature interaction processes (Gaube et al. 2009a; Haberl et al. 2009). In order to allow for a model with the capability to integrate socioeconomic as well as ecological processes we developed a model that is based on a combination of agent-based modelling (ABM) approaches with system-dynamic (stock-flow) modelling approaches and spatial information.

### 3.2.2 Agent-Based Models in LTSER

Agent-based models are based on a formalised representation of social systems consisting of virtual agents and their environment (Epstein and Axtell 1996; Kohler and Gumerman 2000; Parker et al. 2003). Agents interact with other agents and with their environment. The term 'environment' in this case is not limited to the natural environment, but includes the socioeconomic situation of the region as well. Agents possess knowledge of the system they belong to and are able to incorporate information about system states and actions of other agents (Ferber 1999). ABMs can be used to analyse many issues, such as for example predator-prey relationships, supply-chain optimisation, and consumer behaviour or traffic congestion. Among these and many more issues ABMs also have a long tradition of being used for land-use studies. In the context of being used for land-use issues, ABMs can be described as land-user models rather than land-use models, as this approach is focused on modelling decisions of land users, for example farmers. Decisions of individuals or groups are evaluated in terms of their implications for land-use change (Koomen et al. 2007). Decision-making and interaction between agents and their environment constitute the central elements of ABMs. Agents may be individuals (e.g., householders, farmers, developers) or organisations (e.g., NGOs, firms), and an ABM can comprise one or several types of agents. Specifying agents requires defining their state (e.g., preferences) and the rules on which their decision-making is based. Environmental change may follow its own dynamics – often simulated with other modelling approaches such as stock-flow models – but is in turn also influenced by the aggregated agents' behaviour.

Agent-based models (ABM) represent a distinct, new modelling approach that is qualitatively different from other mathematical and statistical approaches (e.g. using differential equations or linear programming). The agents and their interactions involved in those ABMs being used for land-use studies mainly constitute proximate causes of land-use change, such as availability of land and costs of land management (Geist and Lambin 2002), although other causes may be analysed as well. Their main advantage is that ABMs facilitate the analysis and understanding of processes, in particular of decision-making and its effects on land use (DeAngelis and Gross 1992). A lot of work has therefore been undertaken in recent years to develop such models in the context of land-change science (e.g. Berger 2004; Brown and Duh 2003; Parker et al. 2002; Rounsevell et al. 2006; Verburg et al. 2004).

## 3.3 Modelling a Region – Empirical Experiences from Austrian LTSER

In this section we discuss the development of an integrated socio-ecological model in the Eisenwurzen LTSER platform in Austria (see Peterseil et al., Chap. 19 in this volume). The purpose of this study was to evaluate how social and political interventions affect land use as well as socioeconomic conditions in rural regions, and what their ecological implications in terms of land-use change and carbon and nitrogen stocks and flows are (Gaube et al. 2009a). Our aim was to develop the integrated model SERD (Simulation of Ecological Compatibility of Regional Development), intended to help local and regional actors in discussing possible strategies or measures to mitigate potentially adverse effects of such changes. We planned to achieve these targets by developing an agent-based model in combination with a stock-flow model embedded in a participatory project design.

Based on results from previous projects in other regions in Austria (Gaube et al. 2009b; Gaube and Haberl 2012; Newig et al. 2008a), we assumed that participative model development should fit the research requirements of LTSER well by supporting transdisciplinary research right from the beginning. Consequently, we conducted a project located in Reichraming, a municipality situated in the midst of the LTSER platform Eisenwurzen (see Peterseil et al., Chap. 19 in this volume). Reichraming is a mountainous municipality with some higher peaks and flat areas and over 80% forest cover. Most of the forest is managed by Austria´s state-owned forest administration (ÖBf). Most of the 60 farms located in Reichraming raise cattle and produce milk. The 'Kalkalpen' national park (the name of which means 'limestone alps') covers about one-third of Reichraming´s area of approximately 100 km$^2$.

The Eisenwurzen has a long history of metal mining and metallurgy that goes back over 800 years. In pre-industrial times there was a heavy draw on the region's forests stemming from the need to supply fuel wood and to feed a large non-agricultural population of miners, workers in metallurgical facilities and forestry based on the region's low-input/low-output agrarian land-use systems (Gingrich and Krausmann 2008). With the broad use of fossil fuels, most mines were abandoned from the late 1800s onwards and metal smelting is now located in industrial centres outside the region. Like most of the Eisenwurzen, Reichraming is of marginal agricultural productivity and forests are re-growing rapidly. Reichraming experiences problems typical for marginalised rural areas such as declining agriculture, lack of jobs, low incomes and creeping deterioration of infrastructure (Landeskulturdirektion Oberösterreich 1998). One major question underlying the development of the model therefore was to analyse the decision-making process of farmers concerning agricultural production (e.g. intensification vs. extensification, abandonment vs. extension, etc.). Such decisions have direct consequences for Reichraming's land use, which then in turn influences carbon and nitrogen stocks and flows in the ecosystem as well as related GHG emissions.

### 3.3.1 The Model's Conceptual Assumptions

SERD combines an agent-based module used to simulate decisions of farm households with a system-dynamic module that simulates changes in land use coupled with a model capable of simulating stocks and flows of carbon and nitrogen. The model simulates not only socioeconomic flows (e.g. related to fossil-fuel use or food supply), but also ecological flows such as plant growth, carbon accumulation in the soil, etc. – in other words, it is based on a socio-ecological metabolism approach (see Haberl et al., Chap. 2 in this volume). The decisions of the agents are affected by the system in which they are embedded and by external changes fed into the model exogenously. Simultaneously, the dynamics of the whole system depend on the decisions of all agents. Interactions between the agents have direct impacts on the socioeconomic situation, on land use and socio-ecological material flows.

```
         Ecological dimension:
              Land use
              ▲   ▲
             ╱     ╲
            ╱       ╲
           ╱         ╲
          ▼           ▼
Social dimension:  ◄──────►  Economic dimension:
  Labour time                      Income
```

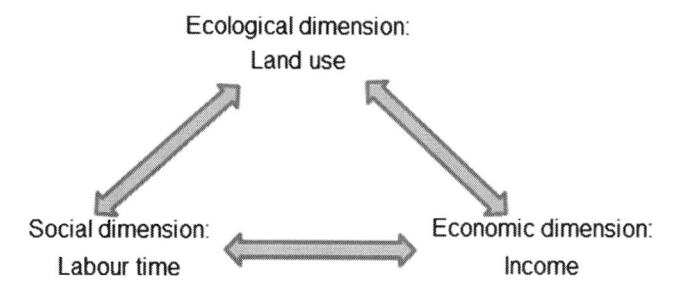

**Fig. 3.2** Sustainability triangle of farmsteads simulated by SERD

The dynamics of the model are driven by assumptions on external conditions in important factors, e.g. agricultural policies and subsidies or prices. Changes in subsidy payments and price relations have direct impacts on family farmsteads; accordingly, the model simulates decisions of farms – in this case, family farmsteads – that are implemented as agents in the model and can take different actions depending on their respective situation. According to the framework conditions at each point in time, each agent evaluates his situation and makes decisions on agricultural production. These decisions in turn affect the socioeconomic situation of the farm and its ecological impact. Consequently, summarising all individual decisions of agents, the model as a whole simulates changes in socioeconomic structures, such as the income and workload of farmsteads, as well as changes in land use and substance flows in the entire study region.

The first question was: How should agents be characterised in order to allow an analysis of socioeconomic actions and dynamics and their link to the biophysical environment? Referring to sustainability's three main dimensions (Fischer-Kowalski 2002; Haberl et al. 2004), we defined each agent in terms of its social, economic and ecological dimensions.

At a local level, one major aspect of the *ecological dimension* is land use, not only because of the direct effects of land use on plant communities, soils, water flows etc., but also because many socioeconomic material flows such as fossil fuels, agricultural products, agrochemicals etc. depend on land-use patterns and intensity. Accordingly, we focused on decisions of farms with respect to land-use type (cropland, grassland and forestry), land-use intensity (conventional, organic, and intensive) and livestock rearing. The *social dimension* may be described as the form and quality of life of a specific social unit in a specific area. In the model this is reflected in various social indicators at the farmstead level, including its family structure and, above all, the use of family labour time required for production and reproduction, with a special focus on the division of labour between men and women. The *economic dimension* is connected with the monetary income of a particular social unit (household, person or community); in our case it was taken into account as the income of all family members living on the farm. The three dimensions are highly interdependent. Trade-offs between these three dimensions are the main criteria taken into account by an agent – that is, a farm – in making its decisions (Fig. 3.2).

According to farm demography – that is, number, age and gender of people living on a farm – only a certain amount of working hours are available and this in turn constrains the extent and intensity of land use. At the same time, as land use requires work, decisions in favour of a certain kind and intensity of land use constrain the time budget that can be used for other activities that are not related to agricultural production, such as working time on the labour market, leisure time or reproduction (e.g., personal care, child-rearing, household work, care for elderly people etc.).

The interdependency between land use and income is implemented in SERD as follows: each agricultural activity generates a certain amount of income, depending on the quality and quantity of the land available and on its respective use. The income, on the other hand, determines and constrains the way in which land can be used: If more income is required, more land can be used, if available, or the land can be used in a different way, e.g. more intensively. Conversely, a high income allows available agricultural land to be extended, for example by leasing additional land. Working time also influences the income. On the one hand, working time constrains income because each activity is assumed to require a certain amount of working hours and the number of working hours that can be mobilised depending on a farm's demography is limited. On the other hand, expectations on minimum income constrain leisure time by requiring more working hours. Every change in the way income is earned – for example, if more time is spent on jobs outside agriculture with a higher hourly income – influences the farm's economic situation.

With respect to all three dimensions, the agents in SERD have in-built dynamic features and are influenced by their environment. For example, time use depends on demography (e.g. how many children and old people must be cared for). It further depends on the socio-cultural system to which a social unit belongs (e.g., social values, traditions and norms, or the infrastructure available). In practice, for example, it is primarily females' working time that is freed up by the availability and usage of an adequate child care system. Income is strongly influenced by the dynamics of economic factors such as markets, prices and subsidies. Finally, land use is constrained by the specific features of the local ecosystem (e.g., rice does not grow in arid areas) as well as by global environmental dynamics such as climate change.

The analysis of decision-making processes within this triangle requires that each agent be defined, including its internal structure in terms of demography, such as family members living on the farm, their gender, age and role on the farm (i.e., agricultural working time). Each farm classified within each respective production type is implemented as one agent. Each agent forms decisions based on its environment as well as on changes in its internal structures (e.g., birth, death, marriage). Farms change their production strategy in terms of intensification or farm size by interacting with their neighbours via a regional land rental market. Beside the extension or reduction of cultivated land area, further reactions are implemented in the model, including, for example, diversification of production, direct marketing of products, increase or decrease of non-agricultural labour time, abandonment of farming, switching to organic agriculture and many more. Decision trees such as those depicted in Fig. 3.3 are used in the model to decide under which conditions and at what time each agent (i.e. farm) decides to change its strategy.

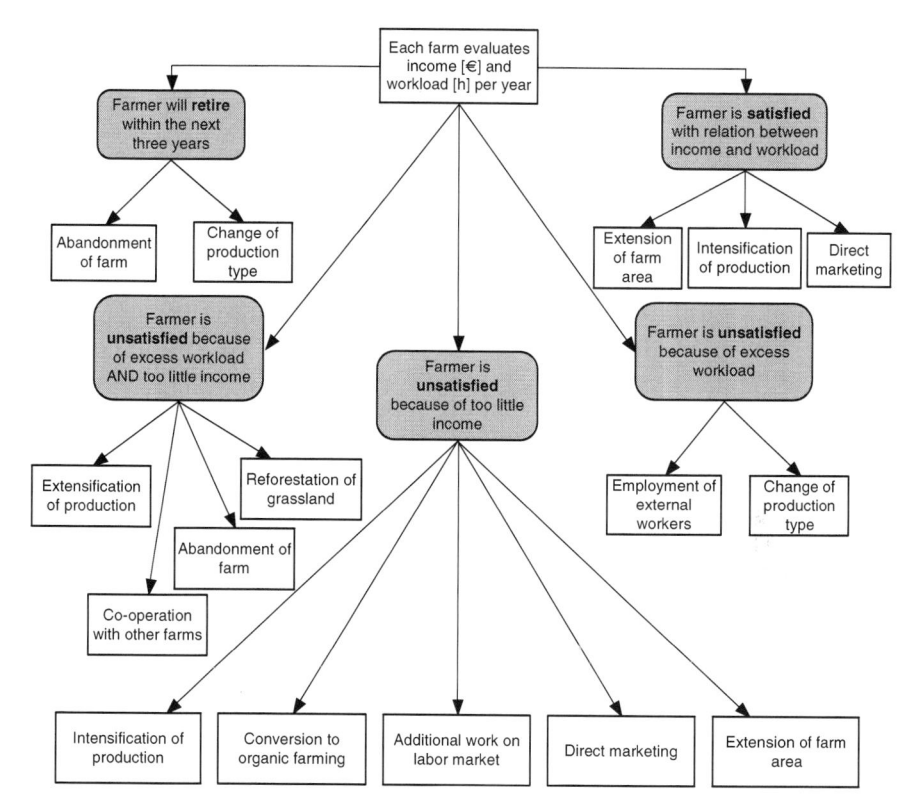

**Fig. 3.3** Decision tree for the 'farm' agent type. *Round light grey boxes*: status of a farm calculated by an automatically induced yearly evaluation of the relation between income and workload. Depending on the status, each farm decides with certain probabilities to take one of the defined actions or to stay passive (that is mostly the case). *Angled boxes*: Options for decisions available in each state. A decision for one of the options changes income, work load, land use, substance flows, etc

Whenever a farm household fails to reach a minimum of income or exceeds a maximum of working hours it seeks to change its situation by choosing an appropriate reaction. Most of the options implemented in SERD are actions that every agent can take without any interactions with other agents in the model. However, since increasing or reducing cultivated land area may affect other agents, a rental market was implemented in the model. Every year, after evaluating its economic and social situation, each farm that wants to rent or lease land 'leaves a message' at the rental market. The model calculates demand and supply for land and allocates it to the single farms. Those farms with requests that cannot be met in a certain year can either take another action or remain passive for the year in question and react again the following year. From the point of modelling requirements, it would have been too complex to implement direct negotiations between the agents. Decision trees are used to determine what kinds of reactions with a specific probability under which circumstances are likely to be made. The design of the decision-making trees as well as all other components of the model has been developed together with local actors by means of a participatory process.

## 3.3.2 Modelling as a Participatory Process

To be able to take into account local actors' requirements and thus derive results or insights that can be implemented in practical measures or policies, many sustainability science approaches employ methods that allow actors to participate in the research, sometimes throughout the entire research process. This entails defining a common research question in a collaborative process between researchers and stakeholder, thus helping to address relevant issues that need to be analysed in order to be able to plan or initiate interventions. Such an approach may enable social actors or social systems to learn throughout the research process. A transdisciplinary research design and a structured participation process are therefore key elements of many approaches in sustainability science (e.g., Jahn 2005; Brand 2000; Hare and Pahl-Wostl 2002; Hirsch Hadorn et al. 2008; Pahl-Wostl 2002a; Rowe and Frewer 2005).

In participatory research approaches, social goals and visions for the future are translated into scientific categories and variables in order to become useful for the research process, that is, they are scientifically operationalised and, in the case of this study, translated into a formalised modelling language. In our study, a considerable number of approximately 15 relevant regional actors were involved in the modelling process right from the beginning. Interviews, focus groups and workshops allowed for research questions, model assumptions and model design to be discussed (Table 3.1).

Throughout the research project, the participatory process contributed to and reflected on the information required for the design of the ABM (for details, see Gaube et al. 2008, 2009a). The ABM was in turn applied as a guiding tool in the participatory process to help local actors reflect on the present and the future as well as in developing strategies and policy priorities. At the beginning of the process, guided interviews with various experts from the provincial government, representatives of the municipality and representatives of the agency of agriculture, the Kalkalpen national park and the federal forestry agency as well as female and male farmers were conducted.

In addition, using a standardised questionnaire, quantitative information on the agricultural production system of each farm (i.e. amount of fodder, harvest and fertiliser) was collected from all 60 farmers in the municipality. Interviews were an effective instrument to establish contact with relevant actors, to gather local knowledge concerning our research questions and to stimulate the interest of actors by making sure that the ongoing research was relevant to them and that they could play an essential role in the research outcome. Another element in the participatory process was a series of workshops and focus groups (Littig and Wallace 1997) with female farmers and representatives of the municipality, aimed at gaining insights into gender-related and land-use affecting issues of Reichraming. Twelve participants

**Table 3.1** Elements of the stakeholder process conducted in Reichraming over a period of 3 years

| Elements of the participatory process | Contents of the meeting | Purpose of the meeting |
| --- | --- | --- |
| Standardised questionnaire for all farmers | Questions on agricultural production, such as fertiliser input, feeding of livestock, number of cuts on grassland, etc. | Standardised data collecting on agricultural management per farm |
| Personal interviews with six farmers (1.5 h each) | Questions on history of the farm, reasons for changes in the production system, influence of family situation on decisions on management, influencing external factors and their impact on decision making, such as subsidies, agricultural product prices, etc. | Gaining in-depth information concerning decision-making on farms |
| Three working group meetings with farmers | Representatives of the three most important production types of the municipality: (1) milk production, (2) fodder production without livestock and (3) farms with horses and game animals discussing future perspectives of their specific production type | Improving system knowledge on both sides (farmers and researchers), improving model assumptions, improving the user interface of the model and visualisation of the scenarios |
| Four stakeholder workshops with representatives of the municipality | Discussion of most relevant challenges the municipality is facing: what are the most important framework conditions influencing socioeconomic parameters such as number of households and farmers, infrastructure such as services, child care facilities, etc. Based on this discussion, future options for Reichraming were developed. | Improving system knowledge on both sides (municipality representatives and researchers), improving model assumptions, improving the user interface of the model and visualisation of the scenarios |
| Three focus groups with women from the municipality | Discussion of the same topics as those in the stakeholder workshops with a specific focus on the women´s perspective: what are the main relevant factors for women influencing their future options in the municipality. | Improving system knowledge on both sides (women and researchers), improving model assumptions, improving the user interface of the model and visualisation of the scenarios |

**Table 3.2** Storylines of the three scenarios trend, globalisation and local policy described by the stakeholders and calculated by the integrated socio-ecological model

| Trend Scenario | Globalisation Scenario | Local policy Scenario |
| --- | --- | --- |
| Strong increase of agricultural prices as well as strongly improved conditions for the use and production of bio-energy will take place. National agricultural subsidies and EU subsidies remain stable. Also internal strategies of farmers and behaviour remain unchanged. | Prices for agricultural products drop substantially. Additionally, agricultural subsidies (national and EU) are cancelled and global conditions for bio-energy becomes less favourable. The behaviour changes to a minimum willingness for cooperation and to a preference for shifting to alternative strategies in the agricultural sector such as direct marketing. | Agricultural subsidies as well as prices decrease (similar to the Globalisation Scenario), but local and regional stakeholders (municipal and provincial policy-makers) try to counteract these conditions by implementing innovative strategies. Consequently, although income decreases, farms are motivated to cooperate and implement innovative strategies (such as organic faming, direct marketing). |

in Reichraming came together to discuss issues and concerns about decision-making processes, time use and workloads in agricultural as well as other households. The intention of the meetings was to test pre-defined hypotheses concerning the ratio between income and the working time of men and women as an influencing factor in the decision-making processes of households.

Several methods for evaluating the agent-based model were applied. One was the discussion of the ABM's preliminary outputs with actors, followed, if necessary, by a redesign and 're-formalisation' of the model (Berger 2004). Another task of the meetings was to discuss the design of the model, where insights that had been generated so far were translated into formalised language and diagrams. Decisions on relevant model outputs (e.g., labour time per farm, cropland area per farm, etc.) and adjustable model parameters required were taken together with the stakeholders. Another important part of the process was the development of 'storyline' scenarios for the municipality (Table 3.2). Inspired by well-known international or even global scenarios (Millenium Ecosystem Assessment 2005a; Nakicenovic and Swart 2000; UNEP 2002), we used the model as a tool to simulate future scenarios depending on changes in (1) external framework conditions, (2) local and regional policies and (3) preferences of individual agents with respect to income and leisure time expectations, willingness to co-operate, etc. Assumptions on local and regional policies as well as on the preferences of local agents were developed in the participatory process described above.

Finally, the thoroughly designed model and the results of impacts of different actions under different framework conditions (scenarios) provided the basis for common discussions among researchers and stakeholders.

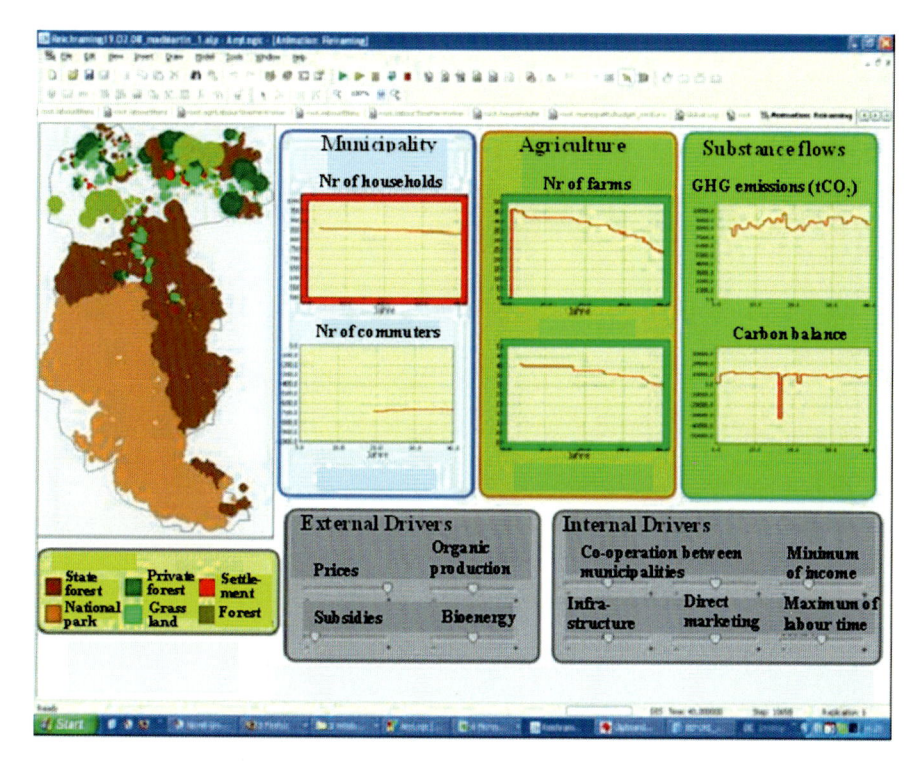

**Fig. 3.4** Interface of the Reichraming model simulating the development of different parameters (socioeconomic and ecological) under different framework conditions (changeable as sliders – *grey boxes* at the *bottom*)

### 3.3.3 Discussion of Results

The main outcome of the project and the transdisciplinary process is a computer simulation model that provides the possibility of changing framework conditions as well as preferences and behaviour of different actors. The design of the interface in terms of parameters that can be changed during each model run and therefore implemented as sliders was an important outcome of the discussions with the stakeholders. Similarly, the decisions on which parameters should be visible as graphs that build up during a model run was discussed and decided together with the stakeholders (Fig. 3.4).

Results of the model runs for the three pre-defined scenarios are summarised in Table 3.3. Some similar changes can be identified in all three scenarios. Grassland areas decrease considerably, coupled with a strong reduction of the number of farms. The proportion of the remaining farms that change their production to an innovative farming type (e.g. rearing of sheep, horses and game animals) grows considerably. The average farm size falls from 40 ha per farm to around 20 ha per farm, including grassland and private forest area. Annual agricultural working time per farmer

**Table 3.3** Results of the three scenarios in comparison with the initial value (all values are in total for the whole region)

| | Initial value | Trend scenario | Globalisation scenario | Local policy scenario |
|---|---|---|---|---|
| Number of farms | 52 | 40 | 36 | 41 |
| Grassland area [ha] | 478 | 383 | 255 | 354 |
| Percent of innovative farms | 54% | 75% | 67% | 68% |
| Average farm size [ha] | 40 | 21 | 18 | 21 |
| Intensive farming area [ha] | 54 | 112 | 0 | 75 |
| Conventional farming area [ha] | 295 | 193 | 76 | 112 |
| Extensive farming area [ha] | 111 | 60 | 157 | 149 |
| GHG emissions of farms [t $CO_2$-eq./ha grassland/year] | 5.19 | 4.97 | 4.05 | 4.90 |
| $N_2O$ emissions of farms [kg N/ha/year] | 2.21 | 2.46 | 1.92 | 2.23 |
| N leaching of farms [kg N/ha/year] | 0.66 | 0.59 | 0.63 | 0.62 |
| $NH_3$ emissions of farms [kg N/ha/year] | 41.49 | 45.69 | 36.51 | 42.29 |

increases in all scenarios, but with variable extent, in line with an increasing annual income per farmer through agricultural labour.

The total number of farms at the end of the simulated period (20 years) is smallest in the globalisation scenario. Many of the remaining farms switch to extensive farming. The agricultural workload per farmer rises to the largest number of working hours of all scenarios, whereas farm incomes decrease. In terms of land use, grassland area is largest in the trend scenario and local policy scenario. In the trend scenario, grasslands are predominantly managed conventionally, however in the local and globalisation scenario, extensively used grasslands play an important role. In the local and trend scenario, farms adopt the most intensive managements styles.

Overall, the different scenarios affect C and N flows in the respective model runs through change in land use and farming intensity. In all three scenarios, forested areas act as a net sink of $CO_2$ from the atmosphere. Greenhouse gas emissions of farms strongly depend on the number of farms and management intensities. For example, agricultural GHG emissions are reduced in the globalisation scenario due to the low number of surviving farms. In terms of nitrogen flows, agriculture is thought to be the main emitter of $NH_3$ and $N_2O$ and is responsible for the biggest part of $NH_3$ leaching due to fertilisers, manure management and animal husbandry (Weiske et al. 2006). The highest effect of the different scenarios on N losses stems from ammonia emissions ($NH_3$), which are almost exclusively caused by agriculture (livestock farming).

Consequently, $NH_3$ emissions are highly dependent on the amount of farmed area and farming intensity and will therefore be lowest in the globalisation scenario.

To summarise, decisions made by farmers are strongly affected by changes in income in terms of subsidies and market prices. However, most of the decisions also depend on the time available for agricultural work and on the preferences of the younger generation regarding how much time they are willing and able to invest in farming. This indicates that the social dimension has great importance for any decision taken on the farm. Constraints upon time availability restrict actions and decisions taken by the farm. Where only one generation is living on the farm, available working time is not sufficient to run the farm as a full-time operation. This is even less of an option where women are engaged in child care. The younger generation will not accept a life without leisure time and without the freedom to make decisions. Finally, infrastructure, such as the child-care system, care for the elderly or the availability of (part-time) jobs in the region, places constraints upon the decisions and actions of farmers. Nearly every action of farmers related to agricultural production either directly or indirectly impact ecological substance flows such as carbon and nitrogen flows.

## 3.4 Outlook and Conclusions

The experiences gathered in the above-discussed project underline that integrated socio-ecological models based on ABM and system-dynamic modelling approaches are a useful tool for integrating social and natural sciences and support sustainability science in many ways. However, integrated modelling also has limitations as a research method. First of all, the advantage of the model in terms of reducing complexity is at the same time a disadvantage in terms of flexibility. Once the concept and framework of the model is decided the tool does not allow further aspects that turn out to be relevant during the research process to be easily integrated. Additionally, developing and programming a simulation model is time and resource consuming. Very specific qualifications are needed to programme the model which can hardly be used by researchers from non-technical disciplines such as social scientists or ecologists. Finally, a limitation of agent-based models is that they are very much site specific. Developing an agent-based model for one LTSER region helps the dynamics in this specific region to be understood. This does not mean that the same model is applicable to many other LTSER regions, especially when their socioeconomic and ecological situation differs clearly. Nevertheless, they are a highly useful tool in LTSER (Haberl et al. 2006) and can of course also be applied in Ecological Economics, Social Ecology, Industrial Ecology (Ayres and Simonis 1994; Fischer-Kowalski 1997), integrated land-change science (GLP 2005; Turner et al. 2007) and integrated sustainability assessment (Pahl-Wostl 2002b).

There are several ways in which integrated socio-ecological models can be used in these contexts. First, in particular in complex systems that are difficult to conceptualise (Young et al. 2006b), the very process of constructing an integrated model is of great help in fostering interdisciplinary integration and mutual learning in interdisciplinary teams (van der Leeuw 2004). The application of the sustainability triangle at the level of individual agents requires that at least three scientific disciplines, in this case ecology, sociology and economics, have to be represented on an equal footing in the model construction process. To define the interdependencies between the three dimensions required extensive discussion between researchers of these disciplines. The need to formalise interrelations through mathematical formulae or decision trees forced researchers to be explicit in formulating assumptions and hypothesised strengths or pathways of interaction between different factors or actors. The functioning of the computer model (or malfunctioning of its earlier versions) built by synthesising the knowledge and data gathered from the different sources provided ample rationale for rejection of failed hypotheses and thus fostered learning and interdisciplinary communication.

Second, modelling provides an opportunity to integrate aspects that are important but often neglected within sustainability research, such as the gender perspective. The fact that we had to define each agent (e.g. farmstead) in terms of its internal structure as represented by family members and their income and time resources facilitated the integration in the model of differences between men and women and their role at the farm. Gender differences were highly important in defining key aspects of agents such as time availability, demography, preferences and many more and their significance is thus naturally reflected in the model structure and modelling outcomes.

Finally, agent-based modelling encourages transdisciplinary research design and helps to structure participative processes, since it requires active and continuous cooperation between researchers and actors over a long period of time. A transdisciplinary research design differs from classical research approaches because it has to be more flexible in various respects (e.g., definition of research goals, selection of actors involved, milestones planned and methods applied). Modelling in cooperation with the actors helps to achieve both structure and flexibility, allowing for regularly reflection upon the research design and outcomes. Furthermore, the active and continuous cooperation over a certain period of time required by the modelling process gives actors the chance to observe and actively shape scientific research, making it more problem-oriented and of greater relevance to 'real life'. The regularity of meetings and the shared goal, that is, building a model that is close to the actors' reality, helps actors to identify with the ongoing research, to understand their own and others' everyday life experience and to formulate possible problem-solving strategies relevant to their own situation. All these aspects together make integrated socio-ecological modelling based on ABM and system-dynamic approaches an instrument able to contribute to local sustainability studies of greater scientific and practical use, and therefore an attractive approach for long-term socio-ecological research.

**Acknowledgments** This research underlying this chapter was funded by the Austrian Ministry of Research and Science within the proVision programme as well as by the EU-FP6 Network of Excellence ALTER-Net (www.alter-net.info) and the EU-FP7 project VOLANTE. It has profited from collaboration within the Global Land Project (http://www.globallandproject.org/), in particular with its Aberdeen Nodal Office on integration and modelling. Our special thanks go to our collaborators, which were, among others: Heidi Adensam, Richard Aspinall, Helene Blanda, Karl-Heinz Erb, Simone Gingrich, Marina Fischer-Kowalski, Peter Fleissner, Thomas Guggenberger, Christina Kaiser, Johannes Kobler, Fridolin Krausmann, Juliana Lutz, Eleanor Milne, Stefan Pietsch, Andreas Richter, Andreas Schaumberger, Jakob Schaumberger, Carol Ann Stannard, Eva Vrzak, Martin Wildenberg, Angelika Wolf, and Michaela Zeitlhofer. All errors are entirely ours.

## References

Agarwal, C., Green, G. M., Grove, M. J., Evans, T. P., & Schweik, C. M. (2002). *A review and assessment of land-use change models: Dynamics of space, time; and human choice*. Bloomington: United States Department of Agriculture.
Ayres, R. U., & Simonis, U. E. (Eds.). (1994). *Industrial metabolism: Restructuring for sustainable development*. Tokyo/New York/Paris: United Nations University Press.
Berger, T. (2001). Agent-based spatial models applied to agriculture: A simulation tool for technology diffusion, resource use changes and policy Analysis. *Agricultural Economics, 25*, 245–260.
Berger, T. (2004). Agentenbasierte Modellierung von Landnutzungsdynamiken und Politikoptionen [Agent-based modeling of land use dynamics and policy options]. *Agrarwirtschaft, 2*, 77–87.
Brand, K.-W. (Ed.). (2000). *Nachhaltige Entwicklung und Transdisziplinarität. Besonderheiten, Probleme und Erfordernisse der Nachhaltigkeitsforschung*. Berlin: Analytica.
Brandt, K. W. (Ed.). (1997). *Nachhaltige Entwicklung, Eine Herausforderung für die Soziologie*. Opladen: Westdeutscher Verlag.
Brown, D. G., & Duh, J.-D. (2003). Spatial simulation for translating from land use to land cover. *International Journal of Geographical Information Science, 17*, 1–26.
Clark, W. C., & Munn, R. E. (Eds.). (1986). *Sustainable development of the biosphere*. Cambridge: Cambridge University Press.
Daly, H. E. (Ed.). (1973). *Toward a steady-state economy*. San Francisco: W.H. Freeman.
DeAngelis, D. L., & Gross, L. J. (Eds.). (1992). *Individual based models and approaches in ecology*. New York: Chapman & Hall.
Dirnböck, T., Bezák, P., Dullinger, S., Haberl, H., Lotze-Campen, H., Mirtl, M., Peterseil, J., Redpath, S., Singh, S. J., Travis, J., & Wijdeven, S. (2008). *Scaling issues in long-term socio-ecological biodiversity research: A review of European cases* (Social Ecology Working Paper 100, pp. 1–39). Vienna: IFF Social Ecology.
Epstein, J. M., & Axtell, R. L. (Eds.). (1996). *Growing artificial societies: Social science from the bottom up*. Washington, DC: The MIT Press.
Erb, K.-H., Haberl, H., Krausmann, F., Lauk, C., Plutzar, C., Steinberger, J. K., Müller, C., Bondeau, A., Waha, K., & Pollack, G. (2009). *Eating the planet: Feeding and fuelling the world sustainably, fairly and humanely – A scoping study* (Social Ecology Working Paper 116, pp. 1–132). Report commissioned by Compassion in World Farming and Friends of the Earth, UK. Vienna/Potsdam: Institute of Social Ecology/PIK Potsdam. 19 Apr 2010.
Ferber, J. (Ed.). (1999). *Multi-agent systems*. Harlow: Addison Wesley Longman.
Fischer-Kowalski, M. (1997). Society's metabolism: On the childhood and adolescence of a rising conceptual star. In M. Redclift & G. R. Woodgate (Eds.), *The international handbook of environmental sociology* (pp. 119–137). Cheltenham/Northhampton: Edward Elgar.
Fischer-Kowalski, M. (2002). Das magische Dreieck von Nachhaltigkeit: Lebensqualität, Wohlstand und ökologische Verträglichkeit. In A. Klotz (Ed.), *Stadt und Nachhaltigkeit* (pp. 25–41). Wien/New York: Springer.

Fischer-Kowalski, M., & Haberl, H. (1997). Tons, joules and money: Modes of production and their sustainability problems. *Society and Natural Resources, 10*, 61–85.

Fischer-Kowalski, M., & Haberl, H. (Eds.). (2007). *Socioecological transitions and global change: Trajectories of social metabolism and land use*. Cheltenham/Northhampton: Edward Elgar.

Fischer-Kowalski, M., & Weisz, H. (1999). Society as a hybrid between material and symbolic realms. Toward a theoretical framework of society-nature interaction. *Advances in Human Ecology, 8*, 215–251.

Fischer-Kowalski, M., Haberl, H., Hüttler, W., Payer, H., Schandl, H., Winiwarter, V., & Zangerl-Weisz, H. (Eds.). (1997). *Gesellschaftlicher Stoffwechsel und Kolonisierung von Natur. Ein Versuch in Sozialer Ökologie*. Amsterdam: Gordon & Breach Fakultas.

Foley, J. A., DeFries, R., Asner, G. P., Barford, C., Bonan, G., Carpenter, S. R., Chapin, F. S., Coe, M. T., Daily, G. C., Gibbs, H. K., Helkowski, J. H., Holloway, T., Howard, E. A., Kucharik, C. J., Monfreda, C., Patz, J. A., Prentice, I. C., Ramankutty, N., & Snyder, P. K. (2005). Global consequences of land use. *Science, 309*, 570–574.

Folke, C., & Gunderson, L. (2006). Facing global change through social-ecological research. *Ecology and Society, 11*, 43.

Gaube, V., & Haberl, H. (2012). Regionale Nachhaltigkeitsmodelle – Partizipativ entwickelt. In H. Egner & M. Schmid (Eds.), *Die Rolle der Wissenschaft in einer vorsorgenden Gesellschaft* (forthcoming). München: Oekom.

Gaube, V., Kaiser, C., Wildenberg, M., Adensam, H., Fleissner, P., Kobler, J., Lutz, J., Smetschka, B., Wolf, A., Richter, A., & Haberl, H. (2008, August 1). *Ein integriertes Modell für Reichraming. Partizipative Entwicklung von Szenarien für die Gemeinde Reichraming (Eisenwurzen) mit Hilfe eines agentenbasierten Landnutzungsmodells* (Social Ecology Working Paper 106, pp. 1–100). Vienna: IFF Social Ecology.

Gaube, V., Kaiser, C., Wildenberg, M., Adensam, H., Fleissner, P., Kobler, J., Lutz, J., Schaumberger, A., Schaumberger, J., Smetschka, B., Wolf, A., Richter, A., & Haberl, H. (2009a). Combining agent-based and stock-flow modelling approaches in a participative analysis of the integrated land system in Reichraming, Austria. *Landscape Ecology, 24*, 1149–1165.

Gaube, V., Reisinger, H., Adensam, H., Aigner, B., Colard, A., Haberl, H., Lutz, J., Maier, R., Punz, W., & Smetschka, B. (2009b). Agentenbasierte Modellierung von Szenarien für Landwirtschaft und Landnutzung im Jahr 2020, Traisental, Niederösterreich. *Verhandlungen der Zoologisch-Botanischen Gesellschaft in Österreich, 146*, 79–101.

Geist, H. J., & Lambin, E. F. (2002). Proximate causes and underlying driving forces of tropical deforestation. *BioScience, 52*, 143–150.

Georgescu-Roegen, N. (Ed.). (1971). *The entropy law and the economic process*. Cambridge, MA: Harvard University Press.

Gibson, C. C., Ostrom, E., & Ahn, T. K. (2000). The concept of scale and the human dimensions of global change: A survey. *Ecological Economics, 32*, 217–239.

Gingrich, S., & Krausmann, F. (2008, October 31). *Der soziale Metabolismus lokaler Produktionssysteme. Reichraming in der oberösterreichischen Eisenwurzen 1830–2000* (Social Ecology Working Paper 107, pp. 1–32). Vienna: IFF Social Ecology.

GLP (Ed.). (2005). *Global land project. Science plan and implementation strategy*. Stockholm: IGBP Secretariat.

Godfray, H. C., Beddington, J. R., Crute, I. R., Haddad, L., Lawrence, D., Muir, J. F., Pretty, J., Robinson, S., Thomas, S. M., & Toulmin, C. (2010). Food security: The challenge of feeding 9 billion people. *Science, 327*, 812–818.

Gunderson, L., & Holling, C. S. (Eds.). (2002). *Panarchy. Understanding transformations in human and natural systems*. Washington, DC: Island Press.

Haberl, H., Erb, K.-H., Krausmann, F., Loibl, W., Schulz, N. B., & Weisz, H. (2001). Changes in ecosystem processes induced by land use: Human appropriation of net primary production and its influence on standing crop in Austria. *Global Biogeochemical Cycles, 15*, 929–942.

Haberl, H., Fischer-Kowalski, M., Krausmann, F., Weisz, H., & Winiwarter, V. (2004). Progress towards sustainability? What the conceptual framework of material and energy flow accounting (MEFA) can offer. *Land Use Policy, 21*, 199–213.

Haberl, H., Winiwarter, V., Andersson, K., Ayres, R. U., Boone, C. G., Castillio, A., Cunfer, G., Fischer-Kowalski, M., Freudenburg, W. R., Furman, E., Kaufmann, R., Krausmann, F., Langthaler, E., Lotze-Campen, H., Mirtl, M., Redman, C. A., Reenberg, A., Wardell, A. D., Warr, B., & Zechmeister, H. (2006). From LTER to LTSER: Conceptualizing the socio-economic dimension of long-term socio-ecological research. *Ecology and Society, 11*, 13. (Online), http://www.ecologyandsociety.org/vol11/iss2/art13/

Haberl, H., Gaube, V., Díaz-Delgado, R., Krauze, K., Neuner, A., Peterseil, J., Plutzar, C., Singh, S. J., & Vadineanu, A. (2009). Towards an integrated model of socioeconomic biodiversity drivers, pressures and impacts. A feasibility study based on three European long-term socio-ecological research platforms. *Ecological Economics, 68*, 1797–1812.

Haberl, H., Fischer-Kowalski, M., Krausmann, F., Martinez-Alier, J., & Winiwarter, V. (2011). A socio-metabolic transition towards sustainability? Challenges for another great transformation. *Sustainable Development, 19*, 1–14.

Hare, M., & Pahl-Wostl, C. (2002). Stakeholder categorisation in participatory integrated assessment processes. *Integrated Assessment, 3*, 50–62.

Hirsch Hadorn, G., Hoffmann-Riem, H., Biber-Klemm, S., Grossenbacher-Mansuy, W., Joye, D., Pohl, C., Wiesmann, U., & Zemp, E. (Eds.). (2008). *Handbook of transdisciplinary research*. Stuttgart/Berlin/New York: Springer.

Holling, C. S. (1986). The resilience of terrestrial ecosystems: Local surprise and global change. In W. C. Clark & R. E. Munn (Eds.), *Sustainable development of the biosphere* (pp. 292–320). Cambridge: Cambridge University Press.

Holling, C. S. (2001). Understanding the complexity of economic, ecological, and social systems. *Ecosystems, 4*, 390–405.

Jahn, T. (2005). Soziale Ökologie, kognitive Integration und Transdisziplinarität. *Technikfolgenabschätzung-Theorie und Praxis, 14*, 32–38.

Janssen, M. A. (2004). Agent-based models. In J. Proops & P. Safonov (Eds.), *Modelling in ecological economics* (pp. 155–172). Cheltenham/Northampton/: Edward Elgar.

Kates, R. W., Clark, W. C., Corell, R., Hall, J. M., Jaeger, C. C., Lowe, I., McCarthy, J. J., Schellnhuber, H. J., Bolin, B., Dickson, N. M., Faucheux, S., Gallopin, G. C., Grübler, A., Huntley, B., Jäger, J., Jodha, N. S., Kasperson, R. E., Mabogunje, A., Matson, P. A., Mooney, H. A., Moore, B., III, O'Riordan, T., & Svedin, U. (2001). Sustainability science. *Science, 292*, 641–642.

Kohler, T. A., & Gumerman, G. J. (Eds.). (2000). *Dynamics in human and primate societies: Agent-based modeling of social and spatial processes*. New York: Oxford University Press.

Koomen, E., Stillwell, J., Bakema, A., & Scholten, H. J. (2007). Modelling land-use change. In E. Koomen, J. Stillwell, A. Bakema, & H. J. Scholten (Eds.), *Modelling land-use change. Progress and applications* (pp. 1–21). Dordrecht: Springer.

Krausmann, F. (2004). Milk, manure and muscular power. Livestock and the industrialization of agriculture. *Human Ecology, 32*, 735–773.

Lambin, E. F., & Geist, H. J. (Eds.). (2006). *Land-use and land-cover change. Local processes and global impacts*. Berlin: Springer.

Lambin, E. F., Turner, B. L. I., Geist, H. J., Agbola, S. B., Angelsen, A., Bruce, J. W., Coomes, O. T., Dirzo, R., Fischer, G., Folke, C., George, P. S., Homewood, K., Imbernon, J., Leemans, R., Li, X., Moran, E. F., Mortimore, M., Ramakrishnan, P. S., Richards, J. F., Skanes, H., Steffen, W., Stone, G. D., Svedin, U., Veldkamp, T. A., Vogel, C., & Xu, J. (2001). The causes of land-use and land-cover change: Moving beyond the myths. *Global Environmental Change, 11*, 261–269.

Landeskulturdirektion Oberösterreich (Ed.). (1998). *Land der Hämmer, Heimat Eisenwurzen, Region Pyhrn – Eisenwurzen*. Salzburg: Residenz-Verlag.

Littig, B., & Wallace, C. (1997). *Möglichkeiten und Grenzen von Fokus-Gruppendiskussionen für die sozialwissenschaftliche Forschung*. Wien: IHS.

Manson, S. M., & Evans, T. (2007). Agent-based modeling of deforestation in southern Yucatán, Mexico, and reforestation in the Midwest United States. *Proceedings of the National Academy of Sciences of the United States of America, 104*, 20678–20683.

Martinez-Alier, J. (Ed.). (1987). *Ecological economics. Energy, environment and society*. Oxford: Blackwell.

Matthews, R. (2006). The people and landscape model (PALM): towards full integration of human decision-making and biophysical simulation models. *Ecological Modelling, 194*, 329–343.

Matthews, R., & Selman, P. (2006). Landscape as a focus for integrating human and environmental processes. *Journal of Agricultural Economics, 57*, 199–212.

Matthews, E., Amann, C., Fischer-Kowalski, M., Bringezu, S., Hüttler, W., Kleijn, R., Moriguchi, Y., Ottke, C., Rodenburg, E., Rogich, D., Schandl, H., Schütz, H., van der Voet, E., & Weisz, H. (Eds.). (2000). *The weight of nations: Material outflows from industrial economies*. Washington, DC: World Resources Institute.

McConnell, W. (Ed.). (2001). *Agent-based models of land-use and land-cover change*. Belgium: LUCC International Project Office.

Meyer, W. B., & Turner, B. L. I. (Eds.). (1994). *Changes in land use and land cover, a global perspective*. Cambridge: Cambridge University Press.

Millennium Ecosystem Assessment (Ed.). (2005a). *Ecosystems and human well-being: Vol. 2. Scenarios*. Washington, DC: Island Press.

Millennium Ecosystem Assessment (Ed.). (2005b). *Ecosystems and human well-being – Our human planet. Summary for decision makers*. Washington, DC: Island Press.

Milne, E., Aspinall, R., & Veldkamp, T. A. (Eds.) (2009). Integrated modelling of natural and social systems in land change science. *Special Issue of Landscape Ecology, 24*(9), 1145–1270. Heidelberg: Springer.

Nakicenovic, N., & Swart, R. (Eds.). (2000). *Special report on emission scenarios*. Cambridge: Intergovernmental Panel on Climate Change (IPCC)/Cambridge University Press.

Newig, J., Gaube, V., Berkhoff, K., Kaldrack, K., Kastens, B., Lutz, J., Schlußmeier, B., Adensam, H., & Haberl, H. (2008a). The role of formalisation, participation and context in the success of public involvement mechanisms in resource management. *Systemic Practice and Action Research, 21*, 423–441.

Newig, J., Haberl, H., Pahl-Wostl, C., & Rothman, D. (Eds.) (2008b). Formalised and non-formalised methods in resource management, knowledge and learning in participatory processes. *Special Issue of Systemic Practice and Action Research, 21*(6), 381–515. Heidelberg/New York: Springer.

Ostrom, E. (2009). A general framework for analyzing sustainability of social-ecological systems. *Science, 325*, 419–422.

Pahl-Wostl, C. (2002a). Participative and stakeholder-based policy design, evaluation and modeling processes. *Integrated Assessment, 3*, 3–14.

Pahl-Wostl, C. (2002b). Towards sustainability in the water sector – The importance of human actors and processes of social learning. *Aquatic Sciences, 64*, 394–411.

Parker, D. C., Berger, T., & Manson, S. M. (Eds.) (2002). *Agent-based models of land-use and land-cover change* (LUCC Report Series No. 6). Louvain-la-Neuve: LUCC International Project Office.

Parker, D. C., Manson, S. M., Janssen, M., Hoffmann, M. J., & Deadman, P. (2003). Multi-agent systems for the simulation of land-use and land-cover change: A review. *Annals of the Association of American Geographers, 93*, 314–337.

Pearce, D., Markandya, A., & Barbier, E. B. (Eds.). (1990). *Blueprint for a green economy*. London: Earthscan.

Redman, C. L., Grove, J. M., & Kuby, L. H. (2004). Integrating social science into the long-term ecological research (LTER) network: Social dimensions of ecological change and ecological dimensions of social change. *Ecosystems, 7*, 161–171.

Reenberg, A., & Lund, C. (2004). Land use and land right dynamics – Determinants for resource management options in Eastern Burkina Faso. *Human Ecology, 26*, 599–620.

Reid, W. V., Chen, D., Goldfarb, L., Hackmann, H., Lee, Y. T., Mokhele, K., Ostrom, E., Raivio, K., Rockström, J., Schellnhuber, H. J., & Whyte, A. (2010). Earth system science for global sustainability: Grand challenges. *Science, 330*, 916–917.

Rounsevell, M. D. A., Reginster, I., Araújo, M. B., Carter, T. R., Dendocker, N., Ewert, F., House, J. I., Kankaanpää, S., Leemans, R., Metzger, M. J., Schmit, C., Smith, P., & Tuck, G. (2006).

A coherent set of future land use change scenarios for Europe. *Agriculture, Ecosystems and Environment, 114*, 57–68.

Rowe, G., & Frewer, L. J. (2005). A typology of public engagement mechanisms. *Science, Technology, and Human Values, 30*, 251–290.

Schaldach, R., & Priess, J. A. (2008). Integrated models of the land system: A review of modelling approaches on the regional to global scale. *Living Reviews in Landscape Research, 2*, 1. (Online) http://www.livingreviews.org/lrlr-2008-1

Sieferle, R. P., Krausmann, F., Schandl, H., & Winiwarter, V. (Eds.). (2006). *Das Ende der Fläche. Zum gesellschaftlichen Stoffwechsel der Industrialisierung*. Köln: Böhlau.

Singh, S. J., Haberl, H., Gaube, V., Grünbühel, C. M., Lisievici, P., Lutz, J., Matthews, R., Mirtl, M., Vadineanu, A., & Wildenberg, M. (2010). Conceptualising long-term socio-ecological research (LTSER): integrating the social dimension. In F. Müller, C. Baessler, H. Schubert, & S. Klotz (Eds.), *Long-term ecological research, between theory and application* (pp. 377–398). Dordrecht/Heidelberg/London/New York: Springer.

Turner, B. L. I., Lambin, E. F., & Reenberg, A. (2007). The emergence of land change science for global environmental change and sustainability. *Proceedings of the National Academy of Sciences of the United States of America, 104*, 20666–20671.

UNEP (Ed.). (2002). *Global environment outlook 3 past, present and future perspectives*. London: United Nations Environment Programme (UNEP)/Earthscan Publications.

van der Leeuw, S. E. (2004). Why model? *Cybernetics and Systems, 35*, 117–128.

Verburg, P. H., Schot, P. P., Dijst, M. J., & Veldkamp, A. (2004). Land use change modelling: Current practice and research priorities. *GeoJournal, 61*, 309–324.

Verburg, P. H., Eickhout, B., & van Meijl, H. (2008). A multi-scale, multi-model approach for analyzing the future dynamics of European land use. *The Annals of Regional Science, 42*, 57–77.

Vitousek, P. M., Ehrlich, P. R., Ehrlich, A. H., & Matson, P. A. (1986). Human appropriation of the products of photosynthesis. *BioScience, 36*, 363–373.

WCED. (1987). *World commission on environment and development: Our common future*. New York: Oxford University Press.

Weiske, A., Vabitsch, A., Olesen, J. E., Schelhaas, M. J., Michel, C., Friedrich, R., & Kaltschmitt, M. (2006). Mitigation of greenhouse gas emissions in European conventional and organic dairy farming. *Agriculture, Ecosystems and Environment, 112*, 221–232.

Weisz, H., Fischer-Kowalski, M., Grünbühel, C. M., Haberl, H., Krausmann, F., & Winiwarter, V. (2001). Global environmental change and historical transitions. *Innovation – The European Journal of Social Sciences, 14*, 117–142.

Wrbka, T., Erb, K.-H., Schulz, N. B., Peterseil, J., Hahn, C., & Haberl, H. (2004). Linking pattern and process in cultural landscapes. An empirical study based on spatially explicit indicators. *Land Use Policy, 21*, 289–306.

Wright, D. H. (1990). Human impacts on the energy flow through natural ecosystems, and implications for species endangerment. *Ambio, 19*, 189–194.

Young, O. R., Berkhout, F., Gallopin, G. C., Janssen, M. A., Ostrom, E., & van der Leeuw, S. (2006a). The globalization of socio-ecological systems: An agenda for scientific research. *Global Environmental Change, 16*, 304–316.

Young, O. R., Lambin, E. F., Alcock, F., Haberl, H., Karlsson, S. I., McConnell, W. J., Mying, T., Pahl-Wostl, C., Polsky, C., Ramakrishnan, P. S., Scouvart, M., Schröder, H., & Verburg, P. (2006b). A portfolio approach to analyzing complex human-environment interactions: Institutions and land change. *Ecology and Society, 11*, 31. (Online) URL: http://www.ecologyandsociety.org/vol11/iss2/art31/

# Chapter 4
# Modelling Transport as a Key Constraint to Urbanisation in Pre-industrial Societies

**Marina Fischer-Kowalski, Fridolin Krausmann, and Barbara Smetschka**

**Abstract** This chapter investigates the constraints for urban growth in pre-industrial societies and focuses on transport as an important component in the functioning of socio-ecological systems. It presents a simple formal model based on sociometabolic relations to investigate the relation between the size of an urban centre, its resource needs and the resulting transport requirements. This model allows, in a very stylised way, light to be shed on some of the physical constraints for urban growth in agrarian societies and a better understanding of how transport shapes the relation between cities and their resource-providing hinterland. The model demonstrates that the growth of urban centres depends upon an extension of the territory and rural population to work the land and generate the supplies cities require. The labour force engaged in urban rural transport rises with the size of urban centre and corresponds to 8–15% of the urban labour force. We find clear indications for a scale limit to agrarian empires, and agrarian centres, due to factors associated with the cost of transport (in terms of human labour time and land). Where this scale limit occurs strongly depends upon agricultural productivity.

**Keywords** Social metabolism • Food and feed transport • Modelling historic socio-ecological systems • Size of urban hinterland • Relations city – rural areas

---

M. Fischer-Kowalski, Ph.D. (✉) • F. Krausmann, Ph.D. • B. Smetschka, M.A.
Institute of Social Ecology Vienna (SEC), Alpen-Adria Universitaet Klagenfurt,
Wien, Graz, Schottenfeldgasse 29, 1070 Vienna, Austria
e-mail: marina.fischer-kowalski@aau.at; fridolin.krausmann@aau.at;
barbara.smetschka@aau.at

## 4.1 Introduction[1]

Urbanisation and spatial concentration are not only key characteristics of industrialisation and industrial societies, but they are also important features of development in pre-industrial times. Urban growth in agrarian societies, however, faced specific biophysical constraints, in particular from low surplus rates in agricultural production systems in the hinterland and from high energy costs of transporting bulk resources like food, feed and fuelwood into the city (Sieferle 2001; Krausmann et al. 2008a). This paper focuses on the role of transport in the functioning and development of pre-industrial socio-ecological systems. We use a simple formal model to investigate the relation between the size of an urban centre, its resource needs and the resulting transport requirements. This model allows us, in a very stylised way, to highlight some of the physical constraints for urban growth in agrarian societies and to better understand how transport shapes the relation between cities and their resource providing hinterland (cf. Gingrich et al. 2012). The paper not only touches important issues of long-term socio-ecological research (Haberl et al. 2006), it also demonstrates how modelling approaches using empirical findings from socio-ecological case studies can be applied in LTSER to investigate the dynamics of socio-ecological systems.

Transportation means moving materials or people across space. Looked at from a technical angle, it deals with physical variables: amounts of matter and people and their location in space, distances to be overcome, technologies to be used, and transportation cost in terms of time and energy. Looked at from a socio-economic angle, it deals with social organisation: populations, territories, urbanisation, food requirements, property, base resources and preciosities; barter and trade, robbery, tribute and taxation; nomadism, migration, flight, expulsion or starvation. Transportation is the decisive link between resources (distributed across the natural environment) and human production and consumption. Using this decisive link, we built a generic formal model building upon the sociometabolic approach (Fischer-Kowalski and Haberl 1993) that allows the reconstruction of social systems of the past and the creation of plausible answers to questions like: How large may an urban centre have been? How many peasants were required to secure its supplies? Which yields per area must be assumed for the urban centre to have a certain number of inhabitants? Could this urban centre subsist on its hinterland, or did it need some additional territory to secure its supplies? In particular, this model allows us to judge whether different data found in historical sources fit together plausibly, or whether one has to introduce additional assumptions (whether plausible or otherwise) to ensure consistency.

In this chapter, we deal first with transport from a technical angle; thereafter, we attempt to demonstrate the close interdependency between technical and social aspects, with a special focus on how the technicalities of transport generate boundary

---

[1] This chapter builds on a previous publication (see Fischer-Kowalski et al. 2004) but uses an improved version of the model together with an extended analysis of scenario results.

conditions for social organisation, and for society-environment relations. However, in contrast to much of the literature on transportation,[2] we shall not concentrate on the *how* of transport, on technologies and infrastructures, but rather on *how much*, and on *what* is to be transported. It is here that a sociometabolic perspective can contribute most.

We use the indicators for the scale and volume of freight transport most common in contemporary transport statistics: "Mass Lifted" (ML in tons) and "Mass Moved" (MM in ton kilometres). Mass Lifted expresses the volume of freight loaded on transport vehicles for a haul, in tons per year. Each time goods are loaded for transportation (irrespective of the distance they are being transported), their weight is counted. The second indicator, Mass Moved, takes distances into account: Mass Moved is equal to Mass Lifted times length of haul, and is given in ton kilometres per year. We deal only with the transportation of persons insofar as this may be a functional alternative to freight transport (e.g. moving people to resources, instead of resources to people).

The systemic relationships generating the necessary background data for the model are based upon the sociometabolic approach as originally developed to describe the social metabolism of contemporary national economies (e.g., Weisz et al. 2006; Fischer-Kowalski et al. 2011) and extended to study historical cases on various spatial scales (Cusso et al. 2006; Barles 2009; Krausmann et al. 2008a, b; Gingrich et al. 2012). This approach regards society as a social system functioning to reproduce a human population within a territory, guided by a specific culture. This definition is sufficiently abstract to be applied to very different historical circumstances and at different scales. The more complex a society becomes, the more it needs to organise energetic and material flows not only to sustain its population biologically, but to maintain a number of intermediate biophysical structures that have a role in social reproduction: animal livestock, built infrastructure, and consumer durables. The flow of material and energy resources required must either be secured from the system's own territory or be imported from other social systems. The composition of these annual flows is commonly termed "metabolic profile" (Fischer-Kowalski and Haberl 2007), and the total materials divided by the number of people in the system is termed "metabolic level". This metabolic level depends on the mode of subsistence and the prosperity of the social system and may vary widely (Weisz et al. 2001; Krausmann et al. 2008a).

Once we know about or can assume a certain metabolic level and certain per capita material quantities, we need to link the respective components by transportation. The first consideration concerns the relation between metabolic input and territory. For the sake of simplification, we assume a human community or society in isolation (that is, a community that sustains itself on its territory, entertaining little or no exchange with other societies). Following this, we need to make some assumptions about the density of critical resources on this territory related to a given technology (such as food for humans); this allows the size of the territory a population must be

---

[2] For a work that is classic in more than a technical sense, see (Ciccantell and Bunker 1998).

able to utilise to be estimated. Or, by the same token, we may know the size of a territory and be able to estimate the size of the population. Finally, population density may be given for a certain area – in which case we could use our toolbox of metabolic profiles to estimate the resource density this implies. Which ever way we turn it, we can learn something about spatial relations and the distances that need to be covered in order to link resources and people.

This rests upon the obvious assumption that material quantities must by some means get from their original location in nature to the human settlements where they are further processed and consumed. So from the information contained in the metabolic profiles, we know the amount and composition of materials to be mobilised from the environment and transported to human settlements, and we have some clues as to the distances that have to be covered. This takes us several steps towards being able to estimate the transport indicators Mass Lifted (ML) and Mass Moved (MM). We further make assumptions about the transport requirements of specific material flows. Thus corn has to be brought to a mill, and from the mill to the household; grass is eaten by the cow on the pasture, and the cow exhales most of the carbon consumed while it walks the milk back home, etc. And finally, we need to make assumptions about transport technology in order to know how much can be transported at a time, over what distance, at what speed and at which energetic and labour time cost.

These considerations are the starting point for our modelling exercises described in more detail below.

## 4.2 Transport in Hunting and Gathering Societies

Even for the most modest metabolic demands, barely extending beyond endosomatic needs, human societies will need a certain amount of transport: natural environments hardly ever continuously offer the amount of resource density continuously that allows for a larger group of humans simply to move themselves around, find appropriate food and eat it on the spot.[3] Deer will have to be hunted at larger distances and will have to be carried home for the rest of the group, and equally nuts, fruits and roots will have to be gathered, put into containers and be transported home, as well as firewood. This need and mode of transportation does not distinguish human hunters and gatherers much from many animals that have to carry food to their young; technologically, the only distinction pertaining to humans may be the use of containers and binding materials. There is a clear ecological/economic limit to such transport: If humans need the same or a higher amount of food for the time period and the effort they invest as they are able to find on their hunting and gathering excursion, or, in other words, if the energetic return upon investment tends towards

---

[3] For human societies, the preparation (cooking) and sharing of food at a common fireplace is a constitutive feature (see Wrangham 2009).

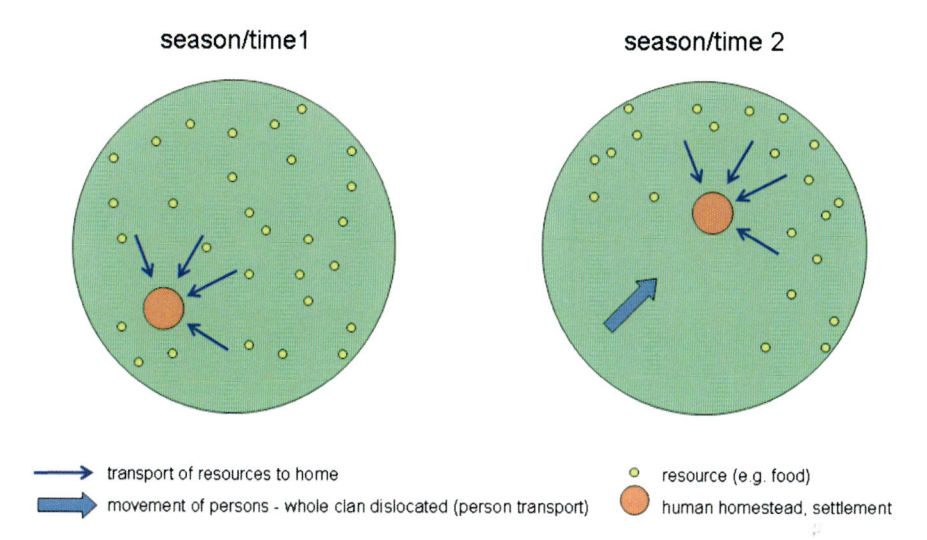

## 4 Modelling Transport as a Key Constraint to Urbanisation in Pre-industrial Societies

**season/time 1**

**season/time 2**

→ transport of resources to home
▬► movement of persons - whole clan dislocated (person transport)
○ resource (e.g. food)
● human homestead, settlement

**Fig. 4.1** Territory and transport in hunting and gathering societies

zero (Krebs and Davies 1984; Hawkes et al. 1982; Goudsblom 1992; Layton et al. 1991), then the whole group has to move to another environment with a higher resource density (see Fig. 4.1).[4] So one may, in the terminology of modern transportation, consider hunting and gathering societies to be sustaining themselves through a sophisticated logistics of freight and person transport (nomadism); the limiting factor being mainly resource density in space at a given time. Since the natural resource density is not intentionally modified through agriculture (or, more generally speaking, through colonising interventions, see footnote 11),[5] and since there are hardly any technologies for storage, resource deficiencies must be balanced by movement in space, both of persons and of resources.

Let us now perform an intellectual experiment and put these general considerations into numbers, using parameters known from socio-metabolic research. If we presuppose a population density of 0.1 person/km², a clan of 50 hunters and gatherers would exploit a territory of 500 km², which, if it were circular, would have a radius of 12.6 km. For each member's nutrition, they would need about 0.5 ton of biomass (3.5 GJ) per year, plus, say, 1 ton of firewood (Krausmann 2011a). The direct material input for this clan ("society") would amount to 75 ton per year (1.5 ton/capita/year). How much transport would this involve? Let us first consider food: the 25 ton/year required for the clan would be, so we assume, randomly distributed

---

[4] Such considerations are captured by "optimal foraging theories" (Harris 1987); usually, the element of transportation is not elaborated explicitly in these theories.

[5] One frequently cited exception from this rule is the controlled use of fire to remove shrubs and trees, thereby increasing the extension of grasslands and the number of herbivores, as well as facilitating the movement of hunters (Lewis 1982).

throughout the territory; part of the food would be eaten where found, and part of it would be carried back home, over an average distance of roughly 2/3 of the radius of the territory, that is in the case of our model 8.2 km. Let us further assume that a person carries 9 kg over this distance, and eats 1 kg on the spot. Accordingly 90% of the food required for the clan has to be transported over an average distance of 8.2 km. Let us now consider firewood: suppose it to be abundant (to be found in an average distance of 0.2 km from home), and all of it, that is 50 ton for the whole clan per year, is to be transported over this short distance, again 9 kg per haul.[6] Put in modern transport statistical indicators, everything taken together would imply a Mass Lifted of 72.5 ton per year (22.5 ton food and 50 ton firewood) and a transportation volume (Mass Moved) of 194.5 tkm, or 3.9 tkm per person. If we now wish to understand what effort this requires, we can translate it into working hours. Again we need a few assumptions: walking speed 4 km/h; each distance has to be covered twice (once without and once with freight); so for each transport of 9 kg of biomass a distance of 5.36 km has to be covered, and this has to happen 8,056 times a year. In total this amounts to a walking distance of 43,178 km a year, or 863 km per person and year. Expressed in time, this is 10,795 h for the whole clan, or 215 h per person, or a little more than half an hour (36 min) per person and day. Even if we assume that only half the population of our clan engages in food collection, we find transportation to consume only a small amount of available working time (approximately 1 h a day), even if such large distances have to be covered.[7]

Thus our model closely reproduces the findings from cultural anthropology that attribute to hunters and gatherers extremely low working hours (Sahlins 1972). Let us now see what happens if we reduce the assumed resource density by a factor of 10, which would allow only a much lower population density (e.g. 0.01 person/km$^2$), not uncommon for hunters and gatherers (Cohen 1977). We would have to assume a tenfold increase of utilised territory for the same group of people: this would imply a more than threefold[8] increase in walking distances for the same amount of resources, and therefore a more than threefold increase in walking time. The resulting average of almost 4 h walking per day and adult person is probably not feasible.[9] At this point, we arrive at the hunters and gatherers complex logistics of person and freight transport: Quite obviously, if resource density is low, the whole clan must periodically move to another (part of the) territory. Thus, within the framework of parameters employed, we would predict hunters and gatherers to be sedentary if the

---

[6] Anyone considering a 9 kg load as small should take into account that female adults usually also carry children, in addition to this load.

[7] Beyond the actual movement in space, no further investment is required: no construction of roads, bridges or carriages, and no breeding and feeding of animals.

[8] If the size of the new territory is ten times as large as the former ($R^2 \pi = 10r^2 \pi$), then the new radius is 3.16 times larger ($R = r \sqrt{10}$).

[9] Lee found !Kung women to walk about 2,400 km a year, which by the above calculation standards would correspond to one and a half hours walking per day (Lee 1980).

resource density of their territory is such as to permanently sustain 50 people within, say, 1,000 km$^2$ (which means 2 h walking per adult per day), but to become increasingly nomadic where resource density is lower than that.[10]

## 4.3 Modelling Transport for Agrarian Societies

The core feature of agriculture consists in colonising terrestrial ecosystems[11] so that their resource density for human purposes would be high enough to allow for a larger human population. This, in principle, should provide relief in terms of transport, as the distances to be overcome to provide food for a given population decrease. On the other hand, new needs and opportunities created by the agrarian regime, such as (seasonal) storage of food, fixed built infrastructure to contain and protect the stored goods and livestock (as food-reserve and working power) generate an overall increase in metabolic level, and thus an increase in the amounts of goods to be transported. At the same time, the territory needs to be restructured: at a close distance to human settlements, resource density is increased (by deforestation, modifying soil structure through ploughing, planting monocultures of highly productive crops and protecting them from competition by weeding and fencing, fertilisation, irrigation, etc.), yet at further distances the land is used more extensively (forests for firewood, grasslands for pasture) and may even be gradually deprived of nutrients in favour of the intensively used core area.[12] This restructuring of territory again tends to reduce transportation needs: the largest amounts are harvested in closer vicinity to the settlement where they are stored[13] and consumed, while the further periphery is in part utilised by livestock who feed by themselves and move around on their own, while only concentrated final products (such as cheese or meat) need to be transported (McNetting 1993; von Thünen 1826; Arnold 1997). The collection of fuelwood, however, may remain an even increasing transportation task as forests become further removed to the periphery.[14] Beyond optimising the use of space,

---

[10] Assuming an average annual food demand of 500 kg per year, this population density can be sustained with an average food yield of 25 kg/km$^2$ (in case of low seasonal variation).

[11] *Colonisation of natural systems* is a concept used in social ecology that refers to society's deliberate interventions into natural systems in order to create or maintain a state of the natural system that renders it more useful for society. Colonisation mainly refers to human labour and the information, technologies and skills involved that make labour effective (Fischer-Kowalski and Haberl 2007; Singh et al. 2010).

[12] How exactly this is done, is of course subject to large geographical and cultural variation; but the principle holds true both for rain-fed and irrigation agriculture (see, for example, Arnold 1997; McNetting 1993).

[13] And in part also have to be transported in the other direction, as seed and dung.

[14] In arid regions, water may also constitute an important transportation issue. In our present approach, we have implicitly assumed water to be ubiquitously available at any time.

**season/time 1**  **season/time 2**

colonization of terrestrial ecosystem (active regulation of resource intensity of land); deforestation and exploitation of forests

⟶ transport of resources to home
○ resource (e.g. food)
● human homestead, settlement
⊛ settlement with stored resources

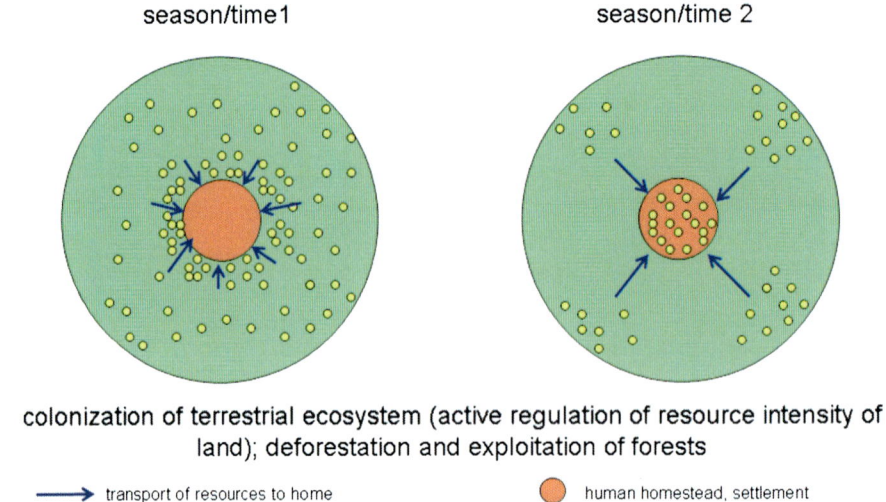

**Fig. 4.2** Territory and transport in agrarian societies, village level

chances to rationalise land-based transport are weak: loads are either frequent or small (vegetables, dung), or larger and seasonal (crop harvest, hay harvest, firewood), and all transports are fairly short distance. It is not economical to provide for elaborate transport infrastructure, either in terms of roads or in terms of vehicles.

On the basis of this rural matrix of human settlements that sustain themselves on a territory they colonise (spatial scale 1), agrarian societies may develop urban centres that accommodate a certain number of people that do not work the land, and that have to be sustained using the surplus production from the rural settlements. This presents the next transportation challenge: to build a centre-periphery network of transport infrastructure and to cover larger distances (with a possibly higher load). Materials for this additional infrastructure, and for the specific infrastructure of urban centres (such as fortifications and public buildings), again raise the overall metabolic level.

Several spatial scales may now be distinguished: scale 1, as we have seen before, consisting of a rural settlement and the territory from which it draws its resources (typically in a seasonal variation, as outlined in Fig. 4.2); scale 2 that consists of an urban centre and the "hinterland" of rural settlements and territories it requires for its sustenance; and scale 3 that may consist of several urban settlements that exchange certain commodities via trade.[15] Each scale builds upon the other and

---

[15] Although very plausible in practical terms, we have not considered the case of a multistage hierarchy of urban centres. Such a case, in principle, does not pose any new challenges to our model: transport loads and distances remain the same whether we deal with one large centre or with a hierarchy of intermediate centres. What matters on this level of abstraction is only the number of people (outside agriculture) to be sustained – the size of the required hinterland results from this (see below for more detail).

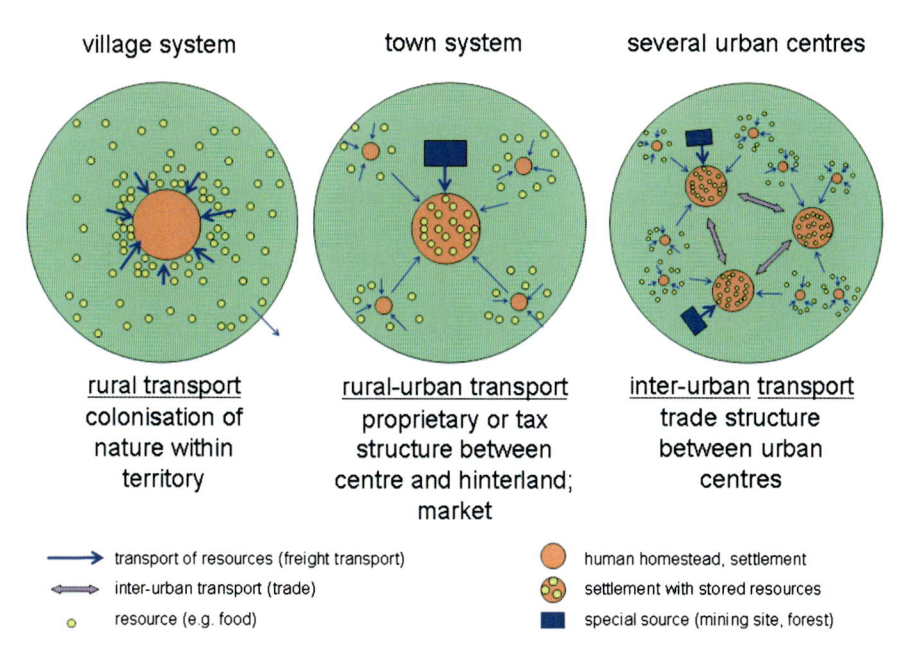

# 4 Modelling Transport as a Key Constraint to Urbanisation in Pre-industrial Societies

**Fig. 4.3** Territory and transport in agrarian civilisations, by spatial scale

implies the full functioning of all transportation processes at the scale below. It was a major achievement of Ester Boserup (1981) to create awareness of the apparently trivial fact that it is not only fertile territory that is required to sustain an urban centre, but that it is a territory sufficiently populated by a rural population that would deliver the necessary labour power to work on it.

According to our knowledge of agrarian systems (McNetting 1981), we felt it was safe to assume that transportation between the rural settlements and the urban centres is largely a one-way process (see Fig. 4.3). Even if produce from the villages is not delivered as income for the (urban) proprietor of the land, tithes or taxes, but to urban markets, the commodities bought in exchange for those (bulk) commodities have a much higher value density (such as metal tools, garments or salt). In terms of weight, they play a negligible role in rural metabolism, and they are merely "carried along" on the way back from the urban centres without posing any transportation problems in themselves.[16]

---

[16] This asymmetric exchange, of course, in the medium term leads to soil depletion in the hinterland: soil nutrients, contained in food and feed, travel to the urban centres, and are there ultimately washed into rivers and the sea, where they may cause pollution through over-fertilisation. This is what – in response to Liebig's (Liebig 1964; Boyden 1987) insights – (Marx 1976) called "metabolic rift" (see also Foster 1999). For some very large centres of agrarian civilizations (such as, for example, Edo in Japan), this problem was managed by the systematic collection of human (and animal) faeces in cities, transporting them back into rural areas to be used as manure (Takashi 1998). See Billen et al. (2009) on nutrient returns from Paris to its rural hinterland and Krausmann (2013) in this volume on similar considerations for the city of Vienna.

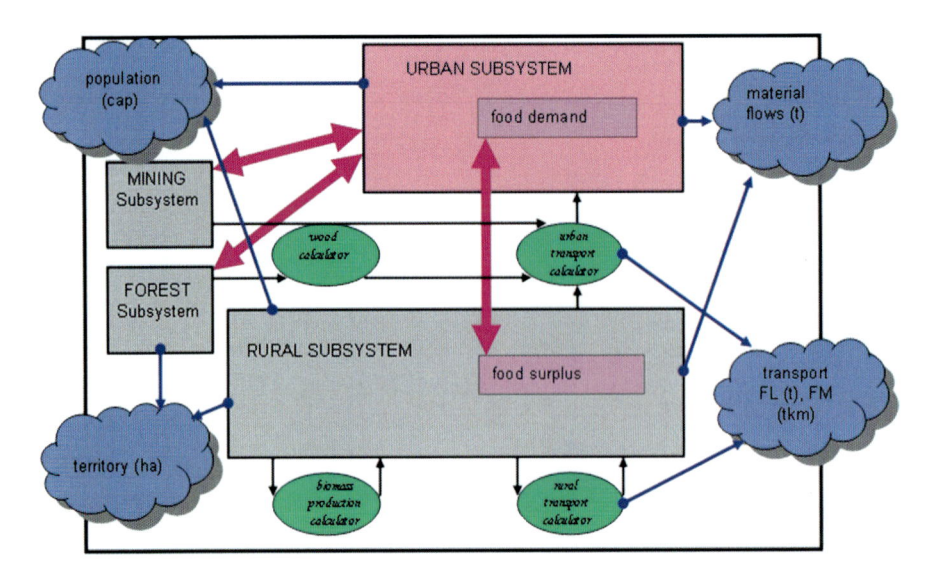

**Fig. 4.4** The overall structure of the model

Such complex relations require a formal model that allows a whole range of variables to be explicitly considered simultaneously, and supports testing the variations in outcome due to different assumptions. Its structure is described in Fig. 4.4. The core of this model consists in two mutually exclusive subsystems: a rural subsystem, comprising the population with a certain metabolic level working in agriculture and reproducing its own subsistence, and eventually a certain food surplus; and an urban subsystem (that can be empty for some model variants), comprising a population not working in agriculture but having a certain level of food demand that has to be satisfied by the surplus from the rural subsystem. In terms of food consumption, we assumed the same metabolic level for urban as for rural people. While for the rural system, we assume all resource requirements for its socioeconomic metabolism (that is food, feed, timber and mineral materials for construction) to be satisfied within the individual village's territory, we add specialised forest and mining compartments to the urban system. This pays tribute to the consideration that the fairly high amounts of timber/firewood required would have to be drawn from special regions that are perhaps remote but offer privileged transport conditions (such as rivers for rafting); the forest system, for simplification, does not comprise people, but only a territory of a size to be calculated by the model. The mining system comprises neither people nor territory, but serves as a source of materials (again calculated by the model according to urban demand) that have to be transported to the urban centres, possibly by special technologies.

There are four blocks of interdependent output variables generated by the model as soon as an urban centre with a certain population size is assumed. From the size of the city follows the size of the rural hinterland in terms of people and territory, the size of the overall material input per year, and transport in terms of Mass Lifted

4 Modelling Transport as a Key Constraint to Urbanisation in Pre-industrial Societies

**Table 4.1** Basic model assumptions of the human food calculator

| Net food requirement | [GJ$_{NV}$/cap/year] | 3.5 |
|---|---|---|
| Adjustment factor for losses and preparation wastes | | 1.3 |
| Average nutritive value of food | [MJ$_{NV}$/kg] | 7.0 |
| Average gross food demand | [kg/cap/year] | 650 |

Based on data given in (Sandgruber 1982; Teuteberg 1986); Krausmann (2004, 2008)

(ML, in tons), and as transport volumes, Mass Moved (MM, in tkm). This "translation" of population size into territory, material flows and transport is mediated by five "calculators", basically technology switches, into which all detailed model assumptions are packed, model assumptions that can be varied according to whatever more specific system features one has in mind.

The calibration of the parameters we use has mainly been informed by in-depth historical studies carried out in local communities in Austria (Krausmann 2004, 2008; Krausmann et al. 2013).

### 4.3.1 Human Food Calculator

The human food calculator translates a certain number of people into a certain net and gross demand for food biomass, in tons/year. Net demand derives from human metabolic need (in joules and in kilos respectively, as a mix of vegetable and animal components), and a certain fraction of losses and unused waste is assumed to arrive at gross demand numbers (see Table 4.1). We assumed food demand to be the same for rural and urban people.

### 4.3.2 Biomass Production Calculator

This calculator relates population and territory (or, in other words, population density) to food output through a "black box" of the agricultural production system. In this "black box" there is a certain distribution between land use categories, their specific yields, and a certain number of livestock per unit area (that produces a certain amount of manure). All this is related to domestic extraction of biomass (DE in tons), out of which a certain fraction amounts to food for humans.[17] From empirical

---

[17] The systemic relations in this "black box" are derived from our research on central European rain-fed agriculture (see Table 4.2). They will be quite unlike, say, South East Asian paddy field agriculture. Such relations result from a long-term learning and optimisation process (for ancient Roman agriculture, see for example Carlsen et al. 1994), and we have not yet managed to formalise them on a general level. So we use the parameters from empirical case studies. These parameters would, of course, be different in an agricultural system that does not use animal traction (such as in China or Japan).

**Table 4.2** Basic model assumptions and characteristic parameters of the biomass production calculator (standard and low productivity scenario), compared with empirical data on nineteenth century Austrian land use systems (three case studies)

|  |  | Empirical data | Standard | Low productivity |
|---|---|---|---|---|
| Food output/rural area | [$GJ_{NV}$/km²/year][a] | 175–280 | 200 | 120 |
| Food output/rural population | [$GJ_{NV}$/cap/year][a] | 5.3–6.2 | 5.7 | 4.8 |
| Potential food surplus | [% of total output] | 14–27 % | 20 % | 5 % |
| Exploitation rate | [% of surplus] | n.d. | 66 % | 66 % |
| Rural population density | [cap/km²] | 30–45 | 35 | 25 |
| Village population | [cap] | 102–129 | 100 | 100 |
| Village territory | [km²] | 2.3–4.2 | 2.9 | 4.0 |

Source: Empirical data based on Krausmann (2004, 2008)
[a]$GJ_{NV}$ refers to $10^9$ J (nutritive value)

**Table 4.3** Basic model assumptions on the relation of food production and agricultural biomass flows (biomass flows per unit of food output) and assumptions on transport stages of different biomass categories

|  | Biomass flows per GJ food output | Transport stages |
|---|---|---|
|  | [$kg_{FW}$/$GJ_{NV}$][a] | [#] |
| DE of primary crops | 171 | 2 |
| DE of straw, litter, hay | 577 | 1 |
| DE by grazing | 161 | 0 |
| DE rural wood | 190 | 1 |
| Manure output | 234 | 1 |

Source: Empirical data based on Krausmann (2004, 2008)
[a]$kg_{FW}$ refers to kg fresh weight (i.e. weight at transport time), $GJ_{NV}$ refers to $10^9$ J (nutritive value)

data on local Central European agricultural production systems of the early nineteenth century, we derive the parameters *rural population density*, *food output per total area* and *total material flows per unit of food output* (see Table 4.2). These exogenic parameters allow us to translate a certain food demand into territory and population as well as into structure and size of material flows (with different transport profiles) (Table 4.3).

### 4.3.3 Rural Transport Calculator

The rural transport calculator translates the manure output and the various types of domestically extracted biomass (that both come out of the biomass production calculator) into Mass Lifted (ML in tons) and transport volumes, Mass Moved (MM in tkm). It works by fractions of materials that have different transportation features: straw, litter and wood are brought from the territory directly to the farmstead; grain

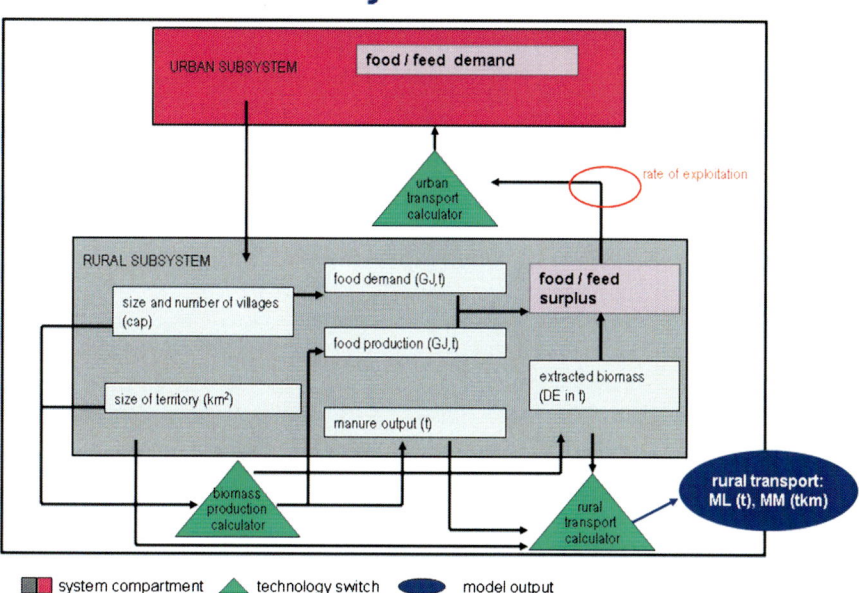

## rural subsystem details

**Fig. 4.5** Detailed model structure of the rural subsystem

is first brought to the local mill, and then to the farmstead; grazed biomass is not transported at all; manure and seed are taken back to the fields. For all these transports, a certain distance, a certain number of transport stages, a certain loading weight, a certain number of draft animals, a certain speed and a certain amount of human labour time for loading and accompanying are estimated.[18]

With the help of these calculators, we can characterise the details of the functioning of the rural subsystem (Fig. 4.5). With either a certain number of population or a certain size of rural territory given (these two are interdependent, as specified in the "biomass production calculator"), we can now calculate the amount of transportation required within the rural system (rural transportation) (as tons ML and tkm MM) and we can calculate the amount of food/feed overhead produced that could be delivered to urban centres as income for urban proprietors, taxes/tithes or to markets. There is still one more important intervening variable: We have denoted it *"rate of exploitation"*, that is the proportion of the surplus over local demand that is actually given or taken away. In the nineteenth century Austrian villages we

---

[18] We had so far no opportunity to calibrate our assumptions empirically. In a discussion with transport technology historians in Dietramszell, Bavaria (organised by the Breuninger Foundation in November 2003), our assumptions survived as plausible (see also Beck 1993; Hitschmann and Hitschmann 1891; Krausmann 2004; Möser 2003).

investigated, the calculated food surplus ranged between hardly anything and roughly one quarter, and from this a certain fraction was exported for purposes outside the village. So in the standard scenarios, the food surplus actually flowing into urban centres amounts to between 11 and 15% of the total food produced in the rural hinterland.[19] In a "low productivity" scenario, the exportable surplus rate of food is only 2%.

### 4.3.4 Wood Calculator

The wood calculator translates a certain urban population into a demand for firewood and timber and then the wood requirement into a corresponding forest area, wood yields presupposed. We assumed an average urban wood demand of 2 t/cap and year and an average (sustainable) wood yield of 2.1 t/ha and year (i.e. 3 $m^3$/ha/year or 300 $m^3$/$km^2$/year). These "urban forests" are added to the rural hinterland, enlarging the overall territorial requirements. The calculation of wood demand for the rural households is included within the systemic mix of the biomass production calculator, following the assumption that part of the territory of each rural community is woodland sufficient to cover the rural household demand for timber and firewood.[20]

### 4.3.5 Urban Transport Calculator

The urban transport calculator translates amounts of materials that are to be transported from the hinterland into the urban centre into Mass Lifted (ML, in tons) and Mass Moved (MM, in tkm), as well as into number of draft animals required and into human labour time. It differentiates between three types of freight: wood, food and feed for urban draft animals. With wood, whatever the size of the hinterland, we assume it is never transported over land more that an average of 5 km, but that the urban wood demand is covered by wood extraction from forests either close to the city or at a close distance from natural or artificial waterways sufficient for floating. Average number of transport stages for wood is assumed to be 2 (land-water-land) and for food and feed to be 1. Average transport distance for food and feed is assumed to be two-thirds of the radius of the total territory of the hinterland.[21] By applying factors for loading time, average transportation speed, weight carried per haul, animals

---

[19] We make no assumptions on how much of this is collected as proprietary income or taxes, and how much is sold on markets. But it is quite obvious that collecting "one tenth" (Zehent) for tithes or taxes is already roughly at the upper limit of what the villages can spare.

[20] This assumption is well warranted for our empirical case studies, but this could of course be very different for other agrarian systems.

[21] It must be borne in mind that this assumption implies an ideal geography that allows the territory to expand evenly in all directions from the centre into the rest of the world. Geometrically, all territories would be circular with the settlement/urban centre in the middle.

4 Modelling Transport as a Key Constraint to Urbanisation in Pre-industrial Societies 91

**Table 4.4** Basic model assumptions to calculate transport indicators

|  |  | Urban food | Urban feed | Urban wood | Rural transport |
|---|---|---|---|---|---|
| Distance per haul |  | 2/3 of radius of hinterland | 2/3 of radius of hinterland | 5 km | 2/3 of radius village territory |
| Transport stages | [#] | 1 | 1 | 2 | 1.2[a] |
| Speed[b] | [km/h] | 3 | 3 | 3 | 2 |
| Weight per haul | [t/haul] | 0.5 | 0.5 | 1 | 0.25 |
| Animals per haul | [#/haul] | 1.5 | 1.5 | 2 | 1 |
| Loading time per stage | [h/haul] | 3 | 3 | 3 | 2 |

[a]See Table 4.3 for details on transport stages per type of material flow
[b]These are the assumptions for the loaded cart. We assumed that speed increases by 25% for the return trip without load

used per haul and hours worked per year by draft animals and humans, we are able to calculate the transport indicators ML, MM and the transport costs in man and animal years for the different material classes and for urban transport (Table 4.4).

The human labour required for rural-urban transport is assumed to be delivered by the urban population.[22] Draft animals required for rural-urban transport are not assumed to be an internal element of the agricultural production system. Their feed requirement is calculated as an additional biomass flow and is added to the biomass demand of the urban system and hence also feeds back into the biomass production model and increases the size of the rural hinterland required (Fig. 4.6).[23]

This is the only dynamic feedback feature built into the otherwise fully linear model. As we intended to use this model to test our hypothesis that transportation functions as a limiting condition for agrarian societies, we have designed it so as to possibly disprove this. The most important background assumptions that work in the direction of facilitating transport are the following:

- we presuppose a homogeneous geography where territories may extend symmetrically in all directions (no shores, no big mountains, no hostile neighbours), taking a circular shape
- we assume territorial homogeneity where all land can be inhabited, utilised and transgressed (no high mountains, no deserts, etc.)

---

[22] By definition, the rural population is fully occupied by agriculture. Practically, of course we may assume that farmers transport some of their produce to markets. But for the bulk goods like corn or flour that make up the overwhelming majority of the mass transported, we may safely assume that some kind of specialised business performs this task.

[23] Feed demand for urban transport was calculated on the assumption of an average daily feed intake of 10 kg per draft animal and day. This already includes the feed demand for the first two live years of the horse during which it is not yet used for draft purposes. Fifty percent of the feed demand was assumed to be grain (oats) and converted with the rural biomass calculator into rural hinterland, population and agricultural biomass flows. The remaining 50 % of the feed demand was assumed to be crop residues and grazing and these were not translated into additional hinterland (see Krausmann 2004).

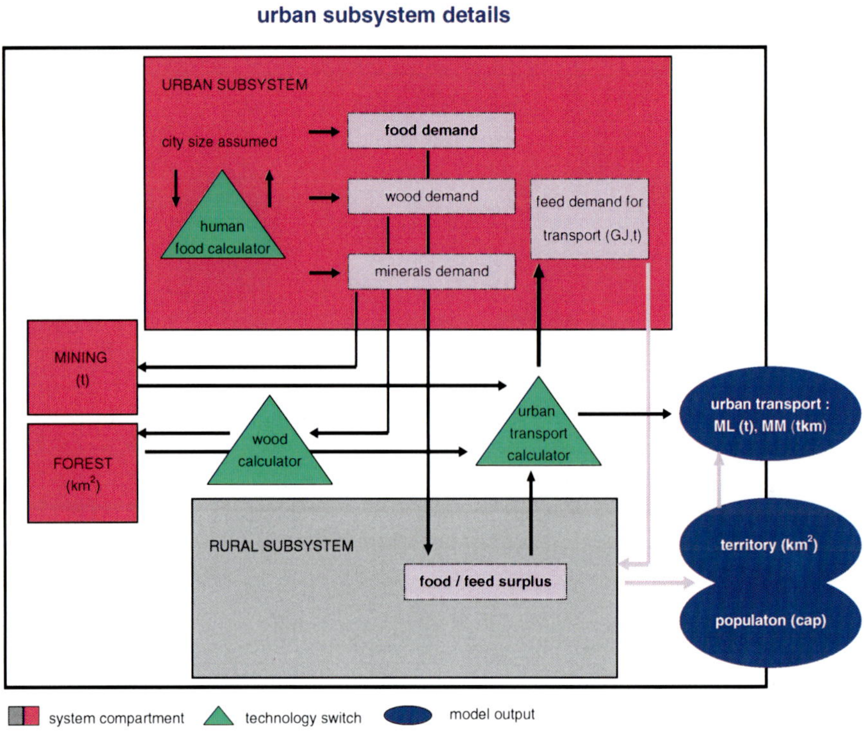

**Fig. 4.6** Urban subsystem details

- we do not include the efforts required for building and maintaining transport infrastructure (roads, bridges, harbours) or transport vehicles
- we do not include the efforts required to secure (politically and militarily) territories and transport infrastructures
- we assume low-price bulk goods (like timber, firewood and construction minerals) to be transported across land for very short distances only, and we ignore the effort needed to transport these goods on waterways or across the sea, and finally
- we assume the population in urban centres to have – on average – the same personal consumption level as people in the hinterland. In other words, we do not contrast rural poverty and urban luxury. This, of course would not preclude an urban elite living by very different standards than the rest – but their overconsumption, on the urban average, would be balanced by urban poor.

Even under such favourable conditions, as we will see in the next chapter, transport is a major task to be mastered by agrarian societies, and this task becomes increasingly difficult to manage as urban centres gain in size.

## 4.4 Centre-Hinterland Relations and Transport Requirements Depending Upon Centre Size: Results from Our Model

Let us first look at the scale 1 characteristics of the system implied by our knowledge of socio-metabolic relations (see Table 4.5). When we assume a population density of 35 people per km$^2$, a reasonable standard for European history,[24] a village of 100 inhabitants utilises a territory of 3 km$^2$. Assuming the settlement to be located in the centre of this territory, it takes a quarter of an hour's walk (1 km) to reach its boundary. The extracted materials within this territory (food, feed and wood) will amount to 5 ton/inhabitant annually (or 500 ton for the village as a whole), which, derived from the composition and processing characteristics of this material, corresponds to a Mass Lifted of 6 ton per inhabitant (or roughly 620 ton for the village annually), with an average length of haul of less than a kilometre. The resulting Mass Moved then amounts to roughly 4 tkm per inhabitant and year, which costs 7,000 human and 4,000 animal working hours a year (corresponding to 5 person and 3 animal years, Table 4.5) or, relative to the estimated total labour power of the village, 7% of human, and 6% of existing animal labour power. This village system can potentially export, per inhabitant, 160 kg (or 1,2 GJ) of food annually, which means that 4 village inhabitants can feed one urban citizen, or one such village of 100 people could feed up to 25 non-agricultural population depending on the exploitation rate of the surplus. Everything said so far assumes that the village functions on a subsistence basis and does not invest any effort into delivering its "exports" anywhere, but that it is a problem of the urban centres to get hold of the supplies they need.

Now we will use our model to understand what happens if the system is increasing in scale. On a next scale, we consider a system with a town of 20,000 inhabitants, sustained by the hinterland of villages required to provide sufficient resources (see Table 4.5).[25] In case of what we assume as standard agricultural productivity (see Table 4.2), a town of 20,000 urban people will require a hinterland of 135,000 rural population, and a territory of 4,100 km$^2$. The radius of this territory would be roughly 36 km, already a day's walking distance. This territory would contain 1,300

---

[24] See, for example, 30–35 cap/km$^2$ in Austria and the UK around 1750 (Krausmann et al. 2013).

[25] This whole comparison, in terms of system characteristics, will be calibrated according to the empirical relations we obtained from the analysis of three nineteenth century Austrian villages, as in the calculation on scale 1. In more general terms, we make our model calculations before the background of a typical central European rain-fed agricultural production system dominated by a traditional three-field rotation and rural subsistence grain production. There is a heavy reliance on draft animals for agricultural labour and transportation, and a limited nutrient availability. The territories are inland with no access to sea, hence no coastal shipping is taken into account. For reasons of simplicity, river transport was taken into account for wood. (Beck 1993; Winiwarter 2002; Krausmann 2004, 2008).

**Table 4.5** Modelling results for material flows and transport in agrarian societies

| Scale | | Village | Town system (20,000) | City system (100,000) |
|---|---|---|---|---|
| Population | [1,000 cap] | 0.1 | 154.7 | 826.5 |
| Area | [1,000 km$^2$] | 0.003 | 4.1 | 21.9 |
| Population density | [cap/km$^2$] | 35 | 38 | 38 |
| total DE = DMC = DMI/cap | [t/cap/year] | 5.1 | 5.3 | 5.2 |
| Mass Lifted rural/cap | [t/cap/year] | 6.2 | 6.4 | 6.4 |
| Mass Lifted urban/cap | [t/cap/year] | | 0.6 | 0.6 |
| Mass Lifted total/cap | [t/cap/year] | 6.2 | 7.0 | 6.9 |
| average distance per haul | [km/haul] | 0.6 | 0.7 | 0.7 |
| Mass Moved rural/cap | [tkm/cap/year] | 4.0 | 4.2 | 4.2 |
| Mass Moved urban/cap | [tkm/cap/year] | – | 3.6 | 6.6 |
| Mass Moved total/cap | [tkm/cap/year] | 4 | 8 | 11 |
| Human transport effort | [1,000 person-years/year] | 0.005 | 9.6 | 55.2 |
| Of which urban transport | [1,000 person-years/year] | 0 | 1.0 | 9.8 |
| Animal transport effort | [1,000 animal-years/year] | 0.003 | 6.6 | 38.1 |
| Of which urban transport | [1,000 animal-years/year] | 0 | 0.9 | 7.8 |
| Rural human transport effort/cap rural | [h/cap/year] | 82 | 82 | 82 |
| Urban human transport effort/cap urban | [h/cap/year] | – | 11 | 18 |
| Share of urban labour force required for transport | | 0% | 8% | 15% |

villages of the size described above that need to be kept motivated to deliver their surplus to the urban centre, by enforcing proprietary relations, taxes and tithes or by market relations. This town would require an annual input of 55,000 ton of biomass materials, of which food would be about one quarter,[26] to be transported on average a distance of roughly 22 km.[27] As we assumed the rural population to be fully occupied with working the land, the labour load of transportation to the urban centre has to be borne by the urban population.[28] The work load of transporting biomass to a city of 20,000 amounts to 1,000 person-years. Assuming two-thirds of the urban population to be between 15 and 65 years of age, this corresponds to about 8% of the potential urban labour force (or 16%, considering males only).

A larger city of 100,000 inhabitants already, under the same conditions, needs a hinterland of 22,000 km$^2$ and a population of 720,000 inhabitants. Its radius would be 84 km (or more, since under ordinary geographical conditions it should not be easy to create circular territories of that size), and within this radius there would be more than 7,300 villages. The work load of transporting biomass to the city amounts to almost 10,000 person-years – under the same assumptions as above, this would already draw 15% of the population, or 30% of the potential urban labour force into the transport sector (see Fig. 4.9).

As we have constructed almost the whole model in a linear fashion, we should not be surprised to see certain constant relations across spatial scales. There are two moments, though, that contribute to non-linearity: one is distances expanding only with a square root function of territory, and therefore of materials extracted, and the other is the feedback effect from increasing distances to increased need for draft animals to again increasing territorial needs to feed those animals. The resulting non-linearity, across the whole chain of causal interlinkages, can be seen with the following parameters (see Figs. 4.7 and 4.8):

---

[26] For the moment, we have ignored construction minerals. Under industrial conditions, construction minerals are of the same order of magnitude (or more) than biomass. Under agrarian conditions, we would presume them to be somewhat less. Calculations based on statistical data for the use of construction minerals in Vienna around the year 1800 and bottom up estimations for the use of construction minerals per building, suggest an annual DMC of construction minerals of 0.5–1.0 ton/c. They would typically not be transported over large distances (see model description above). According to rough estimates including rural and urban demand for construction minerals in our model would increase transport expenditure in terms of human and animal labour by 5–15 %.

[27] Of course, in the longer run the territory around an urban centre would become restructured in a similar way as happens in the villages themselves that organise land use so as to minimise transportation. But there are certainly limits to such a restructuring. For an empirical case see twelfth century Constantinople (Koder 1997).

[28] For our calculation, it does not matter where the labourers in the transport business actually live; by definition, they belong to the urban population, as the rural population works 100% in agriculture.

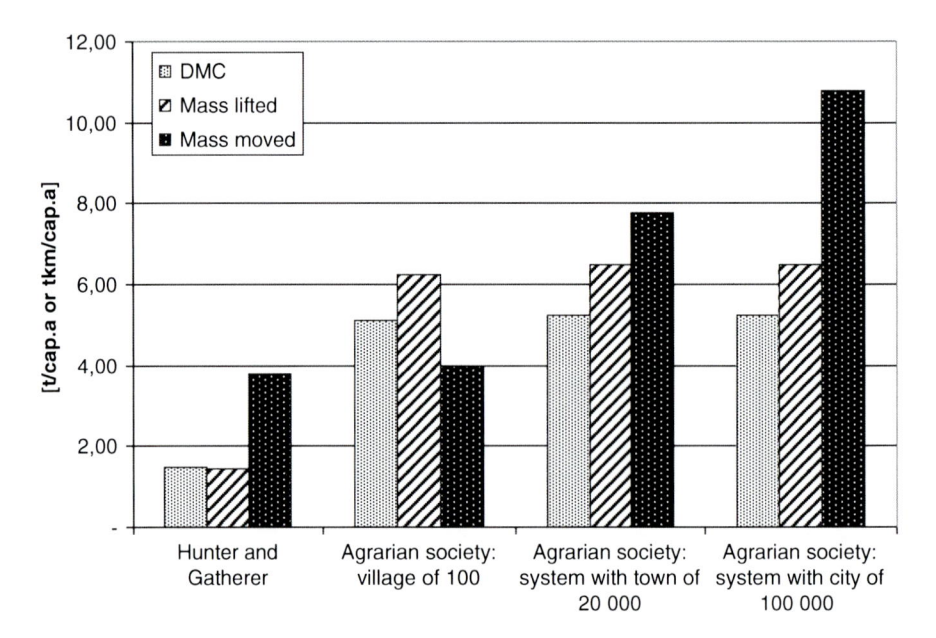

**Fig. 4.7** Per capita material use (*DMC*), and per capita transport effort (*ML, MM*) for hunters & gatherers and for agrarian societies by size of urban centre to be sustained (standard productivity assumptions)

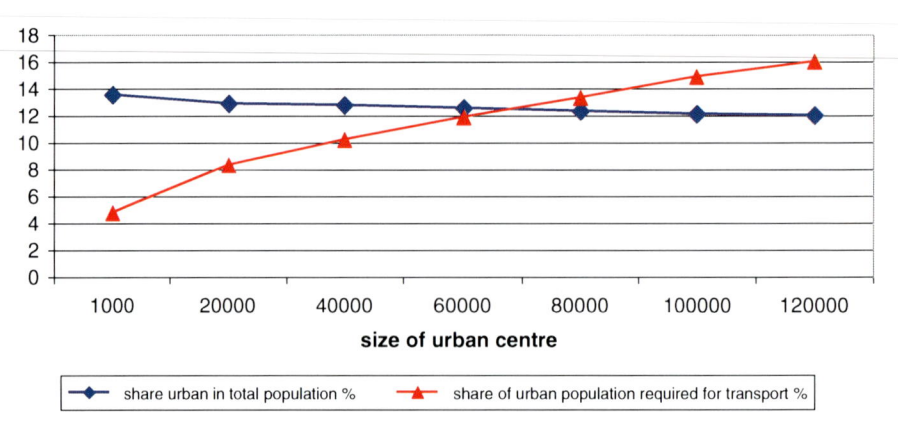

**Fig. 4.8** The share of urban (in contrast to rural, agricultural) population in the total system, and the share of the urban population required for transportation, in relation to size of urban centre (standard productivity assumptions)

- The most dynamic factor is Mass Moved in terms of tkm: if the size of the centre increases, and therefore the territory, Mass Moved increases by a disproportionately large amount (see Fig. 4.7), while per capita material use, and per capita Mass Lifted, remain more or less constant (or even decrease, due to the use of larger vehicles, as we assume).

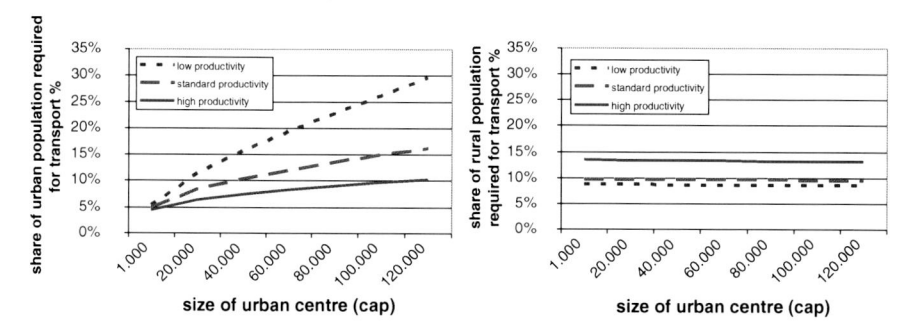

**Fig. 4.9** The relation between agricultural productivity and human transport effort, by size of urban centre. Results from the standard, a low and a high productivity scenario. In the low productivity scenario we assumed a 40 % lower area and 15 % lower labour productivity as compared to the standard scenario (see Table 4.2). In the high productivity scenario we assumed corresponding increases in area and labour productivity

- As a consequence, the need for draft animals rises by a disproportionate amount (much faster than the human labour required, because of the use of larger vehicles with still only one driver).
- And again as a consequence of this, the hinterland needs to increase disproportionately, to allow keeping and feeding all these animals.
- As an interesting and unexpected side-effect of this, the proportion of the urban population (or non-agrarian population) within the total population **declines** again when the size of the urban centre increases beyond a certain point (see Fig. 4.8). So for each agricultural production system (depending on its specific features as described in the "biomass production calculator") there may be an optimum size of urban centres (and of overall territory), a size beyond which the material standard of living starts to decline, because of investments into overcoming distances.[29]

So even under very favourable assumptions, there are clear indications for a scale limit to agrarian empires, and agrarian centres, due to factors associated with the cost of transport (in terms of human labour time and land). Where this scale limit occurs strongly depends upon agricultural productivity. As apparent in Fig. 4.9, which shows model results for scenarios based in different assumptions on land and labour productivity, a city of 100,000 inhabitants with a hinterland of high productivity may have the same transport effort as a city of 20,000 inhabitants in the case of low hinterland productivity (while the transport activity among the rural population is hardly affected). Still, in both cases 10% of the urban population will be engaged in transportation for the supply of the city. If only males perform this task and if you exclude children and old people, this will amount to well over a quarter

---

[29] One should be aware that in our model we keep the material standard of living constant by securing a constant amount of food and other material input per capita.

of the male labour force. Different assumptions on area and labour productivity in the agricultural hinterland do have a large impact on the model results, but whatever the agricultural productivity, such a scale limit will occur.

At the heart of this lies the dilemma that in order to solve the problem of overcoming distances, the problem is structurally aggravated by using additional space to gain the supplies needed. This dilemma, under the conditions of the agrarian socio-ecological regime, where energy and supplies are largely area dependent, is fundamentally insoluble.[30]

## 4.5 Conclusions

Our modelling exercise yields important insights into the functioning of socio-ecological systems. While our analysis focussed on transportation, there surfaced a number of key generic features of agrarian societies and the transition from the agrarian to the industrial socio-ecological regime. As overly simple as it may be, our model allows the intimate interlinkage of territory, labour and subsistence that permeates all social and ecological relations to be grasped. It demonstrates that the growth of urban centres depends upon an extension of the territory and rural population to work the land and generate the supplies that cities require. These changes in the spatial organisation of socio-ecological systems have implications both for ecosystems (e.g. land use intensity, nutrient flows) and social organisation (e.g. division of labour). Our findings shed light upon the causes of the constant competition and struggle over land so prevalent in history, often simply interpreted as resulting from too high ambitions on the part of political and military leaders, while here we are able to show that this may in fact be dynamically driven by the sociometabolic needs of emerging and growing urban centres. Our results also help to explain why even large empires in history that extended over land with low productivity (because of aridity, for example) struggled to maintain urban centres of a still relatively small size. We can also see that raising agricultural productivity was mainly in the interest of urban centres: While it increases the labour burden on peasants (see Boserup 1965) and possibly drives rural population growth to cope with this rising burden, it allows the food surplus required to sustain urban expansion to be increased.[31]

Under agrarian conditions, there were a number of strategies available for urban elites to improve their wellbeing and increase their wealth.

---

[30] On the other hand, by implication the competitive advantage that can be gained from non area-dependent modes of transport, such as downstream rafting or sailing, or using coal/oil for driving engines, becomes obvious.

[31] In the light of this, one may still find it astonishing how little historical urban intelligence was invested into such efforts, perhaps with the notable exception of the Egyptian and Roman Empires, who developed systematic scientific expertise for this purpose.

- increasing the size of the territory and the rural population under their control (as above)
- investing in promoting agricultural productivity
- building a transport infrastructure and improved transport vehicles
- trading with other urban centres
- putting pressure on the rural population by tithes and taxes, thus (in terms of our model) increasing the rate of exploitation
- importing adult slaves to save on the reproduction costs of urban labour.

All these strategies, according to our model, run into constraints from transportation and become at a certain point self-defeating. Accordingly, most major historical urban centres were built in locations where they did not depend upon land transport, but had access to waterways or the sea. Such access did not completely relieve them of these constraints, but helped to push the boundaries a little bit further. So, finally, the sociometabolic approach upon which we base our analysis allows us to understand why, without area-independent sources of transport energy (such as coal or petroleum), urban growth is severely constrained. This also underlines the significance of changes in transport technology and infrastructure (the transport revolution) for the agrarian-industrial transition and the intimate relation between the energy and transport systems (Gingrich et al. 2012).

## References

Arnold, A. (1997). *Allgemeine Agrargeographie*. Gotha/Stuttgart: Klett-Perthes.
Barles, S. (2009). Urban metabolism of Paris and its region. *Journal of Industrial Ecology, 13*, 898–913.
Beck, R. (1993). *Unterfinning. Ländliche Welt vor Anbruch der Moderne*. München: C.H. Beck.
Billen, G., Barles, S., Garnier, J., Rouillard, J., & Benoit, P. (2009). The food-print of Paris: Long-term reconstruction of the nitrogen flows imported into the city from its rural hinterland. *Regional Environmental Change, 9*, 13–24.
Boserup, E. (1965). *The conditions of agricultural growth. The economics of agrarian change under population pressure*. Chicago: Aldine/Earthscan.
Boserup, E. (1981). *Population and technology*. Oxford: Basil Blackwell.
Boyden, S. (1987). *Western civilization in biological perspective*. Oxford: Oxford University Press.
Carlsen, J., Ørsted, P., & Skydsgaard, J. E. (1994). *Landuse in the Roman Empire* (Analecta Romana Instituti Danici: Supplementum 22). Rome: "L'Erma" di Bretschneider.
Ciccantell, P. S., & Bunker, S. G. (1998). *Space and transport in the world-system* (Studies in the political economy of the world-system). Westport/London: Greenwood Press.
Cohen, M. N. (1977). *The food crisis in prehistory. Overpopulation and the origins of agriculture*. New Haven/London: Yale University Press.
Cusso, X., Garrabou, R., & Tello, E. (2006). Social metabolism in an agrarian region of Catalonia (Spain) in 1860 to 1870: Flows, energy balance and land use. *Ecological Economics, 58*, 49–65.
Fischer-Kowalski, M., & Haberl, H. (1993). Metabolism and colonization. Modes of production and the physical exchange between societies and nature. *Innovation – The European Journal of Social Sciences, 6*(4), 415–442.
Fischer-Kowalski, M., & Haberl, H. (2007). *Socioecological transitions and global change: Trajectories of social metabolism and land use*. Cheltenham: Edward Elgar.

Fischer-Kowalski, M., Krausmann, F., & Smetschka, B. (2004). Modelling scenarios of transport across history from a socio-metabolic perspective. *Review Fernand Braudel Center, 27*(4), 307–342.

Fischer-Kowalski, M., Krausmann, F., Giljum, S., Lutter, S., Mayer, A., Bringezu, S., Moriguchi, Y., Schütz, H., Schandl, H., & Weisz, H. (2011). Methodology and indicators of economy wide material flow accounting. State of the art and reliablity across sources. *Journal of Industrial Ecology, 15*(6), 855–876.

Foster, J. B. (1999). Marx's theory of metabolic rift: Classical foundations for environmental sociology. *The American Journal of Sociology, 105*(2), 366–405.

Gingrich, S., Haidvogl, G., & Krausmann, F. (2012). The Danube and Vienna: Urban resource use, transport and land use 1800 to 1910. *Regional Environmental Change, 12*(2), 283–294. doi:10.1007/s10113-010-0201-x.

Goudsblom, J. (1992). *Fire and civilization*. London: Penguin.

Haberl, H., Winiwarter, V., Andersson, K., Ayres, R. U., Boone, C. G., Castillio, A., Cunfer, G., Fischer-Kowalski, M., Freudenburg, W. R., Furman, E., Kaufmann, R., Krausmann, F., Langthaler, E., Lotze-Campen, H., Mirtl, M., Redman, C. A., Reenberg, A., Wardell, A. D., Warr, B., & Zechmeister, H. (2006). From LTER to LTSER: Conceptualizing the socio-economic dimension of long-term socio-ecological research. *Ecology and Society, 11*, 13. (Online), http://www.ecologyandsociety.org/vol11/iss2/art13/

Harris, M. (1987). *Cultural anthropology*. New York: Harper & Collins.

Hawkes, K., Hill, K., & O'Connell, J. F. (1982). Why hunters gather: Optimal foraging and the ache of eastern Paraguay. *American Ethnologist, 9*, 379–398.

Hitschmann, R., & Hitschmann, H. H. (1891). *Vademecum für den Landwirth*. Wien: Moritz Perles.

Koder, J. (1997). Fresh vegetables for the capital. In C. Mango & G. Dagron (Eds.), *Constantinople and its Hinterland* (pp. 49–56). Variourum: Aldershot.

Krausmann, F. (2004). Milk, manure and muscular power. Livestock and the industrialization of agriculture. *Human Ecology, 32*(6), 735–773.

Krausmann, F. (2008). *Land use and socio-economic metabolism in pre-industrial agricultural systems: Four nineteenth-century Austrain Villages in Comparison* (Social Ecology Working Paper; 72). Vienna: IFF Social Ecology.

Krausmann, F. (2011a). The global metabolic transition: a historical overview. In: F. Krausmann (Ed.), *The socio-metabolic transition. Long term historical trends and patterns in global material and energy use* (Social Ecology Working Paper 131, pp. 73–98). Vienna: Institute of Social Ecology.

Krausmann, F. (2013). A city and its Hinterland: Vienna's energy metabolism 1800–2006. In S. Singh, H. Haberl, M. Schmid, M. Mirtl, & M. Chertow (Eds.), *Long term socio ecological research*. New York: Springer.

Krausmann, F., Fischer-Kowalski, M., Schandl, H., & Eisenmenger, N. (2008a). The global socio-metabolic transition: Past and present metabolic profiles and their future trajectories. *Journal of Industrial Ecology, 12*, 637–656.

Krausmann, F., Schandl, H., & Sieferle, R. P. (2008b). Socio-ecological regime transitions in Austria and the United Kingdom. *Ecological Economics, 65*, 187–201.

Krebs, J. R., & Davies, N. B. (1984). *Behavioural ecology: An evolutionary approach*. Oxford: Blackwell.

Layton, R., Foley, R., & Williams, E. (1991). The transition between hunting and gathering and the specialized husbandry of resources: A socio-ecological approach. *Current Anthropology, 32*(3), 255–274.

Lee, R. B. (1980). Lactation, ovulation, infanticide, and women's work: A study of hunter-gatherer population regulation. In M. N. Cohen et al. (Eds.), *Biosocial mechanisms of population regulation* (pp. 321–348). New Haven/London: Yale University Press.

Lewis, H. T. (1982). Fire technology and resource management in aboriginal North America and Australia. In N. M. Williams & E. Hunn (Eds.), *Resource managers: North American and Australian hunter-gatherers* (pp. 45–67). Canberra: Australian Studies Press.

Liebig, J. (1964). *Animal chemistry or organic chemistry in its application to physiology and pathology*. New York: Johnson Reprint. (Original 1842).

Marx, K. (1976). *Capital volume I*. New York: Vintage.

McNetting, R. M. (1981). *Balancing on an Alp. Ecological change and continuity in a Swiss mountain community*. London/New York/New Rochelle/Melbourne/Sydney: Cambridge University Press.

McNetting, R. M. (1993). *Smallholders, householders. Farm families and the ecology of intensive, sustainable agriculture*. Stanford: Stanford University Press.

Möser, K. (2003, November). "Projekt "Der europäische Sonderweg" – Arbeitspapier zur Transportgeschichte". Der Europäische Sonderweg: Die sozialmetabolische Transformation vom Agrarsystem zur Industrialisierung.

Sahlins, M. (1972). *Stone age economics*. New York: Aldine de Gruyter.

Sandgruber, R. (1982). *Die Anfänge der Konsumgesellschaft. Konsumgüterverbrauch, Lebensstandard und Alltagskultur in Österreich im 18. und 19. Jahrhundert*. Wien: Verlag für Geschichte und Politik.

Sieferle, R. P. (2001). *The subterranean forest. Energy systems and the industrial revolution*. Cambridge: The White Horse Press.

Singh, S. J., Haberl, H., Gaube, V., Grünbühel, C. M., Lisievici, P., Lutz, J., Matthews, R., Mirtl, M., Vadineanu, A., & Wildenberg, M. (2010). Conceptualising long-term socio-ecological research (LTSER): Integrating the social dimension. In F. Müller, C. Baessler, H. Schubert, & S. Klotz (Eds.), *Long-term ecological research, between theory and application* (pp. 377–398). Dordrecht/Heidelberg/London/New York: Springer.

Takashi, K. (1998, September 3–5). Edo in the seventeenth century. In *4th international conference on urban history, Venice*.

Teuteberg, H.-J. (1986). Der Verzehr von Lebensmitteln in Deutschland pro Kopf und Jahr seit Beginn der Industrialisierung (1850 bis 1975). Versuch einer quantitativen Langzeitanalyse. In H.-J. Teuteberg & G. Wiegelmann (Eds.), *Unsere tägliche Kost. Geschichte und regionale Prägung* (pp. 225–281). Münster: Coppenrath.

von Thünen, J. H. (1826). *Der isolierte Staat in Beziehung auf Landwirtschaft und Nationalökonomie*. Jena: Fischer.

Weisz, H., Fischer-Kowalski, M., Grünbühel, C. M., Haberl, H., Krausmann, F., & Winiwarter, V. (2001). Global environmental change and historical transitions. *Innovation – The European Journal of Social Sciences, 14*(2), 117–142.

Weisz, H., Krausmann, F., Amann, C., Eisenmenger, N., Erb, K.-H., Hubacek, K., & Fischer-Kowalski, M. (2006). The physical economy of the European Union: Cross-country comparison and determinants of material consumption. *Ecological Economics, 58*, 676–698.

Winiwarter, V. (2002). Der umwelthistorische Beitrag zur Diskussion um nachhaltige Agrar-Entwicklung. *Gaia, 11*(2), 104–112.

Wrangham, R. (2009). *Catching fire. How cooking made us human*. New York: Basic Books.

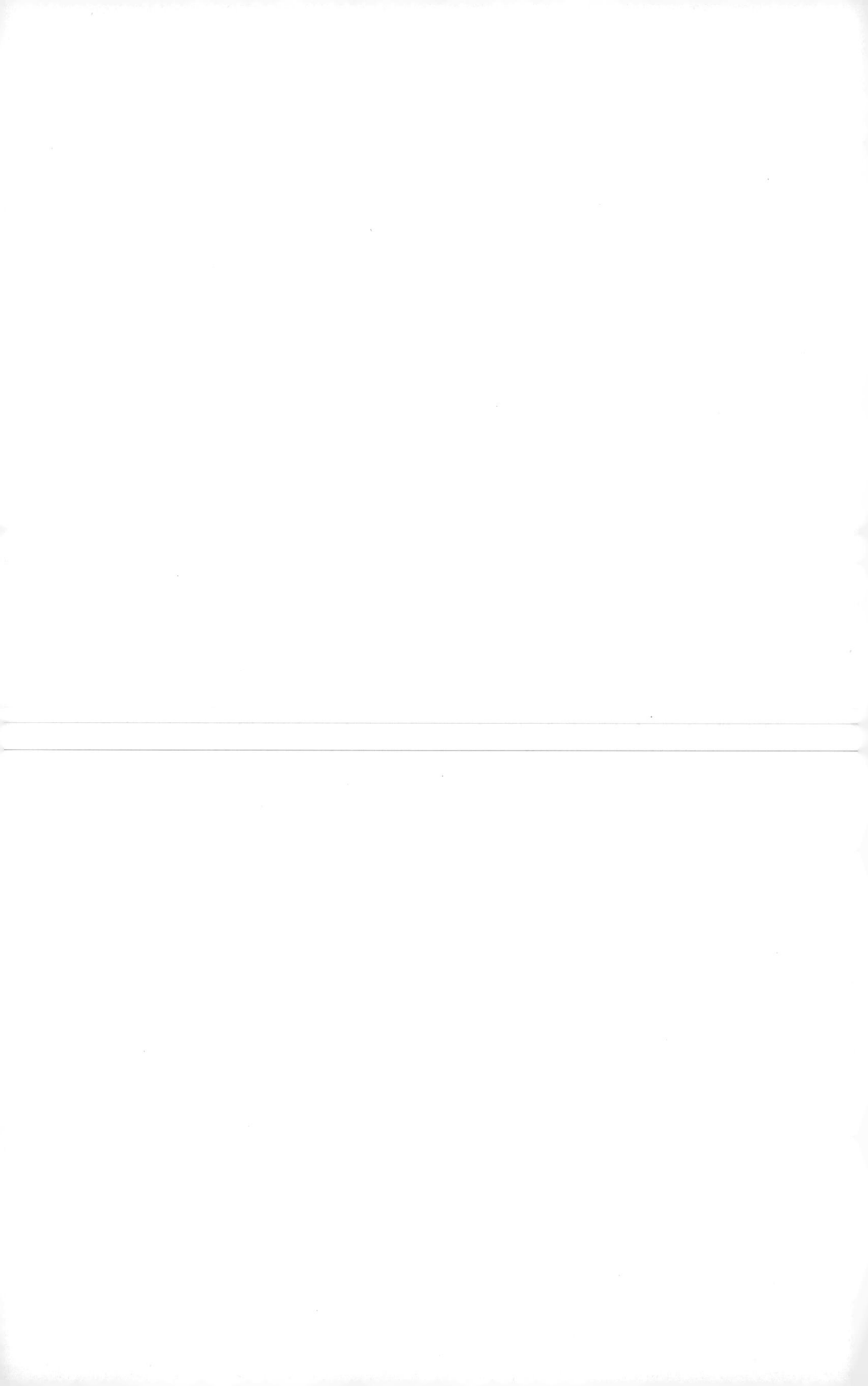

# Chapter 5
# The Environmental History of the Danube River Basin as an Issue of Long-Term Socio-ecological Research

Verena Winiwarter, Martin Schmid, Severin Hohensinner, and Gertrud Haidvogl

**Abstract** Only in a long-term perspective does the profound difference between pre-industrial and industrial society-nature relations become clearly visible. Long-term socio-ecological research (LTSER) extends its temporal scope significantly with contributions from environmental history. This chapter discusses the Danube, Europe's second longest, and the world's most international river, as a long-term case study. We approach the river as a 'socio-natural site', i.e. the nexus of arrangements (such as harbours, bridges, power plants or dams) with practices (such as river regulation, transportation, food- and energy-procuring). Arrangements and practices are both understood as socio-natural hybrids. We discuss how and why practices and arrangements developed over time and which legacies past practices and arrangements had. We emphasise the role of usable energy (so-called exergy) in the transformation of socio-natural sites. Since industrialisation, the amount of exergy harvestable from the Danube's arrangements has increased by orders of magnitude and so have the societal and ecological risks from controlling these exergy-dense arrangements. The arrangements we have inherited from our ancestors determine the scope of options we have in the present when dealing with rivers like the Danube. Current management decisions should therefore be based on the firm ground of historical knowledge.

**Keywords** Environmental history • Danube River Basin • Socio-natural sites • Riverine landscape reconstruction • Historical river geomorphology • River engineering • River-society interaction • Long-term socio-ecological research

V. Winiwarter, Ph.D. (✉) • M. Schmid, Ph.D.
Institute of Social Ecology, Alpen-Adria Universitaet Klagenfurt, Wien, Graz,
Schottenfeldgasse 29, Vienna 1070, Austria
e-mail: verena.winiwarter@uni-klu.ac.at; martin.schmid@aau.at

S. Hohensinner, Ph.D. • G. Haidvogl, Ph.D.
Institute of Hydrobiology and Aquatic Ecosystem Management,
University of Natural Resources and Life Sciences,
Max Emanuel Str. 17, Vienna 1180, Austria
e-mail: severin.hohensinner@boku.ac.at; gertrud.haidvogl@boku.ac.at

## 5.1 Introduction

Environmental history is based on a co-evolutionary concept of relationships between society and nature. The main object of study are those relations in the past. Since the field developed its own contours in the 1970s, pollution, environmentalism, climate, resource use and abuse and its environmental effects, the study of conservation history and, more recently, the environmental effects of war and the human body in polluted environments have been studied (McNeill 2003).

An environmental history of the Danube River Basin (DRB) necessarily has to combine many of these themes, taking the diversity of environments along the river into account. The upper stretch of the Danube connects the Austrian Alps and the Western Carpathian Mountains and comprises the river from its source to the 'Devin Gate' (once called 'Porta Hungarica', near Bratislava). The middle stretch runs from there to the Iron Gate Gorge in the Southern Romanian Carpathians. The lower part reaches from the Iron Gate to the delta-like estuary at the Black Sea (Sommerwerk et al. 2009). Today, hydromorphological change is a major environmental challenge in the upper DRB, pertaining to questions of conservation and resources (water power and flood protection), while pollution is among the biggest concerns in the lower Danube. These together have resulted in the almost total demise of fisheries in the middle DRB. The DRB exhibits a plethora of environmental problems, many of which are likely to be exacerbated by global climate change. None of these problems can be addressed without knowledge about human impacts on the river and the river's impact on humans over time.

The Danube is not only Europe's but the world's most international river. The DRB covers about 800,000 $km^2$ in the territories of currently about 20 states (Wolf et al. 2002). The International Commission for the Protection of the Danube River (ICPDR) was formed in 1994 to monitor the current state of the DRB and to play an active role in conciliating its various stakeholders. The Commission is now responsible for implementing the European Water Framework Directive (WFD).[1] An understanding of historical legacies is required to implement the WFD and for effective integrated river basin management, but the research basis remains very sketchy. Reacting to this research need, the Danube Environmental History Initiative (DEHI) was founded in 2008. The core interest of its interdisciplinary efforts is the comparative study of long-term socio-ecological developments in the DRB.

---

[1] "The Water Framework Directive [WFD] establishes a legal framework to protect and restore clean water across Europe and ensure its long-term, sustainable use. (Its official title is Directive 2000/60/EC of the European Parliament and of the Council of 23 October 2000 establishing a framework for Community action in the field of water policy.) The directive establishes an innovative approach for water management based on river basins, the natural geographical and hydrological units and sets specific deadlines for Member States to protect aquatic ecosystems. The directive addresses inland surface waters, transitional waters, coastal waters and groundwater. It establishes several innovative principles for water management, including public participation in planning and the integration of economic approaches, including the recovery of the cost of water services.", from: http://ec.europa.eu/environment/water/water-framework/index_en.html

As one of the core players within DEHI, the Centre for Environmental History in Vienna concentrates its research on the environmental history of the Danube. Similar to what renowned environmental historian Richard White has suggested for the Columbia River (White 1995), the Danube is conceptualised as an 'Organic Machine', a hybrid between nature and culture. Much of it is, in the words of William Cronon, 'second nature', a nature colonized and changed by humans for millennia (Cronon 1991). Long-term socio-ecological research (LTSER) is based on similar premises (Haberl et al. 2006). Expanding on these concepts, we call the hybrid spaces we investigate 'socio-natural sites'. The first part of this paper explains this conceptual basis, which was developed from the notion of 'social sites' suggested by Theodore Schatzki (2003).

## 5.2 The Danube as a Socio-natural Site

History, the perceivable change of social as well as physical structures like buildings, in short, of everything which relates to humans in this world, can be described as a transformation of socio-natural sites (Winiwarter and Schmid 2008; Schmid 2009). To begin with, humans are able to use their bodies to process the information they have gathered with their senses. How does the body accomplish this feat? By means of energy: Senses react to electromagnetic radiation, sound waves or direct bodily contact, all of which elicit nervous responses in the receptors of the body – we see, hear or feel. How these sensations are communicated between individuals is, in contrast, not a feature of the body but culturally constructed. The different systems of colour distinction utilised by different peoples are a case in point.

A logical follow-up question to this diagnosis could be: How do we construct culturally what our senses produce naturally? This question is misleading, however, because the assumed world on which the question is based is split into the realm of nature and the realm of culture, with the overlap between the two realms inhabited by humankind. While the distinction makes sense for some types of questions as an analytical divide, the natural and social are intertwined, in our bodies as well as in the world, which is a world of hybrids. Rather than investigating the natural and cultural and their interaction, we suggest investigating 'practices' and 'arrangements', both understood as socio-natural hybrids (Fischer-Kowalski and Weisz 1999).

Andreas Reckwitz (2002) has summarised the meaning of 'practices': "A 'practice' (Praktik) is a routinized type of behaviour which consists of several elements, interconnected to one other: forms of bodily activities, forms of mental activities, 'things' and their use, a background knowledge in the form of understanding, know-how, states of emotion and motivational knowledge. A practice – a way of cooking, of consuming, of working, of investigating, of taking care of oneself or of others, etc. – forms so to speak a 'block' whose existence necessarily depends on the existence and specific interconnectedness of these elements, and which cannot be reduced to any one of these single elements. Likewise, a practice represents a pattern which can be filled out by a multitude of single and often unique actions reproducing the practice (a certain way of consuming goods can be filled out by plenty of actual acts of consumption). The single individual – as a bodily and mental agent – then acts as the 'carrier' (Träger) of a practice – and, in fact, of many different practices which need not be coordinated with one another. Thus, she or he is not only a carrier of patterns of bodily behaviour, but also of certain routinized ways of understanding, knowing how and desiring. These conventionalised 'mental' activities of understanding, knowing how and desiring are necessary elements and qualities of a practice in which the single individual participates, not qualities of the individual. Moreover, the practice as a 'nexus of doings and sayings' (Schatzki) is not only understandable to the agent or the agents who [perform these activities], it is likewise understandable to potential observers (at least within the same culture). A practice is thus a routinized way in which bodies are moved, objects are handled, subjects are treated, things are described and the world is understood." (Reckwitz 2002). In our first environmental history micro-study, we called what Reckwitz describes as practices 'Handlungsmuster', that is, 'patterns of action' (Projektgruppe Umweltgeschichte 2000). Practices are based on perception, and practices are impossible without handling material objects. Many of the material objects involved are shaped by practices, and these are called arrangements. Arrangements are the material precipitates of practices. They are maintained by continued interventions, practices of building are followed by practices of maintaining. Practices are constituted by sensual perceptions which lead to representations of them by means of communication; these representations are interpreted and changed into programmes for the intervention into arrangements, and work is needed to actually change arrangements according to the perception-and-interpretation based programmes. The nexus between practices and arrangements is called a socio-natural site.

If one is interested in the nexus of practices and arrangements, physical human interaction with the material world becomes a key issue. Work is the sole possibility of material interaction, although it is based on perception, representation and programmes. Mechanical work is defined as the amount of energy transferred by a force acting through a distance. Hence, the energy involved in the production and maintenance of arrangements becomes a central question of the study of arrangements and practices. Energy is never destroyed during a process; it changes from one form to another. Therefore, one should rather focus on exergy, that is the part of energy that is available to be used (Ayres and Warr 2005). Exergy allows a waterwheel to be moved, allows the water in a pot to be heated, and exergy can, in contrast to energy as such, be used up, as it is changed into less useful forms, the so-called anergy.

Let us turn back from the basics of physics to human work. It could be useful to define as work any activity which harvests exergy or protects arrangements built to harvest exergy from the remaining energy of the system in which harvesting takes place – such as constructing an overflow for a mill creek or preparing for a flood event.

Work, which is available through practices, is an intervention into material arrangements; it changes these arrangements, which leads to changes in practices, which in turn has an impact on where and how which work can be done. Building a bridge facilitates crossing a river, but the same bridge may become an obstacle for shipping transport. Practices and arrangements are transformative with regard to each other; if one changes, the other changes as well, and the socio-natural site transforms.

Environmental history, one could therefore say, investigates the conditions and consequences of interventions into material arrangements through work. The consequences of interventions are visible as changes in arrangements. Following such changes is an important part of environmental history research. Conditions and consequences, however, can only be discerned at a specific point in time, as causes and effects turn into one another over time (Heinz von Foerster 2003). The timing of observation is therefore decisive for the distinction, it is imperative to analyse arrangements not only in terms of consequences, and practices not only in terms of causes.

To summarise the conceptual basis, human beings create, via their practices, arrangements from the material world to harvest exergy. These arrangements deteriorate due to wear and tear. All arrangements are part of the evolutionary setting of humankind, either because of (evolving) humans taking part in them, or because of other living beings which evolve and are part of them. Autopoietic change in arrangements is the norm, not the exception. Life itself is thermodynamically highly improbable and to continue living, exergy is needed, exergy which has to be maximised or at least stabilised via information (cp. the 'maximum power principle' described by Odum and Pinkerton 1955).

Practices of exergy harvest have an impact on arrangements and at the same time they need to account for the autopoietic nature of arrangements, which tend to deteriorate into forms with a lower exergy level. If the exergy of a system is to be maximised, entropy maximisation as the natural course of the world has to be counteracted. Arrangements need continued investment of human labour to stay functional.

Let us turn from the conceptual realm to the river. Rivers are reservoirs of kinetic and potential energy. The harvesting of this energy can take many forms, many different arrangement-practice nexuses can be distinguished: A raft drifting with the flow, a millwheel, a turbine in a power plant, and also the joules contained in the river fish on a human plate constitute some of the ways in which exergy can be harvested from a river.

But, it could be maintained, all physical relations argued for so far only hold true in a closed system. The earth, however, is not a closed system, receiving energy every day in abundance from the sun. As long as the sun continues to do so, exergy remains available. In many cases, exergy is available in excess to what humans want and need. Avalanches or mudslides are effects of such available exergy which endanger the practices and arrangements of human exergy harvest.

This connection leads to the hypothesis that the extent of control via practices which is necessary to harvest exergy is proportional to the exergy density of the arrangement. The higher the amount of exergy which needs to be controlled in an arrangement, the more likely is the deterioration of such an arrangement, and therefore, the more likely is the production of potential harmful legacies and their long-lasting effects. While a small water wheel on the side of a creek has almost no long-lasting effects, even a series of mill weirs surely has (Walter and Merritts 2008), not to mention a large power plant.

We are bound to the maintenance of our arrangements; legacies of earlier interventions (one could think of radioactive waste from nuclear power plants, situated on the Danube's banks and cooled by its waters) have a profound impact on our and our descendants' practices. We need to perceive our environment to construct arrangements; and it depends on our perception as to how we construct them. No work can be pursued without perception of the material layout with which it is supposed to interact. Perception is inescapably tied to motives of the perceiving actor, and hence, inescapably subjective and driven by interests. Political issues are not excluded from the world by focussing on the nexus of practices and arrangements; they are fully integrated via the investigation of perception.

The core question of an environmental history of the DRB, integrating the socio-cultural and the ecological sphere, can thus be reformulated: How has the nexus of arrangements such as harbours, bridges, power plants or dams with practices such as river regulation, transportation, food-procuring and many more, developed over time? Which legacies did these practices and arrangements have? This will be discussed in the following chapter.

## 5.3 The Long History of Interventions into the Riverine Landscape

In his book, *Austria, Hungary and the Habsburgs*, historian Robert Evans describes the role of rivers as contested borders. "River frontiers in fact generated all manner of complex disputes: over water transport and its regulation; over fords, bridges, and their maintenance; over mills, fishing and other riparian rights; over flooding, or conversely over drainage; even – the toughest problems of all – over changing locations of the bed of the stream. The ancient, and in its larger features unquestioned, Austro-Hungarian border was partially riverine, and those sections most gave rise to litigation: protracted arguments about shifting islands at the confluence of the Danube and the March, and elsewhere, and, further south, about the course of the little rivers Leitha and Lafnitz" (Evans 2006, 122).

Seen from an environmental historian's viewpoint, it is the nature of rivers which makes them a source of protracted argument, rivers being dynamic at timescales within human experience. Mountains move, and seas come and go, but do so at geological timescales, normally separated from human perception. Rivers, in contrast, are fast-changing landscape elements and hence, a source of disturbance for societies, in particular for those based on territorial rights such as agrarian, solar-energy-based regimes. Rivers are also, as the quote makes clear, multi-functional elements of landscapes, with many of these potential functions being in conflict with one another. Ship mills could be obstacles for navigation, floating timber threatened fish populations; even the production of fibres was a source of conflict,

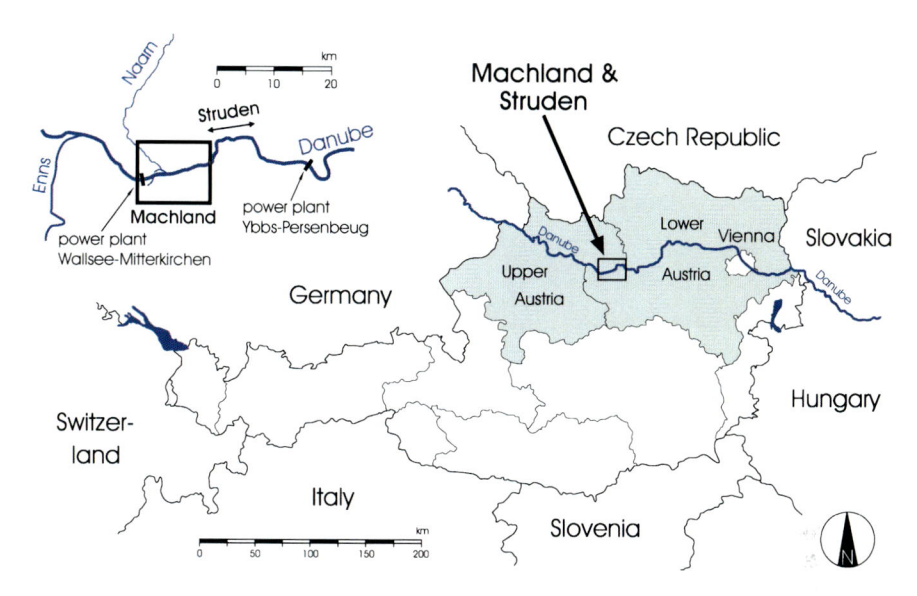

**Fig. 5.1** Location of the Danube sections Machland and Struden, Lower and Upper Austria (Map modified from Hohensinner et al. 2011)

as one can use the river either to condition flax or to catch fish, because decaying flax reduces the water's oxygen content. More examples of such conflicts can be found in historical sources, as archival material for river histories often originated from (environmental) conflicts.

One of the main achievements of Austrian efforts to study the environmental history of the Danube is the reconstruction of a 10 km-long tract of the river in its evolution since 1715, the time for which the earliest accurate map, appropriate for georeferencing, is available. A series of reconstructions clearly show the difference between a largely natural river at the beginning of the period and a channelled, dammed and straightened ship canal in (Fig. 5.1). This series in Fig. 5.2 has been discussed in detail by Hohensinner (2008) and Hohensinner et al. (2004, 2011).

While historians' sources do not allow us longer reconstructions, an environmental history of the Danube could start with the Venus of Willendorf, a c. 25,000 year old artefact found close to the Danube in what is today the province of Lower Austria. The figure's material is of unknown origin, Oolite is not native to the area. The figure or the material must have come from elsewhere, telling a story of movement and transportation in which the Danube might have been involved.

The Fossa Carolina is probably the oldest remnant of an intervention still visible today (Schiller 2008). Charlemagne (724/727-814), King of the Franks and Holy Roman Emperor, wished to connect the Rezat river to the Altmühl river, and hence the Rhine basin to the Danube basin. In 793, Charlemagne gave orders to dig a 3 km-long channel, actually a series of ponds, of which about 500 m are still visible today. Current research claims that the work was intended to facilitate trade between Rhineland and Bavaria. An environmental history of the Danube could

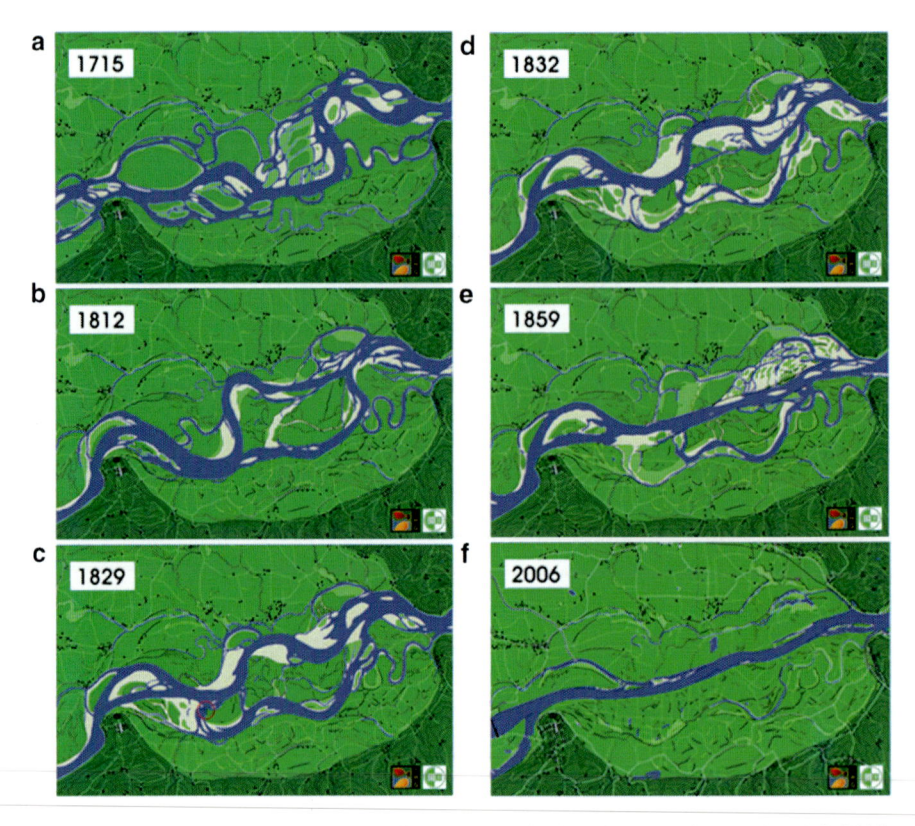

**Fig. 5.2** (**a–f**) Historical development of the Danube in the Machland floodplain 1715–2006: (**a**) and (**b**) prior to channelisation in 1715 and 1812, respectively, (**c**) at the beginning of the channelisation programme in 1829 (*red circle*: first major river engineering measure), (**d**) excavation of the cut-off channel 1832, (**e**) at the end of the channelisation programme in 1859, (**f**) after channelisation and hydropower plant construction in 2006 (Hohensinner 2008)

also start with 'Limes' and 'Ripa', the ancient Roman empire's extensive border zones along the Danube and discuss the river's role in military and civil transportation 2,000 years ago.

Transportation and its facilitation for millennia constituted and still is one of the main driving forces for interventions into the dynamic nature of the Danube. The small Bavarian city of Straubing built a 'Schlacht', a blockage of one river channel to render the other channel, closer to the settlement, more useful as a trading route in the fifteenth century at the latest. Many more such interventions are documented at least for the upper basin in early modern times (Leidel and Franz 1998). Regulation works to prevent the silting of river channels useful for navigation are documented for Vienna from the fourteenth century onwards. It is noteworthy that the operations in Vienna and at other locations were on a large scale, costly, hence conflict-ridden and long-lasting in their effects, as has been shown for the shipping channel at Nussdorf at the northern entrance to Vienna (Thiel 1904, 1906; Mohilla and

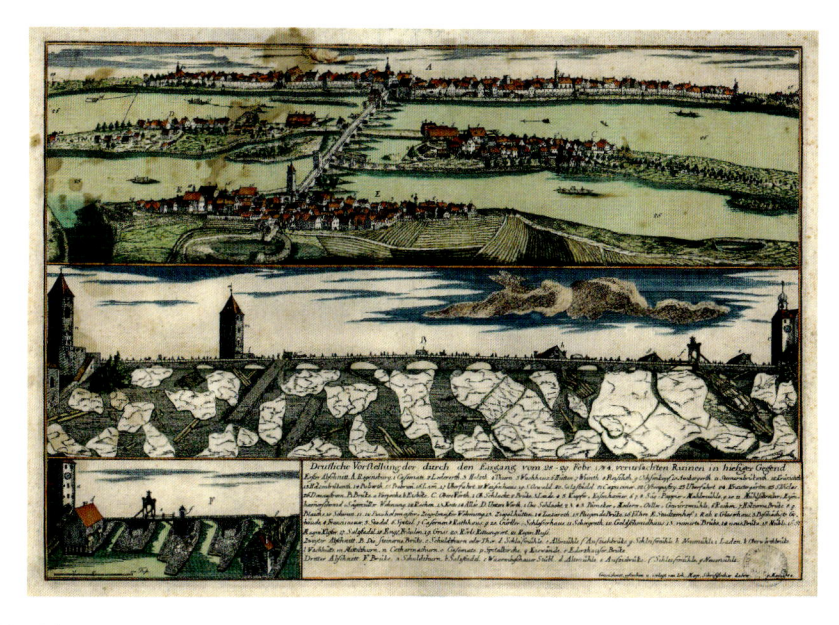

**Fig. 5.3** The flood in Regensburg in 1784 (Angerer 2008) (© Museen der Stadt Regensburg – Historisches Museum)

Michlmayr 1996). Taming rivers for navigation led to widespread interventions also in the pre-industrial period.

Another such intervention concerns bridges. While transportation was much aided by rivers longitudinally, crossing them presented a major hindrance to land transport (although they could be helpful as a barrier to potential enemies). A bridge is usually relatively unproblematic in terms of the flow of water. But bridges are major obstacles to ice floes, which cannot move through them, pile up, creating floods in their aftermath and destroying bridges regularly. The effect of bridges on ice jamming is clearly visible in the image of Regensburg in 1784 on the right, the remains of a wooden bridge, which had been jerked from its anchors and floated 8 km downstream are jamming the flow. The middle tower of the famous medieval stone bridge subsequently had to be removed due to structural damage it had incurred during the jam.

The Danube was an important transport route in pre-industrial times. The main goods transported were wood, which could be made into a carrier of the transported material itself by means of rafts – a perfect arrangement to harvest exergy via the stream's energy – stones from quarries along the river, which were too heavy to be transported over land, produce for urban markets, expensive goods such as salt and – in the many years during which part of the Danube basin was the site of war – war provisions of all kinds. Transportation of people was also regular, if at times quite dangerous. Transport downstream with the flow was, although risky in many stretches, comparatively easy, yet upstream transport was difficult and slow, accomplished by trains of horses pulling the ships against the current (Fig. 5.3).

Examples of hindrances to ship transportation are legion in the environmental history of the Danube, from water levels being too low during autumn to the regular flood events which made navigation very difficult due to two types of dangers: Firstly, static dangers such as Struden (see Fig. 5.1), a spectacular gorge well known for its narrow course with cataracts and a vortex, presented a major obstacle. Attempts to mitigate these obstacles date back at least to the sixteenth century (Slezak 1975). The second type of danger is connected to those stretches where the Danube moves through a wide alluvial valley. Such valleys allowed for rapid shifts of river channels, making navigation very dangerous to inexperienced navigators. Regulation efforts were aimed at providing a deep, stable channel, but the river dynamics worked against this goal as can be seen from a series of failed interventions in the Machland floodplain (see Fig. 5.2).

The first attempt in around 1826 was aimed at closing off one of the two main arms in order to make the other one deeper and hence, less prone to shipwrecking; this measure was also intended to stem further erosion, which was threatening homesteads on the southern bank. Contrary to the hopes of engineers, the remaining channel did not deepen, but instead widened and remained too shallow for safe navigation. And even worse, several farms along the northern banks were now threatened by erosion (see Fig. 5.2b, c). The next intervention, undertaken in 1832, was to excavate a 25 m-wide straight cut-off channel through the large island between the two main channels (see Fig. 5.2d), which would be widened by the river itself due to lateral channel erosion. The engineers were proved right as the river adopted the cut-off channel and widened it considerably, straightening the river course for easier navigation (see Fig. 5.2e). But the bed material removed by the flow was deposited directly downstream of the new channel, creating a maze of gravel bars and shallows which were as dangerous as the upstream area had been before. Shipwrecks added to the problem, as some were nuclei in the formation of new gravel bars and islands, further complicating shipping. Authorities were forced to intensify their efforts and within the next two decades several kilometres of training walls were built, forcing the Danube into a new bed. At the end of the nineteenth century, groynes were added to provide a sufficiently deep waterway during periods of low flow. But it took until the twentieth century with its concrete structures and hydropower plants – made possible due to fossil energy – to restrain the flood dynamic of the Danube (Veichtlbauer 2010). The hydropower plants changed the former river floodplain system up- and downstream of their dams fundamentally and often irreversibly (see Fig. 5.2f).

This history of the Danube in the Machland clearly shows how human practices are shaped by motives, in this case trade for the exchange of material goods and creation of wealth, and how the arrangements they build to support this interest are changed by the river, creating a series of reactions which – in this case – led to massive interventions into river morphology and thus into animal habitat and the ability of the river to serve other purposes. Like much other environmental history, the story of river engineering provides a telling case of the unwanted side-effects of the creation and re-arrangement of socio-natural sites, a case in point to be addressed in future river management.

The second major interest in the river, energy procurement, has already been mentioned in passing. Pre-industrial energy harvest techniques involved wheels of some kind, which translated the flow of the river into the turning of a pivot. Ship mills, which could be used independently of the water table – unless the current was too swift and floating debris endangered them during floods – were the most common facility to use the flow of the Danube. Thousands of these were installed along the river, concentrated where a town or city met its flour needs with grain milled on the ships. The intervention into the river for these mills was very small compared to the elaborate river training detailed above. A bank to secure the ship mill was needed, not unlike the securing of a boat. But they were not without impact: An instance reported in an English newspaper for *Peterwaradin* (Petrovaradin, Serbia) provides information about conflicting uses of the Danube in 1716 and arrangements for different practices: For military purposes, two bridges were to be laid over the Danube there but the action was delayed because violent winds had hindered the towing ashore of ship-mills which lay in the middle of the river in the days before. One of these ship mills was driven down the stream with the wind and damaged the ship bridges, carrying away five ships from one, and 18 from the other.[2]

The harvest of biomass in the fertile floodplains and the transportation infrastructure provided by the river meant that human settlements and cultivated lands were situated in zones where flooding, erosion and sedimentation were not uncommon. The history of floods on the Danube is still underresearched but a wealth of sources exist to reconstruct the interaction between arrangements and river dynamics, and the ensuing activities to restore the arrangements. A remarkable series of major floods in the late eighteenth century seriously affected growing urban agglomerations along the Danube, as has been shown for the case of Budapest (Kiss 2007). A series of reports about the Danube in early British newspapers, which have been analysed for the period from 1687 to 1783, shows a similar picture. During this 96-year period, we read 14 times about ice-induced damage of bridges in Vienna, the longest interval between two incidents being only 17 years. Overall, these newspapers report (noteworthy) cases of destruction every 5 years.

To give but two examples of floods which did not involve ice, in October 1732, the flood on the upper Danube was considered abnormally high as a report from *Ratisbon* (Regensburg) shows, and floats of timber were separated and bridges destroyed. A similarly high flood was reported in the region in June 1737, 'Spoils and broken Furniture are seen every Day floating', writes the paper.[3] In July 1736, we read of great rains which had so swelled the Danube that the inhabitants of the Viennese suburbs of Rossau and Leopoldstadt were obliged to use boats on most of their streets, while the great bridge of Vienna and that of Krems, upriver from the capital, had both suffered considerable damage 'by the Rapidity of the River', and almost all the surroundings of Vienna were reportedly under water.[4]

---

[2] London Gazette (London, England), Saturday, August 18, 1716; Issue 5461; slightly different in Post Man and the Historical Account (London, England), Saturday, August 18, 1716; Issue 11250.

[3] Daily Gazetteer (London Edition) (London, England), Friday, June 10, 1737; Issue 611.

[4] Daily Post (London, England), Wednesday, July 28, 1736; Issue 5265.

Protection from floods was a major imperative of river engineering since its advent. The impact of regulation works on the river was greatest in the middle reaches of the river, in the Hungarian plain, where damming and flood protection was undertaken as early as 1426. In the mid-nineteenth century, Count István Szechényi initiated several extensive engineering projects on the then Hungarian Danube (Harper 2004). But human arrangements were also designed for resilience to dynamic circumstances. By using the flood-prone areas for types of cultivation which needed only low investment and made good use of the prolific new growth after a flood, damage could be minimised. Young willow stands could be used as raw material for basket weaving as well as for early types of regulation devices, fagots, bundles of shoots, tied together and packed into submerged fences along the riverbank. Grazing cattle could be removed in case of a flood. The floods themselves brought nutrients, an important asset in a nutrient-poor world without artificial fertilisers.

Fish, as we pointed out at the beginning, are one way to harvest exergy from the river. The Danube was a source of protein for the longest period of time, with local fish markets being described in a wealth of places. While fisheries were diverse, and many species were commercially interesting, stocks of the most impressive Danube fishes, the Beluga or European sturgeon (*Huso huso*), which could grow to a length of up to 8 m, were already depleted by overfishing by the sixteenth century (Balon 1968; Bartosiewicz et al. 2008). In the eighteenth century, the arrival of a huge "Hausen" in Vienna is already worth a report in a distant newspaper: According to the source, a giant fish reached the Viennese fish market from Hungary in 1732. Its measures are given as 5.5 Ells long and 2.75 broad, the roe weighed 88 lb., the entrails 74 and the body 805, altogether 967 lb. in weight.[5]

One of the great problems of fisheries is their sustainable development. Fish are easily overharvested and can be depleted rapidly. The demise of ocean fisheries in the twentieth century has an earlier parallel in many freshwater fisheries. The catch of large sturgeon was accomplished by structures built into the river (see Fig. 5.4). The arrangement to capture the exergy was a sturdy construction made of wood, often called a fence, but in the case shown here, was more a giant fish trap. These structures were built to last, but the nature of the river led to rapid wear and tear of the wooden beams. Documentation from the river Volga shows how much practices are shaped by arrangements to a degree which could become dangerous for human survival. In an eighteenth century encyclopedia, the life of the divers employed to control the submerged traps is described in some detail.[6] The ice-cold water of the Volga could only be suffered by the divers when they were first heated up in a kind of sauna. They were given spirits to drink, probably in order to increase the blood

---

[5] Daily Journal (London, England), Monday, June 12, 1732; Issue 3569; assuming that the specifications of that fish refer to ancient English units (1 Ell c. 1.143 m; 1 lb (Pound) c. 454 g), it had a length of 6.3 m, a width of 3.1 m and a total weight of c. 440 kg; although this makes this fish a lightweight relative to its size, the measurements fit tolerably with Balon's (1968, 245) calculations on Huso huso's weight-size ratio from historical catch records.

[6] Johann Georg Krünitz in Volume 22 of his *Ökonomische Enzyklopaedie* from 1781; see: http://www.kruenitz1.uni-trier.de/

**Fig. 5.4** Arrangement for catching the Beluga sturgeon (*Huso huso*) at the Iron Gate. (From Luigi Ferdinando Marsigli's '*Danubius Pannonico-Mysicus etc.*', Volume 4, Amsterdam 1726; © Niederösterreichische Landesbibliothek, Topographische Sammlung)

flow through the vessels in their arms and legs. Heated up and slightly intoxicated, they jumped into the river and inspected the underwater arrangement as long as the air in their lungs allowed. The same procedure was repeated several times a day, until they bled from ears and nose and had to be transported ashore. The maintenance work lasted for a week each year, during which the divers worked daily completing up to seven dives. Most divers did not last long through this ordeal, becoming arthritic and suffering from oedema after 3–4 years, and not even the sturdiest could work as divers for more than 10 years, with many dying young.

While this example shows the connection between arrangements and practices, fisheries were usually much worse for fish than for men. Today, commercial river fisheries in Austria no longer exist. A combination of habitat change in the river due to channelisation and power plant building with pollution and changed consumption habits has led to their demise (Haidvogl 2010).

## 5.4 Changes to the Danube River Basin Since Industrialisation

While we have no quantitative information on pre-industrial use of the Danube for drinking water, currently about ten million people on the upper Danube get their drinking water from the river. Protection of its water quality is imperative for the further sustainable development of these urban areas, such as Ulm or Passau. All other uses of the river are to some degree of conflict with its use as a source of potable water.

The most visible difference between pre-industrial and industrial interventions into rivers is the new arrangement of the power plant. An overview of those nations which procure electricity from the Danube renders the political aspect of arrangements immediately visible: Germany, Austria, Slovakia, Serbia and Romania, all of whom control both banks, have built power plants. The Gabcikovo dam was finished by Slovakia in 1996, its Hungarian counterpart Nagymaros was impeded in 1984 by the environmental and anti-communist 'Danube circle' (Duna Kör). Croatia, Bulgaria and Moldova, controlling only one bank, have not built power plants. The Danube's uppermost stretch in Germany is too small to produce sizeable amounts of energy, nevertheless it has also been dammed by several small power plants. We have stated above that the extent of control via practices that is necessary to harvest exergy is proportional to the exergy density of the arrangement. On the upper Danube and its tributaries it was already possible to build power plants in the nineteenth and early twentieth centuries, because the harvestable exergy was low. Austria started building power plants on the Danube between the two world wars. The first plant to be brought on line, Ybbs-Persenbeug, was initiated by the national-socialist regime and was finished only in 1959. To date, about 60 % of Austria's electricity needs are met by hydropower, largely from the Danube and its tributaries. The largest of the Danube's power plants, however, was built at the Iron Gate from 1964 to 1972 (Iron Gate I), and a smaller one (Iron Gate II) was completed in 1984, both in a co-operative effort by the former Yugoslavia and Romania.

Each power plant is a complete blockage of the river and a powerful intervention into its nature. Power plants not only keep migrating fish species such as the anadromous Beluga sturgeon from reaching spawning grounds upstream and hence endanger their survival, they also block sediment transport, which leads to deepening and pothole formation on the bottom. A series of power plants changes the river habitat profoundly. In addition to blocking the connection along the flow, the lower velocity of the river water means that fast-moving water-dependent (rheophilic) fish species lose their habitat. The lower flow velocity also leads to a warming up of the water, which then becomes fit for other (invasive) species, which potentially endanger endemic species. Lastly, the changed river bottom – covered with fine silt rather than with rolling gravel, provides no spawning grounds for the fishes of the unregulated, undammed river of pre-industrial times. What used to be one of the most diverse freshwater habitats is nowadays often dominated by hatched species added to the Danube for the benefit of sport angling (Haidvogl 2010).

Two driving forces dominate the Danube and its further transformation today: Power generation and long-distance mass transport. Interest, and hence perception, practices and arrangements have shifted from the nutrient-carrying abilities of the river and the use of fertile floodplains to the use of its energy in power plants. Banks are steep, secured with stones. Groynes and dikes along the river secure a stabilised difference between land and water, allowing ever bigger ships to use the river for transportation. Harbours are built with concrete. The large harbour structures needed for big ships are built at sizeable distances from each other, so that most of the settlements along the river have lost contact with the Danube, ships pass by rather than land for an exchange of goods (Haidvogl and Gingrich 2010). Meanwhile, tourism along the river flourishes, and boats of various sizes travel up and down the stream, selling views of the few picturesque tracts such as the Austrian Wachau.

Similar stories of profound changes to the river can be told for the Delta region, a European hotspot of biodiversity and a socio-natural site by no means untouched by human practices. The Danube-Black Sea Canal through the Dobruja south of the Delta reduces the distance by boat from Cernavoda to Constanta by almost 350 km. After more than 100 years of planning, international negotiations and modifications of plans due to changed political circumstances, the Romanian government decided in 1949 to build a canal using forced labour. While the labour camps proved disastrous – another incidence of arrangements creating practices beyond human capabilities – the project could not be completed. It was restarted in 1978 by Nicolae Ceausescu. The southern arm was completed in 1984 and the northern arm was inaugurated in 1987 (Turnock 1986). The Canal is also one example in which the Cold War history of the Danube becomes apparent, having provided political motivation for profound changes in the river.

Industrial society, with the force of fossil energy it commands, has led to all-embracing change to the Danube, with consequent losses of species and an ever more pressing need to invest in the stabilisation of the river. More and more infrastructure was built close to the river, floodplains were seen as secure enough to build expensive infrastructure upon them, and only since the 1970s, some change in human interaction with the river has taken place. A case in point is the new Machland dam: The Machland, the alluvial floodplain upstream of one of the gorges in Austria already discussed above, has been fundamentally transformed by two power plants, Wallsee-Mitterkirchen upstream and Ybbs-Persenbeug downstream of the region (see Figs. 5.1 and 5.2f). Today, most parts of the floodplains adjacent to the Danube in the Machland are used as polders for flood retention, with water bodies showing water levels below the backed-up water table of the Danube. Pumping stations are necessary to pump the inflow of tributaries from the polders out into the Danube. This environmental history of the Machland continues to be transformative. In the late twentieth century, flood protection levees were built to shelter settlements in the southeastern former floodplain. Currently a large dam of more than 36 km in length is under construction, several kilometres away from the river to the north; all villages in the floodplain, between the newly built dam and the channelled river, have been razed to the ground and people have been relocated to safer ground. This dam construction is the outcome of a new policy of co-existence: The river is given more room to move and floods can spread out over larger areas.

Vienna, capital of Austria and the former Habsburg Empire and one of nine national capitals in the Danube basin today, is an example for the riverscape's fundamental transformation as part of an urban agglomeration. Vienna exhibited a sharp population increase in particular during the second half of the nineteenth century; population figures grew from some 250,000 around 1820 to half a million in the mid-nineteenth century. In 1910 already more than two million people lived in Vienna (cp. Krausmann, Chap. 11 in this volume). The growing city urgently needed settlement areas and the process commonly termed industrialisation increased the demand for space as industrial and commercial areas had to be made available. After a long and contested debate between river engineers, municipal authorities and the imperial administration, the Great Viennese Danube regulation was accomplished between 1870 and 1875. Similar to other Austrian Danube sections, the main objective of these measures was to improve navigation, as the Danube remained the most important trading route for the city in the late nineteenth century. But flood protection was an additional important motive for reshaping the river. A new channel was excavated, confined by flood protection dykes, and after its completion the river was moved into an artificial bed. New settlement areas became available in the now flood-free former floodplains which had been, with few exceptions such as the villages of Leopoldstadt and Rossau already mentioned, almost unpopulated until the mid-nineteenth century. Since the 1860s, the value of these new land resources had been recognised and discussed by both water engineers and planners and from the 1870s onwards, the former Viennese Danube floodplains were rapidly and densely populated in several phases of urban expansion. Between 1888 and 1918, when Vienna reached its highest recorded population, the districts on the former floodplains showed the highest population growth rates in the metropolis.

This urban expansion of Vienna was only possible due to new types of arrangements such as the new Danube channel, flood protection dykes, a special flood channel and weirs to keep the floodplains free of water. Late-nineteenth century river engineers and urban planners set a course which could not and cannot be diverged from. Although programmes and practices of flood protection changed again in the late twentieth century, the arrangements from the period of promoterism still have to be maintained and adapted by municipal authorities in cities like Vienna. The Great Regulation of the Viennese Danube is but one example of the legacies from past practices in dealing with river dynamics. The arrangements we have inherited from our ancestors determine the scope of options we have in the present when dealing with the river.

## 5.5 Conclusions

Long-term socio-ecological research, LTSER, was developed from long-term ecological research, taking into account that the study of landscapes and sites where human intervention has taken place and continues to take place is important for planning a sustainable future (Haberl et al. 2006). Environmental history, with its conceptual basis and ability to integrate natural sciences and humanities, that is, to integrate research on the impact of human interventions with that on the reasons for such interventions, is ideally suited to provide the LTSER community with long-term case studies which allow management decisions to be based on the firm ground of historical knowledge.

In most cases, the past is not discovered through violent and spectacular events, although these occur and are documented in the written record: A great earthquake happened near Petrovaradin in 1726, which, so an English newspaper paper reported, split a mountain in two, and parts of it fell into the Danube. Vineyards and roads were ruined, and several ship mills which had sunk decades earlier were lifted to the surface.[7] Most often, historians recover the past by putting evidence together bit by bit and carefully scrutinizing it for the biases it might contain. Such work, we argue, is necessary and useful for planning the future. Data on historic sturgeon catches allow estimates to be made of the distribution and stocks of these fishes prior to overexploitation and barrage-induced demise. Such information is needed for management plans. Using the concept of socio-natural sites, the conditionality of possible practices on a sustainable set of arrangements becomes clearly visible and can serve as the basis for planners, who should ask about the fate of humans in the arrangements they propose to harvest the Danube's exergy.

Since industrialisation, the amount of exergy harvestable from arrangements has increased by orders of magnitude, for which the difference between an early modern ship mill and a power plant can serve as an example. But the societal and ecological risks from controlling these exergy-dense arrangements have also increased, as we proposed at the outset of this chapter. The higher the extent of exergy which needs to be controlled in an arrangement, the more likely is the deterioration of such an arrangement, and therefore, the more likely is the production of potentially harmful legacies and their long-lasting effects. None of the arrangements built with fossil fuels can be maintained indefinitely, neither power plants, nor large scale flood protection dykes. A sustainable society based on renewable energy will have to deal with the legacies from exergy-dense arrangements. Only in a long-term perspective, including history and covering at least centuries, does the profound difference between pre-industrial and industrial arrangements become clearly visible.

With contributions from environmental history, LTSER extends its temporal scope significantly. Every historical approach points to the decisive role of politics

---

[7] *Daily Courant* (London, England), Wednesday, November 30, 1726; Issue 7839; *London Journal* (London, England), Saturday, December 3, 1726; Issue CCCLXXXIII.

in the transformation of our sites of research, whether we call them "socio-ecological" or "socio-natural". The Danube has been a theatre of war for centuries or even millennia. We are just starting to assess the effects of war, of war-induced use and abuse, of exploitation and overexploitation of resources in fluvial landscapes. The Danube was a formidable barrier and hence, a battle site. But every river also forms a longitudinal continuum, linking the riparian societies to each other and forcing them to cooperate.

With the European Water Framework Directive, an ecologically sound, natural state of rivers has been defined as the goal for 2015. The Danube has not been natural for at least the past 300 years, although most interventions were of limited scale and only seldom (as in the near extinction of *Huso huso*) as profound as they are today. Historians can point to the choices society has to make. We can try to offer the Danube the space it took in 1715 in the Machland, but this means having to relocate people. The Danube has a 'memory' for previous interventions, and its current behaviour is influenced by them. Knowing this, we should base management decisions on much longer data sets. Knowing the history of side-effects, we might be able to abate them in the future. Looking into the pre-industrial past might allow us a glimpse of the post-fossil fuel age which is inevitable, preparing us for a new regime of interaction with the dynamic Danube.

**Acknowledgments** The research presented in this chapter was funded by the Austrian Science Fund (FWF) within the project 'Environmental History of the Viennese Danube 1500–1890 (ENVIEDAN)' (P22265 G-18). The historical reconstructions of the riverine landscape in the Machland result from the FWF-funded project 'Reconstruction of Danube Habitats in the Austrian Machland 1715–1991' (P14959-B06) at the Institute of Hydrobiology and Aquatic Ecosystem Management, University of Natural Resources and Life Sciences Vienna (BOKU).

# References

Angerer, M. (2008). Regensburg und die Donau – Ansichten aus sechs Jahrhunderten. In C. Ohlig (Ed.), *Historische Wassernutzung an Donau und Hochrhein sowie zwischen Schwarzwald und Vogesen. Schriften der Deutschen Wasserhistorischen Gesellschaft (DWhG) e.V.* (Band 10, pp. 11–27). Norderstedt: Books on Demand.

Ayres, R. U., & Warr, B. (2005). Accounting for growth: The role of physical work. *Structural Change and Economic Dynamics, 16*, 181–209.

Balon, E. K. (1968). Einfluß des Fischfangs auf die Fischgemeinschaften der Donau. *Archiv für Hydrobiologie, Supplement, 34*, 228–249.

Bartosiewicz, L., Bonsall, C., & Sisu, V. (2008). Sturgeon fishing in the middle and lower Danube Region. In C. Bonsall, C. Bonsall, V. Boroneat, & I. Radovanovic (Eds.), *The iron gates in prehistory. New perspectives* (pp. 39–54). Oxford: Archeopress.

Cronon, W. (1991). *Nature's metropolis. Chicago and the Great West*. London/New York: W.W. Norton & Company.

Evans, R. J. W. (2006). *Austria, Hungary, and the Habsburgs: Central Europe c. 1683–1867*. Oxford: Oxford University Press.

Fischer-Kowalski, M., & Weisz, H. (1999). Society as hybrid between material and symbolic realms. Toward a theoretical framework of society-nature interaction. *Advances in Human Ecology, 8*, 215–251.

Haberl, H., Winiwarter, V., Andersson, K., Ayres, R. U., Boone, C. G., Castillio, A., Cunfer, G., Fischer-Kowalski, M., Freudenburg, W. R., Furman, E., Kaufmann, R., Krausmann, F., Langthaler, E., Lotze-Campen, H., Mirtl, M., Redman, C. A., Reenberg, A., Wardell, A. D., Warr, B., & Zechmeister, H. (2006). From LTER to LTSER: Conceptualizing the socio-economic dimension of long-term socio-ecological research. *Ecology and Society, 11*. (Online), http://www.ecologyandsociety.org/vol11/iss2/art13/

Haidvogl, G. (2010). Verschwundene Fische und trockene Auen: Wie Regulierung und Kraftwerksbau das Ökosytem Donau im Machland verändert haben. In V. Winiwarter & M. Schmid (Eds.), *Umwelt Donau: Eine andere Geschichte. Katalog zur Ausstellung des Niederösterreichischen Landesarchivs im ehemaligen Pfarrhof in Ardagger Markt* (pp. 119–135). Sankt Pölten: NÖ Institut für Landeskunde.

Haidvogl, G., & Gingrich, S. (2010). Wasserstraße Donau: Transport und Handel im Machland und auf der Donau im 19. und 20. Jahrhundert. In V. Winiwarter & M. Schmid (Eds.), *Umwelt Donau: Eine andere Geschichte. Katalog zur Ausstellung des Niederösterreichischen Landesarchivs im ehemaligen Pfarrhof in Ardagger Markt* (pp. 91–103). Sankt Pölten: NÖ Institut für Landeskunde.

Harper, K. (2004). Danube river. In S. Krech III, J. R. McNeill, & C. Merchant (Eds.), *Encyclopedia of world environmental history* (pp. 284–285). New York: Routledge.

Hohensinner, S. (2008). *Rekonstruktion ursprünglicher Lebensraumverhältnisse der Fluss-Auen-Biozönose der Donau im Machland auf Basis der morphologischen Entwicklung von 1715–1991*. Dissertation, University of Natural Resources and Life Sciences, Vienna.

Hohensinner, S., Habersack, H., Jungwirth, M., & Zauner, G. (2004). Reconstruction of the characteristics of a natural alluvial river-floodplain system and hydromorphological changes following human modifications: The Danube river (1812–1991). *River Research and Applications, 20*, 25–41.

Hohensinner, S., Jungwirth, M., Muhar, S., & Schmutz, S. (2011). Spatio-temporal habitat dynamics in a changing Danube river landscape: 1812–2006. *River Research and Applications, 27*, 939–955.

Kiss, A. (2007). "Suburbia autem maxima in parte videntur esse deleta" – Danube icefloods and the pitfalls of urban planning: Pest and its suburbs in 1768–1799. In C. Kovács (Ed.), *From villages to cyberspace* (pp. 271–282). Szeged: University Press.

Leidel, G., & Franz, M. R. (1998). *Altbayerische Flusslandschaften an Donau, Lech, Isar und Inn: Handgezeichnete Karten des 16. bis 18. Jahrhunderts aus dem Bayerischen Hauptstaatsarchiv*. Weißenhorn: Anton H. Konrad Verlag.

McNeill, J. R. (2003). Observations on the nature and culture of environmental history. *History and Theory, 42*, 5–43.

Mohilla, P., & Michlmayr, F. (1996). *Donauatlas Wien: Geschichte der Donauregulierung auf Karten und Plänen aus vier Jahrhunderten* [Atlas of the Danube river Vienna. A history of river training on maps and plans of four centuries]. Wien: Österreichischer Kunst- und Kulturverlag.

Odum, H. T., & Pinkerton, R. C. (1955). Time's speed regulator: The optimum efficiency for maximum power output in physical and biological systems. *American Scientist, 43*, 331–343.

Projektgruppe Umweltgeschichte. (2000). *Historische Entwicklung von Wechselwirkungen zwischen Gesellschaft und Natur*. CD-ROM 7 (Forschungsschwerpunkt Kulturlandschaft).

Reckwitz, A. (2002). Toward a theory of social practices. A development in culturalist theorizing. *European Journal of Social Theory, 5*, 243–263.

Schatzki, T. R. (2003). Nature and technology in history. *History and Theory, 42*, 82–93.

Schiller, J. (2008). Von der Fossa Carolina zur Main-Donau-Überleitung. In C. Ohlig (Ed.), *Historische Wassernutzung an Donau und Hochrhein sowie zwischen Schwarzwald und Vogesen. Schriften der Deutschen Wasserhistorischen Gesellschaft (DWhG) e.V.* (Band 10, pp. 61–71). Norderstedt: Books on demand.

Schmid, M. (2009). Die Donau als sozionaturaler Schauplatz: Ein konzeptueller Entwurf für frühneuzeitliche Umweltgeschichte. In A. Steinbrecher & S. Ruppel (Eds.), *Die Natur ist überall bei uns. Mensch und Natur in der Frühen Neuzeit* (pp. 59–79). Basel: Chronos Verlag.

Slezak, F. (1975). Frühe Regulierungsversuche im Donaustrudel bei Grein (1574–1792). *Der Donauraum. Zeitschrift für Donauraum-Forschung, 20*, 58–90.

Sommerwerk, N., Hein, T., Schneider-Jacoby, M., Baumgartner, C., Ostojic, A., Siber, R., Bloesch, J., Paunovic, M., & Tockner, K. (2009). The Danube river basin. In K. Trockner, C. T. Robsinson, & U. Uehlinger (Eds.), *Rivers of Europe* (pp. 59–112). London/Burlington/San Diego: Academic.

Thiel, V. (1904). Geschichte der älteren Donauregulierungsarbeiten bei Wien. Von den älteren Nachrichten bis zum Beginne des XVIII. Jahrhunderts. *Jahrbuch für Landeskunde von Niederösterreich, 1903*, 117–165.

Thiel, V. (1906). Geschichte der Donauregulierungsarbeiten bei Wien II. Vom Anfange des XVIII. bis zur Mitte des XIX. Jahrhunderts. Von der Mitte des XIX. Jahrhunderts bis zur Gegenwart. *Jahrbuch für Landeskunde von Niederösterreich, 1905 und 1906*, 1–102.

Turnock, D. (1986). The Danube-Black Sea Canal and its impact on Southern Romania. *GeoJournal, 12*, 65–79.

Veichtlbauer, O. (2010). Von der Strombaukunst zur Stauseenkette: Die Regulierung der Donau. In V. Winiwarter & M. Schmid (Eds.), *Umwelt Donau: Eine andere Geschichte. Katalog zur Ausstellung des Niederösterreichischen Landesarchivs im ehemaligen Pfarrhof in Ardagger Markt* (pp. 57–73). Sankt Pölten: NÖ Institut für Landeskunde.

von Foerster, H. (2003). *Understanding Understanding. Essays on Cybernetics and Cognition.* New York: Springer

Walter, R. C., & Merritts, D. J. (2008). Natural streams and the legacy of water-powered mills. *Science, 319*, 299–304.

White, R. (1995). *The organic machine. The remaking of the Columbia river.* New York: Hill and Wang.

Winiwarter, V., & Schmid, M. (2008). Umweltgeschichte als Untersuchung sozionaturaler Schauplätze? Ein Versuch, Johannes Colers "Oeconomia" umwelthistorisch zu interpretieren. In T. Knopf (Ed.), *Umweltverhalten in Geschichte und Gegenwart: Vergleichende Ansätze* (pp. 158–173). Göttingen: Attempto.

Wolf, A.T., de Silva, L., & Hatcher, K. (2002). *The program in water conflict management and transformation (PWCMT).* Oregon State University, College of Science. http://www.transboundarywaters.orst.edu/publications/register/tables/IRB_table_6.html

# Chapter 6
# Critical Scales for Long-Term Socio-ecological Biodiversity Research

Thomas Dirnböck, Peter Bezák, Stefan Dullinger, Helmut Haberl, Hermann Lotze-Campen, Michael Mirtl, Johannes Peterseil, Stephan Redpath, Simron Jit Singh, Justin Travis, and Sander M.J. Wijdeven

**Abstract** One challenge in the implementation of Long-Term Socio-Ecological Research (LTSER) is the consideration of relevant spatial and temporal scales. Mismatches between the scale(s) on which biodiversity is monitored and analysed, the scale(s) on which biodiversity is managed, and the scale(s) on which conservation policies are implemented have been identified as major obstacles towards halting or reducing biodiversity loss. Based on a meta-analysis of 18 biodiversity studies and a literature review, we discuss here a set of methods suitable to bridge the various scales of socio-ecological systems. For LTSER, multifunctionality of landscapes provides an inevitable link between natural and social sciences.

T. Dirnböck, Ph.D. (✉)
Environment Agency Austria, Vienna, Austria
e-mail: thomas.dirnboeck@umweltbundesamt.at

P. Bezák, Ph.D.
Institute of Landscape Ecology, Slovak Academy of Sciences, Bratislava, Slovakia
e-mail: peter.bezak@savba.sk

S. Dullinger, Ph.D.
Centre of Biodiversity, University of Vienna, Vienna, Austria
e-mail: stefan.dullinger@univie.ac.at

H. Haberl, Ph.D. • S.J. Singh, Ph.D.
Institute of Social Ecology Vienna (SEC),
Alpen-Adria Universitaet Klagenfurt, Wien, Graz, Schottenfeldgasse 29/5, 1070 Vienna, Austria
e-mail: helmut.haberl@aau.at; simron.singh@aau.at

H. Lotze-Campen, Ph.D.
Potsdam Institute for Climate Impact Research, Potsdam, Germany
e-mail: lotze-campen@pik-potsdam.de

M. Mirtl, Ph.D. • J. Peterseil, Ph.D.
Department of Ecosystem Research and Monitoring, Environment Agency Austria,
Spittelauer Lände 5, 1090 Vienna, Austria
e-mail: michael.mirtl@umweltbundesamt.at; peterseil@umweltbundesamt.at

Upscaling approaches from small-scale domains of classical long-term biodiversity research to the broad landscape scale include landscape metrics and spatial modelling. Multidisciplinary, integrated models are tools not only for linking disciplines but also for bridging scales. Models that are capable of analysing societal impacts on landscapes are particularly suitable for interdisciplinary biodiversity research. The involvement of stakeholders should be an integral part of these methods in order to minimise conflicts over local and regional management interventions implementing broad-scale policies. Participatory approaches allow the linkages between the specific scale domains of biodiversity, its management and policies.

**Keywords** Biodiversity • Conservation • Management • Environmental policy • Long term ecological research • Long term socio-ecological research • Scale • Scale mismatch • Cross-scale interaction

## 6.1 Introduction

Slowing down human-induced biodiversity loss is a prominent target of current sustainability policies. The World Summit on Sustainable Development and the Convention on Biodiversity aimed to significantly reduce the rate of biodiversity loss by the year 2010 (CBD 2003). In its 6th environmental action programme issued in 2002, the European Union formulated the goal to halt the loss of biodiversity by 2010 (EEA 2007). These targets were not achieved. Progress towards reaching new targets will require a better understanding of the pressures on biodiversity and the socioeconomic drivers associated with them, in addition to the ongoing efforts of biologists to document trends and patterns of biological diversity (Haberl et al. 2007). Effective policies to slow down the loss of biodiversity therefore need to be based on an improved understanding of the interactions between socioeconomic and natural systems that result, inter alia, in the currently observed trend of biodiversity loss. Such a knowledge basis can help to realise the proposal to "mainstream biodiversity protection within the political processes by transforming scientific insights regarding pressure sources into criteria applicable in decision-making" (Spangenberg 2007, p. 149) – in other words to redirect social and economic

---

S. Redpath, Ph.D.
Aberdeen Centre for Environmental Sustainability (ACES), University of Aberdeen,
Aberdeen, UK
e-mail: s.redpath@abdn.ac.uk

J. Travis, Ph.D.
Institute of Biological and Environmental Sciences, University of Aberdeen,
Aberdeen, UK
e-mail: justin.travis@abdn.ac.uk

S.M.J. Wijdeven, M.Sc.
ALTERRA, Wageningen, The Netherlands
e-mail: sander.wijdeven@wur.nl

trajectories in a more biodiversity-friendly direction. As many of these processes are slow (e.g. Fischer-Kowalski and Haberl 2007), analyses must cover sufficiently long periods of time to be useful. There is, therefore, a need for long-term socio-ecological research or LTSER (Haberl et al. 2006; Redman et al. 2004; Mirtl et al. 2009; Mirtl 2010; Singh et al. 2010).

One particular challenge within the endeavour of establishing such a long-term research infrastructure is the issue of scale (Redman et al. 2004; Haberl et al. 2006; Mirtl 2010) which, according to Gibson et al. (2000), is defined as "the spatial, temporal, quantitative, or analytical dimensions used to measure and study any phenomenon". The crucial role of spatial and temporal scale for natural processes and patterns has long been recognised (Wiens 1989; Levin 1992; Peterson and Parker 1998). The issue is particularly critical when dealing with biodiversity (Tilman and Kareiva 1997; Yoccoz et al. 2001; Leibold et al. 2004; Rahbek 2005). For that reason this paper uses biodiversity as trigger for analysing the importance of scaling issues for LTSER. The explicit consideration of scaling issues by social scientists is a rather novel phenomenon, but its importance is increasingly recognised (Wilbanks and Kates 1999; Cash and Moser 2000; Gibson et al. 2000; Giampietro 2004; Vermaat et al. 2005). Mismatches between the scale(s) on which ecological processes are observed and analysed, the scale(s) on which these processes are managed, and the scale(s) on which environmental policies are implemented have been identified as major obstacles towards nature conservation (MEA 2005; Carpenter et al. 2006; Cumming et al. 2006).

Most often ecological and social processes operate at a wide variety of scales or levels and cross-scale/level interactions occur frequently (Cash and Moser 2000; MEA 2003; Cash et al. 2006). Therefore, integrated research needs to be conducted at appropriate scales and levels so that efficient and goal-oriented political and management support can be provided. However, ecological and socioeconomic research and monitoring methods often do not match in terms of spatial and temporal scale (Carpenter et al. 2006; Cumming et al. 2006). This chapter proposes a suite of methods which might be useful in bridging the relevant spatial and temporal scales in LTSER, thus complementing previous work aimed at the conceptualisation of this emerging research field (Haberl et al. 2006; Redman et al. 2004; Singh et al. 2010). This review is the result of a workshop (15th–16th May 2006, Vienna, Austria) attended by a team of natural and social scientists of various disciplines during which 18 case studies of European biodiversity research were discussed and evaluated (Dirnböck et al. 2008).

## 6.2 Scale in Interdisciplinary Approaches

The treatment of scale in scientific disciplines has been complicated by differences in the subject of study, conceptual background, and the approaches in data acquisition rather than by different definitions of scale (Gibson et al. 2000; Vermaat et al. 2005).

*Differences in the subject of study*: Political science and economics are primarily concerned with human decision-making on different levels. These disciplines focus on agents and their behaviour, e.g. conservation area managers, local environmental administrators or environmental ministry personnel (Gezon and Paulson 2004). These agents may or may not be directly related to a specific spatial unit of ecological research, like biomes, habitats or species populations (Wiens 1989). On the other hand, ecology and geography are mostly concerned with processes and evolving patterns, the spatial resolution and extent of which may not be directly related to any relevant level of human decision-making (Levin 1992). While the concepts of scales and levels are not necessarily perceived differently, different subjects of study may still lead to mismatches in joint research.

*Differences in the conceptual background*: The problem of scale also arises from different thematic foci in different disciplines. Economics is primarily concerned with aspects of allocative efficiency, i.e. how to allocate available resources in order to maximise some desired output (for instance the maximisation of conserved species subject to a budget constraint, e.g. Ando et al. 1998). While this is, of course, to a certain extent dependent on spatial aspects, the spatial distribution of the allocation itself is not of major interest. In contrast, in ecology the emerging spatial patterns and interactions between processes occurring on different scales are often the primary focus of research, while less attention is often paid to the aggregated outcome (Levin 1992). Only recently have spatial aspects become an important research topic in economics concerned with biodiversity conservation (Wätzold and Drechsler 2005).

*Differences in the data availability*: Some disciplinary preferences for certain scales can also be explained by data availability. For example, human geography, economics and political sciences rely heavily on official statistics and census data referring to areas delineated by defined administrative boundaries such as municipalities, districts, provinces or nation states (Liverman et al. 1998). Their analyses are primarily focused on those spatial units and related levels of decision-making. On the other hand, the spatial resolution of many ecological and biogeophysical research approaches is determined by the observation technology, which can be a field survey technique or the available land cover data (Wiens 1989; Vermaat et al. 2005).

### 6.2.1 What Can We Learn from the Case Studies?

Details of the analysis of the 18 case studies are presented in Dirnböck et al. (2008). We surveyed studies concerned with major threats for biodiversity such as climate and land use changes, eutrophication due to excess nitrogen deposition, and the invasion of alien plants. Scale matches are surveyed between (1) biodiversity, understood as "the variability among living organisms, including diversity within species, between species and of ecosystems" (Article 2 of the Convention on Biological Diversity), (2) biodiversity management, i.e. the various types of local human interventions in the ecosystem, and (3) biodiversity-relevant policy, its goals and targets as well as its specific instruments (see Cash and Moser 2000 for a similar structure). We further highlight monitoring, research and evaluation acting upon each of these parts because these activities are essential to help societies in mitigating their pressures on biodiversity. We then use the criteria and definitions summarised in Table 6.1

**Table 6.1** Definition of scale mismatches between biodiversity, biodiversity management, and biodiversity-relevant policy (upper two rows). Definition of scale mismatches of research, monitoring, and evaluation carried out within each of these parts (third row)

| | Biodiversity | Management | Policy instrument |
|---|---|---|---|
| Management | A mismatch exists if biodiversity is studied at a temporal and spatial scale(s) which does not correspond or only partly corresponds with the particular authoritative reach of the institutional level at which biodiversity is or can be managed | | |
| Policy instrument | A mismatch exists if biodiversity is studied at a temporal and spatial scale(s) which does not correspond or only partly corresponds with the particular authoritative reach of the institutional level at which biodiversity-relevant policy instruments operate | A mismatch exists if biodiversity is managed at an institutional level whose temporal and spatial scale(s) of authoritative reach does not correspond or only partly corresponds with the scale(s) of the authoritative reach of policy instruments | |
| Research/monitoring/ evaluation | A mismatch exists if the scale(s) of research, monitoring and evaluation of biodiversity is carried out does not correspond or only partly corresponds with the temporal and spatial scale at which biodiversity is determined | A mismatch exists if the scale(s) of research, monitoring and evaluation of biodiversity management does not correspond or only partly corresponds with the particular scale of information necessary to effectively manage biodiversity | A mismatch exists if the scale(s) of research, monitoring and evaluation with reference to particular policy instruments do not or only partly correspond with the particular scale of information necessary for political decision making |

to identify scale mismatches. For each case study and for each aspect of biodiversity surveyed, management and policy spatial and temporal scale domains are defined by personal judgement. These domains were then compared with the spatial and temporal scales of the indicators used in the respective study.

Our analysis revealed that scale mismatches result from disregarding the importance of the scale issue. None of the 18 studies succeeded in taking all relevant scales into account. Often the focus was on one end of the scale range. For example, landscape scale processes of plant or animal populations are often neglected when the focus is on local scale dynamics. In these cases, the scales and levels most relevant for management and biodiversity-relevant policies are also disregarded, as they are rather expressed at the landscape and even broader scales. In many case studies, the variety of scales at which biodiversity-relevant policies are implemented are ignored or wrongly addressed. This is often true for local and regional research projects. Owing to their nested design, many European-level research studies succeeded in taking the multi-scale nature of policies into account more carefully by embedding regional case studies in a broader context.

Long-term biodiversity research is still very rare. Knowledge and data on the long-term consequences of management interventions and policies are thus still very limited. The relevant scales of management and of policy are therefore often not addressed.

The following example illustrates a disregard of the highly relevant long-term scale: An extensive and costly experimental study was carried out in the UK, the farm-scale evaluation (FSE), in order to assess the effects of differences in the management – especially the type of herbicides used and the timing of their application – of conventional and genetically modified herbicide-tolerant crops on the diversity and abundance of plants and invertebrates (Firbank et al. 2003). In general, the study demonstrated a series of direct or indirect effects. Notably, various discussion papers appeared with some of them explicitly criticising the limited usefulness of FSE due to neglected temporal and spatial scales: "The most serious limitation of the FSE from the standpoint of public policy is that the study has no predictive component. Forecasts of the likely impacts on biodiversity 10, 20, or even 50 years into the future and at a landscape scale are needed if policy decisions are to be made. However, the FSE was not designed with the goal of estimating parameters for the development of predictive models, but was tied to a rather narrow hypothesis test and constrained to a field scale. Therefore, the current results are inadequate to make long-term policy evaluations" (Freckleton et al. 2003). The case studies highlight that scientific support at relevant scales is particularly limited for biodiversity-relevant policies and, to a lesser extent, for management. Biodiversity research is still dominated by natural sciences with its specific scales of research. In fact, the scale of management is often closer to the one of most biodiversity research. Though exceptions exist, in many case studies, social, economic and political topics are merely accessory matters lacking empirical consideration.

The following example of spatial scale mismatches was particularly striking: The invasion of the alien plant *Rhododendron ponticum* in the UK, which causes considerable conservation problems, provides a good illustration of scale mismatches between the plant's spatial ecology and the methods and policies available to control it. The spatial dynamics of the plant's invasion are driven largely by the scale of seed dispersal and the pattern of habitat available for germination (e.g. Stephenson et al. 2006). Patterns of dispersal and habitat availability at a fine resolution (perhaps measured in metres) can potentially determine the rate of spread of the population at a local or landscape scale. Methods for controlling the plant are well-developed and, when properly implemented, are effective. However, a lack of understanding of the spatial dynamics of the plants means that effective control over a spatial extent is often poor. This is already a problem within an area managed by a single landowner but becomes even more problematic when an infestation of *Rhododendron* occurs over a matrix of different estates owned by numerous individuals. Coordination of control activities at a local or a landscape scale is required for successful spatially-extended control, but this very rarely takes place. A greater understanding of the economic impacts of *Rhododendron* in a spatial context, including external costs associated with the probabilities that the plant spreads from one estate to another (Dehnen-Schmutz et al. 2004) will be helpful in encouraging a change in practice at regional scales. At the policy level, either at county, national or EU scale, a better strategic appreciation of the spatial nature of the problem would help. Currently, much of the funding for control is provided on an ad hoc basis, and there are only infrequent attempts to effectively control *Rhododendron* at a landscape or regional scale.

Last but not least, cross-scale interactions were considered rather superficially in the 18 case studies surveyed.

## 6.3 Towards Scale Explicit LTSER

Much has been written about the concept and treatment of scale in different ecological disciplines (Wiens 1989; Levin 1992; Tilman and Kareiva 1997; Peterson and Parker 1998). Gibson et al. (2000) and Vermaat et al. (2005) presented reviews of the treatment of scale in ecological economics and related fields. LTSER is concerned with the integrated, interdisciplinary analysis of socio-ecological systems.

The case studies analysed (Dirnböck et al. 2008) exemplify that ecological research often misses the broader scales at which social science can be linked to ecological research in order to provide useful support for biodiversity managers and policy makers. For LTSER, the landscape provides an inevitable link between natural and social science, since its structure and processes are the variable outcome of the interplay of nature and society (Haberl et al. 2006). The landscape therefore integrates several scales of this interaction (Farina 2000; Naveh 2000a, b). Biodiversity too, is strongly influenced by processes and structures of landscapes (Leibold et al. 2004). The concept of LTSER advocates long-term research as many ecological and social processes are inherently slow. However, the case studies also illustrate that long-term – particularly interdisciplinary – research is still very rare (Dirnböck et al. 2008). Long-term landscape scale research will be a key focus in most LTSER platforms and innovative meta-analysis methods will be needed to analyse and interpret results across the LTSER network.

### 6.3.1 Up-Scaling Ecological Processes

Long-term ecological experiments and monitoring still provide the most reliable information on ecological processes and are indispensable for evaluating modelling results and improving model structure (Rees et al. 2001; Rastetter et al. 2003). Therefore, they remain a necessary complement within any other strategy for bridging the spatial and temporal scales needed in interdisciplinary LTSER.

One set of methods that can help in scaling ecological processes has emerged in the last decades within landscape ecology, a research field of largely natural-scientific origin. Landscape ecology examines the relationships between landscape patterns and ecological processes (Forman and Godron 1986; Turner 1989; Gustafson 1998), using landscape metrics to quantify and describe characteristics of the landscape structure (Gustafson 1998). Many of the indicators used in landscape-ecological analyses refer to abstract holistic features of the landscape, such as heterogeneity, diversity, complexity, or fragmentation. The purpose of landscape metrics is to obtain sets of quantitative data that allow a more objective comparison of different landscapes (Gustafson 1998; Antrop 2000). The question of scale with reference to resolution and extent is very important for the calculation of many landscape metrics (Meentemeyer and Box 1987; Cullinan and Thomas 1992; O'Neill et al. 1996). Vos et al. (2001) proposed a framework of ecologically scaled landscape indices (ELSIs) that take the different behaviour of species into account. A combination of ELSIs and ecological species profiles is used to facilitate this concept in practice. Other approaches are to identify species-specific and scale-specific thresholds of indices for assessing the effect of, for example, habitat fragmentation on the survival of species (Tischendorf and Fahrig 2000).

Ecological modelling approaches have been developed using theoretical frameworks such as the species-area-relationship, which predicts that species diversity increases with increasing area availability, to model diversity at the landscape scale (Pereira and Daily 2006). Another widely applied technique in this context is habitat distribution modelling, which uses an array of spatial environmental data in order to predict species and/or diversity (Guisan and Zimmermann 2000). Both techniques have been applied successfully. They lack significant determinants of species diversity, however, such as the dispersal of species in fragmented landscapes, which are particularly crucial when effects of environmental change are to be investigated (Ibáñez et al. 2006).

Population Viability Analysis (PVA) is used to gain better insights into the mechanisms which drive diversity by combining life-history data, demographic and sometimes genetic data and data on environmental variability. PVA predicts the probability that a plant or animal population will persist for a given period of time in a given area with a given setting of suitable habitats (habitats where a species can potentially survive). Spatially-explicit PVA's have been used for just over 10 years (Akcakaya et al. 1995), and their use is becoming increasingly common (Akcakaya et al. 2004). Typically, they use GIS technology to create maps of suitable breeding and dispersal habitat for the target species. A stochastic population model then sits on top of the GIS-created matrix and is run to assess the probability that the population persists in a given landscape. Once one moves to a spatial PVA, however, dispersal becomes crucial, and both the rate at which individuals disperse and the spatial scale of movement emerge as vital parameters. Provided that spatial data about the distribution of habitats is available, PVA can be applied to relatively large spatial scales. Interactions can be analysed between different environmental changes. For example range shifts of species resulting from climate change are predicted in that way. The migration process can be limited by a lack of suitable habitats in highly fragmented landscapes so that the species potentially becomes extinct due to the synergetic interaction of both forces (Travis 2003). PVA is thus a promising tool for evaluating such processes at the landscape scale, i.e. at the scale at which it can be linked to policy instruments and management. So far, however, these tools exist only for a limited number of taxonomic groups and species.

## 6.3.2 Developing Multidisciplinary Modelling Approaches for Bridging Scales

A key future task in LTSER will be to combine, or even explicitly couple, spatio-temporal modelling approaches already developed for disciplinary studies. This will not be trivial due to different predominant scale domains of biodiversity, management and policies. Taking for example the conflict between game management and raptor conservation (Thirgood and Redpath 2005; Dirnböck et al. 2008: case study 2). Raptors are scarce and legally protected, but at the same time threaten the livelihoods of gamekeepers and economic returns for private estates. This conflict covers a number of scales and resolutions. An ecological model of the dynamics of the predator and the prey may be built at a resolution set to the size of a single raptor's territory. An economic approach may be set at the resolution of a private estate, which could encompass a number of predator territories. An agent-based model could be set at a regional scale at which the gamekeepers interact. All of these scales and levels are subject to the legal-political framework that operates at a national or even international scale.

Methods to tackle scale issues in integrated land system analyses are still in their infancy (Liverman et al. 1998; Haberl et al. 2006; Young et al. 2006). The most promising approach in model integration – at least for a site-based LTSER network – is a nested structure of various models working at different scales (Schröter et al. 2005; Reidsma et al. 2006). Broad-scale models, e.g. international trade models and macroeconomic models (Edenhofer et al. 2005) should provide the general background against which regional and local decisions and actions are analysed. An international trade model, for instance, would define the regional and local level of food and energy demand, which is a crucial determinant of land-use changes and, finally, landscape structure and processes. To make the processes at different levels more consistent, these models can be used in an iterative way by feeding inputs and outputs back and forth (Root and Schneider 2002). Of course, not all real-world feedback mechanisms can be considered. Depending on the complexity of the models, coupled modelling systems may not always converge on a unique solution and results from different models may not be consistent.

One of the major problems is data availability across different scale domains. As mentioned, many disciplines have, for good reasons, developed tools and models around the available data. Modelling paradigms may also influence the degree of complexity in terms of spatial resolution. For instance, dynamic optimisation models in economics are constructed at a rather aggregated level, in order to allow for solutions at reasonable computational costs. These models are usually less detailed in terms of spatial, temporal and institutional resolution than climate or hydrological models. The properties and the challenges of integrated modelling have been summarised by the SustainabilityA-Test (2010) EU project. Wätzold et al. (2006) and Drechsler et al. (2007) give a general discussion of challenges of ecological-economic modelling including scale issues.

Approaches and models that are capable of analysing society's impacts on landscape structure and processes are particularly relevant for interdisciplinary biodiversity research. The socioeconomic metabolism approach, pioneered in the 1970s (Boulding 1973; Ayres and Kneese 1969), analyses society's stocks and flows of materials, energy or substances (e.g., carbon, nitrogen, lead, copper). These flows are thought to be simultaneously influenced by biogeophysical patterns and processes including climate, geomorphology, soils, and biota on the one hand, and by social interactions and relations such as economic transactions, power relations, legal and political frameworks on the other hand. This "double compatibility" towards ecological and socioeconomic models and data enables the socioeconomic metabolism approach to establish a link between socioeconomic variables and biophysical patterns and processes, with both groups characterised by their predominant scales (Haberl et al. 2004). While some accounting systems derived from the metabolism approach such as Material Flow Analysis (MFA) are mostly applied on national, provincial or municipal scales, others such as the human appropriation of net primary production (HANPP) can also deliver spatially explicit data (maps). GIS techniques enable us to calculate HANPP with the resolution that satellite imagery or aerial photography allow (Haberl et al. 2001; Wrbka et al. 2004). Other land-use related indicators, e.g., indicators relating to carbon stocks or nitrogen flows, could also be calculated using the socioeconomic metabolism approach on any spatial scale for which land-use and land-cover data with sufficient resolution can be generated (Erb 2004). Ultimately, by combining tools to analyse ecological material and energy flows (e.g., biogeochemical process models) with socioeconomic metabolism studies would allow the study of the respective effects of natural and socioeconomic drivers on patterns and processes in integrated socio-ecological systems (Haberl et al. 2006).

Researchers increasingly use models that combine agent-based modules that simulate decisions of and interactions between agents of land use as well as biophysical stocks and flows (Axtell et al. 2002; Janssen 2004; Manson and Evans 2007; Gaube et al. 2009). Agents are not only individuals, but also social or economic units such as farmsteads or households. In these models the behaviour of agents depends not only on natural, social, economic or political factors, but also on the behaviour of other agents. The agent's decisions may have important biophysical effects like changes in land use and these changes may also feed back on agents and modify their behaviour. Currently, these models are being explored mostly on local to regional scales and are thus readily suitable for LTSER platforms. Nevertheless, modellers are aware that processes on broader scales may critically affect trajectories. Such models could be coupled to larger-scale models and help to develop dynamic multi-scale approaches that would allow us to analyse scale interactions much more comprehensively (Janssen and Ostrom 2006).

### 6.3.3 *Implementing the Scales of Decision Making of Management and Policies*

It is widely recognised that key stakeholder groups should participate in decision-making, especially when these decisions have an impact on stakeholder economic or social well-being (Western and Wright 1994; Dirnböck et al. 2008: case studies 2 and 3). The latter case often arises when common goods like biodiversity are to be conserved locally. Participatory approaches have been developed to address this issue. One way of assessing the acceptability of different management options is to quantify the views of stakeholders through the use of a variety of Multicriteria Analyses (Edwards-Jones et al. 2000). Such approaches have been used to assess the management of human-wildlife conflicts (Redpath et al. 2004). In the management of biodiversity conflicts, the general principle of quantifying the perceptions of stakeholders as a means of searching for acceptable solutions has broad relevance (Conover 2002).

## 6.4 Discussion and Conclusions

Policy-relevant interdisciplinary biodiversity research is still in its infancy. The ideal case of long-term inter- and transdisciplinary research tackling all relevant scales of biodiversity, its management and biodiversity-relevant policies has so far been elusive (Dirnböck et al. 2008). A major reason is probably the disciplinary focus in education and research and, hence, the lack of appropriate interdisciplinary theories, methods and expertise (see Furman and Peltola, Chap. 18 in this volume). In biological conservation, to date we are confronted with a patchwork of studies, which have generated a growing number of results covering important issues but often lack usefulness for effective management of biodiversity and goal-oriented political decision making (MEA 2005; Carpenter et al. 2006; Spangenberg 2007).

The future of LTSER in the European LTER network – most probably in analogy with other international efforts towards LTSER – will be based on existing LTER infrastructure and thus confronted with exactly the described situation of many disciplinary studies which can not be integrated straightforward into interdisciplinary efforts (Haberl et al. 2006; Singh et al. 2010; Mirtl et al. 2009; Mirtl 2010). Mismatches in the scales taken into account in disciplinary studies are but one reason. Gaps with regard to relevant scales, which become apparent when screening existing studies in the area of a LTSER platform, should be addressed by a scale-explicit research agenda. Such an agenda will be most effective when including the participation of stakeholders from different scale domains. Studies at the regional scale of LTSER platforms could then be systematically integrated into large-scale modelling exercises or metaanalyses. We hope that the proposed methods and approaches, which are the result of discussions among researchers from a variety of natural and social science disciplines, can pave the way towards scale-explicit LTSER.

**Acknowledgments** The paper was developed within ALTER-Net, a Network of Excellence funded by the EU within its 6th Framework Programme. Apart from the authors, Rehema White, Erik Framstad, Vegar Bakkestuen, Andreas Richter, Clemens Grünbühel, and Norbert Sauberer, also participated in the project. We wish to thank Anke Fischer, Frederic Archeaux and Frank Wätzold for their valuable comments to an earlier draft of the manuscript. This research contributes to the Global Land Project (www.globallandproject.org).

# References

Akcakaya, H. R., McCarthy, M. A., & Pearce, J. L. (1995). Linking landscape data with population viability analysis: Management options for the helmeted honeyeater. *Biological Conservation, 73*, 169–176.

Akcakaya, H. R., Radeloff, V. C., & Mladenhoff, H. S. (2004). Integrating landscape and metapopulation modeling approaches: Viability of the sharp-tailed grouse in a dynamic landscape. *Conservation Biology, 18*, 526–537.

Ando, A., Camm, J., Polasky, S., & Solow, A. (1998). Species distributions, land values, and efficient conservation. *Science, 279*, 2126–2128.

Antrop, M. (2000). Background concepts for integrated landscape analysis. *Agriculture, Ecosystems and Environment, 77*, 17–28.

Axtell, R. L., Andrews, C. J., & Small, M. J. (2002). Agent-based modeling and industrial ecology. *Journal of Industrial Ecology, 5*, 10–13.

Ayres, R. U., & Kneese, A. (1969). Production, consumption and externalities. *The American Economic Review, 59*, 282–297.

Boulding, K. E. (1973). The economics of the coming spaceship earth. In H. E. Daly (Ed.), *Towards a steady state economy* (pp. 3–14). San Francisco: Freeman.

Carpenter, S. R., DeFries, R., Dietz, T., Mooney, H. A., Polasky, S., Reid, W. V., & Scholes, R. J. (2006). Millennium ecosystem assessment: Research needs. *Science, 314*, 257–258.

Cash, D. W., & Moser, S. C. (2000). Linking global and local scales: Designing dynamic assessment and management processes. *Global Environmental Change, 10*, 109–120.

Cash, D. W., Adger, W. N., Berkes, F., Garden, P., Lebel, L., Olsson, P., Pritchard, L., & Young, O. (2006). Scale and cross-scale dynamics: Governance and information in a multi-level world. *Ecology and Society, 11*, 8. (Online) http://www.ecologyandsociety.org/vol11/iss2/art8/

CBD. (2003). *Consideration of the results of the meeting on "2010: The global biodiversity challenge"*. UNEP/CBD/SBSTTA/9/inf/9, Convention on Biological Diversity, Montreal, Canada.

Conover, M. (2002). *Resolving human-wildlife conflicts: The science of wildlife damage management*. Boca Raton: CRC Press.

Cullinan, V. I., & Thomas, J. M. (1992). A comparison of quantitative methods for examining landscape pattern and scale. *Landscape Ecology, 7*, 211–227.

Cumming, G. S., Cumming, D. H., & Redman, C. L. (2006). Scale mismatches in socio-ecological systems: Causes, consequences, and solutions. *Ecology and Society, 11*(2), 14.

Dehnen-Schmutz, K., Perrings, C., & Williamson, W. (2004). Controlling *Rhododendron ponticum* in the British Isles: An economic analysis. *Journal of Environmental Management, 70*, 323–332.

Dirnböck, T., Bezák, P., Dullinger, S., Haberl, H., Lotze-Campen, H., Mirtl, M., Peterseil, J., Redpath, S., Singh, S. J., Travis, J., & Wijdeven, S. (2008). *Scaling issues in long-term socio-ecological biodiversity research. A review of European cases* (Social Ecology Working Paper No. 100). Vienna: IFF Social Ecology. Retrieved from http://www.uni-klu.ac.at/socec/downloads/WP100Webversion.pdf

Drechsler, M., Grimm, V., Mysiak, J., & Wätzold, F. (2007). Differences and similarities between economic and ecological models for biodiversity conservation. *Ecological Economics, 62*, 203–206.

Edenhofer, O., Bauer, N., & Kriegler, E. (2005). The impact of technological change on climate protection and welfare: Insights from the model MIND. *Ecological Economics, 54*, 277–292.

Edwards-Jones, G., Davies, B., & Hussian, S. (2000). *Ecological economics: An introduction*. Oxford: Blackwell Science Ltd.

EEA. (2007). *Europe's Environment. The fourth assessment*. Copenhagen: European Environment Agency.

Erb, K.-H. (2004). Land-use related changes in aboveground carbon stocks of Austria's terrestrial ecosystems. *Ecosystems, 7*, 563–572.

Farina, A. (2000). The cultural landscape as a model for the integration of ecology and economics. *BioScience, 50*, 313–320.

Firbank, L. G., Heard, M. S., Woiwod, I. P., Hawes, C., Haughton, A. J., Champion, G. T., Scott, R. J., Hill, M. O., Dewar, A. M., Squire, G. R., May, M. J., Brooks, D. R., Bohan, A. D., Daniels, R. E., Osborne, J. L., Roy, D. B., Black, H. I. J., Rothery, P., & Perry, J. N. (2003). An introduction to the farm-scale evaluations of genetically modified herbicide tolerant crops. *Journal of Applied Ecology, 40*, 2–16.

Fischer-Kowalski, M., & Haberl, H. (2007). *Socioecological transitions and global change. Trajectories of social metabolism and land use*. Cheltenham/Northampton: Edward Elgar.

Forman, R. T. T., & Godron, M. (1986). *Landscape ecology*. New York: Wiley.

Freckleton, R. P., Sutherland, W. J., & Watkinson, A. R. (2003). Deciding the future of GM crops in Europe. *Science, 302*, 994–996.

Gaube, V., Kaiser, C., Wildenberg, M., Adensam, H., Fleissner, P., Kobler, J., Lutz, J., Schaumberger, A., Schaumberger, J., Smetschka, B., Wolf, A., Richter, A., & Haberl, H. (2009). Combining agent-based and stock-flow modelling approaches in a participative analysis of the integrated land system in Reichraming, Austria. *Landscape Ecology, 24*, 1149–1165.

Gezon, L. L., & Paulson, S. (2004). *Political ecology across spaces, scales and social groups*. New Brunswick: Rutgers University Press.

Giampietro, M. (2004). *Multi-scale integrated analysis of agroecosystems*. Boca Raton: CRC Press.

Gibson, C. C., Ostrom, E., & Ahn, T. K. (2000). The concept of scale and the human dimension of global change: A survey. *Ecological Economics, 32*, 217–239.

Guisan, A., & Zimmermann, N. E. (2000). Predictive habitat distribution models in ecology. *Ecological Modelling, 135*, 147–186.

Gustafson, E. J. (1998). Quantifying landscape spatial pattern: What is the state of art? *Ecosystems, 1*, 143–156.

Haberl, H., Erb, K. H., Krausmann, F., Loibl, W., Schulz, N., & Weisz, H. (2001). Changes in ecosystem processes induced by land use: Human appropriation of net primary production and its influence on standing crop in Austria. *Global Biogeochemical Cycles, 15*, 929–942.

Haberl, H., Fischer-Kowalski, M., Krausmann, F., Weisz, H., & Winiwarter, V. (2004). Progress towards sustainability? What the conceptual framework of material and energy flow accounting (MEFA) can offer. *Land Use Policy, 21*, 199–213.

Haberl, H., Winiwarter, V., Andersson, K., Ayres, R., Boone, C., Castillo, A., Cunfer, G., Fischer-Kowalski, M., Freudenburg, W. R., Furman, E., Kaufmann, R., Krausmann, F., Langthaler, E., Lotze-Campen, H., Mirtl, M., Redman, C. L., Reenberg, A., Wardell, A., Warr, B., & Zechmeister. H. (2006). From LTER to LTSER: Conceptualizing the socio-economic dimension of long-term socio-ecological research. *Ecology and Society*, 11. (Online) http://www.ecologyandsociety.org/vol11/iss2/art13/

Haberl, H., Erb, K.-H., Plutzar, C., Fischer-Kowalski, M., & Krausmann, F. (2007). Human appropriation of net primary production (HANPP) as indicator for pressures on biodiversity. In T. Hak, B. Moldan, & A. L. Dahl (Eds.), *Sustainability indicators. A scientific assessment* (pp. 271–288). Washington, DC/Covelo/London: Island Press.

Ibáñez, I., Clark, J. S., Dietze, M. C., Feeley, K., Hersh, M., LaDeau, S., McBride, A., Welch, N. E., & Wolosin, M. S. (2006). Predicting biodiversity change: Outside the climate envelope, beyond the species–area curve. *Ecology, 87*, 1896–1906.

Janssen, A. M. (2004). Agent-based models. In J. Proops & P. Safonov (Eds.), *Modelling in ecological economics* (pp. 155–172). Cheltenham/Northampton: Edgar Elgar.

Janssen, M. A., & Ostrom, E. (2006). Empirically based, agent-based models. *Ecology and Society*, 11. (Online) http://www.ecologyandsociety.org/vol11/iss2/art37/

Leibold, M. A., Holyoak, M., Mouquet, N., Amarasekare, P., Chase, J. M., Hoopes, M. F., Holt, R. D., Shurin, J. B., Law, R., Tilman, D., Loreau, M., & Gonzalez, A. (2004). The metacommunity concept: A framework or multi-scale community ecology. *Ecology Letters, 7*, 601–613.

Levin, S. A. (1992). The problem of pattern and scale in ecology. *Ecology, 73*, 1943–1967.

Liverman, D., Moran, E. F., Rindfuss, R. R., & Stern, P. C. (1998). *People and pixels, linking remote sensing and social science*. Washington, DC: National Academy Press.

Manson, S. M., & Evans, T. (2007). Agent-based modelling of deforestation in southern Yucatán, Mexico, and reforestation in the Midwest United States. *Proceedings of the National Academy of Sciences of the USA, 104*, 20678–20683.

Meentemeyer, V., & Box, E. O. (1987). Scale effects in landscape studies. In M. G. Turner (Ed.), *Landscape heterogeneity and disturbance* (pp. 15–34). New York: Springer.

Millennium Ecosystem Assessment (MEA). (2003). *Ecosystems and human well-being, a framework for assessment* (Millennium ecosystem assessment series). Washington, DC: Island Press.

Millennium Ecosystem Assessment (MEA). (2005). *Ecosystems and human well-being: Biodiversity synthesis*. Washington, DC: World Resources Institute.

Mirtl, M. (2010). Introducing the next generation of ecosystem research in Europe: LTER-Europe's multi-functional and multi-scale approach. In F. Müller, C. Baessler, H. Schubert, & S. Klotz (Eds.), *Long-term ecological research: Between theory and application* (pp. 75–94). Dordrecht: Springer.

Mirtl, M., Boamrane, M., Braat, L., Furman, E., Krauze, K., Frenzel, M., Gaube, V., Groner, E., Hester, A., Klotz, S., Los, W., Mautz, I., Peterseil, J., Richter, A., Schentz, H., Schleidt, K., Schmid, M., Sier, A., Stadler, J., Uhel, R., Wildenberg, M., & Zacharias, S. (2009). *LTER-Europe design and implementation report – Enabling "next generation ecological science": report on the design and implementation phase of LTER-Europe under ALTER-Net & management plan 2009/2010*. Vienna: Umweltbundesamt, Environment Agency Austria.

Naveh, Z. (2000a). The total human ecosystem: Integrating ecology and economics. *BioScience, 50*, 357–361.

Naveh, Z. (2000b). What is holistic landscape ecology? A conceptual introduction. *Landscape and Urban Planning, 50*, 7–26.

O'Neill, R. V., Hunsaker, C. T., Timmins, S. P., Jackson, B. L., Jones, K. B., Riiters, K. H., & Wickham, J. D. (1996). Scale problems in reporting landscape pattern at the regional scale. *Landscape Ecology, 11*, 169–180.

Pereira, H. M., & Daily, G. D. (2006). Modeling biodiversity dynamics in countryside landscapes. *Ecology, 87*, 1877–1885.

Peterson, D. L., & Parker, V. T. (1998). *Ecological scale: Theory and application*. New York: Columbia University Press.

Rahbek, C. (2005). The role of spatial scale and the perception of large-scale species-richness patterns. *Ecology Letters, 8*, 224–239.

Rastetter, E. B., Aber, J. D., Peters, D. P. C., Ojima, D. S., & Burke, I. C. (2003). Using mechanistic models to scale ecological processes across space and time. *BioScience, 53*, 68–76.

Redman, C. L., Grove, J. M., & Kuby, L. L. H. (2004). Integrating social science into the long-term ecological research (LTER) network: social dimensions of ecological change and ecological dimensions of social change. *Ecosystems, 7*, 161–171.

Redpath, S. M., Arroyo, B. E., Leckie, F. M., Bacon, P., Bayfield, N., Gutiérrez, R. J., & Thirgood, S. J. (2004). Using decision modelling with stakeholders to reduce human-wildlife conflict: A raptor – Grouse case study. *Conservation Biology, 18*, 350–359.

Rees, M., Condit, R., Crawley, M., Pacala, S., & Tilman, D. (2001). Long-term studies of vegetation dynamics. *Science, 293*, 650–655.

Reidsma, P., Tekelenburg, T., van den Berg, M., & Alkemade, R. (2006). Impacts of land-use change on biodiversity: An assessment of agricultural biodiversity in the European Union. *Agriculture, Ecosystems and Environment, 114*, 86–102.

Root, T., & Schneider, S. H. (2002). Strategic cycling scaling: Bridging five orders of magnitude scale gaps in climatic and ecological studies. *Integrated Assessment, 3*, 188–200.

Schröter, D., Cramer, W., Leemans, R., Prentice, I. C., Araújo, M. B., Arnell, N. W., Bondeau, A., Bugmann, H., Carter, T. R., Gracia, A., de la Vega-Leinert, C., Erhard, M., Ewert, F., Glendining, M., House, J. I., Kankaanpää, S., Klein, R. J. T., Lavorel, S., Lindner, M., Metzger, M. J., Meyer, J., Mitchell, T. D., Reginster, I., Rounsevell, M., Sabaté, S., Sitch, S., Smith, B., Smith, J., Smith, P., Sykes, M. T., Thonicke, K., Thuiller, W., Tuck, G., Zaehle, S., & Zierl, B. (2005). Ecosystem service supply and vulnerability to global change in Europe. *Science, 310*, 1333–1337.

Singh, S. J., Haberl, H., Gaube, V., Grünbühel, C. M., Lisievici, P., Lutz, J., Mathews, R., Mirtl, M., Vadineanu, A., & Wildenberg, M. (2010). Conceptualising long-term socio-ecological research (LTSER): Integrating socio-economic dimensions. In F. Müller, H. Schubert, & S. Klotz (Eds.), *Long-term ecological research, between theory and application* (pp. 377–398). Berlin: Springer.

Spangenberg, J. H. (2007). Biodiversity pressure and the driving forces behind. *Ecological Economics, 61*, 146–158.

Stephenson, C. M., MacKenzie, M. L., Edwards, C., & Travis, J. M. J. (2006). Modelling establishment probabilities of an exotic plant, *Rhododendron ponticum*, invading a heterogeneous, woodland landscape using logistic regression with spatial autocorrelation. *Ecological Modelling, 193*, 747–758.

Sustainability A-Test. (2010). Retrieved May 6, 2010, from http://www.sustainabilitya-test.net

Thirgood, S. J., & Redpath, S. M. (2005). Science, politics and human-wildlife conflicts: harriers and grouse in the UK. In R. Woodroffe, S. Thirgood, & A. Rabinowitz (Eds.), *People or wildlife: Conflict or coexistence* (pp. 192–208). London: Cambridge University Press.

Tilman, D., & Kareiva, P. (1997). *Spatial ecology: The role of space in population dynamics and interspecific interaction*. Princeton: Princeton University Press.

Tischendorf, L., & Fahrig, E. (2000). How should we measure landscape connectivity? *Landscape Ecology, 15*, 633–641.

Travis, J. M. J. (2003). Climate change and habitat destruction: A deadly anthropogenic cocktail. *Proceedings of the Royal Society B: Biological Sciences, 270*, 1471–2954.

Turner, M. G. (1989). Landscape ecology: The effect of pattern and process. *Annual Review of Ecology and Systematics, 20*, 171–197.

Vermaat, J. E., Eppink, F., van den Bergh, J. C. M., Barendregt, A., & van Belle, J. (2005). Aggregation and the matching of scales in spatial economics and landscape ecology: Empirical evidence and prospects for integration. *Ecological Economics, 52*, 229–237.

Vos, C. C., Verboom, J., Opdam, P. F. M., & Ter Braak, C. J. F. (2001). Toward ecologically scaled landscape indices. *The American Naturalist, 183*, 24–41.

Wätzold, F., & Drechsler, M. (2005). Spatially uniform versus spatially differentiated compensation payments for biodiversity-enhancing land-use measures. *Environmental and Resource Economics, 31*, 73–93.

Wätzold, F., Drechsler, M., Armstrong, C. W., Baumgärtner, S., Grimm, V., Huth, A., Perrings, C., Possingham, H. P., Shogren, J. F., Skonhoft, A., Verboom-Vasiljev, J., & Wissel, C. (2006). Ecological-economic modeling for biodiversity management: Potential, pitfalls, prospects. *Conservation Biology, 20*, 1034–1041.

Western, D., & Wright, R. M. (1994). *Natural connections: Perspectives in community-based conservation*. Washington, DC: Island Press.

Wiens, J. A. (1989). Spatial scaling in ecology. *Functional Ecology, 3*, 385–397.

Wilbanks, T. J., & Kates, R. W. (1999). Global change in local places: How scale matters. *Climatic Change, 43*, 601–628.

Wrbka, T., Erb, K.-H., Schulz, N. B., Peterseil, J., Hahn, C., & Haberl, H. (2004). Linking pattern and process in cultural landscapes. An empirical study based on spatially explicit indicators. *Land Use Policy, 21*, 289–306.

Yoccoz, N. G., Nichols, J. D., & Boulinier, T. (2001). Monitoring of biological diversity in space and time. *Trends in Ecology & Evolution, 16*, 446–453.

Young, O., Lambin, E. F., Alcock F., Haberl, H., Karlsson, S. I., McConnell, W. J., Myint, T., Pahl-Wostl, C., Polsky, C., Ramakrishnan, P. S., Scouvart, M., Schröder, H., Verburg, P. (2006). A portfolio approach to analyzing complex human-environment interactions: Institutions and land change. *Ecology and Society, 11*. (Online) http://www.ecologyandsociety.org/vol11/iss2/art31/

# Chapter 7
# Human Biohistory

**Stephen Boyden**

**Abstract** Human biohistory is learning about human situations against the background of the story of life on Earth. One of its key features is recognition that the evolutionary emergence of the human capacity for culture was one of the great watersheds in the history of life. Human culture has become a new kind of force in the biosphere – with profound and far reaching impacts not only on humans themselves but also on the rest of the living world. The chapter briefly discusses some important biohistorical principles, including cultural maladaptation and cultural reform, technoaddiction and the evolutionary health principle. Cultural evolution has recently resulted in patterns of human activity across the globe of a magnitude and of a kind that are unsustainable. If present trends continue unabated the ecological collapse of civilisation is inevitable. The future wellbeing of humankind will depend on big changes in the scale, intensity and nature of human activities on Earth. The best hope for the future lies in a rapid transition to a society that is truly in tune with, sensitive to and respectful of the processes of life which underpin our existence. This is referred to as a biosensitive society. However, there will be no transition to biosensitivity unless there come about profound changes in the worldview, assumptions and priorities of our society's dominant culture.

**Keywords** Biohistory • Biosensitivity • Environmental education • Human ecology • Sustainability • Transition

S. Boyden (✉)
Fenner School of Environment and Society, Australian National University,
Canberra, ACT 0200, Australia
e-mail: sboyden@netspeed.com.au

## 7.1 Introduction

There is an approach to learning about human situations which is of immense relevance to every one of us as individuals and to society as a whole. We call it *human biohistory*. Henceforth in this paper I will refer to 'human biohistory' simply as 'biohistory'.

Biohistory is the study of human situations, past and present, in biological and historical perspective – against the background of the story of life on Earth. Biohistory covers the basic principles of evolution, ecology, inheritance, and health and disease, and it pays special attention to the evolutionary background of our own species.

An especially important feature of biohistory is the fact that it recognises the evolutionary emergence of the human capacity for culture as one of the crucial watersheds in the history of life on Earth – of overriding significance not only for humans themselves but also for the rest of the living world.[1] For, as soon as human culture came into existence it began, through its influence on people's behaviour, to have impacts both on humans and on other forms of life. Biohistory is especially concerned with the constant interplay between human culture and biophysical processes.

We argue that basic biohistorical understanding across the community is an essential prerequisite for the future well-being of humankind. However, at present biohistory is not recognised as a *bona fide* subject in academic circles. It does not appear in school curricula and it does not feature in university degree courses or research programmes.

Over recent decades a growing number of writers have emerged who could well be described as leading biohistorians. René Dubos comes first to mind. Others include Hans Zinsser, Jared Diamond, Tim Flannery and Tony McMichael.[2] However, biohistory has yet to be developed systematically as a field of learning in its own right, and it is a long way from occupying the central place it warrants in educational programmes at all levels.

In this chapter I will focus especially on the work of my colleagues and myself at the Australian National University from 1965 until 1990. In my view, our conceptual approach is especially pertinent to the fast developing field which is the theme of this book – namely, long term socio-ecological research (LTSER) (and see Haberl et al. 2006; Singh et al. 2010).

---

[1] The word culture has many rather different meanings. Here it is used to mean the abstract products of the capacity for culture, such as learned language itself and the accumulated knowledge, assumptions, beliefs, values and technological competence of a human population. This use of the term is consistent with the first definition of 'culture' given in Collins Dictionary: 'The total of the inherited ideas, beliefs, values and knowledge, which constitute the shared bases of social action' (Collins Dictionary of the English Language (1979) Collins, Sydney, Auckland and Glasgow).

[2] I mention René Dubos first because his writings capture the essence of biohistory as I see it (e.g. Dubos 1968, 1980). However, he makes no attempt to develop a comprehensive theoretical basis for the subject. The same applies to the other authors mentioned (e.g. Zinsser 1935; Diamond 1997, 2005; McMichael 2001; Flannery 1994).

**Fig. 7.1** Biohistorical pyramid

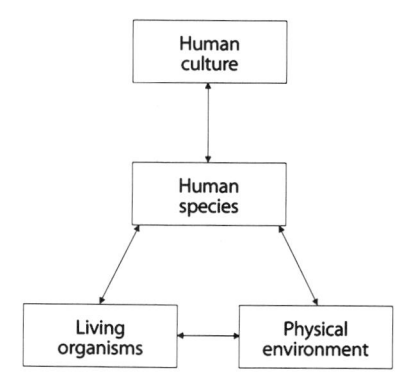

I will also discuss biohistory as an important field of scholarship in its own right as well as its potential contribution to community understanding and social change, and I will explain why I believe that the biohistorical paradigm has an important contribution to make to the transition to ecological sustainability.

## 7.2 Conceptual Starting Point

Our approach to biohistory takes as its starting point the history of life on Earth.

In the beginning there was no life. Only the physical world existed – called the *Physical environment* in Fig. 7.1. Then, perhaps around 4,500 million years ago, the first *Living organisms* came into being.

Eventually, over many millions of years, there evolved an amazing array of different life forms. Among these, emerging some 2,00,000 years ago, was *Homo sapiens*. Because of this animal's special relevance to our studies, it is separated from other living organisms in our conceptual scheme (*Human species* in Fig. 7.1).

Through the processes of biological evolution, the human species had acquired a distinctive and extraordinarily significant biological attribute – the capacity for culture.

The most essential aspect of this capacity is the human ability to invent and learn a symbolic spoken language, and to use it for communicating among ourselves. This linguistic aptitude depends both on characteristics of the human brain and on special anatomical arrangements in the region of the larynx, pharynx and tongue which permit us to utter an amazing range of different sounds.

Another aspect of human behaviour often regarded as an aspect of culture is the ability to invent and learn new technologies and to pass on this technical knowledge from one individual to another and from generation to generation. Some other primates and some birds exhibit a trace of this ability. In humans, this aptitude for technology is greatly enhanced by the use of symbolic language and also by the extraordinary dexterity of our species.

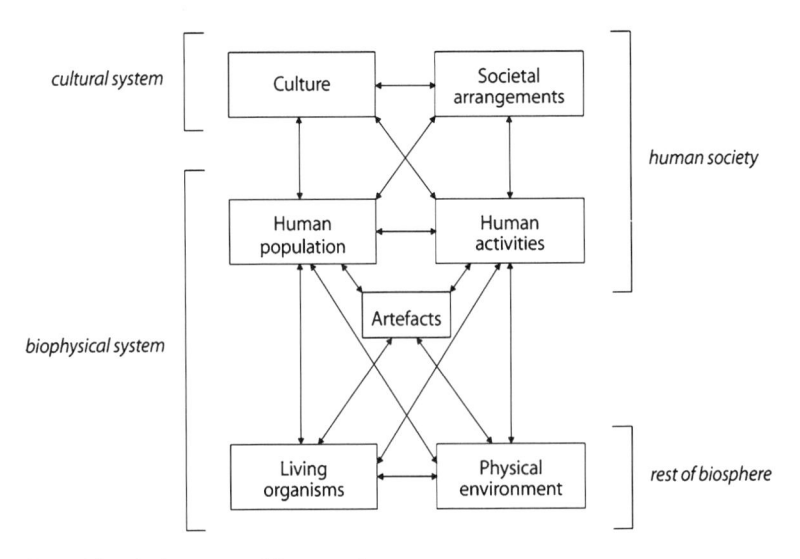

**Fig. 7.2** Biohistorical conceptual framework

As soon as human culture came into existence it began, through its influence on people's behaviour, to have impacts not only on humans themselves but also on other living systems. It evolved as a new kind of force in the biosphere, destined eventually to bring about profound and far-reaching changes across the whole planet.

For the purposes of this discussion, it is useful to complicate the scheme a little. Because we are especially interested in the impacts both on humans and on the environment of what people actually do, it is useful to split the *Human species* into the *Human population* and *Human activities* (Fig. 7.2).

It is also useful to divide *Human culture* into two parts.

The first part is *Culture* itself, which is the information stored in human brains and transmitted through language. Although relatively abstract, human culture is an extraordinarily powerful force in the total system. For a proper understanding of human situations today, it is essential that we take account of the interplay between culture and the processes of life.

In our work the focus has often been on the dominant culture of a society – that is, the culture that largely determines the patterns of human activity in that society.

Culture includes knowledge of language itself, and general knowledge of the environment, history, the arts and technologies, as well as assumptions, priorities and religious beliefs.

The second part is designated *Societal arrangements*, which includes society's economic, regulatory, political and educational arrangements and its institutional structure. Societal arrangements are largely determined by, and to some extent determine, the characteristics of the dominant culture.

In Fig. 7.2 we have added another set of variables – namely *Artefacts*, by which we mean 'things made by humans', including buildings, roads, all kinds of machines and electronic devices, as well as clothes, utensils and works of art.

Although this conceptual framework is based on the sequence of happenings in the history of life on Earth, it can also be applied to the here and now. The same sets of variables are involved. Located at the base of the model are the physical environment and living organisms (the biosphere) – underpinning and supporting the human population, which in turn creates and maintains human culture.

We attempted in our own work to apply an early version of this framework in our study of the ecology of Hong Kong (Boyden et al. 1981). Recently we have adapted it for use as a device to facilitate planning for the future –at the level of individuals and families through to city planning and government policies. It ensures that, in considering different options we take account of their implications both for human wellbeing and for the health of ecosystems (see Appendix).

## 7.3 Why Is Biohistory So Important?

Here in a nutshell are some of the reasons why I believe that biohistory is so crucially important for us all today –at the level of individuals and families and at the level of society as a whole.

### 7.3.1 *Biorealism*

First and foremost, biohistory is important because it constantly reminds us that we are living organisms, products of nature and totally dependent on the processes of life, within us and around us, for our very existence.

It reminds us that life processes underpin, permeate and make possible our whole social system and everything that happens within it. Keeping them healthy must be our first priority – because everything else depends on them.

The dominant culture of our time has lost sight of these fundamental realities – with grave consequences for humankind and the rest of the living world. They are not reflected in governmental policies, political platforms, the structure of educational programmes or the lifestyles of the majority of people.

### 7.3.2 *Human History*

Biohistory tells us that our species has been in existence for some 200,000 years (McDougall et al. 2005).

It shows us that the history of *Homo sapiens* falls into four distinct ecological phases, which differ both in the relationships between human populations and the rest of the living world and in the biological conditions of life and health of humans themselves. Although the dividing lines between the different phases are not always sharp, and occasional societies do not fit neatly into any one of them, the classification is a useful one. The four phases are not mutually exclusive and all of them can exist at the same time.

*Ecological Phase 1: The hunter-gatherer phase*
The hunter-gatherer phase of human existence was by far the longest of the four ecological phases, lasting some 1,80,000 or more years (over 7,000 generations).[3]

*Ecological Phase 2: The early farming phase*
The introduction of farming in some parts of the world around 12,000 years ago (480 generations) marked a turning point in cultural evolution. It was a precondition for all the spectacular developments in human history since that time.

*Ecological Phase 3: The early urban phase*
This phase began around 9,000 years ago (360 generations), when fairly large clusters of people, sometimes consisting of several thousand individuals, began to aggregate together in townships. Many of these people played no part in the gathering or production of food.[4]
Although the new conditions offered protection from most of the hazards of the hunter-gatherer lifestyle, malnutrition and infectious disease became much more important as causes of ill health and death.

*Ecological Phase 4: The high consumption phase*
This phase was ushered in by the industrial revolution, which began a little over 200 years ago (eight generations). It has been associated with profound changes in the ecological relationships between human populations and the rest of the biosphere.

Especially significant ecologically was the introduction of machines for performing different kinds of work and depending on the use of extrasomatic energy, especially fossil fuels. The discoveries and applications of electricity and radioactivity and the spectacular growth of the chemical industry have also been extremely important.

Largely as a consequence of improved living conditions the global population increased from about one billion in 1800 to two billion in the 1930s; and it is now almost seven billion. This population growth, along with the explosive increase in intensity of techno-industrial activities, is resulting in severe ecological disturbances across the whole planet.

The crucially important factor, which should be a central consideration in all government planning and economic deliberations, is the inescapable fact that the days of ecological Phase 4 are numbered. This phase is simply not sustainable ecologically. Either humankind will move into a very different, ecologically

---

[3] In this paper a generation is taken to be 25 years.

[4] One of the most interesting of these very early townships from the socio-ecological standpoint is Çatalhöyük in Anatolia, which was inhabited for approaching 2,000 years from about 9,500 BP (Mellaart 1967; Hodder 2006). The population ranged from 5,000–8,000. There were no apparent social classes. The economy was based primarily on the cultivation of barley, wheat, peas, and lentils, and the breeding of sheep, goats and, later in the period, cattle. Hunting was also still important for meat.

## 7.3.3 Human Culture as a Force in Nature

sustainable and healthy fifth ecological phase of human existence, or human civilisation will collapse.

The rapidity of the evolutionary development of the capacity for culture indicates that, once a rudimentary ability to invent and use symbolic spoken language emerged, it was at once of major biological advantage for its bearers under the prevailing conditions. The nature of this advantage has been the subject of a good deal of speculation among evolutionary biologists (e.g. Dunbar 1997).

In my view, its chief advantage probably lay in its role in the exchange and storage of useful information about the environment. This information was not only communicated within the group, but was also passed on to members of subsequent generations, increasing the likelihood of their good health and successful reproduction. However, it is possible that the aptitude for culture had a number of different biological advantages which collectively contributed to its rapid evolutionary development.[5]

Apart from its practical advantages, culture adds richness to human experience. It did so in the days of our hunter-gatherer ancestors – as in storytelling, musical traditions, dancing and other forms of artistic expression – and it does so today in so many ways. It makes a huge contribution to the sheer enjoyment of life.

However, especially under conditions of civilisation, cultural evolution has often resulted in activities that have caused a great deal of unnecessary distress to humans or damage to ecosystems. Such undesirable culturally-inspired activities are referred to as *cultural maladaptations*.

Biohistory reveals countless examples of cultural maladaptation in human history (Boyden 1987, 2004).

A particularly tragic example of cultural maladaptation was the ancient Chinese custom of foot-binding, which prevented the normal growth of the feet of young girls and caused them excruciating pain. This extraordinary practice well illustrates the propensity of culture to influence people's mind-sets in ways that result in activities that are not only nonsensical in the extreme, but also sometimes very cruel and destructive and contrary to nature. This particular cultural maladaptation was mutely accepted by the mass of the Chinese population for some 40 or more generations.

Throughout the history of civilisation, different cultures, including our own, have come up with a fascinating range of delusions about how social wellbeing, or prosperity, can best be achieved, and some of these delusions have led to blatant examples of cultural maladaptation. Here I will mention only one instance.

---

[5] The fact that the capacity for culture was of biological advantage during the tens of thousands of generations of our species before the advent of agriculture does not mean, of course, that it will necessarily be an advantage under conditions quite different from those of the evolutionary habitat.

According to the dominant culture of the Mayan civilisation, prosperity could best be achieved by pleasing the gods, and the best way to please the gods was to torture, mutilate and then sacrifice human beings. This behaviour can be regarded as a cultural maladaptation, because it certainly caused a great deal of unnecessary human suffering, and it clearly did not do the Mayans' society any good. Their civilisation collapsed suddenly, probably for ecological reasons, around 900 AD.

Again, the point to be emphasised is the fact that while there may well have been a handful of sceptics among the Mayans, the great majority of them really believed that the torture and sacrifice of humans was an entirely appropriate behaviour. Cultural gullibility is indeed a fundamental characteristic of our species.

Biohistory thus alerts us to the need for us to be constantly vigilant – checking that the assumptions of our society's dominant culture are in tune with the processes of life and that they are not leading us to behave in ways that are against nature and against the interests of our species.

## 7.3.4 Cultural Reform

One of the themes of biohistory is human adaptability. Our species shares with all other animals a series of adaptive mechanisms, which include genetic adaptation through natural selection (adaptation of populations over many generations), many kinds of physiological adaptation and adaptation through learning.

However, humans have an extra string to their bow – namely cultural adaptation, which is defined as adaptation through cultural processes.

In the present context we are especially interested in cultural adaptation aimed at overcoming the undesirable consequences of culture itself – that is, adaptation to cultural maladaptations. We refer to this as *cultural reform*.

The processes of cultural reform are often quite complicated, involving prolonged interactions between different interest groups in society. A key role is often played initially by minority groups, occasionally by single individuals, who start the ball rolling by drawing attention to an unsatisfactory state of affairs. We can refer to these people as first-order reformers. A prime example of a first-order reformer is Rachel Carson who, in her ground-breaking book *Silent Spring*, drew attention to the insidious and destructive ecological impacts of certain synthetic pesticides (Carson 1962, see Krausmann and Fischer-Kowalski, Chap. 15 in this volume).

Almost invariably, the expressions of concern coming from first-order reformers are promptly contradicted by others, the *anti-reformers*. This backlash often involves representatives of vested interests who fear that the proposed reforms will be to their disadvantage. They are likely to argue that the problem does not exist or that it has been has been grossly exaggerated, and they try to ridicule the reformers by calling them alarmists, fanatics, scaremongers and prophets of doom. Nowadays some of these anti-reform forces are extraordinarily powerful.

The first-order reformers are, in time, joined by *second-order reformers* who also take up the cause. Eventually, if they are successful, a change comes about in the dominant culture and members of governmental bureaucracies and other organisations set about working out ways and means of achieving the necessary changes. Their efforts may still be hindered to some extent by the stalling tactics of anti-reformers.

Biohistory provides many examples of cultural reform and anti-reform in recent history and at the present time. A well-documented instance of cultural reform from the past is the Public Health Movement of the later part of the nineteenth century (Flinn 1965; Frazer 1950). Other more recent examples include the anti-smoking campaign and the current debates about climate change. In the latter case, the anti-reformers are often referred to as climate change deniers and it is noteworthy that there is often a smattering of scientists among them (Oreskes and Conway 2010).

## 7.3.5 Evolution and Health

Biohistory reminds us that our species has been in existence for some 8,000 generations and that we are basically the same animal as our ancestors who lived long before the advent of farming – that is, an animal genetically adapted through natural selection to the life of the hunter-gatherers.[6] This fact has many important implications – for understanding ourselves and our problems.

One of the outcomes of the processes of evolution is the fact that animals become well adapted in their biological characteristics to the habitat in which they are evolving. In other words, the biological characteristics of any species are such that the individual animals are likely to experience good health in their natural environment.

If an animal is removed from its natural environment, or if its environment changes significantly, then it is likely to be less well adapted to the new conditions, and consequently some signs of physiological or behavioural maladjustment can be expected. This *evolutionary health principle* is a fundamental law of nature (Boyden 1973, 2004).

Ill health or pathological behaviour due to an animal experiencing conditions which deviate from that of its natural environment are referred to as examples of *phylogenetic maladjustment*.

It follows from the evolutionary health principle that if we wish to identify the health needs of any particular kind of animal, the first thing to do is to examine the conditions under which it evolved, because we can be sure that these conditions are capable of providing all the essential ingredients for maintaining and promoting health in that species.

---

[6] This does not mean that evolutionary change in the human species has come to a halt. There has been a relaxation of some selection pressures that were powerful in the hunter-gatherer environment and in the long term this will result in genetic changes in human populations (Rendel 1970). There have also been some new selection pressures associated with the advent of farming that have produced changes in some populations. A well-known example of this is the emergence and spread in European populations of lactase production into adulthood in response to the availability of bovine milk as a food source. For discussion of this change and for other examples, see Cochran and Harpending (2009).

In the case of humankind there is, for example, no diet better for humans than the typical diet of our hunter-gatherer ancestors. Or if we take much more, or much less physical exercise than a typical hunter-gatherer, or if we inhale chemical fumes that were not present in the evolutionary environment, then we are likely to experience ill health.

The evolutionary health principle is of enormous relevance to the health professions, public health policies and personal lifestyle choices. However, it is seldom mentioned in the medical literature.[7]

There are good reasons for believing that the evolutionary health principle applies not only to such physical health needs as clean air and the need for physical exercise, but also to psychosocial aspects of life conditions. For example the lives of hunter-gatherers are usually characterised by the experience of conviviality, effective emotional support networks, incentives and opportunities for creative behaviour and a sense of personal involvement in daily activities. Most of us would agree that such conditions are likely to promote health and well-being in our own society. It is important that we take them into account in assessing the quality of life today and in considering options for the future.[8]

In this context, something must be said about the concept of stressors and meliors. The term 'stressor' is commonly used for an experience that causes anxiety and distress. When stressors are excessive and persistent they can interfere seriously with both mental and physical health. During our work on the ecology of Hong Kong, we became aware of the immense importance of experiences which have the opposite effect to stressors, and which are associated with a sense of enjoyment. We decided to call such experiences *meliors*.

The well-being of individuals at any particular time can be seen to be largely a function of their position on a hypothetical continuum between a state of distress at one extreme and a sense of well-being at the other. While stressors tend to push the individual towards a state of distress, meliors push in the opposite direction, so that a person's position on the continuum is the outcome of the balance between stressors and meliors. Social changes that result in the erosion of meliors are therefore just as undesirable as those that result in an increase in stressors.

There is nothing particularly original about the melior-stressor concept. It is no more than everyday common sense. However, in academic discussion and research, much more emphasis has been placed on stressors than on the opposite kinds of experience. Giving them the name 'meliors' simply serves to remind us to take them properly into account in assessing existing conditions or options for the future.

One of the features of ecological Phase 4 society today is the fact that the achievement of meliors is frequently much more costly, in terms of energy and resources, than it was in the past. The pursuit of meliors makes a substantial contribution to a society's technometabolism (see below).

---

[7] An exception is Cleave and Campbell (1966), who drew attention to the fact that diets containing refined carbohydrates deviated from the natural diet of the human species and consequently gave rise to various forms of maladjustment.

[8] Working lists of the universal health needs of humans, both physical and psychosocial, based on this principle are available on www.biosensitivefutures.org and in Boyden (1987, 2004).

In summary, many cases of ill health in our society today are examples of phylogenetic maladjustment – including most cases of lung cancer, coronary heart disease, obesity and probably much mental depression.

### 7.3.6 Human Behaviour

The fact that humans are basically the same animal today, genetically, as their hunter-gatherer ancestors of, say, 15,000 years ago also has relevance to human behaviour.

The innate behavioural characteristics of our species are the outcome of evolution in an environment very different from that in which we now live. While it can be assumed that these innate behavioural characteristics, such as the capacity for culture, were of biological advantage under the conditions in which they evolved, it is questionable whether this is still the case in the modern setting.

This is an extremely important topic; but because it is complicated and extraordinarily controversial it is not feasible to discuss it further in this short essay (for a discussion see Boyden 2004, Chap. 6).

### 7.3.7 Biometabolism and Technometabolism

An important aspect of biohistory is the study of changing patterns of resource and energy use and waste production by human populations.

Any population of living organisms takes up nutrients and energy from its environment, makes use of them in the processes of life and then discharges wastes and gives off energy in the form of heat. This set of processes is referred to as *population metabolism.*

In the case of the human species, cultural evolution has led to an extra dimension to population metabolism. Thus, in addition to a population's *biometabolism,* which consists of the inputs, internal uses and outputs of energy and materials involved in the biological processes within human bodies, there is also a significant *technometabolism,* which consists of the inputs, uses and outputs of energy and materials resulting from technological processes taking place outside human bodies. Technometabolism is a new phenomenon in the history of life on Earth – of tremendous significance ecologically and in many other ways.

Already in the hunter-gatherer phase of human existence technometabolism became important through the regular use of fire. This development resulted in biologically significant changes in the life conditions of humans, not only by providing them with warmth but also because it led to the consumption of cooked foods, especially meat.

The use of fire by hunter-gatherers sometimes resulted in important ecological changes. In some regions it resulted in the replacement of large areas of woodland with grassland and in big increases in herds of grazing animals, and consequently in the supply of animal protein for humans (Dimbleby 1972; Sands 2005). Fires resulting from human activities had a major impact on vegetation in parts of Australia long before the European invasion of the continent (Jones 1969).

Massive intensification of technometabolism has become an outstanding feature of human society during the fourth high consumption ecological phase of human history, involving a huge surge in resource and energy use and technological waste production. The most evident manifestation of this is anthropogenic climate change – but there are many others.[9]

In 1965, Abel Wolman introduced the concept of urban metabolism and described the metabolism of a hypothetical city of one million inhabitants (Wolman 1965). In the 1970s, studies were carried out on the metabolism of Tokyo (Hanya and Ambe 1976), Brussels (Duvigneaud and Denaeyer-De Smet 1977) and Hong Kong (Newcombe et al. 1978). The last project, which involved a detailed analysis of both technometabolism and biometabolism in an urban system was carried out as part of a broad study on the ecology of Hong Kong and its human population.

In the final report of this work on Hong Kong, attention was drawn to the long-term unsustainability of the ever increasing intensity of resource and energy use and waste production in this city (Boyden et al. 1981). This conclusion is shared by the authors of a more recent study of the metabolism of Hong Kong who write:

> *Per capita* food, water and materials consumption have surged since the 1970s by 20%, 40%, and 149%, respectively. Tremendous pollution has accompanied this growing affluence and materialism, and total air emissions, $CO_2$ outputs, municipal solid wastes, and sewage discharges have risen by 30%, 250%, 245%, and 153%. As a result, systemic overload of land, atmospheric and water systems has occurred. While some strategies to tackle deteriorating environmental quality have succeeded, greater and more far-reaching changes in consumer behaviour and government policy are needed if Hong Kong is to achieve its stated goal of becoming 'a truly sustainable city" in the 21st century. (Warren-Rhodes and Koenig 2001).

Perspectives such as these are crucially significant for our understanding of the true nature of the human predicament today and for planning for sustainability.

Since the 1970s there has been much work on urban metabolism, all of it indicating a progressive increase in the intensity of urban metabolism (Kennedy et al. 2007).

## 7.3.8 Technoaddiction

Another important biohistorical concept is the principle of technoaddiction. In human history it has frequently been the case that new techniques have been introduced simply for curiosity, or sometimes because they have benefited a particular individual or group within society. However, with the passing of time societies have

---

[9] In our work we have described and discussed the technometabolism of Hong Kong (Newcombe et al. 1978; Boyden et al. 1981), Australia (Boyden et al. 1990) and the world (Boyden 1992).

organised themselves around the new techniques and their populations have become progressively more and more dependent on them for the satisfaction of simple, basic needs. Eventually a state of complete dependence is reached.

The dependence of the populations of high-energy societies on fossil fuels is an obvious and extremely serious example. Others include our dependence on electricity and, quite recently, on computer technology.

This insidious form of addiction passes largely unnoticed. It is of immense economic and ecological significance and it explains why our attempts to introduce effective measures to overcome anthropogenic climate change are fraught with so many difficulties.

It is noteworthy that in the present cultural setting the following basic human behaviours usually require significantly more energy and create much more pollution than they did at other times in history: eating; seeking in-group approval; seeking to conform; seeking novelty, excitement and comfort; visiting relatives; being selfish; being greedy and being generous.

## 7.3.9 Cultural Maladaptations Today

Biohistory helps us to appreciate that the worldview and assumptions of our dominant culture today are resulting in cultural maladaptations on a scale and of an intensity never seen before in the history of humankind – maladaptations that are totally incompatible with the survival of civilisation. All the main threats to human wellbeing and survival in the modern world are consequences of cultural maladaptations.

Biohistory also draws attention to the astonishing rate of acceleration in the increase in intensity of humans activities on Earth, and to the fact that very recently *Homo sapiens* has become the first species of animal in the history of life on Earth to bring about significant changes in the ecology of the whole planet.

There are two sets of changes underlying the major ecological difficulties facing humankind today:

- The huge increase in the human population. There are now about 1,000 times as many people on Earth as there were when our ancestors first started farming around 450 generations ago. 70% of this increase has occurred in the past 80 years.
- The massive intensification, especially in the developed countries, of energy and resource use and technological waste production associated with industrialisation, consumerism and economic growth.[10] The human species is now using about 18,000 times as much energy and emitting about 9,000 times as much $CO_2$ as was the case when our ancestors started farming around 10,000 years ago. 90% of this increase has occurred in the past 80 years.

---

[10] Figures for energy use provide a fair indication of the overall impact of humans on the biosphere. People in some of the developed countries today are using around 50 times as much energy per capita as was the case when farming began. Most of this increase has occurred very recently.

Currently the most critical sign of this insensitive over-exploitation of the planet's resources is rapid global climate change. Other areas of serious concern include massive loss of biodiversity on land and in the oceans, thinning of the ozone layer, global pollution of ecosystems with persistent organic pollutants, and various severe forms of land and water degradation – involving distortion of nutrient cycles, loss of topsoil, salinisation, progressive large scale deforestation, biological impoverishment of soil and acidification of the oceans.[11]

The biosphere as a system capable of supporting civilisation will not tolerate this onslaught indefinitely. If present trends in human activity continue unabated the ecological collapse of human civilisation is inevitable.[12]

Apart from these ecological issues, cultural developments during the past 70 years have resulted in the manufacture of weapons of mass destruction which now constitute another horrendous threat to the future of our species and the rest of the biosphere. According to recent estimates, there are around 24,000 nuclear warheads in existence. It would not take many of these to bring an end to civilisation.

Biohistory also shows us how cultural evolution has resulted in the current gross disparities in conditions of life across human populations. Today vast swards of people live in abject poverty, while some individuals have incomes of millions of dollars a year. Such disparities have been common in the early urban and high consumption ecological phases of human history, but they were not a feature of societies in the preceding 190,000 years of human existence.

## 7.4 Hope for the Future

Biohistorical understanding leads to an appreciation that the best hope for humankind lies in a rapid transition to a society that is really in tune with and sensitive to the processes of life – a society that satisfies the health needs of all sections of the human population as well as those of the ecosystems of the biosphere. My colleagues and I call this a *biosensitive society* – that is, a society that in tune with our own biology and in tune with the living systems of the biosphere on which we depend.[13]

---

[11] See Rockström et al. (2009) for an interesting discussion of the full range of interlinked ecological changes resulting from human activities that are causes for serious concern. These authors recognise nine interlinked 'planetary boundaries', three of which have already been transgressed – namely climate change, rate of biodiversity loss and interference with the nitrogen and phosphorus cycles.

[12] In 1992 over 1,500 members of the Union of Concerned Scientists (UCS), including 101 Nobel Prize winners, issued a statement entitled *World's scientists' warning to humanity*. The following extract from the press release that accompanied the publication of this statement summarises their position: "The scientists emphasise the urgency of the problem. As they note in their appeal, 'No more than one or a few decades remain before the chance to avert the threats that we now confront will be lost and the prospects for humanity immeasurably diminished."

[13] For further discussion on biosensitivity see Boyden (2005, 2011) and www.natsoc.org.au/biosensitivefutures.

**Fig. 7.3** Biosensitivity triangle

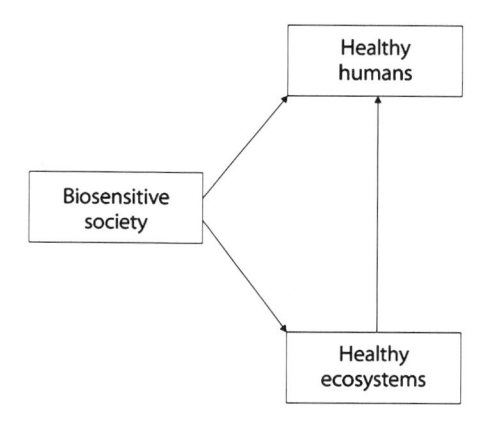

[Diagram: "Biosensitive society" box with arrows pointing to "Healthy humans" and "Healthy ecosystems", which are connected to each other.]

Some explanation is needed of why it was felt necessary to coin the words 'biosensitive' and 'biosensitivity':

The growing concern about our ecological predicament over the past few decades has resulted in a range of important new expressions coming into use. They include, for example, ecological sustainability, environmentalism, carbon footprint and being green.

However, there is a need for a broader, more inclusive term which encompasses both human and ecological wellbeing (Fig. 7.3) and which evokes a positive vision of a society that is based on a real understanding of the living world and the human place in nature and that is truly in tune with the processes of life within us and around us.

The biosensitive society will promote health in all sections of the human population and in the ecosystems of the natural environment.

The transition to a biosensitive society will require sweeping changes in the intensity and nature of human activities, in economic arrangements and in the occupational structure of society. Biosensitivity will be the guiding principle in all spheres of human activity – individual and collective. It will mean biosensitive lifestyles, biosensitive governments, biosensitive technologies and fuel use, biosensitive farming, biosensitive cities, biosensitive design, and a biosensitive economy. Eventually it will also mean moving towards a smaller human population globally (1,000 million?).

Unfortunately the worldview, priorities and assumptions of the dominant cultures that determine patterns of human activity across the world today are totally incompatible with any transition to an ecologically sustainable, healthy and equitable society. They are simply not attuned to ecological realities.

Paramount among the maladaptive assumptions of the dominant culture of our own society is the ideology of 'ever-moreism' – associated with an economic system that results in rampant, continually increasing consumption of resources, use of energy and discharge of technological wastes. This ideology is ecologically absurd. It is leading us faster and faster in the direction of ecological oblivion.

The necessary changes in societal arrangements and patterns of human activity will therefore require revolutionary changes in the dominant culture. That is, our hope for the future lies in the processes of cultural reform. Biosensitivity cannot be achieved until this culture comes to embrace at its heart a sound understanding of the human place in nature and a profound respect for the processes of life. Biosensitivity will be what matters most.

Only then will there be sufficient motivation at all levels of society to make the major changes in societal arrangements and human activities that will be necessary to achieve a sustainable relationship with the ecosystems of which we are a part and on which we depend.

However, in turn, this crucial cultural transformation will not come about until there is widespread understanding right across the community of human situations in biohistorical perspective. Therefore by far the most urgent need is in the realm of learning and education.

Apart from its influence on the cultural worldview, assumptions and priorities, of the dominant culture, biohistory also provides information of enormous practical value for society in its efforts to achieve biosensitivity. It makes clear what 'being in tune with the processes of life' means in practical terms, such as maintaining the biological integrity of soils and natural nutrient cycles, protecting biodiversity and avoiding pollution of the atmosphere. Moreover, it helps us to select lifestyle options that not only promote our own health, but that are also consistent with the health of the natural environment.

So, in summary, shared biohistorical understanding across the whole community is a key prerequisite for the achievement of biosensitivity and hence the survival of civilisation.[14] Until this happens there is unlikely to be any significant change in the dominant culture and therefore no significant move towards sustainability and biosensitivity.

Biohistory should be at the core of every school curriculum – reflecting the reality that we are living beings, products of the processes of life and totally dependent on them for our survival and wellbeing; and by far the most useful role of concerned individuals and community groups at the present time is to actively encourage this kind of understanding in the community and to promote the vision of a biosensitive society.

## 7.5  Biohistory and the Academic Disciplines

Biohistory is, by its very nature, integrative. It requires that we pay attention to the interconnectedness of different parts of the total system – biophysical and cultural – and of the broad classes of variables that determine the overall characteristics of human situations and the life experience of every one of us.

---

[14] Elsewhere, for the sake of brevity, 'biohistorical understanding' has been contracted to 'biounderstanding' (www.natsoc.org.au/biosensitivefutures and Boyden 2011).

Thus biohistory can be said to be 'multidisciplinary' – in that it involves learning about the interplay between different parts of the total system that are conventionally studied by different groups of specialists in the different so-called academic disciplines.[15]

However, it can be argued that biohistory, rather than being seen as 'multidisciplinary', should be regarded as an academic discipline in its own right – but a comprehensive one that one that has crucial links with most, if not all, other disciplines. I suggest that it deserves a place alongside the various fields from the natural and social sciences, such as ecology, ecological anthropology and ecological economics, that are considered to have the potential to make a useful contribution to long-term socio-ecological research (LTSER) (Singh et al. 2010).

The existence of these disciplines, each focusing on a relatively narrow aspect of reality and each with its own set of methods and theory, is an outcome of the vicissitudes of cultural evolution. In fact, of course, the variables and processes studied in these different areas of specialism are interacting parts of a whole, and the interplay between them is of utmost significance for understanding ourselves, our society and our problems. Especially important are the interactions between the culturally inspired human activities and the underpinning processes of life within us and around us.

While there have been increasing calls for multidisciplinarity in academia over recent years, there has not been a great deal of progress in achieving this goal. In my opinion, this is partly due to the fact that it is not sufficient merely to bring together representatives of different disciplines to sit around a table to talk about an issue or topic, only to return afterwards to the security of their own particular academic silos. We need more people who stay at the table and whose full-time professional interest is the interplay in the system between the different sets of components and processes.

Certainly there has been much important work aimed at developing a systems approach to the study of human situations.[16] However, it seems to me that much of this work lacks a sound conceptual base that reflects either the total dependence of human society on the underpinning processes of life or the crucial role of human culture as a determinant of the health and wellbeing of people or of the ecosystems on which they depend.

It is worth noting that social scientists have been very wary of biology ever since they had their fingers burned by social Darwinism. Yet to ignore the life processes which underpin, permeate and make possible all social situations makes no sense. We cannot hope to understand what is going on if we neglect the interplay between this fundamental dimension of the system and the cultural and socio-cultural components.

---

[15] The expressions 'multidisciplinary', 'transdisciplinary' and 'interdisciplinary' are now in common usage, and they have slightly different meanings. However, here 'multidisciplinary' is used to cover all three meanings.

[16] See, for example www.complexsystems.net.au/wiki/Complex_Dynamics_of_Urban_Systems.

In my view the biohistorical framework provides a good starting point for developing a logical conceptual approach to the integrative study of human situations. This is partly because:

- it recognises that the whole social system is life-driven and life-dependent
- it provides a framework for investigating and understanding the interplay between the parts of the system that are conventionally studied by different groups of specialists in the life sciences, social sciences and humanities
- it appreciates the significance of the evolutionary perspective for understanding current situations

# Appendix

## *A Transition Framework*

This Appendix introduces a framework designed to facilitate thinking and communicating about the ecological and health implications of different options for the future. It recognises the crucial role of human culture in the system, and it is based on biohistorical principles discussed in this chapter.

The transition framework is depicted in Fig. 7.4. It is basically an extended version of the 'biosensitivity triangle' (Fig. 7.3) and it also incorporates some of the features of the 'biohistorical framework' depicted in Fig. 7.2.

## *Human Health Needs and Ecosystem Health Needs*

In the biosensitivity triangle (Fig. 7.3), the two boxes on the right-hand side are Healthy people and Healthy ecosystems, which are our ultimate goals in planning for a biosensitive future. However, from the planner's standpoint what is actually more relevant are the immediate requirements for health (e.g. clean air and water for human health, and maintaining biodiversity and soil fertility for ecosystem health). These health requirements are called Human health needs and Ecosystem health needs in Fig. 7.4.

Options for the future must be assessed ultimately in terms of their impacts on these health needs. Boxes 7.1 and 7.2 are working check lists of important health needs of humans and of ecosystems respectively.

## *Biophysical Environment*

This set of factors has been inserted into the triangle because the impacts of human activities on the health needs of people and ecosystems are sometimes indirect, in that the ultimate effect on health is the result of changes brought about in the biophysical environment.

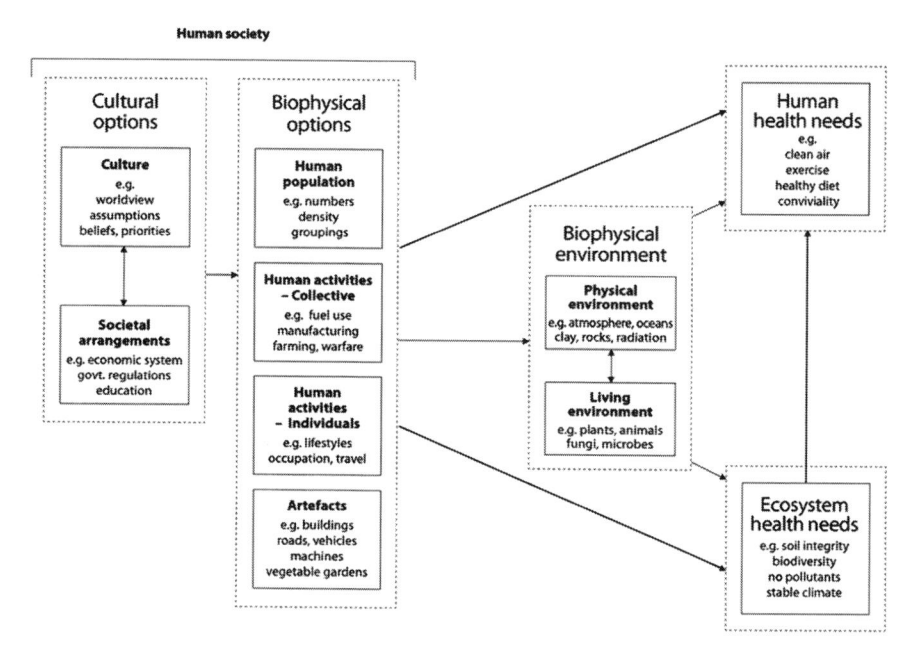

**Fig. 7.4** The transition framework

For example, the human activities that result in the release into the environment of CFCs lead to chemical reactions in the atmosphere and the destruction of ozone in the stratosphere. This change in turn results in an increase in the ultraviolet radiation at the Earth's surface, which interferes with the health both of ecosystems and of humans.

Another example is provided by cases when the application of artificial fertilisers to farmland leads to eutrophication in creeks and rivers. The consequent excessive growth of algae results in anoxia in the aquatic ecosystem and then to loss of biodiversity and also to the production of toxins which can cause illness, even death, in humans and other large animals.

On the other hand, of course, many undesirable impacts of human activities on human and ecosystem health are direct – such as the effects of tobacco smoking on human health and the effects of oil spills on local fauna.

Recognising the crucial role of culture in determining the health both of humans and of the ecosystems on which they depend, *Human society* has been divided into two categories: *Biophysical options* and *Cultural options*.

## Biophysical Options

Biophysical options include the biological and physical aspects of human situations that can be influenced by people's decisions and that directly or indirectly affect the all-important health needs of humans and ecosystems.

**Box 7.1 Human Health Needs**[17]

*Physical*
Clean air
Clean water
Healthy (natural) diet
Healthy (natural) physical activity
Noise levels within the natural range
Minimal contact with microbial or metazoan parasites and pathogens
Natural contact with environmental non-pathogenic microbes
Electromagnetic radiation at natural levels
Protection from extremes of weather

*Psychosocial*
Emotional support networks
Conviviality
Co-operative small-group interaction
Creative behaviour
Learning and practising manual skills
Recreational activities
Variety in daily experience
Sense of personal involvement/purpose
Sense of belonging
Sense of responsibility
Sense of challenge and achievement
Sense of comradeship and love
Sense of security

**Box 7.2 The Health Needs of Ecosystems**

In light of our knowledge of the effects of various human activities on ecosystem health at the present time, we can put together a check list of ecosystem health needs, as follows:

- The absence of polluting gases or particles in the atmosphere which significantly disrupt natural cycles and processes and change the climate
- The absence of polluting gases or particles in the atmosphere which interfere with living processes (e.g. particulate hydrocarbons from combustion of diesel fuel, sulphur oxides)

(continued)

---

[17] This working list of human health needs is based on the evolutionary health principle and our knowledge of the conditions of life in the long natural, or hunter-gatherer phase of human existence.

> **Box 7.2** (continued)
> - The absence of substances in the atmosphere (e.g. CFCs) that result in destruction of the ozone layer in the stratosphere that protects living organisms from the ultraviolet radiation from the sun
> - The absence of chemical compounds in oceans, lakes, rivers and streams in concentrations harmful to living organisms (e.g. persistent organic pollutants – POPs)
> - No ionising radiation that can interfere with the normal processes of life and photosynthesis
> - The absence of chemical compounds in the soil that can interfere with the normal processes of life (e.g. persistent organic pollutants, heavy metals)
> - Soil loss no greater than soil formation (i.e. no soil erosion)
> - No increase in soil salinity and soil sodicity
> - The maintenance of the biological integrity of soil (i.e. maintaining a rich content of organic matter)
> - Intact nutrient cycles in agricultural ecosystems over long periods of time (requiring return of nutrients to farmland)
> - The maintenance of biodiversity in regional ecosystems (including aquatic ecosystems)

Biophysical options are subdivided into four sub-categories:

- *Human population* – such as numbers of people, population density, population age structure
- *Human activities – collective* – such as manufacturing, farming, military activities and transportation.
- *Human activities – individuals* – such as lifestyle options, travel patterns, physical exercise and consumer behaviour
- *Artefacts*[18] – such as buildings, roads, machines, vehicles and furniture.

## Cultural Options

Human activities are to a large extent governed by *Societal arrangements,* such as the prevailing economic system, governmental regulations and the institutional structure of society.

These social arrangements are in turn determined by the worldview, assumptions and priorities of the dominant *Culture*.

---

[18] Artefacts is used to mean 'things made by humans'.

For example, the cultural assumption that the best thing for our society is continuing economic growth, involving ever-increasing use of resources and energy, is a major factor affecting governmental economic policies, and consequently influencing human activities and, ultimately, the health of our planet's ecosystems.

## *Making Use of the Transition Framework*

The transition framework emphasises the fact that the ultimate objective in planning for a biosensitive society is the health both of humans and of the ecosystems on which they depend.

The framework provides a useful starting point for assessing policy options for the future – from the level of individuals and families through to the level of national governments.

In our own work we have made use of the framework to construct a check list of the essential changes that will be necessary in different parts of the total system for the achievement of biosensitivity (see www.natsoc.org.au/biosensitivefutures/vision).

## References

Boyden, S. (1973). Evolution and health. *The Ecologist, 3*, 304–309.
Boyden, S. (1987). *Western civilization in biological perspective: Patterns in biohistory*. Oxford: Oxford University Press.
Boyden, S. (1992). *Biohistory: The interplay between human society and the biosphere – Past and present*. Paris: Parthenon/UNESCO.
Boyden, S. (2004). *The biology of civilisation: Understanding human culture as a force in nature*. Sydney: UNSW Press.
Boyden, S. (2005). *People and nature: The big picture*. Canberra: Nature and Society Forum.
Boyden, S. (2011). *Our place in nature, past present and future*. Canberra: Nature and Society Forum. See also www.natsoc.org.au/biosensitivefutures/vision.
Boyden, S., Millar, S., Newcombe, K., & O'Neill, B. (1981). *The ecology of a city and its people: The case of Hong Kong*. Canberra: Australian National University Press.
Boyden, S., Dovers, S., & Shirlow, M. (1990). *Our biosphere under threat: Ecological realities and Australia's opportunities*. Melbourne: Oxford University Press.
Carson, R. (1962). *Silent spring*. Boston: Houghton Mifflin.
Cleave, D. L., & Campbell, G. R. (1966). *Diabetes, coronary thrombosis and the saccharine disease*. Bristol: John Wright.
Cochran, G., & Harpending, H. (2009). *The 10,000 year explosion: How civilisation accelerated human evolution*. New York: Basic Books.
Diamond, J. (1997). *Guns, germs, and steel: A short history of everybody for the last 13 000 years*. London: Jonathan Cape.
Diamond, J. (2005). *Collapse: how societies choose to fail or survive*. New York: Viking.
Dimbleby, G. W. (1972). Impact of early man on his environment. In P. R. Cox & J. Peel (Eds.), *Population and pollution* (pp. 7–13). London: Academic.

Dubos, R. (1968). *So human an animal*. New York: Charles Scribner's Sons.
Dubos, R. (1980). *Celebrations of life*. New York: McGraw Hill.
Dunbar, R. (1997). *Grooming, gossip, and the evolution of language*. Cambridge, MA: Harvard University Press.
Duvigneaud, P., & Denaeyer-De Smet, S. (1977). L'ecosysteme urbain bruxellois. In P. Duvigneaud & P. Kestemont (Eds.), *Productivité en Belgique* (Traveaux de la Section Belge du Programme Biologique International). Brussels, Paris: Edition Duculot.
Flannery, T. F. (1994). *The future-eaters: An ecological history of the Australian lands and people*. Melbourne: Reed Books.
Flinn, M. W. (Ed.) (1965). *Report on the sanitary conditions of the labouring population of Britain (1842)*. Author E. Chadwick. Edinburgh: Edinburgh University Press.
Frazer, W. M. (1950). *A history of English public health*. London: Balliere, Tindall and Cox.
Haberl, H., Winiwarter, V., Andersson, K., Ayres, R.U., Boone, C., Castillo, A., Cunfer, G., Fischer-Kowalski, M., Freundenberg, W. R., Furman, E., Kaufmann, R., Krausmann, F., Langthaler, E., Lotze-Campen, H. E., Mirtl, M., Redman, C. L., Reenbeg, A., Wardell, A., Warr, B., & Zechmeister, H. (2006). From LTER to LTSER: Conceptualising the socio-economic dimension of long-term socioecological research. *Ecology and society, 11*. (Online) http://www.ecologyandsociety.org/vol11/iss2/art13/
Hanya, T., & Ambe, Y. (1976). A study on the metabolism of cities. In HESC Council of Japan (Ed.), *Science for a better environment* (pp. 228–233). Kyoto: HESC Council of Japan.
Hodder, I. (2006). *The leopard's tale: Revealing the mysteries of Çatalhöyük*. London: Thames and Hudson.
Jones, R. (1969). Fire-stick farming. *Australian Natural History, 16*, 224.
Kennedy, C., Cuddihy, J., & Engel-Yan, J. (2007). The changing metabolism of cities. *Journal of Industrial Ecology, 11*, 43–59.
McDougall, I., Brown, F. H., & Fleagle, J. G. (2005). Stratigraphic placement and age of modern humans from Kibish, Ethiopia. *Nature, 433*, 733–736.
McMichael, A. (2001). *Human frontiers, environments and disease: Past patterns and uncertain futures*. Cambridge: Cambridge University Press.
Mellaart, J. (1967). *Çatal Hüyük: A Neolithic town in Anatolia*. London: Thames and Hudson.
Newcombe, K., Kalma, J. D., & Aston, A. R. (1978). The metabolism of a city: The case of Hong Kong. *Ambio, 7*, 3–13.
Oreskes, N., & Conway, E. M. (2010). *Merchants of doubt: How a handful of scientists obscured the truth on issues from tobacco smoke to global warming*. London: Bloomsbury Press.
Rendel, J. M. (1970). The time scale of genetic change. In S. Boyden (Ed.), *The impact of civilisation on the biology of man* (pp. 27–47). Canberra: Australian National University Press.
Rockström, J. W., Steffen, W., Noorie, K., Persson, Å., Chapin, F. S., Lambin, E. F., Lenton, T. M., Scheffer, M., Folke, C., Schellnhuber, H. J., Nykvist, B., de Wit, C. A., Hughes, T., van der Leeuw, S., Rodhe, H., Sörlin, S., Snyder, P. K., Costanza, R., Svedin, U., Falkenmark, M., Karlberg, L., Corelli, R. W., Fabry, V. J., Hansen, J., Walker, B., Liverman, D., Richardson, K., Crutzen, P., & Foley, J. A. (2009). A safe operating space for humanity. *Nature, 461*, 472–475.
Sands, R. (2005). *Forestry in a global context*. Cambridge, MA: CABI Publishing.
Singh, S. J., Haberl, H., Gaube, V., Grünbühel, C. M., Lisivieveci, P. J., Lutz, P., Matthews, R., Mirtl, M., Vadineanu, A., & Wildenberg, M. (2010). Conceptualising long-term socio-ecological research (LTSER): integrating the social dimension. In F. Müller et al. (Eds.), *Long-term ecological research* (pp. 377–398). Heidelberg: Springer.
Warren-Rhodes, K., & Koenig, A. (2001). Escalating trends in the urban metabolism of Hong Kong: 1971–1997. *Ambio, 30*, 429–443.
Wolman, A. (1965). The metabolism of cities. *Scientific American, 213*, 179–190.
Zinsser, H. (1935). *Rats, lice and history*. London: George Routledge.

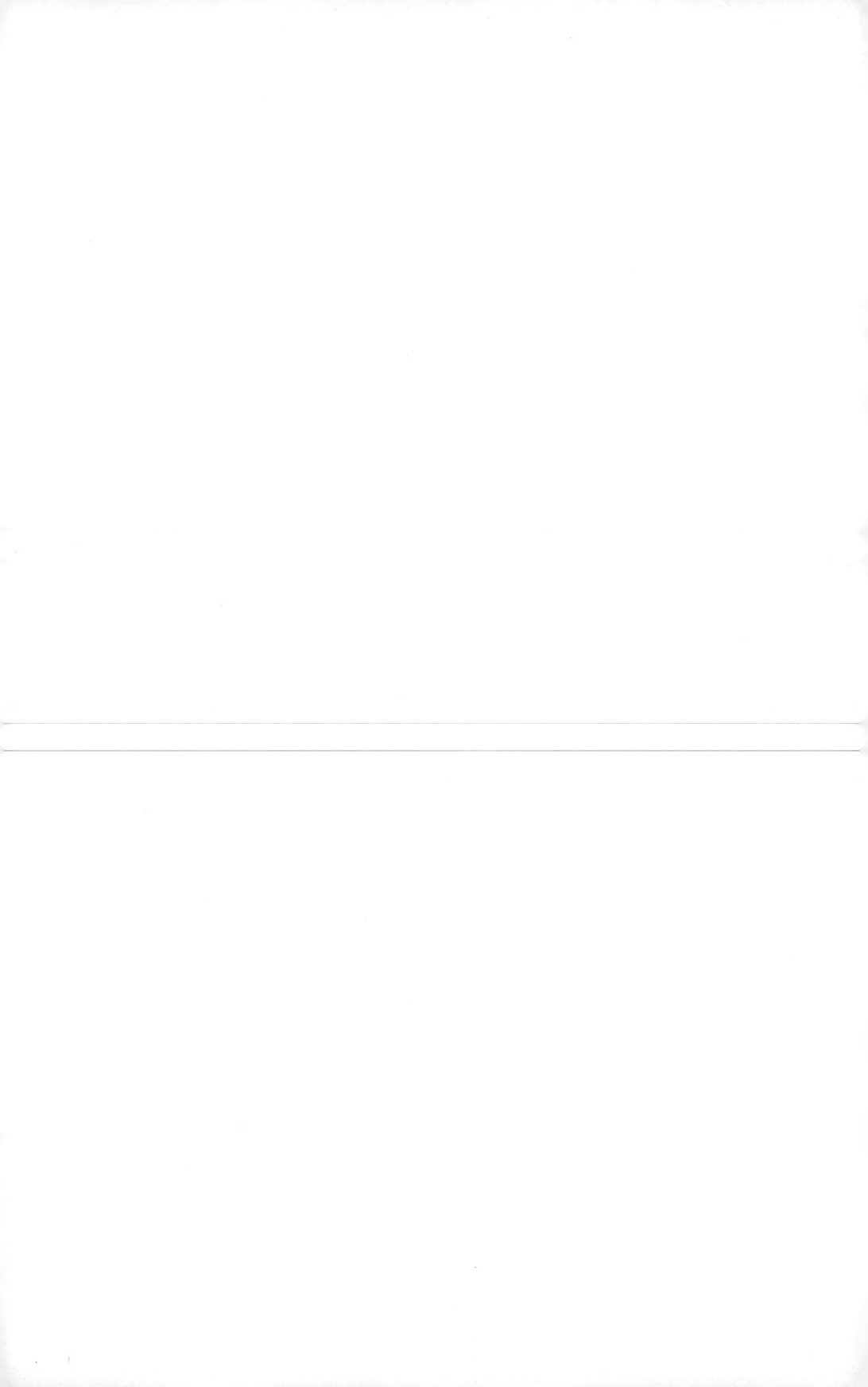

# Chapter 8
# Geographic Approaches to LTSER: Principal Themes and Concepts with a Case Study of Andes-Amazon Watersheds

**Karl S. Zimmerer**

**Abstract** Analysis of current works and trends in geography is conducted, with emphasis on human-environment and nature-society geography (HE-NS) in order to identify areas of conceptual overlap, promising exchange, and potential collaboration with Long-Term Social-Ecological Research (LTSER). HE-NS geography resembles the defining focus of LTSER on the coupled interactions of human societies and environments. Important conceptual connections to LTSER are identified as follows: (i) Coupled Human-Environment Interactions; (ii) Sustainability Science, Social-Ecological Adaptive Capacity, and Vulnerability; (iii) Land-Use and Land-Cover Change (LUCC) and Land Change Science (LCS); (iv) Environmental Governance and Political Ecology; (v) Environmental Landscape History and Ideas; and (vi) Environmental Scientific Concepts in Models, Management, and Policy. Demonstrated promise and potential value of conceptual "points of contact" exist in each of these areas of HE-NS geography and LTSER. Concepts of spatial and temporal scale, human-environment and nature-society interactions, multi-scale and networked spatiotemporal designs, and socio-ecological science theories and methodologies offer specific examples of the bridges between HE-NS geography and LTSER in interdisciplinary environmental studies and policy, a case study of Andean watersheds in the upper Amazon basin, and conclusions.

**Keywords** Human-Environment Geography • Nature-Society Geography • Political Ecology • Environmental Interdisciplinarity • Andes-Amazon Watershed Management • Long-Term Socio-ecological Research • Geographic Monitoring • Land Change Science • Coupled Human-Natural Systems • Human Ecology • Cultural Ecology

K.S. Zimmerer, Ph.D. (✉)
Department of Geography, Earth and Environmental Systems Institute,
Pennsylvania State University, University Park, PA, USA
e-mail: ksz2@psu.edu

## 8.1 Introduction to Geographic Approaches

Long-Term Social-Ecological Research (LTSER) is centred on the changes of coupled socioecological and human-environment systems. In order to understand such changes the approach of LTSER is faced with challenges around four themes – human-environment interaction, scale, spatial generalisability, and social-ecological modelling (Haberl et al. 2006: 3; see also Redman et al. 2004; Singh et al. 2010). In this respect LTSER and its defining challenges are similar to contemporary geography as described, for example, in the recent 155-page report of the U.S. National Resources Council entitled *Understanding the Changing Planet: Strategic directions for the geographical sciences* (NRC 2010). Building on these indications of potentially promising connections, this chapter advances an analysis of LTSER and contemporary geography that argues in favour of their rich and potentially vigorous, albeit previously overlooked collaborations.

Within contemporary geography, the subfield of human-environment and nature-society geography, referred to here as HE-NS, offers a particularly close correspondence to LTSER. HE-NS geography is distinguished by a focus on human-environment interactions and nature-society relations (Zimmerer 2010c). It requires integration of the environmental sciences, blending biogeophysical and social science approaches, in order to examine interconnected and recursive HE-NS interactions and relations, with substantive and often bi-directional influences across the realms of nature and society. As a distinct sub-field, HE-NS is a cornerstone of the quadripartite approach of current four-field geography – a common academic, intellectual, and institutional structure – with the other fields being human geography, physical geography, and GIScience/cartography (Fig. 8.1; Zimmerer 2007).[1] HE-NS geography closely resembles LTSER's defining focus on the "structurally coupled" interactions of human societies and their environments (Haberl et al. 2006: 5) and, also, LTSER's consideration and incorporation of social science concepts that "explain social processes without denying biophysical processes" (Singh et al. 2010: 385). The HE-NS-based frame of this study should be seen as complementary to related geographic perspectives (e.g. spatial data infrastructure and human-environmental observatories of the HERO project; Yarnal et al. 2009) and richly institution-centred approaches (e.g., Vajjhala et al. 2007).

The goal of this study is to undertake the systematic identification of a conceptual "common ground" of LTSER, HE-NS geography, and several related and important

---

[1] This four-field organisation is reflected in the editorial structure of the *Annals of the Association of American Geographers*, the flagship journal of the world's largest geographical association, which distinguishes a quartet of corresponding sections (Zimmerer 2010c). The *Annals* structure grew out of trenchant arguments and debate advancing HE-NS as one of the predominant identities of contemporary geography (Butzer 1990; Kates 1987; Turner 1989, 2002; Zimmerer 2010c). Other geographic fields also promise vitally important contributions to LTSER and allied approaches—and indeed ones that are already recognised—such as the data infrastructure advances and other pioneering contributions of the Human-Environment Regional Observatory (HERO) project and its "collaboratories" (Yarnal et al. 2009).

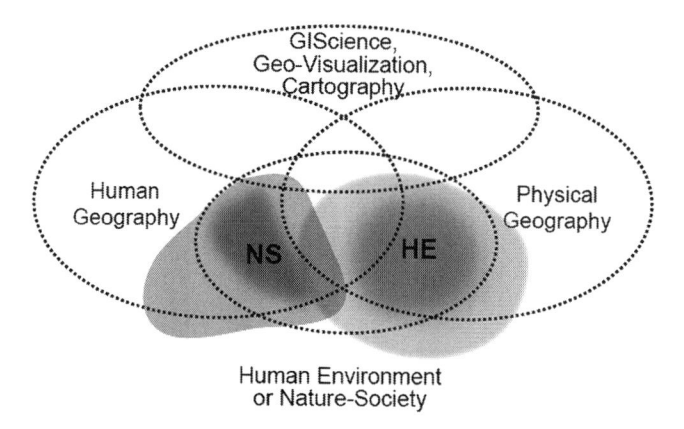

**Fig. 8.1** Visualisation of the principal intellectual spaces of Nature-Society Geography

interdisciplinary approaches. Section 8.2 undertakes the analysis of potential "open points of contact" to LTSER through the drawing of connections to several key concepts, along with concrete examples, that have been developed within the six principal thematic areas of HE-NS: (i) Coupled Human-Environment Interactions; (ii) Sustainability Science, Social-Ecological Adaptive Capacity, and Vulnerability; (iii) Land-Use and Land-Cover Change (LUCC) and Land Change Science (LCS); (iv) Environmental Governance and Political Ecology; (v) Environmental Landscape History and Ideas; and (vi) Environmental Scientific Concepts in Models, Management, and Policy.[1]

At the outset it is important to note that the thematic areas identified above show correspondence to a pair of partially distinct epistemic sub-categories, one associated with human-environment interactions (HE) and the other with nature-society relations (NS) (Zimmerer 2010c), as sketched in Fig. 8.1 and detailed in Table 8.1. Human-environment interactions (HE) is comprised most notably of coupled human-environment interactions; sustainability science, social-ecological adaptive capacity, and vulnerability; and land-use and land-cover change (LUCC)/land change science (LCS). Nature-society relations (NS), as the other sub-category, consists chiefly of political ecology and environmental governance; environmental landscape history and ideas; and environmental scientific concepts in models, management, and policy. The partial distinctness of these sub-categories owes to the noticeable sharing and overlap of certain ideas (e.g., environmental governance). It also owes to the interpretation of potential conceptual co-existence, in addition to scholarly debate and contestation, that has become vital to understanding the increased epistemological range of contemporary HE-NS geography (Zimmerer 2007; Turner 2009; Turner and Robbins 2008).

Analysis in the following sections is focused on a group of key themes and related concepts regarding spatial and temporal scale, human-environment and nature-society interaction, and particular insights and applications stemming from environmental social-ecological science theories and methodologies. The study

**Table 8.1** Core themes of this study and the levels of correspondence to principal areas (Human-environment interactions and nature-society relations) of the geographic sub-field (see also Fig. 8.1)

| | Human-environment interactions | Nature-society relations |
|---|---|---|
| 1. Couple human-environment interaction | Moderate-high | Moderate-high |
| 2. Sustainability science, social-ecological adaptive capacity, and vulnerability | High | Moderate |
| 3. Land-use and land-cover change (LUCC) and land change science (LCS) | High | Moderate |
| 4. Environmental governance and political ecology | Moderate | High |
| 5. Environmental landscape history and ideas | Moderate | Moderate |
| 6. Environmental scientific concepts in models, management, and policy | Moderate | High |

then briefly evaluates the roles of environmental interdisciplinarity and policy (Sect. 8.3), a case study of Andean watersheds in the upper Amazon basin (Sect. 8.4), and conclusions (Sect. 8.5).

## 8.2 Themes and Concepts: Geography and LTSER

### *8.2.1 Coupled Human-Environment Interactions*

Coupled social-ecological systems, a core focus of LTSER (Haberl et al. 2006: 1), is also a sustained emphasis of geography and closely related fields such as ecological and environmental anthropology (Bassett and Zimmerer 2003: 99–101; see also Knight 1971; Nietschmann 1972; Grossman 1977, 1981; Watts 1983; Turner 1989, 1997, 2002; Knapp 1994; Zimmerer 1994, 1996; Walters and Vayda 2009; Singh et al. 2010). In LTSER frameworks these coupled interactions are described as "signals of global environmental change and their impacts on ecosystems across the world" (Haberl et al. 2006: 1). This LTSER conceptualisation resembles a stimulus-response model of human-environment interactions, while also it recognizes that the human component is "complex and cannot be treated as an organism with consistent reactions to external stimuli" (Redman et al. 2004: 163). HE-NS geography has developed an orientation that is broadly related; it is illustrated, for example, in the concept of "decision-making in land management" coupling individual and household-level choices to economic and environmental signals (Blaikie and Brookfield 1987: 70). Understanding these complex couplings has been conceptualised as processes operating at multiple, interconnected spatial scales, typically moving from the local to the global, in both geography (e.g., "chains of explanation" in Blaikie and Brookfield 1987: 27 and "event ecology" in Walters and Vayda 2009) and in LTSER (e.g., "focus on multi-scale approaches;" Redman et al. 2004: 167–168). In geography the "chain of explanation" has contained a pair of principal formulations that vary with respect to the conceptualisation of the initial link (see also Rocheleau 2008). This concept's earlier version posits the initial link of the land or resource manager to local social interactions and environmental change (e.g., soil degradation), then to regional-level ones and finally to national and international political economies (Blaikie and Brookfield 1987: 27). Subsequently, in a revised version, it is biogeophysical processes of human-environmental change that are proposed as the actual point of departure in the multi-level scaling of explanations of land-use change (Blaikie 1994).

Numerous NS-HE examples focus on interactive agricultural and development responses to climate change that are differentiated at multiple *interlinked* social-ecological scales, both spatially and temporally (e.g., across seasonal time spans; Liverman 1990; O'Brien and Leichenko 2003; Polsky 2004). Another cluster of relevant examples stems from geographic studies of herding, livestock, and range ecology in Africa (Bassett 1988; Dougill et al. 1999; Benjaminson et al. 2006; Butt et al. 2009). Fine-grain research reveals that coupled interactions of political and ecological changes occur at interlinked local, regional, and global scales (e.g., through the local gender politics of women herders, for example, that is interlinked to regional markets; Turner 1999).

HE-NS geographic contributions have underscored that the influence of economic and political forces, and their fusion via political economy, can be evident at local scales, even while such forces stem conspicuously from national, international, and global scales. This insight has applied and built on the theory of structuration, indicating the embeddedness of *both* agency in individual, household and firm behaviours *and* their influence on "structures" consisting of such factors as market conditions, social power relations and policy parameters. Household-level land-use decisions and participation in political movements, for example, is mutually embedded in government policies about land use in Peru and Mexico (Zimmerer 1991; Chowdhury and Turner 2006; see also Brenner 2011; Laney 2002). This longstanding HE-NS engagement of the structuration concept has enabled the analysis of agricultural intensification and disintensification within environmentally sensitive tropical and sub-tropical environments. The structuration-based perspective on mutually constitutive, multi-scalar processes in HE-NS change suggests the opportunity for productive dialogue regarding LTSER's emphasis on scale (e.g., Haberl et al. 2006: 6), including the relations of human-environment interactions to non-local politics, economy, and political economy (Redman et al. 2004: 164).

### 8.2.2 *Sustainability Science, Social-Ecological Adaptive Capacity, and Vulnerability*

HE-NS geography is powerfully influenced through the concepts of sustainability science and the related areas of social-ecological adaptive capacity and vulnerability (Adger 2000a, 2006; Kates 1987; Kates et al. 2001; Mustafa 2005; Parris and Kates 2003; Turner et al. 2003; Zimmerer 2010c). Similar to its role in geography, sustainability science is seen as central to the conceptualisation of social-ecological systems in LTSER (Haberl et al. 2006: 2, 7; Singh et al. 2010: 379–380; note LTSER's contributions and collaboration also connect strongly to environmental history, anthropology, archaeology, ecology, resilience science and other key approaches). The concept of adaptive capacity, described as the wherewithal to respond to social-ecological change, is another shared foundation-level interest of both LTSER (e.g. Singh et al. 2010) and geography (e.g., Butzer 1990; Adger 2000a; Eakin 2006; Easterling et al. 2007). One other deeply shared concept is vulnerability – the exposure to social-ecological change – that serves as a shared core tenet in HE-NS geography (e.g., Cutter 2003; O'Brien and Leichenko 2003; Polsky 2004; Mustafa 2005; Barnett et al. 2008), and LTSER (e.g., vulnerability concepts and measures in the HERO project's observatory design; Polsky et al. 2009).

The potential of future geographic contributions to LTSER rests on a number of conceptual areas covered by sustainability science, social-ecological adaptive capacity and vulnerability. Whereas LTSER sees social-ecological dynamics as understood through "the theory of complex adaptive systems" and related concepts (Singh et al. 2010: 382), geography and related environmental social sciences tend to stress the complexity of human-social dynamics and relations, such as the potentially influential and differential role of social and economic power dynamics. So while LTSER is fully aware of multiple "spatiotemporal scales" and the deep chronology of nature-society interaction (Haberl et al. 2006: 6), the tendency is to place emphasis on the commensurability of spatial and temporal scales in human-environmental adjustments and adaptive capacity. One illustration is the ecological theoretical-based concept of temporal scaling via the adaptive cycle (Singh et al. 2010: 382–384), a four-phase sequence of exploitation, consolidation, creative destruction and re-organisation.

HE-NS geography and related fields are engaged and do adopt the above ideas, while they also tend to engage more fully in philosophical perspectives and specific critiques. For example, a productive debate is aimed at examining the extent to which adaptation does, or even can, account for various sorts of HE-NS behaviours and socially related activities (e.g., Watts 1983; Adger 2000b). A fuller perspective is provided also by highlighting the underlying foundations of adaptation and adaptive capacity as time-based concepts intrinsically linked to temporal heterogeneity. New perspectives on temporal heterogeneity of human- and social-environment interactions underscore the role of multiple overlapping scales from extremely short to long-term (Guyer et al. 2007). It can include such temporal pulses as short 5–10-year dynamics of policy shifts, ownership transitions, generational micro-politics, altered crop and cultivation successions, land re-occupation, and frontiers (see also Medley et al. 2003). The importance of these examples – such as forest-policy related impacts on deforestation behaviours and frontier dynamics – is that they help define potential groupings of interaction-based time spans that would be of use to the design and conceptualisation of future LTSER.

## 8.2.3 Land-Use and Land-Cover Change (LUCC) and Land Change Science (LCS)

Geographic analysis of land-use/land-cover change (LUCC) – evolved into land change science (LCS) (Turner et al. 2007) – utilises the concepts and methods of human and cultural ecology, human dimensions of global change, regional science, GIScience, remote sensing, and landscape science and ecology. This approach couples the economics and policy-driven interactions of land-use decision-making and the analysis of land-cover change (typically analysis of remotely sensed images; Rindfuss et al. 2007). It relies on combined empirical and modelling approaches that integrate spatially explicit interactions of socioeconomic signals (including market prices and policy influences) and environmental complexity in establishing predictions of the processes and outcomes of land-cover change (Walker 2003; Walker and Solecki 2004; Caldas et al. 2007). The emphasis of LUCC/LCS on diachronic change bears close relation to a first-order orientation of LTSER, as well as to the scientific and policy community concerned with global environmental change (GEC). Equally significant is a shared emphasis on model building and verification, along with the major incorporation of GIScience, that characterise this approach in geography (e.g., agent-based simulation modelling; Parker et al. 2003) along with pioneering formulations of LTSER (Redman et al. 2004: 168–169; Singh et al. 2010: 390–391).

To-date, LUCC/LCS studies are focused mostly on forest-cover change with a preference for frontier settings. Processes of forest-cover change occur across a spectrum from deforestation (for example in conversions to pastureland and agriculture; Mertens and Lambin 2000; Müller and Munroe 2008) to reforestation in new "secondary forest transitions" (Rudel et al. 2002; Klooster 2006; Farley 2007; Ramankutty et al. 2010). Focus on forest-cover change and frontier settings is both productive and well-placed as well as suggestive of possible re-conceptualisation. On the one hand, it is strategically relevant to major issues of global environmental change such as carbon transfers and forest biodiversity (e.g., on the latter, see Cowell and Dyer 2002; Naughton-Treves 2002; Voeks 2004). On the other hand, LUCC/LCS approaches in geography and LTSER are increasingly in need of understanding landscape interactions in addition to human-forest dynamics (Seto et al. 2012). Transitions of urban and within agricultural areas, such as shifts within multi-species and high-biodiversity agroecosystems, are increasingly recognised as important to biogeochemical cycles and global change (Turner 2010; Zimmerer 2010a, b).

## 8.2.4 Environmental Governance and Political Ecology

Environmental governance is a principal emphasis of political ecology, an approach in HE-NS geography and the environmental sciences that "combines the concerns of ecology and a broadly defined political economy" (Blaikie and Brookfield 1987: 17; see also Bassett 1988; Grossman 1993; Rocheleau and Thomas-Slayter 1996; Zimmerer and Bassett 2003; Robbins 2004; Neumann 2005; McCarthy 2006; Campbell 2007). Similarly environmental governance is proposed as one of the 4–5 essential themes of LTSER (Haberl et al. 2006: 5, 11–13). Both political ecology and LTSER are inclined to adopt a broad definition of governance as "…interventions aiming at changes in environment-related incentives, knowledge, institutions, decision making, and behaviors" (Agrawal and Lemos 2006: 298). The role of multiple ways of knowing about the environment, ranging from scientific know-how to local knowledge, is one example of a governance topic that is central area both in geographic political ecology (Bassett and Zimmerer 2003: 101–102) and in LTSER (Haberl et al. 2006). Another example is the role of human settlements and common resource-use areas as spaces of concentrated socio-ecological systems, which is a focus of political ecology that includes an urban focus (Giordano 2003; Seto et al. 2010; Zimmerer and Bassett 2003; Petts et al. 2008) and the related albeit distinct approach of "urban metabolism" that is considered a potential conceptual foundation of LTSER (Haberl et al. 2006: 8–9; Singh et al. 2010; 379–380).

The theme of environmental governance replete with potentially fruitful intersections to LTSER. One concerns the role of multiple knowledge systems, in addition to environmental science per se. Considered central to LTSER, it is described as the "need to harness indigenous knowledge" (Haberl et al. 2006: 11). Political ecology has unpacked the complex interactions of diverse ways of knowing and learning in combination with Western science and underscored the interconnectedness of these knowledge systems through social power relations, politics, and philosophy (on the role of corridor ideas in East African wildlife conservation as "boundary concepts" connecting local and scientific knowledge see Goldman 2009; see also Robbins 1998; Bassett and Koli Bi 2000; Voeks 2004; Goldman et al. 2011). As a result, multiple systems of environmental knowledge are often socially entwined, frequently form knowledge hybrids and are politically contested This perspective both builds upon and is a general contrast to an earlier ethnoscience approach in geography, anthropology and related fields that, in a now outdated view, conceptualised these twin knowledge systems as parallel and subject to the potential of science to "tap" non-Western ways of knowing.

Future work on environmental governance may promise productive intersections of political ecology with LTSER. These research directions range from the development and application of territorial concepts – focused, for example, on the overly rigid "scalar fixes" of environmental certification in organic coffee production in Mexico that ironically may create sustainable practices on single plots while undermining sustainability at the landscape scale (Mutersbaugh 2002), Thai forest conservation (Roth 2008), and, more generally, protected-area conservation (Zimmerer 1999). One take-home message from these geographic studies is that territorial dynamics are central, both enabling and limiting, to many certification programmes, such as sustainable forestry (Klooster 2006) and payments for ecosystem services (McAfee and Shapiro 2010). As examined broadly in political ecology and related approaches (Bakker 2005; Bebbington 2000; Klooster 2006; Bailey 2007), neoliberal policies have also ushered in an evolving panoply of instruments associated with "market environmentalism" that are widespread and potentially relevant to LTSER site-selection and design criteria (Coomes and Barham 1997).

At the level of local places, these policies often pivot on the roles of non-governmental organisations (NGOs) in the context of local livelihoods, and regional, ethnic, and social power relations including gendered interactions, resource- and land-related social movements and political disparities (Simmons 2004; Wolford 2004). Their programmes and projects offer many important examples for possible LTSER (e.g., land reform, social movement initiatives, sustainable forestry, organic agriculture). While sometimes beneficial, the interventions aimed at improving the livelihoods of certain less powerful social groups, such as those defined by ethnicity or gender, may backfire if their success becomes commandeered by dominant sectors in these societies. Examples include women irrigators in certain parts of Africa, whose increased food-growing capacity, while benefitting from the implementation of pro-women development projects, has been usurped through conjugal relations with male household heads (e.g., Carney 1993; Schroeder 1997). Indeed the time-span of these gender-related interactions may offer insights to LTSER, as well as generally in the broader feminist perspectives of environmental geography (Momsen 2000; Reed and Christie 2009) and specifically in the approach of feminist political ecology (Rocheleau 1995; Rocheleau and Thomas-Slayter 1996; Rocheleau and Edmunds 1997; Nightingale 2003; O'Reilly 2006).

HE-NS contributions have also expanded to focus on the interplay of governance institutions and landscapes in a range of coupled socio-environmental resource systems. Building upon earlier research (Bassett and Zimmerer 2003: 99–101), such geographic advances on the governance of coupled systems encompass: (i) community- and user-based (and nationally and internationally influenced) management of fisheries, marine organisms, and forestry and range resources (Robbins 1998; St. Martin 2001; Young 2001; Mutersbaugh 2002; Mansfield 2004; McCarthy 2006; Campbell 2007); (ii) water resources in urban planning, international relations, and irrigation, including response management and mitigation of climate change (Wescoat 1986; Emel and Roberts 1995; Bakker 2005; Perreault 2008; Petts et al. 2008; Birkenholtz 2009; Feitelson and Fischhendler 2009; Norman and Bakker 2009; Gober et al. 2010); (iii) biodiversity and environmental conservation in utilized landscapes (Zimmerer 1999; Naughton-Treves 2002; Campbell 2007; Roth 2008); (iv) agriculture, land tenure, land change, pesticide use, and agrarian reform and policy institutions, including urban and periurban food production (Grossman 1993; Muldavin 1997; Schroeder 1997; Freidberg 2001; Hovorka 2005; Galt 2010; Jepson et al. 2010); (v) modern environmentalism and popular movements for social and environmental justice (Bowen et al. 1995; Pulido 2000; Liu 2008); (vi) state environmental agencies (Feldman and Jonas 2000); and (vii) industrial and manufacturing regulation (Willems-Braun 1997; Prudham 2003).

## 8.2.5 Environmental Landscape History and Ideas

The theme of environmental landscape history and ideas is a richly active tradition in HE-NS geography. Indeed a historically oriented environmental geography, associated with the so-called Berkeley School of Carl O. Sauer, is one of the principal precursors to contemporary HE-NS geography – it effectively adopted an approach of "cultural-historical ecology" that forged an explicit emphasis on cultural and historical explanation with the use of ecological reasoning (Zimmerer 1996, 2010c). Similarly, historical conceptualisation is central to LTSER where "long term" implies the need for time-based frameworks so as to be able to "monitor change over time and recognize the dynamics and impacts of transitions" (Redman et al. 2004; Haberl et al. 2006: 7). The potential of specifically shared historical orientations include the long-term time frame of human- and paleo-environmental change (Butzer 1990, 1992; Denevan 1996; Redman et al. 2004; Dull 2007); an informed lang durée perspective (Bassett and Zimmerer 2003: 98–99; Haberl et al. 2006: 7); and, in particular, the complex nature of baselines and benchmarks in these chronological comparisons (Etter et al. 2008; Haberl et al. 2006: 7).

Historical concepts of environmental and landscape change are a vitally important arena for the connected area of LTSER and HE-NS geography. Shared interests include such concepts as legacy impacts, the estimation and significance of variable rates of change, and path-dependent trajectories of change (in LTSER see Haberl et al. 2006). Three further intersections can also be highlighted with respect to HE-NS geography. One is the opportunity for multiple temporal scales of human-environmental change to be seen as interlinked to multi-scalar global transformations (Turner 1991). Another is the debunking of the so-called "Pristine Myth" through the use of multi-century frameworks combining environmental historical landscape analysis and complex sociocultural encounters stretching across pre-colonial and colonial periods to present-day settings (Denevan 1992). These works demonstrate the ample, albeit limited, resilience of landscapes amid European colonial conquests and exploitation of varied societies and resource environments. For example, initial colonialism in Latin America (1500s-mid-1600s) often unleashed a pulse of de-vegetation and soil loss, whose characteristics were shaped through pre-existing environmental changes of the pre-European period (e.g., under Inca, Maya, and Aztec imperial concentrations of resources and people). Subsequently, environments regenerated and demonstrated a new fairly stable configuration of landscapes for a couple centuries or potentially longer (mid-1600s–1800s; Butzer 1992; Denevan 1992; Doolittle 1992; Gade 1992; Whitmore and Turner 1992; Endfield and O'Hara 1999; Sluyter 2001; Dunning et al. 2002; see also Butzer and Helgren 2005). This multi-century time scale may be of interest in retrodictive LTSER. The third is the deeply fused social-ecological nature of environmental historical landscape change, whose understanding is being advanced also through concepts of "socionatural hybrids" in geographic works on water resources (Swyngedouw 1999), forest management (Robbins 2001), networks of protected areas designed for sustainable use (Zimmerer 1999), and a range of other resource areas (Goldman et al. 2011).

## 8.2.6 Environmental Scientific Concepts in Models, Management, and Policy

Scientific concepts are similarly at the scholarly centres of HE-NS geography and LTSER alike. One important shared concept concerns equilibria and non-equilibria in human- and social-environmental interactions. It is a foundation of LTSER that "the maintenance of any equilibrium over long time spans is unrealistic" (Singh et al. 2010: 379). LTSER also holds this concept as central to social-ecological dynamics "as 'open systems' operating far from equilibrium" (Singh et al. 2010: 384; see also p. 382). Similarly HE-NS geography has focused on the concept of equilibria and non-equilibria in social-ecological systems, with specific reference to geographic models of adaptation and carrying capacity. For example, the adaptation of crop biodiversity in many agroecosystems is condition through non-equilibrial processes that lead to generalist adaptive capacity within peasant and indigenous food-growing strategies (Zimmerer 1994). Recent geographic works have extended and expanded these ideas on ecological carrying capacity (Sayre 2008), the hydrologic cycle (Linton 2008), biological conservation corridors (Goldman 2009), the science of back-to-nature farming (Ingram 2007), and scientific forestry management (Willems-Braun 1997). These works analyse the power of scientific ideas as deriving from geographic dimensions and their social use in environmental management, such as so-called boundary concepts in place-based sites of interaction, negotiation, and dispute (Goldman 2009). Scientific concepts thus both reflect and also actively influence the interactions of environmental science and scientists with environmental management, policymakers, and stakeholders.

Reflexive practice, ethics, and the promise of participatory approaches are another particularly productive dimension of scientific concepts and their application, taken broadly, that is integral to HE-NS geography and LTSER. Reflexivity, namely the inclusion of scientists and their institutions within models *per se*, can help account for and manage environmental change given its dense social fabric and the extent of scientific complexity and uncertainty (Taylor 2005; Goldman et al. 2011). Issues of reflexivity, ethics and participatory approaches have been considered in cultural and political ecology in the context of development (Bassett 1988; Grossman 1993), applied philosophical pragmatism (Wescoat 1992), and social constructivism (Demeritt 2001). The roles of science, scientists and citizens in responding to global climate change in particular have contributed to various new approaches emphasising reflexivity, ethics, and participation (Shrader-Frechette 1998; Schneider 2001; Easterling et al. 2007).

## 8.3 Geography and LTSER: Environmental Interdisciplinarity and Policy

Above-mentioned concepts of HE-NS geography offer promising interconnections to LTSER also via environmental interdisciplinarity and policy. Similar to many fields, HE-NS geography must translate and communicate its knowledge systems to other disciplines and policymakers in addition to the general public and citizen groups (Bracken and Oughton 2006). This translation may be conceptualised as occurring across geography's "environmental borderlands" (Zimmerer 2007); it occurs via ties with disciplines as diverse as ecology, ecological and environmental anthropology, environmental and resource sociology, environmental history and environmental, agricultural and resource economics. Current environmental interdisciplinarity is seen as posing unprecedented importance and potential opportunities for geography as a consequence of the restructuring of the academy (Turner 2002) and the expansion of interdisciplinary programmes (Baerwald 2010; NRC 2010). Geography's contributions are well developed with respect to the certain interdisciplinary realms of human-environmental research (see details in Zimmerer 2010c), such as earth system science and ecological science (Pitman 2005; Marston 2008); broad environmental social science (Agrawal 2005; Goldman et al. 2011; Rocheleau 2008); and environmental history (White 2004).

Translation of HE-NS geography is equally vital with regard to environmental policy. Such potential and often practical contributions have been underscored in recent works (Wescoat 1992), especially those pertaining to sustainability science (Kates 1995), vulnerability science (Cutter 2003), human dimensions of global environmental change (HDGEC; Turner 1991) and climate change (Easterling et al. 2007), as well as socially sustainable conservation geographies (Zimmerer 1999). Environmental policy-related issues include climate change responses and mitigation; biodiversity and environmental conservation; energy and impact assessment; sustainable resource management and agriculture; urban and industrial environments, planning and design including topics such as "smart growth" and transportation; and environmental economics, justice and social movements (Wilbanks 1994; NRC 1997, 2010; Liverman 1999, 2004; Robbins 2004; Skole 2004; Yarnal and Neff 2004; Neumann 2005; McAfee and Shapiro 2010). More generally these policy contributions reflect the expanded engagement of HE-NS geography with a "normative turn" engaged with environmental values and environmentalism. Geography's directions as outlined in this section thus offer extensive similarities to the concerns, beliefs and values that have been expressed as underlying motivations and commitments of LTSER (Redman et al. 2004; Haberl et al. 2006; Singh et al. 2010).

## 8.4 Andean Watersheds of the Upper Amazon and Networked Sites

Watersheds of the tropical Andes are critical environments well-suited to application of the concepts identified above as promising a "common ground" for HE-NS geography and LTSER. More than 100 million people reside in the mountain watersheds between Venezuela and Argentina. Social-ecological criticality is pronounced in the Central Andes (Ecuador, Peru, and Bolivia). These mountainous headwaters of the Amazon basin are resilient global hotspots of both "wild" biodiversity and

**Table 8.2** Case study-based illustrations of concepts at the intersection of geography and LTSER

| Case study feature | Concept (from Sect. 8.2 with Subsection(s) noted) | Examples of general works cited |
|---|---|---|
| Calicanto irrigation and agricultural land use epochs (see Table 8.2) | Coupled human-environment interactions (8.2.1); temporal scaling and multiple time frames | Chowdury and Turner (2006) |
| | Environmental governance and political ecology (8.2.4); policy, state, and NGO influences | Perreault (2008) Bassett (1988) |
| | Landscape history and ideas (8.2.5); temporal scaling and non-local interactions | Butzer (1990) Doolittle (1992) |
| Calicanto irrigation and social-ecological transitions within land use categories | Land-Use/Cover Change (LUCC) and Land Change Science (8.2.3); extension to non-forest transitions | Naughton-Treves (2002) Rudel et al. (2002) |
| Cochabamba-Chapare-Northern Potosí LTSER Network (see Table 8.3) | Coupled human-environment interactions (8.2.1); complex spatial scaling and geographic network formation | Zimmerer (1999) Campbell (2007) |

complex agroecosystems (Zimmerer 2010a, b). At the same time the Central Andes is subject to multiple trajectories of pronounced social-ecological change. Geographically it is a prime candidate for LTSER efforts. This section draws on my having worked, lived and researched intermittently for nearly 30 years with local farmers, irrigator groups and their communities along with extensive collaborations with local scientists, NGOs, government agencies, and environment-development institutions in Peru, Bolivia, Ecuador, and Colombia. This section is focused primarily on the Cochabamba region of central Bolivia.

Recent research in Cochabamba is used to offer, by way of brief description, illustrations of several of the principal concepts mentioned in Sect. 8.2 (above) where there are notable intersections of geography and LTSER (Table 8.2). The first illustration is the sketch of time frames of combined water-resource and agricultural land use (Table 8.3; see also Sects. 8.2.1 and 8.2.5). This example deploys the concept of differentiated temporal scaling arising through human-environment interactions. It draws on the evidences of the Calicanto irrigated area, nearby uplands, and the surrounding "High Valley" (*Valle Alto*) of Cochabamba. Policies of Bolivia's current government of Evo Morales (2006) are placing a new emphasis on the production and consumption of indigenous food plants with the goals of enhancing food security and reinforcing adaptive capacity with regard to climate change. These Bolivian government policies are beginning to exert influence in the Calicanto area (e.g., increased cultivation and culinary celebration of "ancient foods" or *ñawpaq mikhuna*).

Due to its duration, a more well-defined illustration is the impact on water-resource and agricultural land use of neoliberal multiculturalism during recent decades (1985–2006, especially 1998–2006) (Table 8.3). Policy expressions included the support of indigenous ethnodevelopment based on community resource management

**Table 8.3** Examples of temporal scaling of human-environment interactions and water-resource/agricultural land use in Bolivia (19th and 20th centuries)

| Predominant scaling processes time period | Approximate time period |
|---|---|
| State-peasant coalition (Evo Morales administration), government support of indigenous foods and climate-change policies | 2006 |
| Neoliberalism with increased multiculturalism and ethnodevelopment, leading to legal recognition of indigenous irrigation customs (usos y costumbres, see Perreault 2008) | 1985–2006 |
| Modern state-led agrarian development, with promotion of big dam projects and Green Revolution-style modern development | 1952–1985 |
| Modern private (estate) and technology-centred projects | 1910–1952 |
| State/estate control through taxation and tribute ("tributary state") | 1826–1910 |

With specific reference to the geographic area of the Calicanto irrigated area, nearby uplands, and "High Valley" of Cochabamba during the time period since national independence

and customary water rights (*usos y costumbres*; see Perreault 2008). Development NGOs working in conjunction with the Bolivian government during this period – see Zimmerer (2009) – illustrate a synergy of LTSER and geography at the conceptual intersection of governance issues and political ecology affecting water-resource and agricultural land use (Table 8.2; see also Sect. 8.2.3). These recent time frames show overlap with modern state-led agrarian development (1952–1985) that propelled big dam projects, national land reform, and Green Revolution-style development strategies (Table 8.3). Each of these epochs of water-resource and agricultural land use, distinguished through government policies and human-environment interactions (specifically state-peasant coalition; multicultural neoliberalism, and modern state-led agrarianism and big-dam development), describes a time frame potentially worthy of incorporation into LTSER models and monitoring. Previous time frames can also be identified as potentially of interest (Table 8.3).

Focus on water-resource and agricultural land use in the Calicanto area of the "High Valley" also illustrates the role of important interaction concepts involving human-environment relations and structuration in particular (see Sect. 8.2.1). Modern water-resource development in the form of an internationally financed irrigation project (the LLP), centred on the building of a dam, was not solely an imposition of external forces (Zimmerer 2011a, b; see also Chowdury and Turner 2006; Brenner 2011). Indeed many local irrigators were central agents in the enactment of these changes since they eventually offered crucial support of the LLP. Yet this modern development project led to the eclipse of the irrigators' own pre-existing water-resource management that comprised a landscape technology of spate irrigation or hybrid floodwater-canal farming. Their spate irrigation had benefited from mutually reinforcing bi-directional links with sediment- and nutrient-rich runoff, intensive agriculture, migration, and local economic development (Zimmerer 1993, 2011b). Mounting concerns led the Calicanto irrigators to push for major design adjustments of the LLP and contributed to the transition to community-based resource management at the site (Zimmerer 2009). In short, local resource-users were fully engaged, albeit certainly not as equals, in the playing out of large-scale policy and economic forces (this example of structuration in land-use decision-making is akin to common occurrences in the Central Andes and elsewhere; see Zimmerer 1991; Chowdury and Turner 2006; Brenner 2011).

The case-study of the Cochabamba region can also be used to advance a spatial-network approach informed by the intersection of geography and LTSER. One potential configuration of such sites would be based on spatial-environmental networks that are formed through the movement of people and environmental goods (Table 8.4). The latter is vividly inscribed, for example, through the seed-exchange patterning of high-agrobiodiversity Andean food plants. Most irrigated-agricultural land users in the Calicanto area and the surrounding "High Valley" of central Cochabamba obtain their seeds of high-agrobiodiversity Andean maize and other food plants through local and regional seed networks reaching across valleys and upland Andean areas (e.g., from their Cochabamba communities to the nearby northern Potosí region) (Table 8.4, see also Zimmerer 2010a). At the same time, the movements of people across landscapes, mostly as a result of labour migration, is extremely important due to its high frequency and role in environment-related activities that range from agriculture to forest impacts (whether deforestation/degradation or resilience-enhancing forest management). For example, many land users from the "High Valley" and northern Potosí work seasonally or semi-permanently in lowland tropical areas of the Andean foothills (e.g., the lowland tropical Chapare region) (Table 8.4). The type of network of LTSER sites described here aims to recognize HE-NS processes as tracing multiple designs and scales in extending complexly across landscapes (see also Zimmerer 2000:358–60), rather than primarily or solely as sites based on single parcels, sites, or rigidly nested scalar hierarchies (Table 8.2, see also Section 8.2.1).

## 8.5 Conclusion

Substantive interconnections and promising synergies have been shown to characterise the relations of NS-HE geography and LTSER. This study has systematically identified and examined the interconnections in the following principal themes: (i) Coupled Human-Environment Interactions; (ii) Sustainability Science, Social-Ecological Adaptive Capacity, and Vulnerability; (iii) Land-Use and Land-Cover Change (LUCC) and Land Change Science (LCS); (iv) Environmental Governance and Political Ecology; (v) Environmental Landscape History and Ideas; and (vi) Environmental Scientific Concepts in Models, Management, and Policy. These six themes are core areas within current HE-NS geography, one of the principal divisions of contemporary four-field geography. The study has highlighted interconnections in terms of concepts and theoretical constructs along with specific examples, with secondary emphasis, albeit important, on research design and methods. Study findings detail the richness of numerous well-demonstrated interconnections, as well as ample potential for increased scientific and intellectual interchange and institutional collaboration. In conclusion, geography and LTSER share a significant

**Table 8.4** Principal characteristics of a spatial network of potential LTSER sites (Central Bolivia)

| Geographic sub-unit (place/region) | Area (estimate) | Population (estimate) | Elevation & climate(s) | Biodiversity "hotspot" characteristics | Inter-connecting process | Distance to other regions |
|---|---|---|---|---|---|---|
| Calicanto (C)/"High Valley" (Valle Alto), Cochabamba | 45 km² | 8,000 persons | 2,600–3,100 masl. semi-arid & sub-humid | Andean food plants (mid-elevation) & dry forest | Labour migration (seasonal and semi-permanent), seed flows to VE | 55 km to VE 35 km to SJ |
| Villa (VE) Esperanza/ Chapare | 42 km² | 1,500 person | 300–1,800 masl. tropical humid | Amazonian & Andean foothill food plants, tropical rainforest | Labour migration (high rates) from C and SJ | 55 km to C 90 km to SJ |
| San Juan (SJ) Northern Potosí | 36 km² | 2,200 persons | 2,100–4,000 masl. | Andean food plants (mid- and high-elevation), Andean shrub and tree formations | Labour migration (high rates to C and VE); seed/product flows to C | 35 km to C 90 km to VE |
| Villa (VE) Esperanza/ Chapare | 42 km² | 1,500 person | 300–1,800 masl. tropical humid | Amazonian & Andean foothill food plants, tropical rainforest | Labour migration (high rates) from C and SJ | 55 km to C 90 km to SJ |
| San Juan (SJ) Northern Potosí | 36 km² | 2,200 persons | 2,100–4,000 masl. | Andean food plants (mid- and high-elevation), Andean shrub and tree formations | Labour migration (high rates to C and VE); seed/product flows to C | 35 km to C 90 km to VE |

degree of general similarity, albeit with corresponding distinctness, that promises ample and potentially vital opportunities for future crossover, collaboration and shared directions.[2]

Key conceptual nodes of the potential collaboration of geographic approaches and LTSER show a range from the spatial and temporal scaling of social-environmental processes to the potential design of multi-site social-environmental networks. Conceptualisation of spatial and temporal scaling is especially relevant and important, and points to the usefulness of such concepts as socioenvironmental structuration. One chief insight comes from the idea of historical HE-NS junctures and transitions amid change processes. The case of Bolivia's environmentally vital Andean watersheds illustrates the role of temporal junctures in the water resource landscapes associated with transitions of national agrarianism (1950s), modern irrigation (1960s-early 1990s) and community-based resource management (early 1990s-present). Field studies of the Bolivian Andes are also used to suggest the potential design of multi-site networks for social-environmental analysis and monitoring. Such networks would be seen as comprised of geographic places connected through integrated social-environmental processes (e.g., migration and high-agro-biodiversity seed exchange and food supply) and thus a complement and extension of existing LTSER concepts of network design.

## End Note

1. Each of these themes within HE-NS geography represents a core theme, with ample intersecting and crossing-over among them, and is defined on the basis of a content analysis of more than 90 articles published between 1990 and 2010 (Zimmerer 2010c). Recent HE-NS geography evokes the image of multiple braided streams, in contrast to the earlier dichotomous configuration; on this history of HE-NS geography see Zimmerer (2010c). In addition to the themes mentioned here, HE-NS geography includes emphasis on climate change adaptations and mitigation, conservation geography, cultural and human ecology, theoretical environmental geography, human dimensions of global change, land change science, natural hazards, political ecology, social-ecological resilience, sustainability science, vulnerability science, and several others (Castree et al. 2009).

---

[2] It is important to highlight that HE-NS geography consists of a pair of partially distinct epistemic sub-categories, one associated with human-environment interactions (HE) and the other with nature-society relations (NS) (Zimmerer 2010b), as sketched in Fig. 8.1. Human-environment interactions is comprised most clearly of coupled human-environment interactions; sustainability science, social-ecological adaptive capacity, and vulnerability; and land-use and land-cover change (LUCC)/land change science (LCS). Nature-society relations consists chiefly of political ecology and environmental governance; environmental landscape history and ideas; and environmental scientific concepts in models, management, and policy. The partial distinctness of these sub-categories owes to the noticeable sharing and overlap of certain ideas (e.g., environmental governance) (see Zimmerer 2010b). It also owes to the interpretation of co-existence, in addition to contestation, that has become vital to HE-NS geography (Zimmerer 2007; Turner 2009; Turner and Robbins 2008).

# References

Adger, W. N. (2000a). Institutional adaptation to environmental risk under the transition in Vietnam. *Annals of the Association of American Geographers, 90*, 738–758.

Adger, W. N. (2000b). Social and ecological resilience: Are they related? *Progress in Human Geography, 24*, 347–364.

Adger, W. N. (2006). Vulnerability. *Global Environmental Change – Human and Policy Dimensions, 16*, 268–281.

Agrawal, A. (2005). *Environmentality: Technologies of government and the making of subjects.* Durham: Duke University Press.

Agrawal, A., & Lemos, M. C. (2006). Environmental governance. *Annual Review of Environment and Resources, 31*, 297–325.

Baerwald, T. (2010). Prospects for geography as an interdisciplinary discipline. *Annals of the Association of American Geographers, 100*, 494–501.

Bailey, I. (2007). Market environmentalism, new environmental policy instruments, and climate policy in the United Kingdom and Germany. *Annals of the Association of American Geographers, 97*, 530–550.

Bakker, K. (2005). Neoliberalizing nature? Market environmentalism in water supply in England and Wales. *Annals of the Association of American Geographers, 95*, 542–565.

Barnett, J., Lambert, S., & Fry, I. (2008). The hazards of indicators: Insights from the environmental vulnerability index. *Annals of the Association of American Geographers, 98*, 102–119.

Bassett, T. J. (1988). The political ecology of peasant-herder conflicts in Northern Ivory Coast. *Annals of the Association of American Geographers, 78*, 453–472.

Bassett, T. J., & Koli Bi, Z. (2000). Environmental discourses and the Ivorian Savanna. *Annals of the Association of American Geographers, 90*, 67–95.

Bassett, T. J., & Zimmerer, K. S. (2003). Cultural ecology. In G. L. Gaile & C. J. Wilmott (Eds.), *Geography in America at the dawn of the twenty-first century* (pp. 97–112). Oxford: Oxford University Press.

Bebbington, A. (2000). Reencountering development: Livelihood transitions and place transformations in the Andes. *Annals of the Association of American Geographers, 90*, 495–520.

Benjaminson, T. A., Rohde, R., Sjaastad, E., Wisborg, P., & Lebert, T. (2006). Land reform, range ecology, and carrying capacities in Namaqualand, South Africa. *Annals of the Association of American Geographers, 96*, 524–540.

Birkenholtz, T. (2009). Irrigated landscapes, produced scarcity, and adaptive social institutions in Rajasthan, India. *Annals of the Association of American Geographers, 99*, 118–137.

Blaikie, P. M. (1994). *Political ecology: An evolving view of nature and society* (CASID Distinguished Speaker Series No. 13). East Lansing: Michigan State University.

Blaikie, P., & Brookfield, H. (1987). *Land degradation and society.* London: Methuen.

Bowen, W. M., Salling, M. J., Haynes, K. E., & Cyran, E. J. (1995). Toward environmental justice: Spatial equity in Ohio and Cleveland. *Annals of the Association of American Geographers, 85*, 641–663.

Bracken, L. J., & Oughton, E. A. (2006). 'What do you mean?' The importance of language in developing interdisciplinary research. *Transactions of the Institute of British Geographers New Series, 31*, 371–382.

Brenner, J. C. (2011). Pasture conversion, private ranchers, and the invasive exotic Buffelgrass (*Pennisetum ciliare*) in Mexico's Sonoran desert. *Annals of the Association of American Geographers, 101*, 84–106.

Butt, B., Shortridge, A., & WinklerPrins, A. M. G. A. (2009). Pastoral herd management, drought coping strategies, and cattle mobility in Southern Kenya. *Annals of the Association of American Geographers, 99*, 309–334.

Butzer, K. W. (1990). The realm of cultural-human ecology: Adaptation and change in historical perspective. In B. L. Turner II et al. (Eds.), *The earth as transformed by human action: Global and regional changes in the biosphere over the past 300 years* (pp. 685–701). Cambridge: Cambridge University Press.

Butzer, K. W. (1992). The Americas before and after 1492 – An introduction to current geographical research. *Annals of the Association of American Geographers, 82*, 345–368.

Butzer, K., & Helgren, D. M. (2005). Livestock, land cover, and environmental history: The tablelands of New South Wales, Australia, 1820–1920. *Annals of the Association of American Geographers, 95*, 80–111.

Caldas, M., Walker, R., Arima, E., Perz, S., Aldrich, S., & Simmons, C. (2007). Theorizing land cover and land use change: The peasant economy of Amazonian deforestation. *Annals of the Association of American Geographers, 97*, 86–110.

Campbell, L. M. (2007). Local conservation practice and global discourse: A political ecology of sea turtle conservation. *Annals of the Association of American Geographers, 97*, 313–334.

Carney, J. (1993). Converting the wetlands, engendering the environment – The intersection of gender with agrarian change in the Gambia. *Economic Geography, 69*(329), 348.

Castree, N., Demeritt, D., & Liverman, D. (2009). Introduction: Making sense of environmental geography. In N. Castree, D. Demeritt, D. Liverman, & B. Rhoads (Eds.), *A companion to environmental geography* (pp. 1–16). Malden: Blackwell.

Chowdury, R. R., & Turner, B. L. I. I. (2006). Reconciling agency and structure in empirical analysis: Smallholder land use in the Southern Yucatán, Mexico. *Annals of the Association of American Geographers, 96*, 302–322.

Coomes, O. T., & Barham, B. L. (1997). Rain forest extraction and conservation in Amazonia. *The Geographical Journal, 163*, 180–188.

Cowell, C. M., & Dyer, J. M. (2002). Vegetation development in a modified riparian environment: Human imprints on an Allegheny River wilderness. *Annals of the Association of American Geographers, 92*, 189–202.

Cutter, S. L. (2003). The vulnerability of science and the science of vulnerability. *Annals of the Association of American Geographers, 93*, 1–12.

Demeritt, D. (2001). The construction of global warming and the politics of science. *Annals of the Association of American Geographers, 91*, 307–337.

Denevan, W. M. (1992). The pristine myth – The landscape of the Americas in 1492. *Annals of the Association of American Geographers, 82*, 369–385.

Denevan, W. M. (1996). A bluff model of riverine settlement in prehistoric Amazonia. *Annals of the Association of American Geographers, 86*, 654–681.

Doolittle, W. E. (1992). Agriculture in North America on the eve of contact – A reassessment. *Annals of the Association of American Geographers, 82*, 386–401.

Dougill, A. J., Thomas, D. S. G., & Heathwaite, A. L. (1999). Environmental change in the Kalahari: Integrated land degradation studies for nonequilibrium dryland environments. *Annals of the Association of American Geographers, 89*, 420–442.

Dull, R. A. (2007). Evidence for forest clearance, agriculture, and human-induced erosion in Precolumbian El Salvador. *Annals of the Association of American Geographers, 97*, 127–141.

Dunning, N. P., Luzzadder-Beach, S., Beach, T., Jones, J. G., Scarborough, V., & Culbert, T. P. (2002). Arising from the *bajos*: The evolution of neotropical landscape and the rise of Maya Civilization. *Annals of the Association of American Geographers, 92*, 267–283.

Eakin, H. (2006). *Weathering risk in rural Mexico: Climatic, institutional, and economic change*. Tucson: University of Arizona Press.

Easterling, W. E., Aggarwal, P. K., Batima, P., Brander, K. M., Erda, L., Howden, M., Kirilenko, A., Morton, J., Soussana, J.-F., Schmidhuber, S., & Tubiello, F. (2007). Food, fibre and forest products. In M. L. Parry, O. F. Canziani, J. P. Palutikof, P. J. van der Linden, & C. E. Hanson (Eds.), *Climate change 2007: Impacts, adaptation and vulnerability. Contribution of Working Group II to the fourth assessment report of the Intergovernmental Panel on Climate Change* (pp. 273–313). Cambridge/New York: Cambridge University Press.

Emel, J. L., & Roberts, R. (1995). Institutional form and its effect on environmental change: The case of groundwater in the Southern High Plains. *Annals of the Association of American Geographers, 85*, 664–683.

Endfield, G. H., & O'Hara, S. L. (1999). Degradation, drought, and dissent: An environmental history of colonial Michoacan, west central Mexico. *Annals of the Association of American Geographers, 89*, 402–419.

Etter, A., McAlpine, C., & Possingham, H. (2008). Historical patterns and drivers of landscape change in Colombia since 1500: A regionalized spatial approach. *Annals of the Association of American Geographers, 98*, 2–23.

Farley, K. A. (2007). Grasslands to tree plantations: Forest transition in the Andes of Ecuador. *Annals of the Association of American Geographers, 97*, 755–771.

Feitelson, E., & Fischhendler, I. (2009). Spaces of water governance: The case of Israel and its neighbors. *Annals of the Association of American Geographers, 99*, 728–745.

Feldman, T. D., & Jonas, A. E. G. (2000). Sage scrub revolution? Property rights, political fragmentation, and conservation planning in Southern California under the federal Endangered Species Act. *Annals of the Association of American Geographers, 90*, 256–292.

Freidberg, S. (2001). Gardening on the edge: The social conditions of unsustainability on an African urban periphery. *Annals of the Association of American Geographers, 91*, 349–369.

Gade, D. W. (1992). Landscape, system, and identity in the postconquest Andes. *Annals of the Association of American Geographers, 82*, 460–477.

Galt, R. E. (2010). Scaling up political ecology: The case of illegal pesticides on fresh vegetables imported into the United States, 1996–2006. *Annals of the Association of American Geographers, 100*, 327–355.

Giordano, M. (2003). The geography of the commons: The role of scale and space. *Annals of the Association of American Geographers, 93*, 365–375.

Gober, P. C., Kirkwood, W., Balling, R. C., Ellis, A. W., & Deitrick, S. (2010). Water planning under climatic uncertainty in Phoenix: Why we need a new paradigm. *Annals of the Association of American Geographers, 100*, 356–372.

Goldman, M. (2009). Constructing connectivity: Conservation corridors and conservation politics in East African rangelands. *Annals of the Association of American Geographers, 99*, 335–359.

Goldman, M. J., Nadasdy, P., & Turner, M. D. (Eds.). (2011). *Knowing Nature: Conversations at the intersection of political ecology and science studies*. Chicago: University of Chicago Press.

Grossman, L. S. (1977). Man-environment relationships in anthropology and geography. *Annals of the Association of American Geographers, 67*, 126–144.

Grossman, L. S. (1981). The cultural ecology of economic development. *Annals of the Association of American Geographers, 71*, 220–236.

Grossman, L. S. (1993). The political ecology of banana exports and local food-production in St. Vincent, Eastern Caribbean. *Annals of the Association of American Geographers, 83*, 347–367.

Guyer, J. I., Lambin, E. F., Cliggett, L., Walker, P., Amanor, K., Bassett, T., Colson, E., Hay, R., Homewood, K., Olga, L. O., Pabi, O., Peters, P., Scudder, T., Turner, M., & Unruh, J. (2007). Temporal heterogeneity in the study of African land use. *Human Ecology, 35*, 3–17.

Haberl, H., Winiwarter, V., Andersson, K., Ayres, R. U., Boone, C. G., Castillio, A., Cunfer, G., Fischer-Kowalski, M., Freudenburg, W. R., Furman, E., Kaufmann, R., Krausmann, F., Langthaler, E., Lotze-Campen, H., Mirtl, M., Redman, C. A., Reenberg, A., Wardell, A. D., Warr, B., & Zechmeister, H. (2006). From LTER to LTSER: Conceptualizing the socio-economic dimension of long-term socio-ecological research. *Ecology and Society, 11*, 13. (Online), http://www.ecologyandsociety.org/vol11/iss2/art13/

Hovorka, A. J. (2005). The (re) production of gendered positionality in Botswana's commercial urban agriculture sector. *Annals of the Association of American Geographers, 95*, 294–313.

Ingram, M. (2007). Biology and beyond: The science of "back to nature" farming in The United States. *Annals of the Association of American Geographers, 97*, 298–312.

Jepson, W., Brannstrom, C., & Filippi, A. (2010). Access regimes and regional land change in the Brazilian Cerrado, 1972–2002. *Annals of the Association of American Geographers, 100*, 87–111.

Kates, R. W. (1987). The human environment- the road not taken, the road still beckoning. *Annals of the Association of American Geographers, 77*, 525–534.

Kates, R. (1995). Labnotes from the Jeremiah Experiment: Hope for a sustainable transition. *Annals of the Association of American Geographers, 85*, 623–640.

Kates, R. W., Clark, W. C., Corell, R., Hall, J. M., Jaeger, C. C., Lowe, I., McCarthy, J. J., Schellnhuber, H. J., Bolin, B., Dickson, N. M., Faucheux, S., Gallopin, G. C., Grübler, A., Huntley, B., Jäger, J., Jodha, N. S., Kasperson, R. E., Mabogunje, A., Matson, P., Mooney, H., Moore, B., III, O'Riordan, T., & Svedin, U. (2001). Sustainability science. *Science, 292*, 641–642.

Klooster, D. (2006). Environmental certification of forests in Mexico: The political ecology of a nongovernmental market intervention. *Annals of the Association of American Geographers, 96*, 541–565.

Knapp, G. (1994). Review of smallholders, householders – Farm families and the ecology of intensive, sustainable agriculture by R. M. Netting. *Annals of the Association of American Geographers, 84*, 314–317.

Knight, C. G. (1971). Ecology of African sleeping sickness. *Annals of the Association of American Geographers, 61*, 23–44.

Laney, R. M. (2002). Disaggregating induced intensification for land-change analysis: A case study from Madagascar. *Annals of the Association of American Geographers, 92*, 702–726.

Linton, J. (2008). Is the hydrologic cycle sustainable? A historical-geographical critique of a modern concept. *Annals of the Association of American Geographers, 98*, 630–649.

Liu, L. (2008). Sustainability efforts in China: Reflections on the environmental Kuznets curve through a locational evaluation of "Eco-Communities". *Annals of the Association of American Geographers, 98*, 604–629.

Liverman, D. M. (1990). Drought impacts in Mexico – Climate, agriculture, technology, and land-tenure in Sonora and Puebla. *Annals of the Association of American Geographers, 80*, 49–72.

Liverman, D. M. (1999). Geography and the global environment. *Annals of the Association of American Geographers, 89*, 107–120.

Liverman, D. M. (2004). Who governs, at what scale and at what price? Geography, environmental governance, and the commodification of nature. *Annals of the Association of American Geographers, 94*, 734–738.

Mansfield, B. (2004). Rules of privatization: Contradictions in neoliberal regulation of North Pacific fisheries. *Annals of the Association of American Geographers, 94*, 565–584.

Marston, R. A. (2008). Land, life, and environmental change in mountains. *Annals of the Association of American Geographers, 98*, 507–520.

McAfee, K., & Shapiro, E. (2010). Payments for ecosystem services in Mexico: Nature, neoliberalism, social movements, and the state. *Annals of the Association of American Geographers, 100*, 579–599.

McCarthy, J. (2006). Neoliberalism and the politics of alternatives: Community forestry in British Columbia and the United States. *Annals of the Association of American Geographers, 96*, 84–104.

Medley, K. E., Pobocik, C. M., & Okey, B. W. (2003). Historical changes in forest cover and land ownership in a Midwestern U.S. landscape. *Annals of the Association of American Geographers, 93*, 104–120.

Mertens, B., & Lambin, E. F. (2000). Land-cover-change trajectories in southern Cameroon. *Annals of the Association of American Geographers, 90*, 467–494.

Momsen, J. H. (2000). Gender differences in environmental concern and perception. *Journal of Geography, 99*, 47–56.

Muldavin, J. S. S. (1997). Environmental degradation in Heilongjiang: Policy reform and agrarian dynamics in China's new hybrid economy. *Annals of the Association of American Geographers, 87*, 579–613.

Müller, D., & Munroe, D. (2008). Changing rural landscapes in Albania: Cropland abandonment and forest clearing in the postsocialist transition. *Annals of the Association of American Geographers, 98*, 855–876.

Mustafa, D. (2005). The production of an urban hazardscape in Pakistan: Modernity, vulnerability, and the range of choice. *Annals of the Association of American Geographers, 95*, 566–586.

Mutersbaugh, T. (2002). Building co-ops, constructing cooperation: Spatial strategies and development politics in a Mexican Village. *Annals of the Association of American Geographers, 92*, 756–776.

National Research Council (NRC). (1997). *Rediscovering geography: New relevance for science and society*. Washington, DC: National Academy Press.

National Research Council (NRC). (2010). *Understanding the changing planet: Strategic directions for the geographical sciences*. Washington, DC: National Academy Press.

Naughton-Treves, L. (2002). Wild animals in the garden: Conserving wildlife in Amazonian agroecosystems. *Annals of the Association of American Geographers, 92*, 488–506.

Neumann, R. P. (2005). *Making political ecology*. New York: Oxford University Press.

Nietschmann, B. (1972). Hunting and fishing focus among Miskito Indians, Eastern Nicaragua. *Human Ecology, 1*, 41–467.

Nightingale, A. (2003). Nature-society and development: Social, cultural and ecological change in Nepal. *Geoforum, 34*, 525–540.

Norman, E. S., & Bakker, K. (2009). Transgressing scales: Water governance across the Canada-U.S. borderland. *Annals of the Association of American Geographers, 99*, 99–117.

O'Brien, K. L., & Leichenko, R. (2003). Winners and losers in the context of global change. *Annals of the Association of American Geographers, 93*, 89–103.

O'Reilly, K. (2006). "Traditional" women, "modern" water: Linking gender and commodification in Rajasthan, India. *Geoforum, 37*, 958–972.

Parker, D. C., Manson, S. M., Janssen, M. A., Hoffmann, M. J., & Deadman, P. (2003). Multi-agent systems for the simulation of land-use and land-cover change: A review. *Annals of the Association of American Geographers, 93*, 314–337.

Parris, T. M., & Kates, R. W. (2003). Characterizing a sustainability transition: Goals, targets, trends, and driving forces. *Proceedings of the National Academy of Sciences, 100*, 8068–8073.

Perreault, T. (2008). Custom and contradiction: Rural water governance and the politics of *Usos y Costumbres* in Bolivia's irrigators' movement. *Annals of the Association of American Geographers, 98*, 834–854.

Petts, J., Owens, S., & Bulkeley, H. (2008). Crossing boundaries: Interdisciplinarity in the context of urban environments. *Geoforum, 39*, 593–601.

Pitman, A. J. (2005). On the role of geography in earth system science. *Geoforum, 36*, 137–148.

Polsky, C. (2004). Putting space and time in Ricardian climate change impact studies: Agriculture in the U.S. Great Plains. *Annals of the Association of American Geographers, 94*, 549–564.

Polsky, C., Neff, R., & Yarnal, B. (2009). Establishing vulnerability observatory networks to coordinate the collection and analysis of comparable data. In B. Yarnal, C. Polsky, & J. O'Brien (Eds.), *Sustainable communities on a sustainable planet: The Human-Environment Regional Observatory project* (pp. 83–106). Cambridge: Cambridge University Press.

Prudham, S. (2003). Taming trees: Capital, science, and nature in Pacific Slope tree improvement. *Annals of the Association of American Geographers, 93*, 636–656.

Pulido, L. (2000). Rethinking environmental racism: White privilege and urban development in southern California. *Annals of the Association of American Geographers, 90*, 12–40.

Ramankutty, N., Heller, E., & Rhemtulla, J. (2010). Prevailing myths about agricultural abandonment and forest regrowth in the United States. *Annals of the Association of American Geographers, 100*, 502–512.

Redman, C. L., Grove, J. M., & Kuby, L. H. (2004). Integrating social science into the long-term ecological research (LTER) network: Social dimensions of ecological change and ecological dimensions of social change. *Ecosystems, 7*, 161–171.

Reed, M. G., & Christie, S. (2009). Environmental geography: We're not quite home – Reviewing the gender gap. *Progress in Human Geography, 33*, 246–255.

Rindfuss, R. R., Entwisle, B., Walsh, S. J., Mena, C. F., Erlien, C. M., & Gray, C. L. (2007). Frontier land use change: Synthesis, challenges, and next steps. *Annals of the Association of American Geographers, 97*, 739–754.

Robbins, P. (1998). Authority and environment: Institutional landscapes in Rajasthan, India. *Annals of the Association of American Geographers, 88*, 410–435.

Robbins, P. (2001). Tracking invasive land covers in India or why our landscapes have never been modern. *Annals of the Association of American Geographers, 91*(4), 637–659.

Robbins, P. (2004). *Political ecology: A critical introduction.* Malden: Blackwell.

Rocheleau, D. (1995). Maps, numbers, text, and context – Mixing methods in feminist political ecology. *The Professional Geographer, 47*, 458–466.

Rocheleau, D. E. (2008). Political ecology in the key of policy: From chains of explanation to webs of relation. *Geoforum, 39*, 716–727.

Rocheleau, D. E., & Edmunds, D. (1997). Women, men and trees: Gender, power and property in forest and agrarian landscapes. *World Development, 25*, 1351–1371.

Rocheleau, D. E., & Thomas-Slayter, B. P. (Eds.). (1996). *Feminist political ecology: Global issues and local experiences.* London: Routledge.

Roth, R. J. (2008). 'Fixing' the forest: The spatiality of conservation conflict in Thailand. *Annals of the Association of American Geographers, 98*, 373–391.

Rudel, T. K., Bates, D., & Machinguiashi, R. (2002). A tropical forest transition? Agricultural change, out-migration, and secondary forests in the Ecuadorian Amazon. *Annals of the Association of American Geographers, 92*, 87–102.

Sayre, N. (2008). The genesis, history, and limits of carrying capacity. *Annals of the Association of American Geographers, 98*, 120–134.

Schneider, S. H. (2001). A constructive deconstruction of deconstructionists: A response to Demeritt. *Annals of the Association of American Geographers, 91*, 338–344.

Schrader-Frachette, K. (1998). First things first: Balancing scientific and ethical values in environmental science. *Annals of the Association of American Geographers, 88*, 287–289.

Schroeder, R. A. (1997). "Re-claiming" land in the Gambia: Gendered property rights and environmental intervention. *Annals of the Association of American Geographers, 87*, 487–508.

Seto, K. C., Sánchez-Rodriguez, R., & Fragkias, M. (2010). The new geography of contemporary urbanization and the environment. *Annual Review of Environment and Resources, 35*, 167–194.

Seto, K.C., Reenberg, A., Boone, C. G., Fragkias, M., Haase, D., Langanke, T., Marcotullio, P., Munroe, D. K., Olah, B., & Simon, D. (2012). Urban land teleconnections and sustainability. *Proceedings of the Natonal Academy of Sciences, 109*, 7687–6692.

Simmons, C. S. (2004). The political economy of land conflict in the Eastern Brazilian Amazon. *Annals of the Association of American Geographers, 94*, 183–206.

Singh, S. J., Haberl, H., Gaube, V., Grünbühel, C. M., Lisievici, P., Lutz, J., Matthews, R., Mirtl, M., Vadineanu, A., & Wildenberg, M. (2010). Conceptualising long-term socio-ecological research (LTSER): Integrating the social dimension. In F. Müller, C. Baessler, H. Schubert, & S. Klotz (Eds.), *Long-term ecological research, between theory and application* (pp. 377–398). Dordrecht/Heidelberg/London/New York: Springer.

Skole, D. L. (2004). Geography as a great intellectual melting pot and the preeminent interdisciplinary environmental discipline. *Annals of the Association of American Geographers, 94*, 739–743.

Sluyter, A. (2001). Colonialism and landscape in the Americas: Material/conceptual transformations and continuing consequences. *Annals of the Association of American Geographers, 91*, 410–428.

St. Martin, K. (2001). Making space for community resource management in fisheries. *Annals of the Association of American Geographers, 91*, 122–142.

Swyngedouw, E. (1999). Modernity and hybridity: Nature, regeneracionismo, and the production of the Spanish waterscape, 1890–1930. *Annals of the Association of American Geographers, 89*, 443–465.

Taylor, P. J. (2005). *Unruly complexity: Ecology, interpretation, engagement.* Chicago: University of Chicago Press.

Turner, B. L., II. (1989). The specialist-synthesis approach to the revival of geography—The case of cultural ecology. *Annals of the Association of American Geographers, 79*, 88–100.

Turner, B. L., II. (1991). Thoughts on linking the physical and human sciences in the study of global environmental change. *Research and Exploration, 7*, 133–135.

Turner, B. L., II. (1997). Spirals, bridges and tunnels: Engaging human-environment perspectives in geography. *Ecumene, 4*, 196–217.

Turner, B. L., II. (2002). Contested identities: Human-environment geography and disciplinary implications in a restructuring academy. *Annals of the Association of American Geographers, 92*, 52–74.

Turner, B. L., II, Lambin, E. F., & Reenberg, A. (2007). The emergence of land change sciences for global environmental change and sustainability. *Proceedings of the Natonal Academy of Sciences, 104*, 20666–20671.

Turner, B. L., II, & Robbins, P. (2008). Land-change science and political ecology: Similarities, differences, and implications for sustainability science. *Annual Review of Environment and Resources, 33*, 295–316.

Turner, B. L., II, Matson, P. A., McCarthy, J. J., Corell, R. W., Christensen, L., Eckley, N., Hovelsrud-Broda, G. K., Kasperson, J. X., Kasperson, R. E., Luers, A., Martello, M. L., Mathiesen, S., Naylor, R., Polsky, C., Pulsipher, A., Schiller, A., Selin, H., & Tyler, N. (2003). Illustrating the coupled human-environment system for vulnerability analysis: three case studies. *Proceedings of the National Academy of Sciences, 100*, 8080–8085.

Turner, M. D. (1999). Merging local and regional analyses of land-use change: The case of livestock in the Sahel. *Annals of the Association of American Geographers, 89*, 191–219.

Turner, M. D. (2009). Ecology: Natural and political. In N. Castree, D. Demeritt, D. Liverman, & B. Rhoads (Eds.), *A companion to environmental geography* (pp. 181–197). Malden: Blackwell.

Turner, M. G. (2010). Disturbance and landscape dynamics in a changing world. *Ecology, 91*, 2833–2849.

Vajjhala, S., Krupnick, A., McCormick, E., Grove, M., McDowell, P., Redman, C., Shabman, L., & Small, M. (2007). *Rising to the challenge: Integrating social science into NSF environmental observatories*. Washington, DC: Resources for the Future.

Voeks, R. A. (2004). Disturbance pharmacopoeias: Medicine and myth from the humid tropics. *Annals of the Association of American Geographers, 94*, 868–888.

Walker, R. T. (2003). Mapping process to pattern in the landscape change of the Amazonian frontier. *Annals of the Association of American Geographers, 93*, 376–398.

Walker, R. T., & Solecki, W. D. (2004). Theorizing land-cover and land-use change: The case of the Florida everglades and its degradation. *Annals of the Association of American Geographers, 94*, 311–328.

Walters, B. B., & Vayda, A. P. (2009). Event ecology, causal historical analysis, and human–environment research. *Annals of the Association of American Geographers, 99*, 534–553.

Watts, M. (1983). On the poverty of theory: Natural hazards research in context. In K. Hewitt (Ed.), *Interpretations of calamity: From the viewpoint of human ecology* (pp. 231–262). Boston: Allen and Unwin.

Wescoat, J. L., Jr. (1986). Impacts of federal salinity control on water rights allocation patterns in the Colorado River Basin. *Annals of the Association of American Geographers, 76*, 157–174.

Wescoat, J. L., Jr. (1992). Common themes in the work of White, Gilbert and Dewey, John – A pragmatic appraisal. *Annals of the Association of American Geographers, 82*, 587–607.

White, R. (2004). From wilderness to hybrid landscapes: The cultural turn in environmental history. *Pacific Historian, 66*, 557–564.

Whitmore, T. M., & Turner, B. L., II. (1992). Landscapes of cultivation in Mesoamerica on the eve of the conquest. *Annals of the Association of American Geographers, 82*, 402–425.

Wilbanks, T. J. (1994). Sustainable development in geographic perspective. *Annals of the Association of American Geographers, 84*, 541–556.

Willems-Braun, B. (1997). Buried epistemologies: The politics of nature in (post)colonial British Columbia. *Annals of the Association of American Geographers, 87*, 3–31.

Wolford, W. (2004). This land is ours now: Spatial imaginaries and the struggle for land in Brazil. *Annals of the Association of American Geographers, 94*, 409–424.

Yarnal, B., & Neff, R. (2004). Whither parity? The need for a comprehensive curriculum in human-environment geography. *The Professional Geographer, 56*, 28–36.

Yarnal, B., Polsky, C., & O'Brien, J. (Eds.). (2009). *Sustainable communities on a sustainable planet: The Human-Environment Regional Observatory project.* Cambridge: Cambridge University Press.

Young, E. (2001). State intervention and abuse of the commons: Fisheries development in Baja California Sur, Mexico. *Annals of the Association of American Geographers, 91*, 283–306.

Zimmerer, K. S. (1991). Wetland production and smallholder persistence – Agricultural change in a highland Peruvian region. *Annals of the Association of American Geographers, 81*, 443–463.

Zimmerer, K. S. (1993). Soil erosion and labor shortages in the Andes with special reference to Bolivia, 1953–1991: Implications for conservation-with-development. *World Development, 21*, 1659–1675.

Zimmerer, K. S. (1994). Human geography and the new ecology: The prospect and promise of integration. *Annals of the Association of American Geographers, 84*, 108–125.

Zimmerer, K. S. (1996). Ecology as cornerstone and chimera in human geography. In C. Earle, K. Mathewson, & M. S. Kenzer (Eds.), *Concepts in human geography* (pp. 161–188). London: Rowman and Littlefield.

Zimmerer, K. S. (1999). The reworking of conservation geographies: Nonequilibrium landscapes and nature-society hybrids. *Annals of the Association of American Geographers, 90*, 356–369.

Zimmerer, K. S. (2007). Cultural ecology (and political ecology) in the 'environmental borderlands': Exploring the expanded connectivities within geography. *Progress in Human Geography, 31*, 227–244.

Zimmerer, K. S. (2009). Nature under neoliberalism and beyond in Bolivia: Community-based resource management, environmental conservation, and farmer-and-food movements, 1985-present. In J. Burdick, P. Oxhorn, & K. M. Roberts (Eds.), *Beyond neoliberalism in Latin America? Societies and politics at the crossroads* (pp. 157–174). New York: Palgrave Macmillan.

Zimmerer, K. S. (2010a). Woodlands and agrobiodiversity in irrigation landscapes amidst global change: Bolivia, 1990–2002. *The Professional Geographer, 62*, 335–356.

Zimmerer, K. S. (2010b). Biological diversity in agriculture and global change. *Annual Review of Environment and Resources, 35*, 137–166.

Zimmerer, K. S. (2010c). Retrospective on nature-society geography: Tracing trajectories (1911–2010) and reflecting on translations. *Annals of the Association of American Geographers, 100*, 1076–1094.

Zimmerer, K. S. (2011a). Spatial-geographic models of water scarcity and supply in irrigation engineering and management (Bolivia, 1952–2009). In M. Goldman, P. Nadasdy, & M. D. Turner (Eds.), *Knowing nature: Conversations at the intersection of political ecology and science studies* (pp. 167–185). Chicago: University of Chicago Press.

Zimmerer, K. S. (2011b). The landscape technology of spate irrigation amid development changes: Assembling the links to resources, livelihoods, and agrobiodiversity in the Bolivian Andes. *Global Environmental Change, 21*, 917–934.

Zimmerer, K. S., & Bassett, T. J. (Eds.). (2003). *Political ecology: An integrative approach to geography and environment-development studies.* New York: Guilford Press.

# Chapter 9
# The Contribution of Anthropology to Concepts Guiding LTSER Research

**Ted L Gragson**

**Abstract** Environmental scientists from across the spectrum of physical and biological disciplines are generally agreed that human activities are integral to ecosystems and are organising research networks to identify and address contemporary ecological questions. However, without ready and open access to diverse social areas of expertise and practice, environmental scientists alone will not be able to succeed in carrying out forward-looking, problem-oriented research on simultaneously maintaining Earth's life support systems and meeting human needs. This chapter addresses the specific contribution of anthropology to LTSER research by answering two questions. First, given the numerous calls for interdisciplinary research is there still a role for individual disciplines in contemporary, problem-oriented environmental research? Second, contrary to popular views on what anthropology is and anthropologists do, what is the specific role of anthropology as a discipline in LTSER research? The chapter ends with two case studies from ongoing research in the Coweeta LTER Project in Southern Appalachia that rely on anthropology.

**Keywords** Long-term natural experiment • Place-based research • Interdisciplinary collaboration • Coweeta LTER • Regional anthropology

## 9.1 Introduction

Recognising that pristine systems are rare or non-existent and that the human footprint is global and pervasive (Vitousek et al. 1997; Grimm et al. 2000) serves as a point of departure for examining the contribution of anthropology to LTSER research.

---

T.L. Gragson, Ph.D. (✉)
Department of Anthropology, University of Georgia, Athens, GA, USA
e-mail: tgragson@uga.edu

In terms of the practice of science as usual, there is now general agreement that research should view human activities as integral to ecosystems and that it is important to carry out forward-looking, problem-oriented research on simultaneously maintaining Earth life support systems and meeting human needs (Palmer et al. 2004).

Earth systems are now changing faster than disciplinary research can advance knowledge about their total functioning, and many of the phenomena that underlie global environmental change are nonlinear and cross-scale. In response to this situation, environmental scientists from across the spectrum of physical and biological disciplines, and selected social disciplines, are self-organising into research networks for the purpose of identifying and addressing contemporary ecological questions (Gragson and Grove 2006; Carpenter 2008). Several of these networks are now recognised as environmental observatories – the Long-Term Ecological Research Network (LTER, Hobbie et al. 2003); the National Ecological Observatory Network (NEON, Keller et al. 2008), the Oceans Observatory Initiative (OOI, Isern and Clark 2003), the Global Lake Ecological Observatory Network (GLEON, Kratz et al. 2006), and the Critical Zone Observatory program (CZO, Anderson et al. 2008).

There is no doubt that observations derived from these networks will individually and collectively enhance society's long-term capacity to detect and understand change. However, without the explicit incorporation of social disciplines and the long-term monitoring of social factors it will not be possible to develop models to forecast future conditions in ways that substantively address commonly identified issues about the interaction between humans and biophysical systems. In short, without ready and open access to diverse social areas of expertise and practice, ecologists may not exploit the most cogent or important connections of their research (Boynton et al. 2005).

In the following pages I address the contribution of anthropology as a discipline to LTSER research. I begin by positing two questions that arise with respect to the contribution of any discipline, but in particular a social science discipline, to LTSER research. The numerous calls for interdisciplinary research since Vitousek's article appeared would seem to eclipse the role of disciplines – is there still a role for individual disciplines within contemporary, problem-oriented environmental research? Given the vernacular view of what anthropology is and anthropologists do – what is the role of anthropology in LTSER research? After responding to these questions, I provide two case studies derived from research in the Coweeta LTER Project in Southern Appalachia, a member of the US-LTER Network, that depend on anthropology as a discipline.

## 9.2 Disciplines in an Interdisciplinary World

The questions we confront scientifically and societally defy easy categorisation or solution by traditional disciplinary frameworks. The major challenge in current LTER research, a US research programme that has its roots in the ecosystem ecology of the late 1960s and early 1970s (Coleman 2010), is responding effectively to the call for integration in light of transformations over the last decade in conceptual, fiscal and policy landscapes. The abstract benefits for integrative research at the boundaries between disciplines are indeed exciting (Peters et al. 2008). Nevertheless, it is equally important to understand the conditions by which the promise of interdisciplinarity translates into changes in practice that ensure the vaunted potential of the approach is realised. It should come as no surprise that the constraints on interdisciplinary partnerships increase as the scale and scope of the collaborations increase. A question is whether there continues to be a role for individual disciplines within contemporary, interdisciplinary environmental research.

In the United States, diverse federal agencies including the National Institutes for Health (NIH), the National Science Foundation (NSF), the National Academy of Sciences (NAS) as well as a variety of private organisations such as Ford, MacArthur, Keck, and Heinz have invested heavily in research that transcends single disciplines to focus on a variety of social and environmental problems. While industrial and government laboratories and non-academic settings have largely succeeded in fostering problem-driven research that allows researchers to move easily between working groups, academic institutions are notorious for the administrative and cultural "drag" placed on researchers who would engage in cross-disciplinary research and teaching (CFIR 2004; Whitmer et al. 2010).

State-chartered institutions in particular have emphasised technical advances and the understanding of physical processes as a calculated structural response to funding issues (Jacobs and Frickel 2009; Wainwright 2010). This reflects both the federal funding landscape of the last decade, but also the desire on the part of university officials to avoid conflict with representatives at state and federal levels that might jeopardise institutional funding. Research into social and/or biological processes has been particularly contentious and the National Science Foundation a special target of opportunity in this political arena. Actions include systematic dis-investment in certain programmes and more recently a call to eliminate the Social, Behavioral and Economics Directorate, the principal source of social science research funding in the United States (Coburn 2011). The resulting tension for individuals between the scientific promise of interdisciplinary research and the prospect of tenure and promotion is a true concern at many academic institutions (Rhoten and Parker 2004).

Strong advocates of interdisciplinarity are reacting to the situation in academia when they describe disciplines as disconnected silos that inhibit innovation and stifle inquiry on topics outside the narrow confines of each discipline. It is nevertheless important to recognise that change in academia does occur through the perseverance of individuals. However, there is little empirical research on the processes by which individuals engage in integration and synthesis across disciplines to build explanations and solve problems of relevance to them and the broader society (Hackett and Rhoten 2009; Jacobs and Frickel 2009).

Bibliometric research provides a window on the practices of individuals and a view of science as an interconnected web of scholarship. In a study conducted by the National Science Foundation (NSF 2002), cross-disciplinary citation rates range from highs of 38.3% in biology to lows of 16.8% in earth sciences; the social sciences fall in the middle at 22.7%. Of the 11 broad fields into which the social sciences were grouped, 71.7% of citations in area studies come from journals in other disciplines. Economics is the most insular with only 18.7% of references based on research outside of economics. Anthropology (i.e., ethnography) and archaeology are grouped together, and 47.2% of all citations in these two areas of study come from outside the discipline (Table 6–54, NSF 2002).

Advances in understanding within or beyond disciplines result from the collective action of members of a scientific or intellectual community who organise themselves to address an identified problem (Khun 1962; Jacobs and Frickel 2009). The transformative promise of interdisciplinarity thus lies in its capacity to interpenetrate disciplines. This translates into participants engaging in novel communicative forms and opening channels for renegotiating disciplinary boundaries that generate new epistemic standards (Fuller and Collier 2004; Jacobs and Frickel 2009). The empirical focus of research thus shifts from the structural nature of disciplinary interrelations to questions of process. However, it does not mean the disciplines are no longer important in contemporary interdisciplinary research. According to Abbott (2001), disciplines contribute abstract, theory-driven knowledge that is substantively necessary to the problem-driven knowledge that interdisciplinarity tends to produce.

In the realm of environmental research, learning how to manage feedbacks between ecosystems and humans is vital if we are to move toward a world in which the health of ecosystems and the wellbeing of humans are maintained if not improved. Every ecosystem on Earth is influenced by human actions (Vitousek et al. 1997; Palmer et al. 2004), leading to the realisation that many if not most of today's pressing issues require environments to be viewed as socio-ecological systems (Liu et al. 2007). We need to focus on the dynamic processes of socio-ecological systems, not merely the processes characteristic of de-coupled social or ecological systems, through research that is place-based, long-term, cross-scale, and comparative (Collins et al. 2011).

The focus on patterns that proxy the relation between humans and the environment – e.g., land cover change being a prime example – has precluded significant attention in environmental research to the processes that derive from behavioural decisions at scales from the individual to the institutional level. The nature of the problems we confront, however, require that human behaviour and the institutions they organise themselves into be treated as endogenous elements of ecosystem change, which in turn means dissecting processes of perception, valuation, communication and response of real actors operating within identifiable systems (Sayer 2000; Westley et al. 2002; Haberl et al. 2006; Collins et al. 2011). Disciplines still matter because they make possible the **cross-scale** research necessary to resolve between the agent and system dimensions of socio-ecological research.

Humans self-organise into diverse social systems that exhibit scale dependencies and capabilities that qualitatively change the dynamics of socio-ecological systems. Local actors can modify, ignore, and counteract the influence of public policy and regulator instructions yet remain subject to the influence of national laws and multilateral trade that are in turn responsive to diverse institutional antecedents and path-dependencies (Gibson et al. 2000). Disciplines also matter because they are fundamental to ensuring the study of socio-ecological systems as **long-term** natural experiments. The legacy of past decisions, especially when they involve land-cover change, landscape transformations, or the built environment guide future options by facilitating certain actions and raising barriers to others (Gragson and Bolstad 2006). Different disciplines approach the various dimensions of decision-making in distinct ways that help understand the transitions over time that can guide us toward a sustainable future.

One value frequently emphasised of interdisciplinary research is that it can uncover the full complexity and specificity of concrete reality (Hackett and Rhoten 2009). This value derives from interdisciplinary research being oriented to socially relevant "real-world" problems that no single discipline can resolve (Funtowicz and Ravetz 1993; Klein 2000; Whitmer et al. 2010). "Applied", "use-inspired" or "engaged" research has made interdisciplinary "problem-solving" the focus of knowledge production (Klein 2000; Whitmer et al. 2010), but this implies that research be **place-based**. Place-based research by definition involves local stakeholders – citizens, scientists, public servants, or other individuals belonging to diverse groups varying in self-interest and outlook. Places are not simply backdrops to human activity – they are the outcomes of activity and in turn shape activity (Rodning 2009). It takes time to build places, live in them, and abandon them. Disciplines continue to matter in this context because not all disciplines bring to bear the critical experience of how to establish rapport between discussants within inter-cultural encounters, sample across qualitatively distinct human groups, or have the reflexive awareness necessary to foster "team science" (Stokols et al. 2003; Hackett and Rhoten 2009; Jacobs and Frickel 2009).

Finally, nothing that humans do is ever "natural" in the vernacular meaning of the word. The ultimate grand challenge for social science, top-ten questions not withstanding (Giles 2011), is to account for the diversity of human actions while acknowledging that humans the world over are basically the same biologically. In short, how can we understand individuals and groups without resort to biological essentialism while acknowledging that humans are a special kind of social animal organised into culturally-distinct societies? Place-based research draws attention to the existence of and the necessity to account for human diversity. The risk of place-based research is being interesting but irrelevant, given that environmental problems and their solutions are global in scope. Research that engages this challenge draws on theories and methods that can ensure that both the research and the results are **comparative** and potentially of transcendent value.

## 9.3 More Than Vernacular Anthropology

The previous section gives credence to why disciplines continue to matter in an interdisciplinary world given the demand for contemporary environmental research that is place-based, long-term, cross-scale and comparative. The question remains as to what the potential contribution of anthropology is to LTSER research. To address this question we must first confront what is more than anything a vernacular or popular view as to what anthropology is and anthropologists do. While this discussion has a decidedly American bias, it still serves to render more clearly the disciplinary contribution of anthropology to a global environmental research agenda.

Like all disciplines, anthropology has a disciplinary history linked to founding fathers and mothers. However, it is the history of the discipline during the last half of the twentieth century that is particularly relevant in this context. Clifford Geertz has been characterised as the most influential American cultural anthropologist of the second half of the twentieth century. His writings are viewed by many as defining and giving character to the intellectual agenda of a meaning-centred, non-reductive interpretive social science (Shweder and Good 2005). Geertz argued that cultural anthropology studied people living in out of the way places by participating in their daily lives (Gable 2011). He further believed that the role of anthropology was to interpret the guiding symbols of each culture, which Geertz described as "a system of inherited conceptions expressed in symbolic forms by means of which people communicate, perpetuate, and develop their knowledge about and attitudes toward life" (Geertz 1977: 89).

For Geertz, ethnographic encounters in out–of–the–way places were the basis for illustrating large ideas in philosophical essays that made only passing references to the work of previous authors (Gable 2011). His method of choice was **thick description**. Geertz claimed to have adopted the term from philosopher Gilbert Ryle (Geertz 1977: 3), a follower of Wittgenstein. There is no denying the place of philosophy, meditation, and thick description in advancing certain kinds of knowledge within anthropology. And, Geertz' writing was seductively persuasive in guiding many anthropologists as to the disciplinary means for contributing to the collective understanding of what it means to be human. While having enormous personal influence on the discipline, Geertz advocated an approach emphasising contrasts, which was an extension and transformation of the long Western meditation on the savage going back through Freud, Durkheim, Marx, Rousseau, Montesquieu, and Hobbes.

The savage in this view is a person who is different from yet similar to the Westerner. Each *tribe* – Balinese and Moroccan or ecologists and anthropologists – has its own culture and while similarities abound, it is the contrasts that are emphasised. The most common preconception about anthropology by those outside the discipline is that it consists of anthropologists in search of commonsense understandings of what culture is by reasoning or meditating on contrasts derived from the study of people living in out–of–the–way places. From this vantage point anyone can be, and frequently tries to be, an anthropologist – personal experience is the only prerequisite for explanation. However, commonsense understandings of what culture is that circulate in vernacular discourse, as contrasts between nature vs. culture, developed vs. developing, etc., are not sufficient to the needs of LTSER.

### 9.3.1 Anthropology as a Problem-Oriented Discipline

Anthropology is a non-paradigmatic discipline and its value to LTSER research lies in being a "boundary discipline." There are several dimensions to this characterisation of the discipline, but as with all intensive research, the primary concern lies with what makes things happen in specific cases (Sayer 2000: 20–1). Sol Tax's research (1963) is a counter-example of the anthropological approach advocated by Clifford Geertz, and an interesting case of the circular interdisciplinary flow of information and concepts that advance understanding. Tax worked during the 1930s in the Indian community of Panajachel, Guatemala, and the results of his research and that of those who relied on it has been so thoroughly incorporated into the mainstream of anthropology and economics that it is difficult to appreciate how revolutionary the results were at the time.

From when Tax carried out his research up through its publication it was widely accepted among economists and policymakers that the marginal product of labour in agriculture in developing countries was zero. This meant that labour could be withdrawn from agriculture for industrialisation at no cost to agricultural production (Abler and Sukhatme 2006). It was also widely argued that farmers in developing countries were guided by tradition or culture in the vernacular sense of the word, and therefore did not respond to economic incentives. Tax, who undertook his study prior to the diffusion of general equilibrium theory in economics, sought to demonstrate that the economic life of Indian communities was not unduly dominated by irrational beliefs or religious influences, and that it could be comprehended with the analytical tools of microeconomic theory. For example, he argued that the residents of Panajachel were true representatives of the species Homo economicus.

Tax summarised his research with the claim that he discovered among the Indians not just evidence of individual rationality, but a "money economy organised in single households as both consumption and production units with a strongly developed market which tends to be perfectly competitive" (Tax 1963: 188). To focus only on his conclusion is to miss the significance of Tax's contribution not only to understanding, but the means to understanding (Schweigert 1994; Ball and Pounder 1996; Abler and Sukhatme 2006). *Penny Capitalism* is a true example of a real world economy. Tax literally "discovered" peasant rationality, economic efficiency at the individual enterprise level, and the ubiquity of the price mechanism in a remote, rural community later to be termed an LDC – lesser developed country (Schweigert 1994). Theodore Schultz, who in 1979 would be the first development economist awarded a Nobel Prize, cites the work of Sol Tax as major evidence in support of his efficient-but-poor hypothesis (Schultz 1964) – farmers make efficient use of the resources available to them.

The two major contributions of the work of Schultz (Ball and Pounder 1996) were first, to refute the notion that farmers in developing countries are poor because of their cultural characteristics including lack of a work ethic, failure to understand the idea of saving, or ignorance about the best way to use their resources. The second was the policy implication that followed from this assessment of farmer behaviour. Succinctly, Schultz noted that outside experts such as extension agents and development advisers could not help farmers improve productivity merely by suggesting a reallocation of available factors of production. Investments in education were required to facilitate the diffusion of new factors that could enhance productivity. In short, local knowledge has value in itself and engaging in a dialogue with local practitioners is productive.

Research on poverty by Tucker et al. (2011) is a very contemporary expression of anthropology as a problem-oriented discipline that draws on a disciplinary interest coincident with the origin of modern anthropology in the United States. Franz Boas (1916), frequently referred to as the "Father of American Anthropology," and Alfred Kroeber (1916), a student of Boas famous in his own right for developing the culture area concept, challenged heredity-based explanations for gaps in wealth and achievement in the United States common in the early twentieth century. Tucker et al. also build directly on theoretical, disciplinary and methodological advances that emerge in the wake of the research by Sol Tax at Panajachel.

Poverty reduction is a multi-billion dollar undertaking that each year engages researchers, national governments, and international development organisations from around the globe in discussions and activities that seek to curb the psychosocial suffering, morbidity and diverse problems associated with the condition. Those involved in attempts to reduce poverty hope to demonstrate moral commitment to improving the quality of life while promoting security and stability (Hayami and Godo 2005; Tucker et al. 2011). Nevertheless, poverty is both difficult to define and hard to measure, which accounts in many ways for the range of theoretical vantage points used since the 1960s by anthropologists to examine lives of the poor. These include underdevelopment (Escobar 1995), political economy (Roseberry 1988), modes of production (Siskind 1978), neoliberal policy (Morgen and Maskovsky 2003), consumption (Douglass and Isherwood 1996), and vulnerability (Oliver-Smith 1996).

Placed within environmental and ecological anthropology, however, the work of Tucker et al. (Tucker et al. 2011) seeks to understand the human-environment relation (Haenn and Wilk 2006; Dove and Carpenter 2008) while speaking to policy-relevant issues of social justice and human health. Their research also runs counter to a disturbing trend noted in environmental anthropology (as well as ecology) for publishing qualitative thought and opinion pieces unsubstantiated by empirical evidence or analysis (Charnley and Durham 2010). Tucker et al. ask how individuals labelled as "poor" in non-Western societies with poorly developed markets in which resources of all kinds are more often obtained through non-monetary exchange understand and experience poverty? They create a dialogue between a folk model developed from focus group research among Masikoro (farmers), Vezo (farmers and fishermen), and Mikea (forager-bricoleurs) in southwestern Madagascar and four theoretical models from economics and anthropology.

Tucker et al. clarify that folk models neither validate nor reject theoretical explanations, but help researchers understand the kind of interventions that would be culturally appropriate and effective in specific settings. This is because the folk model is best at revealing community – and regional-level concerns. The Western model that best helps understand the experiences of poverty and wealth in southwestern Madagascar is Mode of Production. This model explains the occurrence of poverty as the consequence of social relations of production that favour the more powerful and create winners and losers (Meillassoux 1981; Wolf 1982; Graeber 2006). The model also helps identify the key to reducing poverty: change property rules to reduce or eliminate exploitation. The significance of this research is that

Tucker et al. (2011) show that the close correspondence between the folk model and the Mode of Production model reveal that pro-growth development will exacerbate rather than reduce poverty under the prevailing conditions in southwestern Madagascar.

## 9.3.2 Anthropology Beyond the Savage

Despite the hold on the public imagination of the Geertzian idea of anthropology as the study of people living in out of the way places, anthropologists also examine the cultures they belong to. In the late 1960s, Laura Nader (1969) suggested that anthropology had done far too much "studying down" and advocated ethnographers to also "study up" by treating as natives upper-level administrators, scientists, and government officials in order to identify the values and beliefs that underlie the decision-making of elites. For Nader, the purpose of "studying up" was to expose how power is naturalised and made normal and invisible.

The study of organisations and institutions is ever more important in contemporary environmental research, in which decision-making depends on integrating scientific understanding to a deliberative process that ensures the science is judged relevant to the decision and credible to the affected parties (Brewer and Stern 2005). This contemporary concern for democracy in many situations challenges us to think about what kinds of rules and forms of enforcement can result in order. In other words, how do congeries of people regardless of political structure come to act as a group and how do its members self-identify as members of a group? A related question is how a particular social structure affects and influences the way thought is ordered and systematised at the personal level. These are fundamentally anthropological questions that have been widely examined at numerous locales under highly diverse circumstances since the middle of the nineteenth century (e.g., Morgan 1877), yet they also show the compounding influence of interdisciplinary discussion over time.

Garret Hardin's (1968) work on the commons equated the commons with tragedy, and argued that humans will inevitably overexploit resources that are open to all. Hardin's solution of "mutual coercion mutually agreed upon" opened the possibility of social control to regulate access (Hardin 1968: 1247). This work helped to side-step prevailing heroic assumptions that individuals either behave identically or that they merely react to external pressure. Vayda and Rappaport built on this idea of social control (Vayda and Rappaport 1968) and pointed out that the tragedy of the commons could be avoided by implementing institutions that promote action in the collective interest. Ciriacy-Wantrup and Bishop described how common property institutions have played socially beneficial roles since prehistory (Ciriacy-Wantrup and Bishop 1975: 713) while Ostrom differentiated between common-pool resources and common property management systems (Ostrom 1990).

Institutions are the "rules of the game" (Acheson 2003) or the "ways of organising activities" (Dietz et al. 2003). They are the social controls that permit the resolution of the collective action dilemma and provide for control over a resource. Crawford and Ostrom (1995) identify three approaches – institutions-as-equilibria, institutions-as-norms, and institutions-as-rules – but argued that the three approaches simply highlight different opportunities and constraints. They suggested that the ideal course of action was to examine what they termed institutional statements (Crawford and Ostrom 1995: 583). Such statements are the shared strategies, norms, and rules that govern behaviour by permitting, forbidding, or prescribing actions. Axelrod (1986) called them norms, March and Olsen (1989) called them rules, and Bourdieu (1977) called them doxic elements of action.

Shared strategies refer to those actions that people generally take or avoid not because there is moral or social pressure to do so or because there are externally-imposed consequences for doing so, but because they feel that it is the most beneficial course of action in itself. Kiser and Ostrom (1982) drew on the diverse and largely disconnected literature regarding the effect institutions have on behaviour to create a meta-theoretical framework for understanding the relationship. Their framework distinguishes between three interrelated yet separate levels of analysis: the operational level, the collective choice level, and the constitutional level. The three levels represent a hierarchy in which the decisions and actions at one level serve to circumscribe those that can be made at lower levels.

Research by Elinor Ostrom has challenged the conventional wisdom that common property is poorly managed and should either be regulated by central authorities or privatised. For this, she received the 2009 Nobel Prize in Economics. These insights are fundamental to the work of anthropologists aided by Nader's reframing of who is a "native" that has resulted in three distinct approaches to organisations: (1) organisations as vessels or containers of cultures (Fiske 1994); (2) organisations as ongoing social processes with a focus on social interactions, relations, events and dynamics involved in the production, reproduction, and alteration of organisational life (Conkling 1984); and (3) the study of policies, policy-makers and the powerful (Heyman 2004). The epistemological distinction between actors and systems expressed in these ideas are fundamental to examining the relations that hold between individuals, institutions and organisations within LTSER research (Haberl et al. 2006).

The functional connectivity of resource-use systems and ecosystems, made more obvious than it already was by globalisation and climate change, presents real challenges for environmental governance (Brondizio et al. 2009). Most resources are linked horizontally to other resources at a similar spatial level, and vertically to larger and smaller systems, with connectivity mediated by levels of social organisational complexity that can range from a local household to an international governance entity (Charnley and Durham 2010). Anthropological research has been particularly important in showing that there is no unique and globally applicable solution for governing ecosystems and their services effectively, efficiently, and equitably on a sustainable basis. As Tucker et al. (2011) show for southwestern Madagascar, the disciplinary strength of anthropology has been the analysis of institutional resource-use systems that facilitate the coproduction, mediation, translation and negotiation of information and knowledge within and across levels (Lansing 1991; Guillet 1998; German et al. 2009; Orlove and Caton 2010).

### 9.3.3 Anthropology of the Mind

The subdiscipline of cognitive anthropology explores the sharing and transmission of culture and knowledge (D'Andrade 1995). It rests on a distributional view of culture as shared understanding within a group of people represented by coherent logical structures termed cultural models (Kempton et al. 1996; Romney and Moore 1998; Bang et al. 2007). The basic building blocks of cultural models are mental models, that are simplified representations of the world held by an individual that allow them to interpret observations, generate novel inferences, and solve problems.

The organisation of knowledge, the perception of environment, and the structure of the environment itself are linked in complex ways (Kaplan and Herbert 1987; Atran et al. 1999; Stern 2000; Dutcher et al. 2007; Hunziker et al. 2008). Understanding this relation, however, is important to understand resources from a macro-social perspective in order to better manage the resources, optimise their allocation, and design policies for their use. This is an issue at the boundary between disciplines to which anthropology and several other disciplines have been major contributors. One of the singular contributors was Harold Conklin, who conducted extensive ethnoecological and linguistic research in Southeast Asia in which he focused on indigenous ways of understanding and knowing the world that developed by "living in the environment." The approach is commonly referred to as ethnoscience (Atran 1993; Ingold 2000).

Individuals organise information about the world into cultural models responsible for constructing the beliefs held by the individual; beliefs inform how individuals evaluate their surroundings, and result in rankings of classes of objects linked to preferences (Slovic 1995; Druckman and Lupia 2000). Preferences in turn provide individuals with inferential structure to their interactions with the world, and it is clear that the formation of preferences reflects both factors internal to the individual as well as beyond their personal control. These include the social and cultural context in which the preferences are formed.

There are references in this research to the ageless problem of the savage as someone both different from and similar to the Westerner. For example, Claude Lévi-Strauss argued that the "savage" mind had the same structures as the "civilised" mind although human characteristics were the same everywhere giving rise to structuralism or the search for the underlying patterns of thought in all forms of human activity (1968). The approach brings *mind* out of the realm of philosophy into a problem-oriented framework providing the basis for understanding decision-making (Purcell et al. 1994; Dutcher et al. 2004; Kaplan and Austin 2004; Hunziker et al. 2008).

Individual decisions are not strictly based on the quantity or quality of resources, but on the comparisons of the relative value of different options and the marginal benefit and cost of these options. The economic value of a good or service reflects the marginal value it contributes to an individual's utility or society's welfare, which depends on many different factors, including the availability of substitutes (Bockstael et al. 2000). In some cases, economic value is fairly easy to identify. For example, the value of timber can be reasonably inferred from timber prices, which reflect many independent individual decisions about the tradeoff between buying or supplying timber and buying or supplying something else. However, services of nature such as water quantity, water quality, and biodiversity are not traded in formal markets and thus do not have a competitive market price from which to infer value. This does not mean these services do not have an economic value only that this value is harder to estimate. Anthropology is a major contributor to the determination of value in such situations since the work of Sol Tax and others has helped to better understand rationality in real world situations.

### 9.3.4 *Anthropology as Cross-Cultural Comparison*

The tendency for human groups to distort their perceptions of the different, another expression of the savage, only became widely appreciated during the twentieth century. Serious efforts to overcome that distortion by inventing ways to understand differences in social life through categories that transcend a single group are also relatively recent in origin. Comparative study helps describe, explain and develop theories about socio-cultural phenomena as they occur in social units (groups, tribes, societies, cultures) that are evidently dissimilar to one another. Comparative social science is not a species of inquiry independent from the remainder of social-scientific inquiry. As noted by Swanson (1971: 145) "Thinking without comparisons is unthinkable." The act of describing a situation – e.g., densely populated or democratic – presupposes a universe of situations that are more or less populated or more or less democratic, and thus rests on the assumption that the situation being described lies somewhere *in comparison with* other like-situations.

Alexis de Tocqueville is widely hailed as a perceptive and brilliant commentator on American society (i.e., *Democracy in America* de Toqueville 2000). Yet his analysis of the condition of America is continually informed by his diagnosis of French society, and rests on an overriding preoccupation with the issue of social equality versus social inequality. de Tocqueville's comparisons and contrasts were made in the context of a partially formulated model of the complex interaction of historical forces (Fig. 9.1, Smelser 1976). While attempting to maintain maximum objectivity, the problems with de Tocqueville's approach are those common to many comparative studies:

1. The use of indirect indicators for comparative variables
2. The selection of comparative cases
3. The imputation of causal relations to comparative associations

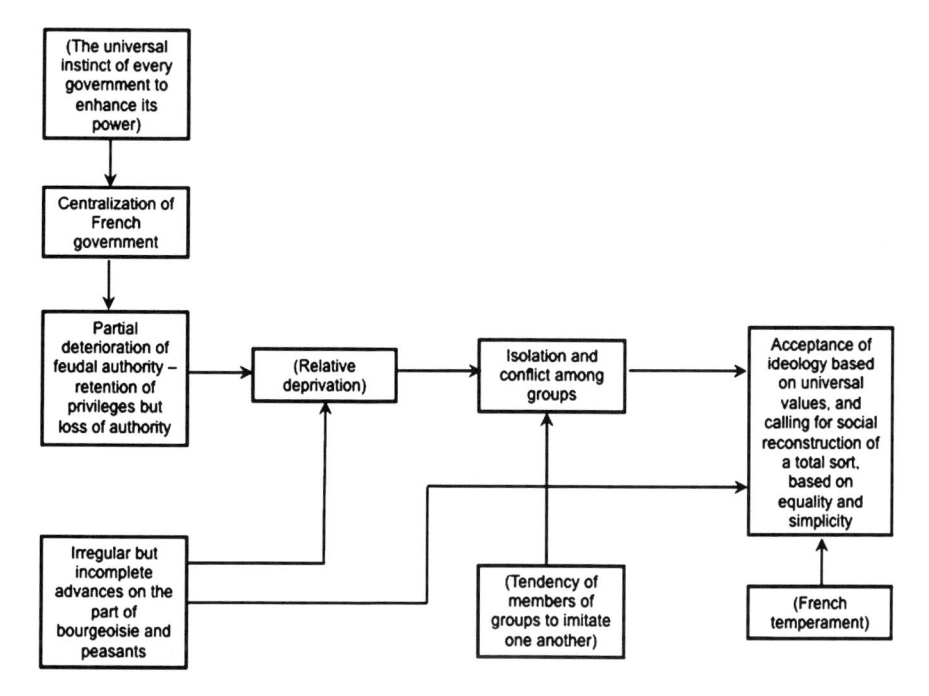

# 9 The Contribution of Anthropology to Concepts Guiding LTSER Research

[Flowchart with the following boxes and connections:

- (The universal instinct of every government to enhance its power) → Centralization of French government → Partial deterioration of feudal authority – retention of privileges but loss of authority
- Partial deterioration of feudal authority → (Relative deprivation) → Isolation and conflict among groups → Acceptance of ideology based on universal values, and calling for social reconstruction of a total sort, based on equality and simplicity
- Irregular but incomplete advances on the part of bourgeoisie and peasants → (Relative deprivation); also connects to the Acceptance of ideology box
- (Tendency of members of groups to imitate one another) → Isolation and conflict among groups
- (French temperament) → Acceptance of ideology box]

**Fig. 9.1** Model of "circumstances remote in time and of a general order" from de Toqueville's analysis of the French Revolution – factors not in parenthesis are features identified by de Toqueville; those in parenthesis are psychological assumptions or assertions (After Smelser 1976)

Comparison is an indispensable scientific technique. However, it is not about drawing sharp distinctions between quantitative and qualitative studies as it is sometimes caricatured. It is best understood as a problem in data reduction and control of variation (Preissle and LeCompte 1981; Bollen et al. 1993) in the course of an investigation.

Most studies can be placed along the continua of four distinct dimensions of the research process. The **inductive-deductive dimension** refers to the place of theory in an investigation. The **generative-verificative dimension** denotes the position of evidence within an investigation and the generalisability of the results. The **constructive-enumerative dimension** refers to the ways in which the units of analysis of a study are formulated and delineated, while the **subjective-objective dimension** refers to the internal vs. external vantage point from which observations are made. Reference back to these dimensions helps with the constant comparison practicing anthropologists rely on in seeking to understand the diversity of human actions without resort to biological essentialism. Nothing done by humans is ever "natural" in the deterministic meaning of the term. The observation is worth making since the developmental trajectory of LTER research in the United States has been directed by the biological and physical sciences exclusive of significant social science input.

## 9.4 LTSER Through an LTER Lens

In LTSER research, individual researchers are called to address issues at the margins of traditional disciplinary ways of examining the world. This follows from the recognition that every ecosystem on Earth is influenced by human actions (Vitousek et al. 1997; Palmer et al. 2004). The consensus view now holds that for many of today's pressing issues the environment is best studied as a socio-ecological system (Liu et al. 2007). So-called pure social and biophysical sciences must continue, but approaches are also needed to understand the dynamic processes unique to socio-ecological systems. While calls for integration are legion (e.g., Palmer et al. 2004; Pickett et al. 2005; Farber et al. 2006; Haberl et al. 2006; Liu et al. 2007), there are relatively few useful roadmaps for implementing integrated, hypothesis-driven research in socio-ecological systems. An exception is the recently proposed Press-Pulse Dynamics Model (Collins et al. 2011) for addressing socio-ecological research needs at US-LTER sites.

LTER sites were initially established as human exclosures isolated from the effects of real world human activities so controlled interventions could be studied. LTER research, however, increasingly dissolves the conceptual boundary between the social and ecological sciences to examine the reality of a world pervasively influenced by humans. Four fundamental, crosscutting questions emerge from an examination of the needs of problem-oriented research at LTER sites:

1. What are the human dimensions of an LTER site?

   - What are the effects of land-use legacies on landscape patterns and processes – past, present, future?
   - How do adjacent and regional land uses influence an LTER site?

2. How do people and organisations influence the spatial and temporal scale of environmental conditions?

   - Which conditions do they influence?
   - Why do they influence these conditions?
   - What are the consequences of their influence?
   - How do social components contribute to the resilience of the system?

3. What affects the distribution of ecological goods and services across spatial and temporal scales?

   - In what quantities?
   - How are they distributed?
   - With what consequences?

4. What is the role of science in environmental decision-making?

   - How is community knowledge represented in LTER data?
   - What happens when site-level data is applied more widely or hierarchically?
   - What factors affect the development and longevity of natural resource decisions, agreements and policy?
   - What should the decision-making process be?

Humans are a biological species that under certain circumstances can be analytically treated like any other organism. However, the practical challenges of conducting socio-ecological research requires a level of sophistication in social science concepts and approaches at least equal to the sophistication in biophysical science concepts and approaches already evident in LTER site-level research. It is neither important nor necessary to convince all biophysical scientists about the inherent value of social science research, nor is the reverse true for social scientists vis-à-vis biophysical science. However, *disciplinary diversity* is required to advance socio-ecological understanding and this diversity must cross the boundary between the biological and the social sciences.

## 9.5 Two Case Studies

Following are two case studies in which anthropology contributed to problem-oriented research in real-world settings that emphasises the dual importance of theory and method for advancing understanding. The case studies derive from ongoing research in the Coweeta LTER Project based in southern Appalachia (Fig. 9.2a, b). The 2008–2014 research objective of the project centres on understanding how the focal ecosystem services of water quantity, water quality, and biodiversity will be impacted by the transition in land uses from wildland to urban and peri-urban, changes in climate, and the interactions between changes in land use and climate. Southern Appalachia is a mountainous region in the Southeastern US characterised by extremely high levels of biodiversity and extensive networks of waterways. Elevations range from 600 to 2,000 masl while vegetation is dominated by temperate deciduous forest. The climate is humid subtropical to marine humid temperate, with a regional average rainfall of 1,400 mm per year.

Evidence for southern Appalachia indicates that humans began extractive use of natural resources about 12,000 years ago with activities concentrated in floodplains and cove sites (Gragson et al. 2008). Land use intensified over time and today approximately 98% of southern Appalachia has been affected through farming, logging, mining, and road building (Davis 2000; Gragson and Bolstad 2006). While population has continually increased since the late 1960s, the economy has shifted from agriculture and manufacturing to tourism and recreation (Gragson and Bolstad 2006), resulting in extensive reforestation of former agricultural land (Wear and Bolstad 1998).

### 9.5.1 *Valuing Ecosystem Services*

Ecosystem services are benefits that humans directly or indirectly receive from the natural environment at different temporal and spatial scales (Farber et al. 2006). The

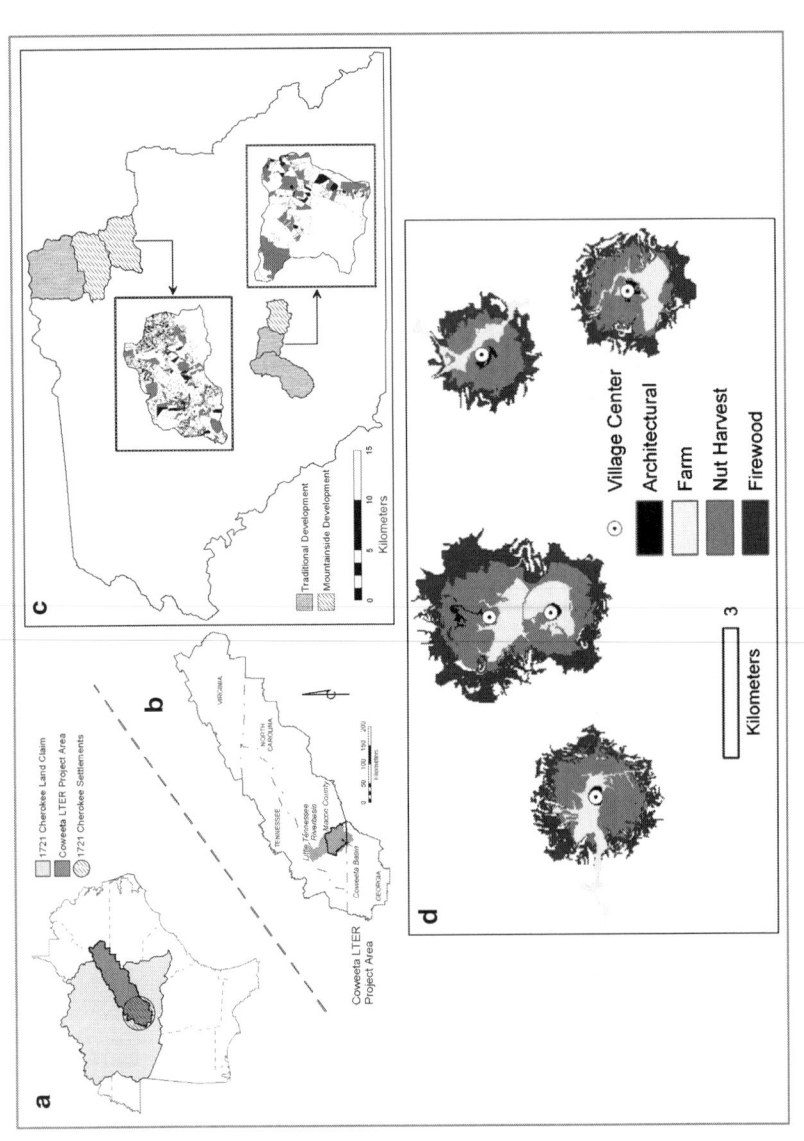

**Fig. 9.2** Coweeta LTER project area scales and research activities. (**a**) SE U.S. showing relation of project areas. (**b**) Coweeta LTER Project area in relation to state boundaries and intensive research sites. (**c**) Macon County NC with research watersheds classified by development type. Insets show watersheds classified by land ownership persistence (*LOPI*) – darker values indicate greater persistence. (**d**) Example of land required for architecture, agriculture, firewood, and mast harvest in proximity to a group of Cherokee villages based on "best" assumptions of per capita requirements (After Bolstad and Gragson 2008)

rapid in-migration to southern Appalachia provides an important opportunity to reconcile the impacts of behaviour at the local scale with the hydrologic processes associated with exurbanisation and development at the watershed scale (Groffman et al. 2003). In this context, connecting patterns on the ground to process scales will link the natural endowment of ecosystem services in southern Appalachia that derive from abundant precipitation and dense stream networks that make the region a water tower (Viviroli et al. 2007) for the Southeastern US. Anthropology in this case helps move the research beyond broad a-priori claims about the universality of human behaviour in relation to water.

Coweeta LTER research has shown the value of distinguishing between land cover and land use (Webster et al. 2012). The objective now is to scale decision-making to the regional level so as to connect physically mediated processes (e.g., climate change) and socially mediated processes (e.g., exurbanisation) in policy-relevant ways. Traditional experiments provide only limited understanding of the complex interactions between services and decisions at regional scales, while correlation-based models omit the relevant processes (Clark et al. 2003; Gragson and Bolstad 2006; Ibáñez et al. 2006). In addition, behavioural scientists routinely publish broad claims about the universality of human psychology and behaviour based on sampling WEIRD societies – Western, Educated, Industrialised, Rich, and Democratic (Henrich et al. 2010). The assumption – implicit or explicit – is that such "standard subjects" are both representative as well as sufficient for understanding the human condition irrespective of geography, economy, culture or any other social dimension.

For all these reasons, the Coweeta LTER research gives great importance to parcel-level decision-making (Fig. 9.2c, e.g., Jurgelski 2004; Gragson and Bolstad 2006; Chamblee et al. 2008). This approach recognises that: (1) observed landscapes produced by actual household decisions are much patchier and have more edge than those predicted by utility-maximisation models; (2) the diversity of actual decisions made by households exceeds the practical allowance of factorial design experiments; and (3) there are many unobserved factors that influence utility maximising decisions that are ignored in many theoretical models (Evans and Moran 2002; Cho et al. 2005; Bolstad and Gragson 2008). Focusing on parcel-level decision-making also helps avoid explanations of human behaviour by resort to the "trickle down theory of neighbourhood effects" in which individual action is understood as merely a response to social, spatial and/or biophysical constraints (Glass and McAtee 2006; Entwisle 2007).

Our approach consists of combining revealed and stated preference methods with face-to-face interviews and representative surveys common to anthropology and many other social fields. This enables us to record attributes about households and its use of the land at a particular moment in time along with knowledge on motivations, incentives, and preferences necessary for explaining actual decisions (Langley 1999; Evans et al. 2006; Grove et al. 2006; Eisenhardt and Graebner 2007). This cross-scale and cross-level approach (Brondizio et al. 2009; Bisaro et al. 2010) lets us capture the perceptions, values and communicative responses of individuals situated within institutions and organisations, embedded within multilevel governance regimes.

Anthropologically, we are bringing together the cognitive and behavioural dimensions of being human. The problem-oriented research is thus designed to seek solutions to the central paradox of exurbanisation: by moving from the city to enjoy the forested and rural landscapes of southern Appalachia, people threaten not only the new place they now cherish but the old places they left behind.

## 9.5.2 A Legacy of Change

Exurbanisation and climate change are global contemporary processes. They are nevertheless related to a common problem in research in establishing the relation between humans and the environment: resolving the relative magnitude of climatic- versus human-induced changes across space and through time. Leigh and Webb (2006) used sedimentation rates to postulate that more frequent flooding occurred in southern Appalachia during the early to middle Holocene interval of global warming, while the late Holocene had a climate regime much like the present. Forest disturbance records for southern Appalachia post-AD 1500 indicate the importance of small forest canopy gaps interrupted by occasional and noticeably higher disturbance peaks (Clinton et al. 1993; Butler 2006). However, the resolution of the sedimentary and biotic archives in these two instances do not allow us to distinguish human from climatic forcing of sedimentation or forest disturbance rates. Some pulses are no doubt due to natural events such as hurricanes while others relate to human events such as agricultural forest clearing.

The nature of the relation between subsistence-based populations and their environments has long been debated. The extremes in this debate are people as agents of landscape degradation versus people as landscape managers. The first position is often advanced by those relying on archaeological and material culture evidence (Kay 1994; Krech 1999; Diamond 2003), while the second is often advanced by those relying on ethnographic and observational evidence (Turner 2005; Berkes 2009). These two types of evidence not only differ in scale, they are often incommensurable in what they reveal about human-environment relations (Brumfiel 1992; Gragson and Bolstad 2006; Lepofsky and Kahn 2011). The challenge is to integrate divergent information to develop a long-term and realistic understanding of human-environmental interaction reflecting the continuum of known human behaviours rather than the *Homo devastans* and Noble Savage caricatures (Balée 1998).

The transition from Native American to EuroAmerican occupation is particularly important for understanding how the past helps define the present and constrains the future of southern Appalachia (Gragson and Bolstad 2006). While parcel ownership serves to link current ecosystem processes and services to contemporary land use, much remains to be known about how landscape structure reflects the historical sequence of property systems imposed over time (Thrower 1966; Price 1995; Russell 1997; Bain and Brush 2004). In a first attempt to realistically portray the continuum of human behaviour in relation to environment in the context of shifting property regimes, we examined Cherokee town placement and population at a regional scale using local analytical procedures (Gragson and Bolstad 2007).

Using the first true census of the Cherokee Nation and the first detailed English map of North America's southern frontier, we modelled the constraints and trade-offs of meeting individual settlement needs for selected resources. In this way we determined the "resource demand footprints" (Fig. 9.2d) for the entire Cherokee territory in the early eighteenth century (Bolstad and Gragson 2008). This provides a critical benchmark for evaluating pre-settlement land use and extends our understanding of the regional historic land-use mosaic. The objective is to explain how the various events of the eighteenth century not only led to changes in the social structure of the Cherokee, but reworked the geography of social relations within and beyond the region they occupied. Individuals in southeast history – Cherokee and others – have been portrayed as mere pawns in ever expanding market relations over which they had little or no control (Dunaway 1996). The anthropological lens that examines the individual as "rational" within the context of a particular cultural, historical and environment setting helps overcome this and other forms of intellectual extremism.

## 9.6 Conclusion

There is no doubt that pristine systems are increasingly rare, even non-existent, on Earth and that the human footprint is global and pervasive. This leads to the realisation that many if not most of today's pressing issues require environments to be viewed as socio-ecological systems. The focus of which has moved beyond description, to examine the dynamic processes of coupled systems through research that is place-based, long-term, cross-scale, and comparative (Collins et al. 2011). Through forward-looking, problem-oriented research, the objective is to simultaneously maintain Earth life support systems and meet human needs (Palmer et al. 2004).

While acknowledging that human activities should be studied as integral to ecosystems, it is even more important to recognise that without drawing on diverse social areas of expertise and practice, the biophysical sciences alone will fail to exploit the most cogent and important connections in socio-ecological systems. Disciplines continue to play an important role in an interdisciplinary world, and anthropology in particular contributes in significant ways to the problem-oriented research that challenges us to understand what it means to be human through the study of culture in place as well as by comparison to other cultures.

LTER research historically examined ecosystem processes divided into five core areas (Gragson and Grove 2006), with human influence circumscribed to controlled experiments (e.g., logging treatments). There is ever greater appreciation that traditional experiments provide only limited understanding of the complex interactions between humans and the environment (Clark et al. 2003, 2011), yet limiting research to human exclosures or pristine settings fails to provide the information necessary to resolve the problems so readily evident in the world today. Ecological research is beginning to be conducted at socially relevant scales, while social research is beginning to recognise that humans are both influenced by and influence the environments they occupy.

Biophysical and social scientists both examine how systems are organised and the roles played by internal versus external influences (Pickett 1991; Pickett et al. 1997). There is as yet no unified theory of a socio-ecological system, but place-based, long-term, cross-scale and comparative research is moving us out of the realm of correlations and associations to a deeper probing of both mechanism and pattern (Collins et al. 2011). Anthropology contributes to the disciplinary diversity required to advance socio-ecological understanding of the world in which we live, and as a boundary discipline is central in bridging between the biological and the social sciences.

**Acknowledgments** This material is based on work supported by the National Science Foundation under Grants DEB-0823293 and DEB-0218001. Any opinions, findings, conclusions, or recommendations expressed in the material are those of the author and do not necessarily reflect the views of the National Science Foundation.

# References

Abbott, A. (2001). *Chaos of disciplines*. Chicago: University of Chicago Press.
Abler, D. G., & Sukhatme, V. A. (2006). The "efficient but poor" hypothesis. *Review of Agricultural Economics, 28*, 338–343.
Acheson, J. M. (2003). *Capturing the commons: Devising institutions to manage the Maine lobster industry*. Hanover: University Press of New England.
Anderson, S. P., Bales, R. C., & Duffy, C. J. (2008). Critical zone observatories: Building a network to advance interdisciplinary study of earth surface processes. *Mineralogical Magazine, 72*, 7–10.
Atran, S. (1993). *Cognitive foundations of natural history: Towards an anthropology of science*. Cambridge: Cambridge University Press.
Atran, S., Medin, D., Ross, N., Lynch, E., Coley, J., Ucan Ek, E., & Vapnarsky, V. (1999). Folkecology and commons management in the Maya lowlands. *Proceedings of the National Academy of Sciences, 96*, 7598–7603.
Axelrod, R. (1986). An evolutionary approach to norms. *The American Political Science Review, 80*, 1095–1111.
Bain, D. J., & Brush, G. S. (2004). Placing the pieces: Reconstructing the original property mosaic in a warrant and patent watershed. *Landscape Ecology, 19*, 843–856.
Balée, W. (1998). Historical ecology: Premises and postulates. In W. Balée (Ed.), *Advances in historical ecology* (pp. 13–29). New York: Columbia University Press.
Ball, R., & Pounder, L. (1996). "Efficient but poor" revisited. *Economic Development and Cultural Change, 44*, 735–760.
Bang, M., Medin, D. M., & Atran, S. (2007). Cultural mosaics and mental models of nature. *Proceedings of the National Academy of Sciences, 104*, 13868–13874.
Berkes, F. (2009). *Sacred ecology: Traditional ecological knowledge and resource management*. Philadelphia: Taylor & Francis.
Bisaro, A., Hinkel, J., & Kranz, N. (2010). Multilevel water, biodiversity and climate adaptation governance: Evaluating adaptive management in Lesotho. *Environmental Science and Policy, 13*, 637–647.

Boas, F. (1916). Eugenics. *The Scientific Monthly, 3*, 471–478.
Bockstael, N., Freeman, A. M., III, Kopp, R. J., Portney, P. R., & Smtih, V. K. (2000). On measuring economic values for nature. *Environmental Science and Technology, 34*, 1384–1389.
Bollen, K. A., Entwisle, B., & Alderson, A. S. (1993). Macrocomparative research methods. *Annual Review of Sociology, 19*, 321–351.
Bolstad, P. V., & Gragson, T. L. (2008). Resource abundance as constraints on early post-contact Cherokee populations. *Journal of Archaeological Science, 35*, 563–576.
Bourdieu, P. (1977). *Outline of a theory of practice*. New York: Cambridge University Press.
Boynton, W., DeVanzo, C., Hornberger, G., Lugo, A., Melillo, J., Pickett, S., & Vaughan, H. (2005). *Report II of the Scientific Task Force's Advisory Committee* (pp. 1–9). Woods Hole, MA.
Brewer, G. D., & Stern, P. C. (Eds.). (2005). *Decision making for the environment: Social and behavioral science research priorities*. Washington, DC: National Academies Press.
Brondizio, E. S., Ostrom, E., & Young, O. R. (2009). Connectivity and the governance of multilevel social-ecological systems: The role of social capital. *Annual Review of Anthropology, 34*, 253–278.
Brumfiel, E. (1992). Breaking and entering the ecosystem: Gender, class, and faction steal the show. *American Anthropologist, 94*, 551–567.
Butler, S. M. (2006). *Forest disturbance history and stand dynamics of the Coweeta Basin, western North Carolina*. MS, University of Maine, Orono, ME.
Carpenter, S. R. (2008). Emergence of ecological networks. *Frontiers in Ecology and the Environment, 6*, 228.
CFIR (Committee on Facilitating Interdisciplinary Research). (2004). *Facilitating interdisciplinary research*. Washington, DC: The National Academies Press.
Chamblee, J. F., Dehring, C. A., & Depken, C. A. (2008). Watershed development restrictions and land prices: Empirical evidence from Southern Appalachia. *Regional Science and Urban Economics, 39*, 287–296.
Charnley, S., & Durham, W. H. (2010). Anthropology and environmental policy: What counts? *American Anthropologist, 112*, 397–415.
Cho, S.-H., et al. (2005). Measuring rural homeowners willingness to pay for land conservation easements. *Forest Policy and Economics, 7*, 757–770.
Ciriacy-Wantrup, S. V., & Bishop, R. (1975). Common property as a concept in natural resource policy. *Natural Resources Journal, 15*, 713–728.
Clark, J. S., Lewis, M., McLachlan, J. S., & HilleRisLambers, J. (2003). Estimating population spread: What can we forecast and how well? *Ecology, 84*, 1979–1988.
Clark, J. S., Hersh, M. H., & Nichols, L. (2011). Climate change vulnerability of forest biodiversity: Climate and competition tracking of demographic rates. *Global Change Biology, 17*, 1834–1849.
Clinton, B. D., Boring, L. R., & Swank, W. T. (1993). Canopy gap characteristics and drought influences in oak forest of the Coweeta Basin. *Ecology, 74*, 1551–1558.
Coburn, T. (2011). *The National Science Foundation: Under the microscope*. Washington, DC: United States Senate.
Coleman, D. C. (2010). *Big ecology: The emergence of ecosystem science*. Berkeley: University of California Press.
Collins, S. L., Carpenter, S. R., Swinton, S. M., Orenstein, D. E., Childers, D. L., Gragson, T. L., Grimm, N. B., Grove, J. M., Harlan, S. L., Kaye, J. P., Knapp, A. K., Kofinas, G. P., Magnuson, J. J., McDowell, W. H., Melack, J. M., Ogden, L. A., Robertson, G. P., Smith, M. D., & Whitmer, A. C. (2011). An integrated conceptual framework for long-term social–ecological research. *Frontiers in Ecology and the Environment, 9*, 351–357.
Conkling, R. (1984). Power and change in an Indonesian government office. *American Ethnologist, 11*, 259–274.
Crawford, S. E. S., & Ostrom, E. (1995). A grammar of institutions. *The American Political Science Review, 89*, 582–600.

D'Andrade, R. G. (1995). *The development of cognitive anthropology*. New York: Cambridge University Press.
Davis, D. E. (2000). *Where there are mountains: An environmental history of the southern Appalachians*. Athens: University of Georgia Press.
de Toqueville, A. (2000). *Democracy in America*. New York: Bantam Books.
Diamond, J. (2003). *Collapse: How societies choose to succeed or fail*. New York: Viking.
Dietz, T., Ostrom, E., & Stern, P. C. (2003). The struggle to govern the commons. *Science, 302*, 1907–1912.
Douglass, M., & Isherwood, B. (1996). *The world of goods: Towards an anthropology of consumption*. New York: Routledge.
Dove, M. R., & Carpenter, C. (2008). *Environmental anthropology: A historical reader*. Malden: Blackwell.
Druckman, J. N., & Lupia, A. (2000). Preference formation. *Annual Review of Political Science, 3*, 1–24.
Dunaway, W. A. (1996). *The first American frontier: Transition to capitalism in Southern Appalachia, 1700–1860*. Chapel Hill: University of North Carolina Press.
Dutcher, D. D., Finley, J. C., Luloff, A. E., & Johnson, J. (2004). Landowner perceptions of protecting and establishing riparian forests: A quantitative analysis. *Society and Natural Resources, 17*, 319–332.
Dutcher, D. D., Finley, J. C., Luloff, A. E., & Johson, J. (2007). Connectivity with nature as a measure of environmental values. *Environment and Behavior, 39*, 474–493.
Eisenhardt, K. M., & Graebner, M. E. (2007). Theory building from cases: Opportunities and challenges. *The Academy of Management Journal, 50*, 25–32.
Entwisle, B. (2007). Putting people into place. *Demography, 44*, 687–703.
Escobar, A. (1995). *Encountering development: The making and unmaking of the third world*. Princeton: Princeton University Press.
Evans, T. P., & Moran, E. F. (2002). Spatial integration of social and biophysical factors related to landcover change. *Population and Development Review, 28*, 165–186.
Evans, T. P., Sun, W., & Kelley, H. (2006). Spatially explicit experiments for the exploration of land-use decision-making dynamics. *International Journal of Geographical Information Science, 20*, 1013–1037.
Farber, S., Costanza, R., Childers, D. L., Erickson, J., Gross, K., Grove, J. M., Hopkinson, C. S., Kahn, J., Pincetl, S., Troy, A., Warren, P., & Wilson, M. (2006). Linking ecology and economics for ecosystem management. *BioScience, 56*, 121–133.
Fiske, S. J. (1994). Federal organizational cultures: Layers and loci. In T. Hamada & W. E. Sibley (Eds.), *Anthropological perspectives on organizational culture* (pp. 95–119). Lanham: University Press of America.
Fuller, S., & Collier, J. H. (2004). *Philosophy, rhetoric, and the end of knowledge*. Mahwah: Lawrence Erlbaum Associates, Inc.
Funtowicz, S. O., & Ravetz, J. (1993). The emergence of post-normal science. In R. von Schomberg (Ed.), *Science, politics, and morality: Scientific uncertainty and decision making* (pp. 85–123). Dordrecht: Kluwer.
Gable, E. (2011). *Anthropology and egalitarianism: Ethnographic encounters from Monticello to Guinea-Bissau*. Bloomington: Indiana University Press.
Geertz, C. (1977). *The interpretation of cultures*. New York: Basic Books.
German, G. L., Karsenty, A., & Tiani, A.-M. (Eds.). (2009). *Governing Africa's forests in a globalized world*. London: Earthscan.
Gibson, C. C., Ostrom, E., & Ahn, T. K. (2000). The concept of scale and the human dimensions of global change: A survey. *Ecological Economics, 32*, 217–239.
Giles, J. (2011). Social science lines up its biggest challenges. *Nature, 470*, 18–19.
Glass, T. A., & McAtee, M. J. (2006). Behavioral science at the crossroads in public health: Extending horizons, envisioning the future. *Social Science and Medicine, 62*, 1650–1671.

Graeber, D. (2006). Turning modes of production inside out or, why capitalism is a transformation of slavery. *Critique of Anthropology, 26,* 61–85.

Gragson, T. L., & Bolstad, P. V. (2006). Land use legacies and the future of Southern Appalachia. *Society and Natural Resources, 19,* 175–190.

Gragson, T. L., & Bolstad, P. V. (2007). A local analysis of early 18th century Cherokee settlement. *Social Science History, 31,* 435–468.

Gragson, T. L., & Grove, M. (2006). Social science in the context of the long term ecological research program. *Society and Natural Resources, 19,* 93–100.

Gragson, T. L., Bolstad, B. V., & Welch-Devine, M. (2008). Agricultural transformation of Southern Appalachia. In C. Redmond & D. Foster (Eds.), *Agricultural transformation of North American landscapes* (pp. 89–121). New York: Oxford University Press.

Grimm, N. B., Grove, J. M., Pickett, S. T. A., & Redman, C. L. (2000). Integrating approaches to long-term studies of urban ecological systems. *BioScience, 50,* 571–584.

Groffman, P. M., Bain, D. J., Band, L. E., Belt, K. T., Brush, G. S., Grove, J. M., Pouyat, R. V., Yesilonis, I. C., & Zipperer, W. C. (2003). Down by the riverside: Urban riparian ecology. *Frontiers in Ecology and the Environment, 1,* 315–321.

Grove, J. M., Troy, A. R., O'Neil-Dunne, J. P. M., Burch, W. R., Cadenasso, M. L., & Pickett, S. T. A. (2006). Characterization of households and its implications for the vegetation of urban ecosystems. *Ecosystems, 9,* 578–597.

Guillet, D. (1998). Rethinking legal pluralism: Local law and state law in the evolution of water property rights in northwestern Spain. *Comparative Studies in Society and History, 40,* 42–70.

Haberl, H., Winiwarter, V., Andersson, K., Ayres, R. U., Boone, C., Castillo, A., Cunfer, G., Fischer-Kowalski, M., Freudenberg, W. R., Furman, E., Kaufmann, R., Krausmann, F., Langthaler, E., Lotze-Campen, H., Mirtl, M., Redman, C. L., Reenberg, A., Wardell, A., Warr, B., & Zechmeister, H. (2006). From LTER to LTSER: Conceptualizing the socioeconomic dimension of long-term socioecological research. *Ecology and Society, 11.* Retrieved from http://www.ecologyandsociety.org/vol11/iss2/art13/ES-2006-1786.pdf

Hackett, E. J., & Rhoten, D. R. (2009). The snowbird charrette: Integrative interdisciplinary collaboration in environmental research design. *Minerva, 47,* 407–440.

Haenn, N., & Wilk, R. R. (2006). *The environment in anthropology.* New York: New York University Press.

Hardin, G. (1968). The tragedy of the commons. *Science, 162,* 1243–1248.

Hayami, Y., & Godo, Y. (2005). *Development economics: From poverty to the wealth of nations.* New York: Oxford University Press.

Henrich, J. H., Heine, S. J., & Norenzayan, A. (2010). The weirdest people in the world? *The Behavioral and Brain Sciences, 33,* 61–135.

Heyman, J. M. C. (2004). The anthropology of power-wielding bureaucracies. *Human Organization, 63,* 487–500.

Hobbie, J. E., Carpenter, S., Grimm, N. B., Gosz, J., & Seastedt, T. (2003). The US long term ecological research program. *BioScience, 53,* 21–32.

Hunziker, M., Felber, P., Gehring, K., Buchecker, M., Nicole Bauer, N., & Kienast, F. (2008). Evaluation of landscape change by different social groups. *Mountain Research and Development, 28,* 140–147.

Ibáñez, I., Clark, J. S., Dietze, M. C., Feeley, K., Hersh, M., LaDeau, S., McBride, A., Welch, N. E., & Wolosin, M. S. (2006). Predicting biodiversity change: Outside the climate envelope, beyond the species-area curve. *Ecology, 87,* 1896–1906.

Ingold, T. (2000). *The perception of the environment: Essays on livelihood, dwelling and skill.* London: Routledge.

Isern, A., & Clark, H. (2003). The ocean observatories initiative: A continued presence for interactive ocean research. *Marine Technology Society Journal, 37,* 26–41.

Jacobs, J. A., & Frickel, S. (2009). Interdisciplinarity: A critical assessment. *Annual Review of Sociology, 35,* 43–65.

Jurgelski, W. (2004). *A new plow in old ground: Cherokees, whites, and land in western North Carolina, 1819–1829*. Ph.D., University of Georgia, Athens, GA.

Kaplan, R., & Austin, M. E. (2004). Out in the country: Sprawl and the quest for nature nearby. *Landscape and Urban Planning, 69*, 235–243.

Kaplan, R., & Herbert, E. J. (1987). Cultural and sub-cultural comparisons in preferences for natural settings. *Landscape and Urban Planning, 14*, 281–293.

Kay, C. (1994). Aboriginal overkill: The role of American Indians in structuring western ecosystems. *Human Nature, 5*, 539–598.

Keller, M., Schimel, D. S., Hargrove, W. W., & Hoffman, F. M. (2008). A continental strategy for the National Ecological Observatory Network. *Frontiers in Ecology and the Environment, 6*, 282–284.

Kempton, W. M., Boster, J. S., & Hatrley, J. A. (1996). *Environmental values in American culture*. Cambridge: The MIT Press.

Khun, T. (1962). *The structure of scientific revolutions*. Chicago: University of Chicago Press.

Kiser, L., & Ostrom, E. (1982). The three worlds of action: A synthesis of institutional approaches. In E. Ostrom (Ed.), *Strategies of political inquiry*. Beverly Hills: SAGE Publications.

Klein, J. T. (2000). A conceptual vocabulary of interdisciplinary science. In P. Weingart & N. Stehr (Eds.), *Practising interdisciplinarity* (pp. 3–24). Toronto: University of Toronto Press.

Kratz, T. K., Arzberger, P., Benson, B. J., Chiu, C.-Y., Chiu, K., Ding, L., Fountain, T., Hamilton, D., Hanson, P. C., Hu, Y. H., Lin, F. P., McMullen, D. F., Tilak, S., & Wu, C. (2006). Toward a global lake ecological observatory network. *Publications of the Karelian Institute, 145*, 51–63.

Krech, S., III. (1999). *The ecological Indian: Myth and history*. New York: Norton.

Kroeber, A. (1916). Inheritance by magic. *American Anthropologist, 18*, 19–40.

Langley, A. (1999). Strategies for theorizing from process data. *The Academy of Management Review, 24*, 691–710.

Lansing, J. S. (1991). *Priests and programmers: Technologies of power in the engineered landscape of Bali*. Princeton: Princeton University Press.

Leigh, D. S., & Webb, P. W. (2006). Holocene erosion, sedimentation, and stratigraphy at raven fork, southern Blue Ridge Mountains, USA. *Geomorphology, 78*, 161–177.

Lepofsky, D., & Kahn, J. (2011). Cultivating an ecological and social balance: Elite demands and commoner knowledge in ancient Maʻohi agriculture, Society Islands. *American Anthropologist, 113*, 319–335.

Lévi-Strauss, C. (1968). *The savage mind*. Chicago: The University of Chicago Press.

Liu, J., et al. (2007). Complexity of coupled human and natural systems. *Science, 317*, 1513–1516.

March, J. G., & Olsen, J. P. (1989). *Rediscovering institutions: The organizational basis of politics*. New York: Free Press.

Meillassoux, C. (1981). *Maidens, meal and money: Capitalism and the domestic community*. Cambridge: Cambridge University Press.

Morgan, L. H. (1877). *Ancient society: Researches in the lines of human progress from savagery through barbarism*. New York: Henry Holt and Company.

Morgen, S., & Maskovsky, J. (2003). The anthropology of welfare "reform": New perspectives on U.S. urban poverty in the post-welfare era. *Annual Review of Anthropology, 32*, 315–338.

Nader, L. (1969). Up the anthropologist – Perspectives gained from studying up. In D. H. Hymes (Ed.), *Reinventing anthropology* (pp. 284–311). Ann Arbor: University of Michigan Press.

NSF (National Science Foundation). (2002). *Integrative graduate education and research traineeship (IGERT) program (NSF 02–145)*. Arlington: National Science Foundation.

Oliver-Smith, A. (1996). Anthropological research on hazards and disasters. *Annual Review of Anthropology, 25*, 303–328.

Orlove, B., & Caton, S. C. (2010). Water sustainability: Anthropological approaches and prospects. *Annual Review of Anthropology, 39*, 401–415.

Ostrom, E. (1990). *Governing the commons: The evolution of institutions for collective action.* Cambridge, MA: Cambridge University Press.
Palmer, M., Bernhardt, E., Chornesky, E., Collins, S., Dobson, A., Duke, C., Gold, B., Jacobson, R., Kingsland, S., Kranz, R., Mappin, M., Martinez, M. L., Micheli, F., Morse, J., Pace, M., Pascual, M., Palumbi, S., Reichman, O. J., Simons, A., Townsend, A., & Turner, M. (2004). Ecology for a crowded planet. *Science, 304*, 1251–1252.
Peters, D. P. C., Groffman, P. M., Nadelhoffer, K. N., Grimm, N. B., Collins, S. L., Michener, W. K., & Huston, M. A. (2008). Living in an increasingly connected world: A framework for continental-scale environmental science. *Frontiers in Ecology and the Environment, 6*, 229–237.
Pickett, S. T. A. (1991). Long-term studies: Past experience and recommendations for the future. In P. G. Risser (Ed.), *Long-term ecological research: An international perspective* (Vol. Scope 47, pp. 71–88). New York: Wiley.
Pickett, S. T. A., Burch, W. R., Jr., Dalton, S., Foresman, T., Grove, J. M., & Rowntree, R. (1997). A conceptual framework for the study of human ecosystems in urban areas. *Journal of Urban Ecosystems, 1*, 185–199.
Pickett, S. T. A., Cadenasso, M. L., & Grove, J. M. (2005). Biocomplexity in coupled natural–human systems: A multidimensional framework. *Ecosystems, 8*, 225–232.
Preissle, J., & LeCompte, M. D. (1981). Ethnographic research and the problem of data reduction. *Anthropology & Education Quarterly, 12*, 51–70.
Price, E. T. (1995). *Dividing the land: Early American beginnings of our private property mosaic.* Chicago: University of Chicago Press.
Purcell, A. T., Lamb, R. J., Mainardi Peron, E., & Falchero, S. (1994). Preference or preferences for landscape. *Journal of Environmental Psychology, 14*, 195–209.
Rhoten, D., & Parker, A. (2004). Risks and rewards of an interdisciplinary research path. *Science, 306*, 2046.
Rodning, C. (2009). Place, landscape, and environment: Anthropological archaeology in 2009. *American Anthropologist, 112*, 180–190.
Romney, A. K., & Moore, C. C. (1998). Toward a theory of culture as shared cognitive structures. *Ethos, 26*, 314–337.
Roseberry, W. (1988). Political economy. *Annual Review of Anthropology, 17*, 161–185.
Russell, E. W. B. (1997). *People and the land through time.* New Haven: Yale University Press.
Sayer, A. (2000). *Realism and social science.* Thousand Oaks: SAGE Publications.
Schultz, T. W. (1964). *Transforming traditional agriculture.* New Haven: Yale University Press.
Schweigert, T. (1994). Penny capitalism: Efficient but poor or inefficient and (less than) second best? *World Development, 22*, 721–735.
Shweder, R. A., & Good, B. (Eds.). (2005). *Clifford Geertz by his colleagues.* Chicago: University of Chicago Press.
Siskind, J. (1978). Kinship and mode of production. *American Anthropologist, 80*, 860–872.
Slovic, P. (1995). The construction of preference. *The American Psychologist, 50*, 364–371.
Smelser, N. J. (1976). *Comparative methods in the social sciences.* Englewood Cliffs: Prentice-Hall Inc.
Stern, P. C. (2000). Toward a coherent theory of environmentally significant behavior. *Journal of Social Issues, 53*, 407–424.
Stokols, D., Fuqua, J., Gress, J., Harvey, R., Phillips, K., Baezconde-Garbanati, L., Unger, J., Palmer, P., Clark, M. A., Colby, S. M., Morgan, G., & Trochim, W. (2003). Evaluating transdisciplinary science. *Nicotine & Tobacco Research, 5*, 21–39.
Swanson, G. E. (1971). Frameworks for comparative research: Structural anthropology and the theory of action. In I. Vallier (Ed.), *Comparative methods in sociology: Essays on trends and applications.* Berkeley: University of California Press.
Tax, S. (1963). *Penny capitalism: A Guatemalan Indian economy.* Chicago: University of Chicago Press.

Thrower, N. J. W. (1966). *Original survey and land subdivision: A comparative study of the form and effect of contrasting cadastral surveys*. Chicago: Rand McNally and Co.

Tucker, B., Huff, A., Tsiazonera, Tombo, J., Hajasoa, P., & Nagnisaha, C. (2011). When the wealthy are poor: Poverty explanations and local perspectives in southwestern Madagascar. *American Anthropologist, 113*, 291–305.

Turner, N. J. (2005). *The earth's blanket: Traditional teachings for sustainable living*. Vancouver: Douglas and McIntyre.

Vayda, A. P., & Rappaport, R. A. (1968). Ecology, cultural and noncultural. In J. A. Clifton (Ed.), *Introduction to cultural anthropology* (pp. 477–497). Boston: Houghton Mifflin.

Vitousek, P. M., Mooney, H. A., Lubchenco, J., & Melillo, J. M. (1997). Human domination of earth's ecosystems. *Science, 277*, 494–499.

Viviroli, D., Dürr, H. H., Messerli, B., Meybeck, M., & Weingartner, R. (2007). Mountains of the world, water towers for humanity: Typology, mapping, and global significance. *Water Resources Research, 43*, 1–13.

Wainwright, J. (2010). Climate change, capitalism, and the challenge of transdisciplinarity. *Annals of the Association of American Geographers, 100*, 983–991.

Wear, D. N., & Bolstad, P. V. (1998). Land-use changes in Southern Appalachian landscapes: Spatial analysis and forecast evaluation. *Ecosystems, 1*, 575–594.

Webster, J. R., Benfield, E. F., Cecala, K. K., Chamblee, J. F., Dehring, C. A., Gragson, T. L., Hepinstall, J. A., Jackson, C. R., Knoepp, J. D., Leigh, D. S., Maerz, J. C., Pringle, C., & Valett, H. M. (2012). Water quality and exurbanization in Southern Appalachian streams. In P. J. Boon & P. J. Raven (Eds.), *River conservation and management* (pp. 89–104). Chichester: Wiley-Blackwell.

Westley, F., et al. (2002). Why systems of people and nature are not just social and ecological systems. In L. H. Gunderson & C. S. Holling (Eds.), *Panarchy: Understanding transformations in human and natural systems* (pp. 103–120). Washington, DC: Island Press.

Whitmer, A., Ogden, L., Lawton, J., Sturner, P., Groffman, P. M., Schneider, L., Hart, D., Halpern, B., Schlesinger, W., Raciti, S., Bettez, N., Ortega, S., Rustad, L., Pickett, S. T. A., & Killilea, M. (2010). The engaged university: Providing a platform for research that transforms society. *Frontiers in Ecology and the Environment, 8*, 314–321.

Wolf, E. (1982). *Europe and the people without history*. Berkeley: University of California Press.

# Part II
# LTSER Applications Across Ecosystems, Time and Space

# Chapter 10
# Viewing the Urban Socio-ecological System Through a Sustainability Lens: Lessons and Prospects from the Central Arizona–Phoenix LTER Programme

Nancy B. Grimm, Charles L. Redman, Christopher G. Boone, Daniel L. Childers, Sharon L. Harlan, and B.L. Turner II

**Abstract** Cities are complex socio-ecological systems (SES). They are focal points of human population, production, and consumption, including the generation of waste and most of the critical emissions to the atmosphere. But they also are centres of human creative activities, and in that capacity may provide platforms for the transition to a more sustainable world. Urban sustainability will require understanding grounded in a theory that incorporates reciprocal, dynamic interactions between societal and ecological components, external driving forces and their impacts, and a multiscalar perspective. In this chapter, we use research from the Central Arizona–Phoenix LTER programme to illustrate how such a conceptual framework can enrich

N.B. Grimm, Ph.D. (✉)
School of Life Sciences, Global Institute of Sustainability,
Arizona State University, Tempe, AZ, USA
e-mail: nbgrimm@asu.edu

C.L. Redman, Ph.D. • D.L. Childers, Ph.D.
School of Sustainability, Arizona State University, Tempe, AZ, USA
e-mail: Charles.redman@asu.edu; dan.childers@asu.edu

C.G. Boone, Ph.D.
School of Sustainability, School of Human Evolution and Social Change,
Arizona State University, Tempe, AZ, USA
e-mail: cgboone@asu.edu

S.L. Harlan, Ph.D.
School of Human Evolution and Social Change, Global Institute
of Sustainability, Arizona State University, Tempe, AZ, USA
e-mail: sharon.harlan@asu.edu

B.L. Turner II, Ph.D.
School of Geographical Sciences and Urban Planning, School of Sustainability,
Arizona State University, Tempe, AZ, USA
e-mail: billie.l.turner@asu.edu

our understanding and lead to surprising conclusions that might not have been reached without the integration inherent in the SES approach. By reviewing research in the broad areas of urban land change, climate, water, biogeochemistry, biodiversity, and organismal interactions, we explore the dynamics of coupled human and ecological systems within an urban SES in arid North America, and discuss what these interactions imply about sustainability.

**Keywords** Urban sustainability • Socio-ecological system • Land-use change • Ecosystem services • Urban heat island • Urban footprint • Urban water dynamics • Urban biogeochemical cycles

## 10.1 Introduction

Cities are focal points of human population, production, and consumption, including the generation of waste and most of the critical emissions to the atmosphere. They also are "places" of diverse economic and social activities. Harnessed appropriately, the economies of scale offered by cities may provide platforms for the transition to a more sustainable world (Bettencourt et al. 2007). Cities are complex socio-ecological systems (SES) that include people as the dominant species, other organisms, and abiotic elements, as well as the social and ecological contexts for these components. The Central Arizona–Phoenix Long-Term Ecological Research programme (CAP LTER) focuses on one such system, providing the science to underlie urban sustainability strategies and approaches that may be applied to cities worldwide. As the urban challenge continues to grow, there is an urgent need to determine what makes cities sustainable or not; whether there is a best configuration, size, shape, or structure for urban SES; and how the impact of cities[1] on the environment outside them can be minimised while the quality of life for their inhabitants is maximised.

To address the urban challenge, CAP LTER science focuses on the broad question of how do the services provided by evolving urban ecosystems affect human outcomes and behaviour, and how does human action (response) alter patterns of ecosystem structure and function and, ultimately, urban sustainability, in a dynamic environment? The question reflects a conviction that ecosystem services—provisioning, regulating, and cultural services,[2] defined as the benefits that people derive from ecosystems (MEA 2005)—are the focal point of interaction between people and their environment. As the urban fabric and infrastructure take shape, modifications to the environment may enhance some services and reduce others,

---

[1] Cities as defined here are synonymous with urban SES and the two terms are use interchangeably throughout the chapter.

[2] In this chapter, we do not include "supporting" ecosystem services as a type of ecosystem service, since these refer to ecosystem processes and indirectly contribute to services (such as nutrient cycling, primary production, and so forth).

sometimes to the point of creating disamenities, and deliberate and unintended tradeoffs often result (Turner 2009; Bennett et al. 2009).

The question of urban sustainability is an important one for our time because of the global increase in urban dwellers and consequent impacts on social and ecological systems worldwide (UNPD 2010). Today, although urban centres cover less than 3% of the Earth's land surface, they are responsible for a disproportionate share of carbon emissions, material extraction, and water use (Brown 2001). The urban footprint, or the total land area required for a city to accommodate these material, energy, and waste-assimilation needs, in some cases cover areas one to two orders of magnitude greater than the cities themselves (Luck et al. 2001; Kennedy et al. 2007). Inputs of both energy (oil, coal, gas, and urban carbon fixation), and materials (water, food, timber, and nutrients [e.g., nitrogen, phosphorus, and carbon]) and their outputs (heat, in the case of energy, and wastes in air, solids, and water) are often described collectively as "urban metabolism" (Wolman 1965; Kennedy et al. 2007), although the analogy with ecosystem metabolism or organismal metabolism is imperfect (Kaye et al. 2006). Instead, Kaye et al. (2006) advocated a mass-balance approach to quantifying inputs and outputs of materials for cities. Ecosystem metabolism of cities, referring to their energy balance, has long been described as an "industrial metabolism" that is highly reliant on ancient carbon, and therefore an important source of atmospheric $CO_2$ (Odum 1997; Collins et al. 2000). The density of cities, the state of urban buildings, and resident consumption patterns each present additional challenges and opportunities in considering the effects of direct (fuel and electricity) and indirect (e.g., transportation to secure consumables for urban residents) energy demands on ecosystem metabolism and urban footprints (Weisz and Steinberger 2010). Studies of ecosystem metabolism and material mass balance of cities reveal the dependence of urban socio-ecosystems on their hinterlands and even ecosystems far removed from them (Collins et al. 2000). Although cities are by nature dependent upon external ecosystems, understanding and perhaps lessening this dependence is key to prospects for urban sustainability (Weisz and Steinberger 2010).

Some ask whether an urban ecosystem can be sustainable at all, particularly one such as Phoenix, which is situated in the harsh environment of a desert and therefore requires extensive environmental modification to ensure "livability." We use the term sustainability in this chapter to refer to the ability of an urban ecosystem to provide comparable levels of services to all its inhabitants, consistent with outcomes that enhance human well-being in broad terms, without threatening the delivery of ecosystem services outside the ecosystem or to future generations. Based upon this definition, urban sustainability will require ever greater understanding of (1) the relationships between human well-being and the special environments of cities, (2) the definition of ecosystem services in built and highly modified environments, (3) the social decisions and processes that drive the distribution of services, and (4) the often hidden impacts of the city ecosystem on the widening landscape from which it draws resources and to which it disperses wastes. These relationships and interactions occur across scales, are irrefutably non-linear and often unpredictable, and are in turn influenced by external forces that are far from constant. Such complexity compels a conceptual framework that incorporates reciprocal, dynamic interactions between societal and ecological components, external driving forces and their impacts, and a multiscalar perspective.

In this chapter, we use research from the CAP LTER to illustrate how such a conceptual framework can enrich our understanding and lead to surprising conclusions that might not have been reached without the integration inherent in the SES approach. We explore the interactions of humans and ecosystems within an urban SES and discuss what these interactions imply about sustainability. First, however, we describe the conceptual framework in some detail.

## 10.2 A Conceptual Framework for Urban SES

In early work, the central Arizona urban SES was studied in the context of a hierarchical, patch-dynamics framework that originated in landscape ecology (Wu and Loucks 1995; Grimm et al. 2000; Wu and David 2002), similar to the approach used by our sibling urban LTER programme, the Baltimore Ecosystem Study (BES; Grimm et al. 2000; Pickett et al. 2011; Grove et al., Chap. 16 in this volume). Spatial heterogeneity and distributions of biophysical and social variables were critical for understanding how metropolitan Phoenix was changing. The scaling of human and ecological phenomena over space and time were featured prominently (e.g., Jenerette et al. 2006; Buyantuyev and Wu 2007; Ruddell and Wentz 2009; Wu et al. 2011). While this modelling platform met some of the requirements described above, CAP scientists were also working with others to conceive new models for integrating human and natural systems (Redman et al. 2004; Haberl et al. 2006; Costanza et al. 2007; Liu et al. 2007a, b; Wu 2008a, b).

The urban SES conceptual framework is dynamic, potentially multiscalar, and describes socio-ecological interactions within parts as well as for the whole heterogeneous system (Fig. 10.1). It builds upon the framework devised by the US LTER network (Collins et al. 2007, 2011) and shares themes and structure with frameworks in sustainability science (MEA 2005; Chapin et al. 2006; Carpenter et al. 2009). Its components are external drivers, space and time scales, press and pulse events, ecosystem structure and function, ecosystem services, and human outcomes and actions.

**Drivers of long-term change**. Global climate change and macroeconomic fluctuations are examples of external forces that can drive long-term change (Fig. 10.1, top). Our approach recognises the interaction between aggregate economic activity, policies to respond to climate change, and changes in local conditions. We also examine internal drivers of change: press events (e.g., air pollution, irrigation, land conversion, urban policies) and pulse events (e.g., flood, housing-market collapse).

**Space and time scales**. Since 1997, our study has centred on a 6,400-km$^2$ rectangular area of central Arizona that includes most of metropolitan Phoenix (Fig. 10.2). This geometrically simple area captures central Arizona's range of landscapes, embedded within a regional matrix of wildlands, urban centres, exurban

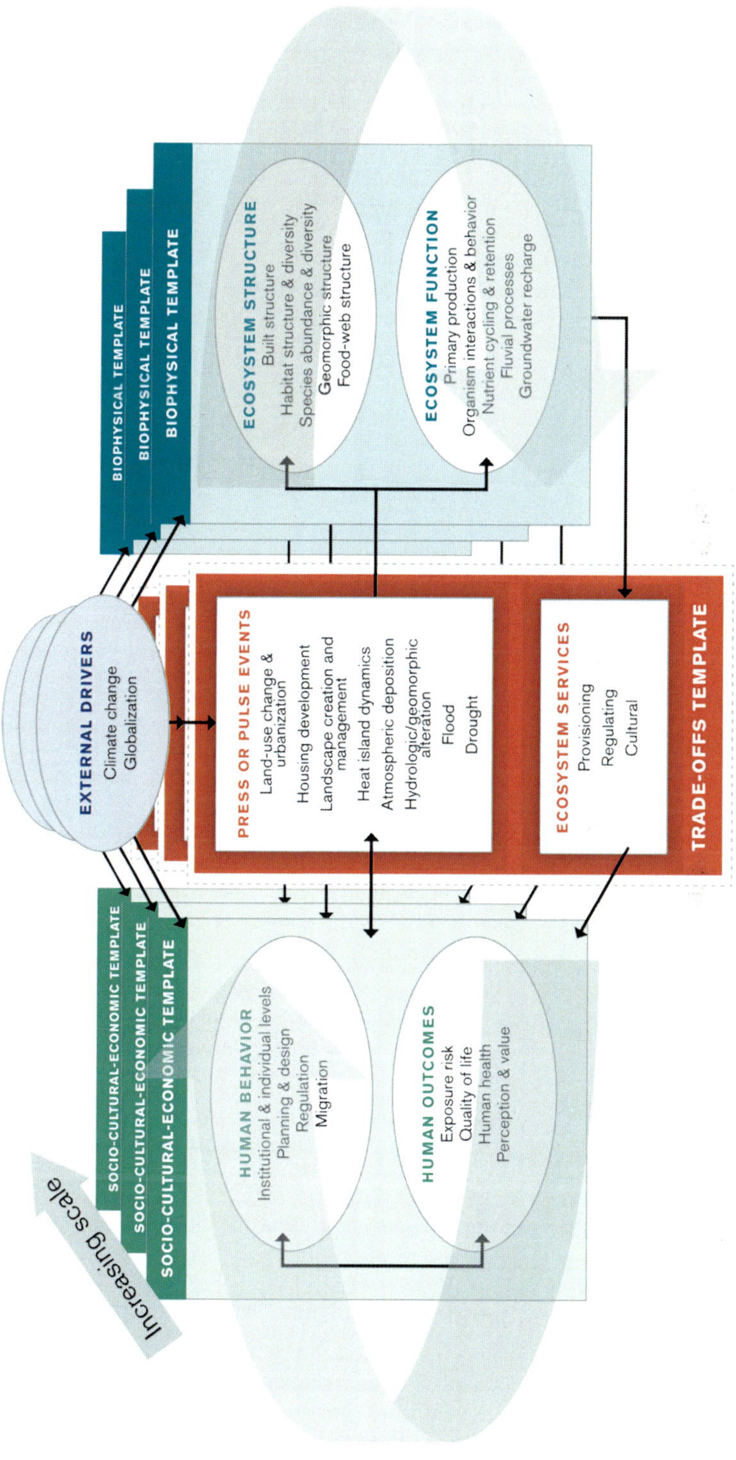

**Fig. 10.1** Conceptual framework for an urban socio-ecological system (*SES*) that can be used to visualize human-environment interactions at multiple scales. These interactions operate continuously in this multiscalar space, as shown by the *faint, gray, elliptical arrows* (Modified for the urban SES based the framework presented in Collins et al. 2011)

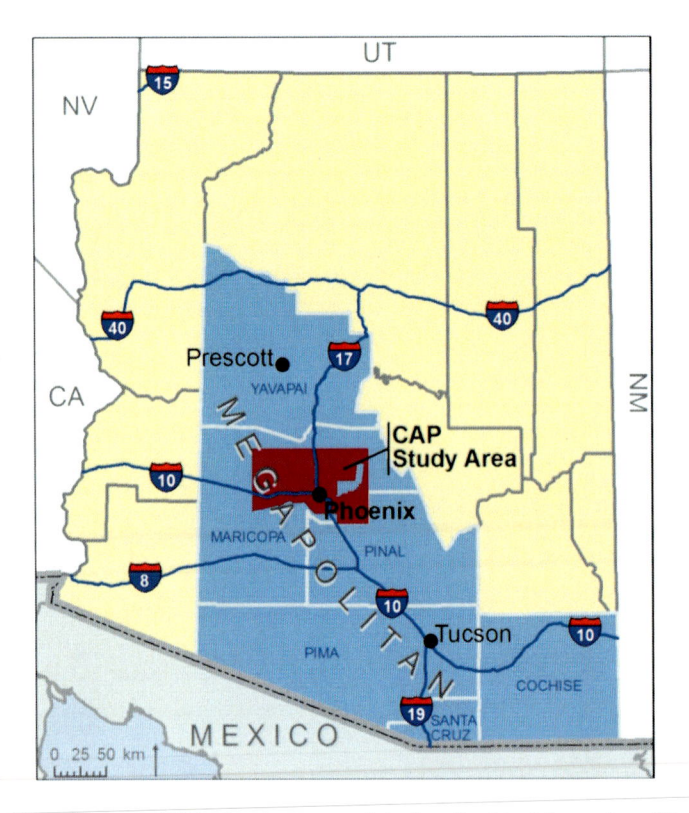

**Fig. 10.2** Map of Arizona, USA, showing extent of the Sun Corridor Megapolitan (*blue shading*) and the CAP study area within it (*red shading*). *Gray lines* are county boundaries

development, and agriculture. We conduct research across the nested hierarchies of landscape scales, ranging from the coarse, urban-agricultural-desert structures to traditional urban land-use categories (e.g., residential, commercial) to differentiated residential landscaping types. Within the socioeconomic realm, we work with units ranging from household to neighbourhood to municipality; within the desert, from plot to site to watershed. Our observational sampling and data-acquisition programmes capture event-based or seasonal time steps for fast variables and annual to 5-year time steps for slower variables, with many of the latter timed to the U.S. Census.

**Ecosystem structures and functions of interest**. In addition to the ecosystem components investigated in any ecosystem study, including soil, nutrient stocks, vegetation, and primary and secondary consumers, we focus on the built environment, including urban infrastructure and designed ecosystems, non-native species, and the human population. These components of the SES interact with and control rates of ecosystem processes and functions, such as primary production and nutrient cycling, which in turn are the "inputs" to ecosystem services (Fig. 10.1, right). Notably, human decisions and management are a major driver of urban ecosystem function.

**Ecosystem services.** Our main foci are (1) regulating ecosystem services, including micro- and meso-climate modulation (largely by vegetation through evapotranspiration and shading), stormwater flow modulation, and air and water quality regulation; (2) the provisioning service of urban food production; and the cultural and aesthetic services arising from biodiversity, and (3) the sense of place provided by natural desert ecosystems (Fig. 10.1, bottom). Recognising that designing and building urban areas with one ecosystem service in mind often degrades another (i.e., produces tradeoffs; Bennett et al. 2009; Turner 2009), we consider multiple services simultaneously.

**Human outcomes and actions.** We evaluate human responses to ecosystem services, such as perceptions and economic preferences for services realised from natural microclimate conditions (e.g., temperature and proximity to rivers or lakes) and those modified or managed using energy and water resources (e.g., swimming pools and irrigated landscaping), and disamenities, such as risks to human health in cities arising from extreme urban climate events and exposure to toxic releases. We study variation in vulnerability within the human population and its implications for environmental equity. We measure human outcomes and actions directly with physical indicators (e.g., incidence of diseases) or indirectly (inferring economic tradeoffs people may make to enhance a valued ecosystem service; for example, from differences in housing price; Klaiber and Smith 2009). When human responses to natural variation in ecosystem services are impossible to observe, stated-preference methods can uncover the choices people might make if given opportunities to change aspects of ecosystem services (Smith 2005). Finally, many of the actions taken by people feed back to the ecosystem, often by changing the pulse or press events that affect ecosystem structure and function (e.g., irrigating residential landscapes affects biodiversity by altering microclimate).

In CAP LTER research, we are asking three focused questions corresponding to our conceptual framework: (1) (urban ecosystem services): How does urbanisation change the structure and function of ecosystems and thereby alter the services they provide (i.e., Fig. 10.1, right)? (2) (human outcomes and actions/responses): How do people perceive and respond to ecosystem services, how are services distributed, and how do individual and collective behaviours further change ecosystem structure and function (i.e., Fig. 10.1, left)? and (3) (urbanisation in a dynamic world): How does the larger context of biophysical drivers and societal drivers influence the interaction and feedbacks between ecological and societal components of the SES (as mediated through ecosystem services) and thereby influence the future of the urban SES (i.e., Fig. 10.1, all)?

In the five research themes that follow, we will address questions 1 and 2, drawing on examples from CAP LTER research. A final section will address question 3, considering the possibility of urban sustainability in a dynamic world. We recognise, however, that a sustainable urban future will require more than research that shows how people and ecosystems interact. Our ultimate goal is to build on this understanding—in collaboration with governmental and nongovernmental partners, other local research groups, and the public—to create scenarios that can guide development of sustainable urban SES. Here we lay the scientific groundwork to begin to meet that challenge.

## 10.3 The Central Arizona–Phoenix SES

The 6,400-km$^2$ CAP LTER study area in central Arizona incorporates metropolitan Phoenix, surrounding Sonoran Desert scrub, and rapidly disappearing agricultural fields (Fig. 10.2). Rapid urbanisation, facilitated by the technological innovation of air conditioning, has been the dominant land change since the 1950s, accompanied by an order-of-magnitude increase in population. Long dominated demographically by White residents, the Hispanic or Latino population of metropolitan Phoenix has risen rapidly, now standing at 31% of the total population (U.S. Census Bureau 2011). Coincident with rapid population growth, the rise of automobile transportation has led to air pollution and other problems, such as the urban heat island, which influence quality of life. Freshwater resources have been appropriated to support first agriculture and later residential development. Native desert vegetation has given way to mostly non-native species maintained by irrigation, affecting biodiversity at higher trophic levels. The context of rapid urbanisation thus has provided fertile ground for research on topics such as climate, water, biogeochemical cycles, and biodiversity, with associated socio-ecological drivers, interactions and feedbacks.

## 10.4 Integrated Projects: Human–Ecosystem Interactions and Feedbacks

Long-term ecological research in the US has been characterised by attention to five "core areas" (see, e.g., Hobbie et al. 2003; Table 10.1), and these themes were early organisers of CAP LTER research (Grimm and Redman 2004). Social scientists within the US LTER network proposed a parallel set of social-science core areas (e.g., Redman et al. 2004), and in CAP LTER, our increased focus on interactions

Table 10.1 LTER ecological core areas, proposed LTER social science core areas, and CAP LTER integrated project areas

| LTER ecological core areas | LTER social science core areas | CAP LTER integrated project areas |
|---|---|---|
| Primary production | Demography | Climate, ecosystems, and people |
| Spatial and temporal distribution of populations | Technological change | Water dynamics in a desert city |
| Organic matter accumulation | Economic growth | Biogeochemical patterns, processes, and human outcomes |
| Inorganic inputs and movements of nutrients | Political and social institutions | Human decisions and biodiversity |
| Site disturbances | Culture | |
| | Knowledge and information exchange | |

and feedbacks between humans and ecosystem processes led to a more specific focus on integrated project-area themes (Table 10.1). At the regional scale, rapid urbanisation can be seen as the central press event (Fig. 10.1). Ensuing changes in ecosystems and the human system in terms of climate, water, biogeochemical processes, and biodiversity have been the specific targets of CAP LTER research. In this section, we highlight examples from this thematic research that reveal the nature of SES interactions.

## 10.4.1 Rapid Urbanisation in Central Arizona: The Press Event

Land and landscape dynamics are pivotal to understanding and assessing SES, especially in intensively built and managed environments that range from the impervious surfaces of the inner city to the open and wildland interfaces of the suburban/peri-urban fringe. The configuration or "architecture" (i.e., kind, amount, distribution and pattern; Turner 2009) of these lands proves critical to the capacity of the ecosystem to deliver services and to the human outcomes resulting from them. Variations in land architecture, such as suburban-wildland patch sizes, movement corridors, proximity to water sources, or locations of introduced vegetation that change habitats, can determine wildlife abundance (Marzluff and Rodewald 2008). The expansion and design of nearby settlements thereby affects the social preferences exhibited in property values at any particular location.

Alterations in patterns of land use and land cover underlie many ecological and social changes in the urban SES and central Arizona. Analysis of remotely sensed data has shown ongoing rapid urbanisation (Buyantuyev and Wu 2007; Buyantuyev et al. 2007; Walker and Briggs 2007; Wu et al. 2011), which is superimposed on centuries of land use. Distinctive silt deposits and associated plant communities along desert washes are legacies of prehistoric agricultural fields of the Hokoham culture over 1,000 years ago (Briggs et al. 2006; Schaafsma and Briggs 2007). Since 1970, rapid urbanisation has led to a decline of arable land and a rise in urban (residential) land uses (Keys et al. 2007; Fig. 10.3). Supporting this spread of housing development, water originally captured and redistributed to irrigated agricultural fields has been shifted to homeowner and municipal use. Legacies of historic (i.e., <150 years) agrarian practices remain (Redman and Foster 2008), however, and can influence contemporary soil biogeochemical pools and fluxes (Lewis et al. 2006; Hall et al. 2009).

Land-use legacies are one of several human influences on the structure and properties of contemporary residential landscapes. Residential landscapes in this desert metropolis fall into distinctive types (Martin et al. 2003; Cook et al. 2004), including "mesic" (irrigated, usually with turf grass and large trees), "oasis" (drip-irrigated, with a small patch of turf within a larger area of decomposed gravel substrate, and often with drought-tolerant trees and shrubs), and "xeric" (desert-mimicking residential landscapes with decomposed gravel substrates and drip-irrigated [or unirrigated], drought-tolerant trees, shrubs, and cacti). The prevalence

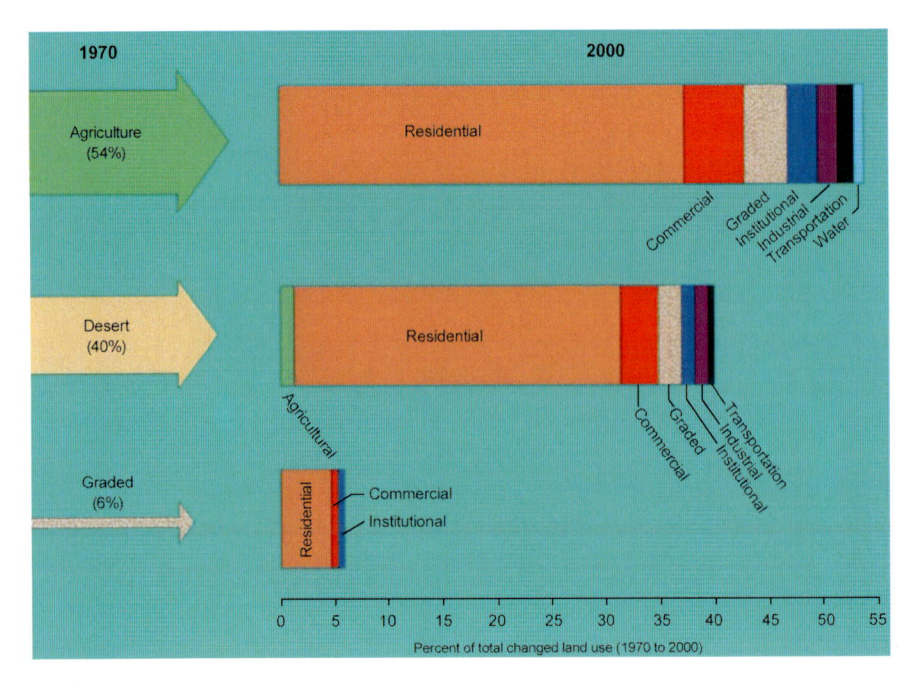

**Fig. 10.3** Analysis of land transitions in central Arizona between 1970 and 2000 (After Keys et al. (2007) based on a graphic designed by B. Trapido-Lurie; used with permission from Taylor & Francis)

of each landscape type varies over time and among locations within the metropolitan area. Furthermore, these residential landscape types may be constructed on lands that were previously desert, or were used for agriculture; the legacy of past land use is a key variable influencing soil properties and nutrient availability (Hope et al. 2005; Kaye et al. 2008). CAP LTER research suggests that desert-like urban landscapes do not resemble the native desert in terms of trophic dynamics, richness, and species composition (Faeth et al. 2005), soil properties (Lewis et al. 2006; Hall et al. 2009), or biogeochemical processes (Hall et al. 2009), although native biotic pollinators can be sustained through native plantings (McIntyre 2000). Myriad decisions, values and norms expressed at the household, neighbourhood and regional scales drive the choices and management of residential landscapes (Larson et al. 2010; Cook et al. 2011). Effects of residential development decisions may last long into the future and become institutionalised by Homeowner Associations' Covenants, Codes and Restrictions (Martin et al. 2003).

At the regional scale, understanding institutional drivers of urban growth is critical because urban sprawl has economic, ecological, and social repercussions. York et al. (in review-b) analysed ballot propositions associated with state trust land and found that conservation and development concerns are ascendant priorities, along with issues of land management and resource use. In a comparison of land fragmentation surrounding urban areas near five LTER sites, researchers asked how urban population dynamics, water provisioning, transportation, amenity-driven growth, and institutional factors influence patterns of land fragmentation. Land-fragmentation patterns around Albuquerque and Las Cruces (both in New Mexico) were similar to those of an earlier era in Phoenix, as suburbs expanded along rivers, but Phoenix today exhibits a monocentric, spreading growth pattern (York et al. 2011), with 'leap-frog' development leaving patches of vacant land (Gober and Burns 2002). Social-survey data reveal that race, gender, political persuasion, and time lived in Greater Phoenix govern perceptions about sprawl and influence support for policy prescriptions (York et al. in review-a).

## 10.4.2 Climate, Ecosystems and People

The goal of this integrated research is to understand interactions among urban and urban-hinterland climate, ecosystems, and social systems. Natural landscape features (terrain) and characteristics of the urban land surface (building and vegetation distribution, irrigation) have modified large-scale atmospheric forcings and substantially altered climate in the region (Brazel et al. 2000; Brazel and Heisler 2009). We expect, therefore, that climate change will play out through the interaction of global drivers with regional presses and pulses (Fig. 10.1). Global climate change is an external driver, while land-use and land-cover changes, driven largely by economic growth or recession, represent the main press and pulse events for local and regional climate change. The Urban Heat Island (UHI) in Phoenix is manifested as an increase of nighttime temperatures of up to 5°C in the past several decades (Hedquist 2005; Hartz et al. 2006a, b; Sun et al. 2009). Temperature changes due to the UHI have already occurred in metropolitan Phoenix and have interfered with historic relationships between local temperature and atmospheric processes in the region (Ruddell et al. in press). Over the last century, there has been a linear downward trend in frost days, an increase in misery days (i.e., days over 100°F [38°C]; especially since 1970), and accelerated warming in heat-wave threshold temperatures (Ruddell et al. in press). Such changes in climate within urbanised central Arizona present microcosms of effects we will likely see elsewhere accompanying global climate change (Grimm et al. 2008b).

Climate is an important driver of ecosystem processes (e.g., primary production) and human outcomes (e.g., health and quality of life). The emergence and intensification of Phoenix's UHI represents an important stressor on humans in the city. CAP LTER researchers found that the UHI varies greatly in space, mirroring the physical heterogeneity of the urban landscape (Myint and Okin 2009). Variations in amounts and distributions of soil, impervious surface, and irrigated vegetation in urban and suburban areas can both exacerbate or ameliorate the UHI, although heat mitigation involves a trade-off with water use (Myint and Okin 2009), and can affect weather patterns (Grossman-Clarke et al. 2008). Superimposed on this are spatially variable demographic characteristics of the human population. Jenerette et al. (2007) found that socioeconomic status of neighbourhoods was the most important social predictor of urban vegetation and thereby indirectly influenced the spatial distribution of temperatures, with higher vegetation cover and cooler temperatures in wealthier neighbourhoods.

The UHI has implications for environmental justice because spatial heat variability affects some segments of the population more than others (Harlan et al. 2006, 2008; Jenerette et al. 2007, 2011). For example, in a July 2006 heat wave, extreme temperatures were variably distributed across Phoenix neighbourhoods. Residents at greatest risk of exposure to heat tended to be minority, low-income, and elderly (Ruddell et al. 2010; Fig. 10.4). Furthermore, Ruddell et al. (2010, 2012) found that respondents to the 2006 Phoenix-Area Social Survey (PASS) were aware of temperature differences in their neighbourhoods relative to others and that their perceptions of hot weather closely tracked measured differences in local temperatures. Respondents answered that they would be willing to pay significantly more for homes comparable to those in which they lived if they were located in neighbourhoods with conditions 5–10°F (~2–6°C) cooler (Harlan et al. 2007).

Urbanisation and the UHI also affect plant phenology. Changes in plant population and community dynamics may result from a significant change in flowering phenology for a small but substantial proportion of the flora (Neil and Wu 2006; Neil 2008). Our urban sites also showed a decoupling of phenology from precipitation, the main driver of phenologic change in the desert (Fig. 10.5). Phenology of urban vegetation instead appears linked to specific ecosystem services, such as food and fodder production, recreation, or cultural aesthetics (Buyantuyev and Wu 2009). This may be a consequence of massive changes in hydrologic systems that have long characterised the urbanisation of central Arizona.

### 10.4.3 Water Dynamics in a Desert City

Redistribution of water may be the single most important effect of urbanisation in arid lands. It is a product of the conversion of natural hydrology to man-made hydrology via modification (local changes), procurement (regional changes), and management (temporal changes; Grimm et al. 2008a; Redman and Kinzig 2008). The Phoenix metropolis now appropriates 100% of the surface flow of the Salt and Verde Rivers and is increasingly exploiting local groundwater and surface water from more distant basins (e.g., the Colorado River). Controlled management and engineering have dramatically shifted the spatiotemporal variability of the hydrologic system. For example, annual sediment transport dropped to low levels in fully urbanised portions of the region, compared to undeveloped and developing areas (El-Ashmawy et al. 2009).

In arid landscapes, water is typically found around rivers with clearly defined, ecologically productive riparian areas. Often located along rivers themselves, cities in arid environments display a similar "oasis" characteristic as humans re-allocate water to grow urban vegetation. We consider the sum of regional hydrologic alterations to

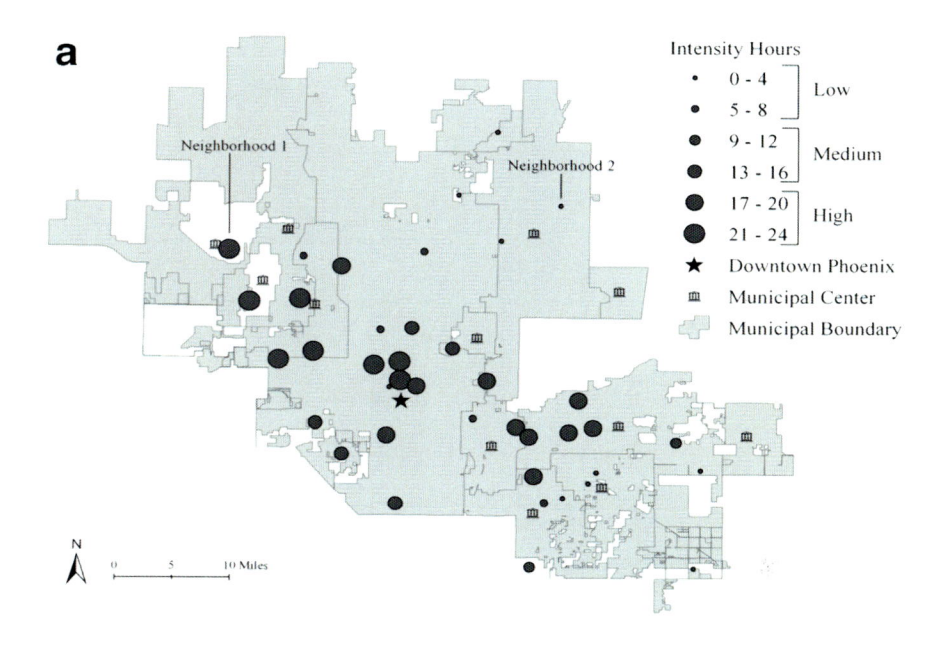

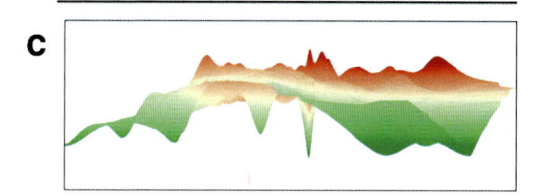

Fig. 10.4 Spatial distribution of heat intensity in Phoenix, AZ, in July 2005. (**a**) Hours in a 4-day period that temperature exceeded 110°F (43 °C), (**b**) demographic characteristics of the population in low-, medium- and high-exposure areas, and (**c**) a graphic representation of the heat exposure "riskscape" for the region (Map in (**a**) and data in (**b**) used with permission from Ruddell et al. 2010)

constitute a "riparianisation" of desert ecosystems, as urbanisation redistributes water more extensively and evenly across the landscape, compared with the pre-human situation. This concept may be more broadly generalised as the "arboreolisation" of grassland ecosystems by urbanisation.

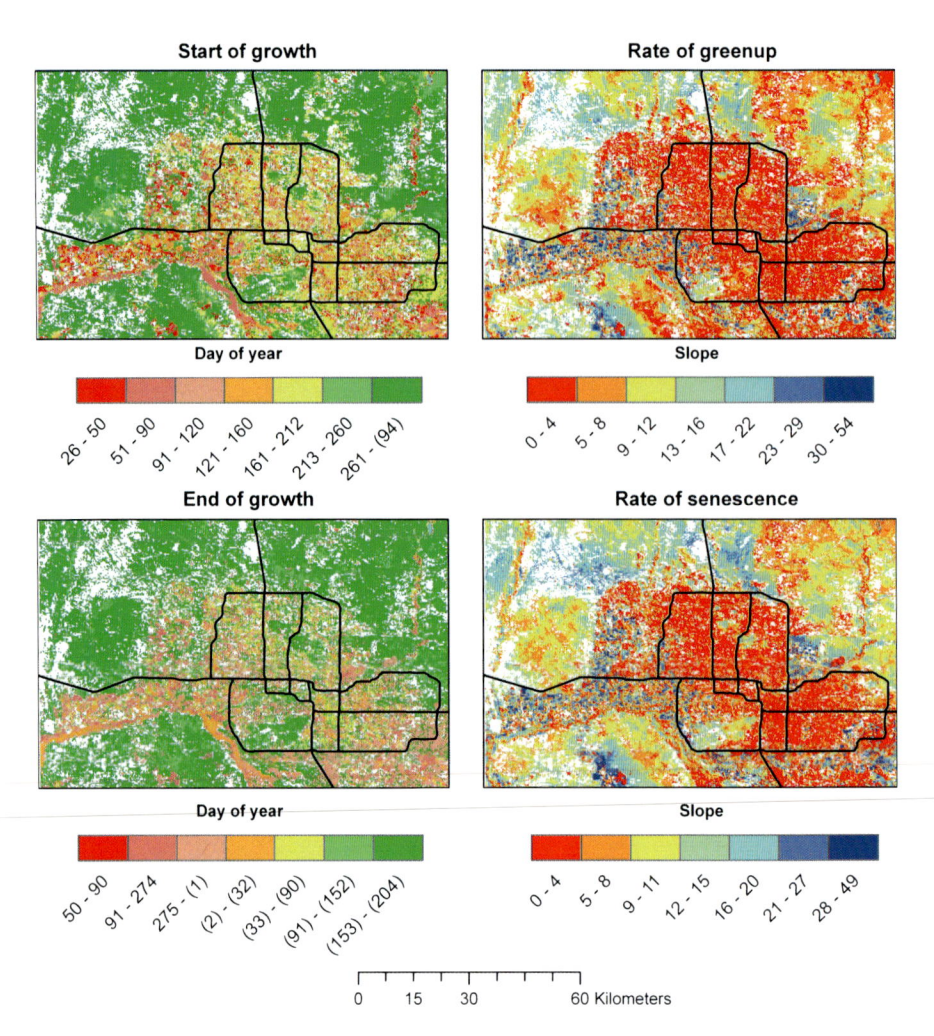

**Fig. 10.5** Maps of the urbanised central Arizona, USA region, showing the spatial distribution of plant phenological variables, start of growth, rate of greenup, end of growth, and rate of senescence. The *thick black lines* are the area freeways, which approximately enclose the urbanised/suburbanised portion of the region, comprising ~24 municipalities of the Phoenix metropolitan area. The large area that is differentiated from the surrounding desert to the north and south of the east-west freeway extending to west from the metropolitan area is an agricultural region that has yet to become urbanised. Seasonal parameters for the initial (spring) growth period were extracted from 2004 to 2005 normalised difference vegetation index (NDVI) data that were filtered with a Savitsky-Golay technique. NDVI data provide an index of greenness from remote imagery that can be correlated with vegetation biomass; changes in NDVI with time reflect growth (*increasing greenness*) or senescence (*decreasing greenness*). Dates are displayed as day of year (year 2005 days are shown in *parentheses*). Rates are calculated as tangent of slope between 20 and 80% levels of NDVI (Data and figures from Buyantuyev 2008)

Water use, vegetation, cooling, and inequitable UHI distribution provides an excellent example of ecosystem-service tradeoffs. Outdoor irrigation accounts for most of the water used by Phoenix area households (an astounding ~800 L person$^{-1}$ day$^{-1}$) and, in turn, domestic water use directly relates to household affluence (Harlan et al. 2009). Lifestyle preferences and priorities embodied in outdoor landscaping help explain the preference for water-intensive lawns and outdoor features (Larsen and Harlan 2006; Yabiku et al. 2008), as do socially constructed ideas about nature and its place in the urban environment (e.g., "I think the desert belongs in the desert"; Larson et al. 2009a). Vegetation helps to ameliorate heat intensity (Stabler et al. 2005; Jenerette et al. 2007; Martin 2008), but this ecosystem service requires water. Unequal access to heat-ameliorating landscapes (supported by irrigation) accounts for spatial variability in vulnerability to the UHI. Jenerette et al. (2011) found that disparity in vegetated cover between neighbourhoods has increased since 1970, and that large increases in regional water use would be required to alleviate current inequalities.

In the desert, winter and summer monsoon storms produce overland flows that generate rapid increases in stream and river discharge, which carries with it high loads of dissolved and particulate materials. Flash floods also are well-known phenomena in urban environments owing to the prevalence of impervious surfaces. Built structures or management may ameliorate or exacerbate these processes. Stormwater management in metropolitan Phoenix features designed systems—retention basins, floodplain parks, and "restored" riparian zones—that may provide a diversity of ecosystem services, some intentional and some not (Larson et al. in press). For example, the amount of nitrogen transported from urban/suburban watersheds during storms depends on a combination of catchment features and storm characteristics (Lewis and Grimm 2007). The type of infrastructure (i.e., designed systems) is hypothesised to be one of the more important catchment features. Nitrogen that is not transported is retained within the catchment, or removed via the process of denitrification (microbial conversion of nitrate to gaseous forms of nitrogen). Since nitrate is a pollutant when it occurs at high concentrations, such as in groundwaters of the region (Xu et al. 2007), nitrogen removal is an important ecosystem service, though it is virtually unknown to the public and scarcely considered in stormwater management.

Indian Bend Wash (IBW) is a designed stream-lake floodway that drains over 200 km$^2$ of Scottsdale (one of the largest municipalities in the Phoenix metropolitan area). A highly manipulated and managed system, IBW is influenced alternately by management and natural hydrologic variation (Roach et al. 2008; Fig. 10.6). The identity of the limiting nutrient (nitrogen or phosphorus) varies temporally in response to deliberate water additions (high in nitrogen) or natural flood inputs (high in phosphorus; Roach and Grimm 2009). In addition, its lakes, stream segments, and floodplains (mostly turf-dominated parks that are fertilised and irrigated using lake water) exhibit high rates of denitrification, especially under storm flows when extensive floodplain areas are inundated (Roach and Grimm 2011). There is great potential to ensure that the design of the Indian Bend Wash floodway, originally intended for flood management only, yields multiple ecosystem services, provided that managers are cognisant of the conditions under which services such as

**Fig. 10.6** Paired aerial photographs showing the change in land cover and use in Indian Bend Wash, Scottsdale and Tempe, Arizona between 1949 (*top*) and 2003 (*bottom*). The 1949 image shows a landscape dominated by farm fields, with country roads (part of the characteristic grid pattern of metropolitan Phoenix) evident as *light gray lines*. These same roads can be seen in the 2003 image, but farm fields have been replaced by housing developments and commercial (*right side of image*) and institutional (*upper left part of image*) land uses. The ephemeral stream (1949) and designed lake chain (2003) can be seen bisecting the images from *top* to *bottom*. Note the wide, shrub and tree-covered channel in the *upper image*; although by 2003 it was replaced by parks, lakes, and streams, the relatively wide channel still contains flash floods (see text for further description)

nitrogen removal, peak flow modulation, recreational amenities, and so forth can be enhanced.

Policies and decisions about water—supply, stormwater, and wastewater—are crucial to the sustainability of a city like Phoenix. A steady weakening of the Groundwater Management Act of 1980, designed to attain "safe-yield" of groundwater, has heightened water insecurity and delayed conservation measures (Hirt et al. 2008; Larson et al. 2009b). At the same time, policymakers are significantly less concerned than the lay public or scientists about regional water use rates; the lay public tend to blame other people for water scarcity and scientists stress the need to control demand (Larson et al. 2009c). Attitudes and actions (or lack thereof) such as these fuel the potential for serious loss of resilience under the threat of increased duration, frequency, and intensity of droughts and other extreme events, such as flash floods, in the US Southwest (Karl et al. 2009).

### *10.4.4 Biogeochemical Patterns, Processes and Human Outcomes*

Human manipulation of biogeochemical cycles through agriculture and energy use has supported societal advances that have increased the carrying capacity of Earth, particularly via the Green Revolution and modern industrial technology. However, these advances have also led to major environmental problems, from local to global scales (Vitousek et al. 1997; Grimm et al. 2008b; Childers et al. 2011), threatening biodiversity, ecosystem integrity, and quality of life. Ecosystem services associated with biogeochemical cycles therefore can be beneficial or harmful. Material fluxes and biogeochemical linkages underlie most ecological processes, but in urban ecosystems they are overwhelmed by human-generated fluxes of nutrients and toxins, and by design and management influences on timing, duration, and magnitude of biogeochemical processes (Groffman et al. 2006; Kaye et al. 2006). Biogeochemical studies in the CAP LTER programme have been conducted from plot/parcel scales to watershed/whole-system scales, including interaction with surrounding ecosystems. In particular, mass balances of nitrogen (Baker et al. 2001) and phosphorus (Metson et al. 2012) have shown very large imports of these elements in food, fuel, and animal feed that are almost entirely human-mediated. Because very little water is transported out of the city, most of these inputs accumulate (in soils, groundwater, and vegetation), although some are exported via atmospheric transport (oxides and aerosols of nitrogen) or in crops (phosphorus).

Our conceptual model of urban biogeochemical processes (Fig. 10.7) identifies four reactive ecosystem compartments (atmosphere, land, surface water, groundwater), any of which may be a source, a recipient system, or a transporting/transforming system for a particular material flux. Toxins and pollutants may become concentrated in urban recipient systems to generate biogeochemical "riskscapes" for urban inhabitants, and nutrients may be transported to low-productivity desert recipient systems where they have a fertilisation effect. Ongoing studies of the

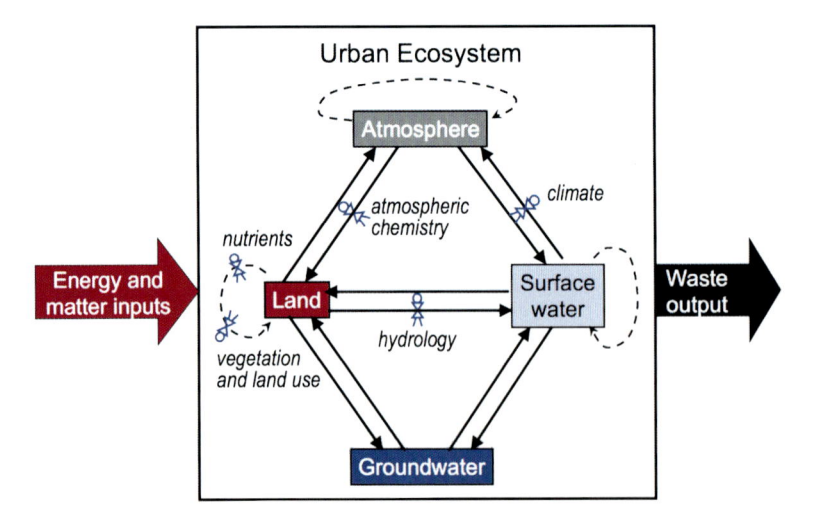

**Fig. 10.7** Model showing major compartments of the biogeochemical cycles (Reprinted from Kaye et al. 2006, with permission from Elsevier)

fates of material fluxes include: (1) desert responses to deposition; (2) soil nutrient distributions; (3) air quality; and (4) water quality.

Our research on atmospheric transport and deposition has found relatively low annual rates of wet and dry nitrogen deposition that did not differ significantly across an area larger than the CAP LTER study region. In contrast, wet and dry deposition of organic carbon was significantly elevated in the urban and downwind desert compared to upwind sites (Lohse et al. 2008). Soil respiration (e.g., microbial activity) showed muted responses to experimental additions of fine particulate organic carbon such as that derived from urban aerosols (Kaye et al. 2011). The low rate of nitrogen deposition is a surprising result, since a mass balance study of Phoenix estimated the generation of nitrogen oxides from automobile use to be nearly 30 kg ha$^{-1}$ year$^{-1}$ (Baker et al. 2001). To determine whether deposition from the urban atmosphere affects desert productivity, CAP LTER researchers established a long-term fertilisation experiment in fifteen locations across a ~100-km transect, including sites upwind, within, and downwind from the city. To date, the perennial desert shrub, creosote bush, has been unaffected by either atmospheric or experimental additions of nutrients, although annuals have shown a response to supplemental N additions when rainfall was sufficient (Hall et al. 2011). The lack of evidence for enhanced nitrogen deposition suggests that the nitrogen is either transported far from the city, or is retained within the city but does not have a fertilising effect on desert plants. Elevated soil nitrogen concentration in desert soils surrounding the city suggests the latter is a plausible mechanism (Zhu et al. 2006).

CAP LTER's extensive soil surveys provide a foundation for understanding controls on and impacts of the spatial distribution of nutrients, organic and inorganic carbon, and metals. Urban soils have significantly higher black carbon content relative to desert soils, and soil concentrations of lead, cadmium, copper, and arsenic correlate with urbanisation (e.g., Fig. 10.8). Urban lead isotopes showed that the

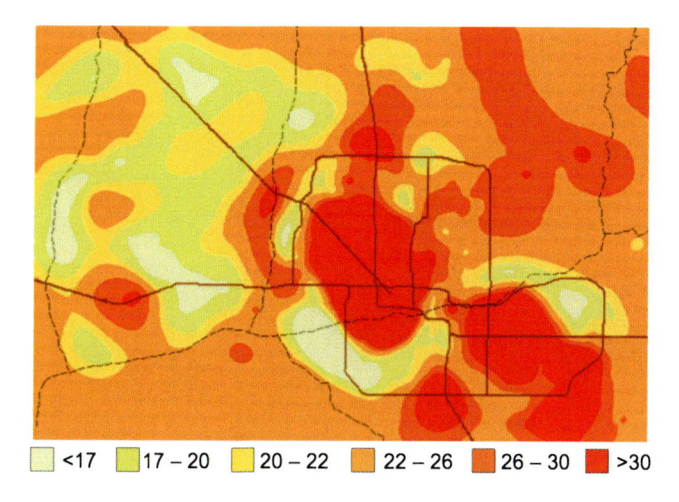

**Fig. 10.8** Lead concentration (μg/kg) measured in 2005 in the surface soil (1–10 cm) across the CAP LTER study area. *Brown lines* show major freeways; the urbanised region is encircled by these roads (Reproduced with permission from Zhuo 2010)

source of this metal was either leaded paint or western coal, but not leaded gasoline (Zhuo 2010; Zhuo et al. in press). We used hierarchical Bayesian models to scale plot data on organic carbon, inorganic carbon, nitrogen and phosphorus to the 6,400-km$^2$ CAP LTER region and estimated that 1,140 Gg of organic carbon and 130 Gg of nitrogen have accumulated in urbanised soils of the region (Kaye et al. 2008; Majumdar et al. 2008), comparable to values estimated previously (Hope et al. 2005; Zhu et al. 2006). This work also confirmed that land-use legacies (i.e., whether a site had ever been farmed) were important determinants of soil-nutrient concentrations.

Distributions of materials also result in uneven distributions of disamenities (environmental factors that negatively affect people) across central Arizona, with ensuing environmental justice implications. For example, Grineski et al. (2007) found distinct sociospatial inequalities in exposure to pollutants; neighbourhoods of lower socioeconomic status, which include higher proportions of renters and Latinos, generally experience higher levels of air pollution (Fig. 10.9). Urban lead distributions also are heterogeneous and higher in poorer neighbourhoods (Zhuo et al. in press). These differential impacts reflect historical patterns of development in Phoenix, with legacies of residential segregation based on class, race, ethnicity, amenities, and disamenities that linger today (Bolin et al. 2005).

## 10.4.5 Human Decisions and Biodiversity

Urbanisation profoundly alters the composition, abundance, and distribution of nonhuman species (McKinney 2002; Schlesinger et al. 2008). Yet biodiversity is key to some ecosystem services (especially cultural services). Reduced access to

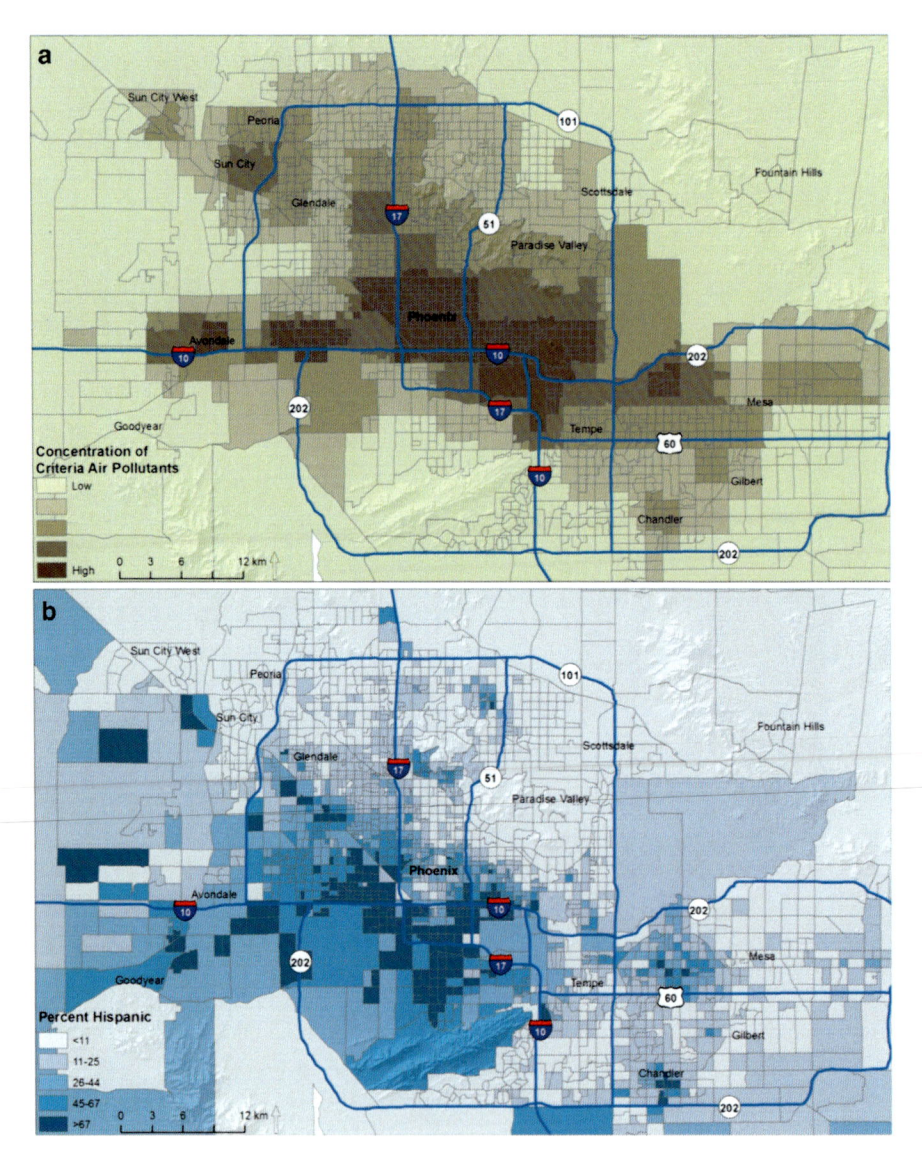

**Fig. 10.9** Spatial distributions of (**a**) (*top*) criteria air pollutants and (**b**) percentage of the population that is in the Hispanic ethnic group, in the central Arizona–Phoenix region (Reproduced with permission from Grineski et al. 2007)

nature, the "extinction of experience" (Pyle 1978), is increasingly thought to be detrimental to human well-being (Shumaker and Taylor 1983; Ryan 2005). CAP LTER research has provided insights into the socioeconomic drivers of urban biodiversity patterns (Kinzig et al. 2005), the functioning of urban food webs (Faeth et al. 2005; Shochat et al. 2006b), and the effects of exposure to native desert landscapes on people (Yabiku et al. 2008).

Ecological approaches to studying human impacts on biodiversity have typically focused on habitat loss and disturbance brought about by human population agglomerations. Our studies have been unique in their focus on mechanisms accounting for changes in species diversity and community composition (Shochat et al. 2006b, 2010). At the metropolitan scale, land-use change and human choice and action have resulted in altered plant, bird, and arthropod communities. Urban plant diversity (influenced most by landscaping aesthetics and socioeconomic status; Martin et al. 2004) is considerably lower and more even (similar numbers of individuals of each species) compared with native desert communities (Hope et al. 2003, 2006; Walker et al. 2009). For birds, community composition mirrors the variation in plant communities associated with landscaping aesthetics and socioeconomics. Irrigation drives ground-arthropod community patterns, with greater abundance and diversity in mesic and oasis (grass with a landscaped gravel border) landscapes (Cook and Faeth 2006). Arthropod species richness has declined over the last decade in desert remnant sites and xeric yards, possibly owing to landscape practices or isolation of these sites from colonist sources (outlying desert; Bang and Faeth 2011). A question that remains is whether species loss occurs due to biotic interactions or differential vulnerability to stress. We do know that some urban birds differ from their desert counterparts in their physiological response to stressors (Fokidis et al. 2009; Deviche et al. 2011; Fokidis and Deviche 2011).

CAP LTER researchers have used experimental and synthetic approaches to determine how urbanisation affects trophic dynamics. Elevated urban habitat productivity and reduced temporal variability contribute to trophic systems that are radically different from their natural counterparts, with a shift to combined bottom-up and top-down control of trophic dynamics (Faeth et al. 2005). Mechanistic, experimental studies of "giving-up density" (a surrogate for how long birds will persist at a foraging patch; Shochat et al. 2004, 2006a, b, 2010) showed that birds tended to feed until remaining food resources were nearly exhausted, indicating that the benefit of continuing to forage outweighed the cost of staying in one place. Even though resource abundance is high in the urban environment, the experiments reveal strong competition among bird species for those food resources in the absence of significant predation risk. Findings from CAP LTER research for diverse groups of biota call into question the "field of dreams" hypothesis (that constructed landscapes meant to imitate the desert are functionally equivalent): trophic dynamics, richness, and species composition in desert-like residential landscapes and desert remnants are not analogous to the native desert.

Human responses to biota—the kinds and forms of vegetation, for example—depend upon a complex set of preferences that we are beginning to unravel with our experimental landscapes work in a single neighbourhood coupled with social survey data (e.g., Larson et al. 2009a). Residents preferred mesic and oasis landscapes (both having some turf and trees) over xeric and desert landscapes. The longer residents had lived in the Phoenix area the less they preferred arid landscapes. Oasis landscapes, with their mixed ground covers of gravel and small patches of turf, have emerged as a compromise, as residents reconcile desires for turf with water scarcity concerns and environmental values (Yabiku et al. 2008). Finally, resident satisfaction with the existing variety of birds in their neighbourhoods was significantly correlated with actual bird diversity and with general levels of neighbourhood satisfaction. Predominantly Hispanic and low-income neighbourhoods in Phoenix had lower amounts and diversity of perennial vegetation (Martin et al. 2004) and lower bird diversity (Kinzig et al. 2005) than middle and higher-income neighbourhoods, suggesting that the aesthetic services associated with biodiversity are inequitably distributed in the region.

## 10.5 Urbanisation in a Dynamic World: Prospects for Sustainability

As a pervasive and accelerating global challenge, urbanisation produces a unique set of environmental, social, and economic problems that desperately need fundamental research as well as action (i.e., a 'problem-based' approach). As a potential solution to the global sustainability challenge, the need for fundamental research uncovering what makes cities sustainable is equally great (i.e., a 'sustainability-based' approach). The two US urban LTER sites, BES and CAP, are exemplars of how studying cities as ecosystems and incorporating SES conceptual frameworks (see Fig. 10.1) can advance understanding of urbanisation. Yet there is a need to understand how patterns, processes, and mechanisms found in Phoenix or Baltimore may contrast with other urban areas, and to examine what the commonalities and differences may reveal about the universal nature of urban SES. For example, comparisons of the "metabolism" or elemental/material mass balances among cities worldwide (Weisz and Steinberger 2010), or for individual cities through time (Krausmann, Chap. 11 in this volume) could reveal what factors may help reduce the dependence of urban ecosystems on their hinterlands. Comparative studies of cities as socio-ecological systems are largely absent from the literature (Grimm et al. 2008b; but see McDonald et al. 2011), and we believe that such analyses and syntheses will be a key to addressing both the 'problem-based' and the 'sustainability-based' approaches posed above.

Examining the research we have described above using a sustainability lens, we ask how do the services provided by evolving urban ecosystems affect human outcomes and behaviour, and how does human action (response) alter patterns of ecosystem structure and function and, ultimately, urban sustainability, in a dynamic environment? Ecosystem services of water provisioning, climate modulation, flood mitigation, nutrient removal, recreation, and aesthetic enjoyment are encompassed in our research on climate, ecosystems and people, water dynamics, biogeochemical patterns, processes, and human outcomes, and human decisions and biodiversity. We have uncovered instances of environmental inequity (e.g., human-health risk from heat and toxic substance distributions) that clearly do not satisfy the justice and equity principles of sustainability. We have identified tradeoffs among services that will have to be considered to ensure sustainability (e.g., water provisioning and climate modulation). Yet we have provided examples wherein built infrastructure can potentially be designed to generate multiple ecosystem services (e.g., floodplains that mitigate high flows, remove nutrients, and provide urban habitat for wildlife and recreation spaces for people). The alteration of ecosystem pattern and process is initially profound when urban ecosystems are created, but their persistence may rely upon a flexibility to design criteria that will allow adjustments under scenarios of climate change (i.e., frequency and magnitude of extreme events) and increased population. Thus, the ultimate answer to the question of sustainability must incorporate a look to the future.

For our studies in CAP LTER, we aim to move beyond the question of what is our relationship with nature and its implications for sustainability, to those of what will be and what ought to be our relationship relative to what is sustainable. The time to raise this question is opportune; Phoenix and Arizona are increasingly posing questions about what their citizens hope for their future (Center for the Future of Arizona and Gallup Poll 2009). Moreover, Phoenix is a metropolitan region at risk from the potential negative impacts of global climate change. Increasing demands for water and land due to the rapidly growing population are on a collision course with high-confidence predictions for a drier, hotter future with reduced water availability in the Southwest (Seager et al. 2007; Barnett and Pierce 2009; Karl et al. 2009; Buizer et al. 2009). At the very least, we are entering a period of non-stationarity, and therefore must learn to plan for uncertainty (Gober et al. 2010). During the current recession, the impacts of higher temperatures, increased energy costs, and water scarcity are already evident in low-income populations.

A decade of research in CAP LTER has generated a strong understanding of socio-ecological dynamics of urban systems. The next generation of research should use that science to inform plausible futures, and in turn provide a basis for scenarios that can guide Phoenix and other metropolitan regions toward a desirable, sustainable future.

**Acknowledgments** This moderately comprehensive and, we hope, forward-looking review is based upon 13 years of research, but especially the most recent 7 years, by a large and talented CAP LTER team. The material presented in this chapter is the work of many current and past CAP LTER investigators, project managers, post-doctoral, graduate, and undergraduate scholars, and countless field teams, technicians, data specialists, and other support staff, and we thank all of them for their contributions over many years to the success of the project. We especially thank Marcia Nation for assembling the many pieces of the story in various reports and drafts. We acknowledge our funding from the National Science Foundation (LTER core grants, supplements, and related projects) and the support of the LTER network. We thank multiple units across Arizona State University for financial and other material support, and for creating an environment that fosters and supports interdisciplinary scholarship. For assistance in preparing the graphics, we thank Bryan Barker and Travis Buckner. Michael Bernstein provided invaluable support in assembling and formatting the literature, obtaining permissions, and copy-editing the manuscript. We acknowledge the reviewers and editors of this volume and thank them for their patience and support. This paper was partially based on work supported by the National Science Foundation while Nancy B. Grimm was working at the Foundation. Any opinions, findings, and conclusions expressed here are those of the authors and do not necessarily reflect the views of the Foundation.

# References

Baker, L. A., Hope, D., Xu, Y., Edmonds, J., & Lauver, L. (2001). Nitrogen balance for the Central Arizona-Phoenix ecosystem. *Ecosystems, 4*, 582–602.

Bang, C., & Faeth, S. H. (2011). Variation in diversity, trophic structure and abundance in response to urbanization: Seven years of arthropod monitoring in a desert city. *Landscape and Urban Planning, 103*, 383–399.

Barnett, T. P., & Pierce, D. (2009). Sustainable water deliveries from the Colorado River in a changing climate. *Proceedings of the National Academy of Sciences, 106*, 7334–7338.

Bennett, E., Peterson, G., & Gordon, L. (2009). Understanding relationships among multiple ecosystem services. *Ecology Letters, 12*, 1394–1404.

Bettencourt, L. M. A., Lobo, J., Helbing, D., Kuhnert, C., & West, G. B. (2007). Growth, innovation, scaling, and the pace of life in cities. *Proceedings of the National Academy of Sciences, 104*, 7301–7306.

Bolin, B., Grineski, S., & Collins, T. (2005). Geography of despair: Environmental racism and the making of south Phoenix, Arizona, USA. *Human Ecology Review, 12*, 155–167.

Brazel, A., & Heisler, G. (2009). Climatology of urban long-term ecological research sites: Baltimore ecosystem study and Central Arizona-Phoenix. *Geography Compass, 3*, 22–44.

Brazel, A. J., Selover, N., Vose, R., & Heisler, G. (2000). The tale of two climates: Baltimore and Phoenix urban LTER sites. *Climate Research, 15*, 123–135.

Briggs, J. M., Spielmann, K. A., Schaafsma, H., Kintigh, K. W., Kruse, M., Morehouse, K., & Schollmeyer, K. (2006). Why ecology needs archaeologists and archaeology needs ecologists. *Frontiers in Ecology and the Environment, 4*, 180–188.

Brown, L. R. (2001). *Eco-economy: Building an economy for the Earth.* New York: Norton.

Buizer, J., Beller-Simms, N., Jacobs, K., Bentzin, B., Shea, E., Anderson, R., & Roy, M. (2009). *Planning integrated research for decision support for climate adaptation and water management: A focus on desert and coastal cities.* From the January NOAA/ASU Workshop Report, Arizona State University, Tempe, AZ.

Buyantuyev, A. (2008). *Effects of urbanization on the landscape pattern and ecosystem function in the Phoenix metropolitan region: A Multi-scale study.* Unpubl. Ph.D. Dissertation, School of Life Sciences, Arizona State University.

Buyantuyev, A., & Wu, J. (2007). Effects of thematic resolution on landscape pattern analysis. *Landscape Ecology, 22*, 7–13.

Buyantuyev, A., & Wu, J. (2009). Urbanization alters spatiotemporal patterns of ecosystem primary production: A case study of the Phoenix metropolitan region, USA. *Journal of Arid Environments, 73*, 512–520.

Buyantuyev, A., Wu, J., & Gries, C. (2007). Estimating vegetation cover in an urban environment based on Landsat ETM+imagery: A case study in Phoenix, USA. *International Journal of Remote Sensing, 28*, 269–291.

Carpenter, S. R., Mooney, H., Agard, J., Capistrano, D., DeFries, R., Diaz, S., Dietz, T., Duraiappah, A. K., Otend-Yeboah, A., Pereira, H. M., Perrings, C., Reid, W. V., Sarukhan, J., Scholes, R. J., & Whyte, A. (2009). Science for managing ecosystem services: Beyond the millennium ecosystem assessment. *Proceedings of the National Academy of Sciences, 106*, 1305–1312.

Center for the Future of Arizona, & Gallup Poll. (2009). *The Arizona we want.* Center for the Future of Arizona. Retrieved from http://www.thearizonawewant.org/

Chapin, F. S., Lovecraft, A. L., Zaveleta, E. S., Nelson, J., Robards, M. D., Kofinas, G. P., Trainor, S. F., Peterson, G. D., Huntington, H. P., & Naylor, R. L. (2006). Policy strategies to address sustainability of Alaskan boreal forests in response to a directionally changing climate. *Proceedings of the National Academy of Sciences, 103*, 16637–16643.

Childers, D. L., Corman, J., Edwards, M., & Elser, J. J. (2011). Sustainability challenges of phosphorus and food: Solutions from closing the human P cycle. *BioScience, 61*, 117–124.

Collins, J., Kinzig, A., Grimm, N. B., Fagan, W., Wu, J., & Borer, E. (2000). A new urban ecology. *American Scientist, 88*, 416–425.

Collins, S. L., Swinton, S. M., Anderson, C. W., Benson, B., Brunt, J., Gragson, T., Grimm, N. B., Grove, J. M., Henshaw, D., Knapp, A. K., Kofinas, G. P., Magnuson, J. J., McDowell, W. H., Melack, J. M., Moore, J., Ogden, L. A., Porter, J., Reichman, O. J., Robertson. G.P., Smith, M. D., Vande Castle, J., & Whitmer, A. C. (2007). *Integrated Science for Society and the Environment: A strategic research initiative* (LTER Network Office Publication No. 23). Retrieved from http://intranet2.lternet.edu/sites/intranet2.lternet.edu/files/documents/LTER_History/Planning_Documents/ISSE_v6.pdf

Collins, S. L., Carpenter, S. R., Swinton, S. M., Orenstein, D. E., Childers, D. L., Gragson, T. L., Grimm, N. B., Grove, J. M., Harlan, S. L., Kaye, J. P., Knapp, A. K., Kofinas, G. P., Magnuson, J. J., McDowell, W. H., Melack, J. M., Ogden, L. A., Robertson, G. P., Smith, M. D., & Whitmer, A. C. (2011). An integrated conceptual framework for long-term social-ecological research. *Frontiers in Ecology and the Environment, 9*, 351–357.

Cook, W. M., & Faeth, S. H. (2006). Irrigation and land use drive ground arthropod community patterns in urban desert. *Environmental Entomology, 35*, 1532–1540.

Cook, W. M., Casagrande, D. G., Hope, D., Groffman, P. M., & Collins, S. L. (2004). Learning to roll with the punches: Adaptive experimentation in human-dominated systems. *Frontiers in Ecology and the Environment, 2*, 467–474.

Cook, E., Hall, S. J., & Larson, K. J. (2011). Residential landscapes in an urban socio-ecological context: Multiscalar drivers and legacies of management practices, ecological structure, and ecosystem services. *Urban Ecosystems.*. doi:10.1007/s11252-011-0197-0.

Costanza, R., Graumlich, L., Steffen, W., Crumley, C., Dearing, J., Hibbard, K., Leemans, R., Redman, C., & Schimel, D. (2007). Sustainability or collapse: What can we learn from integrating the history of humans and the rest of nature? *Ambio, 36*, 522–527.

Deviche, P., Hurley, L. L., & Fokidis, H. B. (2011). Avian testicular structure, function, and regulation. In D. O. Norris & K. H. Lopez (Eds.), *Hormones and reproduction in vertebrates* (Vol. 4, pp. 27–69). Amsterdam: Elsevier/Academic.

El-Ashmawy, L., Toke, N., & Arrowsmith, R. J. (2009). Sedimentary geology of urban environments: An example of measuring sediment production and composition in Tempe, AZ. *The Geological Society of America Abstracts with Programs, 41*, 145.

Faeth, S. H., Warren, P. S., Shochat, E., & Marussich, W. (2005). Trophic dynamics in urban communities. *BioScience, 55*, 399–407.

Fokidis, H. B., & Deviche, P. (2011). Plasma corticosterone of city and desert curve-billed thrashers, *Toxostoma curvirostre*, in response to stress-related peptide administration. *Comparative Biochemistry and Physiology, Part A, 159*, 32–38.

Fokidis, H. B., Orchinik, M., & Deviche, P. (2009). Corticosterone and corticosteroid binding globulin in birds: Relation to urbanization in a desert city. *General and Comparative Endrocrinology, 160*, 259–270.

Gober, P., & Burns, E. K. (2002). The size and shape of Phoenix's urban fringe. *Journal of Planning Education and Research, 21*, 379–390.

Gober, P., Kirkwood, C., Balling, R., Ellis, A., & Deitrick, S. (2010). Water planning under climatic uncertainty in Phoenix: Why we need a new paradigm. *Annals of the Association of American Geographers, 100*, 356–372.

Grimm, N. B., & Redman, C. L. (2004). Approaches to the study of urban ecosystems: The case of central Arizona – Phoenix. *Urban Ecosystems, 7*, 199–213.

Grimm, N. B., Grove, J. M., Pickett, S. T. A., & Redman, C. L. (2000). Integrated approaches to long-term studies of urban ecological systems. *BioScience, 50*, 571–584.

Grimm, N. B., Faeth, S. H., Golubiewski, N. E., Redman, C. R., Wu, J., Bai, X., & Briggs, J. M. (2008a). Global change and the ecology of cities. *Science, 319*, 756–760.

Grimm, N. B., Foster, D., Groffman, P., Grove, J. M., Hopkinson, C. S., Nadelhoffer, K., Peters, D., & Pataki, D. E. (2008b). The changing landscape: Ecosystem responses to urbanization and pollution across climatic and societal gradients. *Frontiers in Ecology and the Environment, 6*, 264–272.

Grineski, S., Bolin, B., & Boone, C. (2007). Criteria air pollution and marginalized populations: Environmental inequity in metropolitan Phoenix, Arizona. *Social Science Quarterly, 88*, 535–554.

Groffman, P. M., Pouyat, R. V., Cadenasso, M. L., Zipperer, W. C., Szlavecz, K., Yesilonis, I. C., Band, L. E., & Brush, G. S. (2006). Land use context and natural soil controls on plant community composition and soil nitrogen and carbon dynamics in urban and rural forests. *Forest Ecology and Management, 236*, 177–192.

Grossman-Clarke, S., Liu, Y., Zehnder, J. A., & Fast, J. D. (2008). Simulations of the urban planetary boundary layer in an arid metropolitan area. *Journal of Applied Meteorology and Climatology, 47*, 752–768.

Haberl, H., Winiwarter, V., Andersson, K., Ayres, R. U., Boone, C., Castillo, A., Cunfer, G., Fischer-Kowalski, M., Freudenberg, W. R., Furman, E., Kaufmann, R., Krausmann, F., Langthaler, E., Lotze-Campen, H., Mirtl, M., Redman, C. L., Reenberg, A., Wardell, A., Warr, B., & Zechmeister, H. (2006). From LTER to LTSER: Conceptualizing the socioeconomic dimension of long-term socioecological research. *Ecology and Society, 11*. Retrieved from http://www.ecologyandsociety.org/vol11/iss2/art13/ES-2006-1786.pdf

Hall, S. J., Ahmed, B., Ortiz, P., Davies, R., Sponseller, R., & Grimm, N. B. (2009). Urbanization alters soil microbial functioning in the Sonoran Desert. *Ecosystems, 12*, 654–671.

Hall, S. J., Sponseller, R. A., Grimm, N. B., Huber, D., Kaye, J. P., Clark, C., & Collins, S. (2011). Ecosystem response to nutrient enrichment across an urban airshed in the Sonoran Desert. *Ecological Applications, 21*, 640–660.

Harlan, S. L., Brazel, A., Prashad, L., Stefanov, W. L., & Larsen, L. (2006). Neighborhood microclimates and vulnerability to heat stress. *Social Science and Medicine, 63*, 2847–2863.

Harlan, S. L., Budruk, M., Gustafson, A., Larson, K., Ruddell, D., Smith, V. K., Wutich, A., & Yabiku, S. (2007). *Phoenix area social survey 2006 highlights: Community and environment in a desert metropolis*. Central Arizona – Phoenix Long-Term Ecological Research Project Contribution No. 4. Global Institute of Sustainability, Arizona State University. Retrieved from http://caplter.asu.edu/docs/contributions/2007_PASS2.pdf

Harlan, S. L., Brazel, A. J., Jenerette, G. D., Jones, N. S., Larsen, L., Prashad, L., & Stefanov, W. L. (2008). In the shade of affluence: The inequitable distribution of the urban heat island. In R. C. Wilkinson & W. R. Freudenburg (Eds.), *Equity and the environment* (Vol. 15, pp. 173–202). Burlington: Emerald Group Publishing Limited.

Harlan, S. L., Yabiku, S., Larsen, L., & Brazel, A. (2009). Household water consumption in an arid city: Affluence, affordance, and attitudes. *Society and Natural Resources, 22*, 691–709.

Hartz, D., Brazel, A. J., & Heisler, G. M. (2006a). A case study in resort climatology of Phoenix, Arizona, USA. *International Journal of Biometeorology, 51*, 73–83.

Hartz, D., Prashad, L., Hedquist, B. C., Golden, J., & Brazel, A. J. (2006b). Linking satellite images and hand-held infrared thermography to observed neighborhood climate conditions. *Remote Sensing of Environment, 104*, 190–200.

Hedquist, B. (2005). Assessment of the urban heat island of Casa Grande, Arizona. *Journal of the Arizona-Nevada Academy of Sciences, 38*, 29–39.

Hirt, P., Gustafson, A., & Larson, K. L. (2008). The mirage in the Valley of the Sun. *Environmental History, 13*, 482–514.

Hobbie, J. E., Carpenter, S., Grimm, N. B., Gosz, J., & Seastedt, T. (2003). The US long term ecological research program. *BioScience, 53*, 21–32.

Hope, D., Gries, C., Zhu, W., Fagan, W. F., Redman, C. L., Grimm, N. B., Nelson, A., Martin, C., & Kinzig, A. (2003). Socioeconomics drive urban plant diversity. *Proceedings of the National Academy of Sciences, 100*, 8788–8792.

Hope, D., Zhu, W., Gries, C., Oleson, J., Kaye, J., Grimm, N. B., & Baker, B. (2005). Spatial variation in soil inorganic nitrogen across an arid urban ecosystem. *Urban Ecosystems, 8*, 251–273.

Hope, D., Gries, C., Casagrande, D., Redman, C. L., Grimm, N. B., & Martin, C. (2006). Drivers of spatial variation in plant diversity across the central Arizona-Phoenix ecosystem. *Society and Natural Resources, 19*, 101–116.

Jenerette, G. D., Wu, J., Grimm, N. B., & Hope, D. (2006). Points, patches and regions: Scaling soil biogeochemical patterns in an urbanized arid ecosystem. *Global Change Biology, 12*, 1532–1544.

Jenerette, G. D., Harlan, S. L., Brazel, A., Jones, N., Larsen, L., & Stefanov, W. L. (2007). Regional relationships between vegetation, surface temperature, and human settlement in a rapidly urbanizing ecosystem. *Landscape Ecology, 22*, 353–365.

Jenerette, G. D., Harlan, S. L., Stefanov, W. L., & Martin, C. A. (2011). Ecosystem services and heat riskscape moderation: Water, green spaces, and social inequality in Phoenix, USA. *Ecological Applications, 21*, 2637–2651.

Karl, T. R., Melillo, J. M., & Peterson, T. C. (Eds.). (2009). *Global climate change impacts in the United States: A state of knowledge report from the U.S. Global Change Research Program*. New York: Cambridge University Press.

Kaye, J. P., Groffman, P. M., Grimm, N. B., Baker, L. A., & Pouyat, R. (2006). A distinct urban biogeochemistry? *Trends in Ecology & Evolution, 21*, 192–199.

Kaye, J. P., Majumdar, A., Gries, C., Buyantuyev, A., Grimm, N. B., Hope, D., Jenerett, G. D., Zhu, W. X., & Baker, L. (2008). Hierarchical Bayesian scaling of soil properties across urban, agricultural, and desert ecosystems. *Ecological Applications, 18*, 132–145.

Kaye, J. P., Eckert, S. E., Gonzales, D. A., Allen, J. O., Hall, S. J., Sponseller, R. A., & Grimm, N. B. (2011). Decomposition of urban atmospheric carbon in Sonoran Desert soils. *Urban Ecosystems*. doi:10.1007/s11252-011-0173-8.

Kennedy, C., Cuddihy, J., & Engel-Yan, J. (2007). The changing metabolism of cities. *Journal of Industrial Ecology, 11*, 43–59.

Keys, E., Wentz, E. A., & Redman, C. L. (2007). The spatial structure of land use from 1970–2000 in the Phoenix, Arizona metropolitan area. *The Professional Geographer, 59*, 131–147.

Kinzig, A. P., Warren, P. S., Gries, C., Hope, D., & Katti, M. (2005). The effects of socioeconomic and cultural characteristics on urban patterns of biodiversity. *Ecology and Society, 10*. Retrieved from http://www.ecologyandsociety.org/vol10/iss1/art23/

Klaiber, H. A., & Smith, V. K. (2009). *Evaluating Rubin's causal model for measuring the capitalization of environmental amenities* (National Bureau of Economic Research Working Paper #14957). Retrieved from http://www.nber.org/papers/w14957

Larsen, L., & Harlan, S. L. (2006). Desert dreamscapes: Landscape preference and behavior. *Landscape and Urban Planning, 78*, 85–100.

Larson, K., Casagrande, D., Harlan, S., & Yabiku, S. (2009a). Residents' yard choices and rationales in a desert city: Social priorities, ecological impacts, and decision tradeoffs. *Environmental Management, 44*, 921–937.

Larson, K. L., Gustafson, A., & Hirt, P. (2009b). Insatiable thirst and a finite supply: Assessing municipal water conservation policy in greater Phoenix, Arizona, 1980–2007. *Journal of Policy History, 21*, 107–137.

Larson, K. L., White, D., Gober, P., Harlan, S., & Wutich, A. (2009c). Divergent perspectives on water resource sustainability in a public-policy-science context. *Environmental Science and Policy, 12*, 1012–1023.

Larson, K. L., Cook, E. M., Strawhacker, C. M., & Hall, S. J. (2010). The influence of diverse values, ecological structure, and geographic context on residents' multifaceted landscaping decisions. *Human Ecology, 38*, 747–761.

Larson, E., Earl, S., Hagen, E., Hale, R., Hartnett, H., McCrackin, M., McHale, M., & Grimm, N. B. (in press). Beyond restoration and into design: Hydrologic alterations in aridland cities. In S. T. A. Pickett., M. Cadenasso, B. McGrath, & K. Hill (Eds.), *Urban ecological heterogeneity and its application to resilient urban design*.

Lewis, D. B., & Grimm, N. B. (2007). Hierarchical regulation of nitrogen export from urban catchments: Interactions of storms and landscapes. *Ecological Applications, 17*, 2347–2364.

Lewis, D. B., Kaye, J. P., Gries, C., Kinzig, A. P., & Redman, C. L. (2006). Agrarian legacy in soil nutrient pools of urbanizing arid lands. *Global Change Biology, 12*, 1–7.

Liu, J., Dietz, T., Carpenter, S. R., Alberti, M., Folke, C., Moran, E., Pell, A. N., Deadman, P., Kratz, T., Lubchenco, J., Ostrom, E., Ouyang, Z., Provencher, W., Redman, C. L., Schneider, S. H., & Taylor, W. W. (2007a). Complexity of coupled human and natural systems. *Science, 317*, 1513–1516.

Liu, J., Dietz, T., Carpenter, S. R., Folke, C., Alberti, M., Redman, C. L., Schneider, S. H., Ostrom, E., Pell, A. N., Taylor, W. W., Ouyang, Z., Deadman, P., Kratz, T., Provencher, W., & Lubchenco, J. (2007b). Coupled human and natural systems. *Ambio, 36*, 639–649.

Lohse, K. A., Hope, D., Sponseller, R. A., Allen, J. O., Baker, L., & Grimm, N. B. (2008). Atmospheric deposition of carbon and nutrients across an arid metropolitan area. *Science of the Total Environment, 402*, 95–105.

Luck, M. A., Jenerette, G. D., Wu, J., & Grimm, N. B. (2001). The urban funnel model and spatially heterogeneous ecological footprint. *Ecosystems, 4*, 782–796.

Majumdar, A., Kaye, J. P., Gries, C., Hope, D., & Grimm, N. B. (2008). Hierarchical spatial modeling and prediction of multiple soil nutrients and carbon concentrations. *Communications in Statistics – Simulation and Computation, 37*, 434–453.

Martin, C. A. (2008). Landscape sustainability in a Sonoran Desert city. *Cities and the Environment, 1*, 1–16.

Martin, C. A., Peterson, K. A., & Stabler, L. B. (2003). Residential landscaping in Phoenix, Arizona, U.S.: Practices and preferences relative to covenants, codes, and restrictions. *Journal of Arboriculture, 29*, 9–17.

Martin, C. A., Warren, P. S., & Kinzig, A. P. (2004). Neighborhood socioeconomic status is a useful predictor of perennial landscape vegetation in residential neighborhoods and embedded small parks of Phoenix, Arizona. *Landscape and Urban Planning, 69*, 355–368.

Marzluff, J. M., & Rodewald, A. D. (2008). Conserving biodiversity in urbanizing areas: Nontraditional views from a bird's perspective. *Cities and the Environment, 1*, 6. Retrieved from http://escholarship.bc.edu/cate/vol1/iss2/6

McDonald, R. I., Douglas, I., Revenga, C., Hale, R., Grimm, N., Grönwall, J., & Fekete, B. (2011). Global urban growth and the geography of water availability, quality, and delivery. *Ambio, 40*, 437–446.

McIntyre, N. E. (2000). Ecology of urban arthropods: A review and a call to action. *Annals of the Entomological Society of America, 93*, 825–835.

McKinney, M. L. (2002). Urbanization, biodiversity, and conservation. *BioScience, 52*, 883–890.

MEA (Millennium Ecosystem Assessment). (2005). *Ecosystems and human well-being: Current state and trends* (Vol. 1). Washington, DC: Island Press.

Metson, G., Hale, R., Iwaniec, D., Cook, E., Corman, J., Galletti, C., & Childers, D. (2012). Phosphorus in Phoenix: A budget and spatial analysis of phosphorus in an urban ecosystem. *Ecological Applications, 22*(2), 702–721. doi:10.1890/11-0865.1.

Myint, S. W., & Okin, G. S. (2009). Modeling land-cover types using multiple endmember spectral mixture analysis in a desert city. *International Journal of Remote Sensing, 30*, 2237–2257.

Neil, K. (2008). *Effects of urbanization on flowering phenology in Phoenix, USA*. Dissertation, Arizona State University.

Neil, K., & Wu, J. (2006). Effects of urbanization on plant flowering phenology: A review. *Urban Ecosystems, 9*, 243–257.

Odum, E. P. (1997). *Ecology: A bridge between science and society*. Sunderland: Sinauer Associates.

Pickett, S. T. A., Cadenasso, M. L., Grove, J. M., Boone, C. G., Groffman, P. M., Irwin, E., Kaushal, S. S., Marshall, V., McGrath, B. P., & Nilon, C. H. (2011). Urban ecological systems: Scientific foundations and a decade of progress. *Journal of Environmental Management, 92*, 331–362.

Pyle, R. M. (1978). The extinction of experience. *Horticulture, 56*, 64–67.

Redman, C. L., & Foster, D. R. (Eds.). (2008). *Agrarian landscapes in transition: A cross-scale approach*. New York: Oxford University Press.

Redman, C. L., & Kinzig, A. P. (2008). Water can flow uphill. In C. L. Redman & D. R. Foster (Eds.), *Agrarian landscapes in transition: A cross-scale approach* (pp. 238–271). New York: Oxford University Press.

Redman, C. L., Grove, J. M., & Kuby, L. (2004). Integrating social science into the long-term ecological research (LTER) network: Social dimensions of ecological change and ecological dimensions of social change. *Ecosystems, 7*, 161–171.

Roach, W. J., & Grimm, N. B. (2009). Nutrient variation in an urban lake chain and its consequences for phytoplankton production. *Journal of Environmental Quality, 38*, 1429–1440.

Roach, W. J., & Grimm, N. B. (2011). Denitrification mitigates N flux through the stream-floodplain complex of a desert city. *Ecological Applications, 21*, 2618–2636.

Roach, W. J., Heffernan, J. B., Grimm, N. B., Arrowsmith, J. R., Eisinger, C., & Rychener, T. (2008). Unintended consequences of urbanization for aquatic ecosystems: A case study from the Arizona desert. *BioScience, 58*, 715–727.

Ruddell, D., & Wentz, E. A. (2009). Multi-tasking: Scale in geography. *Geography Compass, 3*, 681–697.

Ruddell, D. M., Harlan, S. L., Grossman-Clarke, S., & Buyantuyev, A. (2010). Risk and exposure to extreme heat in microclimates of Phoenix, AZ. In P. Showalter & Y. Lu (Eds.), *Geospatial techniques in urban hazard and disaster analysis* (pp. 179–202). New York: Springer.

Ruddell, D., Harlan, S. L., Grossman-Clarke, S., & Chowell, G. (2012). Scales of perception: Public awareness of regional and neighborhood climates. *Climatic Change, 111*(3–4), 581–607. doi:10.1007/s10584-011-0165-y.

Ruddell, D. M., Hoffman, D., Ahmed, O., & Brazel, A. (in press). An analysis of historical threshold temperatures for central Arizona: Phoenix (urban) and Gila Bend (desert). *Climate Research*. doi: 10.3354/cr01130.

Schaafsma, H., & Briggs, J. M. (2007). Hohokam silt capturing technology: Silt fields in the northern Phoenix basin. *Kiva, 72*, 443–469.

Schlesinger, M. D., Manley, P. N., & Holyoak, M. (2008). Distinguishing stressors acting on land bird communities in an urbanizing environment. *Ecology, 9*, 2302–2314.

Seager, R., Ting, M., Held, I., Kushnir, Y., Lu, J., Vecchi, G., Huang, H.-P., Harnik, N., Leetmaa, A., Lau, N.-C., Li, C., Velez, J., & Naik, N. (2007). Model projections of an imminent transition to a more arid climate in Southwestern North America. *Science, 316*, 1181–1184.

Shochat, E., Lerman, S., Katti, M., & Lewis, D. (2004). Linking optimal foraging behavior to bird community structure in an urban-desert landscape: Field experiments with artificial food patches. *The American Naturalist, 164*, 232–243.

Shochat, E., Warren, P. S., & Faeth, S. H. (2006a). Future directions in urban ecology. *Trends in Ecology & Evolution, 21*, 661–662.

Shochat, E., Warren, P. S., Faeth, S. H., McIntyre, N. E., & Hope, D. (2006b). From patterns to emerging processes in mechanistic urban ecology. *Trends in Ecology & Evolution, 21*, 186–191.

Shochat, E., Lerman, S. B., Anderies, J. M., Warren, P. S., Faeth, S. H., & Nilon, C. H. (2010). Invasion, competition and biodiversity loss in urban ecosystems. *BioScience, 60*, 199–208.

Shumaker, S. A., & Taylor, R. B. (1983). Toward a clarification of people-place relationships: A model of attachment to place. In N. R. Feimer & E. S. Geller (Eds.), *Environmental psychology: Directions and perspectives* (pp. 219–251). New York: Praeger.

Smith, V. K. (2005). Fifty years of contingent valuation. In A. Alberini, D. Bjornstad, & J. Kahn (Eds.), *Handbook of contingent valuation*. Cheltenham: Edward Elgar.

Stabler, L. B., Martin, C. A., & Brazel, A. J. (2005). Microclimates in a desert city were related to land use and vegetation index. *Urban Forestry and Urban Greening, 3*, 137–147.

Sun, C.-Y., Brazel, A., Chow, W. T. L., Hedquist, B. C., & Prashad, L. (2009). Desert heat island study in winter by mobile transect and remote sensing techniques. *Theoretical and Applied Climatology, 98*, 323–335.

Turner, B. L., II. (2009). Sustainability and forest transitions in the southern Yucatán: The land architecture approach. *Land Use Policy, 21*, 170–180.

United Nations Population Division (UNPD). (2010). *World urbanization prospects: The 2009 revision*. New York: United Nations Population Division.

U. S. Census Bureau. (2011). *Statistical abstract of the United States*. Retrieved from http://www.census.gov/compendia/statab/2011edition.html

Vitousek, P. M., Aber, J., Howarth, R. W., Likens, G. E., Matson, P. A., Schindler, D. W., Schlesinger, W. H., & Tilman, G. D. (1997). Human alteration of the global nitrogen cycle: Causes and consequences. *Ecological Applications, 7*, 737–750.

Walker, J. S., & Briggs, J. M. (2007). An object-oriented approach to urban forest mapping with high-resolution, true-color aerial photography. *Photogrammetric Engineering and Remote Sensing, 73*, 577–583.

Walker, J. S., Grimm, N. B., Briggs, J. M., Gries, C., & Dugan, L. (2009). Effects of urbanization on plant species diversity in central Arizona. *Frontiers in Ecology and the Environment, 7*, 465–470.

Weisz, H., & Steinberger, J. K. (2010). Reducing energy and material flows in cities. *Current Opinion in Environmental Sustainability, 2*, 1–8.

Wolman, A. (1965). The metabolism of cities. *Scientific American, 213*, 179–190.

Wu, J. (2008a). Toward a landscape ecology of cities: Beyond buildings, trees, and urban forests. In M. M. Carreiro, Y. C. Song, & J. G. Wu (Eds.), *Ecology, planning, and management of urban forests: International perspectives* (Springer series on environmental management, pp. 10–28). New York: Springer.

Wu. J. (2008b). Landscape ecology. In S. E. Jorgensen (Ed.) & B. Fath (Assoc. Ed.), *Encyclopedia of ecology* (pp. 2103–2109). Oxford: Elsevier.

Wu, J., & David, J. L. (2002). A spatially explicit hierarchical approach to modeling complex ecological systems: Theory and applications. *Ecological Modelling, 153*, 7–26.

Wu, J., & Loucks, O. L. (1995). From balance-of-nature to hierarchical patch dynamics: A paradigm shift in ecology. *The Quarterly Review of Biology, 70*, 439–466.

Wu, J., Jenerette, G. D., Buyantuyev, A., & Redman, C. L. (2011). Quantifying spatiotemporal patterns of urbanization: The case of the two fastest growing metropolitan regions in the United States. *Ecological Complexity, 8*, 1–8.

Xu, Y., Baker, L. A., & Johnson, P. A. (2007). Trends in groundwater nitrate contamination in the Phoenix, Arizona region. *Ground Water Monitoring and Remediation, 27*, 49–56.

Yabiku, S., Casagrande, D. G., & Farley-Metzger, E. (2008). Preferences for landscape choice in a southwestern desert city. *Environment and Behavior, 40*, 382–400.

York, A. M., Shrestha, M., Boone, S. G., Zhang, S., Harrington, J. A., Jr., Prebyl, T. J., Swann, A., Agar, M., Antolin, M. F., Nolen, B., Wright, J. B., & Skaggs, R. (2011). Land fragmentation under rapid urbanization: A cross-site analysis of southwestern cities. *Urban Ecosystems, 14*, 429–455.

York, A., Clark, C., Wutich, A., & Harlan, S. (in review-a). What determines public support for development impact fees?. (*Submitted to Land Use Policy* June 2012).

York, A., Conley, S. N., & Helepololei, J. (in review-b). In land we trust: One hundred years of direct democracy on state trust land. (*Submitted to State Politics and Policy* June 2012).

Zhu, W., Hope, D., Gries, C., & Grimm, N. B. (2006). Soil characteristics and the accumulation of inorganic nitrogen in an arid urban ecosystem. *Ecosystems, 9*, 711–724.

Zhuo, X. (2010). *Spatial distributions of toxic elements in urban desert soils: Sources, transport pathways, historical legacies, and environmental justice implications*. Dissertation, Arizona State University.

Zhuo, X., Boone, C. G., & Shock, E. (in press). Soil lead distribution and environmental justice in the Phoenix metropolitan region. *Environmental Justice*.

# Chapter 11
# A City and Its Hinterland: Vienna's Energy Metabolism 1800–2006

**Fridolin Krausmann**

**Abstract** Cities are centres of resource consumption and urban resource use has a considerable influence on both the economy and the environment in the resource-providing hinterland. This chapter looks at cities from a socio-ecological perspective and investigates the evolution of the energy metabolism of the city of Vienna since the beginning of industrialisation. Based on time series data on the size and structure of energy consumption in Vienna in the period from 1800 to 2006, it analyses the energy transition and how it relates to urban growth. It shows that during the last 200 years, a multiplication of energy use and a shift from renewable biomass towards coal and finally oil and natural gas as the dominating energy source have been observed. This energy transition was not a continuous process, but different phases in the energy transition can be distinguished. Also the spatial relations between the city and its resource-supplying hinterland changed. But growth in urban resource use was not simply causing an equal growth of the spatial imprint of urban consumption. Our results show that the size and spatial location of the resource-supplying hinterland is the combined result of various dynamic processes, including transport technology and agricultural productivity.

The paper shows how energy and transport revolution abolished barriers of growth inherent to the old energy regime.

**Keywords** Urban metabolism • Energy consumption • Energy transition • Industrialization • City-hinterland relation

---

F. Krausmann, Ph.D. (✉)
Institute of Social Ecology Vienna (SEC), Alpen-Adria Universitaet Klagenfurt, Wien, Graz, Schottenfeldgasse 29/5, Vienna 1070, Austria
e-mail: fridolin.krausmann@aau.at

## 11.1 Introduction

Cities are centres of consumption. A large and growing fraction of the global population lives in urban centres which draw on vast hinterlands for their supply with water, energy and materials. Urban resource use, therefore, has a considerable influence on both the economy and the environment in the resource-providing hinterland and is a major driver of global environmental change. The role of cities in reducing socioeconomic material and energy flows is increasingly recognized (Weisz and Steinberger 2010). It is commonly agreed that cities and urbanisation are of key importance for sustainable development; however, the particular relationship between urbanisation and sustainability contains many contradictions and is a focus of debate (Grimm et al. 2008; Satterthwaite 2009). A better understanding of cities as socio-ecological systems and, in particular, of urban resource use and the relationship between cities and their resource-providing hinterland is required. This chapter looks at cities from a socio-ecological perspective. It focuses on the biophysical features of urban growth and investigates the emergence of urban patterns of resource use during industrialisation.

Lewis Mumford (1956) was one of the first to address the (social) ecology of cities by examining the relationships between cities and their resource-supplying hinterlands. A decade later, in 1965, Abel Wolman coined the term urban metabolism. He emphasised that cities require physical inputs of materials and energy and produce wastes and emissions and that these metabolic processes are causing environmental pressures. Research into urban metabolism, as a useful way to study cities as socio-ecological systems and to investigate material and energy flows in urban systems more thoroughly, is now growing quickly. Since the early 1980s, a few comprehensive studies on the metabolism of cities have been published (see Boyden et al. 1981) and the pace is now accelerating (Warren-Rhodes and Koenig 2001; Sahely et al. 2003; Niza et al. 2009; Kennedy et al. 2007; Barles 2009). Several authors have explicitly focussed on the spatial imprint of urban consumption, emphasizing why and how urban centres deplete the natural capacities of their hinterlands (Daxbeck et al. 2001; Folke et al. 1997; Luck et al. 2001; Billen et al 2012). Only a few studies, though, have investigated changes in urban resource use over longer periods of time and the emergence of modern industrial patterns of urban metabolism (e.g. Tarr 2002; Schmid-Neset and Lohm 2005; Barles 2005; Hoffmann 2007; Billen et al. 2009; Gingrich et al. 2011; Marull et al. 2010).

This chapter takes up the urban metabolism approach and applies it to a historical case. It investigates the evolution of the energy metabolism of the city of Vienna since the beginning of industrialisation. Based on time series data on the size and structure of energy consumption in Vienna in the period from 1800 to 2006, it analyses the energy transition (Grübler 2004) and how it relates to urban growth. It shows from which regions and over which distances materials and energy had to be transported into the growing city. The spatial dimension of urban metabolism is complemented by an estimate of the changing size of the hinterland[1] required to supply a city with sufficient resources.

## 11.2 Methods, Data and Sources

To analyze the long-term development of the socioeconomic energy system of the City of Vienna, methods of material and energy flow accounting (MEFA) have been applied as they are used in industrial ecology (Ayres and Ayres 1998; Daniels and Moore 2001). The energy flow accounting used here extends conventional approaches of energy analysis, which tend to focus on technical energy (fossil fuels, hydro power or fuelwood), by including traditional energy carriers (Haberl 2001). In particular, it includes food for humans and feed for draught animals, which are important types of final energy in pre-industrial societies. This methodology has already been successfully applied for historical cases and in long-term socio-ecological research (Sieferle et al. 2006).

Based on these methods the indicator domestic energy consumption (DEC) has been calculated. It is defined as the sum of extraction of energy (carriers) within the boundaries of the considered socio-economic system plus all imports and minus exports of energy. DEC is a measure of apparent energy consumption. Only a small share of a city's energy needs is extracted within the city, so that demand is largely met by imports of energy. As a consequence, a considerable part of the primary to final energy conversion does not take place within the city, but final energy ready for consumption is imported in forms such as electricity, transport fuel, food or feed for draught animals. Therefore, in the case of cities, DEC measures a mix of primary and final energy.[2]

All energy flows are given in Joules. Flows measured in mass or volume in original sources were converted into Joules by applying material specific gross calorific values. To capture changes in the relation of the city to its hinterland we combine energy flow accounting with tools to investigate spatial aspects of social metabolism, such as the actual area demand (cf. Haberl et al. 2001; Wackernagel et al. 2004) or

---

[1] In this chapter, the notion of 'hinterland' is understood in a broader sense and is not restricted to the immediate rural, comparatively infrastructure-poor areas surrounding urban centres. Instead, from a socio-ecological perspective, the urban hinterland encompasses the full extent of regions supplying the urban centre with natural resources (cf. Jones 1955; Fischer-Kowalski et al. 1997). In an abstract sense, hinterland is understood as the environmental space or ecological footprint required to sustain the city with material and energy. From this perspective, the extent of the hinterland and the intensity of the relation between centre and hinterland changes over time. During industrialisation, the direct spatial relation between a city and its hinterland has increasingly vanished as the hinterland of the modern industrial city spreads across the globe (Mumford 1956).

[2] In energy accounting, this problem is sometimes overcome by calculating primary energy equivalents of imported final energy (e.g. coal required to produce imported electricity). This procedure has not been applied for this chapter.

the virtual forest approach (Sieferle et al. 2006). By using time- and region-specific biomass yields we assign production areas to individual energy/material flows. Details on the assumptions behind these conversions are provided in Sect. 11.5.

Statistical records and sources provide information on Vienna's energy and raw material input since the mid-eighteenth century, but meaningful annual time series can be reconstructed only from around 1800. For the nineteenth century, various sources can be used to quantify energy inputs into the city. First, a consumption tax (*Verzehrungssteuer*) was collected at the fortification surrounding the city (*Linienwall*, see below). Tax records since 1829 list goods imported into the city in both physical and monetary units (Hauer 2010). Additionally, transport statistics are available that record wood and coal delivered to the city by boat or float on the river Danube and later on by rail (Gingrich et al. 2011). Digests of these data have been published in annual statistical series since the early nineteenth century.[3] Several of these sources have already been evaluated and for some periods edited data compilations are available especially for the annual supply of fuelwood and coal.

Pioneering compilations of data concerning the use of different energy carriers in the City of Vienna covering the time period 1760 to the early twentieth century have been published by the Austrian economic historian Roman Sandgruber (Sandgruber 1978, 1983, 1987). Data on energy use in the twentieth century are available from the statistical yearbook of Vienna (MSW 1885, 2008) which has been published since 1885. For the period 1800–1860, this study draws on the statistical data compiled by Sandgruber (1987); from 1860 to1883, on data published in the statistical periodical *Statistische Monatschrift* (Pizzala 1884); and from 1883 to 1921, on Sandgruber (1983) and various years of the statistical yearbook of Vienna. Since 1921, data have been compiled on the basis of information provided in various volumes of the statistical yearbook of Vienna and a number of special studies dealing with Vienna's energy situation (Nagl 1966; Wiener Stadtwerke 1975, 1978, 1983, 1994; Stenitzer et al. 1997). Important sources of information on energy use in Vienna, transport and regional supply include the forestry yearbooks edited by Joseph Wessely (1880, 1882) and a statistical compendium edited by the Viennese Chamber of Commerce (Handels- und Gewerbekammer in Wien (Ed.) 1867). Price series for wood and coal are available from Mühlpeck et al. (1979) and Sandgruber (1983).

Annual food consumption was extrapolated from estimates of per capita intake for different points in time for the nineteenth and twentieth centuries (derived from Mühlpeck et al. 1979; Statistik Austria 2008; BMLF – Bundesministerium für Land- und Forstwirtschaft 1997). Feed demand for draught animals was calculated on the basis of numbers of draught animals in Vienna (Sandgruber 1983; MSW various years) and assumptions of average daily feed intake per head (Krausmann 2004).

---

[3] The most significant annual statistical publications are: Tafeln zur Statistik (1828–1865); Statistisches Jahrbuch der Stadt Wien (SJB, 1883-today); Ergebnisse der Verzehrungssteuer im Verwaltungsjahr (1860–1891); Statistische Ausweise über die Preise der Lebensmittel und der Approvisionierung in Wien (1879ff). See Sandgruber (1978 and 1986) and Hauer (2010) for a more detailed discussion of available sources.

## 11.3 The City of Vienna: Location, Population and Spatial Expansion

The City of Vienna was the capital and administrative and economic centre of the Austrian part of the Habsburg Empire (*Cisleithanien*), a state which encompassed roughly 300,000 km$^2$ and 28 million inhabitants in 1911. Following the collapse of the Empire after World War I in 1919, Vienna became the capital of the newly-formed and much smaller Republic of Austria (83,900 km$^2$ and slightly above 6.4 million inhabitants in 1922).

The city is located on the banks of the Danube River between the foothills of the Alps in the west and fertile lowlands of the Danube basin in the south and east, stretching into the agricultural regions of Lower Austria and Hungary. The preconditions for supplying a pre-industrial urban centre with sufficient resources were favourable: the extensive woodlands and fertile agricultural land in the immediate hinterland and along the Danube provided the city with raw materials, fuelwood and staple food. The Danube and its feeders allowed for water transport of bulk materials right into the city centre (Gingrich et al. 2012).

Vienna never had a pronounced industrial character. At times it had a reputation as a centre of the textile industry and manufacturing in the nineteenth century, but never attracted considerable energy-intensive heavy industries. This did not change substantially throughout the observed period. The numbers of steam engines in operation in Vienna indicate the low significance of Vienna as an industrial site. According to data derived from Sandgruber (1983), in 1852 only 0.16 steam engines were installed per 1,000 inhabitants. By 1890, the number grew to 0.8, corresponding to an increase in installed power from 6 to 22 hp per 1,000 inhabitants. In industrialised Great Britain, for example, the installed steam power was much higher and in the same period rose from 37 to 191 hp per 1,000 inhabitants (Castaldi and Nuvolari 2003).

At around 1800, approximately 250,000 people lived in the capital of the Habsburg Empire. Throughout the nineteenth century until World War I, the city grew rapidly, experiencing waves of immigration from other provinces of the Empire (see Fig. 11.1). Population grew at exponential rates and reached a peak before World War I (Juraschek 1896; MSW various years). According to the census of 1910, slightly more than two million people, or 7% of the population of *Cisleithanien*, lived in the city at that time. The collapse of the Habsburg Empire along with the economic crises of the 1930s and World War II brought urban expansion to a halt with population even declining between 1910 and 1950. Since then, Vienna's population has fluctuated around 1.6 million inhabitants or about one-fifth of the Austrian population.

For an investigation of urban development and urban metabolism it is essential to keep in mind that the administrative boundaries that define the city as a political (and statistical) entity are not static but change over time: Until the first expansion of the city's territory in 1850, the City of Vienna was formally limited to the area within the city wall, approximately the current first district (2.8 km$^2$). Since 1704, however, a fortification circle has surrounded the City and its suburbs, spanning around the cur-

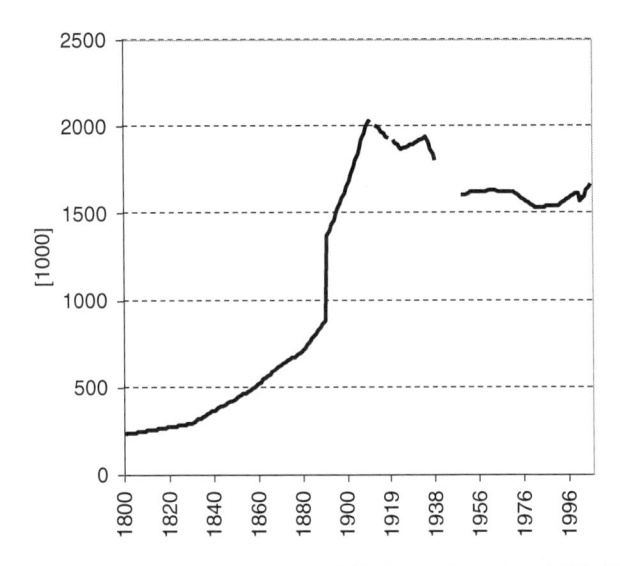

**Fig. 11.1** Population development, Vienna 1800–2000. System boundary: 1800–1890 territory of districts 1–9; from 1890 on the respective administrative boundaries, see text (Source: MSW various years, Juraschek 1896)

rent districts numbered 1–9 and parts of district 10. Above all, this fortification, the so called "Linienwall", served as a tax-border, at which a total of 220 goods had to be declared and tax was collected (Peterson 2005; Buchmann 1979). The expansion of the city limits of 1850 formally absorbed the villages, residential and commercial areas which had been growing between city centre and the Linienwall into the city and increased the administrative territory to approximately 60 km$^2$. In 1892 and 1905, the city limits were again moved outward and former suburbs were incorporated, extending the city to 178 and 278 km$^2$, respectively. After WWII, the city reached its current extension, stretching over an area of 414 km$^2$. This area is by no means equally urbanised. A considerable part of the current territory is used agriculturally (23%) or covered with protected woodlands (17%) (Eigner and Schneider 2005; Juraschek 1896). Accordingly, population density within the city varied in 2001 from 1,300 to 24,000 cap/km$^2$ across the 23 districts (MSW 2002).

For the period 1800–1890, all quantitative information on population and energy use presented in this chapter refers to the area within the Linienwall (i.e., districts 1–9, partly 10, 55.4 km$^2$); for the period 1890–1939 to an area of 278 km$^2$ and from 1948 to 2000 to the current administrative territory (414 km$^2$). The periods 1912–1918 and 1939–1947 were excluded from our data series because of irregularities and incomplete data records. It has to be noted that the shift in system boundaries, in particular the expansion of the system boundaries in 1890, results in a considerable statistical break. The expansion of the city limits in 1890 when the city absorbed industrial and residential areas outside the old Linienwall increased the territory of the city by a factor of 3.2. As a consequence, Vienna's population grew by a factor of 1.6 (see Fig. 11.1) and coal consumption by a factor of 1.4 (see also Fig. 11.2a).

As population and energy flow data consistently refer to the same territorial system boundaries, these statistical breaks are diminished when per capita figures are presented.

## 11.4 Changes in Energy Use During Industrialisation

### 11.4.1 The Energy Transition

Figure 11.2 shows the development of energy consumption in the City of Vienna from 1800 until 2006. During the first half of the nineteenth century, energy use was growing slowly and the first doubling of energy consumption took almost 70 years. Growth accelerated mid-century and from 1870, it took less than 20 years for the next doubling to occur. Another 20 years later, in 1910, energy consumption was already 10 times larger than at the beginning of the observed period 100 years earlier. After a slump in the aftermath of World War I, energy consumption quickly recovered. It reached a new peak before the economic crisis of the 1930s and entered a phase of unprecedented growth after World War II – quite similar to the development observed at the national scale (Krausmann and Haberl 2007). Growth lasted until the early 1970s, when the oil price shocks of 1973 and 1979 slowed down the pace of growth (see also Krausmann and Fischer-Kowalski, Chap. 15 in this volume). Since then, energy inputs have remained high but subject to considerable annual fluctuations stemming mostly from ups and downs in the consumption of transport fuel and natural gas.[4] At the beginning of the twenty-first century, energy consumption amounted to 180 PJ or 25 times the energy consumed in 1800. Vienna's share in Austria's final consumption grew from 5% in 1800 to almost one-fifth two centuries later.[5]

Vienna's energy use not only multiplied during industrialisation, but also its composition changed: Fig. 11.2 shows that until the 1850s, the city's energy demand was almost completely met from renewable biomass: Most of it was wood for the provision of heat for households and manufacturing. Food for urban dwellers and feed for draught animals accounted for only one-fifth of urban energy demand.[6]

---

[4] The reasons for these fluctuations are not fully clear. They partly reflect actual ups and downs in energy consumption (due to e.g. fluctuations in winter temperatures) but they might be partly a result of the peculiarities of the compilation of energy statistics (in particular transport fuel) for urban systems.

[5] A large fraction of the energy used in Vienna (as in many other cities) is imported as final energy ready for consumption (e.g. food, fuels, gas for heating, electricity). In practical terms, the energy consumption (DEC) calculated for Vienna is closer to final energy use than to primary energy supply – and for this reason Austrian final energy consumption is used as a reference measure.

[6] Feed for draught animals for urban transport is most likely underestimated, as this estimate only includes draught animals reported within the city limits. These numbers seem to be rather low. A significant share of transport services may have been provided by carrying trade located outside the city.

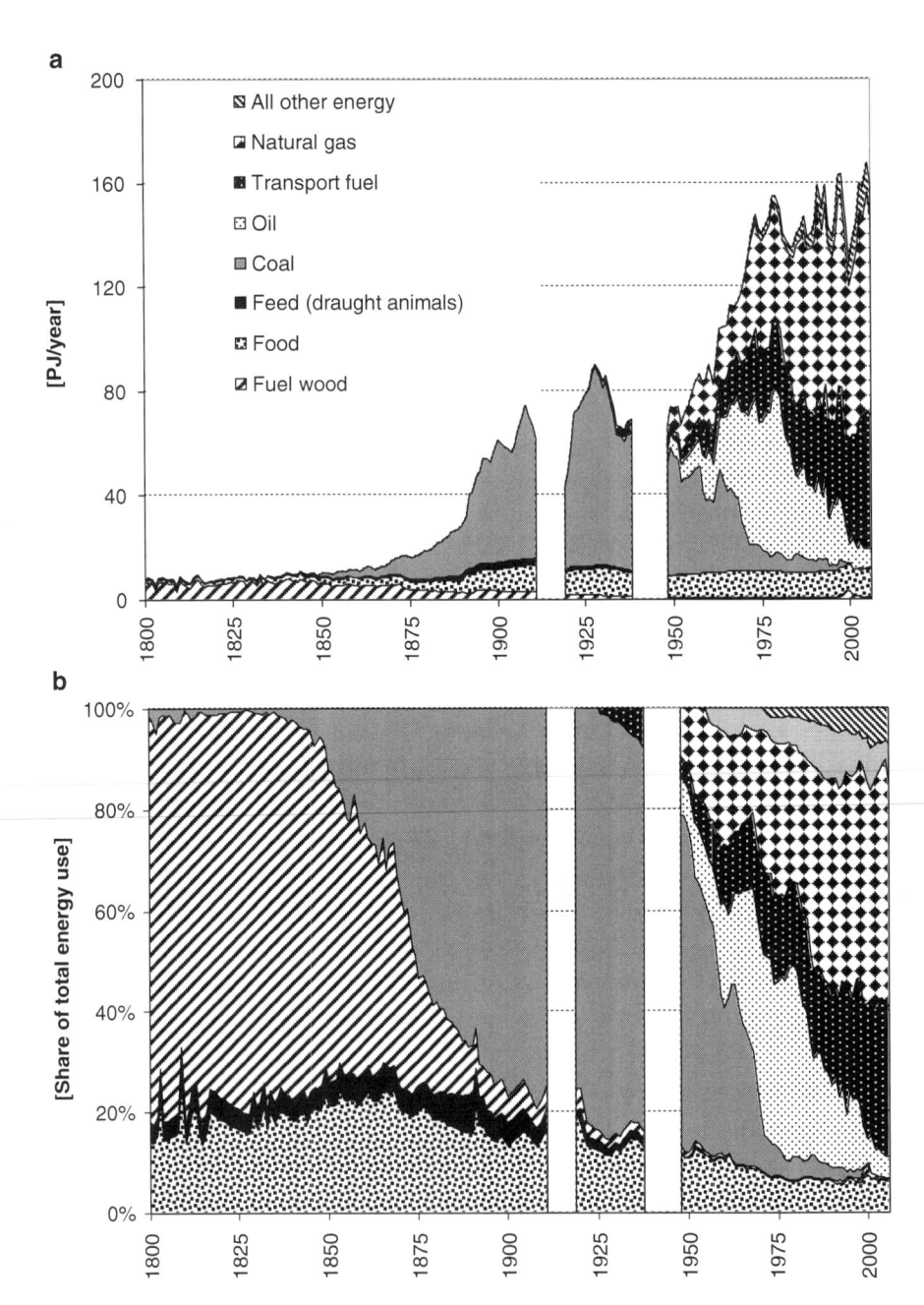

**Fig. 11.2** Energy consumption (DEC) in the city of Vienna, 1800–2000: DEC in PJ/year (**a**) and share of energy carriers in DEC (**b**)

For the first decades of the nineteenth century, coal was of very minor importance. Its share began to increase in the late 1830s and only in the 1850s exceeded a benchmark of 5%. From then on, the first energy transition progressed at a very rapid pace: in 1870, coal already accounted for one-third of total energy input and by 1910, its share had risen to more than three-quarters. The share of coal reached its peak before the economic crises of the 1930s. From then on, petroleum products and natural gas increasingly substituted for coal. After WWII, the significance of coal rapidly diminished. In only two decades, its share in energy supply fell from 80% to around 15%. From the 1980s, natural gas, with a share of 45% of DEC in 2000, emerged as the dominant energy carrier, followed by heating oil and transport fuel, accounting for about one-third of supply. Over the 200-year period, the share of biomass declined to somewhat less than 10% and has remained at this level since the mid-1970s. Even though renewable or alternative commercial energy sources such as hydropower and district heat have gained importance since the early 1990s, their share of total energy supply is still less than 10% in Vienna. This is considerably lower than the Austrian average, where renewables account for more than one-fifth of primary energy supply.

The rising importance of electricity is only partly visible in Fig. 11.2b, which only shows hydropower and imported electricity but ignores the electricity produced in the city's thermal power plants. Electricity is the most universal form of useful work and key to industrial energy systems. The electrification of the city began as early as the 1880s, when the first thermal power stations (*Dampfzentralen*) were installed. Two decades later, the exploitation of water power for electricity generation also began. By the 1920s, around 20% of the electricity supply of Vienna was provided by small-scale hydroelectric plants within the city. However, the use of electricity was confined to a few applications and overall electricity consumption remained low, amounting to only a few PJ throughout the first half of the twentieth century. This changed after WWII, when the city rapidly became electrified: electricity consumption grew steadily to more than 40 PJ/year at the beginning of the twenty-first century and continues to rise. Also, the sources for electricity supply changed. While in the 1920s, 15–20% of all coal consumed in Vienna was burnt at very low efficiency in thermal power plants in the city to produce 80% of the electricity demand, current net imports of electricity account for over 50% of total supply (Fig. 11.3). Most of the remainder is produced from thermal plants in Vienna, which have shifted from using coal to oil and, since the 1980s, mostly towards natural gas as the main feedstock. Thermal conversion efficiency has been improved significantly, reaching 45%. Only a minor fraction of the city's electricity supply is produced in urban hydropower plants.

### 11.4.2 Energy and Population Growth: From Scarcity to Abundance

Growth of energy consumption is not surprising for a rapidly growing city. But how exactly was growth in energy use related to urban growth? Figure 11.4 compares the development of population and energy use in Vienna. From Fig. 11.4a, it seems that

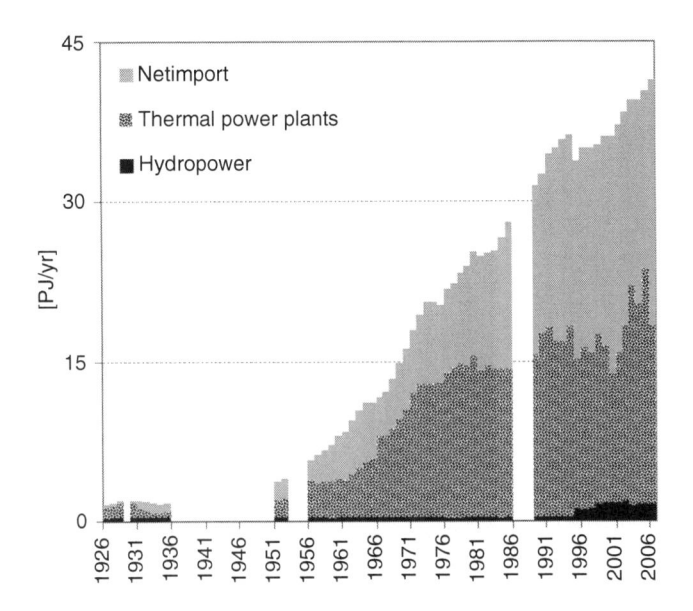

**Fig. 11.3** Vienna's electricity supply by source 1926–2006

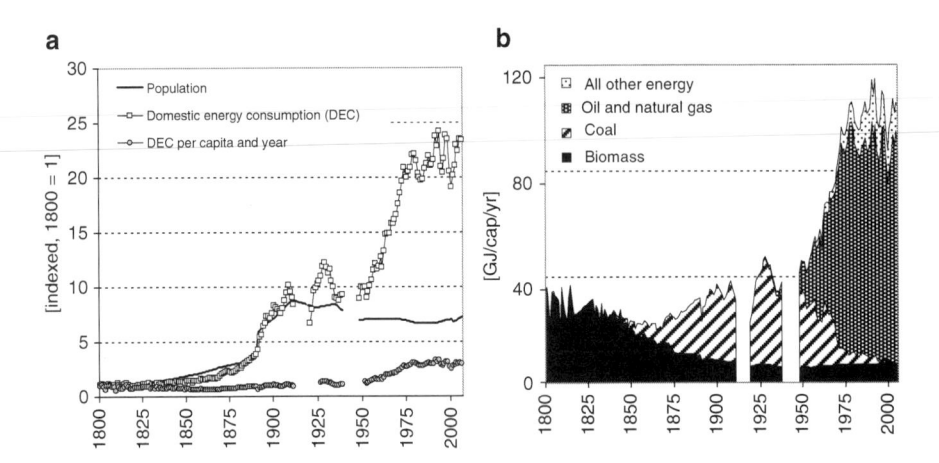

**Fig. 11.4** Population and energy consumption 1800–2000. (**a**) Development of population, domestic energy consumption (DEC) and DEC per capita and year (indexed 1800=1); (**b**) DEC per capita and year by main energy carriers in GJ

throughout the nineteenth century, energy use grew by and large in proportion to population growth and only in the twentieth century did energy consumption outpace population growth and per capita energy consumption began to rise. Figure 11.4b provides a more detailed picture of the development of per capita energy use. It reveals that, in the period from 1800 to the 1860s, population indeed grew faster than energy supply. Energy available per inhabitant declined and by the 1860s was at only half of the value of 1800. At the same time, energy prices went up: the price of fuelwood increased from the 1830s until the mid 1870s, roughly doubling in this period. Only then did prices for wood begin to decline in response to the beginning of the dominance of coal.

The development of per capita energy use in this early period of rapid urbanisation is a product of several underlying factors. Firstly, the demand for fuel wood increased along with population and gradually required more distant forests and woodlands to be exploited to feed the growing city (Johann 2005). This increased transport costs and, hence, energy prices (see Sect. 11.5). Secondly, declining per capita consumption also reflects a certain impoverishment of the population: Population growth was driven by massive immigration of people from the rural provinces and the relative growth of a poor working class. Thirdly, efficiency gains, especially in household stoves, may have compensated at least partly for the decline in fuel supply and available useful energy probably did not decline at the same pace as primary energy input (see Sandgruber 1987; Radkau 1989). Nevertheless, declining per capita consumption and rising prices of energy in the middle of the century are a strong indication that supplying the city with sufficient energy became increasingly difficult in this period. It can be assumed that throughout the wood and the early coal period of industrialisation, household energy consumption and most likely also energy services did decline – in particular as industry and manufacturing were using growing amounts of energy.

Coal was used in Vienna from the 1830s, but only with the expansion of the railroad system[7] and improved river transportation did the increasing substitution of coal for wood begin to mitigate the tight situation with respect to urban energy supply, although a monopolistic tariff policy of the railway companies kept coal prices high. Energy consumption per capita only began to grow again in the mid-1870s. By 1910, it had reached the level typical for the early 1800s. Again, this is reflected in energy prices: Between 1870 and 1910, the prices for coal declined by 20%. Per capita energy consumption in the coal period of industrialisation peaked before the World Economic Crisis in 1928. This marked the beginning of a new phase of the urban energy transition.

In particular the period following WWII was characterised by a surge in per capita energy use, which almost tripled in less than 30 years: In this period, urban growth came to a halt and the population stabilised while energy use continued to grow. The new growth dynamic of the 1950s and 1960s was related to the next phase of the energy transition, in which petroleum products and natural gas rapidly substituted for coal as the prevailing energy carrier and electricity became a universal form of energy. Among the underlying factors were the relative decline in energy prices in that period (Pfister 2003) and the rapid penetration of new energy (conversion) technologies, such as the internal combustion engine (see Krausmann and Fischer-Kowalski, Chap. 15 in this volume).

---

[7] The northern and southern railway stations were opened in the late 1830s. It took several decades, however, until Vienna was fully connected to the industrial centres of Moravia, Bohemia, Silesia and Prussia in the north (1848) and Styria and the Adriatic harbour of Trieste (1859) in the south.

Technological change was supported by massive political and economic efforts. The "Marshall plan" for European recovery after the war, programmes for electrification and the extension of transport and communication networks helped to rapidly provide the infrastructure basis for this transition. This facilitated the spread of individual transport, central heating and electric household appliances, all contributing to a surge in energy consumption, in particular of households. This process, which has been termed the *1950s syndrome* by the Swiss environmental historian Christian Pfister (1996, 2003), changed Vienna's energy metabolism: during the period from 1950 to 1979, per capita energy use in Vienna almost tripled, reaching 100 GJ/cap/year; electricity use grew fourfold from 3.5 to 15.4 GJ/cap/year (Fig. 11.4a). After the oil price shocks in 1973 and 1979, growth slowed down considerably and eventually came to a halt. During the last decades, per capita energy use in Vienna fluctuated between 105 and 110 GJ/cap/year. At the beginning of the twenty-first century, the provision of transport services is responsible for one-third of the total final energy consumption in the city, heating and cooling in households and offices demands almost 40%, while industry consumes only one-fifth (MSW 2008).

## 11.5 The Spatial Imprint of Urban Consumption

### 11.5.1 Where Did Wood, Food and Coal Consumed in Nineteenth Century Vienna Come From?

Urban centres are centres of concentrated consumption. They require large amounts of material and energy inputs to sustain their population, infrastructures and production. These resources are supplied from their immediate or distant hinterlands and urban consumption patterns shape the use of land and resources in these regions. Based on the sparse empirical evidence available from sources, the following section will briefly highlight some of the spatial aspects of Vienna's supply with energy resources in the nineteenth century. It explores from which regions the city's demand for energy resources was met in the nineteenth century and how the first transport revolution and the expansion of the railroad network changed the spatial imprint of the city.

According to data reported in Wessely (1882), around 60% of the wood delivered to the city in the 1860s originated from Lower Austria, partly from the woodlands surrounding Vienna towards the Alps in the West. The remainder came from distant woodlands as far west as Bavaria, Salzburg and Tyrol. In contrast, extensive woodlands located much closer to the west and south of Vienna in Styria and Lower Austria could not be exploited for urban supply because of the lack of transportation infrastructure.

Wood felled in the abundant woodlands in the immediate surroundings of Vienna (*k.k. Wienerwald*, see Johann 2005) was mostly transported several kilometres over land on horse carts. Around 20% of the wood delivered to Vienna in the 1860s was transported as far as several hundred kilometres on the rivers Danube and Inn from distant places in Upper Austria, Salzburg and Bavaria. Another 15% was brought to Vienna from the north from Moravia and Bohemia (today's Czech Republic) by rail. In total, 60% of the more than 600,000 m$^3$ that were shipped to Vienna in the 1860s was floated down the Danube, one-fifth was delivered on horse carts and only 16% by rail. Throughout the eighteenth and nineteenth centuries, the network of flumes, artificial waterways and log slides to transport wood from the forests alongside the major floating rivers was extended and changed forest ecosystems along the major water transport routes (Johann 2005). With increasing demand, more and more distant woodlands were exploited. This increased transportation costs and contributed to growing fuelwood prices in Vienna (see above) and the increasing shift towards coal.

Coal deposits are scarce in the surroundings of Vienna. Large deposits were located 100 or more kilometres away to the south (Styria) and north (Bohemia and Moravia) (Lorenz von Liburnau 1878). In the pre-railroad era, several minor coal deposits within a distance of 50–100 km of the city were exploited. This coal, totalling a few thousand tonnes per year, was delivered to the city on rafts or horse carts. From the second half of the nineteenth century, the northern railroad connected the city to the rich deposits in the north and supplied the city with increasing amounts of coal. In 1860, more than half of the more than 100,000 t of coal consumed in the city came from mines in Bohemia and Moravia and 16% from Prussia on this line and only 22% from the Austrian provinces of the Empire, mainly from Lower Austria.

The spatial patterns of food supply for nineteenth-century Vienna are more complex. In the pre-railroad period, not only transportation distances mattered, but also the limited possibilities to conserve food products shaped the spatial patterns of resource supply. In particular, refrigeration was limited and perishable foods such as vegetables and milk were generally produced in the immediate surroundings or even within the city limits. The main staple crops were produced on farms in the fertile agricultural regions stretching to the west and east of Vienna in the Danube basin and transported overland on horse carts or on the Danube and its feeders. A significant fraction of the annual cereal supply came from large holdings in Moravia and Hungary and from smaller farms in Lower Austria. But cereals were shipped from as far away as present-day Romania and transported upstream with the support of horses. Cattle were driven to Vienna in large herds from the Hungarian plains and slaughtered in the city (Peterson 2005).

The spatial patterns of Vienna's supply of wood, coal and food highlight the fact that for the development of urban centres in the early industrial period, the spatial distribution of key natural resources was crucial because the cost of overland transport was prohibitive for bulk materials.[8] Until the expansion of the railroad network,

---

[8] According to Braudel (in Sieferle et al. 2006), transport costs on horse carriages are by a factor 9 higher than those on natural water ways. Sandgruber (1987) quotes figures for transport costs of fuelwood in 1855 that indicate that transportation on horse carriages is by a factor 10–20 more costly than transport on the Danube per unit of area.

the most important means to transport bulk materials such as fuelwood or cereals to Vienna were the Danube and its feeders and a modest network of channels which expanded the catchment area of the city. These low-density transport networks determined the relationship of the city with its resource-supplying hinterlands: Bulk resources like wood or mineral materials originated mostly from a more or less narrow corridor along water routes to minimise expensive animal-driven overland transport. Scarcity or abundance, in particular of wood, was not merely a question of overall resource availability in a region, but at least equally of accessibility and transport possibilities.

Based on the information available concerning the supply of wood, coal and food in the nineteenth century, a very rough quantification of the actual size of the hinterland upon which Vienna drew for its energy supply is possible. Bulk materials such as wood, coal and cereals were transported into the city from a hinterland with a radius of perhaps 200–300 km, corresponding to an area of 125,000–280,000 $km^2$. But the urban utilisation of the different regions forming this area was of varying intensity. The actual spatial relations were shaped by their distance to the prevailing transport networks rather than by distance to the city alone (von Thünen 1826), and while some far away regions were well connected to the city via the Danube, others in the close vicinity were hardly affected at all by the urban need for energy resources. The emergence of the railroad system radically changed the spatial relations between city and hinterland in the mid-nineteenth century. Within only a few decades, the railway lines connected Vienna to regions in the north and south that had hitherto had little function for Vienna's resource supply, thus extending the urban hinterland (see Gingrich et al. 2012). In the twentieth century, the spatial imprint of the city changed again. The road system provided a transport network with an area density of more than one order of magnitude beyond that of the railroad network.[9] In addition, during the second phase of the energy transition, energy was increasingly transported in grid-bound transportation systems (pipelines, supra-regional electric grids) over several thousand km from the location of extraction to conversion and consumption. At the beginning of the twenty-first century, the hinterland of the city is global: While some of the food (vegetables, staple crops) are still produced in the vicinity of Vienna or even within the city, most of the energy originates from distant regions.

### 11.5.2 What Area Was Required to Provide the City with Sufficient Energy?

In this next section we turn from the extension of the resource-supplying hinterland to a more abstract measure of the size of the urban footprint: It presents an attempt to calculate how much productive land was actually required to provide the city

---

[9] The area density of transportation systems changes dramatically with technology: Typical natural and artificial waterways under optimum conditions may reach 10–15 $m/km^2$, railroad systems 80–100 $m/km^2$, and modern road systems up to 2,000 $m/km^2$ (Central European averages, author's own calculations).

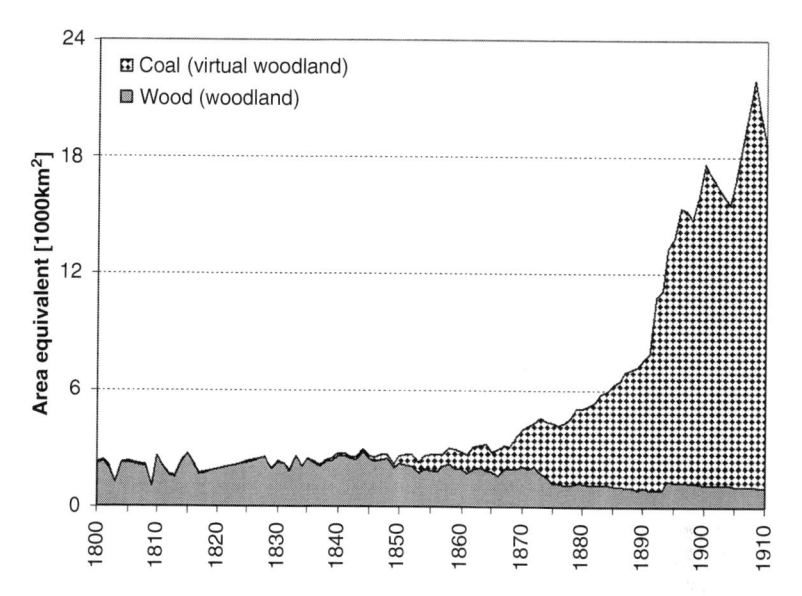

Fig. 11.5 Actual and virtual forest area required to supply Vienna with wood and coal (Source: Author's own calculations, see text)

with sufficient wood fuel and food and how the size of this urban footprint changed during industrialisation and with the energy transition (see Fig. 11.5):

A rough estimation of the forest area required to supply the city with fuelwood can be made by applying average nineteenth-century wood yields. According to cadastral records of Lower Austria (where a large part of the wood was harvested) from the 1840s, around 2.5–3.5 solid m$^3$ of wood could be harvested per hectare (ha) woodland per year in a sustainable way, that is without diminishing standing timber stocks. Using an average of 3 m$^3$/ha/year, the fuelwood consumed per year in Vienna in the first half of the nineteenth century corresponds to a total forest area of 2,000–3,000 km$^2$, an area equal to roughly 50% of the total forest area of Lower Austria (or 6% of the Austrian forests) at that time (Fig. 11.5). As losses during logging, hauling and floating were considerable and could amount to 25% or more of the felled wood, this calculation probably underestimates the total required forest area. From the 1850s onwards, when wood consumption began to decline, the required forest area sank to 1,000 km$^2$ and below.

Environmental historian Rolf Peter Sieferle (2001) has coined the term "subterranean forest" to illustrate the impact of coal on the area-based energy system (Krausmann and Fischer-Kowalski, Chap. 15 in this volume). He argues that the energy supplied by coal far exceeds the theoretical capacity of land-based energy systems and was thus a major precondition for industrial growth. This also holds true with respect to the impact of the growing use of coal in Vienna: to calculate the size of the subterranean forest or the "virtual forest area" corresponding to the amount of burnt coal, we assume that the energy contained in coal can be substituted one to one by an equal amount of calorific energy contained in fuelwood. The amount of fuelwood required to substitute coal can than be converted into a corresponding forest area by applying the same procedure as described above. This calculation shows that the annual fuelwood supply of Vienna was equivalent to an area of roughly 3,000 km$^2$ of woodlands in the hinterland (not considering losses, see above), an area which remained stable until the mid-nineteenth century. In contrast, the virtual forest area corresponding to Vienna's coal consumption rapidly increased to an area of more than 20,000 km$^2$ before the outbreak of World War I. This area equals 60% of the total Austrian forest area. The fossil fuels burnt annually in Vienna at the turn of the twenty-first century equates to a forest area of 50,000 km$^2$ (greater than the entire Austrian forest at nineteenth-century yields). These calculations once more underline the fact that coal was not merely a substitute for wood but a precondition for urban growth in the nineteenth century.

The quantification of the agricultural area required to provide enough (staple) food for urban supply[10] is intricate. Some rough estimates, which help to highlight the area intensity of urban food supply, are possible, however. In 1800, some 140,000 tonnes of food, consisting of cereals, potatoes, meat and dairy products, beer and wine with an overall energy content 1.4 PJ/year, were consumed per year in Vienna. To calculate the agricultural area corresponding to this amount of food, it is not sufficient simply to apply crop-specific net yields or food-specific conversion ratios. Low-input agriculture, characteristic of the early nineteenth century, was based on complex land-use systems maintaining productive capacity by non-uniform land use, biomass and nutrient transfers and recycling processes (Loomis and Connor 1992). That is, the production of a specific crop was not only related to the plot of land where it was grown, but was also part of a multifunctional land-use system integrating cropland, grassland, woodlands and livestock husbandry at the farm or village scale (Krausmann 2004). This makes it difficult to calculate crop yields and it seems more plausible to use information of net aggregate food output per ha of agricultural land instead. Such information was derived from a number of case studies on food output in rural production systems in nineteenth-century Lower Austria and from average Austrian values (Krausmann 2004). According to these studies, an average of 3.9 GJ of food [11] was produced per average hectare of agricultural land and year in the mid-nineteenth century.

During the first agricultural revolution, which increased the area and labour productivity of traditional agricultural production systems, average food output increased to 7.2 GJ/ha/year in 1910. The fossil fuel-powered industrialisation of agriculture in the second half of the twentieth century tripled this value to more than 23 GJ of food per ha agricultural area and year in the late twentieth century. Applying these average conversion ratios to the amount of food consumed in Vienna results in a demand of 3,700 km$^2$ of net agricultural area to meet the annual food requirements

---

[10] The following calculations only refer to major domestic staple foods (e.g. cereals, potatoes, vegetables, meat, and dairy products). The inclusion of areas for special cultivars, such as tea, coffee, tropical fruits, olive oil etc., may have a significant impact on the outcome and increase in particular the food footprint for the contemporary period (see Erb et al. 2001).

[11] Calculated as the net output of plant and animal based food per total agricultural area.

of 250,000 urban dwellers by around 1800. This area is equivalent to 10% of the agricultural area available in Austria (in its current boundaries). By 1910, urban growth had pushed the size of this area to 14,000 km², despite significant increases in area productivity. Only the fossil fuel-powered industrialisation of agriculture reversed the trend and from the mid-twentieth century, the net agricultural area required to feed the city declined considerably: In the 1990s, less area was required to provide 1.6 million urban dwellers with sufficient food than for 0.25 million in 1800.

The above calculations implicitly assumed that all the food produced in an agricultural production system was available for urban consumption. This was, of course, not the case and a quite different picture emerges if changes in labour efficiency and agricultural surplus are also considered. The labour efficiency of pre-industrial agriculture was low and only a fraction of the 3.9 GJ/ha/year that were produced in the hinterland could be exported to urban markets. The larger part of the produce was still required to meet the subsistence needs of the producing agricultural population (see also Fischer-Kowalski et al. 2004). Using information on labour productivity and surplus rates derived from the abovementioned case studies, we can assume that in the early nineteenth century, an average of 80% of the food output per ha was consumed by the local agricultural population and that only 20% was eligible for export. By 1910, the surplus rate had increased to 60% and in modern industrialised agriculture almost 100% of the net food output can be exported. Applying these coefficients to the numbers on area productivity presented above, we arrive at a total agricultural hinterland of 22,000 km² in 1800, but only slightly more than that (24,000 km²) in 1910. This illustrates the tremendous significance of the improvements to traditional farming methods and the gains in both area and labour productivity for urban growth in the nineteenth century: Despite a multiplication of the urban population, the agricultural footprint hardly increased. Industrialisation of agriculture and the unprecedented surge in yields and labour productivity further diminished the area required for urban food supply: At the turn of the twenty-first century, only 3,500 km² of productive land were required to feed a city of 1.6 million people.

Another important aspect relating urban consumption to land use systems in the hinterland is the urban draw on the scarce nutrient reservoirs of the agricultural regions, supplying the centre with biomass. By exporting agricultural produce to urban centres, producing regions are deprived of large amounts of plant nutrients (see Schmid-Neset and Lohm 2005; Barles 2007). Nitrogen may serve as an example here: Based on data on food consumption in Vienna and average nitrogen content of food products, it is possible to estimate the amount of nitrogen that was withdrawn every year from agricultural areas by urban consumption. According to this calculation, the amount of nitrogen contained in staple food consumed in the city increased from 2,000 t/year in 1,800–18,000 t/year in 1910; practically none of this nitrogen was returned to the producing regions. This was a massive loss of nutrients even on a larger scale: by 1910, the nitrogen drainage caused by Vienna's food demand was roughly equal to the total amount of nitrogen fixed by leguminous crops or to almost one-third of all nitrogen returned to fields by manuring in Austrian agriculture (on the current territory of Austria) (Krausmann 2004). In the nineteenth century, urban centres increasingly emerged as sinks of crucial plant nutrients which ended in urban soils, waste water or air emissions. For pre-industrial land-use systems, where the maintenance of soil fertility depended on low level natural inputs and a complex system of transfers and recycling of essential plant nutrients, these losses were an important factor. This is also reflected in the increasing efforts to recycle plant nutrients contained in nightsoil from urban centres in the nineteenth century (Barles 2007). Only with the availability of artificial fertiliser and massive nutrient replacement in the first half of the twentieth century was this problem relieved.

## 11.6 Conclusions

The case of Vienna is a textbook example of the energy transition. During the last 200 years, a multiplication of energy use and a shift from renewable biomass towards coal and finally oil and natural gas as the dominating energy source have been observed. Furthermore, the spatial relations between city and its resource-supplying hinterland changed with the energy system and corresponding technologies. These changes were not a continuous process, but different phases in the energy transition can be distinguished (Table 11.1): The biomass-based energy system prevailed until the 1860s. In this phase, population grew faster than energy supply and the amount of energy available per urban dweller declined. The growing city increasingly faced constraints inherent to the traditional solar energy system (see Krausmann and Fischer-Kowalski, Chap. 15 in this volume). But rather than an absolute scarcity of energy resources in the surroundings, the spatial location of wood and food in relation to navigable waterways was the bottleneck for supplying the growing city with sufficient energy. In the 1860s, the expansion of the railroad network and the shift from biomass to coal began to relieve the tight supply situation and abolished major limitations of the solar-based energy system. For roughly 65 years, coal became the dominant energy source and triggered a multiplication of the size of the city and urban resource needs. In this phase, both population and energy supply grew at high rates and per capita energy availability increased only modestly (Table 11.1). This changed during the third phase of the energy transition, when after World War II coal was rapidly replaced by oil, natural gas and electricity. In this period of oil-driven growth, urban population

**Table 11.1** Phases of the urban energy transition: Average annual growth rates of population and energy consumption (DEC) and the share of biomass and coal in total energy consumption

|  |  | Biomass phase <1865 | Coal phase 1865–1928 | Oil driven growth 1934–1973 | Industrial metabolism 1973–2006 |
|---|---|---|---|---|---|
| Population growth | % | 1.4 | 1.9 | −0.5 | 0.1 |
| DEC growth | % | 0.6 | 3.2 | 2.1 | 0.3 |
| DEC per capita growth | % | −0.8 | 1.3 | 2.5 | 0.2 |
| Share of biomass | % | 100–70 | 70–15 | 15–7 | 7 |
| Share of coal | % | 0–30 | 30–85 | 85–6 | <1 |
| DEC range | GJ/cap/year | 35–20 | 20–47 | 41–100 | 95–105 |

began to decline while energy consumption continued to rise. Energy consumption per capita multiplied and drove overall urban energy demand. The oil price shocks in the 1970s brought an abrupt end to a century of growth in urban energy consumption. From then, both population and energy use grew at very modest rates and per capita energy consumption remained at a high level.

Growth in urban resource use was not simply causing an equal growth of the spatial imprint of urban consumption. It shows that the size and spatial location of the resource-supplying hinterland is rather the combined result of various dynamic processes. Changes in transport technology were an important underlying factor: The transport revolutions from water transport to railroad and finally road- and grid-based transport networks have radically altered the spatial relation between Vienna and its hinterland. In addition, increases in agricultural area and labour productivity, another effect of industrialisation, had a large impact on the urban footprint. Growth in output per unit of area and surplus rates contributed to the stabilisation and even decline of urban demand for agricultural areas, in spite of a multiplication of urban population and food demand.

The urban demand upon the resources of its hinterland is neither new nor is it a sustainability problem per se. Urban centres never have been nor will they be self-sufficient with respect to their resource needs. But the Viennese case underlines that it is of key importance to understand the centre-hinterland relations in order to minimise inefficient patterns of resource supply and use as well as negative environmental impacts of urban consumption in distant regions, where they are not visible to the urban consumer. This is of particular significance as the extension of the urban hinterland has grown to a global scale and soon the majority of the global population will live in cities.

These findings show that urban growth is intrinsically linked to the emergence of a fossil fuel-based energy system. Only the shift from renewable biomass towards fossil energy carriers abolished barriers of growth inherent to the old energy regime. Modern cities consume large amounts of energy, both in absolute terms and per urban dweller; these patterns of urban industrial metabolism are intrinsically linked to the functioning of urban socio-ecological systems. Reducing material and energy use in urban centres will, therefore, require far-reaching changes in the functioning and spatial organisation of cities, of mobility and housing.

**Acknowledgments** The research for this paper was supported by the Austrian Science Fund (Project No. P21012 G11). I want to thank Rolf Peter Sieferle, Verena Winiwarter, Marina Fischer-Kowalski and Simone Gingrich for their support of this research and Marian Chertow and Helmut Haberl for a critical review of the manuscript.

# References

Ayres, R. U., & Ayres, L. W. (1998). *Accounting for resources, 1, economy-wide applications of mass-balance principles to materials and waste*. Cheltenham/Lyme: Edward Elgar.
Barles, S. (2005). A metabolic approach to the city: Nineteenth and twentieth century Paris. In D. Schott, B. Luckin, & G. Massard-Guilbaud (Eds.), *Resources of the city. Contributions to an environmental history of modern Europe* (pp. 28–47). Aldershot: Ashgate.

Barles, S. (2007). Feeding the city: Food consumption and flow of nitrogen, Paris, 1801–1914. *The Science of the Total Environment, 375*, 48–58.

Barles, S. (2009). Urban metabolism of Paris and its region. *Journal of Industrial Ecology, 13*, 898–913.

Billen, G., Barles, S., Garnier, J., Rouillard, J., & Benoit, P. (2009). The food-print of Paris: Long-term reconstruction of the nitrogen flows imported into the city from its rural hinterland. *Regional Environmental Change, 9*, 13–24.

Billen, G., Garnier, J., Barles, S. (2012). History of the urban environmental imprint: introduction to a multidisciplinary approach to the long-term relationships between Western cities and their hinterland. *Regional Environmental Change, 12*, 249–253.

BMLF – Bundesministerium für Land- und Forstwirtschaft. (1997). *Lebensmittelbericht Österreich*. Vienna: BMLF.

Boyden, S., Millar, S., Newcombe, K., & O'Neill, B. J. (1981). *The ecology of a city and its people: The case of Hong Kong*. Canberra: ANU Press.

Buchmann, B. M. (1979). Die Verzehrungssteuer. *Wiener Geschichtsblätter, 1979*(1), 20–29.

Castaldi, C., & Nuvolari, A. (2003). *Technological revolutions and economic growth: The "age of steam" reconsidered* (Eindhoven Centre for Innovation Studies Working Paper 03.25). Eindhoven: Eindhoven Centre for Innovation Studies.

Daniels, P. L., & Moore, S. (2001). Approaches for quantifying the metabolism of physical economies, part I: Methodological overview. *Journal of Industrial Ecology, 5*, 69–93.

Daxbeck, H., Kisliakova, A., & Obernosterer, R. (2001). *Der ökologische Fußabdruck der Stadt Wien*. Vienna: Magistrat der Stadt Wien (MA22).

Eigner, P., & Schneider, P. (2005). Das Wachstum von Wien. In K. Brunner & P. Schneider (Eds.), *Umwelt Wien. Geschichte des Natur- und Lebensraumes Wien* (pp. 22–53). Vienna: Böhlau.

Erb, K.-H., Krausmann, F., & Schulz, N. B. (2001) *Der ökologische Fußabdruck des österreichischen Außenhandels* (Social Ecology Working Paper 62). Vienna: IFF Social Ecology.

Fischer-Kowalski, M., Haberl, H., Hüttler, W., Payer, H., Schandl, H., Winiwarter, V., & Zangerl-Weisz, H. (1997). *Gesellschaftlicher Stoffwechsel und Kolonisierung von Natur. Ein Versuch in Sozialer Ökologie*. Amsterdam: Gordon & Breach Fakultas.

Fischer-Kowalski, M., Krausmann, F., & Smetschka, B. (2004). Modelling scenarios of transport across history from a socio-metabolic perspective. *Review Fernand Braudel Center, 27*, 307–342.

Folke, C., Jansson, A., Larsson, J., & Costanza, R. (1997). Ecosystem appropriation by cities. *Ambio, 26*, 167–172.

Gingrich, S., Haidvogl, G., & Krausmann, F. (2012). The Danube and Vienna: Urban resource use, transport and land use 1800 to 1910. *Regional Environmental Change, 12*, 283–294. doi:10.1007/s10113-010-0201.

Grimm, N. B., Faeth, S. H., Golubiewski, N. E., Redman, C. L., Wu, J. G., Bai, X. M., & Briggs, J. M. (2008). Global change and the ecology of cities. *Science, 319*, 756–760.

Grübler, A. (2004). Transitions in energy use. In C. J. Cleveland (Ed.), *Encyclopedia of energy* (pp. 163–177). Amsterdam: Elsevier.

Haberl, H. (2001). The energetic metabolism of societies, part I: Accounting concepts. *Journal of Industrial Ecology, 5*, 11–33.

Haberl, H., Erb, K.-H., & Krausmann, F. (2001). How to calculate and interpret ecological footprints for long periods of time: The case of Austria 1926–1995. *Ecological Economics, 38*, 25–45.

Handels- und Gewerbekammer in Wien (Ed.). (1867). *Statistik der Volkswirtschaft in Nieder-Oesterreich 1855–1866*. Vienna: Leopold Sommer.

Hauer, F. (2010). *Die Verzehrungssteuer 1829–1913 als Grundlage einer umwelthistorischen Untersuchung des Metabolismus der Stadt Wien* (Social Ecology Working Paper 129). Vienna: IFF Social Ecology.

Hoffmann, R. C. (2007). Footprint metaphor and metabolic realities. Environmental impacts of medieval European cities. In P. Squatriti (Ed.), *Natures past. The environment and human history* (pp. 288–325). Ann Arbor: The University of Michigan Press.

Johann, E. (2005). Die städtische Holzversorgung vom 17. bis zum 19. Jahrhundert. In K. Brunner & P. Schneider (Eds.), *Umwelt Wien. Geschichte des Natur- und Lebensraumes Wien* (pp. 170–179). Vienna: Böhlau.

Jones, L. W. (1955). The hinterland reconsidered. *American Sociological Review, 20,* 40–44.
Juraschek, F. (1896). Das Wachsthum des Territoriums, der Bevölkerung und des Verkehers von Wien 1857–1894. *Statistische Monatsschrift, 22,* 328–344.
Kennedy, C. A., Cuddihy, J., & Engel-Yan, J. (2007). The changing metabolism of cities. *Journal of Industrial Ecology, 11,* 1–17.
Krausmann, F. (2004). Milk, manure and muscular power. Livestock and the industrialization of agriculture. *Human Ecology, 32,* 735–773.
Krausmann, F., & Haberl, H. (2007). Land-use change and socio-economic metabolism. A macro view of Austria 1830–2000. In M. Fischer-Kowalski & H. Haberl (Eds.), *Socioecological transitions and global change: Trajectories of social metabolism and land use* (pp. 31–59). Cheltenham/Northampton: Edward Elgar.
Loomis, R. S., & Connor, D. J. (1992). *Crop ecology: Productivity and management in agricultural systems.* Cambridge: Cambridge University Press.
Lorenz von Liburnau, J. R. (1878). *Atlas der Urproduction Oesterreichs.* Vienna: R. von Waldheim.
Luck, M. A., Jenerette, G. D., Wu, J., & Grimm, N. B. (2001). The urban funnel model and the spatially heterogeneous ecological footprint. *Ecosystems, 4,* 782–796.
Marull, J., Pino, J., Tello, E., & Cordobilla, M. J. (2010). Social metabolism, landscape change and land-use planning in the Barcelona metropolitan region. *Land Use Policy, 27,* 497–510.
Mühlpeck, V., Sandgruber, R., & Woitek, H. (1979). Index der Verbraucherpreise 1800 bis 1914. Eine Rückberechnung für Wien und den Gebietsstand des heutigen Österreich. In Anonymous (Ed.), *Geschichte und Ergebnisse der zentralen amtlichen Statistik in Österreich 1829–1979* (pp. 649–687). Vienna: Kommissionsverlag.
Mumford, L. (1956). The natural history of urbanization. In W. L. Thomas Jr. (Ed.), *Man's role in changing the face of the Earth* (pp. 382–398). Chicago: The University of Chicago Press.
Nagl, H. (1966). *Die Energiewirtschaft Wiens.* Dissertation, University of Vienna, Vienna.
Niza, S., Rosado, L., & Ferrao, P. (2009). Urban metabolism: Methodological advances in urban material flow accounting based on the Lisbon case study. *Journal of Industrial Ecology, 13,* 384–405.
Peterson, B. (2005). Die Lebensmittelversorgung der Stadt. In K. Brunner & P. Schneider (Eds.), *Umwelt Wien. Geschichte des Natur- und Lebensraumes Wien* (pp. 207–221). Vienna: Böhlau.
Pfister, C. (1996). *Das 1950er Syndrom: Der Weg in die Konsumgesellschaft.* Bern/Vienna: Haupt.
Pfister, C. (2003). Energiepreis und Umweltbelastung. Zum Stand der Diskussion über das "1950er Syndrom". In W. Siemann (Ed.), *Umweltgeschichte Themen und Perspektiven* (pp. 61–86). Munich: C.H. Beck.
Pizzala, J. (1884). Der Brennstoffverbrauch Wiens in den Jahren 1860 bis 1882. *Statistische Monatsschrift, 10,* 323–326.
Radkau, J. (1989). *Technik in Deutschland. Vom 18. Jahrhundert bis zur Gegenwart.* Frankfurt am Main: Edition Suhrkamp.
Sahely, H. R., Dudding, S., & Kennedy, C. A. (2003). Estimating the urban metabolism of Canadian cities: Greater Toronto Area case study. *Canadian Journal for Civil Engineering, 30,* 468–483.
Sandgruber, R. (1978). Wirtschaftswachstum und Energie in Österreich 1840–1913. In H. Kellenbenz (Ed.), *Wirtschaftswachstum, Energie und Verkehr vom Mittelalter bis ins 19. Jahrhundert* (pp. 67–95). Stuttgart/New York: Fischer Verlag.
Sandgruber, R. (1983). *Wiens Energieverbrauch und Energieversorgung in der Phase der Industrialisierung.* Vienna: Magistrat der Stadt Wien.
Sandgruber, R. (1987). Die Energieversorgung Wiens im 18. und 19. Jahrhundert. In A. Kusternig (Ed.), *Bergbau in Niederösterreich* (pp. 459–491). Vienna: NÖ Institut für Landeskunde.
Satterthwaite, D. (2009). The implications of population growth and urbanization for climate change. *Environment and Urbanization, 21,* 545–567.
Schmid-Neset, T.-S., & Lohm, U. (2005). Spatial imprint of food consumption. A historical analysis for Sweden, 1870–2000. *Human Ecology, 33,* 565–580.
Sieferle, R. P. (2001). *The subterranean forest. Energy systems and the industrial revolution.* Cambridge: The White Horse Press.
Sieferle, R. P., Krausmann, F., Schandl, H., & Winiwarter, V. (2006). *Das Ende der Fläche. Zum gesellschaftlichen Stoffwechsel der Industrialisierung.* Köln: Böhlau.

Statistik Austria. (2008). *Online database ISIS of Statistik Austria.* www.statistik.at

Stenitzer, M., Fickl, S., Papousek, B., & Cerveny, M. (1997). *Energieeinsatz und CO2-Emissionen in Wien.* Vienna: Magistrat der Stadt Wien, MA 22.

Tarr, J. A. (2002). The metabolism of the industrial city. The case of Pittsburgh. *Journal of Urban History, 28,* 511–545.

Thünen, J. Hv. (1826). *Der isolierte Staat in Beziehung auf Landwirtschaft und Nationalökonomie.* Jena: Fischer.

Wackernagel, M., Monfreda, C., Schulz, N. B., Erb, K.-H., Haberl, H., & Krausmann, F. (2004). Calculating national and global ecological footprint time series: Resolving conceptual challenges. *Land Use Policy, 21,* 271–278.

Warren-Rhodes, K., & Koenig, A. (2001). Escalating trends in the urban metabolism of Hong Kong: 1971–1997. *Ambio, 30,* 429–438.

Weisz, H., & Steinberger, J. K. (2010). Reducing energy and materials flows in cities. *Current Opinion in Environmental Sustainability, 2,* 185–192.

Wessely, J. (1880). *Forstliches Jahrbuch für Oesterreich-Ungarn.* Vienna: Carl Fromme.

Wessely, J. (1882). *Forstliches Jahrbuch für Oesterreich – Ungarn. Oesterreichs Donauländer. II. Theil: Spezial-Gemälde der Donauländer.* Vienna: Carl Fromme.

Wien. Magistrat der Stadt Wien (MSW). (1885). *Statistisches Jahrbuch der Stadt Wien* (various years 1885 to 2008).

Wien. Magistrat der Stadt Wien (MSW). (2002). *Statistisches Jahrbuch der Stadt Wien für das Jahr 2000.*

Wien. Magistrat der Stadt Wien (MSW). (2008). *Statistisches Jahrbuch der Stadt Wien für das Jahr 2006.*

Wiener Stadtwerke. (1975). *Energiekonzept der Stadt Wien.* Vienna: Wiener Stadtwerke Generaldirektion.

Wiener Stadtwerke. (1978). *Energiekonzept der Stadt Wien.* Vienna: Wiener Stadtwerke Generaldirektion.

Wiener Stadtwerke. (1983). *Energie für Wien. Energiekonzept der Stadt Wien. 1. Fortschreibung.* Vienna: Wiener Stadtwerke.

Wiener Stadtwerke. (1994). *Energie in Wien.* Wien: Wiener Stadtwerke.

# Chapter 12
# Sustaining Agricultural Systems in the Old and New Worlds: A Long-Term Socio-Ecological Comparison

**Geoff Cunfer and Fridolin Krausmann**

**Abstract** During the late nineteenth and early twentieth centuries, tens of millions of migrants left Europe for the Americas. Using case studies from Austria and Kansas, this chapter compares the socio-ecological structures of the agricultural communities immigrants left to those they created on the other side of the Atlantic. It employs material and energy flow accounting (MEFA) methods to examine the social metabolic similarities and differences between Old World and New World farm systems at either end of the migration chain. Nine indicators reveal significant differences in land use strategy, labour deployment and the role of livestock. Whereas Old World farms had abundant human and animal labour but a shortage of land, Great Plains farms had excess land and a shortage of labour and livestock. Austrian farmers returned 90% of extracted nitrogen to cropland, sustaining soils over many generations, but they produced little marketable surplus. A key difference was livestock density. Old World communities kept more animals than needed for food and labour to supply manure that maintained cropland fertility. Great Plains farmers used few animals to exploit rich grassland soils, returning less than half of the nitrogen they extracted each year. Relying on a stockpiled endowment of nitrogen, they produced stupendous surpluses for market export, but watched crop yields decline between 1880 and 1940. Austrian immigrants to Kansas saw their return on labour increase 20-fold. Both farm systems were efficient in their own way, one

G. Cunfer, Ph.D. (✉)
Department of History, School of Environment and Sustainability,
University of Saskatchewan, Saskatoon, Saskatchewan, Canada
e-mail: geoff.cunfer@usask.ca

F. Krausmann, Ph.D.
Institute of Social Ecology Vienna (SEC), Alpen-Adria Universitaet Klagenfurt,
Wien, Graz, Schottenfeldgasse 29/5, Vienna 1070, Austria
e-mail: fridolin.krausmann@aau.at

producing long-term stability, the other remarkable commercial exports. Kansas farmers faced a soil nutrient crisis by the 1940s, one that they solved in the second half of the twentieth century by importing fossil fuels. Austrian and Great Plains agriculture converged thereafter, with dramatically increased productivity based on oil, diesel fuel, petroleum-based pesticides and synthetic nitrogen fertilisers manufactured from natural gas.

**Keywords** Historical agro-ecosystems • Socio-Ecological metabolism • Agricultural frontier • Material and energy flow accounting • Agricultural land use • Biophysical economy • Soil sustainability • Austro-Hungarian agriculture • Great Plains sustainability • Grassland ecosystem

## 12.1 Migration

George Thir had a busy year in 1884.[1] Along with his parents, George and Theresia Thir, he emigrated from the corner of central Europe where today Austria, Hungary, and Slovakia meet. He travelled to the United States, made his way to the far edge of agricultural settlement in western Kansas, and selected a farm that would become his home for the remainder of his life. Kansas had organised its western territory just 6 years earlier, including the Thirs' new home of Decatur County. By the time the Thirs arrived, the gently undulating mixed-grass prairie of western Kansas was filling up with farmers. Most came from eastern parts of the United States, but a significant number came directly from Germany, Austria-Hungary, Sweden and other countries. The Thirs most likely immigrated from Gols, in what is now Austria, where most of their Kansas neighbours originated. They certainly came from somewhere in the German-speaking portion of the Austro-Hungarian empire. Over the course of his life, various official documents identified the younger George as German, Hungarian, Austro-Hungarian, and Austrian. The Austro-Hungarians who settled in the northwest corner of Decatur County, Kansas came from a cluster of farming villages within 25 km of one another, including Gols and Zurndorf in what is now Austria, and Ragendorf and Kaltenstein in present-day Hungary.[2] Born in May 1865, George Thir was 19 when he travelled to Kansas. Within a few months of arrival he chose suitable farmland in Section 18 of Finley Township and, on 9 October 1884, filed a Homestead claim on 65 ha of grass (Decatur County Historical Book Committee 1983, 25–31; Homestead records from Kansas GenWeb 2009;

---

[1] This study is supported by U.S. National Institute of Child Health and Human Development grant nos. HD044889 and HD033554. An earlier version of this text appeared as Cunfer and Krausmann (2009).

[2] For details on the emigration from this region of the Austro-Hungarian Empire see Dujmovits (1992) and Antoni (1992).

The reconstruction of Thir and Demmer family history comes from the following sources: U.S. Population Census manuscript schedules, Decatur County, Kansas, 1880, 1900, 1910, 1920, 1930; Kansas State Board of Agriculture, population census manuscripts, Decatur County, Kansas, Kansas State Board of Agriculture 1885, 1895, 1905, 1915, 1925, held at Kansas State Historical Society, Topeka, hereafter cited as KSHS).

Turning raw prairie into a farm was slow, hard work. In March 1885, the new homestead, valued at $50, had no cropland, no livestock, no fences and no house. Thir worked as a blacksmith and boarded with neighbours. He had not really started farming his new land yet when the census-taker recorded his presence in the spring of 1885, but the next 10 years would see considerable progress on the Thir farm (Kansas State Board of Agriculture 1895 held at KSHS).

In 1888, when George Thir was 23 years old, he married Elizabeth Demmer, aged 20. Born in Gols in 1868, at the age of 13 she and her family had joined the chain migration to far western Kansas. Between the 1870s and 1890s, dozens of families left Gols, Ragendorf, Zurndorf and Kaltenstein for the United States, travelling by ship across the Atlantic, then by train to Nebraska. Many settled near Crete, Nebraska, where a community of Austro-Hungarian immigrants welcomed new arrivals. The motivations for migration varied. Most migrants sought free agricultural land and an opportunity for economic improvement, while some fled the military draft. In 1983, for example, Carl Resch recalled his grandfather's reason for leaving: "In 1883 John Resch Sr. immigrated to America with his wife and children to escape conscription into the army of Francis Joseph, Emperor of Austria-Hungary, and in search of good land and a better life—free from militarism that ravaged Europe periodically." Another Gols native, Andreas Wurm, had already been drafted and discharged by the age of 17 when, in 1878, he joined two friends travelling to Nebraska. Like many others, they found Crete already full, and moved southwest to Decatur County, Kansas, where free land was still available. Not yet old enough to file a homestead claim, Wurm brought his parents from Austria-Hungary to Kansas so that they could file a claim for him (*Decatur County, Kansas*, cit., 152, 204, 333–334, 351–2, 374, 425, 428–433).

George's new wife, Elizabeth Demmer, was also part of a multi-generational migration. She was one of five children born to Mathias and Maria Ecker Demmer. In 1881 the whole family moved to Crete, Nebraska, and then on to Decatur County, Kansas. Several other branches of the Demmer family made the move between the late 1870s and mid-1880s to the United States, where they found (and often intermarried with) former neighbours from Austria-Hungary. Families from Gols, Ragendorf, and Kaltenstein selected homesteads all around Finley Township, where George and Elizabeth Thir made their new farm (Fig. 12.1). Elizabeth gave birth to a daughter, Susie M. Thir, in January 1889. A second daughter, born in May 1892, took her mother's name. Their third and final child, George Jr., was born in May 1895. By that year the farm, now worth $800, was thriving. It boasted cropland planted to corn, spring wheat, sorghum and potatoes, plus hay and grazing land for three horses, one milk cow, and one hog (*Decatur County, Kansas*, cit., 152, 184, 374, 430; *Standard Atlas of Decatur County, Kansas* (1905), held at KSHS, Kansas

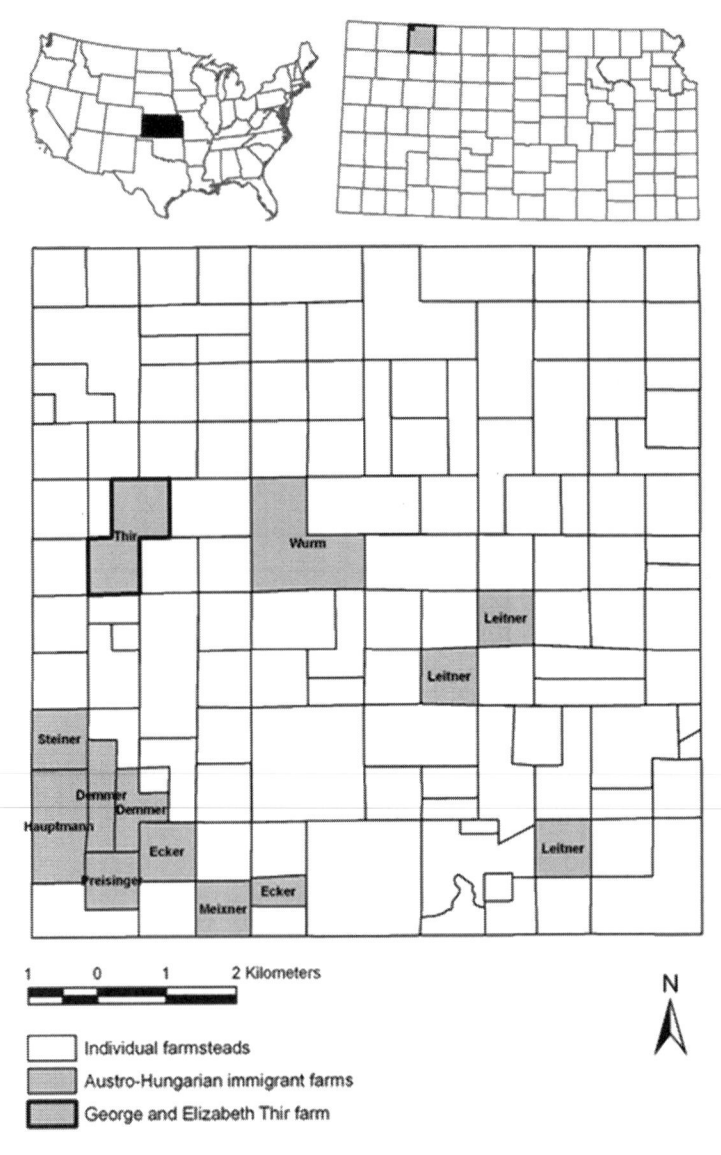

**Fig. 12.1** Austro-Hungarian immigrant farms, including the Thir farm, situated within Finley Township. Small locator maps show the location of Kansas within the United States and of Decatur County and Finley Township within the state of Kansas

State Board of Agriculture 1905, held at KSHS; Kansas State Board of Agriculture, population and agricultural census manuscripts, Decatur County, Kansas, 1895).

Over the next few decades, as the Thir children grew up, the farm expanded. By 1905 it had doubled in size to 130 ha, with buildings, implements, a dozen milk cows, 10 beef cattle, 4 horses, 11 hogs, and a variety of cropland, hay land, and pasture, all together worth $2,000. Ten years later, the farm had doubled in size again to 259 ha—one square mile of fertile Kansas farmland. The daughters moved out of the family home in their early twenties to join new husbands. George Jr. remained single, continued to live with his parents and farmed in partnership with his father into the 1940s. George Sr. died in 1949 and Elizabeth in 1953 (Kansas State Board of Agriculture 1905, 1915, 1920, 1925, 1930, 1935, 1940; U.S. Population Census 1900, 1910, 1920, 1930; Herndon Union Cemetery records, Rawlins County, Kansas).

When George and Elizabeth Thir migrated across the ocean, they left behind an agro-ecological system in Central Europe where farmland supported high populations on smallholdings, where rainfall was reliable, where nutrients and energy flowed through tightly bound pathways linking soil, plants, animals and people into a complex and highly evolved system. For centuries, farmers had pushed the land to produce as much food as possible to support growing populations, but in a way that could be sustained over many generations. In Austria-Hungary, land was scarce, labour (and hungry mouths) abundant. Livestock were a crucial component of the system, providing food and clothing, but also physical labour and manure to fertilise cropland (Krausmann 2004).

They arrived in an agro-ecological setting in Kansas that had immense potential but little existing structure. There fertile soil was abundant and cheap, labour hard to come by, and rainfall uncertain. Population density was low, and even livestock were in short supply and expensive. George and Elizabeth spent their lives creating a new agro-ecological system where none had existed. They brought labour to bear: their own strong backs plus those of three children and a barnyard full of animals. They tapped into a rich stockpile of soil nutrients accumulated under native grassland over geological time. They organised a new farm system alongside neighbours from home and from many different parts of the world, one that meshed their cultural inheritance with a semi-arid plains environment. The result was very different from the agricultural world they had left behind.

In order to understand the environmental history of farming communities like those the Thirs inhabited, it is important to recognise that agriculture is a coupled human-environment system (Haberl et al. 2006; Liu et al. 2007). Borrowing methods from sustainability science, this chapter employs a long-term socio-ecological perspective to focus on biophysical relations between society and the natural environment (Ayres and Simonis 1994; Fischer-Kowalski 1998). Recognising that all economic activity is based on a throughput of materials and energy, social metabolism links socioeconomic activity to ecosystem processes. The corresponding set of methods—material and energy flow analysis (MEFA)—allows one to trace material and energy flows through socioeconomic systems and provides a quantitative picture of the physical exchanges between societies and their environment. This approach has been applied in historical studies of local rural systems to investigate the relationship between land, humans, livestock, and the flows of materials and energy related to production and reproduction in agricultural systems (Sieferle et al. 2006; Krausmann 2004; Cusso et al. 2006; Guzman Casado and Gonzalez de Molina 2009; Cunfer 2004; Marull et al. 2008). George and Elizabeth Thir were not just farming—they were also manipulating energy and nitrogen, shifting them across the landscape and directing them into and out of particular soils, biota, crops, and animals. MEFA methods take us beneath the surface to understand the ecological implications of socioeconomic activities.

## 12.2 Comparative Old World and New World Farm Systems

How did the farm system that immigrants left behind compare with that which they found (and created) on the Great Plains frontier? This chapter uses a long-term socio-ecological approach to explore similarities and differences in land use at either end of the migration chain (Fischer-Kowalski and Haberl 2007).[3] It employs two community case studies, one in Austria and the other in Kansas, to compare the ways that people turned the raw materials of soil, climate, and biota into the finished products of food, field, and culture.

Theyern, Austria, as it existed around 1830, serves as the first case study. Theyern is about 100 km northwest of Gols. A pre-existing dataset makes it possible to model Theyern's land-use history in great detail. Although regional differences between farming systems in the nineteenth century are considerable, the basic socio-ecological characteristics of pre-industrial agriculture in Theyern and the Gols-Ragendorf-Kaltenstein region that fed Finley Township's nineteenth century population boom are comparable. Theyern was a typical lowland farming system with an area of 2.3 $km^2$ and a population of 102 in 1829. The village lay in the low, rolling countryside of northeastern Austria. A loess soil over conglomerate rock with a high lime content provides good conditions for cultivation. With an average annual temperature of 10 °C and 521 mm of precipitation, Theyern has favourable climatic conditions for cereal production. The village has been cultivated for many centuries (Sonnlechner 2001). By the early nineteenth century, more than half of Theyern's area was cropland (Fig. 12.2a). Despite a rather large livestock herd, only 3% of the village was in grassland, but woodland commons provided additional grazing. Woodlands covered roughly one-third of the territory, but only prevailed on soils unsuitable for cultivation. They served not only as a source for fuel and timber but also provided grazing and litter for animal bedding (Krausmann 2004, cit. 735–773). Theyern, like Gols, was on the edge of a wine-growing region and, although there are no vineyards in Theyern itself, farmers had access to vines in neighbouring villages. Population density was high: 45 persons per $km^2$. In 1829, Theyern was home to 17 families who farmed an average of 8 ha each (Cadastral Schätzungs Elaborat der Steuergemeinde Theyern, held at Landesarchiv St. Pölten). However, three of the farms were larger (13–19 ha), while four had very small holdings of under 4 ha, probably producing barely enough for subsistence.

---

[3] For an early discussion of agro-ecology as a central subject for environmental history see Worster (1990).

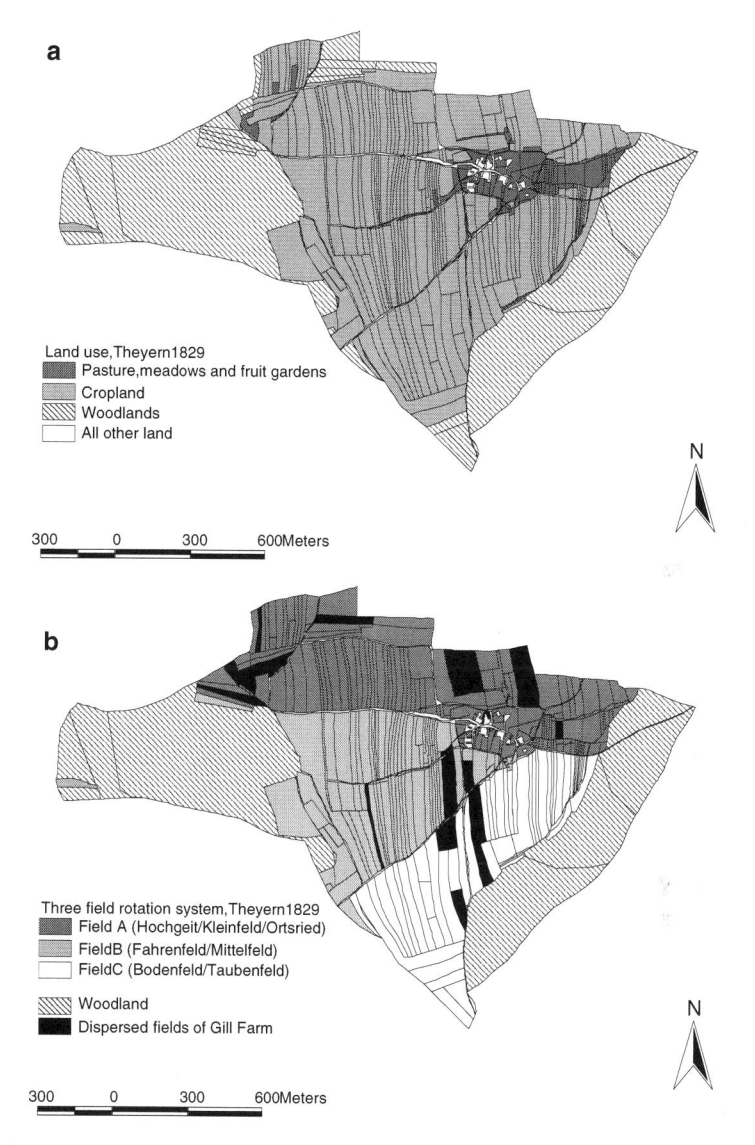

**Fig. 12.2** Theyern land management; (**a**) Small meadows and orchards clustered closely around residential house lots, while cropland surrounded the village. On the outskirts of the community, woodlands prevailed on poor soils not suitable for cropping; (**b**) The cropland portion of the agro-ecosystem rotated annually through a three-field sequence. Family farms consisted of scattered plots distributed across all parts of the village, as illustrated here for the Gill family, one of the larger holdings (ca. 13 ha farmland)

Until the mid-nineteenth century, land did not belong to the peasants but to the local manor, which assigned it to particular families. In the case of Theyern, the nearby Benedictine monastery of Göttweig served this function, and also collected tithes and taxes (in the form of money, compulsory human and animal labour, or a share of agricultural produce). Beside the peasant families and the manor, the village itself was an important institution of land-use decision-making. The village managed its woodlands collectively as commons. Also, the village as a whole determined the temporal rhythm of cultivation and crop choice. Each family tended numerous small plots of land scattered across the municipality. A three-field rotation system necessitated joint decisions and efforts with respect to ploughing and harvesting of crops (Fig. 12.2b) (Cadastral Schätzungs Elaborat der Steuergemeinde Theyern, held at Landesarchiv St. Pölten).

The main source for the reconstruction of Theyern's land-use and farming systems is the Franciscan Cadastre (Franziszeischer or Stabiler Kataster; Moritsch 1972; Sandgruber 1979). This tax survey dates to the first half of the nineteenth century (1817–1856) and covered most of the territory of the Austro-Hungarian Empire, some 3,00,000 km$^2$. It included a geodetic survey of the territory, estimations of crop yields for all land-use classes and a report of monetary outputs (Lego 1968; Finanz-Ministerium 1858).

Up to 39 different land-use classes plus up to four distinct quality designators appear on the maps. The Cadastral Summary (Catastral Schätzungs Elaborat) is the basic data source for the reconstruction of land-use practice and biomass and nutrient flows. This handwritten text exists for each map and offers an extensive description of topography, demography and the farming system. It contains detailed information on land use and land cover, yields, population, livestock and farming practices, as well as livestock feeding practices, soil manuring standards, general information on the number of farms, wealth of the community, use of animals and markets. In addition to the data provided by the cadastre, we used a wide variety of sources and literature about local, regional and general aspects of the structure and functioning of pre-industrial farming systems.[4] Furthermore, from previous research projects, published and unpublished data and analyses relating to the environmental history of the case study regions are available.[5]

Theyern's Cadastral survey dates to 1829. Rather than reflecting specific conditions during any single year, the cadastre reports long-term averages. A reconstruction of the agro-ecosystem on the basis of these data represents a valid average for the first half of the nineteenth century. While this restricts the direct comparability of the farming system that the Thirs left behind in Austria when they emigrated in the 1880s and their Kansas farm, the data still allows for a comparison of the general socio-ecological characteristics of different types of nineteenth century farming systems, which is the main goal of this chapter (Sandgruber 1978).

At the other end of the migration lay Decatur County, Kansas. George and Elizabeth ended their separate travels on the Great Plains, a flat to gently undulating grassland environment, slowly rising in elevation from east to west. Recently

---

[4] For a detailed description see Krausmann (2004, 2008).

[5] This material includes digitised versions of the original cadastral maps of the village, specific evaluations of parcel protocols (e.g., the quantification of the extent of external land use, land use data, and factor costs at the farm level). See Projektgruppe Umweltgeschichte (1997, 1999) and Winiwarter and Sonnlechner (2001).

buffalo range controlled by Cheyenne, Pawnee, and Arapaho horse cultures, Decatur County sat at the transition zone between dry mixed-grass prairie and very dry short-grass steppe (Fig. 12.1). Rainfall averaged 475 mm, and the dominant native vegetation was little bluestem, grama, and buffalo grasses. Trees were rare—less than 5% of ground cover—and appeared only in narrow bands along rivers and streams. Here soils were quite rich, but rainfall was unreliable, reeling between wet years with 800 mm or more and droughts when less than 250 mm fell.[6] To the Thirs and their neighbours the land promised a prosperous future.

The reconstruction of Decatur County's agro-ecosystem comes mainly from agricultural censuses compiled periodically by the State of Kansas and the U.S. federal government. Census descriptions for individual farms in this part of Kansas are available for 1885, 1895, 1905, 1915, 1920, 1925, 1930, 1935 and 1940. These nine snapshots describe land-use activity over 55 years, from the beginning of frontier farm-making to the establishment of a fully developed, modern agricultural system. Censuses report the acreage and yield of various crops on each farm, the number of livestock, the amount of irrigation, fencing, and agricultural implements owned. With these data we can follow the progress of the Thir homestead from raw prairie to integrated farm. Identical data exist for every farm in Finley Township, allowing a comparison between the Thir farm and the several dozen that surrounded it. Aggregated county level data are more readily available, existing for each year between 1880 and 1940. Thus it is possible to study the land-use history of the region at nested scales, from the individual farm to the rural neighbourhood of the township, to the entire 230,000-ha county, and, indeed, for all 105 counties in the state of Kansas.

Population censuses reveal important elements of the social side of farm systems. Manuscript population schedules are available for 1885, 1895, 1900, 1905, 1910, 1915, 1920, 1925, and 1930. These data reveal the life cycles of families, as couples married and had children, as children grew up and left home, as people aged and died. Again, we can observe these changes at various scales, from individual people and families to aggregated townships and counties. Together, the population and agricultural censuses provide basic data about the social metabolism of Kansas farmsteads (Sylvester et al. 2006).

---

[6] Climate data come from two sources. The first is Karl, T.R., Williams, C.N. Jr., Quinlan, F.T., and Boden, T.A. (1990). United States Historical Climatology Network serial temperature and precipitation data. Environmental Science Division. Publication No. 3404. Oak Ridge, Tenn.; Carbon Dioxide Information and Analysis Center, Oak Ridge National Laboratory. The historical climatology data are stored as point data for weather stations at monthly intervals for 1,221 stations in the United States. The second source is National Climatic Data Center, Arizona State University, and Oak Ridge National Laboratory, Global Historical Climatology Network. This data set includes comprehensive monthly global surface baseline climate data. The Great Plains Population and Environment Project (www.icpsr.umich.edu/plains) interpolated data from 394 weather stations in the Great Plains to counties for each month between 1895 and 1993.

## 12.3 A Long-Term Socio-Ecological Approach to Agricultural Systems

This chapter uses a simple conceptual model of agriculture as a coupled socioeconomic and natural system (Fig. 12.3). It builds on basic assumptions about the relation of population, land use and agricultural production formulated by Ester Boserup, but extends this perspective by explicitly including flows of material and energy (Boserup 1965, 1981). It is specific about the interactions of socioeconomic systems and ecosystems, allowing one to capture important technological developments related to the industrialisation of agriculture. In its most general form, the model defines the main biophysical relations in terms of flows of energy and materials between (and within) a natural system (i.e. the agro-ecosystem, characterised by biogeographic conditions and land use types) and a socioeconomic system, consisting of a population subsystem (characterised by demographic attributes) and an economic production subsystem (including infrastructure, farm technology and livestock).[7] The model describes a farming unit (here a farm, township or village) as an agro-ecosystem managed by a local population investing labour and energy, applying a certain mix of technology, and generating a certain return of agricultural produce. It maintains exchange processes with other demographic, socioeconomic, and ecological systems. On a more detailed level, the model specifies the relation of land use and land cover with the extraction of biomass, different types of conversion and consumption processes within the local production system, and the flows into and out of the local environment. Such a systemic perspective allows one to analyse

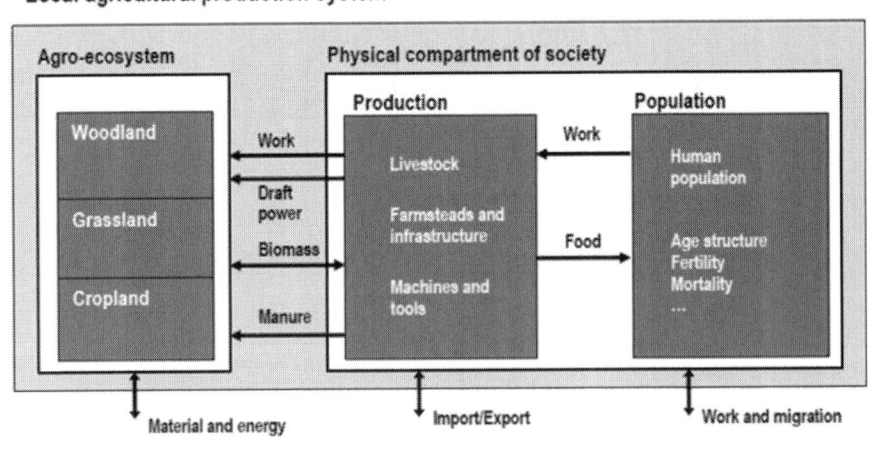

**Fig. 12.3** A conceptual model of agriculture as a coupled socioeconomic and natural system (See text for explanation)

---

[7] This version of the model focuses on biophysical relations between society and nature and thus reduces the socioeconomic system to its physical components, i.e. the population and the production subsystem. See Fischer-Kowalski and Weisz (1999).

all biomass and energy flows and their interrelations within the farming unit, and to link them to land use, ecosystem processes and the demographic system.

The Austrian cadastral records and the Kansas agricultural and population censuses can be used to quantify the flows of nutrients, materials and energy through the various subsystems described in this model. This technique allows one to cross-check the validity of historical data and to fill gaps in the data when omissions or flaws occur in the original sources. For example, even though only fragmentary quantitative data on feed supply and livestock may be available from the cadastral record, knowledge about the reproductive patterns of livestock as well as species-specific feed demand make it possible to generate a picture of feed requirements compared to available supply.[8]

This study identifies nine key socio-ecological indicators that describe the physical stocks and flows of the two farm systems. Those indicators fit into three categories: people and space, farm productivity and livestock, and nutrient management. This text includes graphic figures to represent the most important indicators; the complete data behind those figures are available in Tables 12.1, 12.2, 12.3, 12.4, 12.5, and 12.6.

People and Space

- **population density**: census population divided by land area (people/km$^2$)
- **average farm size**: agricultural area[9] divided by number of farms (ha/farm)
- **land availability**: agricultural area divided by number of farm labourers reported in the Kansas census or estimated based on Theyern's age structure (ha/person)

Annual Farm Productivity

- **grain yield**: cereal production (including grain returned as seed) divided by total area planted, excluding fallow (kg/ha)
- **area productivity**: plant and animal produce for human nutrition, including edible produce available for export, converted into food energy and divided by agricultural area (GJ/ha)[10]
- **labour productivity**: plant and animal produce for human nutrition, including edible produce available for export, converted into food energy and divided by number of farm labourers reported in the Kansas census or estimated based on Theyern's age structure (GJ/person)[11]

---

[8] See, for example, Schüle (1989).

[9] Throughout the paper we define "agricultural area" as not only cultivated and intensively used land such as cropland, meadows or fruit gardens but also uncultivated prairie and woodlands. Uncultivated prairie in Kansas and woodlands in Theyern were integral components of both agricultural systems, as they were used for grazing or to extract bedding materials and also served as sources of biomass and plant nutrients transported to intensively used cropland (Cf. Krausmann 2004; Cunfer 2004).

[10] One Giga Joule (GJ) corresponds to $10^9$ J or 239 Mega calories (Mcal). Food output is measured in Joules of nutritional value according to standard nutrition tables.

[11] We use "area productivity" and "labour productivity" in conformity with their usage in the long-term socio-ecological literature. Readers should be aware that economists have different definitions for these terms.

**Table 12.1** Population, land use, livestock and crop production in Finley Township, 1895–1940

| Variable | Unit | 1895 | 1905 | 1915 | 1920 | 1925 | 1930 | 1935 | 1940 |
|---|---|---|---|---|---|---|---|---|---|
| Population | Persons | 227 | 389 | 341 | 392 | 373 | 379 | n.d. | n.d. |
| Agricultural population | Persons | 169 | 332 | 260 | 286 | 230 | 255 | 287 | 259 |
| Farms | Number | 32 | 64 | 58 | 65 | 63 | 64 | 72 | 65 |
| Total area (land in farms) | ha | 2,939 | 8,320 | 7,376 | 10,006 | 9,487 | 8,792 | 9,233 | 8,761 |
| Cropland | ha | 1,341 | 3,545 | 5,048 | 4,643 | 5,344 | 4,780 | 4,938 | 5,142 |
| Corn | ha | 830 | 1,095 | 1,079 | 783 | 1,778 | 1,571 | 1,784 | 1,383 |
| Wheat | ha | 291 | 1,509 | 3,148 | 3,186 | 2,957 | 2,738 | 2,116 | 1,416 |
| Barley | ha | n.d. | 373 | 177 | 212 | 128 | 181 | 219 | 639 |
| All other crops | ha | 220 | 568 | 645 | 462 | 482 | 290 | 819 | 1,704 |
| Grassland | ha | 1,598 | 4,775 | 2,328 | 5,364 | 4,142 | 4,012 | 4,295 | 3,619 |
| All other land | ha | 44 | 125 | 111 | 150 | 142 | 132 | 139 | 131 |
| Cattle | Head | 161 | 1,541 | 557 | 1,035 | 1,244 | 432 | 1,548 | 701 |
| Horses (and mules) | Head | 136 | 435 | 497 | 656 | 556 | 299 | 257 | n.d. |
| Pigs | Head | 167 | 1,749 | 531 | 335 | 1,114 | 344 | 222 | 30 |
| Corn (harvest) | t | 1,173 | 2,580 | 2,203 | 1,476 | 1,676 | 3,085 | 420 | 565 |
| Wheat (harvest) | t | 117 | 1,825 | 2,961 | 3,853 | 1,788 | 2,392 | 995 | 447 |
| Barley (harvest) | t | n.d. | 663 | 314 | 319 | 117 | 263 | 106 | 299 |

Sources: See text

**Table 12.2** Population, land use, livestock and crop production on the Thir farm, 1895–1940

| Variable | Unit | 1895 | 1905 | 1915 | 1920 | 1925 | 1930 | 1935 | 1940 |
|---|---|---|---|---|---|---|---|---|---|
| Population | Persons | 4 | 5 | 4 | 3 | 3 | 3 | 3 | 3 |
| Agricultural population | Persons | 4 | 5 | 4 | 3 | 3 | 3 | 3 | 3 |
| Farms | Number | 1 | 1 | 1 | 1 | 1 | 1 | 1 | 1 |
| Total area | ha | 65 | 130 | 259 | 162 | 162 | 162 | 227 | 227 |
| Cropland | ha | 25 | 52 | 118 | 59 | 75 | 80 | 88 | 134 |
| Corn | ha | 20 | 8 | 8 | 12 | 16 | 26 | 24 | 28 |
| Wheat | ha | 3 | 32 | 81 | 40 | 49 | 51 | 57 | 34 |
| Barley | ha | 0 | 5 | 0 | 2 | 6 | 0 | 4 | 0 |
| All other crops | ha | 1 | 7 | 29 | 4 | 4 | 3 | 3 | 71 |
| Grassland | ha | 40 | 77 | 141 | 103 | 87 | 82 | 138 | 93 |
| All other land | ha | 1 | 2 | 4 | 2 | 2 | 2 | 3 | 3 |
| Cattle | Head | 1 | 22 | 26 | 30 | 21 | 11 | 25 | 4 |
| Horses (and mules) | Head | 3 | 5 | 8 | 9 | 9 | 8 | 5 | n.d. |
| Pigs | Head | 1 | 11 | 5 | 7 | 10 | 3 | 1 | 9 |
| Corn (harvest) | t | 29 | 19 | 17 | 23 | 15 | 52 | 6 | 12 |
| Wheat (harvest) | t | 1 | 39 | 76 | 49 | 29 | 44 | 27 | 11 |
| Barley (harvest) | t | 0 | 9 | 0 | 3 | 6 | 0 | 2 | 0 |

Sources: See text

**Table 12.3** Socio-ecological characteristics, Finley Township, 1895–1940

| Variable | Unit | 1895 | 1905 | 1915 | 1920 | 1925 | 1930 | 1935 | 1940 |
|---|---|---|---|---|---|---|---|---|---|
| Population density | cap/km² | 2.5 | 4.2 | 3.7 | 4.2 | 4.0 | 4.1 | n.d. | n.d. |
| Farm size | ha per farm | 92 | 130 | 127 | 154 | 151 | 137 | 128 | 135 |
| Land availability | ha per agric. labourer | 36 | 45 | 47 | 58 | 69 | 59 | 55 | 58 |
| Grain yield | kg/ha/year | 1,141 | 1,687 | 1,244 | 1,351 | 736 | 1,278 | 370 | 378 |
| Area productivity | GJ/ha/year | 4.6 | 4.9 | 7.0 | 5.1 | 1.7 | 6.5 | 0.4 | 1.6 |
| Labour productivity | GJ/labourer/year | 168 | 220 | 327 | 293 | 114 | 385 | 19 | 92 |
| Marketable crop production | % of total production | 74% | 53% | 69% | 66% | 26% | 72% | −14% | 43% |
| Livestock density | animal per km² | 4.2 | 22.9 | 13.2 | 14.5 | 17.8 | 7.5 | 14.9 | 4.9 |
| Nitrogen return on cropland | % of total extraction | 27% | 30% | 30% | 22% | 38% | 21% | 68% | 51% |

Sources: See text

**Table 12.4** Socio-ecological characteristics, Thir farm, 1895–1940

| Variable | Unit | 1895 | 1905 | 1915 | 1920 | 1925 | 1930 | 1935 | 1940 |
|---|---|---|---|---|---|---|---|---|---|
| Population density | cap/km² | 6.2 | 3.9 | 1.5 | 1.9 | 1.9 | 1.9 | 1.3 | 1.3 |
| Farm size | ha per farm | 65 | 130 | 259 | 162 | 162 | 162 | 227 | 227 |
| Land availability | ha per agric. labourer | 32 | 43 | 86 | 54 | 54 | 54 | 76 | 76 |
| Grain yield | kg/ha/year | 1,274 | 1,427 | 1,041 | 1,371 | 709 | 1,246 | 406 | 369 |
| Area productivity | GJ/ha/year | 4.9 | 4.8 | 3.1 | 3.7 | 1.3 | 5.4 | 0.5 | 0.7 |
| Labour productivity | GJ/labourer/year | 159 | 209 | 267 | 198 | 68 | 293 | 34 | 55 |
| Marketable crop production | % of total production | 75% | 59% | 59% | 54% | 23% | 65% | 6% | 33% |
| Livestock density | animal per km² | 4.2 | 17.7 | 10.5 | 20.0 | 16.3 | 10.1 | 11.1 | 2.1 |
| Nitrogen return on cropland | % of total extraction | 20% | 22% | 58% | 25% | 39% | 21% | 58% | 47% |

Sources: See text

**Table 12.5** Population, land use, livestock and crop production in Theyern municipality, 1829

| Variable | Unit | 1829 |
|---|---|---|
| Population | Persons | 102 |
| Agricultural population | Persons | 102 |
| Farms | Number | 17 |
| Total area | ha | 225 |
| Cropland | ha | 135 |
| Rye | ha | 41 |
| Cereal mix | ha | 41 |
| All other crops | ha | 13 |
| Fallow | ha | 28 |
| Grassland | ha | 7 |
| Woodland | ha | 79 |
| All other land | ha | 4 |
| Cattle | Head | 85 |
| Horses and mules | Head | 5 |
| Pigs | Head | 42 |
| Sheep | Head | 77 |
| Rye (harvest) | t | 35 |
| Cereal mix (Linsgetreide) (harvest) | t | 32 |

Sources: See text

**Table 12.6** Socio-ecological characteristics, Theyern municipality, 1829

| Variable | Unit | 1829 |
|---|---|---|
| Population density | cap/km² | 45.3 |
| Farm size | ha per farm | 13 |
| Land availability | ha per agr. labourer | 3 |
| Grain yield | kg/ha/year | 819 |
| Area productivity | GJ/ha/year | 4.4 |
| Labour productivity | GJ/labourer/year | 9 |
| Marketable production | % of total production | 25% |
| Livestock density | animal per km² | 24 |
| Nitrogen return on cropland | % of total extraction | 92% |

Sources: See text

- **marketable crop production**: cereal production minus grains required for feed, seed and subsistence (percentage of extracted biomass as tons dry matter)

Livestock and Nutrient Management

- **livestock density**: large animal units of 500 kg live weight divided by agricultural area (animals/km²)[12]

---

[12] We converted livestock numbers into large animal units of 500 kg live weight by using species and region-specific data on average live weight in the observed period. See Krausmann (2004); 735–773 and Krausmann (2008), 56.

- **nitrogen return**: N inputs from natural deposition, free fixation, manure and leguminous crops divided by N contained in harvested biomass (percentage of extracted N returned to soil)[13]

## 12.4 People and Space

Theyern, Austria was typical of European agro-ecological systems. With episodic agricultural occupation dating at least as early as 1000 B.C., we know that population expansion during the late Middle Ages led to a gradual re-colonisation of the area for agriculture. By 1830, Theyern had existed as a discrete community for hundreds of years and its cropland, hay meadows, grazing commons and surrounding forests had been producing food, feed and shelter, year in and year out, for a very long time. Most members of the community lived nearly at the subsistence level, producing as much food and supporting as many people as possible, given current cultivation practices, technology and energy availability. The fully populated land achieved its peak productive potential. Theyern's population density in 1830 was 45 people per $km^2$ (Fig. 12.4a). The average family farmed 13 ha of land, and there were 2 ha of agricultural land per person in the community (1 ha/cap if woodland is excluded; Fig. 12.4b, c). Over centuries, the people of Theyern had learned how to use their land intensively, supporting the highest number of people possible, and sustaining those populations for multiple generations.

The situation in Decatur County, Kansas, when Elizabeth Demmer, George Thir, and their compatriots arrived, was just the opposite. Here was land that had never known widespread agricultural use. For 10,000 years since the end of the last ice age, the Great Plains had been steppe grassland, home to wild grazers—bison—and browsers—pronghorn—but few other large animals. The indigenous people were mobile hunters and gatherers, travelling on foot over wide distances. Native agriculture expanded on the plains only after 1000 A.D. and only over a very small area. Occasional patches of maize, beans, and squash dotted the narrow river valleys

---

[13] This estimate of nitrogen return to soils is only approximate. This analysis does not include a full soil nutrient balance. For one thing, it does not consider N losses due to volatilisation and leaching. Furthermore, a comprehensive assessment of soil fertility would need to include phosphorus, potassium, and organic matter, plus the structural properties of soils. Given the limitations of historical data, this paper focuses on those N inputs and extractions that farmers control most directly. For further details concerning the procedure used to estimate nitrogen flows see Krausmann (2004, 2008, 17–20) and Cunfer (2004). On soil nutrient balances more broadly, see Loomis (1978, 1984), Campbell and Overton (1991), Loomis and Connor (1992), and Shiel (2006a, b).

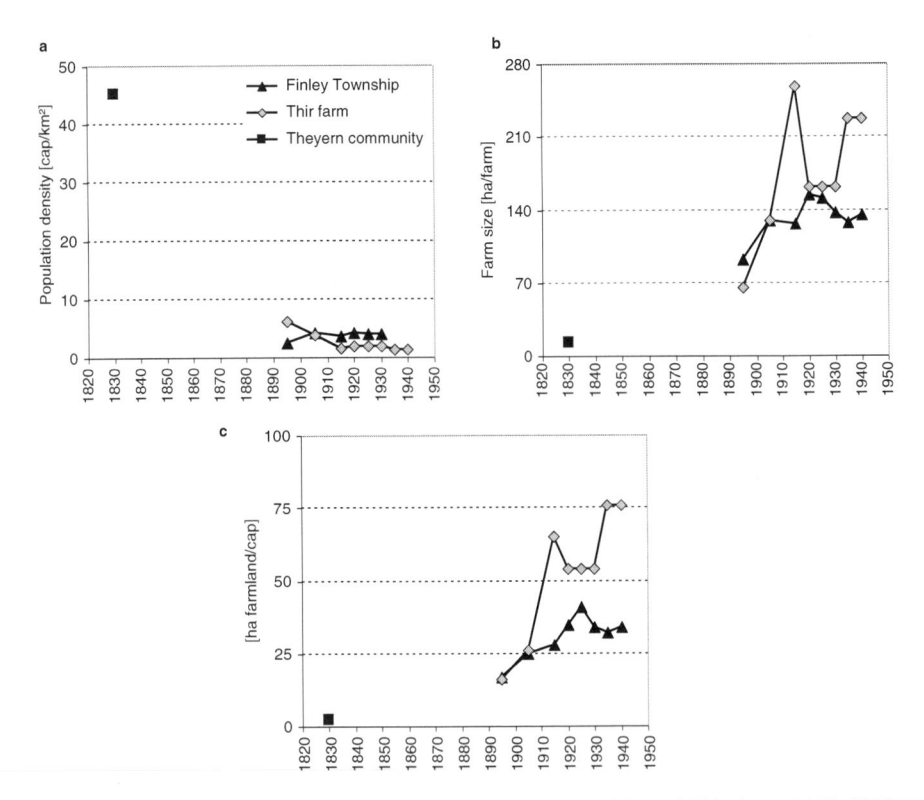

Fig. 12.4 People and space, Theyern, 1829 and Finley Township and Thir farm, 1895–1940; (a) population density; (b) average farm size; (c) land availability (Sources: see text)

winding through vast uncultivated upland grasslands.[14] At their greatest extent, Indian crop fields never reached even 1% of the area of the Great Plains. After the seventeenth century, many natives adopted horse-based hunting and gathering, and some moved in the direction of horse pastoralism.

European farmers who moved into the region in the late nineteenth century entered an agricultural vacuum. Importing livestock with them, and thus increasing their ability to work the soil by 100-fold, American, German, and Austro-Hungarian settlers began the enormous task of agricultural colonisation, plowing sod that had lain intact for thousands of years. The contrast with European agricultural villages

---

[14] Farming Indians maintained soil fertility by swidden, moving their villages wholesale every 5–10 years when soil nutrients failed and crop yields declined. The most notable difference between New World and Old World agriculture was the presence of domesticated animals in the latter. Indian farmers had no domesticated livestock. Women tilled the soil entirely through human labour. Thus Indian agriculturalists never farmed the widespread uplands of the Great Plains. Both population densities and the area of arable land remained very low. See Hurt (1987, 57–64) and Wedel (1978).

could not have been greater. The population density in Finley Township, where George and Elizabeth Thir made their new farm, was only two people per km$^2$ in 1895, an order of magnitude lower than in Theyern. The average farm size was an incredible 92 ha, so large that for the first several decades, few farmers could make use of all of their land and a considerable fraction of the available land was used only for extensive grazing. There were 17 ha of land in the township for every man, woman, and child. The amount of land available to be worked per agricultural labourer was huge and increased from 36 ha in 1895 to almost 70 ha in 1925, when the first tractors appeared in the township. Given the shortage of labour on this agricultural frontier, much of the land remained unused. On the Thir homestead, 65 ha supported and employed two adults and three children. Compared to the community as a whole, the Thir farm was nearly representative, with a population density of six people per km$^2$ and about 16 ha of land per person.

The pioneer era in Decatur County lasted about 50 years, from 1870 to 1920. During that time farmers filled the land, adjusted their farming practices to fit local soils, climate and topography, and moved toward an agricultural equilibrium. Population density in Finley Township increased during the initial period of homesteading and then stabilised at between 4 and 5 people per km$^2$. During the same period, average farm sizes rose rapidly, from 92 ha in 1895 to a peak at 154 ha in 1920, then dropped slightly to settle at around 130 ha for the next few decades. Land per person followed a similar curve, rising from 17 ha in 1895 to 35 in 1920, and thereafter floating between about 30 and 40 through the early twentieth century. On the Thir farm, rapid acquisition of additional land pushed these numbers higher for the family. In 1915, 30 years after immigration from Austria, the Thirs owned 259 ha of land, a whopping 65 ha for each person in the family. While farmers on the Kansas frontier went through a period of adaptation and adjustment, they did not move toward an Old World style farm system of high population densities on intensively used land; If anything, they moved away from that model.

## 12.5 Annual Farm Productivity

Theyern farmers maximized their grain yields, but within the bounds of long-term sustainability. They grew as much food as possible without undermining the ability of the land to support people for indefinite generations into the future. Theyern farms in 1830 produced 819 kg of grain per hectare, which, together with animal products, were enough to provide 9 GJ of nutritional energy for every farm labourer (Fig. 12.5a). Area productivity was 2.9 GJ of food per hectare (Fig. 12.5b). The highly integrated subsistence system supported a lot of people, but surplus above local demand was low and for the smaller farms production accomplished bare survival only. Here farmers had been re-using soils over centuries for agricultural production. The population density matched agricultural production, given local climate and available technology. The largest share of farm output went toward local consumption.

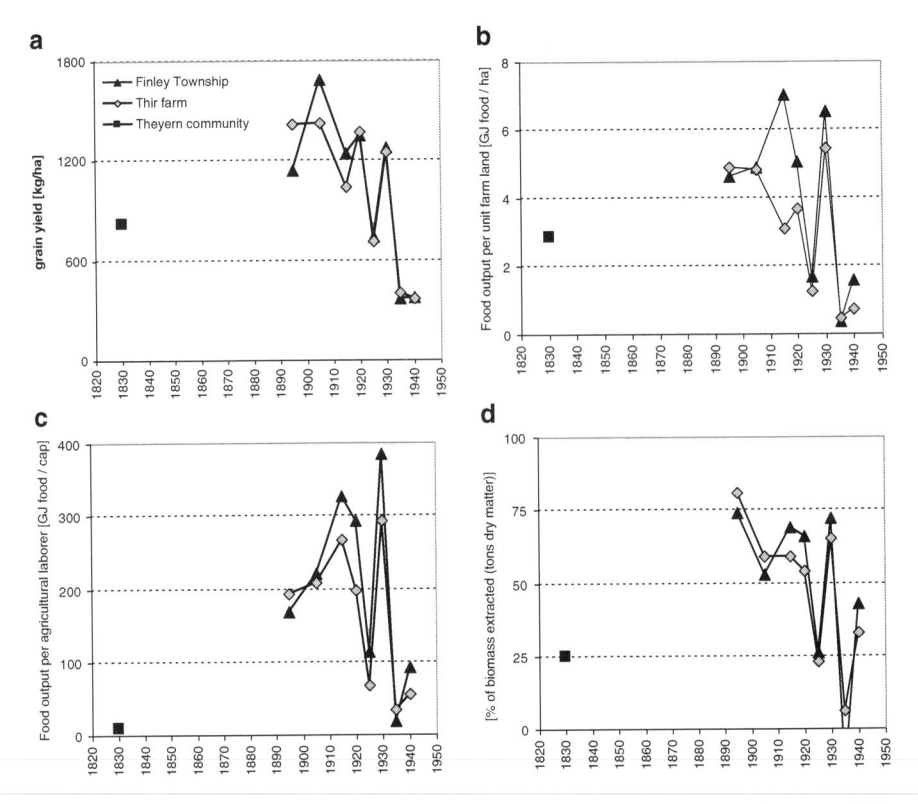

**Fig. 12.5** Annual farm productivity, Theyern 1829 and Finley Township and Thir farm, 1895–1940; (**a**) grain yield; (**b**) area productivity; (**c**) labour productivity; (**d**) marketable crop production (Sources: see text)

Theyern exported from the local system no more than 25% of its agricultural produce through sales in nearby markets or rent paid to the landlord (Fig. 12.5d). This profile provides a long-term average of the community's typical productivity throughout the first half of the nineteenth century.

In western Kansas the freshly ploughed soils produced much higher yields in the first couple of decades. Taking advantage of 10,000 years of stockpiled soil nutrients, the Thir farm produced 1,274 kg of grain per hectare in 1895, 56% higher than Theyern's yield, while Finley Township as a whole averaged 1,141 kg, a 39% surplus over the Austrian case. The township's area productivity in 1895 was significantly higher than in Theyern, at 4.6 GJ/ha, and because there were fewer people on the land in Kansas, nutritional energy production per farm labourer was 168 GJ in Finley Township (Fig. 12.5b, c). Such return on labour—nearly 20 times Theyern's rate—was stupendous. Whereas one Theyern farm labourer grew enough food to feed about 2.5 people, one agricultural labourer in Finley Township could feed nearly 50. No person could reasonably consume so much food. Rather, the excess production beyond subsistence needs went into market exports. Agriculture in the

Great Plains was from the beginning oriented towards commercial production and was reliant on the expanding railroad network to transport grain to urban markets. Three-quarters of the grain grown in Finley Township was in excess of local food and feed needs, and instead found national and international markets. At harvest farmers bagged their wheat, hauled it to grain elevators on the railroad line and shipped their produce east. Cities grew rapidly in the late nineteenth century as other immigrants poured in to take factory jobs in the United States' industrialising economy (Prickler 2003). Kansas wheat farmers fed not only themselves but those distant urban workers too.

The exploitation of stockpiled soil nutrients could not continue indefinitely. Through the early twentieth century, cereal yields in western Kansas fell, plummeting to less than a quarter of their peak levels. As farmers ploughed up fresh land in the first two decades of agricultural settlement, yields remained high, rising from 1,141 kg/ha in 1895 to 1,687 kg 10 years later. Thereafter, once most of the new land was already in production, yields began to fall, down to 1,244 kg in 1915 and 736 kg in 1925. By the 1920s, in the fourth decade of agricultural settlement, grain yields dropped to levels similar to those Theyern farmers had produced a century earlier. Still, yields continued to fall, to below 400 kg during the 1930s drought. The Thir farm closely followed community-wide trends.

The decline in yields was unmistakably downward over half a century, but from year to year there were sharp upturns and downturns. For example, 1925 saw township-wide yields of only 736 kg/ha, but 1930 produced a bumper crop at 1,278 kg. Five years later, in 1935, production was down sharply again. Area productivity likewise varied widely, fluctuating between 4 and 7 GJ/ha, then dropping to less than 2 in 1925 and again in the 1930s. Crop yields in Kansas derived not only from soil fertility, but also from soil moisture. The extreme annual variation in rainfall at the centre of the continent hovered just above or just below the minimum precipitation necessary to sustain wheat, corn, and other cereals. Unlike in Theyern, rainfall controlled yields as much as soil quality did. Thus the extremely low yields in 1935 and 1940 resulted more from the deep drought of those years than from depleted soils. The downward trend in yields over the long term reveals a combination of declining rainfall and soil mining in western Kansas during the pioneer era. Newly-arrived farmers produced stupendous food excesses and sold those crops into the cash market. In the process, they exploited the stockpiled soil fertility that had accumulated century by century under native grass.

None of the primary sources report actual exports of farm produce. Instead, we estimate marketable crop production by calculating how much of the harvest was needed for feeding the people and livestock in the community and for seeding next year's crop. Any surplus would have been available for sale on the market. The marketable production in excess of subsistence needs moved downward in Finley Township, along with yields, from 74% in 1895 to just 26% in 1925. It bounced back with strong rainfall in 1930 to 72%, but then fell with the arrival of drought in the 1930s. By 1935, cereal production actually fell 14% below what was needed for bare subsistence, but was up again to more than 40% of total production just 5 years later.

## 12.6 Livestock and Nutrient Management

In addition to high human population density, Old World farm systems had high densities of livestock. The menagerie of European agriculture included oxen, beef cattle, milk cows, draft horses, mules, donkeys, hogs and pigs, goats, and an array of birds, including chickens, ducks and geese. Theyern, for example, had 24 large animals (500 kg equivalent) per km$^2$ around 1830 (Fig. 12.6a). The impact of livestock cannot be understated. Most obviously, farm animals provided food (beef, pork, poultry, milk, eggs, lard, butter) and clothing (leather, wool). They also provided labour for ploughing soil, cultivating weeds, harvesting crops and transporting farm produce over short and long distances.[15] More subtle, but no less significant, was the impact of manure produced by livestock. Rich in nitrogen, organic carbon, and other soil nutrients, livestock manure was a vector by which people could redirect nutrients from biomass that humans cannot digest (grass, brush, stubble, litter) to agricultural crops. Livestock also functioned as a means to move fertility from place to place across the landscape. For example, cattle grazing grass or brush growing on steep hillsides, in forests or over non-arable soils, accumulated nutrients that they brought back to the farm yard and deposited on the ground. When farmers applied manure to their crop fields, they essentially transported soil nutrients from untillable land to arable land, subsidising fertility in the infields with nutrients transported by livestock from the outfields. Theyern farmers maintained significantly more livestock than they needed for food and labour; they kept additional animals because of their manure production (Allen 2008; Frissel 1978; Cusso et al. 2006).

Every year, Theyern farmers returned to the soil more than 90% of the nitrogen that they extracted from it in crops (Fig. 12.6b). Much of that restored nitrogen flowed through livestock and their manure. Collecting, processing and properly applying manure was labour-intensive work. The whole system was intricately interrelated: Feeding a dense population required maintaining animals that produced manure, which in turn required a significant labour force and thus dense populations. Domesticated animals enabled the soil restoration necessary for continuous cropping into the indefinite future. The presence of these animals distinguished Old World farming from that of Native Americans. In the Americas, natives had no livestock, and managed soil fertility by moving to new farm fields every 5–20 years as soil fertility declined.

---

[15] The most common draft animals used in Theyern around 1830 were oxen. Only the larger farms kept horses, while in small holdings cows were also used for labour (working fields and fallow areas) and transport (moving harvest from dispersed fields), fuelwood from the community forests, and manure back to the fields. Krausmann (2004) estimates that installed power amounted to 0.17 kW per ha of cropland. According to Schaschl (2007), who quantified the monthly supply of and demand for human and animal labour during the course of a year for individual farms in Theyern, the supply of animal labour exceeded demand even during peak seasons in March and April. In Finley Township, horses were the only animals used to provide work until the first tractors appeared in the 1920s. According to our estimate, installed power per unit of cropland was similar to that in Theyern.

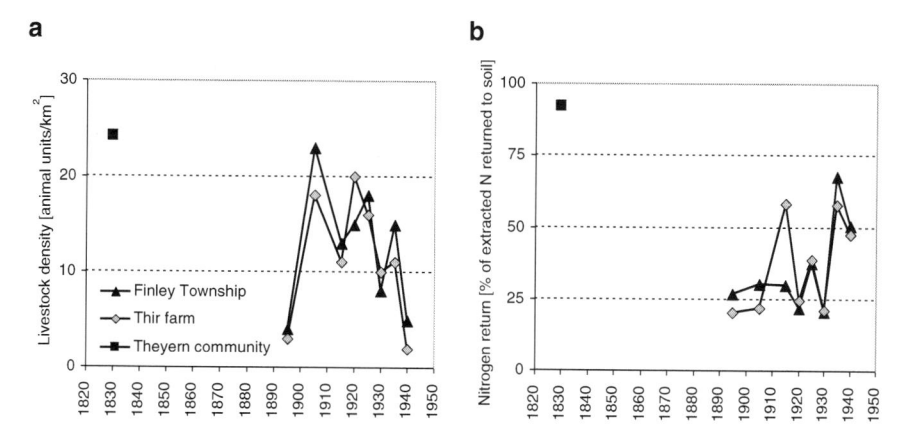

**Fig. 12.6** Livestock and nutrient management, Theyern, 1829 and Finley Township and Thir farm, 1895–1940; (**a**) livestock density; (**b**) nitrogen return (Sources: see text)

Another mechanism for the maintenance of soil nitrogen in the European system was fallow rotation. In 1829, cropland in Theyern was still cultivated in the traditional three-field rotation. A crop of winter cereal in the first year and a summer cereal in the second year was followed by a year of fallow. During the fallow period, the land was manured and vegetation regrowth was ploughed into the soil. Mineralised nutrients from organic matter accumulated for the benefit of crops in subsequent years. Natural ecosystem processes also provided additions of soil nitrogen, including free fixation by soil microorganisms and nitrogen deposited from the atmosphere in rain, snow or dust. At the turn of the nineteenth century, Austrian farmers were only beginning to include nitrogen-fixing legume fodder crops such as clover or alfalfa fodder into their crop rotations, but in the coming decades legumes gradually replaced fallow in the crop rotation system, emerging as a crucial element in the management of soil fertility. In Theyern in 1829, roughly one-fifth of the fallow field was planted with clover, already providing a considerable contribution to soil nitrogen stocks. Thus, by a combination of means Theyern farmers were essentially in balance, replacing about as much soil nitrogen as they extracted each year.

Finley Township, for its part, was decidedly out of balance with the nitrogen system. The initial plough-up accelerated the decomposition of accumulated organic matter and spiked nitrogen into the soil for the first several years (Parton et al. 2005). But ongoing ploughing and cultivation soon generated nitrogen declines through both chemical and biological processes (Hass et al. 1957). Exposure of soils to the atmosphere initiated ammonia volatilisation by which stored nitrogen escaped into the air. Tillage also encouraged bacterial denitrification, in which soil bacteria converted nitrate to nitrogen gases by means of digestion, returning soil nitrogen to the atmosphere. Ploughing could accelerate leaching of nitrogen via rainwater deep into the soil, plus additional losses from water and wind erosion (Stevenson 1982; Cunfer 2004; Burke et al. 2002). Thus it is not surprising that crop yields began at remarkably high levels, then dropped throughout the next 50 years after settlement.

In addition to these natural nitrogen losses, Kansas farmers extracted more nitrogen from their soils than they returned each year, in large part because they put little manure back onto the fields. Finley Township had a low livestock density of only four large animals per hectare in 1895, far below Theyern's 24. That number rose to 23 animals per km$^2$ in 1905 (mostly beef cattle, horses, and milk cows), and then dropped steadily over the next 40 years, down to just five again by 1940. The relative shortage of livestock on Kansas farms meant that farmers had correspondingly less manure with which to return nitrogen to cropland soils. Farmers there returned only 27% of the nitrogen they extracted in 1895, and that number remained below 40% through the 1920s. The 1930s saw an increase in nitrogen return to between 50 and 70% only because significant crop failures during drought years prevented farmers from extracting much nitrogen from their land.[16] With natural soil fertility that far exceeded subsistence needs and that produced large, exportable surpluses for two decades, farmers did not feel the need to husband large numbers of livestock for the purpose of manure accumulation. They needed horses for labour and used cattle and pigs for household food and to create added value to uncultivated prairie. But beyond that, they did not maintain additional animals simply for their soil fertility benefits, as in Theyern.

As George and Elizabeth Thir and their neighbours took more nitrogen than they returned every year, crop yields fell. It took a couple of generations before crisis loomed, and in the 1930s several regional problems converged. Low and declining soil fertility began to pressure farms just as a 9-year drought devastated the region and a world-wide economic depression further challenged farm sustainability. The eventual solution came, not in adopting Old World-style farm management, but from the importation of fossil fuel energy. The decline in livestock density in Finley Township after 1905 went hand-in-hand with the advent of fossil fuel energy deployment. When farmers adopted tractors, trucks, and other internal combustion engines in the early twentieth century, they decreased their horse populations, simultaneously decreasing their manure supply. After World War II, farmers addressed their soil fertility problem by applying synthetic fertiliser in place of the missing manure. Nitrogen fertiliser also represents a fossil fuel import, since its production requires large amounts of natural gas. Thus twentieth century farmers substituted fossil fuel-driven tractors for the labour function of livestock, and substituted fossil fuel-derived fertilizers for the manure function of livestock. In multiple ways, fossil fuels provided substitutes for the missing livestock in the Kansas farm system.

---

[16] While the peaks in the rate of nitrogen return in Finley Township and at the Thir farm in the 1940s are due to harvest failures and consequent low nitrogen extraction rather than to increases in nitrogen input, leguminous crops contributed to the high return rate (above 50%) which can be observed for the George Thir farm in 1915. This was the only year when Thir planted a considerable fraction of his cropland with alfalfa.

## 12.7 Conclusion

This chapter presents a detailed picture of the social ecology and metabolic characteristics of farming systems in Decatur County, Kansas and their development over time. The Austrian case, the rural village of Theyern, serves as a reference point to contrast the Kansas farm system and highlight defining socio-ecological characteristics. Even though direct comparability may be hampered by differences in time period, environmental context, and institutional settings, some conclusions about factors that determine the socio-ecological characteristics of farming systems and their development over time are possible.

In some respects, the two farm systems were similar. Both were mixed farming communities that integrated cereal production with domesticated livestock. Area productivity, the amount of food produced per area of farmland, was similar. In 1830, 1 ha of farmland in Theyern produced about 2.9 GJ of food; in 1895, 1 ha in Finley Township, Kansas produced 4.6 GJ. Area productivity fluctuated with rainfall in Kansas, between highs of 7 GJ and lows of less than 1, but both farm systems were at the same order of magnitude.

The same was not true for labour productivity. Theyern produced about 9 GJ of food per farm labourer while those in Decatur County produced 200 GJ, 20 times their cross-Atlantic counterparts. The Theyern farm system coaxed food from the soil through intensive applications of labour, both human and animal. Maintaining area productivity meant high population densities of both people and livestock to sustain soil fertility. In Kansas, farmers needed (or invested) very little labour to produce large amounts of food. Consequently, population and livestock densities were lower, and declined between 1905 and 1940.

The two farm systems had different optimisation goals. The long history of subsistence farming, the tight social networks of village, manor and church in Theyern aimed not at peak production but at risk minimisation and long-term sustainability.[17] Theyern's greatest resource was a high labour supply, which it employed to maintain soil fertility. The tiny, scattered village fields, managed collectively, did not encourage peak production, but rather diversified holdings for all families and reduced the risk of catastrophic failures.

Finley Township, Kansas, followed a different strategy aimed at taking advantage of new commercial grain markets in the industrialising cities, new transportation opportunities as railroads spread across North America, and a rich endowment of fertile soils. Here were economies of scale with large, consolidated farms. Kansas was short of labour, but instead exploited its chief resource: abundant soil nitrogen and organic carbon, accumulated through millennia and mined in the first 50 years after settlement. The two systems were both efficient in their own way. Theyern supported the most people possible over long periods of time, usually producing enough food to keep them alive but rarely enough to make them wealthy. Finley

---

[17] For a discussion of risk minimisation strategies see McCloskey (1976).

Township maximized productivity, dramatically raising the standard of living for immigrants and their descendents. The nine socio-ecological indicators discussed in this study define and frame the two strategies.

But agricultural systems never remain static, and the social metabolic systems in both Austria and the Great Plains changed through the nineteenth and early twentieth centuries. In some ways their trajectories crossed paths. Austria as a whole moved steadily upward from relatively low yields and labour productivity in the early nineteenth century to higher production and increasing labour productivity by the century's end. Yields doubled over 75 years (Sieferle et al. 2006). Finley Township, for its part, began with high yields and labour productivity in 1895, and drifted downward over the decades, to a nadir in the 1930s. Kansas had reached a crisis of soil fertility by World War II. Thus through the nineteenth century and early twentieth century, the two farm systems moved in different directions.

After World War II, the application of fossil fuels to agricultural systems transformed both locations and began a transformation of productivity never seen before in the history of agriculture. The import of energy—diesel fuel for tractors, natural gas for nitrogen fertiliser, petroleum for pesticides, and gasoline and electricity for a multitude of farm machinery—presented a new solution to the ancient problem of maintaining soil fertility. With fossil fuels, Austrian farmers no longer needed to invest enormous amounts of labour in demanding livestock to provide power and manure. With fossil fuels, Kansas farmers could continue farming their depleted prairie soils by applying synthetic nitrogen every year as they watched crop yields rebound, match pioneer-era levels, and then exceed any previous production levels. It was not clear at the time, but the solution to the age-old problem of agricultural sustainability—soil maintenance—created a different one: unsustainable external energy inputs. But in the gap between the soil crisis and the oil crisis, Austrian and Kansas agricultural metabolism converged, with each moving toward high output commercial farming. By the end of the twentieth century, average cereal yields in Austria and Kansas were at a similar level and ranged between 6.5 and 7.5 t/ha (Sieferle et al. 2006; Kansas State Board of Agriculture. *Biennial Reports*. Topeka, Kans.).

Pioneer farms are rarely in equilibrium with their environment. By definition, settlers undertake the task of transforming their environment and inevitably undergo an adaptation process as they learn the limits of their new home, its climates, soils, plants, animals, and microorganisms. The Thir family liberated themselves from conservative Old World institutions and constrained Old World agro-ecosystems. But the farm they built on the Kansas frontier was unsustainable. The soil mining enterprise played out over several generations, between 1880 and 1930, but by then a soil fertility crisis loomed. It is no coincidence that the 1930s stand out in American memory as a time of rural crisis, population turmoil, and transformation in government agricultural policy. The drought, dust storms and global economic depression certainly contributed, but frontier farming in the Great Plains would have faced a dramatic change even without those forces. The application of fossil fuel energy saved the region for commercial agriculture, and allowed farmers to sustain their land-use practices for another 75 years.

In a broader global context, the stories of Old World and New World agriculture are intimately connected. Even as nitrogen flowed through local human, livestock, and cropland systems, broader flows across the Atlantic tethered these places to one another. The New World agricultural frontier provided novel opportunities for European farmers escaping subsistence lifestyles, and millions followed the Thirs and Demmers across the ocean. The grain and beef they produced flowed the other way, flooding Europe with cheap American food that undermined farm villages across the continent. It was that economic pressure on traditional European agriculture that forced innovation and led to Austria's steadily increasing yields in the late nineteenth century. Economists have argued that highly efficient New World farmers pressured backward and inefficient Old World people to improve agriculture (which some did) or to abandon it for industrialising cities (which most did) (Hayami and Ruttan 1985; Persson 1999; Williamson 2006; Van Zanden 1991; Koning 1994). This chapter points out an ecological component to the story that economists have missed or downplayed. One of the key reasons why New World farmers were so efficient and able to produce such stupendous crop surpluses for export between 1870 and 1930 was their endowment of stockpiled soil nutrients. For half a century, Great Plains farmers mined their rich soils and dumped those nutrients on the world market, disrupting risk-averse, long-lasting agricultural systems across the ocean. New World farming could not be sustained over the long term yet it undermined Old World systems that had been in place for centuries. Then, as the mid-twentieth century soil depletion crisis loomed, fossil fuel fertilisers and other high energy inputs rescued farmers, as the developed world substituted oil for soil.

## References

Allen, R. C. (2008). The nitrogen hypothesis and the English agricultural revolution. A biological analysis. *The Journal of Economic History, 68*, 182–210.

Antoni, M. (1992). *Nach Amerika… Materialien zur Landesausstellung in Güssing*. Eisenstadt: Pädagogisches Institut des Bundes für Burgenland.

Ayres, R. U., & Simonis, U. E. (1994). *Industrial metabolism: Restructuring for sustainable development*. Tokyo/New York/Paris: United Nations University Press.

Boserup, E. (1965). *The conditions of agricultural growth. The economics of agrarian change under population pressure*. Chicago: Aldine/Earthscan.

Boserup, E. (1981). *Population and technological change – A study of long-term trends*. Chicago: The University of Chicago Press.

Burke, I. C., Lauenroth, W. K., Cunfer, G. A., Barrett, J., Mosier, A., & Lowe, P. (2002). Nitrogen in the Central Grasslands Region of the United States. *BioScience, 52*, 813–823.

*Cadastral Schätzungs Elaborat der Steuergemeinde Theyern*, held at Landesarchiv St. Pölten.

Campbell, B. M. S., & Overton, M. (Eds.). (1991). *Land, labour and livestock: Historical studies in European agricultural productivity*. Manchester: Manchester University Press.

Cunfer, G. A. (2004). Manure matters on the great plains frontier. *The Journal of Interdisciplinary History, 34*, 539–567.

Cunfer, G., & Krausmann, F. (2009). Sustaining soil fertility. Agricultural practice in the old and new worlds. *Global Environment, 4*, 8–47.

Cusso, X., Garrabou, R., & Tello, E. (2006). Social metabolism in an agrarian region of Catalonia (Spain) in 1860 to 1870: Flows, energy balance and land use. *Ecological Economics, 58*, 49–65.

Decatur County Historical Book Committee. (1983). *Decatur County, Kansas*. Lubbock: Craftsman Printers, Inc.

Dujmovits, W. (1992). *Die Amerikawanderung der Burgenländer*. Pinkafeld: Desch-Drechsler.

Finanz-Ministerium, K. K. (Ed.). (1858). *Tafeln zur Statistik des Steuerwesens im österreichischen Kaiserstaate mit besonderer Berücksichtigung der directen Steuern und des Grundsteuerkatasters*. Wien.

Fischer-Kowalski, M. (1998). Society's metabolism. The intellectual history of material flow analysis, part I: 1860–1970. *Journal of Industrial Ecology, 2*, 61–78.

Fischer-Kowalski, M., & Haberl, H. (2007). *Socioecological transitions and global change: Trajectories of social metabolism and land use*. Cheltenham/Northhampton: Edward Elgar.

Fischer-Kowalski, M., & Weisz, H. (1999). Society as a hybrid between material and symbolic realms. Toward a theoretical framework of society-nature interaction. *Advances in Human Ecology, 8*, 215–251.

Frissel, M. J. (Ed.). (1978). *Cycling of mineral nutrients in agricultural ecosystems*. Amsterdam/Oxford/New York: Elsevier.

Guzman Casado, G. I., & Gonzalez de Molina, M. (2009). Preindustrial agriculture versus organic agriculture: The land cost of sustainability. *Land Use Policy, 26*, 502–510.

Haberl, H., Winiwarter, V., Andersson, K., Ayres, R. U., Boone, C. G., Castillio, A., Cunfer, G., Fischer-Kowalski, M., Freudenburg, W. R., Furman, E., Kaufmann, R., Krausmann, F., Langthaler, E., Lotze-Campen, H., Mirtl, M., Redman, C. A., Reenberg, A., Wardell, A. D., Warr, B., & Zechmeister, H. (2006). From LTER to LTSER: Conceptualizing the socio-economic dimension of long-term socio-ecological research. *Ecology and Society, 11*, 13. (Online), www.ecologyandsociety.org/vol11/iss2/art13/

Hass, H. J., Evans, C. E., & Miles, E. F. (1957). *Nitrogen and carbon changes in Great Plains soils as influenced by cropping and soil treatments*. Washington, DC: GPO.

Hayami, Y., & Ruttan, V. W. (1985). *Agricultural development. An international perspective*. Baltimore: John Hopkins University Press.

Herndon Union Cemetery. *Herndon Union Cemetery Records, Rawlins County, Kansas*.

Hurt, R. D. (1987). *Indian agriculture in America: Prehistory to the present*. Lawrence: University Press of Kansas.

ICPSR. (2011). *The Great Plains Population and Environment project*. Inter-University Consortium for Political and Social Research. www.icpsr.umich.edu/plains. Accessed 28 June 2011.

Kansas GenWeb. (2009). *Homestead records*. http://skyways.lib.ks.us/genweb/decatur/Land%20 Records/finley_homesteading.htm. Accessed 16 Feb 2009.

Kansas State Board of Agriculture. (1885, 1895, 1905, 1915, 1920, 1925, 1930, 1935, 1940). *Population census manuscripts*, Decatur County, Kansas.

Karl, T. R., Williams, C. N. Jr., Quinlan, F. T., & Boden, T. A. (1990). *United States Historical Climatology Network (HCN) serial temperature and precipitation data. No. 3404* (pp. 1–389). Oak Ridge: Carbon Dioxide Information and Analysis Center; Oak Ridge National Laboratory.

Koning, N. (1994). *The failure of agrarian capitalism: Agrarian politics in the U.K., Germany, Netherlands, and the U.S.A., 1846–1919*. New York: Routledge.

Krausmann, F. (2004). Milk, manure and muscular power. Livestock and the industrialization of agriculture. *Human Ecology, 32*, 735–773.

Krausmann, F. (2008). *Land use and socio-economic metabolism in pre-industrial agricultural systems: Four nineteenth-century Austrain villages in comparison* (Social Ecology Working Paper, 72). Vienna: IFF Social Ecology.

Lego, K. (1968). *Geschichte des österreichischen Grundkatasters*. Wien: Bundesamt für Eich- und Vermessungswesen.

Liu, J. G., Dietz, T., Carpenter, S. R., Folke, C., Alberti, M., Redman, C. L., Schneider, S. H., Ostrom, E., Pell, A. N., Lubchenco, J., Taylor, W. W., Ouyang, Z. Y., Deadman, P., Kratz, T., & Provencher, W. (2007). Coupled human and natural systems. *Ambio, 36*, 639–649.

Loomis, R. S. (1978). Ecological Dimensions of Medieval Agrarian Systems. An Ecologist Responds. *Agricultural History, 52*, 478–484.

Loomis, R. S. (1984). Traditional agriculture in America. *Annual Review of Ecology and Systematics, 15*, 449–478.

Loomis, R. S., & Connor, D. J. (1992). *Crop ecology: Productivity and management in agricultural systems*. Cambridge: Cambridge University Press.

Marull, J., Pino, J., & Tello, E. (2008). The loss of landscape efficiency: An ecological analysis of land use changes in western Mediterranean agriculture (Valles County, Catalonia, 1853–2004). *Global Environment, 2*, 112–150.

McCloskey, D. (2001). English open fields as behavior toward risk. In Deirdre N. McCloskey and Stephen Ziliak (Eds.), *Measurement and meaning in economics: The essential Dierdre McCloskey* (pp. 17–63). Cheltenham: E. Elgar.

Moritsch, A. (1972). Der Franziszeische Grundsteuerkataster Quelle für die Wirtschaftsgeschichte und historische Volkskunde. *East European Quarterly, 3*, 438–448.

Parton, W. J., Gutmann, M. P., Williams, S. A., Easter, M., & Ojima, D. (2005). Ecological impact of historical land-use patterns in the Great Plains: A methodological assessment. *Ecological Applications, 15*, 1915–1928.

Persson, K. G. (1999). *Grain markets in Europe, 1500–1900: Integration and deregulation*. Cambridge: Cambridge University Press.

Prickler, L. (2003). Ebene im Osten: Der Seewinkel im Bezirk Neusiedl am See. In E. Bruckmüller, E. Hanisch, & R. Sandgruber (Eds.), *Geschichte der österreichischen Land- und Forstwirtschaft im 20. Jahrhundert. Regionen, Betriebe, Menschen* (pp. 741–794). Wien: Ueberreuter.

Projektgruppe Umweltgeschichte. (1997). *Historische und ökologische Prozesse in einer Kulturlandschaft*. Studie im Auftrag des BMWVK, Endbericht. Wien.

Projektgruppe Umweltgeschichte. (1999). *Kulturlandschaftsforschung: Historische Entwicklung von Wechselwirkungen zwischen Gesellschaft und Natur*. Wien: CD-ROM, Bundesministerium für Wissenschaft und Verkehr.

Sandgruber, R. (1978). Die Agrarrevolution in Österreich. Ertragssteigerung und Kommerzialisierung der landwirtschaftlichen Produktion im 18. und 19. Jahrhundert. In A. Hoffmann (Ed.), *Österreich-Ungarn als Agrarstaat. Wirtschaftliches Wachstum und Agrarverhältnisse in Österreich im 19. Jahrhundert* (pp. 195–271). Wien: Verlag für Geschichte und Politik.

Sandgruber, R. (1979). Der Franziszeische Kataster und die dazugehörigen Steuerschätzungsoperate als wirtschafts- und sozialhistorische Quellen. *Mitteilungen aus dem niederösterreichischen Landesarchiv, 3*, 16–28.

Schaschl, E. (2007). *Rekonstruktion der Arbeitszeit in der Landwirtschaft im 19. Jahrhundert am Beispiel von Theyern in Niederösterreich* (Social Ecology Working Paper, 96, pp. 1–174). Vienna: IFF Soziale Ökologie.

Schüle, H. (1989). *Raum-zeitliche Modelle – ein neuer methodischer Ansatz in der Agrargeschichte. Das Beispiel der bernischen Viehwirtschaft als Träger und Indikator der Agrarmodernisierung 1790 – 1915*. Bern: Lizensiatsarbeit, historisches Institut der Universität Bern.

Shiel, R. (2006a). An introduction to soil nutrient flows. In J. R. McNeill & V. Winiwarter (Eds.), *Soils and societies: Perspectives from environmental history* (pp. 7–12). Isle of Harris: White Horse Press.

Shiel, R. (2006b). Nutrient flows in pre-modern agriculture in Europe. In J. R. McNeill & V. Winiwarter (Eds.), *Soils and societies: Perspectives from environmental history* (pp. 216–242). Isle of Harris: White Horse Press.

Sieferle, R. P., Krausmann, F., Schandl, H., & Winiwarter, V. (2006). *Das Ende der Fläche. Zum gesellschaftlichen Stoffwechsel der Industrialisierung*. Köln: Böhlau.

Sonnlechner, C. (2001). Umweltgeschichte und Siedlungsgeschichte. Das Waldviertel. *Zeitschrift für Heimat- und Regionalkunde des Waldviertels und der Wachau, 50*, 361–382.

*Standard Atlas of Decatur County, Kansas*. (1905). Chicago: George A. Ogle & Co.
Stevenson, F. J. (Ed.). (1982). *Nitrogen in agricultural soils* (Agronomy Series No. 22). Madison: American Society of Agronomy, Crop Science Society of American, and Soil Science Society of America.
Sylvester, K. M., Leonard, S. H., Gutmann, M. P., & Cunfer, G. (2006). Demography and environment in grassland settlement: Using linked longitudinal and cross-sectional data to explore household and agricultural systems. *History and Computing, 14*, 31–60.
U.S. Population Census. (1880, 1900, 1910, 1920, 1930). *U.S. Population Census manuscript schedules, Decatur County, Kansas*.
Van Zanden, J. L. (1991). The first green revolution: The growth of production and productivity in European agriculture, 1870–1914. *The Economic History Review, 44*, 215–239.
Wedel, W. R. (1978). The prehistoric plains. In J. D. Jennings (Ed.), *Ancient Native Americans*. San Francisco: W.H. Freeman and Company.
Williamson, J. G. (2006). *Globalization and the poor periphery before 1950*. Cambridge: MIT Press.
Winiwarter, V., & Sonnlechner, C. (2001). *Der soziale Metabolismus der vorindustriellen Landwirtschaft in Europa*. Stuttgart: Breuninger Stiftung.
Worster, D. (1990). Transformations of the earth: Toward an agroecological perspective in history. *The Journal of American History, 76*, 1087–1106.

# Chapter 13
# How Material and Energy Flows Change Socio-natural Arrangements: The Transformation of Agriculture in the Eisenwurzen Region, 1860–2000

Simone Gingrich, Martin Schmid, Markus Gradwohl, and Fridolin Krausmann

**Abstract** This contribution presents empirical results on changes in socio-ecological metabolism and land use in agriculture in two regions in and around the Upper Austrian Eisenwurzen LTSER Platform from the late nineteenth century to the turn of the twenty-first century. Based on local and regional statistical records, changes in the agricultural production systems are traced and it is shown how industrialisation (marked e.g. by a strong increase in use of machinery and output of yields) shaped two very distinct patterns of change in two biogeographically different regions. While this investigation contributes to two major themes of LTSER, i.e. socio-ecological metabolism and land use, the systemic and quantitative perspective does not per se address LTSER's third major theme, governance and decision making. We suggest that concepts from environmental history have the potential to fill this gap. Using the concept of socio-natural sites, we explore how the systems perspective can benefit from an actors' perspective along three examples which could merit empirical research.

**Keywords** Socio-ecological Metabolism • Land use change • Agriculture • Industrialisation • Socio-natural arrangements • Eisenwurzen region • Local case study

---

S. Gingrich, Ph.D. (✉) • M. Schmid, Ph.D. • F. Krausmann, Ph.D.
Institute of Social Ecology Vienna (SEC), Alpen-Adria Universitaet Klagenfurt,
Wien, Graz, Schottenfeldgasse 29/5, Vienna 1070, Austria
e-mail: simone.gingrich@aau.at; fridolin.krausmann@aau.at; martin.schmid@aau.at

M. Gradwohl, M.Sc.
Centre for the Study of Agriculture, Food and Environment, University of Otago,
Dunedin, New Zealand
e-mail: grama999@student.otago.ac.nz

## 13.1 Introduction

Long-term socio-ecological research (LTSER) aims at an integrated understanding of temporal patterns in society-nature interaction with the goal of providing environmental policy with relevant information to foster sustainable regional development. It requires interdisciplinary cooperation between scholars from the natural sciences and social sciences (Redman et al. 2004), as well as the humanities (Haberl et al. 2006). Three major themes in LTSER include (1) socio-ecological metabolism, (2) land use and landscapes, and (3) governance and decision making (Haberl et al. 2006). This chapter empirically addresses two of these three themes, i.e. socio-ecological metabolism and land use and landscapes, namely in two case study regions situated in and around Austria's Eisenwurzen LTSER Platform. We trace changes in agricultural systems from the late nineteenth to the turn of the twenty-first century, using statistical publications on land use, agricultural production, livestock and other agricultural structural parameters, and conducting energy flow analyses for selected years. Such an analysis provides useful information on the biophysical functioning of agricultural systems and its change over time. In order to generate results which contribute to LTSER's third theme, governance and decision making, we complement this perspective by considerations at smaller scales. We address the role of actors, their agency, perception and motives in making changes to the land in their care. While we do not present empirical results in this respect, we aim to establish the conceptual basis for further research, combining a birds-eye or systems perspective with a close-up view on human actors.

Recently, some research has been carried out concerned with the role of actors in the current functioning of socio-ecological systems. Many of these studies involve local stakeholders in participatory processes (e.g. Singh et al. 2010a), others apply participatory modelling to understand interrelations between different actors and the local or regional ecosystems (Gaube et al. 2009; Gaube and Haberl, Chap. 3 in this volume). Rural historians try to bridge the gap between systemic and actors-centred approaches by using concepts such as 'farming styles' from rural sociology (e.g. Langthaler 2006). To our knowledge, little effort has been made so far to integrate an actor's perspective into LTSER (Peterseil et al., Chap. 19 in this volume). We suggest that environmental history provides fruitful potential in this endeavour.

For this purpose, we adopt the concept of socio-natural sites which has recently been developed in environmental history (Winiwarter and Schmid 2008; Schmid 2009; Winiwarter et al., Chap. 5 in this volume). In short, this concept offers a conceptual basis to understand and to explain social, cultural and ecological changes over long periods of time in specific places. It focuses on 'practices' (what humans did and how) and 'arrangements', the material requirements (e.g. machines) and outcomes (e.g. buildings) of human practices. Both practices and arrangements are seen as socio-natural hybrids; they are social and natural at the same time. Thus in line with LTSER, the concept points to the fact that phenomena we usually regard as 'natural' are also a result of societal decisions and human labour – one could think of agricultural crops which are both the result of biological evolution and the product of breeding or genetic engineering – and what we conventionally regard as purely 'social' is also shaped by natural processes – patterns of time use on a farm linked to seasons may serve as an example. A socio-natural site is defined as the nexus of arrangements with practices; if one changes, the other will change too and the site transforms (for a more detailed introduction to the concept see Winiwarter et al., Chap. 5 in this volume).

Because the concept of socio-natural sites focuses on arrangements and practices and with the latter on an important and widespread term in the complex landscape of contemporary social and cultural theories (Reckwitz 2002), it can be linked rather easily to a variety of branches of research beyond the natural sciences. Generally, LTSER should experiment with different theoretical frameworks to foster its interdisciplinary integration and to bridge the great divide between natural sciences, social sciences and humanities. This chapter is one such experiment; we will reconstruct flows of material and energy through agricultural systems in the nineteenth and twentieth centuries and ask how the fundamental transformation we observed in the agro-ecological systems may have affected the socio-natural sites involved.

The chapter is structured as follows: after an introduction to the case study regions and the used data and methods, we present the empirical findings in three time slots, showing the state of the agricultural systems in biophysical terms at specific points in time. In the conclusions we return to the concept of socio-natural sites and re-interpret our empirical results within that framework. While our empirical results allow us to grasp historical changes in agricultural arrangements, a specific focus on actors and their changing forms of agriculture (practices) requires further interdisciplinary empirical work, including history and anthropology among other disciplines.

## 13.2 Case Study Regions

This study presents two key features of LTSER on agricultural systems, data on the socio-ecological metabolism and land use change for two regions in (and bordering, respectively) the Austrian Eisenwurzen LTSER Platform (see Peterseil et al., Chap. 19 in this volume): Sankt Florian and Grünburg (Fig. 13.1). The focus of analysis is the biophysical functioning of the regional agricultural system and its change over time.[1] The regions were chosen on the grounds of exceptional data availability for the late nineteenth century (Lorenz 1866, see below). They are situated only 30 km from each other and represent two different Central European cultural landscapes. Each region encompasses 6,000 ha or ten villages.

---

[1] The empirical results presented in this contribution are based largely on two academic theses concerned with (1) the biophysical functioning of pre-industrial agricultural systems in the two regions (Gradwohl 2004), based on a particularly detailed source of the late nineteenth century (Lorenz 1866), and (2) the temporal patterns of change in these regions during industrialisation, i.e. from the late nineteenth century until the turn of the twenty-first century (Gingrich 2004), based on various statistical publications.

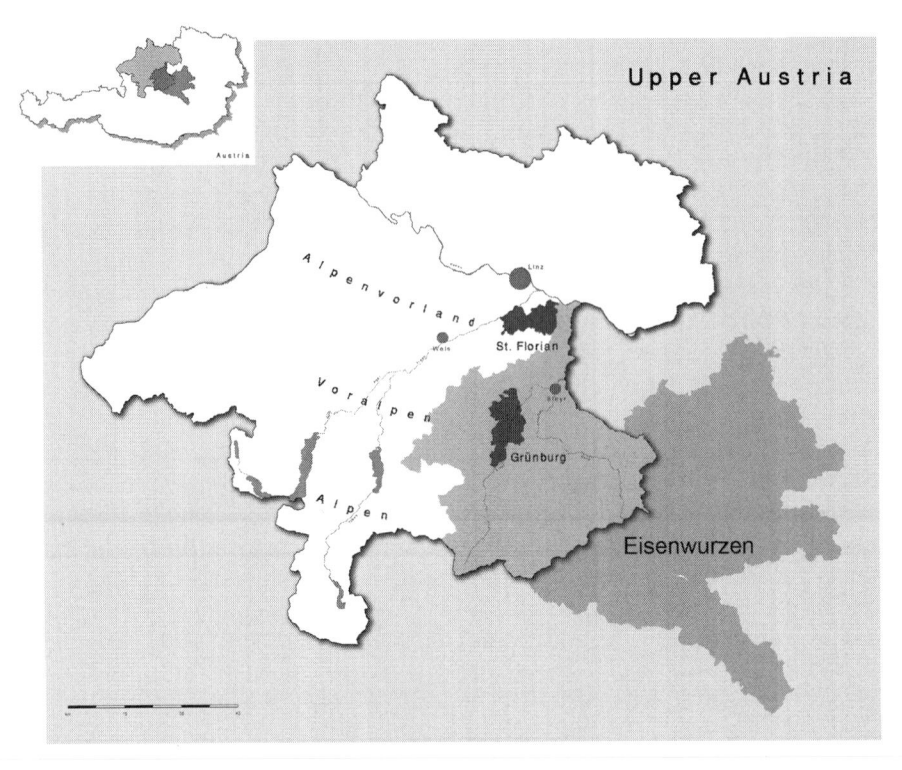

**Fig. 13.1** Geographic position of the two case study regions

Sankt Florian is located in the fertile lowland of central Upper Austria, in the geographical region known as Alpenvorland (alpine foreland). Soils and climate are favourable for crop agriculture and hills covered with fields are characteristic for this region. Sankt Florian has traditionally served as an urban hinterland, providing the nearby cities of Linz and Wels with agricultural products, and, since the opening of the Westbahn railway in 1860, even larger urban centres further away, such as Salzburg and Vienna. In recent times Sankt Florian has been strongly affected by suburbanisation (see also Sandgruber 2003).

Grünburg, in contrast, is more remote and rather inhomogeneous in terms of topography and soils, as it cuts across the border of two topographical regions. In this heterogeneity, it is representative of the pre-alpine landscape (Voralpen). Its northern part extends to the Traun-Enns-Platte similar to Sankt Florian, but the southern part, separated by a thin belt of sandstone hills, reaches into the Limestone Alps and includes peaks up to 900 m a.s.l.. Livestock farming and grassland cultures have dominated agriculture here for centuries. Grünburg's knife manufacturing works were involved in the Eisenwurzen region's iron industry (Schuh and Sieghartsleitner 1997; Landeskulturdirektion Oberösterreich 1998), and since the final collapse of this sector in the late nineteenth century,

Grünburg has suffered from marginalisation. Recently the region has become more and more engaged in tourism, fostered by the nearby Kalkalpen (Limestone Alps) National Park.

## 13.3 The Empirical Basis: Data and Methods

The data used are mainly quantitative and come from different statistical publications from the late nineteenth century onwards. The earliest source is a very detailed account of statistics describing agricultural production in Sankt Florian and Grünburg in 1864 (Lorenz 1866). Lorenz, who was interested in the development of a statistical monitoring framework for agriculture, has chosen these regions (around 6,000 ha each) as model areas for landscapes in the province of Upper Austria. The boundaries of the regions Lorenz studied are not identical with administrative borders, neither in his nor in later times. Statistical data for the period after 1864 are available only for administrative units. For Sankt Florian we chose three municipalities (overall area: 8,400 ha), and in Grünburg four municipalities (11,500 ha) to represent the region. Both areas contain landscape types and topography in roughly the same proportions as the regions represented in Lorenz' study of the 1860s. However, due to these distortions in the reference systems, we compare only relative numbers throughout the time series, generally dividing total numbers by total area or agricultural area (arable land plus grassland).

Data were compiled for land use and other agricultural structure parameters, such as population and agricultural work force, livestock, farm size, agricultural production and agricultural machinery. Table 13.1 presents the statistical sources used.

Energy flow analysis provides a tool to understand the dimensions of biophysical exchange between a socioeconomic system (the agricultural production system), its natural environment and other social systems. We quantified energy flows (Haberl 2001; Schandl et al. 2002) at three points in time: 1864, 1950 and 2000. The framework has been used for historical analyses in agricultural production systems (Cusso et al. 2006; Winiwarter and Sonnlechner 2000; Krausmann 2004; Sieferle et al. 2006). It allows for an integrated analysis of the biophysical exchange processes between the agro-ecosystem of a region and the agriculturally active social system, defined in our case as the regional 'agricultural production system' (see Fig. 13.2). The 'agricultural production system' comprises all persons involved in agriculture, as well as agricultural machinery and livestock. The 'domestic environment' refers to the territory used for agriculture. Other socioeconomic systems include households and non-agricultural sectors in the region, as well as all socioeconomic systems outside of the region (cf. Singh et al. 2010b).

Fundamentally, the assessment of energy flows relies on empirical data from statistical publications as described above: We converted mass values reported in statistics into energy units using conversion factors provided by Haberl (1995). Some flows not reported in statistics were estimated based on model assumptions. In Domestic Extraction (DE) we comprise biomass harvest and grazing.

**Table 13.1** Data sources

| Data | Year(s) of reference | Data source |
|---|---|---|
| Population, agricultural population | 1864 | Lorenz (1866) |
| | 1934 | Bundesamt für Statistik (1935) |
| | 1951 | Österreichisches statistisches Zentralamt (1952b) |
| | 1949–2000 | ISIS-Database[a] |
| Land use | 1864 | Lorenz (1866) |
| | 1878 | Foltz (1878) |
| | 1900 | K.k.statistische Zentralkomission (1903) |
| | 1949–2000 | ISIS-Database |
| Plant production | 1864 | Lorenz (1866) |
| | 1949–2000 | ISIS-Database (3-year averages) |
| Livestock, slaughter rate, animal production | 1864 | Lorenz (1866) |
| | 1950 | Österreichisches statistisches Zentralamt (1950) |
| | 1951 | Österreichisches statistisches Zentralamt (1952a) |
| | 1960–2000 | ISIS-Database |
| Farm numbers, farm size | 1878 | Foltz (1878) |
| | 2000 | ISIS-Database |
| Agricultural machinery | 1953 | Österreichisches statistisches Zentralamt (1954) |
| | 1960–2000 | ISIS-Database |

[a]ISIS-database provided by Statistik Austria (www.statistikaustria.at)

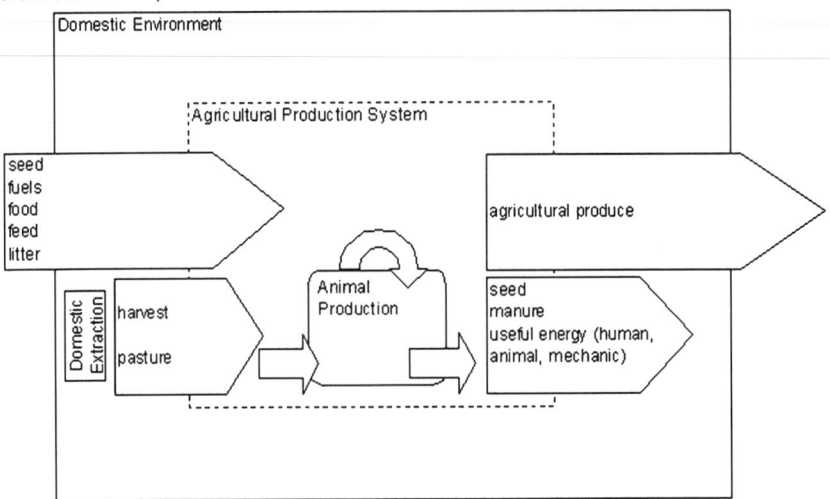

**Fig. 13.2** Conceptual scheme of the applied system boundaries and the energy flows calculated in this study

For grazing, we assumed that in 1864, all agricultural land except cropland was used for grazing (Krausmann 2004). In the twentieth century, only areas designated as pastures were considered as actual grazing areas. We applied typical productivities to grazing areas (Krausmann 2008).

Imports were calculated as the difference between DE and domestic demand for different fractions of biomass. For the twentieth century, we assessed increasing market integration of agricultural production on the basis of expert interviews (K. Grammer, Interview with the mayor of Grünburg, Grünburg, personal communication, April 14, 2004; G. Seiser, Interview with a social anthropologist specialist on Upper Austrian agriculture, personal communication, August 4, 2004; K. Zarzer, Interview with a specialist on grasslands and fodder cropping, Linz: Upper Austrian Chamber of Agriculture, Linz, personal communication, April 14, 2004). Feed demand was calculated using demand factors based on livestock numbers and live weight (Löhr 1952; Hohenecker 1980). Feed imports were assumed to have increased from the difference between demand and supply in 1864 and 1950 to one-third of feed demand in 2000 (K. Grammer, Interview with the mayor of Grünburg, Grünburg, personal communication, April 14, 2004; G. Seiser, Interview with a social anthropologist specialist on Upper Austrian agriculture, personal communication, August 4, 2004; K. Zarzer, Interview with a specialist on grasslands and fodder cropping, Linz: Upper Austrian Chamber of Agriculture, Linz, personal communication, April 14, 2004). Litter demand was accounted for according to litter demand factors from literature (Beer et al. 1990; Dissemond and Zauchinger 1994). Food imports were calculated from the number of agricultural work force, work hours and percentage of food provided within the agricultural production system (Darge 2002). We assumed 20% of seed demand to be imported in 1950 (K. Grammer, Interview with the mayor of Grünburg, Grünburg, personal communication, April 14, 2004; G. Seiser, Interview with a social anthropologist specialist on Upper Austrian agriculture, personal communication, August 4, 2004; K. Zarzer, Interview with a specialist on grasslands and fodder cropping, Linz: Upper Austrian Chamber of Agriculture, Linz, personal communication, April 14, 2004), while in 2000 all seed demand was categorised as imports. All fossil fuel use in tractors was considered as import and was calculated on the basis of energy use per hour of operation (Darge 2002; Leach 1976). All agricultural products which are not used as food, feed, litter, or seed within the agricultural production system were considered as exports.[2]

In outputs to the domestic environment, we considered useful energy, seed and manure. In the calculation of muscular power we relied on the number of individuals, average "working hours" and average power (Darge 2002; Smil 2001; Löhr 1952). The useful energy delivered by tractors was estimated by multiplying the demand for fossil fuels by an average efficiency factor of 30%. Output of seed was calculated as a fraction of the vegetal produce from agricultural output (Löhr 1983). We assumed that all manure was dissipated to agricultural areas. Manure production was calculated as percentage of live weight (Vetter and Steffens 1986) and converted to gross calorific energy values using factors by Darge (2002). Mineral fertiliser was not considered, since it does not represent a direct energy flow.

---

[2] The calculation of exports excludes losses during storage, as we assume that most of these would occur outside the agricultural production system during further processing and transport. Similarly losses of agricultural products between harvest and local consumption (e.g. during storage on farm) were not considered in the calculation of imports, assuming the products most imported were not significantly affected.

## 13.4 Empirical Results: Biophysical Changes of the Agricultural System

Let us now turn to the empirical results of our analysis. Since the aim of this article is to complement such a biophysical (or systems) perspective with a humanities (or actors) perspective, we will roughly organise our results along key terms of the concept of socio-natural sites. We select 3 years of reference (1864, c. 1950 and c. 2000) and discuss for each time slot those arrangements and practices about which we can infer from our empirical data. Compilations of the presented data are given in Tables 13.2 and 13.3.

Table 13.2 Structural parameters characterising the case studies: Number of population, farms, and livestock; land availability and use; average yields and productivity

| Parameter | [unit] | Sankt Florian 1864 | 1950 | 2000 | Grünburg 1864 | 1950 | 2000 |
|---|---|---|---|---|---|---|---|
| Population | [cap] | 4,718 | 14,692 | 26,344 | 5,214 | 8,865 | 9,677 |
| Population density | [cap/km$^2$] | 78 | 175 | 313 | 85 | 77 | 84 |
| Agricultural labour force | [%] | 67 | 12 | 3 | 53 | 26 | 13 |
| average farm size | [ha$_{agr}$/farm] | 3 | n.a. | 12 | 4 | n.a. | 10 |
| Land use | | | | | | | |
| Total area | [ha] | 6,014 | 8,410 | 8,410 | 6,139 | 11,521 | 11,510 |
| Arable land | [%] | 60 | 53 | 61 | 33 | 26 | 23 |
| Gardens; permanent cultures | [%] | 2 | 1 | 1 | 1 | 0 | 1 |
| Grassland | [%] | 16 | 24 | 4 | 36 | 41 | 32 |
| Forests | [%] | 15 | 17 | 14 | 27 | 20 | 23 |
| Other areas | [%] | 7 | 5 | 20 | 3 | 13 | 21 |
| Livestock | | | | | | | |
| Livestock density | [LSU/ha$_{agr}$][a] | 62 | 106 | 104 | 46 | 122 | 242 |
| Horses | [%] | 23% | 0% | 0% | 5% | 1% | 0% |
| Cattle | [%] | 61% | 43% | 3% | 83% | 70% | 53% |
| Pigs | [%] | 12% | 32% | 14% | 8% | 26% | 45% |
| Poultry | [%] | 1% | 24% | 83% | 0% | 2% | 0% |
| Others | [%] | 3% | 0% | 0% | 3% | 0% | 1% |
| Tractors | [tractors/km$^2$] | – | 2.7 | 6.2 | – | 0.4 | 7.2 |
| Productivity | | | | | | | |
| Grain yield (3 year average) | [t$_{FW}$/ha] | 1.6 | 1.7 | 6.5 | 1.1 | 1.3 | 5.5 |
| Animal production | [t$_{FW}$/ha$_{agr}$] | 1.1 | 0.7 | 0.6 | 0.8 | 0.8 | 2.2 |

Source: Authors' own calculations, see text
[a]Livestock Units (LSU), corresponding to 500 kg live weight

**Table 13.3** Biomass flows in the agricultural sector of the studied land use systems: Domestic extraction (DE), Imports (Im), Direct Input (DI), Exports (Ex) and Domestic Consumption (DC) of biomass; per capita (GJ/cap) and per agricultural area (GJ/ha$_{agr}$)

| Parameter | [unit] | Sankt Florian 1864 | 1950 | 1999 | Grünburg 1864 | 1950 | 1999 |
|---|---|---|---|---|---|---|---|
| Domestic extraction | [GJ/cap] | 73 | 28 | 34 | 38 | 49 | 71 |
| Imports | [GJ/cap] | 1 | 7 | 8 | 2 | 10 | 53 |
| Direct input | [GJ/cap] | 74 | 35 | 42 | 40 | 60 | 124 |
| Exports | [GJ/cap] | 5 | 8 | 26 | 1 | 6 | 19 |
| Domestic consumption | [GJ/cap] | 70 | 27 | 16 | 39 | 54 | 105 |
| Domestic extraction | [GJ/ha$_{agr}$] | 74 | 65 | 150 | 47 | 58 | 105 |
| Imports | [GJ/ha$_{agr}$] | 1 | 15 | 35 | 2 | 12 | 77 |
| Direct input | [GJ/ha$_{agr}$] | 75 | 80 | 185 | 49 | 70 | 182 |
| Exports | [GJ/ha$_{agr}$] | 5 | 18 | 109 | 1 | 7 | 28 |
| Domestic consumption | [GJ/ha$_{agr}$] | 70 | 62 | 76 | 48 | 63 | 154 |
| Imports | [% of DE] | 1 | 24 | 23 | 5 | 21 | 74 |
| Exports | [% of DE] | 6 | 28 | 73 | 3 | 11 | 27 |

Source: Authors' own calculations, see text

## 13.4.1 Labour Intensive Practices in Fairly Closed Energy Cycles: The Late Nineteenth Century

In 1864, farmers in Sankt Florian and Grünburg practiced a relatively intensive form of agriculture, still limited largely by the restrictions of a solar energy-based or agrarian mode of production (Krausmann et al. 2008; see also Cunfer and Krausmann, Chap. 12 in this volume). Compared to the Austrian average at that time (42 cap/km$^2$), population density was relatively high in both regions with 78 cap/km$^2$ in Sankt Florian and even 85 cap/km$^2$ in the then still iron processing region of Grünburg, where just over half of the population worked in agriculture (in Sankt Florian this share was at around two-thirds). Agriculture was organised on a small scale and based mainly on the resources available on the land: Both regions were home to around 1,200 farms with an average agricultural area per farm of around 5 ha. Signs of agricultural industrialisation were first visible in Sankt Florian in the late nineteenth century (Sandgruber 2003; Hoffmann 1974): Steam-driven threshing machines and steam ploughs supported farmers in their work (Lorenz 1866). But machines had by no means already replaced muscle power. Horses still served as the most important draught animals in Sankt Florian. In Grünburg, on the other hand, draught power was provided mainly by oxen, which were better suited to the hilly terrain.

Land use in both regions was relatively similar, since both Sankt Florian and Grünburg were characterised by mixed farming, but with a different distribution of land use types (Fig. 13.3a, b). In Sankt Florian, 60% of the total area was used as cropland. Grassland and forests each made up only about 15% of the total. In Grünburg, the area was divided almost equally between arable land (33%), grassland (33%) and forests (26%). The composition of livestock in the two regions also shows considerable differences, although in both regions draught animals represented an important share of the livestock. In Sankt Florian, cattle and pigs were raised in similar numbers (0.55 cattle/$ha_{agr}$ and 0.61 pigs/$ha_{agr}$). In Grünburg, more cattle (0.61 cattle/$ha_{agr}$) and less pigs (0.30 pigs/$ha_{agr}$) were kept. The higher number of cattle correlates with the higher share of grassland.

Looking at the energy flows of the two regional agricultural production systems, we find that both regions operated in fairly closed cycles. Local plant production fed much of the agricultural population and livestock, and outputs to nature in terms of useful work, manure and seed were relatively small (2 GJ/$ha_{agr}$/year in both regions). But we also find considerable differences: Domestic Extraction (DE) of biomass per unit of agricultural area in Sankt Florian was about 50% higher than in Grünburg (74 GJ/$ha_{agr}$/year and 47 GJ/$ha_{agr}$/year, respectively), while the share of DE used as feed was much higher in Grünburg (87%, compared to 69% in Sankt Florian). Thus in Sankt Florian a considerably higher amount of biomass (mainly cereals) was available as output to other socioeconomic systems (23 GJ/$ha_{agr}$/year in Sankt Florian versus 6 GJ/$ha_{agr}$/year in Grünburg).

Despite similar population density and similar farm size, agriculture in Sankt Florian was much more productive in terms of surplus production available to the non-agricultural population in the late nineteenth century. The differences can be related to both more intensive land use (more agricultural workers, more machinery in use), but also to the different basic biogeographical conditions of the two regions.

### 13.4.2 A Time of Crisis: Disruptions After World War II

Around 1950, agricultural production had been clearly affected by war. The share of cropland had decreased considerably in both regions, supposedly a direct effect of the war when general manpower was lacking. Since 1864, population density in Sankt Florian had risen rapidly due to suburbanisation processes of the growing industrial city of Linz. In Grünburg in contrast, population density had decreased with the decline of iron industries since the late nineteenth century (Kropf 1997; Sandgruber 1997). Hence, individual farm size had increased while their total number had decreased in both regions. Mechanisation of agriculture was clearly underway by World War II, though to different extents: In 1953, while Sankt Florian's tractor density was 2.7 tractors/$km^2$, Grünburg showed a far lesser degree of agricultural mechanisation (0.4 tractors/$km^2$).

The biophysical organisation of the two agricultural production systems had not changed significantly until immediately after World War II. In both regions, cereal yields went up slightly between 1864 and 1950: from 1.6 to 1.7 t/ha/year in Sankt Florian and from 1.1 to 1.3 t/ha/year in Grünburg. In Sankt Florian, overall area productivity dropped to 65 GJ/$ha_{agr}$/year, most likely as a result of lacking infrastructure and work force, resulting in the above-mentioned reduction of arable land due to war. In Grünburg, area productivity rose to 58 GJ/$ha_{agr}$/year. In 1950

Grünburg's livestock, which had grown in numbers, could not be fed sufficiently by local production, but (in theory) relied on imports; more likely, these animals experienced temporary malnourishment. In both regions, the supply of livestock with feed and litter made up for almost 90% of biomass import. Possibly, the assumption that feed and litter consumption equalled demand in the post-war years implies an overestimation of feed and litter imports. As in 1864, more output to other socioeconomic systems was available in Sankt Florian (18 GJ/ha$_{agr}$/year) than in Grünburg (7 GJ/ha$_{agr}$/year). However, outputs in Sankt Florian were lower after World War II than they had been in the late nineteenth century.

### 13.4.3 Specialisation and Disintegration: Two Paths of the Fundamental Transformation After World War II

Between 1950 and 2000, Austrian agriculture underwent a very rapid industrialisation process (Bruckmüller et al. 2002, 2004), which is also visible in Sankt Florian and Grünburg, albeit in very different ways. In Sankt Florian, population density almost doubled during this period as a result of suburbanisation. Only 3% of all inhabitants still worked in agriculture in 2000. In Grünburg, where population had remained stable, 13% of the population were classified as agricultural in 2000, a relatively high value compared to the average of 8% in the province of Upper Austria. The average farm size had grown to 12 ha in Sankt Florian and 10 ha in Grünburg.

The agro-ecosystems looked fundamentally different around 2000: While the share of arable land stayed constant in Sankt Florian at around 60%, grasslands decreased significantly. In Grünburg arable land declined continuously, but grasslands stayed constant. In both regions, 'other areas' went up significantly in the late twentieth century (Fig. 13.3a, b). In Sankt Florian, the high increase in population and also in buildings hints at growing settlement areas. In Grünburg on the other hand, the cadastral data indicate that most 'other areas' are likely to be forests managed by non-residents, which are not distinguished in communal land use statistics. Since 1950, average cereal yields more than tripled to 6.5 t/ha/year in Sankt Florian and 5.5 t/ha/year in Grünburg. The decrease in relative difference of cereal yields since the nineteenth century indicates that in Grünburg, lower-yielding plots of cropland were taken out of use. With the introduction of industrial feed, industrial fertiliser and tractors to replace draught animals, the livestock composition in the two regions changed dramatically from the 1950s. In Sankt Florian, cattle stocks decreased to only 15% of the late nineteenth century density, while pig numbers more than doubled and chicken stocks reached extremely high densities. Nevertheless, the overall livestock density stayed at a similar level. By contrast, in Grünburg, the overall density of livestock tripled from 1864 to 2000. Cattle stocks doubled between 1864 and 2000, and pig stocks even increased tenfold. Most of this increase took place only in the second half of the twentieth century. Tractor densities more than doubled from the mid-twentieth century and in 1999 they reached 6.2 tractors/km$^2$ in St. Florian and even 7.2 tractors/km$^2$ in Grünburg.

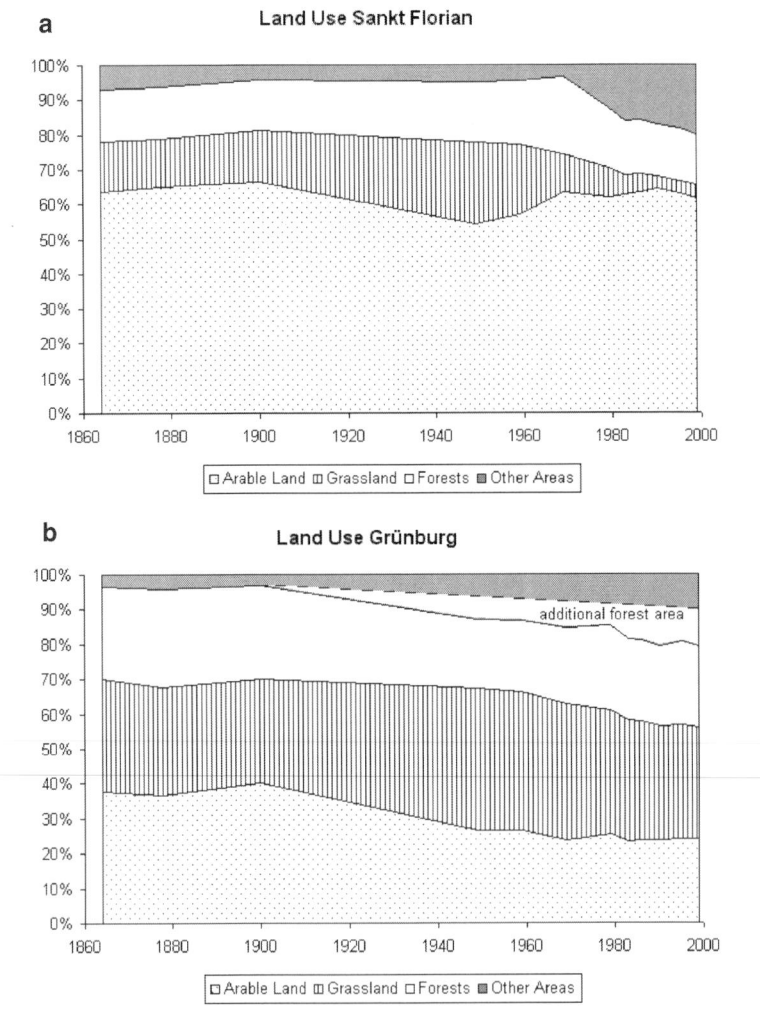

**Fig. 13.3** (a) Land use in Sankt Florian, 1864–2000; (b) Land use in Grünburg, 1864–2000 (Source: Authors' own calculations, see text). "Additional forest area" in Grünburg region was modelled based on the differences between cadastral forest records available only in 1995 and the statistical data used in all other points in time

The energy flows through the agricultural production systems in Sankt Florian and Grünburg had changed dramatically by the year 2000. Both systems now relied on high inputs from other socioeconomic systems (including fossil fuels for tractors), and produced much more surplus than in all earlier years. Additionally, output and domestic production of plant biomass had reached high levels in both regions. Aside from these general developments we can distinguish two very different strategies for using the land: In Sankt Florian, area productivity had risen to an extremely high level with 150 GJ/ha$_{agr}$/year, due to the abandoning of less extensive land uses in favour of intensive crops (cereals, maize and sugar beet). In Grünburg on the other hand, area productivity was about one-third lower (105 GJ/ha$_{agr}$/year), due to the higher significance of grassland production. In Grünburg, livestock density had reached a level that could no longer be fed by the local area and depended highly on feed imports. In 2000, feed plus litter imports still accounted for about 75% of Grünburg's imports. Because of the high imports for livestock husbandry, imports were higher in Grünburg (77 GJ/ha$_{agr}$/year) than in Sankt Florian (35 GJ/ha$_{agr}$/year). However, as livestock "transforms" energy from feed to animal products with relatively low conversion rates, cattle production being the least efficient, the energetic value of animal production in Grünburg was low as compared to feed and litter demand. In Sankt Florian, where animal production was dominated by pork and chicken production, the feed/animal production ratio was slightly higher. Output to other socioeconomic systems and total agricultural production was much higher in Sankt Florian.

## 13.5 Conclusion and Outlook

At the turn of the twenty-first century, we observe two very different agricultural systems: Sankt Florian has become a suburban area and agriculture is specialised on high-yielding crops, as well as chicken and pig farming. Grünburg on the other hand has been marginalised and now practices less intensive farming based largely on cattle breeding, and some pig and maize farming (for data compilations see Tables 13.2 and 13.3). The disintegration of land use systems thus happened on a very small regional scale: only 30 km apart, the two regions appear fundamentally different today. Additionally, the data presented in this study also show that this transformation did not happen continuously from the nineteenth century onwards. Most of these changes occurred only from the mid-twentieth century. The study thus complements national level analyses for Austria (Krausmann et al. 2003; Haberl and Krausmann 2007) on a local scale, i.e. on the level of municipalities.

While industrialisation led to very distinct changes in agricultural production in the two regions, on a more abstract level, there are some important similarities: mechanisation and intensification can be observed in both regions. Agriculture in both Sankt Florian and Grünburg today relies on large exchange flows with external partners, providing mainly feed and fuels, and consuming the high surplus production.

The concept of socio-natural sites focuses on the nexus between human practices and arrangements; the latter are seen as the biophysical and material precipitates and prerequisites of practices (cp. Winiwarter et al, Chap. 5 in this volume). This study's empirical results allow changes in the arrangements in both sites between 1864, 1950 and in particular after WW II until 2000 to be reconstructed. These changes in arrangements include the partly dramatic transformation of the landscapes as such, with e.g. increasing forests in Grünburg and the spread of built-up areas in Sankt Florian (Fig. 13.3a, b), they include changes in livestock and its composition (Table 13.2) and, among other features, the overall increase of agricultural machinery. These changing arrangements also hint at changes in human practices. To illustrate that general idea with only one example: the yoke and other equipment required to harness oxen as draught animal may still be found in a local museum, but these have become quaint relics, detached from the once embodied tacit knowledge of their users, isolated from the practices, from the other things and creatures once involved in their usage. Nevertheless, such objects refer to a specific past state of the socio-natural site in which agriculture had to operate in fairly closed cycles and resulted in appropriate diversified landscapes – just as much as today's tractor stands for agriculture's dependency on energy inputs from other places around the world.

The second point we wish to stress here concerns LTSER as research across spatial scales. Systemic approaches such as the study of social metabolism inevitably have to define the exact boundaries of their systems due to methodological reasons (Singh et al. 2010b). But to explain the transformation of these systems, the changing biophysical and communicative relationships to other sites have to be taken into account. These other sites may be close or very far away. Sankt Florian's path in particular after WW II is obviously affected by the process of suburbanisation in the sprawl of urban agglomerations like Linz. Over centuries, Grünburg's socio-ecological development cannot be explained without reference to economic cycles of the region's iron industries. Increasingly fossil fuel dependent arrangements in both sites after WW II make the agricultural systems' productivity more and more dependent on the performance of distant sites, even sites on other continents. The specialisation in agriculture after 1950 we observe in both places required constant flows of material, energy and people (if we think of work migration) from and to other places. A concept such as that of socio-natural sites is more flexible in defining spatial borders according to the questions and phenomena under consideration, because its application does not first and foremost depend on statistical data from distinct administrative units but rather on qualitative information. Thus it may help to follow such mutual socio-ecological relationships between different sites over long time spans.

How did people in the two regions experience these fundamental transformations? Was there a common experience of industrialisation, or did farmers sense the regional divergence? A concept such as that of socio-natural sites raises questions on these issues and thus opens the way for future LTSER including and giving humanities and social sciences a more important role. Such LTSER would aim at a better understanding of how actors and social organisations affect and are affected by the transformation of socio-ecological systems (cp. Gragson, Chap. 9 in this volume). Integrative research of this kind would deal with the general question of human agency in the transition of socio-ecological systems: how much scope was (and is) there for conscious human choice and action? Are patterns of material and energy flows, are metabolic regimes an iron cage in which (historical) actors are imprisoned? (cp. Brewer 2010; see also Cunfer and Krausmann, Chap. 12 in this volume).

The environmental histories of Sankt Florian and Grünburg may not be as spectacular as those of other places in the world. In many respects they rather stand for the often gradual but nevertheless fundamental transformation of rural worlds that took place mainly during the twentieth century. A conceptual framework like socio-natural sites reminds us that data on social metabolism and land use are also expressions of a history in which society and nature have been and still are indispensably intertwined, a history in which actors changed their environments partly dramatically through their decisions and thus their ways of living substantially. LTSER needs both empirically sound data to assess and explain the transformation of socio-ecological systems and distinct histories of human actors and social organisations. There are human life worlds behind our figures waiting to be discovered. Regionalised LTSER as a common effort of scientists and humanists has the potential to make those discoveries.

## References

Beer, K., Kotiath, H., & Podlesak, W. (1990). *Organische und mineralische Düngung*. Berlin: Deutscher Landwirtschaftsverlag.

Brewer, J. (2010). *Microhistory and the histories of everyday life. eseries* 5. Edited by the Center for Advanced Studies of Ludwig-Maximilians-Universität München. www.cas.lmu.de/publikationen/eseries

Bruckmüller, E., Hanisch, E., Sandgruber, R., & Weigl, N. (2002). *Geschichte der österreichischen Land- und Forstwirtschaft im 20. Jahrhundert. Politik Gesellschaft Wirtschaft*. Wien: C. Ueberreuter.

Bruckmüller, E., Hanisch, E., & Sandgruber, R. (Eds.). (2004). *Geschichte der österreichischen Land- und Forstwirtschaft im 20. Jahrhundert. Regionen Betriebe Menschen*. Wien: C. Ueberreuter.

Bundesamt für Statistik. (1935). *Ergebnisse der österreichischen Volkszählung vom 22. März 1934*. Wien: Österreichische Staatsdruckerei.

Cusso, X., Garrabou, R., & Tello, E. (2006). Social metabolism in an agrarian region of Catalonia (Spain) in 1860 to 1870: Flows, energy balance and land use. *Ecological Economics, 58*, 49–65.

Darge, E. (2002). *Energieflüsse im österreichischen Landwirtschaftssektor 1950–1995, Eine humanökologische Untersuchung* (Social Ecology Working Paper 64). Wien: IFF Social Ecology.

Dissemond, H., & Zauchinger, A. (1994). *Strohaufkommen in Österreich. Erhebung und Analyse des Überschußstrohaufkommens und seiner Eignung für energetische Nutzung*. Wien: Bundesministerium für Wissenschaft und Verkehr.

Foltz, C. (1878). *Statistik der Bodenproduction von Oberösterreich*. Wien: Faesn & Frick.

Gaube, V., Kaiser, C., Wildenberg, M., Adensam, H., Fleissner, P., Kobler, J., Lutz, J., Schaumberger, A., Schaumberger, J., Smetschka, B., Wolf, A., Richter, A., & Haberl, H. (2009). Combining agent-based and stock-flow modelling approaches in a participative analysis of the integrated land system in Reichraming, Austria. *Landscape Ecology, 24*, 1149–1165.

Gingrich, S. (2004). *Veränderungen von Landnutzung und Energieflüssen in ausgewählten Agrarökosystemen Oberösterreichs 1866–2000*. Diploma thesis, University of Vienna, Vienna.

Gradwohl, M. (2004). *Biomasse- und Energieflüsse in vorindustriellen Agrarökosystemen. Vergleich zweier Gebiete des Alpenvorlandes und der Donauebene 1864*. Diploma thesis, University of Vienna, Vienna.

Haberl, H. (1995). *Menschliche Eingriffe in den natürlichen Energiefluß von Ökosystemen: Sozio-ökonomische Aneignung von Nettoprimärproduktion in den Bezirken Österreichs* (Social Ecology Working Paper 43). Wien: IFF Social Ecology.

Haberl, H. (2001). The energetic metabolism of societies, part I: Accounting concepts. *Journal of Industrial Ecology, 5,* 11–33.

Haberl, H., & Krausmann, F. (2007). The local base of the historical agrarian-industrial transition, and the interaction between scales. In M. Fischer-Kowalski & H. Haberl (Eds.), *Socio-ecological transitions and global change: Trajectories of social metabolism and land use* (pp. 116–138). Cheltenham/Northampton: Edward Elgar.

Haberl, H., Winiwarter, V., Andersson, K., Ayres, R. U., Boone, C. G., Castillio, A., Cunfer, G., Fischer-Kowalski, M., Freudenburg, W. R., Furman, E., Kaufmann, R., Krausmann, F., Langthaler, E., Lotze-Campen, H., Mirtl, M., Redman, C. A., Reenberg, A., Wardell, A. D., Warr, B., & Zechmeister, H. (2006). From LTER to LTSER: Conceptualizing the socio-economic dimension of long-term socio-ecological research. *Ecology and Society, 11.* (Online), http://www.ecologyandsociety.org/vol11/iss2/art13/

Hoffmann, A. (1974). *Bauernland Oberösterreich: Entwicklungsgeschichte seiner Land- und Forstwirtschaft.* Linz: Trauner Verlag.

Hohenecker, J. (1980). *Ernährungswirtschaftsplanung für Krisenzeiten in Österreich. Vierter Teilbericht. Futtermittelbilanzen für Österreich. Schema und Berechnungen für die Wirtschaftsjahre 1972/73 bis 1976/77.* Wien.

K.k.statistische Zentralkomission. (1903). *Gemeindelexikon der im Reichsrate vertretenen Königreiche und Länder. Bearbeitet aufgrund der Ergebnisse der Volkszählung vom 31. Dezember 1900.* Wien: Hölder.

Krausmann, F. (2004). Milk, manure and muscular power. Livestock and the industrialization of agriculture. *Human Ecology, 32*(6), 735–773.

Krausmann, F. (2008). *Land use and socio-economic metabolism in pre-industrial agricultural systems: Four nineteenth-century Austrian villages in comparison* (Social Ecology Working Paper 72). Wien: IFF Social Ecology.

Krausmann, F., Haberl, H., Schulz, N. B., Erb, K.-H., Darge, E., & Gaube, V. (2003). Land-use change and socio-economic metabolism in Austria. Part I: Driving forces of land-use change: 1950–1995. *Land Use Policy, 20*(1), 1–20.

Krausmann, F., Schandl, H., & Sieferle, R. P. (2008). Socio-ecological regime transitions in Austria and the United Kingdom. *Ecological Economics, 65*(1), 187–201.

Kropf, R. (1997). Die Krise der Kleineisenindustrie in der oberösterreichischen Eisenwurzen im 19. Jahrhundert. In Anonymous (Ed.), *Heimat Eisenwurzen. Beiträge zum Eisenstraßensymposion Weyer* (pp. 114–154). Steyr: Ennsthaler Verlag.

Landeskulturdirektion Oberösterreich. (1998). *Land der Hämmer, Heimat Eisenwurzen, Region Pyhrn – Eisenwurzen.* Salzburg: Residenz-Verlag.

Langthaler, E. (2006). Agrarsysteme ohne Akteure? Sozialökonomische und sozialökologische Modelle in der Agrargeschichte. In A. Dix & E. Langthaler (Eds.), *Jahrbuch für Geschichte des ländlichen Raumes 3: Grüne Revolutionen* (pp. 216–238). Innsbruck/Wien/München/Bozen: Studien Verlag.

Leach, G. (1976). *Energy and food production.* Guildford: IPC Science and Technology Press.

Löhr, L. (1952). *Faustzahlen für den Landwirt.* Graz: Leopold Stocker Verlag.

Löhr, L. (1983). *Faustzahlen für den Landwirt.* Graz: Leopold Stocker Verlag.

Lorenz, J. R. V. (1866). *Statistik der Bodenproduction von zwei Gebietsabschnitten Oberösterreichs (Umgebung von St. Florian und von Grünburg).* Wien: k.k. Ministerium für Handel und Volkswirthschaft.

Österreichisches statistisches Zentralamt. (1950). *Ergebnisse der landwirtschaftlichen Statistik in den Jahren 1946–1949.* Wien: Österreichische Staatsdruckerei in Wien.

Österreichisches statistisches Zentralamt. (1952a). *Ergebnisse der Landwirtschaftlichen Statistik im Jahre 1951.* Wien: Kommissionsverlag der Österreichischen Staatsdruckerei.

Österreichisches statistisches Zentralamt. (1952b). *Ergebnisse der Volkszählung vom 1. Juni 1951 nach Gemeinden.* Wien: Verlag Carl Ueberreuter.

Österreichisches statistisches Zentralamt. (1954). *Ergebnisse der Erhebung des Bestandes an landwirtschaftlichen Maschinen und Geräten im Jahre 1953.* Wien: Druck und Kommissionsverlag Carl Ueberreuter.

Reckwitz, A. (2002). Toward a theory of social practices. A development in culturalist theorizing. *European Journal of Social Theory, 5*, 243–263.

Redman, C. L., Grove, J. M., & Kuby, L. H. (2004). Integrating social science into the long-term ecological research (LTER) network: Social dimensions of ecological change and ecological dimensions of social change. *Ecosystems, 7,* 161–171.

Sandgruber, R. (1997). Eine Einleitung. In Anonymous (Ed.), *Heimat Eisenwurzen. Beiträge zum Eisenstraßensymposion Weyer* (pp. 9–24). Steyr: Ennsthaler Verlag.

Sandgruber, R. (2003). Im Viertel der Vierkanter. Landwirtschaft im oberösterreichischen Zentralraum. In E. Bruckmüller, E. Hanisch, & R. Sandgruber (Eds.), *Geschichte der österreichischen Land- und Forstwirtschaft im 20. Jahrhundert. Regionen Betriebe Menschen* (pp. 439–490). Wien: Ueberreuter.

Schandl, H., Grünbühel, C. M., Haberl, H., & Weisz, H. (2002). *Handbook of physical accounting. Measuring bio-physical dimensions of socio-economic activities. MFA – EFA – HANPP.* Vienna: Federal Ministry of Agriculture and Forestry, Environment and Water Management.

Schmid, M. (2009). Die Donau als sozionaturaler Schauplatz: Ein konzeptueller Entwurf für frühneuzeitliche Umweltgeschichte. In A. Steinbrecher & S. Ruppel (Eds.), *Die Natur ist überall bei uns. Mensch und Natur in der Frühen Neuzeit* (pp. 59–79). Basel: Chronos Verlag.

Schuh, G., & Sieghartsleitner, F. (1997). *Heimat Eisenwurzen. Beiträge zum Eisenstraßensymposium Weyer*. Steyr: Ennstaler Verlag.

Sieferle, R. P., Krausmann, F., Schandl, H., & Winiwarter, V. (2006). *Das Ende der Fläche. Zum gesellschaftlichen Stoffwechsel der Industrialisierung.* Köln: Böhlau.

Singh, S. J., Haberl, H., Gaube, V., Grünbühel, C. M., Lisievici, P., Lutz, J., Matthews, R., Mirtl, M., Vadineanu, A., & Wildenberg, M. (2010a). Conceptualising long-term socio-ecological research (LTSER): Integrating the social dimension. In F. Müller, C. Baessler, H. Schubert, & S. Klotz (Eds.), *Long-term ecological research, between theory and application* (pp. 377–398). Dordrecht/Heidelberg/London/New York: Springer.

Singh, S. J., Ringhofer, L., Haas, W., Krausmann, F., & Fischer-Kowalski, M. (2010b). *A researcher's guide for investigating the social metabolism of local rural systems* (Social Ecology Working Paper 120). Wien: IFF Social Ecology.

Smil, V. (2001). *Enriching the Earth. Fritz Haber, Carl Bosch, and the transformation of world food production.* Cambridge, MA: MIT Press.

Vetter, H., & Steffens, G. (1986). *Wirtschafteigene Düngung: umweltschonend – bodenpflegend – wirtschaftlich.* Frankfurt/M: Verlagsunion Agrar.

Winiwarter, V., & Schmid, M. (2008). Umweltgeschichte als Untersuchung sozionaturaler Schauplätze? Ein Versuch, Johannes Colers "Oeconomia" umwelthistorisch zu interpretieren. In T. Knopf (Ed.), *Umweltverhalten in Geschichte und Gegenwart: Vergleichende Ansätze* (pp. 158–173). Göttingen: Attempto.

Winiwarter, V., & Sonnlechner, C. (2000). *Modellorientierte Rekonstruktion vorindustrieller Landwirtschaft* (Schriftenreihe "Der Europäische Sonderweg", Band 2). Stuttgart: Breuninger-Stiftung.

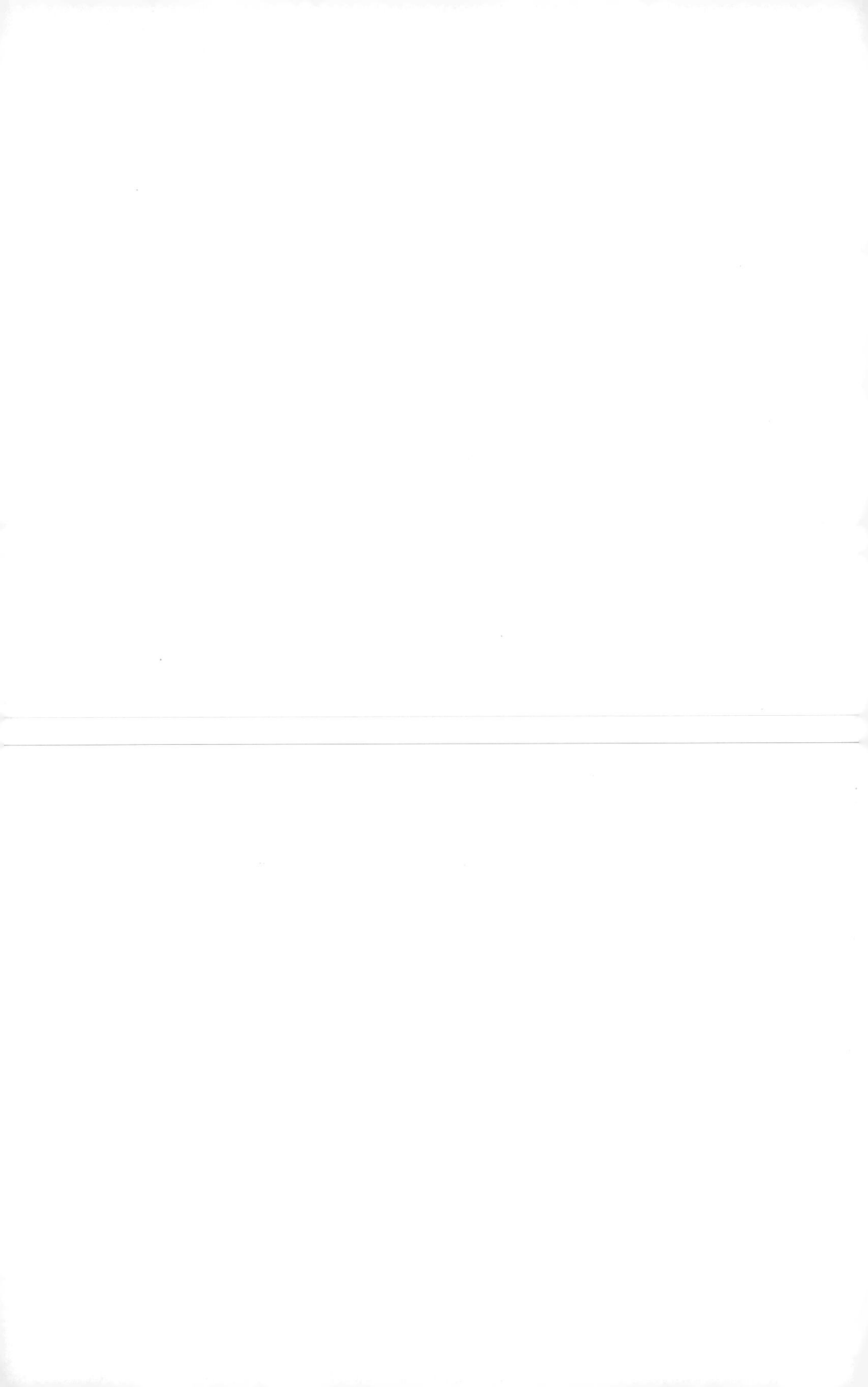

# Chapter 14
# The Intimacy of Human-Nature Interactions on Islands

Marian Chertow, Ezekiel Fugate, and Weslynne Ashton

**Abstract** Islands provide a place to conceptualise human-nature interactions in socio-ecological systems and to explore how such phenomena occur within decisive boundaries. Isolation, vulnerability to disruption, and constraints on the availability of natural resources add urgency to island sustainability questions with limited solution sets. This chapter presents findings that contribute to the larger issues of resiliency and vulnerability on islands. Cross-cutting reflections are offered based on studies conducted over the last 10 years at the Yale Center for Industrial Ecology of four diverse islands: Singapore, a highly developed island city-state; Puerto Rico, an island rich with nature and industry; O'ahu, a high density, tourism-dependent island, home to Honolulu, Hawai'i; and Hawai'i Island, also known as "The Big Island", with a larger land area and a lower population density than O'ahu. Over the course of the twentieth century, each of these islands became heavily dependent on imports such as water, food, or fuel to sustain basic human needs and modern economic functions. Within the last decade, each has consciously sought to restructure its socio-ecological configurations by using more locally available resources in one or more of its metabolic linkages. This pattern has the potential to reconnect island economies with their natural systems while simultaneously enhancing relationships and increasing resilience.

---

M. Chertow, Ph.D. (✉)
Center for Industrial Ecology, Yale School of Forestry and Environmental Studies,
Yale University, 195 Prospect Street, New Haven, CT 06511, USA
e-mail: marian.chertow@yale.edu

E. Fugate, M.Sc.
Renaissance School and Agro-Ecology, CHEC, Charlottesville, VA, USA
e-mail: ezekiel.fugate@gmail.com

W. Ashton, Ph.D.
Illinois Institute of Technology, Chicago, IL, USA
e-mail: washton@iit.edu

**Keywords** Island socio-ecological systems • Industrial ecology • Industrial metabolism • Material and energy flow analysis - MEFA • Hawaii • Oahu • Puerto Rico • Singapore

## 14.1 Introduction

Islands provide a place not only to conceptualise human-nature interactions in socio-ecological systems[1] but also to explore and document how such phenomena occur within decisive boundaries. Isolation, vulnerability to disruption, and constraints on the availability of natural resources add urgency to island sustainability questions with limited solution sets. This chapter presents findings that contribute to the larger questions of resiliency and vulnerability on islands, rooted in the idea that human and natural systems are intimately interwoven and that knowledge heavily weighted to either system is not a sufficient guide to action, especially when considering sustainability issues. Drawing from Singh et al. (2010), Haberl et al. (2004, 2006) and predecessors, the types of changes to natural ecosystems caused by human activities and the underlying socioeconomic driving forces of these changes are discussed in this chapter. Considering the special case of islands, four themes are explored:

1. Tighter and looser coupling of socio-ecological activity on islands;
2. Isolation from and connectivity to the global economy;
3. Targeting dependence and self-sufficiency of natural resource use;
4. Dynamics of socio-ecological change.

In exploring these themes, this chapter provides cross-cutting reflections based on studies conducted over the last 10 years at the Yale Center for Industrial Ecology of four diverse islands:

1. Singapore, a highly developed island city-state;
2. Puerto Rico, an island rich with nature and industry;
3. O'ahu, a high density, tourism-dependent island, home to Honolulu, Hawai'i;
4. Hawai'i Island, also known as "The Big Island", with a larger land area and a lower population density than O'ahu.

Like many islands, these were once largely self-contained units after first human settlement. Eventually, the resource dependence of these island societies increased beyond the physical boundaries of the islands. Each experienced substantial transformations in the period around 1960: Hawai'i became the 49th state of the United States in 1959; Singapore claimed its independence from Malaysia in 1965; and Puerto Rico initiated a programme in the 1950s for "bootstrapping" its economy

---

[1] Although we acknowledge disputes over the terms 'socio-ecological system', 'social-ecological system', 'coupled human and natural system', and 'coupled human-environment system', we have chosen here to use these terms synonymously. For a more in-depth analysis, see for example, Young et al. (2006), Gallopín et al. (1989), Berkes and Folke (1998), or Turner et al. (2003).

upward through industrialisation and land reform. Many years later, these islands have now become heavily dependent on imports such as water, food or fuel to sustain basic human needs and modern economic functions. Each island, however, has consciously sought to restructure its socio-ecological configurations by using more locally available resources in one or more of its metabolic linkages. This pattern has the potential to reconnect island economies with various aspects of their natural systems as discussed herein.

## 14.2 Studying Islands and Human-Natural Systems

While an island is generally perceived as a small unit of land surrounded by water, there is little consensus in the literature on island studies as to what precisely defines an island, including the upper bounds of its area and population (Deschenes and Chertow 2004). Geographers generally consider Greenland, with 2.2 million $km^2$ of land area, to be the largest island, while Australia, with a land area of approximately 7.6 million $km^2$, is the smallest continent. Human population does little to clarify this distinction, as Taiwan, with a 2010 population of 23 million and a land area of 36,000 $km^2$, is more populous than many continental countries. Frequently, islands are viewed as microcosmic in that they display the dynamics of competition for scarce resources and, increasingly, the pressures and impacts of humans on the environment. Yet, islands are distinct in their role as small systems affected by global-scale forces. Many unique problems arise when comparatively closed, fragile island environments are coupled with open, global economic systems. Nauru Island in the central Pacific Ocean offers a tragic example. An important source of phosphates for the global market in the twentieth century, extensive mining depleted Nauru's phosphate reserves and devastated many of its natural systems (McDaniel and Gowdy 2000).

Since the time of Darwin, islands have served as "model systems" for the transformative study of numerous phenomena including evolution, nutrient cycling, and speciation (Baldacchio 2004). Islands are particularly useful model systems for ecological science because they are closed in many regards, with clear physical boundaries, relatively small geographic areas, and simplified driving forces that can be separated and experimentally controlled (Bateson 1972; Vitousek 2006). Each island under study is characterised by a unique combination of ecological, physical, social, cultural and economic factors. Hawai'i Island, for example, the most thoroughly studied island from an ecological perspective in our group of four, is distinguished by well-defined biogeochemical gradients, orthogonal variation in climatic conditions, extreme geographic isolation, and a diverse composition of endemic and indigenous biota (Vitousek 2006). These conditions have inspired researchers to create a deep wealth of scientific knowledge in areas such as ecosystem ecology, biogeochemistry, evolutionary biology and anthropology (e.g., Vitousek 1995, 2002, 2004, 2006; Chadwick and Chorover 2001; Ladefoged et al. 2009; Kirch 1986; Kirch et al. 2006). Natural histories of islands describing flora and fauna are also common (Carlquist 1965; Kingdon 1990).

Less common, however, are socio-ecological studies that build on the base of physical and natural sciences and are strongly influenced by an increased understanding of the pervasive and growing influence of humans on ecosystem functioning. UNESCO's Man and the Biosphere programme in the 1970s was the first to examine islands by investigating the dynamics of modern human societies and the natural systems in which they exist. Over time, these issues have prompted a fundamental restructuring in the conceptualisation and study of ecosystems to include a human dimension, and this new paradigm is gaining broader acceptance (Grimm et al. 2000, 2008; Alberti et al. 2003; Redman et al. 2004; Gragson and Grove 2006; Haberl et al. 2006; Pickett and Grove 2009). With respect to islands, in the modern, interdependent world, deeper insights into socio-ecological conditions "present island populations with the challenges of limited resource availability, tenuous resource security, and limited natural carrying capacity," making islands excellent focal points for studies that systematically analyse the interactions between human/industrial activities and the natural environment (Deschenes and Chertow 2004; Graedel and Allenby 2002).

## 14.3 The Islands Under Study: Descriptions and Methods

The island studies of the Yale Center for Industrial Ecology follow the theme of socio-ecological research, but they are particularly focused on identifying and studying social, economic, and cultural factors that drive resource use and environmental change. Table 14.1 provides comparative physical and economic information about each of the four islands.

The Yale Center for Industrial Ecology relies on flagship analytical tools of industrial ecology to understand socio-ecological systems; these tools facilitate the study of flows of materials, energy, and water through human/industrial systems at different levels and scales. When presented holistically within the methodological framework of industrial metabolism (also referred to as social or socioeconomic metabolism), these tools provide a systems-level view of the natural resource dynamics of society (Ayres 1989; Ayres and Simonis 1994; Fischer-Kowalski and Haberl 1993; Fischer-Kowalski and Hüttler 1999; Daniels and Moore 2001).[2] The quantitative insights gleaned from industrial metabolic studies are especially important for managing resource use and environmental impacts in coupled human-natural systems on islands given their spatial and natural resource constraints. Indeed, just as the ecological studies referenced above have contributed substantially to a deepened understanding of natural island systems, we posit that, by analogy, studying material and energy flows can provide insight into socio-ecological interactions (see also Haberl

---

[2] When the concept of industrial metabolism is applied to a city or specific geographic region, it is increasingly referred to as urban metabolism (Wolman 1965; Baccini and Brunner 1991; Kennedy et al. 2007).

**Table 14.1** Characteristics of the selected islands

|  | Puerto Rico | Singapore | Oahu | Hawaii |
|---|---|---|---|---|
| Land area (km$^2$) | 8,870 | 687 | 1,545 | 10,432 |
| Agricultural land (2007) | 24.7% | 1.10% | 15.8% | 26.5% |
| Population | 3,978,702 | 4,701,069 | 907,574 | 177,835 |
| GDP PPP ($ billion) | 86.9 | 243.2 | 48.1 (2008) | 5.1 (2007) |
| GDP/capita (PPP) | $17,100 | $52,200 | $42,423 (2008) | $29,702 (2007) |
| Electricity production (billion kWh) | 23.7 | 41.7 | 8.2 | 1.2 |
| Per capita (kWh) | 5,962 | 8,875 | 9,068 | 7,023 |
| Exports ($ billion) | 20.9 | 274.5 | 0.55 (2008) | NA |
| Imports ($ billion) | 14.9 | 240.5 | 3.9 | NA (state only) |
| Internet users | 1 million | 3.37 million | 819,000 | NA (state only) |
| Visitors | 3,551,000 | 7,488,000 | 4,024,888 | 1,215,256 |
| Tourism expenditures/visitor | $978[11] | $1,802[12] | $1,269[13] | $1,029[13] |
| Visitors/population | 0.89 | 1.59 | 4.43 | 6.83 |
| Airport passenger movement | 9,265,713[14] | 37,203,978[15] | 11,157,524 | 996,620 |

Sources:
1. *The World Factbook 2010*. Washington, DC: Central Intelligence Agency, 2010. https://www.cia.gov/library/publications/the-world-factbook/index.html
2. *The State of Hawaii Data Book*. Honolulu, HI: State of Hawaii, Department of Business, Economic Development & Tourism, 2010
3. *2007 Census of Agriculture*. Washington, DC: National Agricultural Statistics Service, United States Department of Agriculture, 2009
4. *World Development Indicators*. The World Bank, 2010
5. *Regional Economic Accounts*. Bureau of Economic Analysis, 2010
6. *County of Hawaii Data Book*. Department of Research and Development, County of Hawaii, 2010
7. *State and County QuickFacts*. U.S. Census Bureau, 2010. http://quickfacts.census.gov/qfd/index.html
8. *Exports from U.S. Metropolitan Areas*. The International Trade Administration, U.S. Department of Commerce, 2009. http://www.trade.gov/mas/ian/Metro/index.html
9. State Trade Data. Foreign Trade Division, U.S. Census Bureau, 2010. http://www.census.gov/foreign-trade/statistics/state/
10. *Current Population Survey*. Bureau of Labor Statistics, 2010. http://www.bls.gov/cps/
11. *UNdata*. United Nations Statistics Division, 2010. http://data.un.org/Default.aspx
12. Global Market Information Database. Euromonitor International, 2010
13. *2009 Annual Visitor Research Report*. Hawaii Tourism Authority, 2010
14. *Passenger Traffic 2009*. Autoridad de los Puertos de Puerto Rico, 2010
15. *World Airport Traffic Report for 2009*. Airports Council International, Montreal, 2010

Notes
(a) NA not available
(b) Urban population calculation methods may differ between sources
(c) Hawai'i GDP not published at the county level, only personal income is calculated by Bureau of Economic Analysis and Census Bureau
(d) Only available as state total, not published by island/county for Hawaii
(e) Honolulu accounts for 89% of exports in state
(f) Data is for 2009 unless otherwise noted

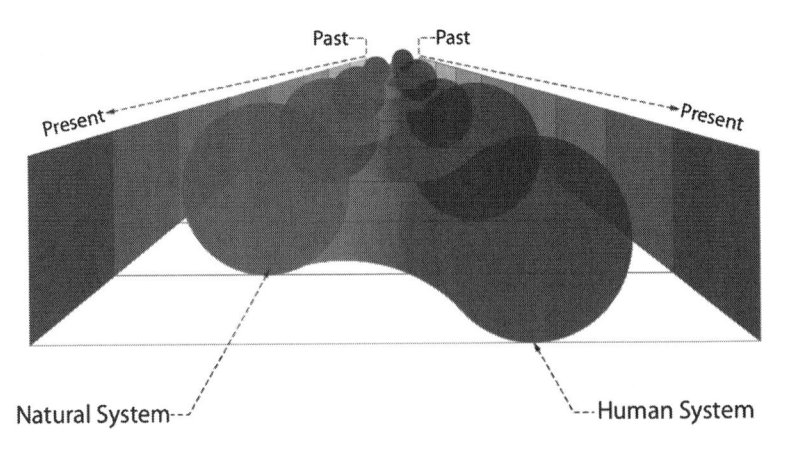

**Fig. 14.1** Line of sight along the dynamic metabolic interface between natural and human systems where quantitative measurements of socio-ecological variables are made

et al., Chap. 2 in this volume). The neutral, materialist perspective offered by these studies attempts to understand both ecological integrity and sustainable livelihoods by providing a quantitative line of sight along the permeable boundary that links natural and human systems (Fig. 14.1).

The industrial metabolism framework, with some variations in the particular tools and methods employed based upon the specific questions being investigated, was used to study all four islands. In studies of Puerto Rico from 2001 to 2008, we concentrated on the metabolism of the island's extensive industrial parks. Using the parks as a unit of analysis helped to advance the sub-field of industrial symbiosis through exploration of agglomeration economics, social network analysis, and embeddedness theory (Chertow et al. 2008; Ashton 2008; Chertow and Ashton 2009). Research in Singapore has been more concentrated on material flow analysis (MFA) than on energy and water flows. We have reviewed and expanded the scope of an island-wide metabolism study that focused on direct material flows (Schulz 2007) to include both direct and indirect flows (Chertow et al. 2011), and we are now gathering data at the household and district levels to explore multi-level material flows (Chertow et al. 2010).

The City and County of Honolulu is coterminous with the Island of O'ahu in Hawai'i. Here, we have extensively studied one symbiotic industrial cluster (Chertow and Miyata 2011) and conducted an island-wide material flow analysis to recommend new solutions to material and waste flow problems (Eckelman and Chertow 2009a, b). In both O'ahu and the larger-in-area, smaller-in-population Hawai'i Island, we have investigated resource use and environmental change across recent transitions associated with the growth and development of the islands by constructing material and energy flow analyses (MEFA). MEFA, similar to MFA but with an added focus on energy, is a powerful tool used to quantify the metabolic framework by tracking the input, output, conversion and accumulation of materials, energy or selected substances. While primarily used as a research tool within industrial ecology, it is also used to inform natural resource, pollution control, and waste management policy making at the local, regional and national levels. All of the metabolic studies rely on the following equation (Fischer-Kowalski and Hüttler 1999):

$$\Sigma(I_T) = \Sigma(O_T) + \Delta(S_T)$$

where $\Sigma(I_T)$ = sum of material/energetic inputs into the system; $\Sigma(O_T)$ = sum of outputs from the system; and $\Delta(S_T)$ = total changes in stock within the system.

Over time, metabolic information can be used to identify and monitor patterns of material, water and energy throughput and cycling as areas undergo socioeconomic and ecological change (Haberl et al. 2004). Ongoing work for Hawai'i Island includes an island-wide historical MEFA dating back to the 1800s. To conduct this longer-term MEFA and, thus, to better understand socio-ecological transitions in Hawai'i, we are drawing on an extensive database of imports and exports that flowed through the major ports of the islands. In addition, comprehensive metabolic studies of Hawai'i Island's two urban areas – Kailua-Kona on the east side and Hilo on the west side of the island – are being performed to provide a comparative analysis of the structure and function of two socio-ecological systems related through resource exchanges, geographic proximity, and historical and contemporary cultural configurations.

## 14.4 Themes

To draw a portrait of human-nature interactions on these islands, the four themes introduced above are discussed in detail using context-specific examples. The studies of the four islands were conducted separately; thus, this chapter provides a first opportunity to derive a broader, synthesised understanding across the different island settings and to propose lessons learned. We draw specific examples from our research rather than a complete picture of each island to help define these themes and to tease out their usefulness in island long-term socio-ecological research (LTSER) studies.

### 14.4.1 Tighter and Looser Coupling of Socio-ecological Activity on Islands

There is considerable ambiguity in the human-natural systems literature about the occurrence of metabolic coupling and the conditions under which this coupling becomes tighter or looser, or synonymously, stronger or weaker. Take, for example, an island whose people are dependent on food delivered by a ship with a port of origin 1,000 miles away. Is this island tightly coupled to the shipping network because the sustenance of its human population is dependent on it? Or, is it loosely coupled because the food was not locally grown in the first place? To clarify this relationship, we first isolate the social and ecological systems of interest and then examine both the temporal and spatial dimensions of metabolic couplings involving a physical or energetic exchange. In the food example, the configuration can shift over time from a coupling in which only the humans and the natural systems of an island are involved (i.e., food grown on the island is consumed by humans on the island) to a situation in which the humans of the island are coupled to ecological systems elsewhere for the majority of their food supply.

In a well-defined example of the dynamics of spatial and temporal coupling, the first group of Polynesian settlers who arrived in modern-day Hawai'i in approximately 800 AD represented an essentially closed human system penetrating a previously closed (i.e., uninhabited by humans) natural system. Until the time of European contact in 1778, everything consumed by the entire lineage of humans, except for the provisions brought by the initial settlers, was derived from the islands or from near-shore waters. Further, all physical objects that exited the human system, such as excrement and agricultural waste, entered some natural system on the islands. Indeed, the social and ecological systems that evolved on the Hawaiian Islands after the arrival of the first settlers were inextricably coupled within the spatial extent of the islands. Such tight couplings between natural and human systems were characteristic of the co-evolutionary process that unfolded on islands prior to the industrial revolution and, indeed, are still characteristic of more traditional cultures in which development has not followed a strict path of globalisation and vital linkage to distant markets.

By the mid-nineteenth century, Hawai'i had developed a robust global trade network. In 1860, imports valued at over $1 million were registered in Honolulu that year. Concurrently, some 32 domestic exports, including sugar, hides, coffee and whale oil, were shipped from Hawaiian ports to destinations such as New Zealand, Australia, Great Britain and China. The modern development of transportation and communication systems has enabled remote island communities to obtain an ever-increasing portion of their goods and services from distant shores. In O'ahu in 2008, we found that islanders depended on 15.1 million tonnes of imports and produced only 3.4 million tonnes of goods domestically (Eckelman and Chertow 2009a). Today's ships contain a remarkably complex assemblage of globally produced goods, as witnessed by reviewing the freight records of "big box" stores. Thus, both temporal factors, which characterise the stage of development of an economy in terms of its industrialisation, and spatial factors, which characterise a society's connection to outside, possibly global markets, determine the composition, extent and dynamics of coupling.

In general, the spatial extent of modern couplings and their characterisation in relation to "simpler" times are quite difficult to describe. We assert here, however, that identifying instances of tighter and looser coupling on islands is relatively less complicated than doing so in the middle of a large land mass because the number of pathways and ports of embarkation and debarkation are geographically restricted. In contrast, because a mainland city's boundaries are crossed by multiple train lines, highways, electric lines, water pipelines, and foot paths, metabolic linkages are challenging to identify, characterise, and track.

Often, modern couplings are dictated not only by geographical proximity but also by the extent to which services are mediated by technology. Energy consumption presents a useful example. An energy consumer on an island typically purchases power from the electricity grid, which can produce energy from local sources such as geothermal wells or from imported fossil fuels. Thus, the social system consuming energy on an island can be coupled with local primary energy sources or with sources from other parts of the world. Consider an example in which tourists are vacationing in a contemporary island hotel: are they tightly or loosely coupled to any particular natural systems on the island? For the purpose of this example, the tourists consume a certain amount of energy while they are in the hotel, and the hotel produces all of its own electricity from an on-site hydroelectric generator. In this example, we consider the tourists to be *tightly coupled* to the natural hydrologic system that generates the energy consumed in the hotel.

Next, imagine that instead of producing its own electricity, the hotel is connected to an island-wide energy grid. The energy flowing through the grid is generated on the island by geothermal and hydroelectric power plants, but both of these are located on the other side of the island in a different watershed. Here, we still consider the tourists to be coupled to the natural volcanic and hydrologic systems on the island that generate the energy consumed in the hotel, but we consider this coupling to be *looser*. The coupling now spans greater distances and is mediated by a network of pipes, pumps, and tanks as part of a more extensive human-constructed infrastructure.

Finally, imagine that the grid energy is generated by an electric power plant burning petroleum products. These fossil fuels are not produced on the island; rather, they are extracted from a forested ecosystem in Asia, processed nearby, and imported in an oil tanker. Following our previous logic, we consider the tourists to be coupled to the forested ecosystem from which the fuels were extracted. We describe the tourists in this situation, however, as being *loosely coupled* to the forested ecosystem because their coupling is mediated by global markets, governmental institutions, technology, and physical infrastructure.

A third lesson from the islands suggests more broadly that non-material relationships including culture, laws, market prices, and marketing messages influence changes in the organisation and regulation of material and energy flows (Costanza et al. 2001; Moran and Ostrom 2005; Reenberg et al. 2008; Ostrom and Cox 2010). In particular, coupling on the islands we have studied is notably influenced by governance structures. This extends well beyond the activities of government; indeed, we have measured the weight of the US military presence in O'ahu in tonnes of housing materials for military families and the quantity of fuel imported for military operations (Eckelman and Chertow 2009b). The point of interest here, however, is not the activities themselves, but the power of a non-material change to substantially influence underlying material relationships amid other social and economic ones.

Examples from the islands under study illustrate this point. Hawai'i became the 49th state of the US in 1959, and with this change, a new level of banking and investment security emerged. This change prompted rapid real estate development and exponential growth in the tourism industry, increased the demand for imports, and substantially transformed Hawai'i's metabolic impacts. Further, Hawai'i's continued reliance on tourism contributes to the disproportionate use of resources by visitors. Data collected by the State of Hawai'i show that compared to residents, tourists consume 0.26 more cubic metres of water, use 22.0 more kWh of electricity, and generate 0.8 more kilograms of waste on a daily basis (Hawaii DBEDT 2005).

Singapore's independence in 1965 followed British rule starting in 1819, Japanese occupation during WWII, and a brief alliance with Malaysia in 1961. After its independence, Singapore lacked the natural resources needed to support many traditional industries and it experienced a high unemployment rate, a housing crisis, and difficulties associated with a fledgling educational system. The establishment of a massive programme of modernisation focused on industrial development and education substantially changed land use and metabolic couplings. In the first 10 years, the majority of the population was moved into new government housing units, and financial and development institutions were established. Singapore also created its own military during this time. The enormous social and economic successes that Singapore achieved have distinct physical manifestations. For example, numerous land reclamation projects have increased the area of Singapore from 582 km$^2$ in 1962 to 641 km$^2$ in 1992 and 687 km$^2$ in 2010 (Goudie and Cuff 2008). The rapid transition, described by Singapore's famous leader, Lee Kuan Yew, as a move "from Third World to First," hinged on the substantial metabolic reconfiguration associated with the government's heavily planned and meticulously executed development programme.

Although Puerto Rico was ceded by Spain to the US in 1898, another 50 years passed before the US Congress gave Puerto Ricans the right to their own constitution and the ability to vote for their own governor. The first democratically elected governor began service in 1949, and launched an ambitious programme of industrialisation called "Operation Bootstrap" that included policies and tax incentives to convince US-based manufacturers to locate in Puerto Rico while maintaining access to US markets. As US citizens, Puerto Ricans gained the option to migrate freely to the mainland US, which eased employment and resource consumption demands on the island. With regard to island metabolism, these programmes, coupled with land reform to limit the holdings of large sugarcane interests, initiated the conversion from an agricultural to an industrial economy. These examples illustrate and identify the relationship between non-material changes and changes in the organisation and regulation of material and energy flows and, thus, in the underlying socio-ecological configurations of these islands.

In summary, studies of these islands show that the co-evolutionary process between natural and human systems is maintained by specific, organised flows of materials, energy, information and money across appropriate interfaces. This process of flux, in which resources move from one system to another and sometimes back again (e.g., wastewater discharge, waste heat, and bioengineered crops), is regulated by a complex combination of both material factors (e.g., physical infrastructure and technology) and non-material factors (e.g., governance structures and access to information). The heavy dependence of these islands on imports of food, energy, and manufactured goods suggests reduced material coupling with island ecosystems and increased coupling with more distant ecosystems. In this regard, a human system on one island can become coupled not only to local natural systems but also to natural systems across the globe. Thus, not ONE coupled socio-ecological system, but MANY socio-ecological systems are anchored around a local human system and a local natural system. The boundaries around the system change as we move from more tightly coupled socio-ecological systems on islands to more loosely coupled, globalised socio-ecological systems.

## 14.4.2 Isolation from and Connectivity to the Global Economy

The extent to which the social and ecological systems of islands are connected via the global economy to social and ecological systems elsewhere varies widely. Both Singapore and Puerto Rico are highly connected to the global economy, and there we find characteristic open economies in which the available goods and services parallel those available in nearby mainland systems. Both islands are also located much closer to their mainlands than Hawai'i, facilitating resource movement. In fact, Singapore has a reputation as a place where a consumer can purchase almost anything, including highly diverse foods, electronics, and exotic building materials. Puerto Rico is a key connecting point for Americans travelling to other Caribbean islands by air and sea, especially on cruise ships. Conversely, islanders from the region visit Puerto Rico to engage in an American shopping experience and to gain entry into the US.

Both Puerto Rico and Singapore are also important hubs in global manufacturing supply chains. For many years, multinational corporations have sited pharmaceutical manufacturing operations in Puerto Rico, using intermediate products sourced from around the world and taking advantage of tax code exemptions. As shown in Table 14.2 below, the governance changes that provided incentives to develop the pharmaceutical industry in Puerto Rico greatly altered material flows. Today, material exports from Puerto Rico are dominated by five industrial products, four of which are tied to the pharmaceutical industry, which is valued at over US $13 billion and is Puerto Rico's leading industry (Table 14.2). Socially, Puerto Rican leaders adapted the education system to respond to the needs of the growing manufacturing sector and now produce a highly skilled technical workforce in key areas. Perhaps the largest ecosystem impact associated with the growth of the pharmaceutical industry is the industry's extensive consumption of the island's limited freshwater supply. Groundwater levels in Barceloneta, the most concentrated cluster of pharmaceutical plants, fell by as much as 45 m in the first 20 years after industrial withdrawals commenced; this figure is much greater than anywhere else on the island (Renken et al. 2002).

Singapore's container shipping port, which facilitates the agglomeration and subsequent distribution of a wide array of goods, has been the world leader in container traffic since 2006 (Port of Hamburg 2011). The country's educated labour force and strategic location on one of the world's major shipping routes make it a

**Table 14.2** Puerto Rico's top five export Commodities, USD million

| Description | 2009 value | 2009% share |
|---|---|---|
| Medicaments (HS 300490) | 7,006 | 33.5 |
| Antisera and blood fractions | 4,626 | 22.1 |
| Compounds with an unfused pyridine ring | 1,251 | 6.0 |
| Parts for mineral processing machines | 679 | 3.2 |
| Polypeptide protein, glycoprotein hormones, and derivatives | 480 | 2.3 |

Source: Foreign Trade Division, U.S. Census Bureau, State Import Series. http://www.census.gov/foreign-trade/statistics/state/data/imports/pr.html
As classified by *Schedule B: Statistical Classification of Domestic and Foreign Commodities Exported from the United States*, US Census Bureau, 2010

**Table 14.3** Singapore's top five import commodities, USD million

| Description | 2009 value | 2009% share |
|---|---|---|
| Petroleum, petroleum products | 58,841 | 23.4% |
| Electrical machinery, apparatus, and appliances | 52,462 | 20.9% |
| General industrial machinery and equipment | 18,737 | 7.5% |
| Office machines and automatic data-processing machines | 14,722 | 5.9% |
| Other transport equipment (not passenger cars, commercial trucks, motorcycles, or trailers) | 9,040 | 3.6% |

Source: Euromonitor GMID database. As classified by the Standard International Trade Classification, Rev.3, United Nations, Department of Economic and Social Affairs

highly connected and desirable location for manufacturing operations that supply goods and services to Asian markets. The fact that four of Singapore's top five import commodities are also four of its top five export commodities is a testament to the island's role as a highly connected trans-shipment or *entrepôt* port, a port that specialises in the storing, processing, and reshipping of goods moving between other countries. Trans-shipment, a physical form of connectedness and flow in the economy, is so common in Singapore that most of the economy's imports and exports, in terms of both currency and material volume, are dominated by the movement of these four commodities (Tables 14.3 and 14.4).

Some of the colonised Pacific Islands are highly connected to the global economy, while others are largely self-contained. The Hawaiian archipelago is no exception to this pattern. O'ahu is the most globally connected of the chain, and the other islands are dependent on O'ahu as the principal port of the state. Goods first arrive in Honolulu from other states and countries and are then transshipped to the outer islands. Tourism is the dominant industry, but the prevalence of tourism varies across islands. Maui and Kauai are highly connected and have the highest ratio of tourists to total population.[3] Hawai'i Island, with the largest resident population outside of O'ahu, exhibits great socioeconomic variation. The island of Moloka'i,

---

[3] Maui County had a population of 145,157 in 2009 and a tourist count of 2,639,929 in 2007. Kaua'i had a population of 63,689 in 2008 and a tourist count of 1,271,000 in 2007.

## 14 The Intimacy of Human-Nature Interactions on Islands

**Table 14.4** Singapore's top five export commodities, USD million

| Description | 2009 value | 2009% share |
|---|---|---|
| Electrical machinery, apparatus, and appliances | 73,624 | 26.3% |
| Petroleum, petroleum products | 40,857 | 14.6% |
| Office machines and automatic data-processing machines | 24,824 | 8.9% |
| General industrial machinery and equipment | 16,637 | 5.9% |
| Organic and inorganic chemicals | 11,139 | 4.0% |

Source: Euromonitor GMID database. As classified by the Standard International Trade Classification, Rev.3, United Nations, Department of Economic and Social Affairs

a small Hawaiian island near O'ahu, presents an interesting contrast because it is currently undergoing economic reconfiguration after its largest employer left the island amid a dispute over expanded tourist infrastructure. Some observers speculate that it is transitioning toward more local livelihoods based on sustainable agriculture and away from greater global connectivity (Hamabata 2009, personal communication). It is interesting to note that the top five countries from which the State of Hawai'i imports over $2 billion of goods (Vietnam, Saudi Arabia, Indonesia, Thailand, and China) are entirely different from the countries to which it sends most of its exports (Japan, Singapore, South Korea, Netherlands, and Australia), excluding the US.

Even on a single island, the level of connectedness to the global economy and the accompanying levels of metabolic flows can vary greatly. Hawai'i Island has two urban areas, Kailua-Kona and Hilo, which are similar in population and size. Yet, the two have markedly different socioeconomic and biophysical characteristics. Hilo, with a wetter and cooler climate, has struggled with the demise of the sugarcane industry and the transition from a plantation economy to a more diversified base, while Kailua-Kona, both hotter and drier, has experienced explosive growth fueled by its attractiveness as an international tourist and second-home destination. Although Hilo's airport was established in 1928 and Kailua-Kona's current airport was opened in 1970, nearly one million international passengers passed through the Kona airport in 2009, while only 257 passed through the Hilo airport. Further, Kailua-Kona has 6,296 hotel rooms, while Hilo has only 549.

### 14.4.3 Targeting Dependence and Self-sufficiency of Natural Resource Use

The combined characteristics of boundedness, isolation and size limit resource availability on islands. The trajectory of all the islands under study over the last 50 years has been growth in population and greater use of three resources: water, energy, and waste assimilation capacity. The islands are far from self-sufficient, and Fig. 14.2, portraying resource dependence on O'ahu, is typical for all. As shown in the figure, O'ahu's dependence on foreign materials ranges from 10 % for construction minerals to 100 % for fossil fuels. Crude oil alone represented two-thirds of the annual value of commodities shipped to Hawaii in 2009 reinforcing the fact that the

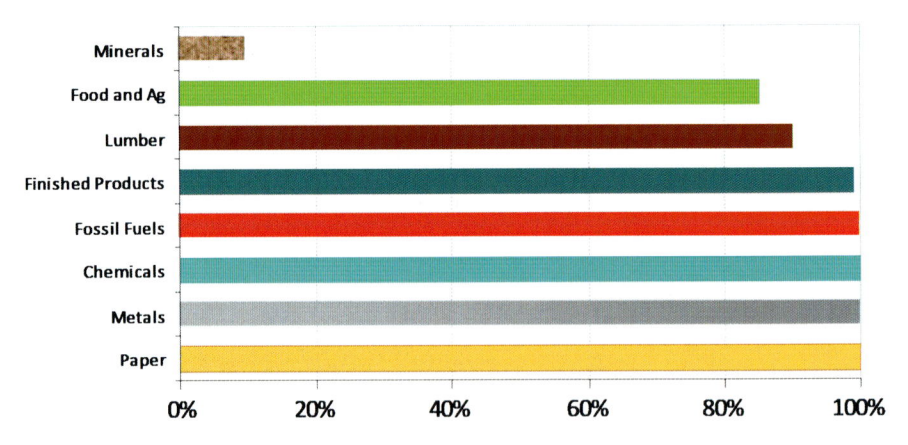

**Fig. 14.2** Oahu's dependence on various types of imported goods by percentage (Eckelman and Chertow 2009b)

State of Hawaii relies on imported fossil fuels for more than 80 % of electricity production. O'ahu, Hawai'i Island, and Puerto Rico are water self-sufficient (Eckelman and Chertow 2009b; Fugate 2008; Molina-Rivera 2005), whereas Singapore is not.

When an isolated area is dependent on outside resources for vital services, it is highly vulnerable to risks. Hawai'i, with a 10-day food supply, is vulnerable to shipping disruptions. Even if it were to grow all of its own food, it would be subject to other risks, such as crop failures due to inclement weather or the introduction of invasive species. For an island such as Singapore, the decision to take land out of commercial development to produce food faces high opportunity costs that suggest the continued importation of food may be preferable. Still, an interesting finding is that each island has been striving to decrease its external dependence in at least one major resource area. Singapore is much more water self-sufficient today than it was 20 years ago. Puerto Rico is seeking less dependence on fossil fuels for energy. O'ahu is struggling to handle its waste materials so as to avoid the export of waste. Hawai'i Island is attempting to increase its self-sufficiency of both food and energy. These trends are discussed below.

The story of water in Singapore is one of increasing self-sufficiency through investments in research and technology. Heavily dependent on neighbouring Malaysia for its water supply, Singapore foresaw the potential threats of increasing prices and supply disruption as urbanisation and industrialisation increased the demands of a small area with few domestic fresh water resources. Until 2003, Singapore relied on two water sources: the continued use of imported water from Malaysia and the increased use of catchment areas. Now, Singapore touts its Four National Taps programme. The 3rd tap, potable water reclaimed from secondary treated sewage, began when the first plant opened in 2003. Known as NEWater, reclaimed water now makes up 30% of the domestic water supply. The 4th tap, desalinised seawater, was established in 2005 and can supply 136,000 $m^3$ per day (PUB 2011). Singapore has also been vigilant about demand management; per capita domestic water consumption has decreased by several percentage points, and water system losses have been reduced from 11% in the 1980s to 5% today, one of the lowest leakage levels in the world. Singapore's goal is to achieve water self-sufficiency by 2061, 100 years after the first agreement with Malaysia (The Straits Times 2010).

Prior to 2003, Puerto Rico's power generation system was based almost completely on fuel oil, with only 1% of energy produced by hydropower sources. Since then, there has been strong interest in reducing the island's dependence on fossil fuels. Initially, two power plants, a combined cycle natural gas facility and a coal fluidised bed cogeneration facility, were commissioned to diversify the fossil energy mix and increase island-wide electricity output and efficiency. Each new plant was slated to contribute approximately 15% of the island's total power generation. In addition, these plants were highly efficient, using waste heat to replace nearby aging industrial steam boilers and to support a desalinisation plant. Despite many efforts, implementation of renewable energy projects has proven extremely difficult because of a combination of politics, environmental concerns and union opposition. Some signals of progress appeared at the end of 2010, as the Energy Diversification through Renewable, Sustainable and Alternate Energy Sources Law was passed. This law requires the government to achieve specific targets of energy production from renewable sources (12% by 2015, 15% by 2020, and 20% by 2035) and to implement a tradable renewable energy certificates programme on the island. The first two 250-kW wind turbines were installed on the island in 2010 at the Bacardi Rum distillery in San Juan Bay. Work on a 30-MW wind farm, which has been in the planning stages for more than 10 years, and a 100-MW wind farm were expected to begin in 2011 in Guayanilla and Santa Isabel, respectively (Energy Business Daily 2010).

In 2000, the City and County of Honolulu completed a study of new approaches for waste management technologies, and in 2005, representatives began updating the solid waste management plan. With only one public landfill and one waste-to-energy plant on O'ahu, the government has been forced to address the need for new capacity despite local opposition to increased landfill space and the anticipated expansion of the waste-to-energy plant by an additional 275,000 tonnes of waste per year by 2012. To address the most immediate problem, in September 2009, an agreement was reached to allow O'ahu to ship 90,000 tonnes of solid waste to Washington State each year (AP News 2009). The export solution failed when the contractor was unable to obtain the proper permits, and thousands of tonnes of waste accumulated on the island (Star Advertiser 2010). City advocates are taking a more positive approach by looking closely both at the waste stream and at the material import stream to assess the potential of creating green jobs and by determining the substitutability of imports with local products, including agricultural materials and inorganic minerals. When the waste plant expansion is complete, the total amount of landfill diversion to generate energy will be 33%; this is in addition to a current material recycling rate of 35%. If plans for ash and residue recycling are approved, the city could reach landfill diversion rates of approximately 80% (City and County of Honolulu n.d.).

In 2008, Hawai'i Island was approximately 80% fossil fuel dependent, with 68% and 99.9% of electricity generation and transportation energy, respectively, coming from fossil fuels. As a result, roughly 16% of the Gross County Product is spent on fossil fuels; this is twice the amount spent in the mainland US (Hawai'i County R&D 2011). According to the County's Sustainable Energy Plan, the County could move from 80 to 31% fossil fuel dependency by 2030, based on 66 recommendations that are gradually being adopted by the County. First and foremost, the policy goals are "to minimize energy use to the greatest extent possible and to meet remaining demand with energy generated from locally generated renewable resources." Hawai'i Island's determination to couple its energy requirements more strongly with local resources is demonstrated by the County's commitment to more cost effective and less polluting operations.

### *14.4.4 Dynamics of Socio-ecological Change*

On each of the islands under study, traditional land uses have given way to higher density development; visitors to any of the islands in 1965 would barely recognise them a generation later. Singapore is the most extreme example, as teachers and taxi drivers still recollect the locations of family farms and local herds where highways and housing blocks now stand. The speed and thoroughness of Singapore's socio-ecological change is the most advanced, based on the volume of construction and monetary expenditures following independence.

Hawai'i Island has had several eras of rapid change since the arrival of canoes from Tahiti and Micronesia over 1,000 years ago. Shortly after European contact began, a revolutionary period from 1795 to 1820, during which King Kamehameha I eagerly accepted metal use, began trading for guns, and then forcefully proceeded to unite all of the Hawaiian Islands under one banner for the first time prior to his death, drastically changed socio-ecological systems on the islands. Those who know little else of O'ahu know that 7 December 1941, Pearl Harbor Day, was another extreme example of the compression of space and time in an island setting. As Clarke (2001) stated, "on continents, economic and political changes evolve over decades; on islands, a ship appears on the horizon, a seaplane lands in a harbour, a European explorer arrives, and a single day changes everything forever."

Scholars studying change dynamics are seeking patterns in human-nature interactions. How is a natural system affected by the humans who enter it and under what circumstances does a human system revert back to nature? In Puerto Rico, for example, researchers have noted the spontaneous regeneration of forests following the abandonment of agricultural land in favour of industrialisation and urbanisation (Rudel et al. 2000; Grau et al. 2004). Before the 1940s, the majority of Puerto Ricans worked in the agricultural sector, many on sugar and coffee plantations scattered across the island. In 1948, with a new constitution ensuring "commonwealth" status, the government focused on attracting manufacturers from the mainland US to create new employment opportunities, particularly in cities. As a result of the movement of workers from farms to factories within Puerto Rican cities and across to the mainland US, forest cover increased from 9% of land area in 1950 to 37% by 1990, and farm workers as a percentage of the island's total labour decreased from 35 to 3.7% (Rudel et al. 2000). The Puerto Rican Farmers Association now reports that the island is self-sufficient only in bananas and milk, while producing 15–20% of the food supply locally (Fox News Latino 2011).

A similar pattern of the re-establishment of forest land was documented in Austria. The regrowth of forests in Austria, however, was attributed to the increased use of imported fossil fuels and the consequent decreased use of domestic biomass as the primary energy source (Krausmann and Haberl 2007; Erb et al. 2008). Puerto Rico's service sector, and especially the tourism industry, relies on the health of the island's coastal and forest ecosystems to attract domestic and foreign visitors. While the forests have recovered, however, locations such as Barceloneta, where manufacturing activities have clustered since the early1970s, have born a heavier environmental burden, receiving elevated levels of air and water pollution (Ashton 2009).

This pattern of change has been described as following C.S. Holling's model of succession in complex adaptive systems. In this model, some series of events triggers the collapse of a system, freeing resources for uptake by many diverse species or actors who can then colonise the system. Over time, another set of species or actors come to dominate the system because they use its resources well and out-compete others. These actors then begin to conserve energy and material resources for their own benefit, developing organisational structures to store the materials and further limit entry by potential competitors. Such structures, however, can be vulnerable to change if they do not have the capacity to adapt; this may result in another system collapse given a certain confluence of events. In Barceloneta, Puerto Rico, the middle of the twentieth century saw the collapse of the agricultural sector and the subsequent establishment of many diverse industries. The chemical sector was dominant from the 1970s into the 2000s due to abundant groundwater resources and favourable tax incentives, but all manufacturing, especially chemical manufacturing, began a slow decline starting in the 1990s (Ashton 2009). Figure 14.3 illustrates Holling's model by revealing a succession of changes in the industry mix in the Barceloneta region. This model is applicable at a conceptual level to the other islands in this study.

Another alarming example of disruptive socio-ecological change occurred on Hawai'i Island. In 1793, George Vancouver, a captain who had sailed with James Cook, introduced cattle to Hawai'i Island as a gift to King Kamehameha I (Tomich 1986). The King then prohibited the killing of cattle, and by 1802, the free roaming animals were regularly destroying Hawaiian agricultural plots (Barrera and Kelly 1974). Over the next several decades, land reform, zoning, tax laws, and the wandering ungulates accelerated the conversion of native vegetation to ranchland (Gagne 1988). By 1851, the cattle population on the island had grown to over 20,000 and the detrimental effects of cattle were evidenced by the destruction of forests and the conversion of once green areas to open plains overrun with invasive grasses (Influence of Cattle 1856; Henke 1929; Daehler and Goergen 2005). Today, a small number of feral cattle still exist on the island, but the majority are kept on expansive ranchlands (Cuddihy and Stone 1990). Our preliminary research suggests that the

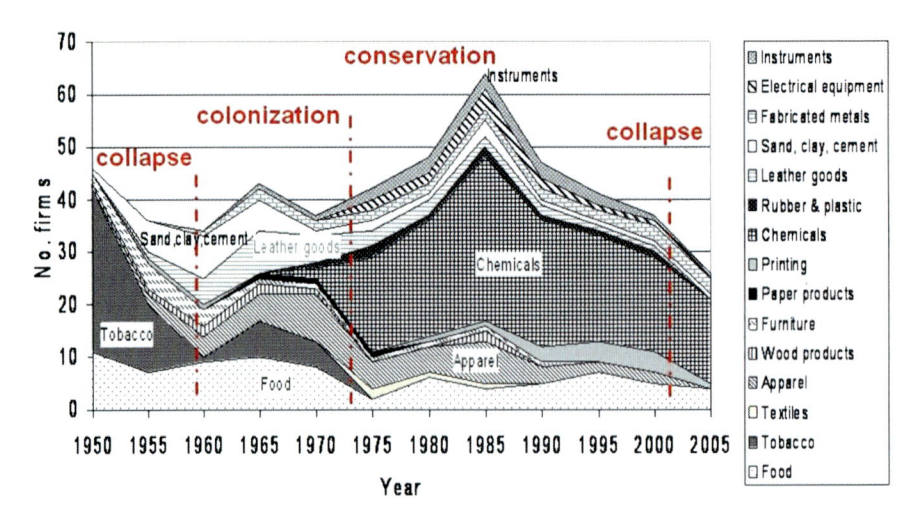

**Fig. 14.3** Number of manufacturing enterprises by sector and successional stage in Barceloneta, PR, 1950–2005 (Based on Ashton 2009)

cattle on Hawai'i Island are more influential than humans in the cycling of phosphorus and nitrogen (Siart and Skeldon 2009). The introduction of cattle serves as a cautionary tale regarding the unintended and potentially transformative dynamics of socio-ecological change.

## 14.5 Conclusion: The Need for Multi-level, Multi-scale Metabolic Analysis Within the Island Context

In a recent overview in *Science* of what has been learned from the current wave of coupled natural-human system research and theory, the authors concluded that (1) most prior work had been theoretical, (2) much more empirical work was needed, and (3) earlier studies focused on "interactions within the system, rather than interactions among different coupled systems" (Liu et al. 2007). Many of the examples assembled in this article demonstrate the usefulness of MFA for measuring changes in the metabolism of the human system, which, in turn, induced changes in coupled human and natural systems. By discussing individual islands, we have attempted to look beyond one system toward defining and illustrating metabolic couplings among several systems. In the cases of Puerto Rico and Hawai'i, we were also able to analyse regional systems within and among the islands. Thus, the call of Liu et al. for studies of the interactions among different coupled systems has been partially heeded by this study of human-nature interactions on islands.

Multi-level and multi-scale metabolic analyses are needed to expand upon and improve this work and to understand the co-evolution of coupling in the globalised world. Since the beginning of the industrial revolution, natural and human system interactions have changed dramatically in space and time. Islands with endogenously coupled systems and relatively closed boundaries have transitioned into exogenously coupled systems with more open boundaries. In industrialised societies, individuals, and households are simply less dependent on the direct production of material, energy and resources and are, therefore, less likely to be inclined to monitor and maintain ecological balance in the local environment. In place of local coupling, links to natural systems that provide material, energy and food resources are established at the regional, national and even global levels. This has led to substantial heterogeneity in both perceived and actual interactions with natural systems. These links are mediated by a complex combination of culture, market structures, businesses, institutions, and physical infrastructures that can be more thoroughly investigated with a multilevel, multi-scale view.

One barrier to multi-level studies is the difficulty of data collection at levels below the national economy. Data for Singapore, a separate country, and Puerto Rico, a separate commonwealth, were usually easier to obtain than data for individual counties within the State of Hawai'i. Our work in Hawai'i brings in the temporal scale of socio-ecological change by investigating material and energy flows over a period of 150 years. The acquisition of metabolic data across the years and the association of these data with natural system changes raises complicated questions, but this historical perspective also permits a richer understanding of the impacts of material and non-material changes during socio-ecological transitions.

Just as islands have been a useful focus for many scientific tasks, their value for the study of socio-ecological systems is also apparent. In each of the islands examined, the physical, social, economic, and environmental changes have been so extensive and rapid that visitors of even 50 years ago would barely recognise the islands today. In the light of these transitions, some of which have been more detrimental than others, tools to measure metabolism can be used to illuminate our understanding of past human-nature interactions and to chart a more sustainable course for the future.

# References

Alberti, M., Marzluff, J., Shulenberger, E., Bradley, B., Ryan, C., & Zumbrunnen, C. (2003). Integrating humans into ecology: Opportunities and challenges for urban ecology. *BioScience, 53*, 1169–1179.
AP News. (2009). Hawaii trash headed for Northwest landfill. Reported in *The Oregonian*, Portland, Oregon.
Ashton, W. (2008). Understanding the organization of industrial ecosystems: A social network approach. *Journal of Industrial Ecology, 12*, 34–51.
Ashton, W. (2009). The structure, function and evolution of a regional industrial ecosystem. *Journal of Industrial Ecology, 13*, 228–246.
Ayres, R. (1989). Industrial metabolism. In J. Ausubel & H. Sladovich (Eds.), *Technology and environment* (pp. 23–49). Washington, DC: National Academy Press.
Ayres, R., & Simonis, U. (1994). *Industrial metabolism: Restructuring for sustainable development*. New York: United Nations University Press.
Baccini, P., & Brunner, P. (1991). *Metabolism of the anthroposphere*. New York: Springer.

Baldacchio, G. (2004). The coming of age of island studies. *Tijdschrift voor Economische en Sociale Geografie, 95*, 272–283.

Barrera, W. Jr., & Kelly, M. (1974). *Archeological and historical surveys of the Waimea to Kawaihae Road corridor, Island of Hawaii* (Number 74–1 in Departmental Report Series). Honolulu: Department of Anthropology, Bishop Museum.

Bateson, G. (1972). *Steps to an ecology of mind: Collected essays in anthropology, psychiatry, evolution, and epistemology*. San Francisco: Chandler Publishing Co.

Berkes, F., & Folke, C. (Eds.). (1998). *Linking social and ecological systems: Management practices and social mechanisms for building resilience*. Cambridge, MA: Cambridge University Press.

Carlquist, S. (1965). *Island life: A natural history of the islands of the world*. Garden City: Natural History Press.

Chadwick, O. A., & Chorover, J. (2001). The chemistry of pedogenic thresholds. *Geoderma, 100*, 321–353.

Chertow, M., & Ashton, W. (2009). The social embeddedness of industrial symbiosis linkages in Puerto Rican industrial regions. In F. A. Boons & J. Howard-Grenville (Eds.), *The social embeddedness of industrial ecology* (pp. 128–151). Cheltenham: Edward Elgar Publishers.

Chertow, M., & Miyata, Y. (2011). Assessing collective firm behavior: Comparing industrial symbiosis with possible alternatives for individual companies in Oahu, Hawaii. *Business Strategy and the Environment, 20*, 266–280.

Chertow, M., Ashton, W., & Espinosa, J. (2008). Industrial symbiosis in Puerto Rico: Environmentally related agglomeration economies. *Regional Studies, 42*, 1299–1312.

Chertow, M., Kua, H. W., Ashton, W., & Jiang, B. B. (2010, November). *Multi-level material flows in a city-state: Implications for policy and planning*. Paper presented at the International Society of Industrial Ecology – Asia Pacific meeting, Tokyo, Japan.

Chertow, M., Choi, E., & Lee, K. (2011). The material consumption of Singapore's economy: An industrial ecology approach. In V. Savage & L. H. Lye (Eds.), *Environment and climate change in Asia: Ecological footprints and green prospects*. Pearson: Prentice Hall.

City and County of Honolulu, Department of Environmental Services (n.d.). Retrieved October 1, 2010, from http://www.opala.org/solid_waste/archive/Future_Plans.html

Clarke, T. (2001). *Searching for Crusoe: A journey among the last real islands*. New York: Ballantine Books.

Costanza, R., Low, B., Ostrom, E., & Wilson, J. (Eds.). (2001). *Institutions, ecology and sustainability*. Boca Raton/London: CRC Press.

Cuddihy, L., & Stone, C. (1990). *Alteration of native Hawaiian vegetation* (pp. 40–99). Honolulu: University of Hawai'i Press.

Daehler, C., & Goergen, E. (2005). Experimental restoration of an indigenous Hawaiian grassland after invasion by buffel grass (*Cenchrus ciliaris*). *Restoration Ecology, 13*, 380–389.

Daniels, P., & Moore, S. (2001). Approaches for quantifying the metabolism of physical economies: Part I – Methodological overview. *Journal of Industrial Ecology, 5*, 69–93.

Deschenes, P., & Chertow, M. (2004). An island approach to industrial ecology: Toward sustainability in the island context. *Journal of Environmental Planning and Management, 47*, 201–217.

Eckelman, M., & Chertow, M. (2009a). Using material flow analysis to illuminate long-term waste management solutions in Oahu, HI, USA. *Journal of Industrial Ecology, 13*, 758–774.

Eckelman, M., & Chertow, M. (2009b). *Linking waste and material flows on the Island of Oahu, Hawaii: The search for sustainable solutions* (Report Number 21). Sponsored by the Hawai'i Community Foundation. New Haven: Yale School of Forestry & Environmental Studies.

Energy Business Daily. (2010). *Big wind project could spur renewable energy revolution in Puerto Rico*. December 13th, 2010. Available at: http://energybusinessdaily.com/renewables/big-wind-project-could-spur-renewable-energy-revolution-in-puerto-rico/

Erb, K.-H., Gingrich, S., Krausmann, F., & Haberl, H. (2008). Industrialization, fossil fuels and the transformation of land use: An integrated analysis of carbon flows in Austria 1830–2000. *Journal of Industrial Ecology, 12*, 686–703.

Fischer-Kowalski, M., & Haberl, H. (1993). *Metabolism and colonization: Modes of production and the physical exchange between societies and nature* (Discussion paper series, DP5).

London: Centre for the Study of Global Governance, London School of Economics and Political Science.

Fischer-Kowalski, M., & Hüttler, W. (1999). Society's metabolism: The intellectual history of materials flow analysis: Part II, 1970–1998. *Journal of Industrial Ecology, 2*, 107–136.

Fox News Latino. (2011, February 3). Puerto Rican farmers warn of coming food shortages. *Fox News Latino.* http://latino.foxnews.com/latino/money/2011/02/03/puerto-rican-farmers-warn-coming-food-shortages/#ixzz1PlbYMLVB

Fugate, E. (2008, August). *Life cycle energy analysis of Hawai'i Island water systems.* Paper presented at Gordon research conference on industrial ecology, New London, NH.

Gagne, J. (1988). Conservation priorities in Hawaiian natural systems. *BioScience, 38*, 264–271.

Gallopín, G., Gutman, P., & Maletta, H. (1989). Global impoverishment, sustainable development and the environment: A conceptual approach. *International Social Science Journal, 121*, 375–397.

Goudie, A., & Cuff, D. (Eds.). (2008). *The Oxford companion to global change.* Oxford: Oxford University Press.

Graedel, T., & Allenby, B. (2002). *Industrial ecology* (2nd ed., pp. 268–272). Upper Saddle River: Prentice-Hall.

Gragson, T., & Grove, M. (2006). Social science in the context of the long-term ecological research program. *Society and Natural Research Programs, 19*, 93–100.

Grau, H. R., Aide, T. M., Zimmerman, J. K., & Thomlinson, J. R. (2004). Trends and scenarios of the carbon budget in postagricultural Puerto Rico (1936–2060). *Global Change Biology, 10*, 1163–1179.

Grimm, N., Grove, J., Pickett, S., & Redman, C. (2000). Integrated approaches to long-term studies of urban ecological systems. *BioScience, 50*, 571–584.

Grimm, N., Faeth, S., & Golubiewski, N. (2008). Global change and the ecology of cities. *Science, 319*, 756–760.

Haberl, H., Fischer-Kowalski, M., Krausmann, F., Weisz, H., & Winiwarter, V. (2004). Progress towards sustainability? What the conceptual framework of material and energy flow accounting (MEFA) can offer. *Land Use Policy, 21*, 199–213.

Haberl, H., Winiwarter, V., Andersson, K., Krister, P., & Ayres, R. (2006). From LTER to LTSER: Conceptualizing the socioeconomic dimension of long-term socioecological research. *Ecology and Society, 11.* (Online) http://www.ecologyandsociety.org/vol11/iss2/art13/

Hawaii County Department of Research and Development. (2011). *Energy.* Retrieved January 1, 2011, from http://www.hawaiicountyrandd.net/energy

Hawaii Department of Business, Economic Development and Tourism. (2005). *Planning for sustainable tourism* in Hawaii, Modeling Study Report, October.

Henke, L. (1929). *A survey of livestock in Hawaii* (Research Publication No. 5). Honolulu: University of Hawaii.

Influence of cattle on the climate of Waimea and Kawaihae, Hawaii. (1856). *Sandwich Islands' Monthly Magazine.*

Kennedy, C., Cuddihy, J., & Engel Yan, J. (2007). The changing metabolism of cities. *Journal of Industrial Ecology, 11*, 43–59.

Kingdon, J. (1990). *Island Africa.* London: Collins.

Kirch, P. (1986). *Island societies: Archaeological approaches to evolution and transformation.* Cambridge: Cambridge University Press.

Kirch, P., Chadwick, O., Tuljapurkar, S., Ladefoged, T., Graves, M., Hotchkiss, S., & Vitousek, P. (2006). Human Ecodynamics in the Hawaiian ecosystem, from 1200–200 yr B.P. In T. Kohler (Ed.), *Modeling long-term cultural change.* Santa Fe: Santa Fe Institute.

Krausmann, F., & Haberl, H. (2007). Land-use change and socioeconomic metabolism: A macro view of Austria 1830–2000. In M. Fischer-Kowalski & H. Haberl (Eds.), *Socioecological transitions and global change* (pp. 31–59). Cheltenham: Edward Elgar.

Ladefoged, T., Kirch, P., Gon, S., III, Chadwick, O., Hartshorn, A., & Vitousek, P. (2009). Opportunities and constraints for intensive agriculture in the Hawaiian archipelago prior to European contact. *Journal of Archaeological Science, 36*, 2374–2383.

Liu, J., Dietz, T., Carpenter, S., Alberti, M., Folke, C., Moran, E., Pell, A., Deadman, P., Kratz, T., Lubchenco, J., Ostrom, E., Ouyang, Z., Provencher, W., Redman, C., Schneider, S., & Taylor, W. (2007). Complexity of coupled human and natural systems. *Science, 317*, 1513–1516.

McDaniel, C., & Gowdy, J. (2000). *Paradise for sale: A parable of nature*. Berkeley: University of California Press.

Molina-Rivera, W. (2005). *Estimated water use in Puerto Rico, 2000* (Open File Report 2005–1201). Reston: United States Geological Survey.

Moran, E., & Ostrom, E. (2005). *Seeing the forest and the trees*. Cambridge, MA: MIT Press.

Ostrom, E., & Cox, M. (2010). Moving beyond panaceas: A multi-tiered diagnostic approach for social-ecological analysis. *Environmental Conservation, 37*, 451–463.

Pickett, S., & Grove, M. (2009). Urban ecosystems: What would Tansley do? *Urban Ecosystems, 12*, 1–8.

Port of Hamburg. (2011). Retrieved December 2, 2011, from http://www.hafen-hamburg.de/en/content/container-port-throughput-global-comparison

PUB. (2011). *Singapore's National Water Agency*. Retrieved January 2011, from http://www.pub.gov.sg/water/Pages/default.aspx

Redman, C., Grove, M., & Kuby, L. (2004). Integrating social science into the long-term ecological research (LTER) network: Social dimensions of ecological change and ecological dimensions of social change. *Ecosystems, 7*, 161–171.

Reenberg, A., Birch-Thomsen, T., Mertz, O., Fog, B., & Christiansen, S. (2008). Adaptation of human coping strategies in a small island society in the SW Pacific—50 years of change in the coupled human–environment system on Bellona, Solomon Islands. *Human Ecology, 36*, 807–819.

Renken, R. A., Ward, W. C., Gill, I. P., Gómez-Gómez, F., & Rodriguez-Martinez, J. (2002). *Geology and hydrogeology of the Caribbean Islands aquifer system of the commonwealth of Puerto Rico and the U.S. Virgin Islands* (United States Geological Survey No. 11). Reston: United States Geological Survey.

Rudel, T., Perez-Lugo, M., & Zichal, H. (2000). When fields revert to forest: Development and spontaneous reforestation in post-war Puerto Rico. *The Professional Geographer, 53*, 386–397.

Schulz, N. (2007). The direct material inputs into Singapore's development. *Journal of Industrial Ecology, 11*, 117–131.

Siart, S., & Skeldon, M. (2009). *Soil and ecological conditions on Hawai'i Island over 1200 years: A review from human settlement to 1950*. Waimea: The Kohala Center and the Yale Center for Industrial Ecology.

Singh, S. J., Haberl, H., Gaube, V., Grünbühel, C., Lisivieveci, P., Lutz, J., Matthews, R., Mirtl, M., Vadineanu, A., & Wildenberg, M. (2010). Conceptualising long-term socio-ecological research (LTSER): Integrating the social dimension. In F. Müller, C. Baessler, H. Schubert, & S. Klotz (Eds.), *Long term ecological research* (pp. 377–398). Dordrecht: Springer.

Star Advertiser. (2010). Retrieved December 30, 2010, from http://www.staradvertiser.com/columnists/20101230_Garbage_slated_for_shipping_now_being_burned_at_HPOWER.html

The Straits Times. (2010). Singapore's fifth and largest Newater factory in Changi opens. *The Straits Times*. Retrieved January 2011, from http://www.home-in-singapore.sg/sgp/cms.www/content.aspx?sid=1172

Tomich, P. (1986). *Mammals in Hawaii: A synopsis and notational bibliography* (Rev ed.). Honolulu: Bishop Museum.

Turner, B., Jr., Matson, P., McCarthy, J., Corell, R., Christensen, L., Eckley, N., Hovelsrud-Broda, G., Kasperson, J., Kasperson, R., Luers, A., Martello, M., Mathiesen, S., Naylor, R., Polsky, C., Pulsipher, A., Schiller, A., Selin, H., & Tyler, N. (2003). Illustrating the coupled human–environment system for vulnerability analysis: Three case studies. *Proceedings of the National Academy of Sciences, 100*, 8080–8808.

Vitousek, P. M. (1995). The Hawaiian Islands as a model system for ecosystem studies. *Pacific Science, 49*, 2–16.

Vitousek, P. M. (2002). Oceanic islands as model systems for ecological studies. *Journal of Biogeography, 29*, 573–582.

Vitousek, P. M. (2004). *Nutrient cycling and limitation: Hawai'i as a model system*. Princeton: Princeton University Press.
Vitousek, P. M. (2006). Ecosystem science and human-environment interactions in the Hawaiian archipelago. *Journal of Ecology, 94*, 510–521.
Wolman, A. (1965). The metabolism of cities. *Scientific American, 213*, 179–190.
Young, O., Berkhout, F., Gallopín, G., Janssen, M., Ostrom, E., & van der Leeuw, S. (2006). The globalization of socio–ecological systems: An agenda for scientific research. *Global Environmental Change, 16*, 304–316.

# Chapter 15
# Global Socio-metabolic Transitions

Fridolin Krausmann and Marina Fischer-Kowalski

**Abstract** This chapter provides a macro-perspective on the evolution of society-nature interactions during industrialisation. It explores the emergence of the industrial metabolic regime and investigates the links between economic development, population growth, resource use and environmental change. It discusses the constraints that the environment imposes upon socioeconomic development and the role of technology in both alleviating these constraints and altering the natural environment. Starting from a discussion of the sociometabolic characteristics of the agrarian socio-ecological regime, the paper develops a socio-ecological perspective of global industrialisation taking the development in different world regions into account. It shows how a shift from a solar energy system tapping into flows of renewable biomass towards a fossil fuel powered energy system based on the exploitation of large stocks of energy resources allowed for an emancipation of the energy system from land use and abolished traditional limits of growth. This metabolic transition facilitated unprecedented population growth and triggered a surge in the per capita use of material and energy. The paper argues that industrial society's high demand for material and energy resources is structurally determined and cannot be reduced simply by a more frugal or efficient use of resources.

**Keywords** Social metabolism • Metabolic regimes • Decoupling • Resource use • Industrialization

F. Krausmann, Ph.D. (✉) • M. Fischer-Kowalski, Ph.D.
Institute of Social Ecology Vienna (SEC), Alpen-Adria Universitaet Klagenfurt, Wien, Graz
e-mail: fridolin.krausmann@aau.at; marina.fischer-kowalski@aau.at

## 15.1 Introduction

Long term socio-ecological research, such as many of the case studies presented in this volume, often has a local focus and investigates the interplay of ecosystems and societal dynamics in specific regional settings. Technological change or economic developments at the national or global scale, however, have a decisive impact on society-nature interactions in specific regions and it is important to understand how local and regional socio-ecological systems are embedded in developments at larger scales (Haberl et al. 2006). In this chapter we provide a macro-perspective on the evolution of society-nature interactions during industrialisation. We are, above all, interested in the socio-ecological significance of technological change and investigate how it has influenced the interplay between societies and their natural environment. With this analysis we aim to contribute to a better understanding of the constraints that the environment imposes upon socioeconomic development and of the significance of technology in both alleviating these constraints and altering the natural environment.

To arrive at a socio-ecological understanding of industrialisation, we draw upon concepts of societal metabolism and the colonisation of nature (Baccini and Brunner 1991; Fischer-Kowalski and Haberl 1997). The use of metabolism as a concept in socio-economic studies was originally formulated by Karl Marx, who used it to denote the need that humans have to obtain their means of subsistence through an exchange with nature, in a process that is socially organised and connected with labour (Fischer-Kowalski 1998; see also Singh et al. 2010). This concept has since been further differentiated, implemented statistically in parallel with macroeconomic accounting, and historically specified: It is not merely the 'human being', which in terms of its metabolism is reliant on and has an impact upon nature, but rather the respective forms of societal production and consumption that generate its qualitative and quantitative characteristics.

Energy is a determining dimension in the metabolism of a society. The availability of energy plays a crucial role in defining relationships with nature by placing limits on the capacity of humans to alter nature and to extract, transport and process resources. Thus the question of how much energy a society has available, and from what sources, makes a great difference – not only to relationships with nature but also to relationships within society. In connection with this, it is possible to identify several major "sociometabolic regimes" that have existed during human history to date, between which there are significant transitions that are in general referred to as 'revolutions': the neolithic revolution, which marks the transition between the hunter-gatherer regime and agrarian society, and the industrial revolution, which marks the transition from the agrarian to the industrial regime (Sieferle 2003).

We focus on the agrarian-industrial transition, which is still ongoing at the global scale. Thus far, what is described here is not particularly new, yet in our contribution we shall attempt to show that using such an expanded socio-ecological perspective allows technological change to be observed and understood in a novel way. If one takes not only human actors and their societal relationships into account but also the natural preconditions and consequences of their activities, one may arrive at an understanding of the requirements, limitations and causal relationships that can also be described in quantitative terms and that make it possible to avoid representing these developments as stories of progress or decline. It is a common feature of both progress and decline narratives that they pursue an understanding of nature that is at least implicitly a magical interpretation rather than a realistic one, tested in the context of natural sciences. The magical phrase "Faith can move mountains" would be answered by the realist thus: "That may be, but it will certainly require plenty of energy to do so too."

## 15.2 Society-Nature Relationships Prior to the Industrial Revolution – The Metabolism of Agrarian Societies

The metabolism of all pre-industrial societies is based on the use of biomass and thus upon the ability of plants to utilise solar energy via photosynthesis to create energy-rich material from carbon dioxide, water and mineral compounds. In the form of nutrition and animal feed, biomass provides the energetic basis for sustaining the existence of humans and their livestock and can be converted by these into mechanical energy. Combustion (burning of fuelwood, for example) provides space and process heating for domestic households (cooking) and manufacturing (metal smelting) as well as light. The conversion of heat into mechanical energy was not possible prior to the invention of the steam engine and thus the availability of mechanical energy is subject to strict limitations. Water and wind power play an important but nonetheless in terms of quantity, rather subordinate role. With very few exceptions,[1] biomass is by far the most important energy source until the industrial revolution, generally accounting for 99% of all available primary energy sources. By far the largest proportion of biomass was utilised as nutrition for people and livestock. The proportion used as fuelwood was subject to significant regional variations related to the local availability of wood and climatic conditions and only a modest amount of extracted biomass was used for non-energetic purposes.

With the use of biomass, humans intervene in renewable energy flows. Nature is transformed through agrarian economy in a way that enables societal benefits in the form of utilisable biomass to be increased. The German environmental historian Rolf Peter Sieferle thus speaks of "the controlled solar energy system" of agrarian societies (Sieferle 2001). At a global level, this controlled solar energy system characterises societal relationships with nature for most of humankind until the

---

[1]In the seventeenth century Netherlands, for example the exploitation of large peat deposits, intensive use of wind energy and a dense network of waterways suitable for shipping formed the energetic basis for an exceptional economic development, the *Dutch Golden Age*. It is estimated that during this period up to 1.5 million tonnes of peat were dug annually, involving the excavation of 700 ha of peatland each year. Peat is naturally a source – albeit not one of the oldest – of fossil energy, (see De Zeeuw 1978).

twenty-first century. We define this fundamental form of societal relationship with nature as the 'agrarian sociometabolic regime'. In all its regionally specific variants, which are dependent upon a variety of biogeographical and societal factors,[2] this regime has a range of common features that clearly distinguish it from other sociometabolic regimes (such as the 'hunter-gatherer' or indeed the 'industrial regime'). The production of available energy is based upon the controlled transformation of ecosystems with the aim of increasing the utilisable yield of biomass, i.e. upon the colonisation of nature. Labour is invested in redesigning ecosystems and increasing the yield of utilisable biomass that can be harvested per unit area. The basic precondition for this form of subsistence is that a positive energy yield (*energy return on investment, EROI*[3]) is obtained from agrarian activity: through agrarian land-use, significantly more energy in the form of biomass must be produced than is expended in the form of human labour (and prerequisite energetic expenditure such as nutrition in particular). It has been estimated that in Central Europe before the beginning of industrialisation in the agrarian economy, an EROI of c. 10 to 1 was achieved (Leach 1976; Krausmann 2004). Any surplus may be used to supply the non-agrarian sectors of society – that is, to provide nutrition and fuelwood for the inhabitants of cities and those inhabitants not taking part in agrarian activities, as well as feed for draught animals that have to transport all this material.

The higher the surplus, the more complex the possible societal structures become. However, this surplus is never particularly high, since a system must be very well organised for the work of ten farm families to be able to sustain more than 1–2 other households (such as aristocratic landowners, craftspeople or officials). The reaction under the agrarian regime to an increase in food demand, which is usually caused under agrarian conditions by population growth, initially involves expanding the area dedicated to agrarian production – and this may often lead to attempts to capture new territories. As a last resort, where land is scarce and territory limited, the option remains to apply a greater investment of labour to the same land area with the aim of achieving a greater yield, in other words, the intensification of land-area use. However, the yield per invested hour of labour declines as intensity of use increases and asymptotically approaches a physical limitation, from which point there is no benefit to be achieved by further intensification. In other words, growth is possible, but leads to a diminishing marginal utility of labour. When this limit is reached, we find the 'typical' picture of agrarian societies, in which the majority of the population, including children, performs demanding physical work on a continuous basis, while still suffering from shortages of essential resources. This logic, which the anthropologist Esther Boserup (1965, 1981) has studied on a worldwide basis and of which she provides a detailed description, represents a fundamental

---

[2] The local characteristics of agrarian subsistence types depend partly on the distribution of precipitation and temperature through the year, on population density and the available labour resources, as well as on forms of governance and land ownership. Thus the appearance of pre-industrial agrarian societies differs widely, ranging from simple shifting cultivation and nomadic herding to complex and differentiated societies based on farming with and without livestock, irrigation or crop rotation.

[3] On the concept of EROI, see Hall et al. (1986, 28).

limitation upon societal development based on agrarian regimes: as a rule, growth in this regime eventually leads, despite progress made regarding methods of husbandry and plant cultivation, to the stagnating or even diminishing availability of per capita material and energy resources.

An additional limit to growth results from constraints on transportation. Overland transportation relies on the physical work of humans or animals,[4] is costly in energy terms and is only profitable for bulk raw material over distances of a few kilometres. Biomass is a decentral raw material with low energy density and is thus particularly affected by this transport limitation. Bulk raw materials can only be transported for longer distances where waterways are available. In the agrarian regime, the growth of cities is therefore subject to strict limitations and larger urban centres can only develop along rivers or coastal areas with a fertile agrarian hinterland. Furthermore, the lack of possibilities to transform heat energy into mechanical work limits the degrees of freedom: mechanical work can only be performed through the physical work of humans, animals and water/wind energy and the productivity that was thereby attainable remained relatively low.[5] Altogether, the size and structure of societal metabolism and its spatial differentiation were subject to limitation through the controlled solar energy system: in Europe before the beginning of the Industrial Revolution, between two and four tonnes of raw material and 30–70 gigajoules (GJ)[6] of primary energy were appropriated per capita and year, whereby biomass accounted for over 80% of all material and 95% of all energy inputs: as food for human population, livestock feed and wood for construction and fuel. Regional differences in metabolism were related in particular to the varying relevance of holding livestock and climatic conditions.[7]

Although agrarian societies have the potential to be ecologically sustainable in energetic terms, since they make use of renewable flows and do not consume exhaustible resources, the reliance of the agrarian regime upon a massive transformation of nature is associated with risks and leads to a range of specific environmental

---

[4]It is important to bear in mind that it makes no energetic or economic sense for draught animals and the people working with them to require more foodstuff for the transport route (and return journey) than they can carry. They can thus only transport either very valuable goods, which can be exchanged in terms of weight equivalent for large quantities of nutrients, or foodstuff for short journeys. It must also be considered that additional cultivated land is required for these animals and people in order to feed them, which also means the distances which have to be travelled increase too (see Sieferle 1997, 87), (Fischer-Kowalski et al. 2013, Chap. 4 in this volume).

[5]One should imagine that a Pharaoh with 2,000 labourers to build the pyramids had little more capacity at his disposal than a worker would today using a larger road construction machine.

[6]One Joule represents 0.24 cal and is a very small unit. Commonly derived units like megajoule $(MJ) = 10^6$ J, gigajoule $(GJ) = 10^9$ J and exajoule $(EJ) = 10^{18}$ J are used. Adequately feeding a human being requires approximately 10 million Joules (MJ) per day. The energy content (calorific value) of wood is roughly 15 MJ/kg, that of coal 20–30 MJ/kg and that of petroleum 45 MJ/kg.

[7]The highest biomass conversion rates are seen in pastoral societies with a very high per capita livestock holding and the lowest are recorded in societies whose means of subsistence relies predominantly on human physical work and plant-based diets (for example in the rice-cultivating societies of south and southeast Asia).

problems. In most regions, deforestation was a precondition for the spread of agriculture: In England, for example, only a small percentage of land area was still forested before the industrial revolution and in central Europe more than 50% of the forested areas were cleared between 900 and 1900 (Bork et al. 1998; Darby 1956). Changes in land cover and use bring with them changes in water and nutrient cycles and are often connected with soil degradation and erosion (see for example the extensive anthropogenic, i.e. man-made, karst formation in the Mediterranean region). The transformation of the ecosystem leads to changes in fauna and flora and the human-induced transfer of plants, livestock and parasites has many unwanted side-effects (for example, see Crosby 1986). The close contact with livestock encourages the spread of parasites and infectious diseases and in cities, water and air become polluted. However, these environmental problems only had a regional character and they were frequently triggered or enhanced by natural processes such as extreme weather phenomena. Thus societal policies aimed at avoiding such problems included portfolio strategies, i.e. relying on diversity instead of specialisation, and underexploitation of available resources (Müller-Herold and Sieferle 1998).

However, a decisive factor for the sustainability of agrarian societies was whether they managed to obtain a balance between population and soil fertility and with this, the relative stability of agricultural yields in the long term. With regard to population, the cultural (and legal) regulation of families and reproduction served to further this aim. Restrictions on marriage and sexual taboos (e.g. strict penalties applied to pre- and extra-marital sexual activity of women in particular) are characteristic of all agrarian societies (see for example Harris and Ross 1987). Merely stabilising, not to mention increasing per capita food yields in a form of agriculture that is entirely dependent on internal and biological means of production is a difficult undertaking and one that has not always been successful: soil degradation, desertification and, in some cases, the collapse of social structures, were all outcomes of failed attempts to operate agricultural systems or of an imbalance between population and the capacity of an agrarian system (Diamond 2005; Tainter 1988). In the three-field crop rotation system, which was widespread in central Europe at the beginning of the nineteenth century, the stabilisation of levels of important plant nutrients was achieved through a complex and labour-intensive system of field crop rotation that provided for fallow land, collection and spreading of animal manure and foodstuff transfers from woodland and grassland to arable land (Mazoyer et al. 2006, see also Cusso et al. 2006; Krausmann 2004).

A further global sustainability problem, however, remained completely unnoticed by agrarian societies and indeed in a regional sense they even benefited from this. The metabolism of agrarian societies is essentially based upon carbon: hydrocarbons, proteins and vegetable oils constitute the basis for nutrition and energy supply. Globally speaking, this metabolism remains within the framework of existing biogeochemical cycles, since the carbon that is released into the atmosphere through digestion and combustion processes ($CO_2$ and other compounds) will be reabsorbed in the course of new vegetation growth. However, in practical terms this is only partly the case. Deforestation of original woodland vegetation releases large quantities of carbon, whereas the plants that are preferred in agriculture (largely grasses) store much less carbon in their plant mass and often also in soil than forests. Thus the spread of agrarian societies involving the loss of forested land has led to an accumulation of $CO_2$ in the atmosphere that is not insignificant. It is estimated that 30–50% of the $CO_2$ enrichment of the atmosphere today can be traced back to changes in vegetation.[8]

## 15.3 The Coal Phase of the Metabolic Transition, or the English Success Story from the Mid-Seventeenth Century

A process began in England in the seventeenth century, whereby increasing use of coal led to the development of a new energy system. This energy transition was characterised by a shift from the use of energy flows with low power density in the form of biomass that is regrown annually to the exploitation of large-scale energy deposits that had accumulated over geological eras and which existed in concentrated form as coal, with a high power density (Smil 2003). Initially, coal was used solely as an often quite unpopular fuel for stoves in the households of manufacturing workers in urban centres, whose increasing requirements could not be supplied by fuelwood alone. In England coal supplies were to be found close to these centres and coal could also be transported at low cost via waterways. These densely populated manufacturing centres had come into existence because, as early as the seventeenth century, the English owners of large estates found it more profitable to use their land for the production of raw materials for the textile industry than to produce foodstuff for a rural population which, in their eyes at least, was seen as partly expendable.

By 1800, 900 kg of coal per capita and year were already being used in England (Fig. 15.1a). This was a completely new development path worldwide and – as England's rapid economic upsurge showed – one that promised great success. England's share of global coal extraction in 1800 was about 90%, of which a not insignificant amount was exported to other European countries, which in turn soon began to recognise the advantages of a coal-based economy. In quantitative terms, however, large parts of the rest of Europe, the USA and Japan, together with all other regions of the world, remained dependent to the greatest possible extent upon the agrarian sociometabolic regime and relied almost entirely on biomass as raw material and energy source. It was not until 1850 that the energy transition appeared in other European countries and per capita coal consumption also increased rapidly in Germany, France and the USA (Fig. 15.1a). In leading industrial nations such as Germany and the USA, a consumption level of 1,000 kg/capita and year was already surpassed by 1870, whereas in most other European countries, such as France and Austria, this occurred at a significantly later date. Late developers such as Japan and Russia/USSR only began to use larger quantities of coal after the beginning of the

---

[8] There has even been speculation that this has hindered the statistically foreseeable development of a new ice age (see Ruddiman 2003; Prentice et al. 2001).

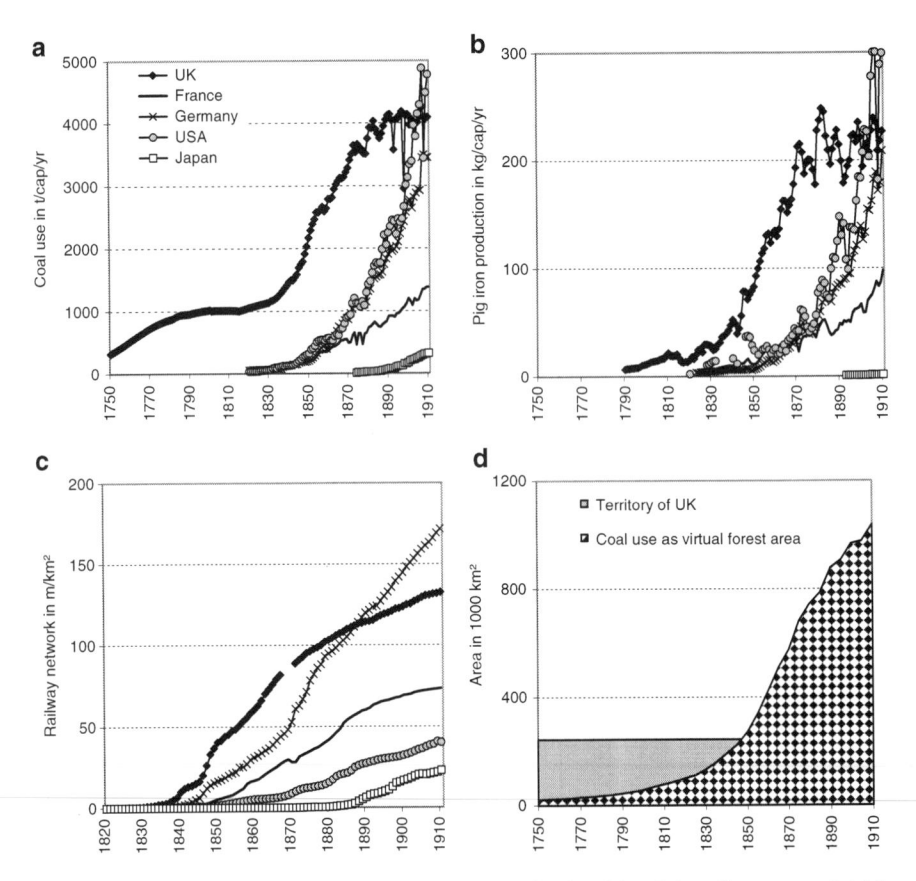

**Fig. 15.1** The development of coal use (**a**), pig-iron production (**b**) and the railway network (**c**) in selected countries from 1750/1830 to 1910 and coal use in the United Kingdom (UK) as virtual forest area (**d**) (Datasources: Authors' calculations based on Mitchell 2003; Maddison 2008; Schandl and Krausmann 2007). To convert coal use into virtual forest area (**d**), it was assumed that a quantity of fuelwood with the equivalent energy content to the coal used can be provided through sustainable forest management (i.e. through the use of annual growth and not standing timber mass). The forest area required to produce this volume of fuelwood is presented as a virtual forest area. Accordingly, by 1900, coal use in the United Kingdom represented a forest area five times the size of the entire country

twentieth century. During this phase, the metabolic transition was largely limited to Europe, Japan and the USA. In 1900, over 70% of coal extracted globally was used by only four countries: England, France, Germany and the USA. In nearly all other world regions, by contrast, regional urban-industrial centres at most were affected by this metabolic transition. Accordingly, the average per capita coal use remained negligible in countries such as India, China or Brazil even at the beginning of the twentieth century, comprising far less than 100 kg per capita and year. Indeed, the European countries that were in the process of industrialisation had an active interest in using colonialism to ensure that other world regions played a role as suppliers of cheap agricultural products and other raw materials, as well as outlet markets for growing industrial production and certainly not in allowing them to participate in industrial development themselves (see further discussion on this theme below).

### 15.3.1 Coal, Steam Engines, Steel and Railways

The Industrial Revolution and the dominance of the new energy system were very closely linked to the establishment of a new technology complex, characterised by the cooperation and positive feedbacks between coal, steam engine, iron and steel production, and the railway (Grübler 1998, 207). The stationary steam engine was first used as a pump in coal-mines and enabled the exploitation of deeper coal reserves and reduced the costs of coal extraction. Conversely, the use of coal enabled iron production to be greatly increased and, from 1870, high-quality steel to be manufactured. The steam engine together with large quantities of iron and later steel made a transport revolution possible, by means of the railways and steamship transport.

The development of coal consumption, pig-iron production and the railway networks during the nineteenth century (Fig. 15.1) underlined the leading position of the United Kingdom and the process of catch-up experienced by latecomers such as Germany and the USA – who only achieved the same degree of industrialisation as the United Kingdom by the beginning of the twentieth century. Between 1840 and 1860, a rapid expansion of railway networks began in several countries. Rail and, to an equal degree, steamship transportation made the large-scale separation of population segments producing foodstuffs and increasingly large population segments requiring these foodstuffs as the basis for other, i.e. industrial, production processes possible for the first time in human history. This meant that for the first time ever, there was no immediate limitation upon the growth of urban centres (see Krausmann 2013, Chap. 11 in this volume).

Steam engines enabled the conversion of coal into mechanical power. This led to a dramatic increase in the available power compared to the previous regime. The possibility of extracting, transporting, processing and consuming materials underwent radical change and as a result an entirely new form of societal metabolism came into being: in addition to biomass, huge quantities of coal, construction materials and ores were extracted and processed. In the United Kingdom, materials used, for example, increased between 1750 and 1900 from 60 to 400 million tonnes per year. Population growth during this phase happened at a somewhat slower pace as the increase in material and energy use. For the first time in history, there was a rapidly growing demand for non-agricultural workers: The mechanical performance of large coal-powered machines created the conditions that produced immense numbers of jobs required for final manufacturing. During this phase, although there was a rise in per capita material and energy consumption, this did not produce an increase in mass prosperity but was channelled instead into the expansion of the factory system and into exports. Meanwhile, the environmental conditions experienced by city dwellers worsened noticeably.

In his study of global environmental history in the twentieth century, John McNeill refers to Charles Dickens in describing this phase as the *Coketown* period (McNeill 2000, 296): distinguishing features were growing urban industrial regions with smoking chimneys, acrid smog, contaminated watercourses, grim working-class districts and slums. The growing material and energetic input into the economic system was accompanied by equal increases in poisonous gas and soot emissions and in the formation of effluents and waste materials. This led to new types of environmental problems and, above all in the high-density industrial and urban centres with their concentrated use of resources, produced a degree of pressure upon the surrounding ecosystems and quality of life that had never been experienced before. The extreme smog phenomena that occurred in London during the nineteenth and twentieth centuries and that involved extraordinarily high levels of soot and sulphur dioxide with immediate danger to health have been well-documented (Brimblecombe 1987). During this phase, air and water pollution and the hygiene and health problems associated with this developed into serious and in part trans-regional environmental problems with a stark effect upon the health of city dwellers. Measures such as the construction of high chimneys and canal networks diverted or diluted the problematic substances, but were only able to lessen their local impact somewhat.

## *15.3.2 The Emancipation of the Energy System from Land Area*

Coal represented a first important step towards emancipating the energy system from the land area and removing traditional limitations on economic growth. Rolf Peter Sieferle (2001) coined the vivid phrase of the "subterranean forest" for this phenomenon. He showed that by 1850 the energy (calorific value) contained in the amount of coal that was combusted annually in the United Kingdom had already reached the equivalent of the fuelwood that could be produced from a virtual forest area the size of the entire country. By 1900, this had risen to the area equivalent of a subterranean forest covering four times the land area of the entire country (see Fig. 15.1d). This means that in order to maintain a societal metabolism at the same level, the United Kingdom would have required a territory four times greater than its actual land area, and one that was entirely covered with forest.

However, coal use did not remove all the limitations of the solar energy system. A very profound reliance upon the area-dependent resource of biomass remained in place: the need for nutrition. The early industrialisation period was connected with a marked increase in population. In England, for example, the population more than doubled between 1750 and 1900 and inhabitants, including women and children, were also employed in non-agricultural production. Access to more (technical) energy had not in any sense replaced human physical work, but in fact had increased the demand for this. In a similar way, the railway did not replace the need for draught and working animals but instead the opposite was true: the wide-meshed network of railway lines in combination with the increase in transported goods and people led to an increased demand for working animals for distribution services and regional transport. Stocks of draught animals grew continuously into the twentieth century.

While coal did indeed provide a substitute for fuelwood, more timber than ever before was required for building the railways and for the emerging paper industry. Altogether, the demand for biomass grew paradoxically alongside the transformation of the energy system, in order to feed people and animals and to supply new industries with raw material. At the same time, the potential for expanding the cultivatable area was largely exhausted and the means of raising area productivity were limited. The most important limitation was the chronic shortage of fertiliser. Although mineral fertilisers such as guano, Chile saltpetre or superphosphate were increasingly used in agriculture by the end of the nineteenth century, the volumes employed were low and limited to special crops such as oranges or tobacco, and the supply of plant nutrients for most of the cultivated land still had to rely on farm-internal means (manure, leguminous crops, etc.). Thus a fundamental limitation upon traditional agriculture remained in place, which, in spite of successful biological innovations such as new cultivated plants and new land-use practices, for example in England, where these innovations were certainly implemented early on, led to stagnating grain yields in the nineteenth century. The import of foodstuffs from overseas colonies or from the USA (which had after all shaken off its colonial status) and Russia was therefore necessary (Krausmann et al. 2008b).

In the USA a completely different development took place: a rapidly growing population, but an extremely low population density of only two persons per km$^2$, meant that with the expansion of the railway system huge swathes of fertile prairie land could be cultivated for food production. Within a few decades of homesteading, over 100 million hectares of high quality agricultural land were gained in the midwestern USA between 1850 and 1920, after the indigenous peoples with their extensive land-use practices had been violently expelled (Cunfer 2005). The nutrient-rich soils of the Great Plains allowed for high initial yields to be achieved with little labour input. The labour productivity of this system of agriculture was extraordinarily high and enabled a small rural population to supply the densely-populated urban centres on the coasts as well as to export large quantities of foodstuffs to Europe. By around 1880, the USA was already exporting over four million tonnes of grain, providing basic nutrition for over 20 million people (Krausmann and Cunfer 2009).

This development too was closely associated with the technology cluster of the *Coketown* era. It depended upon the expansion of the railway system and steamship transportation, the supply of the local population in the treeless plains with their harsh climate with energy in the form of coal and the opportunities represented by high-quality machines produced from steel for the mechanisation of working processes in agriculture.

## 15.4 Oil and the Car: The USA's Success Story from 1900

Not only the agrarian productivity of a pioneer country but also another resource – oil – positioned the USA to become the leading nation during the next phase of the industrial transformation. Like coal, oil was not a new resource. It had been used by humans for many centuries, but only in very small quantities, for example for lighting purposes in lamps. But the great oil boom began around only 1900 with the discovery of large oil fields in Spindletop, Texas. Oil production in the USA rose during the first three decades of the new century from under 10 to 140 million tonnes per year. Before the world economic crisis at the beginning of the 1930s, more than 1.2 tonnes of oil per capita and year was being extracted. The USA dominated global oil production throughout the first half of the twentieth century, much as England had previously achieved with coal. Only after the Second World War did the exploitation of huge reserves in western Asia begin, forcing the USA out of its previously dominant position on the world markets. Oil has a still higher energy density than coal, is cheaper to extract and, given the right infrastructure, can be easily and cheaply transported. Of all the various options, oil is thus in many ways an ideal energy resource.

However, in contrast to coal, reserves of oil (and natural gas) are distributed very unevenly across the planet. The industrial countries in Europe that had been so successful thus far possessed only minor exploitable oil reserves and had to first build up capital-intensive distribution networks, such as pipelines, oil tankers and refineries, in order to be able to use this resource, whereby part of the profits of industrial production were channelled to other world regions (initially to the USA in particular). In this way, the use of oil brought with it new geopolitical power relations that were quite new.

A new technology cluster emerged with the use of oil as an energy resource. McNeill termed the combination that was created from oil together with the combustion engine, automobile and airplane, the (petro)chemical industry and finally electricity the *Motown Cluster*, after the centre of the US automobile industry (the *Motor Town* Detroit) (McNeill 2000, 297). The mobile combustion engine, employed in cars and airplanes, brought with it the individualisation and speeding up of human and goods transport, which triggered a new transport revolution. And with electricity, a new and universally applicable form of energy became available, which enabled the mechanisation of numerous technological processes via the electric motor. As with biomass during the first phase of transformation, during the second phase coal was also not completely replaced as a source of energy but remained the basis for steel production and thermal electricity generation. Nonetheless, coal consumption in the USA had already reached its historical peak by 1920 and in European countries several decades later, with consumption beginning to sink rapidly thereafter. In contrast, the share of total energy flow worldwide represented by oil rose within just a few decades to nearly 50% (see Fig. 15.2).

The economic context for the establishment of the new energy system in the USA constituted a combination of cheap energy, assembly line production and rising incomes among the working class, which is defined in the literature as *Fordism* and which heralded the era of mass production and mass consumption (von Gottl-Ottlilienfeld 1924; Grübler 1998). The new technologies found application in affordable goods suitable for the masses and now households too benefited from rising levels of energy and material consumption, with their material well-being rising dramatically. The key material- and energy-intensive products during this phase

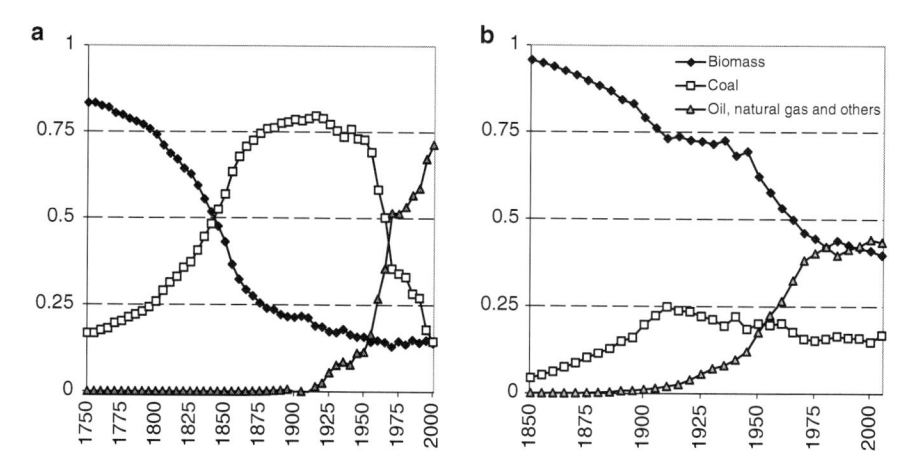

**Fig. 15.2** The establishment of new energy sources in the United Kingdom (1750–2000) (**a**) and worldwide (1850–2005). (**b**) In this diagram, the share of total primary energy supply represented by the three fractions biomass, coal and oil/natural gas (including other energy forms) is depicted. The biomass fraction includes all biomass used as food for humans and livestock and biomass used for all other purposes, together with fuelwood (Data sources: Authors' calculations based on Schandl and Krausmann 2007, 97 (United Kingdom) and Krausmann et al. 2009; Podobnik 1999; IEA 2007 (World))

were the automobile, central heating, electrical household equipment and meat. These products became affordable and their consumption expanded comprehensively within a few decades across all social strata. The *American way of life* thus came into being. In Europe and Japan, this dynamic (partly driven by American economic aid) established itself fully only after the Second World War and then led to a per-capita doubling of annual energy and material consumption (and of course of the concomitant waste products and emissions too) within 25 years. The Swiss environmental historian Christian Pfister coined the term '1950s' Syndrome' to describe this unprecedented growth dynamic in societal metabolism. Pfister showed that in Europe in the decades following World War Two until the oil crisis of the 1970s, a fundamental transformation of society-nature relationships took place (Pfister 2003). Three socioeconomic factors made a decisive contribution to the rapid establishment of the new sociometabolic regime: on the one hand, energy prices sank significantly relative to the price of other goods, so the importance of energy as a cost factor diminished. On the other hand, state-run infrastructure programmes and intervention measures drove forward the expansion of the pipelines and electricity networks and helped to create the transport infrastructure that was required.[9] Thirdly, the new broad-based, state-run welfare system (introduced by Roosevelt in the USA in the

---

[9] In the context of the *New Deal*, a million km of highways and 77,000 bridges were constructed in the USA in the 1930s and the country was eventually covered by a comprehensive motorway system from 1956, with help from the Federal Highway Aid Program.

framework of the *New Deal* in the 1930s and in the rest of Europe after the Second World War) helped to secure the income of the majority (Lutz 1989). In Europe, it was the reconstruction after the Second World War and the *European Recovery Program* (Marshall Plan) that drove this rapid establishment forward. Altogether, the 1950s Syndrome produced a catch-up development and the dissemination of the *American way of life* in Western Europe, Canada, Australia and Japan. Most other regions of the world and thus the majority of the world's population remained initially unaffected.

## 15.4.1 The Automobile

The automobile is one of the most important factors in the transformation of societal metabolism in the twentieth century. It provided the basis for a further transport revolution: At several 1,000 m per $km^2$, the density of the road network was greater than that of the rail network by one or two orders of magnitude; draught animals thus became completely redundant as delivery agents to centralised networks. After the Second World War, the fleet of motorised vehicles in industrialised countries grew rapidly. Already by 1970, there were between 250 and 350 vehicles for every 1000 inhabitants in many European countries, an in the USA, this figure was even twice as high (Fig. 15.3a). The affordable automobile made comprehensive access to individual transport possible for the first time and automobile production became the most important sector of industry and thus crucial for the economic establishment of the new system. This system of transport causes both directly and indirectly enormous material and energy flows in its production and daily operations (Freund and Martin 1993): per automobile, up to 30 tonnes of materials are used for the manufacturing process. As late as the 1990s in the USA, 10–30% of all metals used and two-thirds of rubber production were used by the automobile industry. Furthermore, the construction and maintenance of the requisite transport infrastructure consumes material and energy. Per kilometre of motorway, 40,000 tonnes of cement, steel, sand and gravel are required and the area used for road building is 10–15 times as great as that required by the railway. In this phase, the transport sector replaces industry as the greatest direct consumer of energy. The fuel consumption of the vehicle fleet becomes, along with the energy requirement for space heating, the greatest single factor in the societal energy consumption of industrialised countries.

## 15.4.2 Electrification

Electricity had already been used commercially since the late nineteenth century. Electricity generation is not dependent on a specific primary energy source. It was

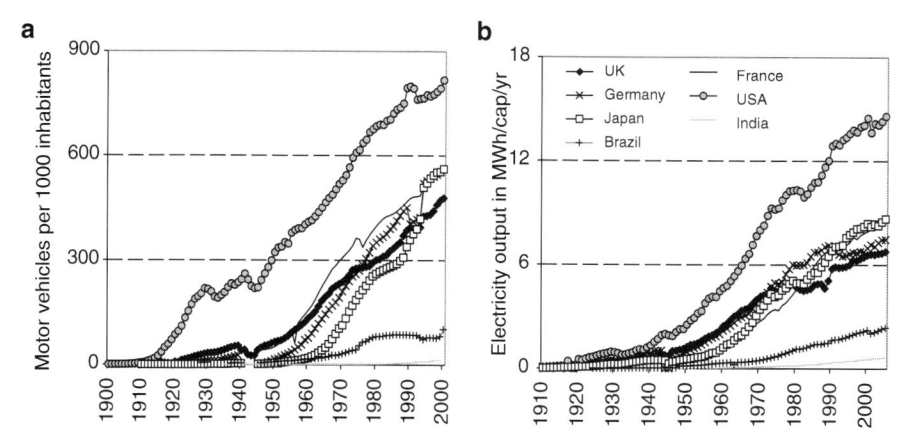

**Fig. 15.3** Motor vehicle stocks (**a**) and electricity generation (**b**) in the 20th century (Data sources: Authors' calculations based on Mitchell 2003; Maddison 2008)

first generated using hydro-power, then in thermal power stations using coal and later oil, gas or waste products, and from the 1960s onwards, also in nuclear power stations. Starting in the USA, comprehensive electrification became one of the fundamental prerequisites for industrial development and high general standards of living in the twentieth century. The demand for electricity is constantly on the rise, with the increase being coupled directly with economic growth. In all industrialised countries, a continual increase in demand can be observed to the present levels of 8–10 MWh per capita and year (see Fig. 15.3b); in the USA, however, electricity consumption is double this level. Electricity generation requires huge amounts of energy, whereby up to 60% of the primary energy is lost in conversion and transmission. In industrialised countries, 20–25% of the total primary energy consumption is used for electricity generation. Individual countries take different paths in this respect and according to their resource endowment rely on hydro-power (Austria, Sweden), nuclear energy (France) and most frequently on coal (this applies particularly in the case of newly-industrialised countries such as China or India). Two-thirds of the current global coal extraction is used for thermal power stations. All technologies have their own specific negative impact on the environment: Hydro-power represents an intervention in ecosystems and, at least in industrial countries, there are only a few remaining river systems that have retained their natural condition; thermal power stations make a significant contribution to global $CO_2$ emissions; nuclear energy is associated with significant risks (major accidents in Three Mile Island, 1979, in the USA; Chernobyl, 1986, in Ukraine) and the unsolved problem of storing long-term radioactive waste. Today, approximately 15% of electricity is produced using nuclear power, with three nations – the USA, France and Japan – accounting for 56% of the total capacity.

Electricity is universal, convenient to employ and can be used to create lighting, heating or to perform mechanical activity. The electric motor allows the mechanisation of extremely difficult processes. It has revolutionised the time expended in household activities through the spread of equipment such as washing machines, dishwashers and vacuum cleaners and has allowed a far-reaching decoupling of physical work from production processes in industry. Finally, transistors and computer chips have created a revolution in information and communication technology (telephone, television and computer technology).

## 15.4.3 The Green Revolution

As shown, in the nineteenth century the USA was able, thanks to its highly productive form of agriculture, to effectively compensate the weaknesses of the English transformation model (i.e., difficulties in producing sufficient food for a high-density and growing population) and to turn this to its advantage. However, it became clear that this level of agricultural productivity had no long-term potential and indeed after only a few decades, it ran up against massive ecological limitations: the combination of large land area with a low investment of labour was only possible because the prairie soil then being ploughed up for the first time contained huge reservoirs of plant nutrients accumulated over a long historical period. These reservoirs, however, quickly began to deplete in the first decades of ploughing. The yields began to decrease and enormous problems with erosion appeared (Cunfer 2005). In a situation where oil could be obtained cheaply, it was possible, with a bundle of technologies coupled to this new energy source, to create a new and successful agricultural model. The tractor allowed for the substitution of all animal and a large proportion of human labour in agriculture, much as the motor saw raised the speed of tree-felling in comparison with the axe by a factor of 100–1,000 (and thus enabled the rapid deforestation of the rainforests). On the other hand, the agrochemical industry, based on petroleum and natural gas, helped to lift the chronic limitations on plant nutrients from which agriculture was suffering. From the 1920s onwards, huge amounts of atmospheric nitrogen were made available for agricultural use using the Haber-Bosch process, which requires a high energy input (Smil 2001). The average nitrogen application in crop farming increased to several 100 kg per hectare as a result. Together with industrial potassium and phosphate fertilisers, pesticides and successes in plant and livestock breeding area yields and labour productivity in agriculture were multiplied within a very short space of time (Grigg 1992).

Starting in the USA and disseminated by agricultural companies active on the global market, these new agricultural methods were distributed around the world under the term 'the green revolution'.[10] They found application in Europe after the Second World War, as a result of which the proportion of the population engaged in

---

[10] The term *green revolution* was first coined in 1968 by William S. Gaud, the director of the United States Agency for International Development USAID (see also Leaf 2004).

agriculture fell to 5% or less. The 'green revolution' also took hold in large sectors of agriculture in the southern hemisphere and helped create the conditions in which global food production was able to keep pace with the quadrupling of the world population in the twentieth century.

The industrialisation of agriculture led to a massive transformation of the agrarian landscape, which had to be rendered suitable for machine activity, and this led to a range of specific environmental problems: Large-scale fields were created from which all hedgerows and even many geo-morphological aspects of the terrain that presented obstacles for mechanisation were removed. Farming with heavy machinery encouraged the compaction and erosion of the soil, and large-scale monocultures required a high level of agrochemical use, which had a negative impact on soil and groundwater. Furthermore, the position of agriculture in societal metabolism and the energy system changed fundamentally. Industrialised agriculture requires a high energy input and today, more energy is invested in agricultural production than is thereby obtained in the form of food. This is partly due to the large quantity of high-quality agricultural produce that is fed to livestock. In general, agriculture has been altered during the course of the sociometabolic transformation from being the most important source of useful energy to becoming an energy sink (Pimentel and Pimentel 1979). With the industrial transformation, society has made itself dependent on abundant external energy sources for the most important part of its metabolism, namely the feeding of its population.

In spite of this, the 'green revolution' created the conditions for a new relationship between the industrial centres and the global periphery. It shrank the need of densely-populated European countries for colonial territories (a need that had anyway never formed part of the development model emanating from the USA). Under the clear political and military leadership of the USA, the first priority was to head off any danger of a planned economy, socialism or state capitalism developing and to establish the worldwide dominance of the western capitalist economic model.

## 15.5 What Next? The Limits of the Industrial Transformation's Dynamic Since the 1970s: Present and Futures

### 15.5.1 The Beginning of the End for the US-Dominated Oil Regime?

At the beginning of the 1970s in the USA, the world's first civilian mainframe computer produced a complex simulation model for the global relationship between societal and economic dynamics with its natural causes and consequences, under the provocative title *The Limits to Growth* (Meadows et al. 1972). The plausibility of the results of this study by Dennis and Donella Meadows was underlined the following year by the so-called first oil price shock for the world economy. In reaction to the Yom Kippur war, the OPEC countries restricted their oil supplies and from one day to the next, the price of oil rose from 3 to 5 dollars per barrel. Further oil price crises followed in 1979 (first Gulf War) and 1990 (second Gulf war). Where the USA was concerned, the country had already reached *peak oil* i.e. the maximum in terms of domestic oil extraction by 1970/1971, a fact already predicted in 1956 by M. K. Hubbert (Deffeyes 2001). Thus the USA was increasingly becoming a net purchaser of oil instead of creating considerable revenues from oil exports as in previous decades – and on top of this, the costly wars waged in the Middle East brought no success in terms of securing long-term American control over oil supplies. Furthermore, nuclear power, which had been supported with large-scale investment as the new hope for energy supply, failed to match expectations. Not only did nuclear energy prove more expensive and its technological development slower to develop than expected, it also experienced a set-back at this time with the serious accidents first at the reactor on Three Mile Island (1979) and a few years later in Chernobyl (1986).

Worldwide, a range of measures were implemented that reduced oil use and slowed down the increase in energy consumption significantly. An abrupt (but short term) fall-off in the material and energy consumption of industrialised countries and the cessation of the rapid growth of societal per-capita material and energy consumption took place (Fig. 15.3a, b). From that time on through the following three decades, the energy and material expenditure of industrialised countries stabilised at a high level and no further convergence between Europe/Japan and the level of the USA, where this expenditure level was twice as high, took place. Starting in Japan, there was a re-orientation of the key industry of the oil regime, the car industry, towards producing smaller and more energy-efficient vehicles. The US car industry, however, continued to adhere to its traditional and resource-intensive course until the collapse of large parts of this industry during the crisis of 2008.

One can interpret the early 1970s, heralded culturally by the worldwide student protests of 1968 and militarily by defeat in Vietnam, as a turning point in which a long-term sociometabolic regime promoted by the USA and entailing a wasteful approach to natural resources began to move towards an end. At the same time, a new regime, that of information and communication technologies, was just beginning that brought with it the opportunity to satisfy important human needs while using fewer resources. If we interpret this period thus, we would have to conclude that at the least the USA failed to understand the signs sufficiently and despite all its power, attempted for several decades thereafter to continue along the *business as usual* path of its previous regime. While the USA had the satisfaction along the way of enjoying the triumph that came with the collapse of the Soviet empire,[11] the oil regime led to the creation of a new and potentially hostile periphery of countries that were pre-industrial but that had suddenly become extremely wealthy due to their oil reserves,

---

[11] From a sociometabolic perspective, the Soviet Union and the countries economically and politically connected with it still largely represented the English coal-steel-railway regime, but were nonetheless distinguished by a particularly high use of resources together with a low income.

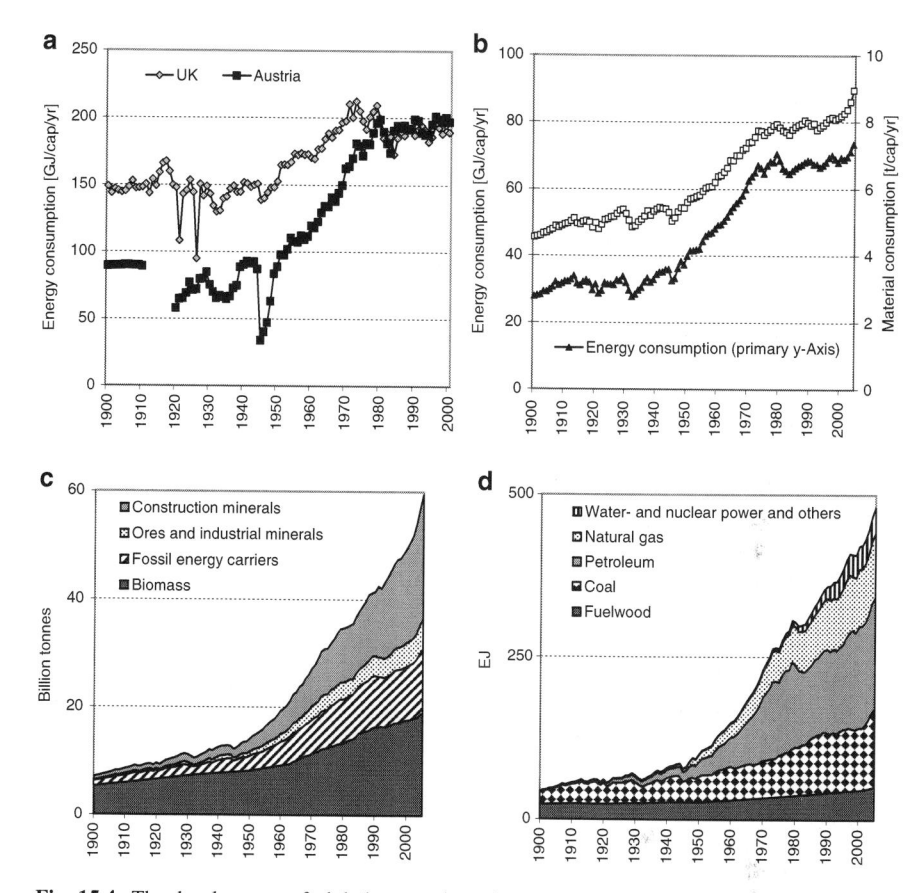

**Fig. 15.4** The development of global energetic and material societal metabolism. (**a**) Per capita energy consumption in the UK and Austria, (**b**) Global per capita energy and material consumption, (**c**) Global primary energy consumption by technology, and (**d**) Global material consumption (Data sources: based on Krausmann et al. 2008b, 2009)

and many of whom – an historical coincidence – shared a cultural tradition of proselytising and aggressive religious conviction that turned them against the USA.

## 15.5.2 Metabolism and Environment in the Twenty-First Century

Altogether, the second phase of the metabolic transition on a global scale during the last 100 years has led to an increase in yearly material flows from 8 to 60 billion tonnes, while primary energy consumption has increased from 50 to 480 EJ/year (Fig. 15.4c, d). The fact that societal metabolism has become so much larger is partly driven by the pronounced increase in global human population, which roughly quadrupled during the same period. A significant contribution was made, however,

**Table 15.1** Sociometabolic profile of selected countries in 2000

|  | GDP per capita (income) | Material use | Energy use | $CO_2$ emissions | Electricity use | Motor vehicles | Ecological footprint |
|---|---|---|---|---|---|---|---|
|  | $/cap/ year | t/cap/ year | GJ/cap/ year | tC/cap/ year | GJ/cap/ year | #/1,000 Inhab. | ha/cap |
| USA | 31,618 | 28 | 440 | 5.6 | 52 | 761 | 9.6 |
| Japan | 23,804 | 16 | 202 | 2.5 | 31 | 551 | 4.4 |
| France | 23,735 | 17 | 252 | 1.6 | 29 | 548 | 5.6 |
| Germany | 23,391 | 20 | 225 | 2.7 | 25 | 553 | 4.5 |
| UK | 22,560 | 12 | 214 | 2.6 | 24 | 418 | 5.6 |
| Korea | 14,010 | 15 | 208 | 2.5 | 20 | 223 | 4.1 |
| Argentina | 11,012 | 22 | 227 | 1.0 | 9 | 204 | 2.3 |
| Mexico | 8,231 | 15 | 117 | 1.0 | 7 | 144 | 2.6 |
| Brazil | 6,646 | 16 | 139 | 0.5 | 8 | 92 | 2.1 |
| China | 3,491 | 7 | 55 | 0.6 | 4 | 10 | 1.6 |
| India | 2,234 | 6 | 37 | 0.3 | 2 | 11 | 0.8 |

Data sources: The World Bank Group (2007) (Gross domestic product in const. USD for the year 2000, adjusted for purchasing power parity (PPP)); Krausmann et al. (2008a) (Material and energy use); Marland et al. (2007) ($CO_2$ Emissions); IEA (2007) (electricity use); United Nations Department of Economic and Social Affairs (2004) (motor vehicle stocks); Global Footprint Network (2006) (ecological footprint)

by the increase in the volume of per capita material and energy consumption in industrialised countries. In mature industrial economies, the average annual material flow per capita generally amounts to between 15 and 30 tonnes/year and the energy flow to between 200 and 450 GJ/year, while the share of biomass in material expenditure is under 30% (Table 15.1). The metabolic transition also brought with it an enormous diversification in the materials used.[12]

Industrial metabolism produced a large number of regional and a range of global environmental problems and raised societal dominance over natural systems to an entirely new level. At the beginning of the twenty-first century, there are no ecosystems that remain untouched by human influence and many species and ecosystems have vanished altogether or are threatened with extinction (Millennium Ecosystem Assessment 2005). In the case of many environmental problems, mostly classical pollution problems caused by industrialisation and rapid physical growth, technological solutions could indeed be found, thanks to the rising wealth of industrialised nations. Waste disposal systems were installed and problematic substances and toxins were removed from circulation or rendered harmless in a controlled way through the use of filters, decomposition processes and similar measures. In her book *Silent Spring*, published in 1962, Rachel Carson had revealed the consequences for health and ecology of the widespread use of DDT, a highly efficient insecticide, in agriculture (Carson 1962). At the beginning of the 1970s, the use of this toxin was finally banned

---

[12] Up to 60 different metals, including extremely rare types, are used in electronic equipment such as PCs or mobile telephones. Recycling these components, given the tiny amounts and fine distribution, is impossible in many cases (see Hilty 2008, 168).

(at least in industrialised countries). A further example is acid rain and the forest dieback that occurs as a result: sulphur dioxide is released in large quantities by the combustion of fossil fuels and leads via the accumulation of sulphuric acid to acid rain. In the 1970s, this phenomenon, which was already long-known, played a central role in the discourse of environmental politics and it was feared that it would cause the large-scale decline of forests and the acidification of waterways. This led eventually to rigid air purity regulations: the use of filtration technologies and de-sulphurisation of energy sources became established practice so that emissions were vastly reduced. Finally, the hole in the ozone layer became a matter for discussion: only 5 years after its discovery in 1985, an extensive ban on the use of chlorofluorocarbons was agreed and implemented internationally, since when the hole in the ozone layer has slowly been reduced. In this way, some major environmental problems were successfully contained or entirely eliminated – with the newly-developed politics of the environment playing a major role in this respect. However, there are certain fundamental problems caused by the metabolic transition that environmental politics alone are unable to deal with.

Altogether, the material flows created by human activity have attained a similar scale to the material volumes that are converted by the biogeochemical processes of the planet itself: it is estimated that, for example, humankind at the beginning of the twenty-first century is appropriating nearly 30% of annual biomass regrowth (net primary production) and thus a large portion of the means of existence for all heterotrophic organisms (Haberl et al. 2007). Similarly, annual anthropogenic reactive nitrogen emissions meanwhile contribute more to fixed N to terrestrial ecosystems then all natural contributions (Galloway et al. 2008). In response to the increasing domination of the earth system by human activities, scientists have begun to speak about the era of the Anthropocene (Steffen et al. 2007).

This is happening in a situation and at a time in which roughly a billion people live according to an industrial metabolism profile, while the remaining five billion aspire to do so and in part exist in conditions of extreme poverty. Table 15.1 shows the great differences that exist on a global scale between the sociometabolic profiles of different countries and exhibits the close coupling between aggregate income (measured in per capita GDP) and the indicators given for resource and environmental consumption. Between the wealthy industrialised countries and the poorest countries in the world there is a huge difference of between one and two orders of magnitude. In general it is the case that the higher the per capita GDP of a country is, the greater is its environmental consumption.

It was not only the USA that failed to recognise the signs of the times in the early 1970s – the same may be said of all the other industrialised countries. It has not been possible to substitute the old resource-exploiting structures of the industrial regime with a new metabolic profile of an information and communications society, instead the innovations were merely added on top of the existing system. In this sense, the chance to demonstrate a new model of technological and economic development to emerging economies that could have offered a high degree of quality of life at a far lower ecological cost was missed. Now the dynamic of the relationship between society and nature threatens to lead to major crises, if not catastrophes.

If the process of catching up, in which currently two-thirds of the world population are engaged, involves reaching the present metabolic level of developed industrialised countries, even assuming significant gains in efficiency, this would mean almost the tripling of annual resource extraction by 2050. The provision of a growing and increasingly wealthy population with sufficient food is likely to increase pressure on the few remaining untouched ecosystems and will raise land-use intensity on cultivated land. With respect to technical energy, a development of this kind means a return to coal (see e.g., EIA 2010), a trend that is already ongoing globally and that will create high costs or even lead to catastrophe in terms of climate change. Concerning freshwater and drinking water reserves, the situation in many regions of the world is already dire, with the future hanging by the slender threads of climate change. In relation to other resources such as metals and rare earths, competition would increase to such an extent that the danger of this leading to military conflict would rise. The world currently finds itself following precisely this development path – and yet also in a global economic crisis that perhaps may enforce a reversal of the trend.

A political move towards just such a change of direction was attempted with the Kyoto Protocol and with the measure to restrict the emissions of the major outflow of industrial metabolism, i.e. carbon dioxide, on a global scale. $CO_2$ and other so-called greenhouse gases are responsible for producing an alteration to biogeochemical cycles that is unprecedented in human history and that has grave consequences for the global climate system. Rising emissions of greenhouse gases are a direct consequence of increasing combustion of fossil fuels. Currently a global total of over 8 Giga tonnes (Gt) of carbon are emitted, which represents a global per capita rate of 1.5 tonnes per year. Figure 15.5 shows the development of $CO_2$ emissions in a number of selected industrial and developing countries. The trend in the emission rates very closely mirrors the phases of global metabolic transition: the emission rate in England in 1750, i.e. in the early phase of the industrial revolution, was c. 0.25 t of carbon (C) per capita and year.[13] This had doubled by 1800 to 0.5 t/cap/year and by 1850 had doubled again to 1 t/cap/year. The European latecomers and the USA reached the 0.25 t/cap/year threshold only by the middle of the nineteenth century, but then required far less time to double this to 0.5 and then to 1 t/cap/year.

A further race to catch up began in the first half of the twentieth century, when the aspiring industrial nations in Europe (USSR), East Asia (Japan and later also South Korea) and Latin America (Mexico) increased their $CO_2$ emissions within only two or three decades from 0.25 tC per capita and year to over 1 t/cap/year. For the great majority of the world's countries, however, this process has yet to happen: at the beginning of the twenty-first century, large economies such as India or Brazil still present very low rates of emissions of under 0.5 t per capita and year. Industrialised countries today emit on average 3.5 tonnes per capita and year, while all other countries emit less than 0.5 tonnes per capita and year on average.

---

[13] Carbon dioxide ($CO_2$) emissions are often recorded in tonnes of carbon (C). One tonne of C represents 3.67 tonnes of $CO_2$.

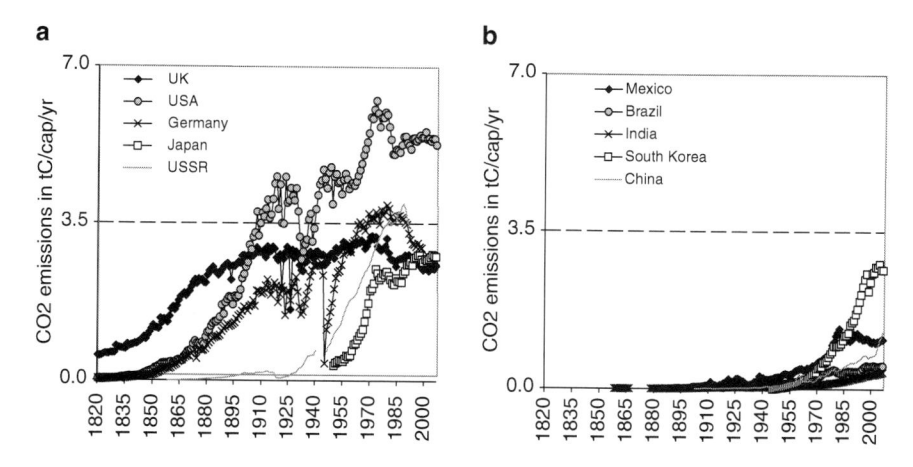

**Fig. 15.5** $CO_2$ emissions resulting from combustion of fossil energy sources and cement production in selected countries. Data given in tonnes of carbon (C) per capita and year. (**a**) Industrialised countries (**b**) Southern hemisphere countries (Source: Authors' own calculations based on Marland et al. 2007 ($CO_2$ emissions); Maddison 2008 (population))

To restrict global warming to the required upper limit of plus 2°, it would be necessary to reduce average emissions rates per capita to 1.3 t/cap/year (IPCC 2007).

## 15.6 Conclusion

At the beginning of the twenty-first century, there are completely different conceptions of how to deal with the global ecological crisis – or in some cases the very existence of such a crisis is even doubted. Then there is the hope placed in technological solutions that might make it possible to continue more or less undisturbed along the path of economic growth taken so far. Many people rely on the notion that, as in the past, technological *end of pipe* solutions will take effect and hope to find a way of tackling the problem of climate change through what is known as *Geo-engineering* (National Academy of Sciences 1992): Examples include experiments to increase the capacity of the oceans to absorb greenhouse gases through large-scale fertilisation, to increase the reflection of incoming solar radiation through artificial cloud formation, or to capture and store underground emitted carbon dioxide through *Carbon Capture and Storage* (CCS) (IPCC 2005).

Others count on efficiency gains that enable products and services to be provided with significantly lower use of energy and materials, which should in turn allow societal metabolism to be decoupled from economic development (Weizsäcker et al. 1995). In connection with a way of life that entails lower consumption and the use of new and carbon-free sources of energy, from wind power to solar electricity and nuclear energy, it should remain possible to have economic growth while simultaneously reducing the pressure on the global climate and the environment in general. Finally, there are also more radical ideas that envision a complete renunciation of the established paradigm of growth (*de-growth*) and thus call for a new model of society, since this would be the only way to sustain the physical conditions for the living requirements of a growing world population in the longer term Jackson (2009).[14]

A long-term socio-ecological perspective delivers no clear answer here, but it does provide important insights. It becomes clear that the industrial society's high demand for material and energy resources is structurally determined and cannot be reduced simply by a more frugal use of resources. In the industrial sociometabolic regime, economic development and metabolism are as closely interlinked as it is possible to be, and gains in efficiency, although sometimes enormous, have never in the past led to a reduction in metabolism but have rather driven further growth (Ayres and Warr 2009). Although the historical perspective shows that technological solutions have often come into play in the past, it also reveals that these very solutions create new types of problems and that a spiral of risk continues to turn. Finally, society will have to recognise that physical growth is limited and that it is thus more important to decouple the quality of human life from further material and energy use. This will not be achieved by means of technological solutions alone but rather requires far-reaching changes to be made in society. Such changes will occur, irrespective of whether the relevant political and economic actors wish them to do so. Those who advocate the concept of sustainable development believe that it would be wiser to organise such a change proactively.

**Acknowledgments** This paper draws on research funded by the Austrian Science Fund FWF (Projects P21012-G11 and P20812-G11). We would like to express our heartfelt gratitude to Rolf Peter Sieferle, who was kind enough to look through earlier versions of this text and made helpful comments and to Helmut Haberl and Martin Schmid for their critical review, which helped to improve the paper. We would also like to thank Michael Neundlinger, who supported us in collecting and preparing the data. Ursula Lindenberg supplied a thoughtful translation of the German original into English.

## References

Ayres, R. U., & Warr, B. (2009). *The economic growth engine: How energy and work drive material prosperity*. Cheltenham/Northhampton: Edward Elgar.
Baccini, P., & Brunner, P. H. (1991). *The metabolism of the anthroposphere*. Berlin: Springer.
Bork, H. R., Bork, H., Dalchow, C., Faust, B., Piorr, H.-P., & Schatz, T. (1998). *Landschaftsentwicklung in Mitteleuropa*. Gotha/Stuttgart: Klett-Perthes.
Boserup, E. (1965). *The conditions of agricultural growth. The economics of agrarian change under population pressure*. Chicago: Aldine/Earthscan.

---

[14] In April 2008 the first scientific conference on the theme of *degrowth* "Economic De-growth for ecological sustainability and social equity" took place in Paris. See http://events.it-sudparis.eu/degrowthconference/en/

Boserup, E. (1981). *Population and technological change – A study of long-term trends*. Chicago: The University of Chicago Press.

Brimblecombe, P. (1987). *The big smoke. A history of air pollution in London since medieval times*. London: Methuen.

Carson, R. (1962). *Silent spring*. Boston: Houghton Mifflin Company.

Crosby, A. W. (1986). *Ecological imperialism. The biological expansion of Europe, 900–1900*. Cambridge: Cambridge University Press.

Cunfer, G. (2005). *On the Great Plains: Agriculture and environment*. College Station: Texas A&M University Press.

Cusso, X., Garrabou, R., & Tello, E. (2006). Social metabolism in an agrarian region of Catalonia (Spain) in 1860 to 1870: Flows, energy balance and land use. *Ecological Economics, 58*(1), 49–65.

Darby, H. C. (1956). The clearing of the woodland in Europe. In W. L. Thomas Jr. (Ed.), *Man's role in changing the face of the Earth* (pp. 183–216). Chicago: The University of Chicago Press.

De Zeeuw, J. W. (1978). Peat and the Dutch golden age (A.A.G. Bijdragen 21, pp. 3–31). Wageningen: Afdeling Agrarische Geschiedenis Landbouwhogeschool.

Deffeyes, K. S. (2001). *Hubbert's peak, the impending world oil shortage*. Princeton: Princeton University Press.

Diamond, J. M. (2005). *Collapse: How societies choose to fail or succeed*. New York: Viking.

EIA. (2010). *International energy outlook 2010*. Washington, DC: Energy Information Administration, Department of Energy (DOE).

Fischer-Kowalski, M. (1998). Society's metabolism. The intellectual history of material flow analysis, part I: 1860–1970. *Journal of Industrial Ecology, 2*(1), 61–78.

Fischer-Kowalski, M., & Haberl, H. (1997). Stoffwechsel und Kolonisierung: Konzepte zur Beschreibung des Verhältnisses von Gesellschaft und Natur. In M. Fischer-Kowalski et al. (Eds.), *Gesellschaftlicher Stoffwechsel und Kolonisierung von Natur* (pp. 3–12). Amsterdam: Gordon & Breach Fakultas.

Freund, P., & Martin, G. T. (1993). *The ecology of the automobile*. Montreal: Black Rose Publishing.

Galloway, J. N., Townsend, A. R., Erisman, J. W., Bekunda, M., Cai, Z., Cai, J., Freney, L. A., Martinelli, S. P., Seitzinger, M., & Sutton, A. (2008). Transformation of the nitrogen cycle: Recent trends, questions and potential solutions. *Science, 320*, 889–892.

Global Footprint Network. (2006). *Ecological footprint and biocapacity data*. http://www.footprintnetwork.org/en/index.php/GFN/page/footprint_for_nations/

Grigg, D. B. (1992). *The transformation of agriculture in the west*. Oxford: Blackwell.

Grübler, A. (1998). *Technology and global change*. Cambridge: Cambridge University Press.

Haberl, H., Winiwarter, V., Andersson, K., Ayres, R. U., Boone, C. G., Castillio, A., Cunfer, G., Fischer-Kowalski, M., Freudenburg, W. R., Furman, E., Kaufmann, R., Krausmann, F., Langthaler, E., Lotze-Campen, H., Mirtl, M., Redman, C. A., Reenberg, A., Wardell, A. D., Warr, B., & Zechmeister H. (2006). From LTER to LTSER: Conceptualizing the socio-economic dimension of long-term socio-ecological research. *Ecology and Society, 11*(2), 13. (Online) http://www.ecologyandsociety.org/vol11/iss2/art13/

Haberl, H., Erb, K.-H., Krausmann, F., Gaube, V., Bondeau, A., Plutzar, C., Gingrich, S., Lucht, W., & Fischer-Kowalski, M. (2007). Quantifying and mapping the human appropriation of net primary production in earth's terrestrial ecosystems. *Proceedings of the National Academy of Sciences of the United States of America, 104*(31), 12942–12947.

Hall, C. A. S., Cleveland, C. J., & Kaufmann, R. K. (1986). *Energy and resource quality. The ecology of the economic process*. New York: Wiley Interscience.

Harris, M., & Ross, E. (1987). *Death, sex & fertility: Population regulation in pre-industrial & developing societies*. New York: Columbia University Press.

Hilty, L. M. (2008). *Information technology and sustainability. Essays on the relationship between information technology and sustainable development*. Norderstedt: Books on Demand.

IEA. (2007). *Energy statistics of non-OECD countries, 2004–2005* (2007 ed.) CD-ROM. Paris: International Energy Agency (IEA), Organisation of Economic Co-Operation and Development (OECD).

IPCC. (2005). *IPCC special report on carbon dioxide capture and storage. Prepared by Working Group III of the Intergovernmental Panel on Climate Change.* Cambridge/New York: Cambridge University Press.
IPCC. (2007). *Climate Change 2007 – Impacts, adaptation and vulnerability. Contribution of the Working Group II to the fourth assessment report of the IPCC.* Cambridge/New York/Melbourne: Cambridge University Press.
Jackson, T. (2009). Prosperity without growth: Economics for a finite planet. London: Earthscan.
Krausmann, F. (2004). Milk, manure and muscular power. Livestock and the industrialization of agriculture. *Human Ecology, 32*(6), 735–773.
Krausmann, F., & Cunfer, G. (2009, August 4–8). *Agroecosystems on the American frontier: Material and energy systems and sustainability.* Presentation at the World Congress of Environmental History, Copenhagen.
Krausmann, F., Fischer-Kowalski, M., Schandl, H., & Eisenmenger, N. (2008a). The global sociometabolic transition: Past and present metabolic profiles and their future trajectories. *Journal of Industrial Ecology, 12*(5/6), 637–656.
Krausmann, F., Schandl, H., & Sieferle, R. P. (2008b). Socio-ecological regime transitions in Austria and the United Kingdom. *Ecological Economics, 65*(1), 187–201.
Krausmann, F., Gingrich, S., Eisenmenger, N., Erb, K.-H., Haberl, H., & Fischer-Kowalski, M. (2009). Growth in global materials use, GDP and population during the 20th century. *Ecological Economics, 68*(10), 2696–2705.
Leach, G. (1976). *Energy and food production.* Guildford: IPC Science and Technology Press.
Leaf, M. J. (2004). Green revolution. In S. Krech III et al. (Eds.), *Encyclopedia of world environmental history* (pp. 615–619). London/New York: Routledge.
Lutz, B. (1989). *Der kurze Traum immerwährender Prosperität. Eine Neuinterpretation der industriellkapitalistischen Entwicklung im Europa des 20. Jahrhunderts.* Frankfurt am Main/New York: Campus Verlag.
Maddison, A. (2008). *Historical statistics for the world economy: 1–2006 AD.* http://www.ggdc.net/maddison/
Marland, G., Boden, T. A., & Andres, R. J. (2007). Global, regional, and national $CO_2$ emissions. In Oak Ridge National Laboratory, U.S. Department of Energy (Ed.), *Trends: A compendium of data on global change.* Oak Ridge: Carbon Dioxide Information Analysis Center (CDIAC).
Mazoyer, M., Roudart, L., & Membrez, J. H. (2006). *A history of world agriculture: From the neolithic age to the current crisis.* London: Earth Scan.
McNeill, J. R. (2000). *Something new under the sun. An environmental history of the twentieth century.* London: Allen Lane.
Meadows, D. L., Meadows, D. H., & Randers, J. (1972). *Die Grenzen des Wachstums. Bericht an den Club of Rome.* Stuttgart: DVA.
Millennium Ecosystem Assessment. (2005). *Ecosystems and human well-being: Vol. 1. Current state and trends.* Washington/Covelo/London: Island Press.
Mitchell, B. R. (2003). *International historical statistics.* New York: Palgrave Mcmillan.
Müller-Herold, U., & Sieferle, R. P. (1998). Surplus and survival: Risk, ruin and luxury in the evolution of early forms of subsistence. *Advances in Human Ecology, 6*, 201–220.
National Academy of Sciences. (1992). Geoengineering. In Committee on Science Engineering and Public Policy (U.S.). Panel on Policy Implications of Greenhouse Warming (Ed.), *Policy implications of greenhouse warming: Mitigation, adaptation, and the science base* (pp. 433–464). Washington, DC: National Academy Press.
Pfister, C. (2003). Energiepreis und Umweltbelastung. Zum Stand der Diskussion über das "1950er Syndrom". In W. Siemann (Ed.), *Umweltgeschichte Themen und Perspektiven* (pp. 61–86). München: C.H. Beck.
Pimentel, D., & Pimentel, M. (1979). *Food, energy and society.* London: Edward Arnold.
Podobnik, B. (1999). Toward a sustainable energy regime, a long-wave interpretation of global energy shifts. *Technological Forecasting and Social Change, 62*(3), 155–172.
Prentice, I. C., Heimann, M., & Sitch, S. (2001). Contribution of Working Group I to the third assessment report of the Intergovernmental Panel on Climate Change. In J. T. Houghton et al.

(Eds.), *Climate change 2001: The scientific basis* (pp. 183–237). Cambridge, MA: Cambridge University Press.

Ruddiman, W. F. (2003). The anthropogenic greenhouse era began thousands of years ago. *Climatic Change, 61*(3), 261–293.

Schandl, H., & Krausmann, F. (2007). The great transformation: A socio-metabolic reading of the industrialization of the United Kingdom. In M. Fischer-Kowalski & H. Haberl (Eds.), *Socioecological transitions and global change: Trajectories of social metabolism and land use* (pp. 83–115). Cheltenham/Northampton: Edward Elgar.

Sieferle, R. P. (1997). *Rückblick auf die Natur: Eine Geschichte des Menschen und seiner Umwelt*. München: Luchterhand.

Sieferle, R. P. (2001). *The subterranean forest. Energy systems and the industrial revolution*. Cambridge: The White Horse Press.

Sieferle, R. P. (2003). Nachhaltigkeit in universalhistorischer Perspektive. In W. Siemann (Ed.), *Umweltgeschichte Themen und Perspektiven* (pp. 39–60). München: C.H. Beck.

Singh, S. J., Haberl, H., Gaube, V., Grünbühel, C. M., Lisievici, P., Lutz, J., Matthews, R., Mirtl, M., Vadineanu, A., & Wildenberg, M. (2010). Conceptualising Long-term socio-ecological research (LTSER): Integrating the social dimension. In F. Müller et al. (Eds.), *Long-term ecological research, between theory and application* (pp. 377–398). Dordrecht/Heidelberg/London/New York: Springer.

Smil, V. (2001). *Enriching the earth. Fritz Haber, Carl Bosch, and the transformation of world food production*. Cambridge, MA: MIT Press.

Smil, V. (2003). *Energy at the crossroads. Global perspectives and uncertainties*. Cambridge, MA/London/England: MIT Press.

Steffen, W., Crutzen, P. J., & McNeill, J. R. (2007). The anthropocene: Are humans now overwhelming the great forces of nature. *Ambio, 36*(8), 614–621.

Tainter, J. A. (1988). *The collapse of complex societies*. Cambridge: Cambridge University Press.

The World Bank Group. (2007). *World development indicators 2007*. CD-ROM. Washington, DC: The World Bank.

United Nations, D. o. E. a. S. A. (2004). *Statistical yearbook* (Forty Eight issue). New York: United Nations.

von Gottl-Ottlilienfeld, F. (1924). *Fordismus. Über Industrie und Technische Vernunft*. Jena: Fischer.

Weizsäcker, E. U., Lovins, A. B., & Lovins, H. L. (1995). *Faktor Vier – Doppelter Wohlstand, halbierter Naturverbrauch. Der neue Bericht an den Club of Rome*. München: Droemer Knaur.

# Part III
# LTSER Formations and the Transdisciplinary Challenge

# Chapter 16
# Building an Urban LTSER: The Case of the Baltimore Ecosystem Study and the D.C./B.C. ULTRA-Ex Project

J. Morgan Grove, Steward T.A. Pickett, Ali Whitmer, and Mary L. Cadenasso

**Abstract** There is growing scientific interest, practical need, and substantial support for understanding urban and urbanising areas in terms of their long-term social and ecological trajectories: past, present, and future. Long-term social-ecological research (LTSER) platforms and programmes in urban areas are needed to meet these interests and needs. We describe our experiences as a point of reference for other ecologists and social scientists embarking on or consolidating LTSER research in hopes of sharing what we have learned and stimulating comparisons and collaborations in urban, agricultural, and forested systems. Our experiences emerge from work with two urban LTSERs: the Baltimore Ecosystem Study (BES) and the District of Columbia-Baltimore City Urban Long-Term Ecological Research Area-Exploratory DC-BC ULTRA-Ex project. We use the architectural metaphor of constructing and maintaining a building to frame the description of our experience with these two urban LTSERs. Considering each project to be represented as a building gives the following structure to the chapter: (1) building site context; (2) building structure; and (3) building process and maintenance.

---

J.M. Grove, Ph.D. (✉)
Northern Research Station, US Department of Agriculture Forest Service,
Baltimore, MD, USA
e-mail: mgrove@fs.fed.us

S.T.A. Pickett, Ph.D.
Cary Institute of Ecosystem Studies, Millbrook, NY, USA
e-mail: picketts@caryinstitute.org

A. Whitmer, Ph.D.
Georgetown University, Washington, DC, USA
e-mail: whitmer@georgetown.edu

M.L. Cadenasso, Ph.D.
Department of Plant Sciences, University of California, Davis, CA, USA
e-mail: mlcadenasso@ucdavis.edu

**Keywords** Urban Ecology • LTSER • Social Ecology • Baltimore • Sustainability

## 16.1 Introduction: A Building Tour

In this chapter, we address the challenge of building long-term social-ecological research (LTSER) platforms and programmes in urban areas. The motivation for addressing this challenge is that there is growing scientific interest, practical need, and substantial support for understanding urban and urbanising areas in terms of their long-term social and ecological trajectories: past, present, and future.

We present this overview of our experiences as a point of reference for other ecologists and social scientists embarking on or consolidating LTSER research in hopes of sharing what we have learned and stimulating comparisons and collaborations in urban, agricultural, and forested systems. Our experience emerges from work with two urban LTSERs: the Baltimore Ecosystem Study (BES) and the District of Columbia-Baltimore City Urban Long-Term Ecological Research Area-Exploratory – DC-BC ULTRA-Ex – project. The two projects partially overlap in their geography, but are motivated and structured differently. Hence, this chapter benefits from both similar and contrasting experiences.

We use the architectural metaphor of constructing and maintaining a building to frame the description of our experience with two these urban LTSERs. Considering each project to be represented as a building gives the following structure to the chapter (1) building site context, (2) building structure, and (3) building process and maintenance.

## 16.2 Building Site Context: Historical Origins of the Baltimore Ecosystem Study

The history of BES is an important thread in the emergence of urban ecological science in the United States. Although ecological thinking had been applied to American cities by specialists in social sciences, geography, planners, and urban designers, these important strands did not have very much empirical input from scientists trained in ecology. Urban wildlife ecology had been well developed, and there were also empirical studies to assess effects of urban contaminants (Vandruff et al. 1994), yard management (Loucks 1994), or urban metabolism (Boyden 1979). Calls for concerted action in the 1970s were cogent and forward looking, but ultimately they did not fundamentally expand the focus of ecological science to cities (Stearns 1970; Stearns and Montag 1974). In this relative empirical desert BES was established.

The roots of BES are in comparative ecological science conducted in the New York metropolitan region. Dr. Mark McDonnell, then of the Cary Institute of Ecosystem Studies in his role as forest ecologist for the New York Botanical Garden (NYBG) in the Bronx, New York, planted a seed which prepared the way for BES and which we believe was crucial in establishing contemporary urban ecological science in the US. When he and Dr. Carl White attempted to compare the nitrogen metabolism in the old growth forest on the grounds of the NYBG in 1985, they discovered the soils to be hydrophobic. Although this phenomenon was known from other cities, the finding stimulated McDonnell to broaden his comparison between the urban forest and other oak forests on similar substrates but located at greater distances from the New York City urban core. Ultimately, this comparison became known as the Urban-rural Gradient Ecology (URGE) project, and was advanced by interactions with Dr. Richard Pouyat, and an increasingly broad group of ecological researchers such as Dr. Margaret Carreiro.

With support from the Cary Institute of Ecosystem Studies under the leadership of Dr. Gene Likens, a postdoctoral researcher in geography, Dr. Kimberly Medley was hired as the first expert in social structures and processes to join the collaboration. The URGE project could soon claim a plethora of findings concerning soil contamination, nitrogen loading, denitrification, the role of exotic earthworms and different fungi, and forest structure along the gradient, for example (McDonnell et al. 1997; Pouyat et al. 2009). Attempts to increase the scope of the study by adding social science and economic collaborators beyond the expertise provided by Medley met with little success, due perhaps to limited interactions at the time between the social sciences and ecological sciences in general, and to the high level of prior commitment that characterised the social scientists McDonnell and colleagues approached.

In 1993, McDonnell became director of the Bartlett Arboretum, the Connecticut State Arboretum. Research on the New York metropolitan URGE project began to be carried out in diverse institutional homes of the established collaborators and as graduate students and post-doctoral associates moved on to other positions. Continued efforts to establish working relationships with social scientists bore fruit when Pouyat introduced McDonnell and Steward Pickett to his colleague in the USDA Forest Service, Dr. J. Morgan Grove, and through him to the social ecologist Dr. William R. Burch, Jr., from Yale University's School of Forestry and Environmental Studies and to their decade long research project in Baltimore, Maryland (Grove and Hohmann 1992). The desire of Grove and Burch to familiarise the ecologists with their social science research and community engagement resulted in a field trip to Baltimore. It became clear that these social scientists had established extensive social capital, including "street cred" in Baltimore. Their social networks with communities, action-oriented Non-Governmental Organisations (NGOs), and key environmentally relevant Baltimore city agencies were a precious and site-specific resource. If the desire to increase integration between biophysical science and social science were to be fulfilled, Baltimore seemed to be an ideal place to realise that goal. Pickett's position permitted him the freedom to pursue funding opportunities using Baltimore as a research arena. Over the next few years, colleagues interested in Baltimore, some from Baltimore area institutions like the University of Maryland, Baltimore County, and Johns Hopkins University were courted and became contributors to an emerging intellectual framework to support integrated biophysical and social research and outreach in Baltimore.

These efforts positioned the informal network of researchers to respond forcefully to the National Science Foundation's (NSF) call for proposals for up to two urban Long-Term Ecological Research sites in the United States. These urban sites would complement some two dozen other LTER sites that had been established across the United States to understand the structure and functioning of forested, grassland, agricultural, coastal, lake, river, and tundra ecosystems. The programme officers at NSF, especially the late Dr. Thomas Callahan, were convinced that ignoring urban systems left a gap in the understanding of America's diversity of ecological systems.

The initial team that produced the proposal included soil scientists, vegetation ecologists, ecological economists, social scientists, educators, landscape ecologists, spatial analysts, paleoecologists, hydrologists, microbial ecologists, climatologists, geomorphologists, and wildlife ecologists. The team included graduate and postdoctoral students, leaders in the NGO Parks & People Foundation, senior researchers from academic institutions and federal agencies including the USDA Forest Service and the US Geological Survey. Indeed, as budget planning proceeded, it became clear that the in-kind support for research and staff time from the USDA Forest Service under the leadership of Dr. Robert Lewis of the (then) Northeastern Research Station would exceed the funds available from NSF. Partnerships were sought with managers and policy makers from Baltimore City, Baltimore County and the State of Maryland, including environmental officers and school officials.

Various options for a Baltimore home, given that the proposed grantee institution would be the Cary Institute of Ecosystem Studies in Millbrook, NY, were sought. Ultimately the enthusiastic support of Dr. Freeman Hrabowski, III, President of the University of Maryland, Baltimore County to host the Baltimore offices and labs of the new urban LTER led to the establishment of a convenient and intellectually engaging home for the project. The Baltimore Ecosystem Study, named analogously to the Hubbard Brook Ecosystem Study LTER from which we drew inspiration for the experimental watershed approach, was established in November of 1997. As a result of the same competition, the NSF also funded the Central Arizona-Phoenix LTER, headquartered at Arizona State University and led by stream ecosystem ecologist Dr. Nancy Grimm, and archaeologist Dr. Charles Redman.

## 16.3 Building Structure

### 16.3.1 Why We Seek to Know

#### 16.3.1.1 Practical Motivations

There are several motivations for developing LTSER research focused on urban and urbanising areas. From a practical perspective, it is essential to recognise that *the Earth is an urban planet*. In 1800, about 3% of the world's human population lived in urban areas. By 1900, this proportion rose to approximately 14% and exceeded 50% by 2008. Every week nearly 1.3 million additional people arrive in the world's cities, amounting to a total of about 70 million a year (Brand 2006; Chan 2007). The urban population of the globe is projected by the UN to climb to 61% by 2030 and eventually reach a dynamic equilibrium of approximately 80%

urban to 20% rural dwellers that will persist for the foreseeable future (Brand 2006; Johnson 2006). This transition from 3% urban population to the projected 80% urban is a massive change in the social-ecological dynamics of the planet (Brand 2009; Seto et al. 2010).

The spatial extent of urban areas is growing as well. In industrialised nations the conversion of land from wild and agricultural uses to urban and suburban settlement is growing at a faster rate than the growth in urban population. Cities are no longer compact (Pickett et al. 2001); they sprawl in fractal or spider-like configurations (Makse et al. 1995) and increasingly intermingle with wild lands. Even for many rapidly growing metropolitan areas, suburban zones are growing faster than other zones (Katz and Bradley 1999). The resulting new forms of urban development include edge cities (Garreau 1991) and a wildland-urban interface in which housing is interspersed in forests, shrublands, and desert habitats.

An important consequence of these trends in urban growth is that cities have become *the dominant global human habitat* of this century in terms of geography, experience, constituency and influence. Accompanying the spatial changes are changes in perspectives and constituencies. Although these urbanising habitats were formerly dominated by agriculturists, foresters and conservationists, they are now increasingly dominated by people possessing resources from urban systems, drawing upon urban experiences and expressing urban habits. This reality has important consequences for social and ecological systems at global, regional and local scales, as well as for natural resource organisations attempting to integrate ecological function with human desires, behaviours and quality of life.

## From Local to Global, Cities Play a Critical Role in Climate Change Vulnerabilities, Mitigation, and Adaptation

Urbanisation creates both ecological vulnerabilities and efficiencies. For instance, coastal areas, where many of the world's largest cities occur, are home to a wealth of natural resources that are rich with diverse species, habitat types and productive potential. They are also vulnerable to land conversion, changes in hydrologic flows, outflows of sediment and waste and sea level rise (Grimm et al. 2008). In the US, 10 of the 15 most populous cities are located in coastal counties (NOAA 2004) and 23 of the 25 most densely populated US counties are in coastal areas. These areas have already experienced ecological disruptions (Couzin 2008).

While ecological vulnerabilities are significantly associated with urban areas, urbanisation also fosters ecological efficiencies. The ecological footprint of a city, i.e., the land area required to support it, is quite large (Folke et al. 1997; Johnson 2006; Grimm et al. 2008). Cities consume enormous amounts of natural resources, while the assimilation of their wastes – from sewage to the gases that cause global warming – are also distributed over large areas. London, for example, occupies 1,70,000 ha and has an ecological footprint of 21 million hectares – 125 times its size (Toepfer 2005). In Baltic cities, the area needed from forest, agriculture, and marine ecosystems corresponds to approximately 200 times the area of the cities themselves (Folke et al. 1997).

Ecological footprint analysis can be misleading, however, for numerous reasons (Deutsch et al. 2000). It ignores the more important question of efficiency, defined here as persons-to-area: how much land area (occupied area and footprint area) is needed to support a certain number of persons? From this perspective, it becomes clear that urbanisation is critical to delivering a more ecologically sustainable and resource-efficient world because the per-person environmental impact of city dwellers is generally lower than people in the countryside (Brand 2006; Johnson 2006; Grimm et al. 2008). For instance, the average New York City resident generates about 29 % of the carbon dioxide emissions of the average American. By attracting 9,00,000 more residents to New York City by 2030, New York City can actually save 15.6 million metric tons of carbon dioxide per year relative to the emissions of a more dispersed population (Chan 2007). The effects of urbanisation on ecological efficiency may mean that social-ecological pressures on natural systems can be dramatically reduced in terms of resources used, wastes produced, and land occupied. This may mean that cities can provide essential solutions of mitigation and adaptation to the long-term social-ecological viability of the planet, given current population trends for this century.

Current global demographic trends are paralleled by changing conceptions of cities and urbanisation. In very broad historical terms we have begun a new paradigm for cities. Since the 1880s, a great deal of focus has centred on the "Sanitary City," with concern for policies, plans and practices that promoted public health (Melosi 2000). While retaining the fundamental concern for the Sanitary City, we have begun to envelop the Sanitary City paradigm with a concern for the "Sustainable City," which places urbanisation in a social-ecological context at local, regional and global scales (Pincetl 2010).

Urban ecology and long-term studies have a significant role to play in this context. Already, urban ecology has an important applied dimension as an approach used in urban planning, especially in Europe. Carried out in city and regional agencies, the approach combines ecological information with planning methodologies (Hough 1984; Spirn 1984; Schaaf et al. 1995; Thompson and Steiner 1997; Pickett et al. 2004; Pickett and Cadenasso 2008).

Major investments in urban ecology theory, data, and practices are required to meet the needs of cities and urbanising areas. Cities face challenges that are increasingly complex and uncertain. Many of these complexities are associated with changes in climate, demography, economy, and energy at multiple scales. Because of these complex, interrelated changes, concepts such as resilience (Gunderson 2000), vulnerability (Turner et al. 2003), and ecosystem services (Bolund and Hunhammar 1999) may be particularly useful for addressing both current issues and preparing for future scenarios requiring long-term, and frequently capital-intensive, change.

Cities have already begun to address these challenges and opportunities in terms of policies, plans, and management. For example, on June 5th, 2005, mayors from around the globe took the historic step of signing the Urban Environmental Accords – Green City Declaration with the intent of building ecologically sustainable, economically dynamic and socially equitable futures for their urban citizens. The Accords covered seven environmental categories to enable sustainable urban living

Applied and Basic Research

| Quest for fundamental understanding? | | Considerations of use? | |
|---|---|---|---|
| | | No | Yes |
| | Yes | Pure basic research (Physicist Bohr) | Use-inspired basic research (Biologist Pasteur) |
| | No | -- | Pure applied research (Inventor Edison) |

**Fig. 16.1** In *Pasteur's Quadrant*, Stokes categorises four different types of research. Most research associated with BES would be located in Pasteur's quadrant: Use-inspired basic research (Adapted from Stokes 1997)

and improve the quality of life for urban dwellers: (1) energy, (2) waste reduction, (3) urban design, (4) urban nature, (5) transportation, (6) environmental health, and (7) water (http://www.citymayors.com/environment/environment_day.html). International associations such as *ICLEI-Local Governments for Sustainability* (http://www.iclei.org/) are developing and sharing resources to address these issues. The ability to address these seven categories will require numerous, interrelated strategies and scientific domains.

### 16.3.1.2 Scientific Motivations

There are diverse scientific motivations for examining urban areas as LTSERs. Stokes (1997) offers a useful heuristic, *Pasteur's Quadrant* (Fig. 16.1), for different motivations or categories of scientific research. Three of these quadrants are of particular interest for urban LTSERs. The two most familiar quadrants may be the first and third quadrants. Stokes defines the first quadrant, Pure Basic Research, as science performed without concern for practical ends. This quadrant is labelled Bohr's Quadrant since physicist Nils Bohr had no immediate concern for use as he worked to develop a structural understanding of the atom. In this quadrant LTSERs work to understand physical, biological, and social laws that advance our fundamental understanding of the world. For instance, urban systems can be useful end members for understanding the effects of altered climates, organismal components, substrates and land forms (Zipperer et al. 1997; Carreiro et al. 2009), or changes in livelihoods and lifestyles on consumption, social institutions, identity and status. The third quadrant, Pure Applied Research, is defined as science performed to solve a social problem without regard for advancing fundamental theory or scientific method. Stokes labelled this Edison's Quadrant, since inventor Thomas Edison never considered the underlying implications of his discoveries in his pursuit of commercial illumination. In this quadrant LTSERs work to develop solutions to specific problems, such as bio-retention systems for removing pollutants from stormwater or social marketing to increase household participation in tree planting programmes.

Stokes defines the second quadrant, Use-inspired Basic, as science that is designed to both enhance fundamental understanding and address a practical issue. This quadrant is labelled Pasteur's Quadrant, because biologist Louis Pasteur's work on immunology and vaccination both advanced our fundamental understanding of biology and saved countless lives. In this quadrant LTSERs work to advance scientific theories and methods while addressing practical problems; for example, how do households' locational choices affect ecosystem services and vulnerabilty to climate change or how do ecological structures and social institutions interact over the long term to affect urban resilience and sustainability? While BES research can be located in each of these quadrants, most BES research is Use-inspired Basic located in Pasteur's Quadrant and addresses many of the practical motivations identified earlier (Pickett et al. 2011).

### *16.3.2 What We Seek to Know: From an Ecology in Cities to an Ecology of Cities*

A driving idea for the design of the BES has been to promote the transition from an "Ecology *in* Cities" to an "Ecology *of* Cities." The study of social-ecological systems in general, and urban systems in particular has been an emerging area of significant attention over the past 15 years. During this time, a body of research and applications emerged that may be labelled "The Ecology *of* Cities". This corpus of work may be best described as the transition in the study of urban systems from an "Ecology *in* Cities" to an "Ecology *of* Cities," (Pickett et al. 1997a; Grimm et al. 2000) where the study of the "Ecology *in* Cities" focused historically on ecologically familiar places and compared urban and non-urban areas: parks as analogues of rural forests (e.g. Attorre et al. 1997; Kent et al. 1999) and vacant lots as analogues of fields or prairies (Vincent and Bergeron 1985; Cilliers and Bredenkamp 1999). Urban streams and remnant wetlands were the object of ecological studies similar in scope and method to those conducted in non-urban landscapes.

An "Ecology *of* Cities" in its current form incorporates new approaches from ecology in general and ecosystem ecology in particular. An "Ecology *of* Cities" builds upon but is very different from an "Ecology *in* Cities." First, the "Ecology *of* Cities" addresses the complete mosaic of land uses and management in metropolitan systems, not just the green spaces as rural analogues that were the focus of "Ecology *in* Cities". Second, spatial heterogeneity, expressed as gradients or mosaics, is critical for explaining interactions and changes in the city. Third, humans and their institutions are a part of the ecosystem, not simply external, allegedly negative influences. Finally, the role of humans at multiple scales of social organisation, from individuals through to households and neighbourhoods, and to complex and persistent agencies, is linked to the biophysical scales of urban systems. Thus, an Ecology of Cities opens the way towards understanding feedbacks among the biophysical and human components of the system, towards placing them in their dynamic spatial and temporal contexts, and towards examining their effects on ecosystem inputs and outputs at various social scales, including individuals, households, neighbourhoods, municipalities, and regions (Grove and Burch 1997).

The shift to addressing the complete mosaic of land uses and management in metropolitan systems is also important to the practical needs of enhancing urban sustainability. Most of the land in urban areas is not in "urban-rural" analogues. For example, the division between public and private ownership in the City of Philadelphia is 33% public and 67% private. Of those private lands, 85% are residential lands, with 459,524 individual parcels. Thus, in many cases, the new "forest owner" in urban areas is most often a residential homeowner. The shift to an "Ecology *of* Cities" provides a much better scientific understanding of the social and ecological characteristics of diverse ownerships and the dominant ownership type, which is required in order to enhance urban sustainability.

### *16.3.3 How We Seek to Know: Integrative Tools*

To facilitate the transition from an Ecology *in* Cities to an Ecology *of* Cities, we employ a suite of integrative tools in our LTSER toolbox: (1) An Ecosystem Approach and LTSER Frameworks, (2) Patch Dynamics, (3) Complexity in Social-Ecological Systems, (4) Scalable Data Platforms, and (5) Watersheds. Many research projects have attempted to bring mainstream ecology and crucial social sciences more closely together (Pickett et al. 1997b; Grimm et al. 2000; Alberti et al. 2003; Redman et al. 2004).[1]

#### 16.3.3.1 An Ecosystem Approach and LTSER Frameworks

We employ the ecosystem concept because of its utility for integrating the physical, biological, and social sciences (Pickett and Grove 2009) and addressing (1) differences among land uses and variations in management within and among land uses in terms of fluxes of individuals, energy, nutrients, materials, information and capital, and (2) ecological structures and social institutions that may regulate those fluxes.

The ecosystem concept owes its origin to Tansley (1935), who noted that ecosystems can be of any size, as long as the concern was with the interaction of organisms and their environment in a specified area. Further, the boundaries of an ecosystem

---

[1] When we began to contribute to this research agenda through the Baltimore Ecosystem Study in 1997, it was important to employ familiar concepts that each discipline could embrace. Hence, we began with ecosystems, watersheds, and patch dynamics as tools to organize research and conceptualize an urban area as an interdisciplinary research topic (Cadenasso et al. 2006). The sequence of integrative tools we present in this chapter is different from their historical development in BES. However, we chose the sequence presented here because watersheds are a particular application of the preceding four tools.

**The Human Ecosystem**

```
┌─────────────────────────────────────────────────────┐
│  ┌ ─ ─ ─ ─ ─ ─ ─ ─ ─ ─ ─ ─ ─ ─ ─ ─ ─ ─ ─ ─ ─ ─ ┐    │
│  │      The Bioecological Ecosystem            │    │
│  │  ┌──────────────┐      ┌──────────────────┐ │    │
│  │  │Biotic Complex│◄────►│ Physical Complex │ │    │
│  │  └──────────────┘      └──────────────────┘ │    │
│  └ ─ ─ ─ ─ ─ ─ ─ ─ ─ ─ ─ ─ ─ ─ ─ ─ ─ ─ ─ ─ ─ ─ ┘    │
│          ▲  ▲           ▲  ▲                         │
│          │   ╲         ╱   │                         │
│          │    ╲       ╱    │                         │
│          ▼     ╲     ╱     ▼                         │
│  ┌──────────────┐      ┌──────────────────┐          │
│  │Social Complex│◄────►│   Built Complex  │          │
│  └──────────────┘      └──────────────────┘          │
└─────────────────────────────────────────────────────┘
```

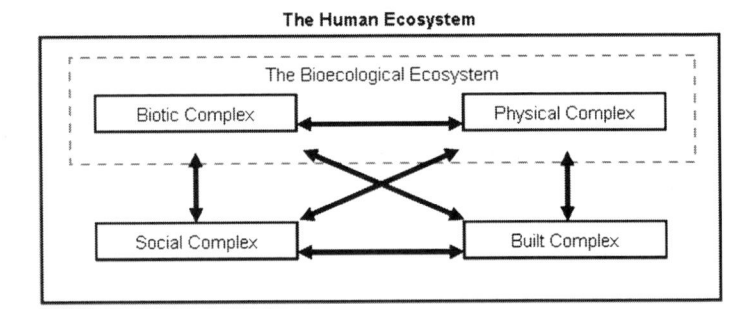

**Fig. 16.2** The human ecosystem concept, bounded by the *bold line*, showing its expansion from the bioecological concept of the ecosystem as proposed originally by Tansley (1935) in the *dashed line*. The expansion incorporates a social complex, which consists of the social components and a built complex, which includes land modifications, buildings, infrastructure, and other artefacts. Both the biotic and the physical environmental complexes of urban systems are expected to differ from those in non-urban ecosystems (Figure copyright BES LTER and used by permission (Pickett and Grove 2009))

are drawn to answer a particular question. Thus, there is no set scale or way to bound an ecosystem. Rather, the choice of scale and boundary for defining any ecosystem depends upon the question asked and is the choice of the investigators. In addition, each investigator or team may place more or less emphasis on the chemical transformations and pools of materials drawn on or created by organisms; or on the flow, assimilation, and dissipation of biologically metabolisable energy; or on the role of individual species or groups of species on flows and stocks of energy and matter. The fact that there is so much choice in the scales and boundaries of ecosystems, and how to study and relate the processes within them, indicates the profound degree to which the ecosystem represents a research approach rather than a fixed scale or type of analysis (Allen and Hoekstra 1992; Pickett and Cadenasso 2002).

The Human Ecosystem Framework

When Tansley (1935) originated the term 'ecosystem', he noted carefully that "… *ecology must be applied to conditions brought about by human activity. The "natural" entities and the anthropogenic derivatives alike must be analyzed in terms of the most appropriate concepts we can find.*" An explicit conception of the human ecosystem brings all the resilient ideas in Tansley's original, core ecosystem concept together. Tansley's core definition of ecosystem was focused on the main ecological topics of his day: organisms and the physical environment. That way of conceiving of the ecosystem is outlined in the inner, dashed box in Fig. 16.2. However, if ecologists are to account for all the kinds of patterns, processes, and interactions that have been identified for social-ecological research (Machlis et al. 1997; Redman et al. 2004; Collins et al. 2011), then it is useful to include two additional "complexes" within the idea of the ecosystem appropriate for the twenty-first century (Fig. 16.2).

It is important to note that each complex shown in Fig. 16.2 can be disaggregated. In the case of the social complex, social scientists have focused on interactions between humans and their environments since the origin of their disciplines. Further, social scientists have focused specifically on an expanded view of the ecosystem approach that includes humans along a continuum from urban areas to wilderness since the 1950s (Hawley 1950; Schnore 1958; Duncan 1961, 1964; Burch and DeLuca 1984; Machlis et al. 1997). Recently, the social and medical sciences have focused increasingly on the ecosystem concept because of its usefulness for natural resource policy and management (Rebele 1994) and public health (Northridge et al. 2003).

Frameworks for urban ecosystems need to recognise the reciprocal relationships of biological structures and processes, socioeconomic structures and processes, slowly changing historical or evolutionary templates, and global or regional external drivers. Furthermore, they need to acknowledge the role of social differentiation and the perception by individuals or institutions as mediators of the interactions between biophysical and socioeconomic patterns and processes. Feedbacks, often with time lags and indirect effects, are a part of the conceptual frameworks of urban ecosystems.

BES has used the Human Ecosystem Framework (Fig. 16.3) as the "disaggregated" version of Fig. 16.2 (Burch and DeLuca 1984; Machlis et al. 1997; Pickett et al. 1997b). Originally proposed by social ecologists Bill Burch, Gary Machlis and colleagues (Burch and DeLuca 1984; Machlis et al. 1997), this analytical framework identifies and describes the various structures and kinds of interactions that are important for including humans as components of ecosystems. The framework identifies the resource base of the ecosystem, which includes biophysical and social resources and the kinds of ways in which people organise themselves to exploit and manage those resources in order to accomplish the various functions of life (Fig. 16.3). The framework also recognises that individuals and institutions change over time, based on inherent human physiological rhythms and institutional 'timing cycles' (Fig. 16.3).

This analytical framework is not a theory or model in and of itself. Rather we have used it in BES to identify the specific kinds of variables and interactions to be included in our urban ecological research and applications. Machlis et al. (1997) note that some features of the framework are "orthodox to specific disciplines and not new". They also indicate, however, that the framework contains some less commonplace features such as myths as cultural resources, or justice as a critical institution. We adopt their view that the human ecosystem framework is a coherent entity that is useful in structuring the study of human ecosystems.

The Human Ecosystem Framework is crucially important for reminding all participants in BES that they are studying, explaining, or contributing to the management of a complex, human inhabited ecosystem (Machlis et al. 1997). This has been especially important in linking our biophysical and social scientists, engineers, urban designers, and decision makers (Grove and Burch 1997; Pickett et al. 2001; Grove et al. 2005). There are several elements that are critical to the successful application of this framework. First, it is important to recognise that the primary drivers of human ecosystem dynamics are both biophysical and social. Second, there is no single,

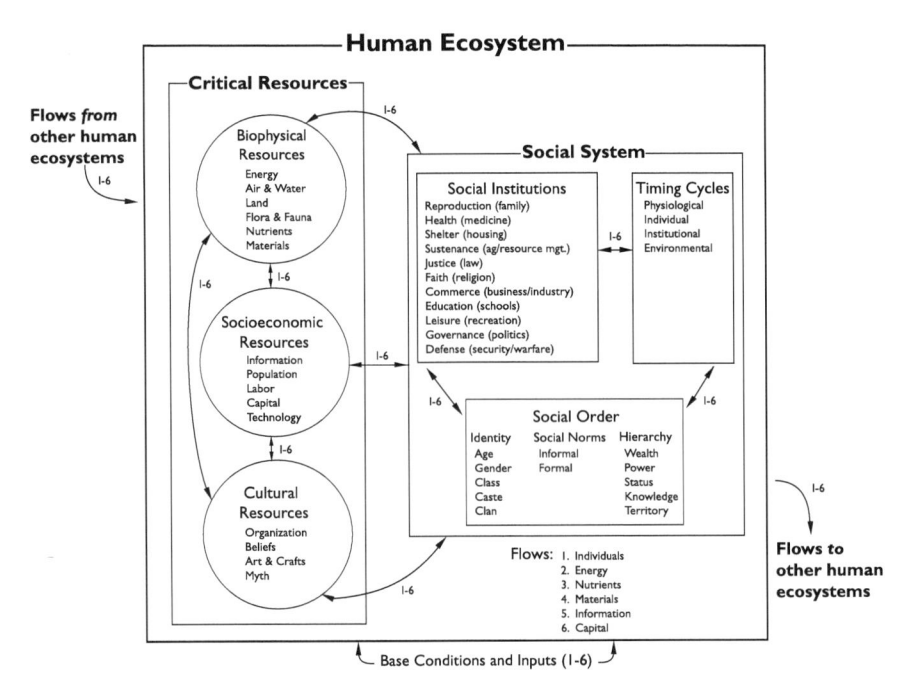

**Fig. 16.3** The human ecosystem framework. This conceptual framework identifies the components of the resource and human social systems required by inhabited ecosystems. The resource system is comprised of both biophysical and social resources. The human social system includes social institutions, cycles, and the factors that generate social order. This is a framework from which models and testable hypotheses suitable for a particular situation can be developed. It is used to organise thinking and research and is a valuable integrating tool for the BES (Re-drawn from Machlis et al. 1997)

determining driver of anthropogenic ecosystems. Third, the relative significance of drivers may vary over time. Fourth, components and their interactions with each other need to be examined simultaneously (Machlis et al. 1997). Finally, researchers need to examine how dynamic biological and social allocation mechanisms such as ecological constraints, economic exchange, authority, tradition and knowledge, affect the distribution of critical resources including energy, materials, nutrients, population, genetic and non-genetic information, labour, capital, organisations, beliefs and myths, within any human ecosystem (Parker and Burch 1992).

Press-Pulse Dynamics Framework

We include the Press-Pulse Dynamics Framework (PPD) as a complimentary, interactive framework to the Human Ecosystem Framework (HEF). The PPD was developed over a 3-year period by members from ecological and social science

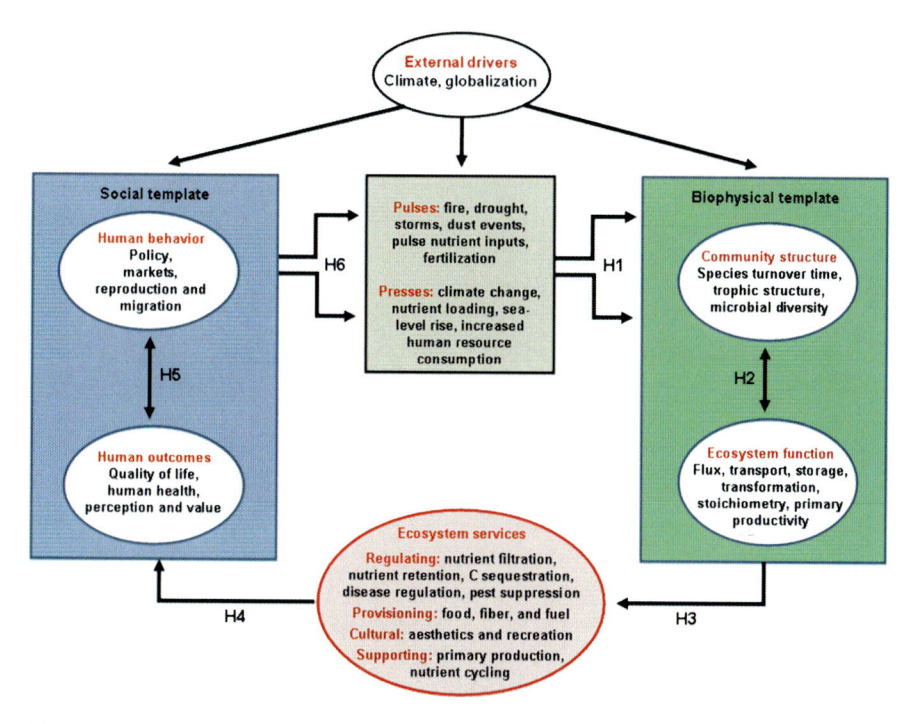

Fig. 16.4 Press–pulse dynamics framework (*PPD*). The PPD framework provides a basis for long-term, integrated, socio–ecological research. The *right-hand side* represents the domain of traditional ecological research; the *left-hand side* represents traditional social research associated with environmental change; the two are linked by pulse and press events influenced or caused by human behaviour and by ecosystem services, top and bottom, respectively (Collins et al. 2011). Individual items shown in the diagram are illustrative and not exhaustive

communities in the United States to promote long-term social ecological research (Collins et al. 2007, 2011). Like the Human Ecosystem Framework, the PPD (Fig. 16.4) is not a theory nor a model in and of itself. And like the Human Ecosystem Framework, the PPD incorporates methods and data from the geophysical, biological, social, and engineering sciences. The PPD differs from the Human Ecosystem Framework because of its focus on (1) "press and pulse events" that may drive socio-ecological systems and (2) the linkages between social and ecological templates in terms of changes in the quantity and quality of ecosystem services. The PPD adds to the traditional topics of existing long-term ecological research – i.e. the biophysical template (structure and function) and regulating and provisioning ecosystem services shown in Fig. 16.4 – by including topics such as cultural and supporting ecosystem services; the social template; social pulse and press drivers; and the relationships between these topics. The PPD and the HEF are complementary because the PPD provides a template for flows, interactions and connections of the entities and processes identified by the Human Ecosystem Framework.

The intention of the PPD is to provide a generalisable, scalar, mechanistic, and hypothesis-driven framework to promote socio-ecological research within existing long-term ecological research projects, the development of new long-term socio-ecological research, and comparisons among existing and new projects. The PPD can be used to focus a long-term socio-ecological research agenda through the identification of and connections among six strategic research questions (Collins et al. 2011):

1. How do long-term press disturbances and short-term pulse disturbances interact to alter ecosystem structure and function (H1)?
2. How can biotic structure, including built structure, be both a cause and consequence of ecological fluxes of energy and matter (H2)?
3. How do altered ecosystem dynamics affect ecosystem services (H3)?
4. How do changes in vital ecosystem services alter human outcomes (H4)?
5. How do changes in human perceptions and outcomes affect human behaviours and institutions (H5)?
6. Which human actions influence the frequency, magnitude or form of press and pulse disturbance regimes across ecosystems and what determines these actions (H6)?

Because the PPD framework focuses on press and pulse types of disturbance, we feel it is important to define the term. Disturbance is a technical term when used in socio-ecological research. It was originally used to refer to events with sharp onset and short duration, and with the ability to affect the physical structure of any ecological system. The term was provocative when first introduced because such events, although they disrupted some aspects of ecological systems, generated results that were positive for some other features of ecological systems. For example, disturbance often created opportunities for disadvantaged species to persist or enter an ecosystem, or provided locations in which resource conversion rates increased, facilitating access by suppressed or newly establishing organisms. Disturbance as originally introduced was distinct from ecological stress, which was often a longer lasting event that directly affected the function or metabolism of a system. Both disturbance and stress can be unified as perturbations, and this term reminds researchers that the effects on any specific system component or entire system may be positive, negative, or neutral. Since its introduction in the mid-1980s, the concept of disturbance has been refined, and has led to a new consideration of ecological events in general. Events are now considered to be complex occurrences characterised by an onset, a duration in time, and potentially a later decline. Ecologists recognise that the complexities of onset, duration, and demise of events will result in different effects. A short flood may not kill many plant species on a floodplain, while an unusually long flood may cause mortality and may even remove sensitive species from the system.

The complexity of ecological events can be abstracted in the contrast of pulses and presses. While this contrast does not consider all possible combinations of sharpness of onset, length of duration, or existence or rate of decline (Pickett and Cadenasso 2009), it focuses attention on two end members of that rich array of events: those that are transient and those that are persistent, at least for a relatively long time. Pulse events have sharp attack and quick demise, though they may have substantial effects on ecological systems. An earthquake is a good example of a pulse event. Press events alter the conditions in a system over a long time period, and may in fact be more akin to stresses. A shift in a climate regime from wet to dry, or the injection of a new level of resource supply through pollution are examples of biological pulses. Social pulses and presses are also important. New investment in a neighbourhood may be a pulse. A shift in demographic composition in a district of a city would illustrate a press. Presses and pulses are raw material for advancing integration between bioecological and social structures and processes in human ecosystems. The human ecosystem (Fig. 16.2) is a general concept that can be made operational by using a framework of potential components in the form of the Human Ecosystem Framework (Fig. 16.3), and a hypothetical model of connections and interactions in the form of the PPD (Fig. 16.4). How such models can be applied to spatially heterogeneous ecosystems such as urban areas is the concern of the next major framework used by BES.

### 16.3.3.2 Patch Dynamics

Patch dynamics is the second integrative tool employed by BES. It is important to the transition from an Ecology *in* to an Ecology *of* Cities because it can be used to account for the spatial heterogeneity of different lands uses and variations in management within land uses in terms of fluxes of (1) individuals, (2) energy, (3) nutrients, (4) materials, (5) information, and (6) capital and the ecological mechanisms and social institutions that control the flux of those resources across space.

Patch dynamics recognises that spatial heterogeneity is a key attribute of ecological systems. Emerging in the late 1970s, and originating from much the same impetus as the spatially focused discipline of landscape ecology, patch dynamics was used to describe the spatial structure of areas, the flows of materials, energy, and information across spatial mosaics, and the changes in individual spatial components of the mosaic as well as in the mosaic as a whole (Pickett et al. 2001). In other words, patch dynamics is concerned with the spatial structure, function and change of mosaic systems. At a particular scale, the heterogeneity can be resolved into patches that differ from each other. Although the patches may be heterogeneous at finer scales, at the scale of interest they are internally homogeneous relative to one another (Cadenasso et al. 2003). Examples include forest and field patches discriminated at the scale of km, or, at the scale of metres, tree fall pits and mounds in old growth forests (Pickett and White 1985). It is important to note that the mosaics can comprise discrete, bounded patches, or can be conceived of as gradients or fields defining continuous surfaces of differentiation. Patch dynamics can also be applied within a hierarchical structure, with different types of patches identified with processes at nested scales (Wu and Loucks 1995; Grove and Burch 1997).

Patch dynamics is important in urban systems as well because urban social-ecological systems are notoriously heterogeneous or patchy (Jacobs 1961; Clay 1973). Biophysical patches are a conspicuous layer of heterogeneity in cities. The basic topography, although sometimes highly modified, continues to govern important processes in the city (Spirn 1984). The watershed approach to urban areas has highlighted the importance of slopes, and of patchiness along slopes, in water flow and quality (Band et al. 2006). Steep areas are often the sites of remnant or successional forest and grassland in and around cities. Soil and drainage differ with the underlying topography. Vegetation, both volunteer and planted, is an important aspect of biophysical patchiness. The contrast in microclimate between leafy, green neighbourhoods versus those lacking a tree canopy is a striking example of biotic heterogeneity (Nowak 1994). Additional functions that may be influenced by such patchiness include carbon storage (Jenkins and Riemann 2001) and animal biodiversity (Adams 1994; Hostetler 1999; Niemela 1999).

Social differentiation and its spatial manifestation in terms of patches is also pronounced in and around cities and is not limited to categories such as land use (Grove and Burch 1997). Social patchiness can exist in such phenomena as economic activity and livelihoods, family structure and size, age distribution of the human population, wealth, educational level, social status, and lifestyle preferences (Burch and DeLuca 1984; Field et al. 2003). Social differentiation is important for human ecological systems because it affects both locational choices and the allocation of critical resources, including natural, socioeconomic, and cultural resources. In essence, social differentiation determines "who gets what, when, how and why" (Lenski 1966; Parker and Burch 1992). This allocation of critical resources is rarely equitable, but instead results in rank hierarchies. Unequal access to and control over critical resources is a consistent fact within and between households, communities, regions, nations and societies (Machlis et al. 1997). Environmental justice scholarship is a particular application of social differentiation research. Environmental justice research has demonstrated that disadvantaged groups, especially racial and ethnic minorities, are disproportionately burdened with environmental disamenities and enjoy fewer amenities compared to the privileged majority (Mohai and Saha 2007). These inequitable outcomes are typically the result of unjust procedures that burden racial and ethnic minorities with the most polluted and hazardous environments close to where they live (Bolin et al. 2005; Lord and Norquist 2010; but see Boone et al 2009).

Temporal dynamics are just as important as spatial pattern, since socio-ecological patterns are not fixed in time. For example, a city patch possessing a tree canopy will change as the trees mature and senesce, reducing canopy extent. Patches can also exhibit social dynamics, as when a neighbourhood of predominantly older residents shifts to dominance by young families. In these contrasting states, the patch makes different demands on the infrastructure and government. For example, young families may want playgrounds and access to schools while elderly residents may demand access to health services and passive recreation. The social requirements of specific patches will thus shift through time.

It is also critical to include the built nature of cities as a component of patch dynamics as well. Most people, and indeed most architects and designers, assume that the built environment is a permanent fixture. However, buildings and infrastructure change, as do their built and biophysical context. This elasticity in the urban system suggests a powerful way to re-conceptualise urban design as an adaptive, contextualised pursuit (Pickett et al. 2004; Shane 2005; Colding 2007; McGrath 2007). Such dynamism combines with the growing recognition of the role of urban design in improving the ecological efficiencies and processes in cities, particularly as existing structures may offer challenges to realising these efficiencies. Although this application of patch dynamics is quite new, it has great promise to promote the interdisciplinary melding of ecology and design and to generate novel designs with enhanced environmental benefit (McGrath et al. 2007). Thus, patches in urban systems can be characterised by physical structures, biological structures, social structures, built structures, or a combination of the four over time (Cadenasso et al. 2006) (Fig. 16.2).

### 16.3.3.3 Complexity in Socio-ecological Systems: Types and Levels

A third integrative tool in BES is a framework for examining complexity in terms of three types of complexity and levels of complexity within each type (Cadenasso et al. 2006). This complexity framework is useful in making the transition from an Ecology *in* Cities to an Ecology *of* Cities because it builds upon our patch dynamics approach by further developing its spatial and temporal dimensions and by adding a multi-scalar dimension to the types and levels of understanding social-ecological systems. In essence, this structural approach to socio-ecological systems can be used to ask what pieces are there and how are they arranged, how do the pieces interact, and how do they change through time (Fig. 16.5)? The complexity framework thus permits researchers to examine three realms: (1) the spatial heterogeneity of the urban system, (2) the organisation and connectivity of the spatially arrayed components, and (3) the role of history, contingency, and path dependency on the dynamics of urban ecosystems (Cadenasso et al. 2006).

The first type of complexity is spatial heterogeneity. Complexity of spatial heterogeneity increases as the perspective moves from patch type and the number of each type, to spatial configuration, and to the change in the mosaic through time (Wiens 1995; Li and Reynolds 1995) (Fig. 16.5). At the simplest structural end of the spatial axis, systems can be described as consisting of a roster of patch types. Richness of patch types summarises the number of patch types making up the roster. Structural complexity is increased as the number of each patch type is quantified. This measurement is expressed as patch frequency. How those patches are arranged in space relative to each other increases the complexity of understanding the heterogeneity or structure of the system (Li and Reynolds 1993). Finally, each patch can change through time. Which patches change, and how they change and shift identity constitutes a higher level of spatial complexity. The most complex understanding of system heterogeneity is acquired when the system can be quantified as a shifting mosaic of patches, or when the patch dynamics of the system is spatially explicit and quantified (Fig. 16.5). Although the passage of time is an element at the highest level of spatial complexity, this is distinct from historical complexity, where the function of such phenomena as lags and legacies is the concern (Cadenasso et al. 2006).

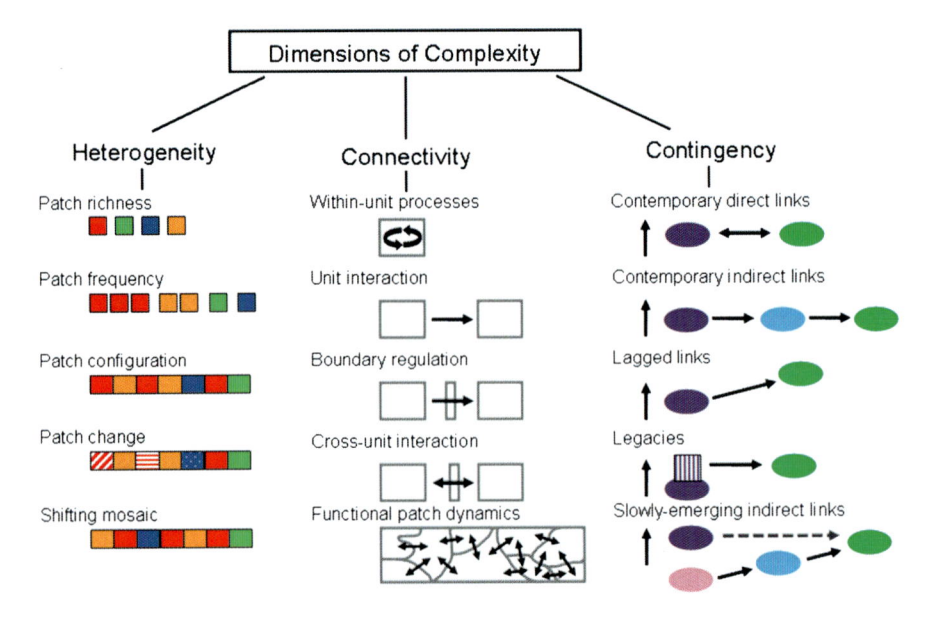

**Fig. 16.5** Framework for complexity of socio-ecological systems. The three dimensions of complexity are spatial heterogeneity, organisational connectivity and temporal contingencies. Components of the framework are arrayed along each axis increasing in complexity. For example, a more complex understanding of spatial heterogeneity is achieved as quantification moves from patch richness, frequency and configuration to patch change and the shift in the patch mosaic. Complexity in organisational connectivity increases from within unit process to the interaction of units and the regulation of that interaction to functional patch dynamics. Finally, historical contingencies increase in complexity from contemporary direct effect through lags and legacies to slowly emerging indirect effects. The *arrows* on the *left* of each illustration of contingency represent time. While not shown in the figure, connectivity can be assessed within and between levels of organisation (Cadenasso et al. 2006)

The second type of complexity is organisational connectivity. The organisational axis reflects the increasing connectivity of the basic units that control system dynamics within and between levels of organisation. Within organisational hierarchies, causality can move upward or downward (Ahl and Allen 1996). Organisational complexity drives system resilience, or the capacity to adjust to shifting external conditions or internal feedbacks (Holling and Gunderson 2002). Following our structural approach, we can return to the patch as an example of the basic functional unit of a system to explain this axis more fully. In this case, the simplest end of the connectivity axis is within-patch processes. As the interaction between patches is incorporated, complexity increases. Understanding how that interaction may be regulated by the boundary between patches constitutes a higher level of complexity. The organisational complexity axis continues to increase with recognition that patch interaction may be controlled by features of the patches themselves in addition to the boundary. Finally, the highest level of structural complexity on the organisational axis is the functional significance of patch connectivity for patch dynamics, both of a single patch and of the entire patch mosaic (Fig. 16.5). Note that from the perspective of complex behaviour, each range of this axis would be considered a structure whose complex behaviour could be evaluated and compared to other ranges of the gradient (Cadenasso et al. 2006).

The third type of complexity is historical contingency. Historical contingency refers to relationships that extend beyond direct, contemporary ones. Therefore, the influence of indirect effects, legacies or apparent memory of past states of the system, the existence of lagged effects, and the presence of slowly appearing indirect effects constitute increasing historical complexity (Fig. 16.5). To explain the steps of this axis we start with the simple or contemporary ones. Contemporary interactions includes those interactions where element A influences element B directly. Indirect contemporary interactions involve a third component, C, to transmit the effect of A on B. An interaction is lagged if the influence of element A on element B is not immediate but manifested over some time period. A higher level of temporal complexity is invoked by legacies. Legacies are created when element A modifies the environment and that modification, whether it be structural or functional, eventually influences element B. At the high end of the temporal complexity axis are slowly emerging indirect effects. These types of interaction occur when the apparent interaction of elements A and B is illusory and element B is actually influenced by some earlier state of element A and that influence is mediated through an additional element, C (Fig. 16.5) (Cadenasso et al. 2006).

### 16.3.3.4 Scalable Data Platform

The spatial, temporal, and hierarchical dimensions of socio-ecological systems require a scalable data platform to integrate biophysical and social patterns and processes in urban regions. The BES and DC-BC ULTRA-Ex use a parcel-relevant sampling approach that combines both plot-based and pixel-based data in an extensive-intensive sampling framework, with pixels as a type of extensive data, plots as a type of intensive data, and parcels as a scale of analysis.

There are several motivations for using parcels as a crucial unit of sampling and analysis. First, parcels are a complete census of an entire urban area and all ownership types. Since most of the land in urban areas is not in "urban-rural" analogs, the use of parcels is critical for the shift to an "Ecology *of* Cities." Second, parcels are a basic unit of decision-making associated with household and firm locational choices and behaviours. Third, parcels and their owners have social and ecological histories, and their geographies and attributes can be documented and described over time through a variety of sources (Boone et al. 2009; Lord and Norquist 2010; Buckley and Boone 2011). Fourth, parcels can be aggregated into other units of analysis, such as patches, neighbourhoods, watersheds, and municipalities (Grove et al. 2006a). Disaggregation is also possible, for example, to investigate differences between biophysical and social features of front and back residential yards (Loucks 1994).

An extensive – intensive sampling framework provides the capacity for linking pattern and process (Figs. 16.6 and 16.7). In general, extensive sampling may be

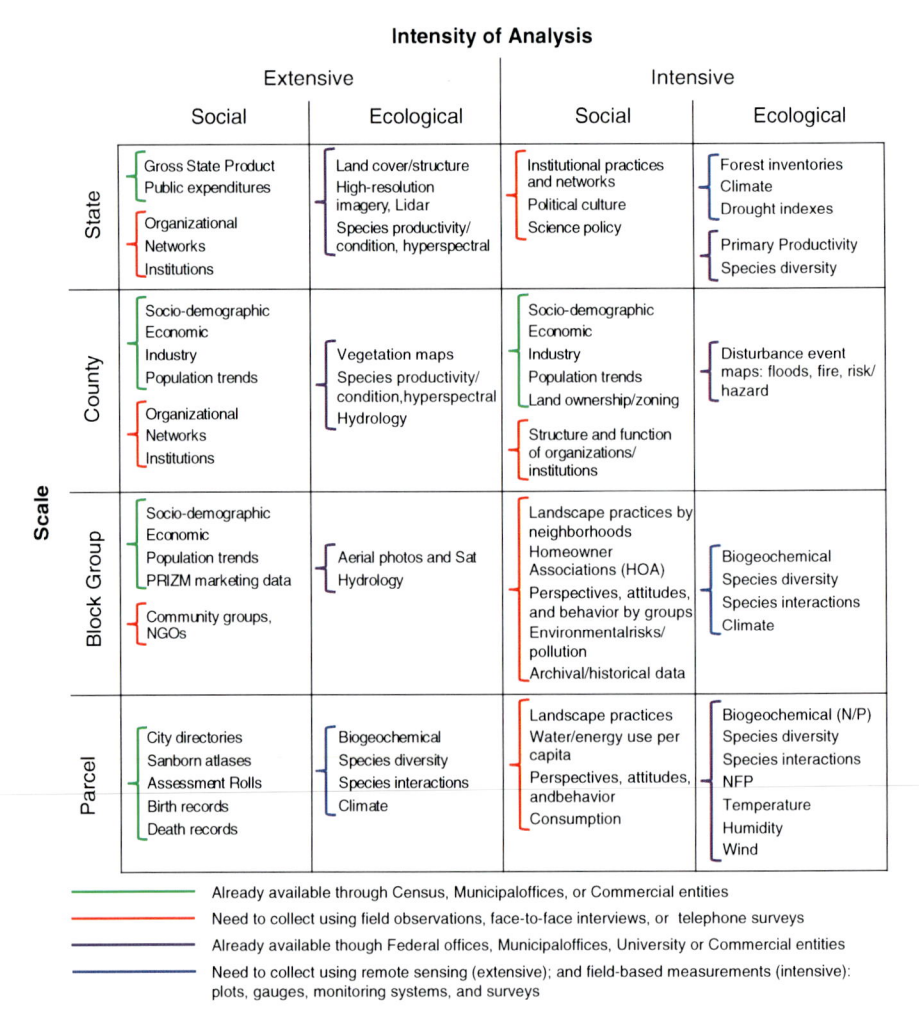

|  | | Intensity of Analysis | | | |
|---|---|---|---|---|---|
|  | | Extensive | | Intensive | |
|  | | Social | Ecological | Social | Ecological |
| **Scale** | State | Gross State Product<br>Public expenditures<br>Organizational<br>Networks<br>Institutions | Land cover/structure<br>High-resolution imagery, Lidar<br>Species productivity/condition, hyperspectral | Institutional practices and networks<br>Political culture<br>Science policy | Forest inventories<br>Climate<br>Drought indexes<br>Primary Productivity<br>Species diversity |
|  | County | Socio-demographic<br>Economic<br>Industry<br>Population trends<br>Organizational<br>Networks<br>Institutions | Vegetation maps<br>Species productivity/condition, hyperspectral<br>Hydrology | Socio-demographic<br>Economic<br>Industry<br>Population trends<br>Land ownership/zoning<br>Structure and function of organizations/institutions | Disturbance event maps: floods, fire, risk/hazard |
|  | Block Group | Socio-demographic<br>Economic<br>Population trends<br>PRIZM marketing data<br>Community groups,<br>NGOs | Aerial photos and Sat<br>Hydrology | Landscape practices by neighborhoods<br>Homeowner Associations (HOA)<br>Perspectives, attitudes, and behavior by groups<br>Environmental risks/pollution<br>Archival/historical data | Biogeochemical<br>Species diversity<br>Species interactions<br>Climate |
|  | Parcel | City directories<br>Sanborn atlases<br>Assessment Rolls<br>Birth records<br>Death records | Biogeochemical<br>Species diversity<br>Species interactions<br>Climate | Landscape practices<br>Water/energy use per capita<br>Perspectives, attitudes, and behavior<br>Consumption | Biogeochemical (N/P)<br>Species diversity<br>Species interactions<br>NFP<br>Temperature<br>Humidity<br>Wind |

─────── Already available through Census, Municipal offices, or Commercial entities
─────── Need to collect using field observations, face-to-face interviews, or telephone surveys
─────── Already available though Federal offices, Municipal offices, University or Commercial entities
─────── Need to collect using remote sensing (extensive); and field-based measurements (intensive): plots, gauges, monitoring systems, and surveys

**Fig. 16.6** Examples of socio-ecological data types organized by scale and intensity of analysis. Data types marked in *green* are data that LTSER sites must typically acquire, document, and archive. Data types marked in *red* are typically collected by LTSER sites (Zimmerman et al. 2009)

more useful for measuring pattern and inferring process, while intensive sampling may be more appropriate over time for the direct measurement and quantification of process and mechanism, particularly the motivations of social actors. An extensive – intensive data framework provides complementary sampling opportunities. Extensive sampling provides a basis for developing strategies for more intensive sampling, including the formulation of stratified sampling plans across space and time. Intensive sampling can be used to validate extensive data because the same phenomenon can be measured both extensively and intensively. For instance, vegetation productivity can be measured using both remote sensing and field-based measurements. Intensive

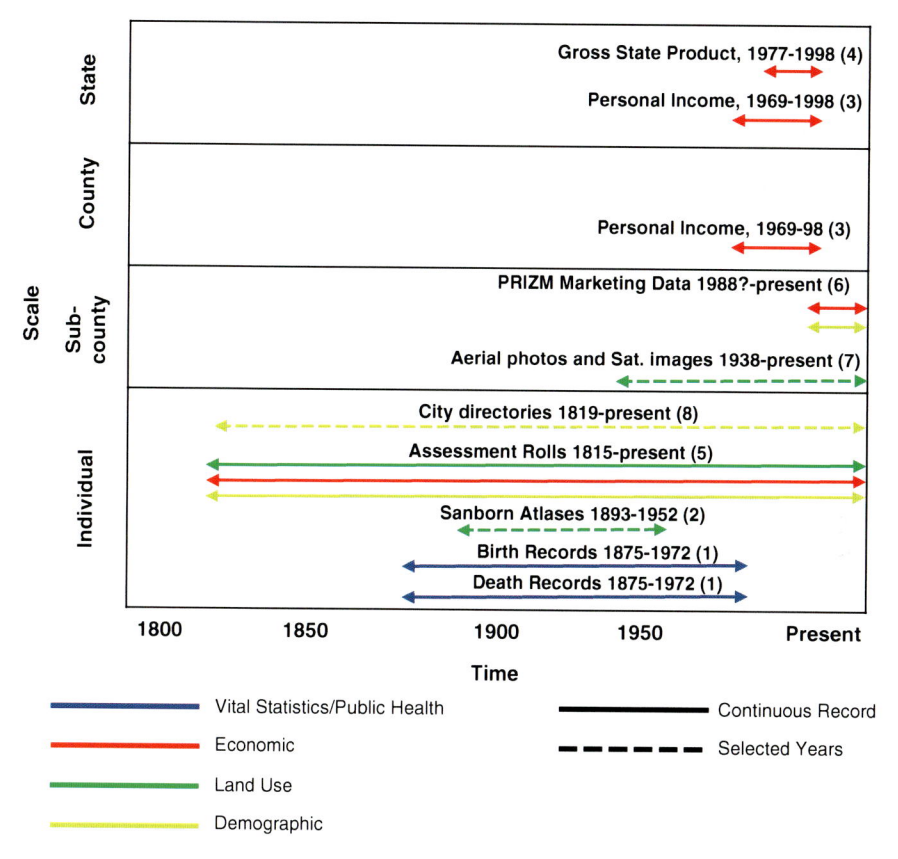

**Fig. 16.7** Example of non-census data sets with spatial reference to Baltimore City, 1800–2000 (Figure developed for Baltimore LTER Figure copyright BES LTER and used by permission from Boone)

sampling can be used to generate more detailed, process-based and mechanistic models. Subsequently, extensive sampling provides the basis for generalising these process-based models across space and time.

The empirical ability to examine and integrate social and ecological characteristics in an extensive-intensive framework is relatively new in the United States. The widespread adoption of Geographic Information Systems (GIS) by federal, state, and local governments and recent advances in remote sensing have greatly increased the availability of high-resolution geospatial data (<1 m). In particular, cadastral information maintained by local governments in hardcopy format is increasingly available digitally (Troy et al. 2007). These cadastral maps include a variety of information such as the boundaries and ownership of land parcels and infrastructure such as streets, storm drains, and retention ponds. High resolution imagery can be used to derive vegetation cover and combined with cadastral data and digital surface water data to distinguish vegetation extent, structure, and productivity between private property and public rights-of-way, including along streets. These parcel data also include attributes such as building type, building age, and building characteristics such as the number of bedrooms and bathrooms, building condition, transacted value, owner, land use and zoning (Zhou et al. 2009a, b).

These empirical advances provide a foundation for combining traditional social and ecological data. For instance, different types of social and ecological surveys can be linked to these data by both a unique address and by latitude and longitude. In the case of telephone surveys, telephone lists can include address and spatial location, while field observations and interviews can record both address and Global Positioning System (GPS) location as surveys are conducted. All of these data can be linked to Census geographies, which provide the basis for including demographic and socioeconomic data from the Census and commercial marketing data that are available at the Block Group level. Some of these marketing data include residential land management behaviours such as household expenditures on lawncare supplies and services (Zhou et al. 2009b). These data can be further combined with a variety of present and archival data that are address based, including real estate transactions, business directories, legal and health records, biographies and diaries, photographs and neighbourhood association minutes, for example (Merse et al. 2009).

### 16.3.3.5 Watersheds

We use the watershed approach in Baltimore because of its proven success and general application in ecology (e.g. Bormann and Likens 1979), its integrative role as a particular application of an ecosystem and patch dynamics approach (Band and Moore 1995; Law et al. 2004), and its practical relevance to decision making.

Hydrologists examine how the abiotic attributes of different patches within a watershed, such as temperature and physical characteristics including topography, soil properties, water table depth and antecedent soil moisture, contribute variable amounts of water and nutrients to streamflow, depending upon their spatial location in the watershed (Black 1991). Hydrologists have summarised mosaics of the characteristics listed above using the Variable Source Area (VSA) approach. The VSA approach can be integrated with a delineation of patches based upon the biotic attributes of the watershed such as vegetation structure and species composition (Bormann and Likens 1979), and the social attributes of the watershed such as indirect effects from land-use change, forest/vegetation management and direct effects from inputs of fertilisers, pesticides, and toxins to examine how the abiotic, biotic, and social attributes of different patches within a watershed contribute variable amounts of water and nutrients to streamflow (Grove 1996). This integrated approach builds on the VSA approach introduced by hydrologists to combine nested hierarchies of land use and land cover, socio-political structures and hydrological heterogeneity. By dividing watersheds into areas that differ in their ability to absorb or yield water, a more mechanistic understanding of the water yield from a watershed can be achieved.

One of the powers of the watershed approach is that large catchments can be divided into smaller catchments, or aggregated into still larger drainages. In other words, the source areas can be subdivided or grouped together. Therefore, the watershed approach can be scaled to match the extent or grain of the research question, or of the model or theory used to link with another discipline. This approach resonates with the hierarchical patch dynamics approach (Wu and Loucks 1995; Grove and Burch 1997).

A focus on watersheds also facilities interactions between scientists and decision makers, which we discuss later in this chapter. In Baltimore, watersheds are increasingly a management focus. The Chesapeake Bay, on which Baltimore is located, is the largest estuary in the United States, drawing on a watershed that intersects seven states. Although it is vast, the Bay is shallow, and pollution has reduced its water and habitat quality (Kennedy and Mountford 2001). Policy and management decisions are promulgated by public agencies and NGOs such as watershed associations and neighbourhood groups in order to reduce nitrogen pollution and sedimentation to the Bay.

In BES, we focus on a diverse set of watersheds draining the City of Baltimore and much of adjacent Baltimore County. These watersheds present a range in size, condition, use and history: from industrial and commercial lands along the Inner Harbor; to established residential patches of varying densities, structures, and ages; to commercial strips and zones; to stable agriculture; rural forests preserves; and agricultural land actively being converted into suburban housing and business uses. Small catchments have been selected in each of these areas, and the cumulative effect of urbanisation on water quality and the pattern of water flow has been sampled and monitored to assess the ecological structure and function throughout the metropolis (Fig. 16.8).

The Gwynns Falls is the largest of our intensively sampled watersheds, covering 17,150 ha. Gwynns Falls is sampled by three stations on the main stem of the stream, focusing on headwaters, middle and upper reaches, and the downstream reach. The sampling stations represent a gradient of urbanisation, and the downstream reach represents the net output of the watershed. The contribution of the upper reaches can be assessed by subtraction. In addition, four tributary subwatersheds of Gwynns Falls, representing contrasting land covers, are sampled. The tributary watersheds represent (i) dense urban, with industrial and commercial as well as residential land; (ii) early twentieth century rowhouse suburbs; (iii) agricultural land in a suburban matrix; and (iv) recent low density suburban development. All Gwynns Falls stations are sampled weekly for water flow and quality.

Pond Branch, a tributary of the Baisman Run, serves as the forested reference watershed for comparison to the more built-up watersheds. This gauged catchment has been sampled weekly for flow and water chemistry. Smaller watersheds were added to the sampling network to address specific situations of land cover and management. The remainder of Baisman Run represents recent large-lot, suburban development. Moore's Run is the location of the atmospheric eddy flux tower, and drains heavily wooded older suburbs. Mine Bank Run is the site of a restoration

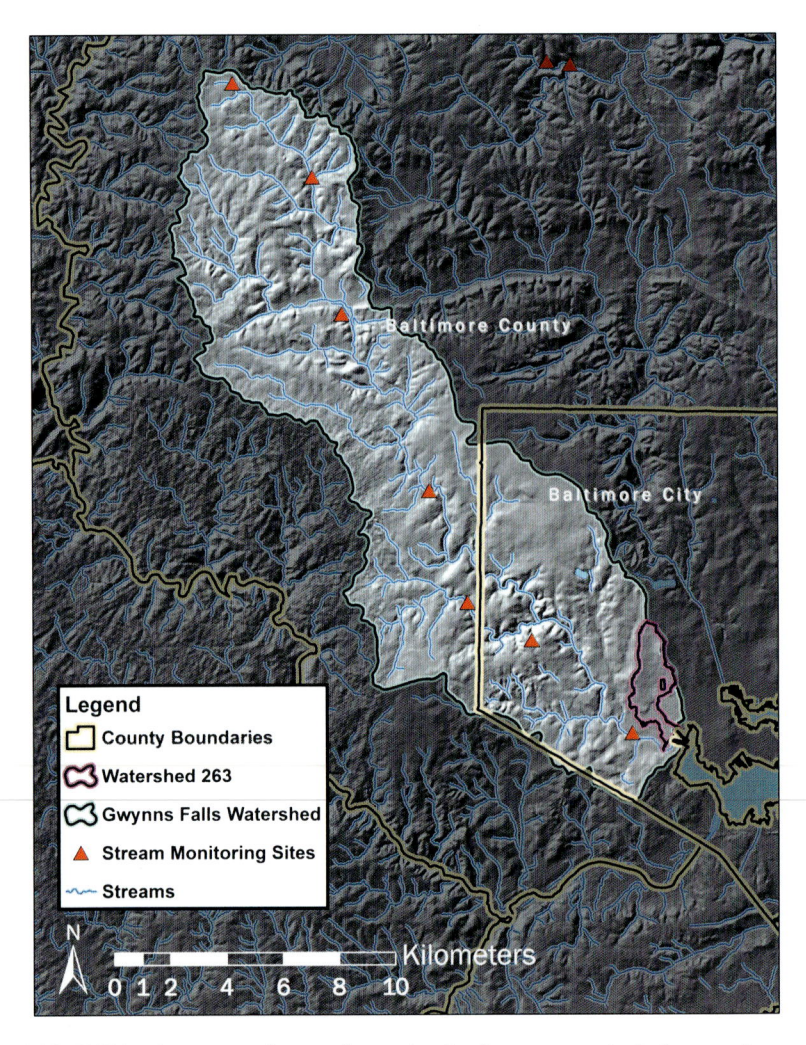

**Fig. 16.8** BES has instrumented a set of nested and reference watersheds that vary in current, historical, and future land use and condition (Figure copyright BES LTER and used by permission from O'Neil-Dunne)

project conducted by Baltimore County, in which BES scientists measure variables that can assess the success of the restoration. The Minebank Run project seeks to restore stream channel geomorphology, riparian function, and in-stream nutrient and organic matter retention. Finally, a 367-ha storm drain catchment, Watershed 263, in Baltimore City has recently been added to the network. Sampling is conducted in the storm drain pipes in two subwatersheds on the same schedule as the major surface drainages. Watershed 263 is an urban restoration and greening project to test the impact of extensive tree planting and removal of impervious surfaces on storm water quality.

## 16.4 Ways of Knowing: Analytical Strategies

We employ three analytical strategies, or "ways of knowing" in combination with our integrative tools. These three analytical strategies are: (1) Carpenter's Table, (2) Environmental History, and (3) Linkages between Scientists and Decision makers.

### *16.4.1 Carpenter's Table: Long-Term Monitoring, Experimentation, Comparative Analyses, and Modelling*

Research in BES is organised around the idea that long-term social-ecological research can be viewed as a table with four legs: long-term monitoring, experiments, comparative analyses, and modelling. This strategy is modified from an analysis by Carpenter (1998). The table metaphor suggests that the largest goal of the scientific enterprise is understanding or theory, represented by the table top. For complete understanding of a topic, such as socio-ecological systems in the long term, all four activities must be conducted. To the extent that one or more of the activities are absent or poorly developed, understanding will be incomplete (Fig. 16.9).

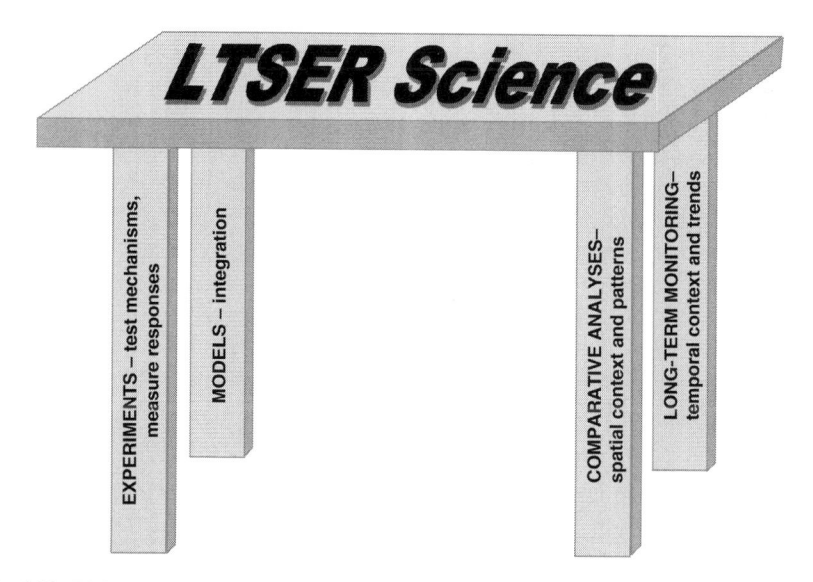

**Fig. 16.9** LTSER Platforms are similar to a table with four legs essential to the integrity of the whole: long-term monitoring, experimentation, comparative analyses, and modelling (Figure adapted from Carpenter 1998)

Long-term monitoring. In BES, long-term monitoring includes a variety of social and biophysical data that are organised within our scalable data framework, including both extensive and intensive data. Long-term monitoring is intended to continue for long periods into the future. Consistency of method, overlapping of methods when it is necessary to change instruments or approaches, regularity of sampling, and continual quality assurance and quality control are features of successful long-term monitoring (Likens 1989). The collection of long-term data include those from historical sources, such as archives and published records. Paleoecological approaches also extend long-term data into the past.

Experiments. Traditional manipulative experiments are difficult to carry out in urban watersheds due to concerns about environmental justice and constraints on human subjects research (Grove and Burch 1997; Cook et al. 2004). However, spatial variation in the nature and extent of land cover, i.e. the urban-rural gradient, provides numerous opportunities for experimental variation of factors controlling biophysical and social parameters. In addition, management initiatives such as efforts to improve sanitary sewer infrastructure, watershed restoration projects such as W263 (Fig. 16.8), Baltimore City's Urban Tree Canopy Goal (Fig. 16.13), and conversion of abandoned lots to community-managed open space represent experimental opportunities, which are sometimes called "natural" experiments in the social sciences. These management efforts provide strong opportunities for integration of biophysical and social sciences and for education and outreach.

Comparative analyses. Comparative analyses occur between social and ecological geographies and periods of time within the BES research area. Comparative analyses can also be made with other LTSER projects.

Modelling. A long-term goal of social and biogeophysical modelling activities in BES is to establish an "end-to-end system" of models and observational instruments to gather and synthesise information on social and biogeophysical components of the human ecosystem represented in the PPD framework (Fig. 16.4). The objective of this synthesis is to understand how individual and institutional behaviours; the urban landscape and infrastructure; ecosystem services, other push/pull factors; and climate interact to affect water and biogeochemical storage and flux in the urban hydrologic cycle; terrestrial and aquatic carbon and nutrient cycling and storage; and the regulation of surface energy budgets. BES uses biogeophysical models to simulate water, carbon, and nitrogen cycle processes and econometric and structural models to simulate locational choices and patterns of land development and change at multiple scales across the region. Coupling of these models is intended to provide predictive understanding of the feedbacks between environmental quality, ecosystem services, locational choice, and land development and redevelopment. BES uses specific policy scenarios aimed at enhancing sustainability in the Baltimore region related to water quality and carbon sequestration to motivate its coupled modelling and synthesis activities. These modelling activities are crucial for formalising our existing knowledge of how the system functions over time as well as identifying gaps in our current theory and observational systems. These models can also be used with decision makers to test future scenarios comparing current conditions, future trends, and possible policy interventions.

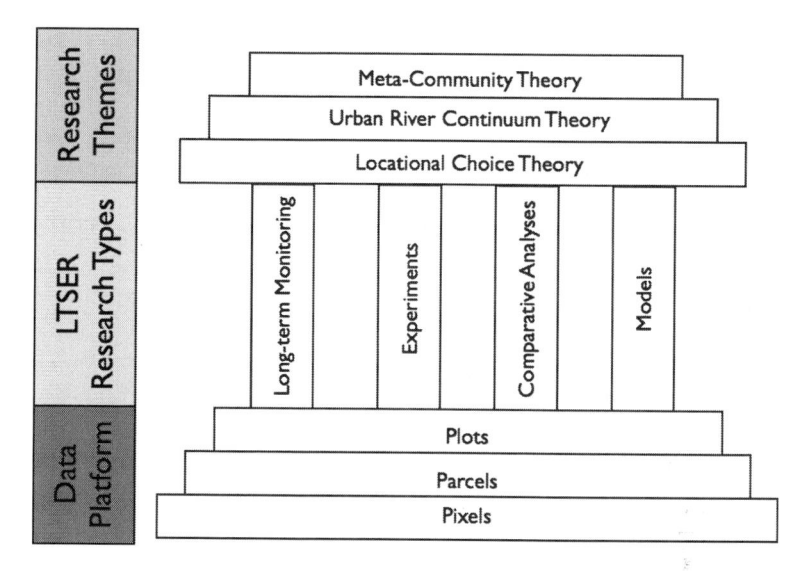

**Fig. 16.10** The LTSER data temple, with specific BES research themes included (Figure copyright BES LTER and used by permission from Grove)

#### 16.4.1.1 Putting It All Together: From Carpenter's Table to BES Temple

In BES we have taken the structural legs of Carpenter's table and converted them into columns (Fig. 16.10). We make this change in order to represent more fully the emerging and essential structure for BES research, which combines our integrative tools, ways of knowing, and research themes. The visual representation of this framework is a classic Greek temple composed of three primary elements. The first element is the foundation of the temple, or scalable data framework, which provides the base for all BES activities. This base is made up of pixels, parcels, and plots that are organised using our extensive-intensive approach (Figs. 16.6 and 16.7). The second element of our temple is the four columns to the temple, or research types (Fig. 16.9). These research types are supported by the data framework and, in turn, support the roof of the temple, or research themes from our current research foci. Each type of research uses our patch dynamics approach and pursues different types and levels of complexity (Fig. 16.5) appropriate to the research theme. The roof of the temple is made up of our BES research themes. Each theme can be located in one of Pasteur's Quadrants (Fig. 16.1) and detailed using the PPD (Fig. 16.4).

### *16.4.2 Environmental History*

Environmental history is a second way of knowing used by BES. Employing historical methods has several benefits (see Winiwarter et al., Chap. 5 in this volume; Cunfer and Krausmann, Chap. 12 in this volume; Gingrich et al., Chap. 13 in this volume).

First, it is good practice for researchers focused on the present and future. We have found in our research that historical analyses provide a conducive environment for researchers who focus on contemporary systems to work together in understanding "why" and "how" socio-ecological processes have operated to produce present-day patterns (van der Leeuw 1998; Pickett et al. 1999; Foster et al. 2002). The time spent in collaborating on background issues helps to build trust among researchers – an intangible but crucial and constructive element of integrated research. Second, historical analyses help build skills for understanding key dimensions of socio-ecological systems because the past is already integrated (interdisciplinary) and phenomena can be located in space and time, and attributed to different social and ecological scales. Finally, current and future conditions are often influenced by lags, legacies, and path dependencies (cp. Figs. 16.5 and 16.7). Thus, it is critical to understand the past because it informs the present and the future.

## 16.4.3 Linking Decision Making and Science

A third way of knowing is based upon linking decision making and science. In this case, science located in Pasteur's and Edison's quadrants are most relevant to this discussion. We have found that there are two parts to linking decision making and science: a) a framework for identifying linkages between decision making, science, and monitoring and assessment, and b) understanding the dynamic feedbacks between decision making and science. We note that decision makers are potentially made up of a diverse set of actors, including government agencies, NGOs, community groups, or individual citizens of all ages.

### 16.4.3.1 Linkages Between Decision Making, Science, and Monitoring and Assessment

There are several types of linkages between decision making, science, and long-term monitoring (Fig. 16.11). We describe these linkages with illustrations from BES.

Decision making and Science (A). There are numerous opportunities for decision making and research to intersect, and these intersections will be either use-inspired basic research or pure applied research. Some examples of activities that are part of this intersection include research to understand the ability of riparian areas, forests and lawns to take up nitrogen (Groffman et al. 2003), examining the relationship between residential land management and crime (Troy et al. 2007), or studying the relationships among climate change, vector-borne diseases and public health (LaDeau et al. 2011).

Decision making and Monitoring and Assessment (B). Decision makers rely upon a variety of data to monitor and assess the effectiveness, efficiency, and equity of their activities. These data can be associated with each of the human ecosystem complexes

**Fig. 16.11** Linkages between decision making, science, and monitoring & assessment (Figure copyright BES LTER and used by permission from Grove)

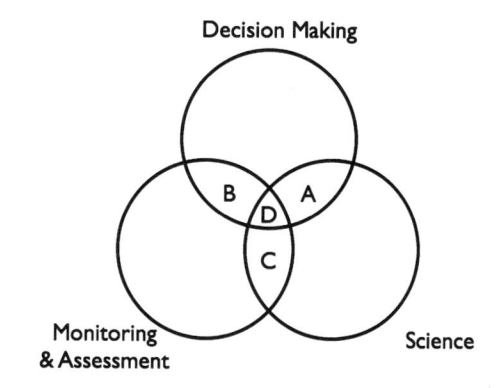

described earlier (Fig. 16.2), such as climate, flooding and air quality (physical), landcover, tree species and tree health (biological), public health, crime, employment and ownership (social), and the distribution and condition of buildings, roads, and sanitary and stormwater pipes (built).

Increasingly, local governments are making their data publicly available. In the case of Baltimore, OpenBaltimore has been developed to provide access to City data in order to support government transparency, openness and innovative uses that will help improve the lives of Baltimore residents, visitors and businesses. A goal of OpenBaltimore is to enable local software developer communities to develop applications that will help solve city problems (http://data.baltimorecity.gov/).

Science and Monitoring and Assessment (C). Scientists in BES contribute to long-term monitoring and assessments. Like (B) above, these data are associated with each of the human ecosystem complexes. Further, BES data are structured using the scalable data framework described previously. Data are documented with metadata and publicly available (http://www.beslter.org). Because BES collects data at parcel, neighbourhood, and county levels, and over the long term, comparisons can be made among these geographies for a specific point in time, or in terms of trends over time.

Intersection of A+B+C (D). The intersection of A, B, C occurs in D. Activities in D primarily involve coordinating activities among government agencies, NGOs, BES scientists and citizens. For instance, BES participates in and helps support a technical committee and workshops that include mid-level managers from government agencies and NGOs focused on urban sustainability issues such as land management, storm water, and urban agriculture.

An important opportunity for decision makers and scientists is that decision makers' policies, plans, and management represent important changes to the socio-ecological system. In Baltimore, scientists can help monitor and assess past, current and future activities. Important lessons can be learned about the effectiveness, efficiency and equity of decisions and the underlying social and ecological dynamics of the region. Thus, coordinating activities in D can be helpful to alert scientists of decision makers' plans and provide scientists and decision makers with time to initiate monitoring activities before decision makers begin to implement changes

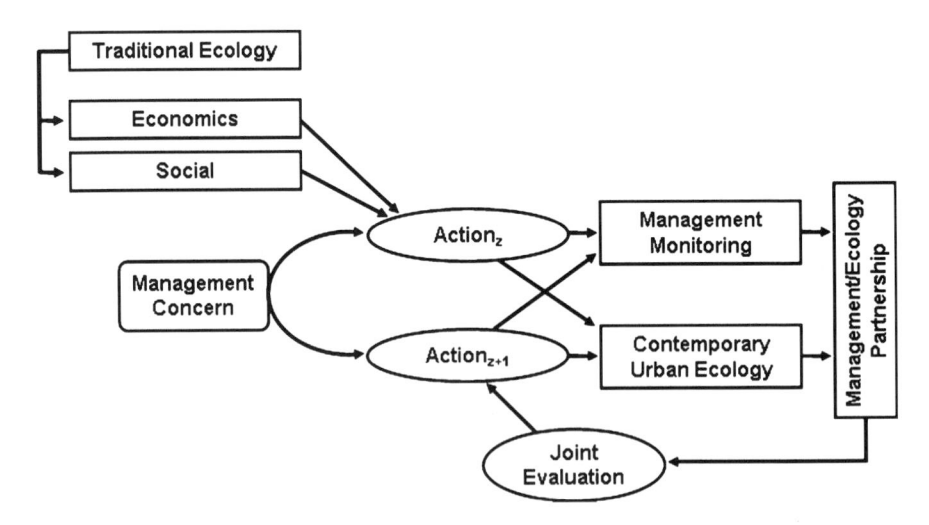

**Fig. 16.12** An abstracted cycle of interaction between research and management. The cycle begins with the separate disciplines of ecology, economics and social sciences interacting with a management or policy concern. In the past, ecology has neglected the urban realm as a subject of study, leaving other disciplines to interpret how ecological understanding would apply to an urban setting. A management or policy action (Action$_z$) results. Management monitors the results of the action to determine whether the motivating concern was satisfied. Contemporary urban ecology, which integrates with economics and social sciences, is now available to conduct research that recognises the meshing of natural processes with management and policy actions. Combining this broad, human ecosystem and landscape perspective with the concerns of managers can generate a partnership to enhance the evaluation of management actions. New or alternative management actions can result (Actions$_{z+1}$) (Pickett et al. 2007)

in policies, plans, or management. Some examples of the intersection between decision making, science, and monitoring and assessment include research related to and technical assistance with the development of Baltimore City's Urban Tree Canopy (UTC) policies, plans and management (Grove et al. 2006a, b; Galvin et al. 2006; Troy et al. 2007; Locke et al. 2011) and with urban watershed reclamation projects, such as W263 (Fig. 16.8 and http://www.parksandpeople.org/greening/greening-for-water-quality/watershed-263/). In both cases, BES assists in the monitoring and assessment of the projects' social, economic and ecological costs and benefits.

### 16.4.3.2 Dynamic Feedbacks Between Decision Making and Science

LTSERs can lead to a dynamic coupling between scientists, interested parties and decision makers (Fig. 16.12). An example from BES illustrates this opportunity. The Baltimore region is characterised by ecologically functional watersheds and stream valleys that have contributed to Baltimore's economic and cultural history.

An early test of the BES LTER project was to apply and demonstrate the utility of forested, watershed studies from the Coweeta, H. J. Andrews, and Hubbard Brook Experimental Forests/LTERs in the United States (Bormann and Likens 1979) to an urban watershed system. One of the initial questions that BES asked, using a watershed approach, was "do riparian zones, thought to be an important sink for N in many non-urban watersheds, provide a similar function in urban and suburban watersheds?"

Somewhat surprisingly, BES analyses found that rather than sinks, riparian areas had the potential to be sources of nitrogen in urban and suburban watersheds. This finding could be explained by the observation that hydrologic changes in urban watersheds, particularly incision of stream channels and reductions in infiltration in uplands due to stormwater infrastructure, led to lower groundwater tables in riparian zones. This "hydrologic drought" created aerobic conditions in urban riparian soils which decreased denitrification, an anaerobic microbial process that converts reactive nitrogen into nitrogen gases and removes it from the terrestrial system (Groffman et al. 2002, 2003; Groffman and Crawford 2003).

Based upon these results, the Chesapeake Bay Program re-assessed their goals for riparian forest restoration in urban areas (Pickett et al. 2007). Given that riparian zones in deeply incised urban channels were not likely to be functionally important for nitrate attenuation in urban watersheds, the programme focused instead on establishing broader urban tree canopy goals for entire urban areas (Fig. 16.13), with the idea that increases in canopy cover across the city would have important hydrologic and nutrient cycling benefits to the Bay (Raciti et al. 2006).

This science-decision making cycle is dynamic and iterative. The Urban Tree Canopy (UTC) example has already progressed through four cycles. After the establishment of Baltimore's UTC goal, analyses of the relationship between property regimes and urban tree canopy found that an "All Lands, All People" approach would be critical for achieving the City of Baltimore's urban tree canopy goal (Action$_{z+2}$). Private lands under the control of households are a critical component to achieving any vegetation management goal in the City. Total existing canopy cover is 20%, with 90% of that cover located on private lands. Likewise, about 85% of the unplanted land area, where potential planting could occur in the future, is on private land as compared to under 15% on public rights of way (Galvin et al. 2006).

The importance of residential households to achieving Baltimore's UTC goal led to research addressing the relationships between households, their lifestyle behaviours, and their ecologies (Grove et al. 2006a, b; Troy et al. 2007; Boone et al. 2009; Zhou et al. 2009b). A critical finding from this body of research was that although lifestyle factors such as family size, life stage and ethnicity may be weakly correlated with socioeconomic status, these lifestyle factors play a critical role in determining how households manage the ecological structure and processes of their properties. These findings suggested the need for novel marketing campaigns that differentiated between and promoted UTC efforts to different types of neighbourhoods (Action$_{z+3}$). The need to "market" to different neighbourhoods led to the need to understand existing and potential gaps in stewardship networks (Dalton 2001; Svendsen and Campbell 2008; Romolini and Grove 2010) – both functional and

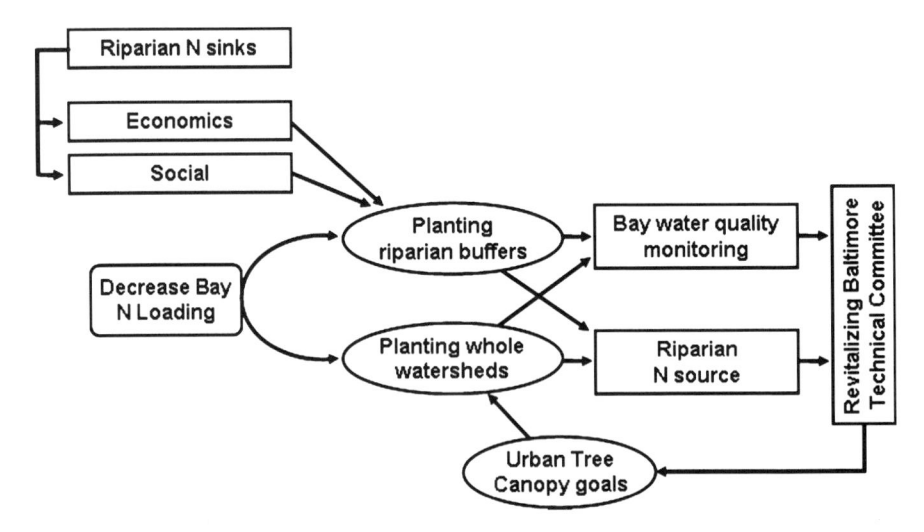

**Fig. 16.13** An example of the management-research interaction in Baltimore City watersheds. Traditional ecological information indicated that riparian zones are nitrate sinks. The management concern was to decrease nitrate loading into the Chesapeake Bay. In an effort to achieve that goal, an action of planting trees in riparian zones was proposed. Management monitoring indicated that progress toward decreasing Bay nitrate loadings was slow. Results from BES research suggested that stream channel incision in urban areas has resulted in riparian zones functioning as nitrate sources rather than sinks. In partnership with managers and policy makers in Baltimore City and the Maryland Department of Natural Resources, a re-evaluation of strategies to mitigate nitrate loading was conducted. This led to a decision to increase tree canopy throughout the entire Chesapeake Bay watershed. Baltimore City adopted an Urban Tree Canopy goal, recognising both the storm water mitigation and other ecological services such canopy would provide (Pickett et al. 2007)

spatial dimensions of the network as a mechanism to communicate and organise local, private stewardship ($Action_{z+4}$).

Practical benefits from urban LTSER projects are not limited to a LTSER site. The findings and methods developed in Baltimore through these successive science-decision making cycles have had widespread utility in other urban areas. For instance, the tools developed in Baltimore to assess and evaluate existing and possible UTC have been disseminated through existing Forest Service networks and applied to more than 70 urban areas in the United States and Canada (http://nrs.fs.fed.us/urban/utc/).

## 16.5 Building Process and Maintenance: Platform and Programme

Platform and programme are like building structure and function. They emphasise two related parts of a research "design". We do not talk much about platforms for relatively simple or focused research efforts, because the platform is such a familiar part of the culture that it hardly seems remarkable. For a single or small group of disciplinary investigators, the platform is the two or three bay university laboratory, or the experimental field plot or the glasshouse. Each of these familiar facilities has a standard set of fittings and supporting infrastructure. Labs may have benches, gas, 220 V electric service, fume hoods and the like. The simple field plot might consist of several hectares a few miles from the university campus on the edge of town, and equipped with a chain link fence, a padlocked gate and a small storage shed for some field gear.

In contrast, for large teams or multidisciplinary groups, or work requiring special technology, the platform becomes something quite complex and sometimes massive. Examples include the particle accelerators of physics, the telescopes of astronomy and the research vessels of oceanography. Even a university field station or an experimental forest is relatively complex. The largest of these are separate properties, often with administrative offices, laboratories, motor pools, class and meeting rooms and dormitories. Field stations sometimes house sophisticated laboratory and analytic equipment that facilitate the work of research on site. The fittings and support infrastructure for each of these platforms are an order of magnitude greater than those of a university lab, greenhouse, or field plot. A platform serves one or more complex questions, or a question that requires extraordinary technology. Platforms often, indeed most likely must, serve questions and researchers well beyond the identity of the founding investigators or questions.

A programme is the suite of research questions, experiments, comparisons, and models. Such activities are decided upon by the deliberations of a community of researchers, often in collaboration with a funding entity. Complex programmes have a number of characteristics. They may bind different disciplines together, reach across scales, or take the lead in employing radically different approaches to a research question or mission-oriented research problem. As the empirical base of the field(s) expands, and as data are generated through the use of the platform itself, the research questions will very likely evolve. New questions will be posed, and old ones will be answered or deemed uninteresting in light of advancing understanding.

Socio-ecological programmes have many characteristics that suggest the need for complex research and engagement platforms. Note that engagement with the communities, decision leaders, institutions of governance, government agencies in abutting or overlapping jurisdictions, and property holders and managers are requirements of successful socio-ecological research. Therefore, research platforms for socio-ecological projects must have the capacity and mandate for engagement with constituents and stakeholders.

Platforms consist of many elements. There are field sites and laboratories, offices and meeting space, dedicated vehicles, instruments for measurement, facilities for data management, storage and dissemination, web-based communication and collaboration technologies, and analytic software. It is important to recognise that many aspects of these platforms are intangible and are more likely to be grounded in collaboration science than in engineering. The specific form these components take reflects the specific research needs, the funding availability, and the intellectual network assembled to address the research questions.

This last point about platforms introduces the idea that a programme is inhabited by or emerges from a community of collaborators. While it may not be appropriate to speak about people as a part of the programme, the latter hardly exists separate from them. Hence, a community and its culture are essential to a programme. The culture and maintenance of the programme includes such features as the social network of communication, a shared vocabulary, the habits of credit and collaboration, the recruitment and assimilation of new scientists and students, and the schedule and scope of formal and informal meetings within the project and between the programme and interested parties beyond. Crucial to success in socio-ecological research is the willingness, training or ability of participants to listen and communicate effectively across disciplinary boundaries, and to understand and respect the different expertises and approaches members represent (Pickett et al. 1999).

In conclusion, programmes and platforms are inextricably linked as a research and engagement system. The platforms employed by vigorous research programmes become generators of new questions, new transdisciplinary knowledge, and new generations of researchers and scholars equipped for work at a farther frontier. Part of the challenge in developing an urban ecology research platform is the need to combine physical instruments (e.g., Hubble telescope or a NOAA ship) with people. The scientific community is used to researchers making the case for building something physical. The fact that building an urban ecology research platform sounds too much like "community development" to many represents a challenge for this work.

## 16.6 Conclusion

We have offered a working blueprint for the design of urban LTSERs based upon our experiences since 1997 in Baltimore and more recently in Washington, D.C.. We note, however, that the approach to urban long-term socio-ecological research described here can apply to any such system. Our architectural metaphor of siting, constructing, and maintaining a "building" highlights processes that can support research in any socio-ecological system. Using common approaches to LTSER platforms and programmes can help us develop overarching theories for understanding differences among urban areas, and among urban, agricultural, and forested systems. The need for and challenge to LTSERs for all types of systems will only grow as societies and ecologies become increasingly intertwined.

**Acknowledgments** This material is based on work supported by the USDA Forest Service Northern Research Station and the National Science Foundation under DEB 1027188 (Baltimore Ecosystem Study) and DEB 0948947 (Washington D.C./Baltimore ULTRA-Ex). We gratefully acknowledge additional support from the Center for Urban Environmental Research and Education (CUERE) at the University of Maryland, Baltimore County and Georgetown University. Partnerships with the US Geological Survey, the City of Baltimore Department of Public Works and Department of Recreation and Parks, the Baltimore County Department of Environmental Protection and Resource Management and Department of Recreation and Parks, the Maryland Department of Natural Resources, Forest Service, The Parks & People Foundation, and the Casey

Trees Foundation have been instrumental in the lessons reported here. We are grateful to our colleagues at these institutions for their intellectual contributions and insights into the environment and environmental management in the Baltimore region.

We would like to thank Cherie L. Fisher, Jarlath O'Neil-Dunne, Chris Boone and Gary Machlis for their assistance with figures for this chapter and helpful suggestions from Chris Boone, Dan Childers and two anonymous reviewers.

# References

Adams, L. W. (1994). *Urban wildlife habitats: A landscape perspective*. Minneapolis: University of Minnesota Press.

Ahl, V., & Allen, T. F. H. (1996). *Hierarchy theory: A vision, vocabulary, and epistemology* (p. 206). New York: Columbia University Press.

Alberti, M., Marzluff, J. M., Shulenberger, E., Bradley, G., Ryan, C., & Zumbrunnen, C. (2003). Integrating humans into ecology: Opportunities and challenges for studying urban ecosystems. *BioScience, 53*, 1169e–1179e.

Allen, T. F. H., & Hoekstra, T. W. (1992). *Toward a unified ecology*. New York: Columbia University Press.

Attorre, F., Stanisci, A., & Bruno, F. (1997). The urban woods of Rome. *Plant Biosystems, 131*, 113–135.

Band, L. E., & Moore, I. D. (1995). Scale: Landscape attributes and geographical information systems. *Hydrological Processes, 9*, 401–422.

Band, L. E., Cadenasso, M. L., Grimmond, S., & Grove, J. M. (2006). Heterogeneity in urban ecosystems: Pattern and process. In G. Lovett, C. G. Jones, M. G. Turner, & K. C. Weathers (Eds.), *Ecosystem function in heterogeneous landscapes* (pp. 257–278). New York: Springer.

Black, P. E. (1991). *Watershed hydrology*. Prentice Hall: Englewood Cliffs.

Bolin, B., Grineski, S., & Collins, T. (2005). The geography of despair: Environmental racism and the making of south Phoenix, Arizona, USA. *Human Ecology Review, 12*, 156–168.

Bolund, P., & Hunhammar, S. (1999). Ecosystem services in urban areas. *Ecological Economics, 29*, 293–301.

Boone, C., Cadenasso, M., Grove, J., Schwarz, K., & Buckley, G. (2009). Landscape, vegetation characteristics, and group identity in an urban and suburban watershed: Why the 60s matter. *Urban Ecosystems, 13*, 255–271.

Bormann, F. H., & Likens, G. (1979). *Patterns and processes in a forested ecosystem*. New York: Springer.

Boyden, S. (1979). *An integrative ecological approach to the study of human settlements*. Paris: UNESCO.

Brand, S. (2006). City planet. *Strategy + Business, 42*. Retrieved from: http://www.strategy-business.com/press/16635507/16606109

Brand, S. (2009). *Whole earth discipline: An ecopragmatist manifesto*. London: Viking/Penguin.

Buckley, G. L., & Boone, C. G. (2011). "To promote the material and moral welfare of the community": Neighborhood improvement associations in Baltimore, Maryland, 1900–1945. In R. Rodger & G. Massard-Guilbaud (Eds.), *Environmental and social justice in the city: Historical perspectives* (pp. 43–65). Cambridge: White Horse Press.

Burch, W. R., Jr., & DeLuca, D. R. (1984). *Measuring the social impact of natural resource policies*. Albuquerque: New Mexico University Press.

Cadenasso, M. L., Pickett, S. T. A., Weathers, K. C., Bell, S. S., Benning, T. L., Carreiro, M. M., & Dawson, T. E. (2003). An interdisciplinary and synthetic approach to ecological boundaries. *BioScience, 53*, 717–722.

Cadenasso, M. L., Pickett, S. T. A., & Grove, J. M. (2006). Dimensions of ecosystem complexity: Heterogeneity, connectivity, and history. *Ecological Complexity, 3*, 1–12.

Carpenter, S. R. (1998). The need for large-scale experiments to assess and predict the response of ecosystems to perturbation. In M. L. Pace & P. M. Groffman (Eds.), *Successes, limitations, and frontiers in ecosystem science* (pp. 287–312). New York: Springer.

Carreiro, M. M., Pouyat, R. V., Tripler, C. E., & Zhu, W. (2009). Carbon and nitrogen cycling in soils of remnant forests along urbanerural gradients: case studies in New York City and Louisville, Kentucky. In M. J. McDonnell, A. Hahs, & J. Breuste (Eds.), *Comparative ecology of cities and towns* (pp. 308–328). New York: Cambridge University Press.

Chan, S. (2007, December 4). Considering the urban planet 2050. *New York Times*, New York.

Cilliers, S. S., & Bredenkamp, G. J. (1999). Analysis of the spontaneous vegetation of intensively managed open spaces in the Potchefstroom Municipal Area, North West Province, South Africa. *South African Journal of Botany, 65*, 59–68.

Clay, G. (1973). *Close up: How to read the American City*. New York: Praeger Publishers.

Colding, J. (2007). Ecological land-use complementation' for building resilience in urban ecosystems. *Landscape and Urban Planning, 81*, 46–55.

Collins, S., Swinton, M, Anderson, C. W., Benson, B. J., Brunt J., Gragson, T. L., Grimm, N., Grove, J. M., Henshaw, D., Knapp, A. K., Kofinas, G., Magnuson, J. J., McDowell, W., Melack, J., Moore, J. C., Ogden, L., Porter, L., Reichman, J., Robertson, G. P., Smith, M. D., vande Castle, J., & Whitmer, A. C. (2007). *Integrated science for society and the environment: A strategic research initiative* (LTER Network Office Publication No 23). Albuquerque: LTER Network Office.

Collins, S., Carpenter, S. R., Swinton, S. M., Orenstein, D. E., Childers, D. L., Gragson, T. L., Grimm, N. B., Grove, J. M., Harlan, S. L., Kaye, J. P., Knapp, A. K., Kofinas, G. P., Magnuson, J. J., McDowell, W. H., Melack, J. M., Ogden, L. A., Robertson, G. P., Smith, M. D., & Whitmer, A. C. (2011). An integrated conceptual framework for long-term social–ecological research. *Frontiers in Ecology and the Environment, 9*, 351–357.

Cook, W. M., Casagrande, D. G., Hope, D., Groffman, P. M., & Collins, S. L. (2004). Learning to roll with the punches: Adaptive experimentation in human-dominated systems. *Frontiers in Ecology and the Environment, 2*, 467–474.

Couzin, J. (2008). Living in the danger zone. *Science, 19*, 748–749.

Dalton, S. E. (2001). *The Gwynns Falls watershed: A case 3 study of public and non-profit sector behavior in natural resource management*. Published doctoral dissertation, Johns Hopkins University, Baltimore.

Deutsch, L., Jansson, A., Troell, M., Ronnback, P., Folke, C., & Kautsky, N. (2000). The "ecological footprint": communicating human dependence on nature's work. *Ecological Economics, 32*, 351–355.

Duncan, O. D. (1961). From social system to ecosystem. *Sociological Inquiry, 31*, 140–149.

Duncan, O. D. (1964). Social organization and the ecosystem. In R. E. L. Faris (Ed.), *Handbook of modern sociology* (pp. 37–82). Chicago: Rand McNally & Co., Il printing.

Field, D. R., Voss, P. R., Kuczenski, T. K., Hammer, R. B., & Radeloff, V. C. (2003). Reaffirming social landscape analysis in landscape ecology: A conceptual framework. *Society and Natural Resources, 16*, 349–361.

Folke, C., Jansson, A., Larsson, J., & Costanza, R. (1997). Ecosystem appropriation by cities. *Ambio, 26*, 167–172.

Foster, D. R., Swanson, F., Aber, J., Tilman, D., Bropakw, N., Burke, I., & Knapp, A. (2002). The importance of land-use legacies to ecology and conservation. *BioScience, 53*(1), 77–88.

Galvin, M. F., Grove, J. M., & O'Neil-Dunne, J. P. M. (2006). *A report on Baltimore City's present and potential urban tree canopy*. Annapolis: Maryland Department of Natural Resources, Forest Service.

Garreau, J. (1991). *Edge city: Life on the new frontier*. New York: Doubleday.

Grimm, N., Grove, J. M., Pickett, S. T. A., & Redman, C. L. (2000). Integrated approaches to long-term studies of urban ecological systems. *BioScience, 50*, 571–584.

Grimm, N. B., Faeth, S. H., Golubiewski, N. E., Redman, C. L., Wu, J., Bai, X., & Briggs, J. M. (2008). Global change and the ecology of cities. *Science, 319*, 756–760.

Groffman, P. M., & Crawford, M. K. (2003). Denitrification potential in urban riparian zones. *Journal of Environmental Quality, 32*, 1144–1149.

Groffman, P. M., Boulware, N. J., Zipperer, W. C., Pouyat, R. V., Band, L. E., & Colosimo, M. F. (2002). Soil nitrogen cycling processes in urban riparian zones. *Environmental Science and Technology, 36*, 4547–4552.

Groffman, P. M., Bain, D. J., Band, L. E., Belt, K. T., Brush, G. S., Grove, J. M., Pouyat, R. V., Yesilonis, I. C., & Zipperer, W. C. (2003). Down by the riverside: Urban riparian ecology. *Frontiers in Ecology and the Environment, 1*, 315–321.

Grove, J. M. (1996). *The relationship between patterns and processes of social stratification and vegetation of an urban-rural watershed*. Published doctoral dissertation, Yale University, New Haven.

Grove, J. M., & Burch, W. R., Jr. (1997). A social ecology approach and applications of urban ecosystem and landscape analyses: A case study of Baltimore, Maryland. *Journal of Urban Ecosystems, 1*, 259–275.

Grove, J. M., & Hohmann, M. (1992). GIS and social forestry. *Journal of Forestry, 90*, 10–15.

Grove, J. M., Burch, W. R., & Pickett, S. T. A. (2005). Social mosaics and urban forestry in Baltimore, Maryland. In R. G. Lee & D. R. Field (Eds.), *Communities and forests: Where people meet the land* (pp. 248–273). Corvalis: Oregon State University Press.

Grove, J. M., Cadenasso, M. L., Burch, W. R., Jr., Pickett, S. T. A., O'Neil-Dunne, J. P. M., Schwarz, K., Wilson, M., Troy, A. R., & Boone, C. (2006a). Data and methods comparing social structure and vegetation structure of urban neighborhoods in Baltimore, Maryland. *Society and Natural Resources, 19*, 117–136.

Grove, J. M., Troy, A. R., O'Neil-Dunne, J. P. M., Burch, W. R., Cadenasso, M. L., & Pickett, S. T. A. (2006b). Characterization of households and its implications for the vegetation of urban ecosystems. *Ecosystems, 9*, 578–597.

Gunderson, L. H. (2000). Ecological resilience – In theory and application. *Annual Review of Ecology and Systematics, 31*, 425–439.

Hawley, A. H. (1950). *Human ecology: A theory of community structure*. New York: Ronald Press.

Holling, C. S., & Gunderson, L. H. (2002). Resilience and adaptive cycles. In L. H. Gunderson & C. S. Holling (Eds.), *Panarchy: Understanding transformations in human and natural systems* (pp. 25–62). Washington, DC: Island Press.

Hostetler, M. (1999). Scale, birds, and human decisions: A potential for integrative research in urban ecosystems. *Landscape and Urban Planning, 45*, 15–19.

Hough, M. (1984). *City form and natural process: Towards a new urban vernacular*. New York: Van Norstrand Reinhod Company.

Jacobs, J. (1961). *The death and life of great American cities*. New York: Vintage Books.

Jenkins, J. C., & Riemann, R. (2001). What does nonforest land contribute to the global C balance? In R. E. McRoberts, G. A. Reams, P. C. Van Deusen, & J. W. Moser (Eds.), *Proceedings of the third annual forest inventory and analysis symposium, general technical report* (pp. 173–179). St Paul: Department of Agriculture, Forest Service, North Central Research Station.

Johnson, S. (2006). *The ghost map: The story of London's most terrifying epidemic – and How It changed science, cities, and the modern world*. New York: Riverhead Books.

Katz, B., & Bradley, J. (1999, December). Divided we sprawl. *Atlantic Monthly*, pp. 26–42.

Kennedy, V. S., & Mountford, K. (2001). Human influences on aquatic resources in the Chesapeake Bay watershed. In P. D. Curtin, G. S. Brush, & G. W. Fisher (Eds.), *Discovering the Chesapeake: The history of an ecosystem* (pp. 191–219). Baltimore: The Johns Hopkins University Press.

Kent, M., Stevens, R. A., & Zhang, L. (1999). Urban plant ecology patterns and processes: A case study of the flora of the city of Plymouth, Devon, UK. *Journal of Biogeography, 26*, 1281–1298.

LaDeau, S. L., Calder, C. A., Doran, P. J., & Marra, P. P. (2011). West Nile virus impacts in American crow populations are associated with human land use and climate. *Ecological Research, 26*, 909–916.

Law, N. L., Band, L. E., & Grove, J. M. (2004). Nutrient input from residential lawn care practices. *Journal of Environmental Planning and Management, 47*, 737–755.

Lenski, G. E. (1966). *Power and privilege: A theory of social stratification*. New York: McGraw-Hill.

Li, H., & Reynolds, J. F. (1993). A new contagion index to quantify spatial patterns of landscapes. *Landscape Ecology, 8*, 155–162.

Li, H., & Reynolds, J. F. (1995). On definition and quantification of heterogeneity. *Oikos, 73*, 280–284.

Likens, G. E. (Ed.). (1989). *Long-term studies in ecology: Approaches and alternatives*. New York: Springer.

Locke, D. H, Grove, J. M., Lu, J. W. T., Troy, A., O'Neil-Dunne, J. P. M., & Beck, B. (2011). The 3Ps March on: Prioritizing potential and preferable locations for increasing urban tree canopy in New York City. *Cities and the Environment, 3, article 14*. Retrieved from http://escholarship.bc.edu/cate/vol3/iss1/4

Lord, C., & Norquist, K. (2010). Cities as emergent systems: Race as a rule in organized complexity. *Environmental Law, 40*, 551–597.

Loucks, O. L. (1994). Sustainability in urban ecosystems: Beyond an object of study. In R. H. Platt, R. A. Rowntree, & P. C. Muick (Eds.), *The ecological city: Preserving and restoring urban biodiversity* (pp. 48–65). Amherst: University of Massachusetts Press.

Machlis, G. E., Force, J. E., & Burch, W. R., Jr. (1997). The human ecosystem part I: The human ecosystem as an organizing concept in ecosystem management. *Society and Natural Resources, 10*, 347–367.

Makse, H. A., Havlin, S., & Stanley, H. E. (1995). Modeling urban growth patterns. *Nature, 377*, 608–612.

McDonnell, M. J., Pickett, S. T. A., Pouyat, R. V., Parmelee, R. W., & Carreiro, M. M. (1997). Ecosystem processes along an urban-to-rural gradient. *Urban Ecosystems, 1*, 21–36.

McGrath, B., Marshall, V., Cadenasso, M. L., Grove, J. M., Pickett, S. T. A., Plunz, R., & Towers, J. (Eds.). (2007). *Designing urban patch dynamics*. New York: Columbia University Graduate School of Architecture, Planning and Preservation, Columbia University.

Melosi, M. V. (2000). *The sanitary city: Urban infrastructure in America from colonial times to the present*. Baltimore: Johns Hopkins University Press.

Merse, C. L., Buckley, G. L., & Boone, C. G. (2009). Street trees and urban renewal: A Baltimore case study. *The Geographical Bulletin, 50*, 65–81.

Mohai, P., & Saha, R. (2007). Racial inequality in the distribution of hazardous waste: A national-level reassessment. *Social Problems, 54*, 343–370.

Niemela, J. (1999). Ecology and urban planning. *Biodiversity and Conservation, 8*, 118–131.

NOAA. (2004). *Population trends along the coastal United States: 1980–2008*. Washington, DC: U.S. Department of Commerce.

Northridge, M. E., Sclar, E. D., & Biswas, P. (2003). Sorting out the connection between the built environment and health: A conceptual framework for navigating pathways and planning healthy cities. *Journal of Urban Health: Bulletin of the New York Academy of Medicine, 80*, 556–568.

Nowak, D. J. (1994). Atmospheric carbon dioxide reduction by Chicago's urban forest. In E. G. McPherson (Ed.), *Chicago's urban forest ecosystem: Results of the Chicago Urban Climate Project* (pp. 83–94). Radnor: Northeastern Research Station, USDA Forest Service.

Parker, J. K., & Burch, W. R., Jr. (1992). Toward a social ecology for agroforestry in Asia. In W. R. Burch Jr. & J. K. Parker (Eds.), *Social science applications in Asian agroforestry* (pp. 60–84). New Delhi: IBH Publishing.

Pickett, S. T. A., & Cadenasso, M. L. (2002). The ecosystem as a multidimensional concept: Meaning, model, and metaphor. *Ecosystems, 5*, 1–10.

Pickett, S. T. A., & Cadenasso, M. L. (2008). Linking ecological and built components of urban mosaics: An open cycle of ecological design. *Journal of Ecology, 96*, 8–12.

Pickett, S. T. A., & Cadenasso, M. L. (2009). Altered resources, disturbance, and heterogeneity: A framework for comparing urban and non-urban soils. *Urban Ecosystems, 12*, 23–44.

Pickett, S. T. A., & Grove, J. M. (2009). Urban ecosystems: What would Tansley do? *Urban Ecosystems, 12*, 1–8.

Pickett, S. T. A., & White, P. S. (Eds.). (1985). *The ecology of natural disturbance and patch dynamics*. Orlando: Academic.

Pickett, S. T. A., Burch, W. R., Jr., & Dalton, S. (1997a). Integrated urban ecosystem research. *Urban Ecosystems, 1*, 183–184.

Pickett, S. T. A., Burch, W. R., Jr., Dalton, S., Foresman, T., Grove, J. M., & Rowntree, R. (1997b). A conceptual framework for the study of human ecosystems in urban areas. *Journal of Urban Ecosystems, 1*, 185–199.

Pickett, S. T. A., Burch, W. R., Jr., & Grove, J. M. (1999). Interdisciplinary research: Maintaining the constructive impulse in a culture of criticism. *Ecosystems, 22*, 302–307.

Pickett, S. T. A., Cadenasso, M. L., Grove, J. M., Nilon, C. H., Pouyat, R. V., Zipperer, W. C., & Costanza, R. (2001). Urban ecological systems: Linking terrestrial ecological, physical, and socioeconomic components of metropolitan areas. *Annual Review of Ecology and Systematics, 32*, 127–157.

Pickett, S. T. A., Cadenasso, M. L., & Grove, J. M. (2004). Resilient cities: Meaning, models and metaphor for integrating the ecological, socio-economic, and planning realms. *Landscape and Urban Planning, 69*, 369–384.

Pickett, S. T. A., Kolasa, J., & Jones, X. (2007). *Ecological understanding: The nature of theory and the theory of nature* (2nd ed.). New York: Springer.

Pickett, S. T. A., Cadenasso, M. L., Grove, J. M., Boone, C. G., Groffman, P. E., Irwin, E., Kaushal, S. S., Marshall, V., McGrath, B. P., Nilon, C. H., Pouyat, R. V., Szlavecz, K., Troy, A. R., & Warne, P. (2011). Urban ecological systems: Scientific foundations and a decade of progress. *Journal of Environmental Management, 92*, 331–362.

Pincetl, S. (2010). From the sanitary to the sustainable city: Challenges to institutionalizing biogenic (nature's services) infrastructure. *Local Environment, 15*, 43–58.

Pouyat, R. V., Carreiro, M. M., Groffman, P. M., & Pavao-Zuckerman, M. A. (2009). Investigative approaches to urban biogeochemical cycles: New York metropolitan area and Baltimore as case studies. In M. J. McDonnell, A. Hahs, & J. Breuste (Eds.), *Ecology of cities and towns: A comparative approach* (pp. 329–351). New York: Cambridge University Press.

Raciti, S., Galvin, M. F., Grove, J. M., O'Neil-Dunne, J., & Todd, A. (2006). *Urban tree canopy goal setting: A guide for communities*. Annapolis: USDA Forest Service, Chesapeake Bay Program Office, Northeastern Area, State and Private Forestry.

Rebele, F. (1994). Urban ecology and special features of urban ecosystems. *Global Ecology and Biogeographical Letters, 4*, 173–187.

Redman, C. L., Grove, J. M., & Kuby, L. H. (2004). Integrating social science into the long-term ecological research (LTER) network: Social dimensions of ecological change and ecological dimensions of social change. *Ecosystems, 7*, 161–171.

Romolini, M., & Grove, J. M. (2010, August 3). *Polycentric networks and resilience in urban systems: A comparison of Baltimore & Seattle*. Ecological Society of America annual meeting, Pittsburgh, PA.

Schaaf, T., Zhao, X., & Keil, G. (1995). *Towards a sustainable city: Methods of urban ecological planning and its application in Tianjin, China*. Berlin: Urban System Consult GmbH.

Schnore, L. F. (1958). Social morphology and human ecology. *The American Journal of Sociology, 63*, 620–634.

Seto, K., Sanchez-Rodriguez, R., & Fragkias, M. (2010). The new geography of contemporary urbanization and the environment. *Annual Review of Environmental Resources, 35*, 167–194.

Shane, G. D. (2005). *Recombinant urbanism: Conceptual modeling in architecture, urban design, and city theory*. Chichester: Wiley.

Spirn, A. W. (1984). *The granite garden: Urban nature and human design*. New York: Basic Books, Inc.

Stearns, F. (1970). Urban ecology today. *Science, 170*, 1006–1007.

Stearns, F., & Montag, T. (Eds.). (1974). *The urban ecosystem: A holistic approach*. Stroudsburg: Dowden, Hutchinson and Ross, Inc.

Stokes, D. E. (1997). *Pasteur's Quadrant – Basic science and technological innovation*. Washington, DC: Brookings Institution Press.

Svendsen, E. S., & Campbell, L. K. (2008). Urban ecological stewardship: Understanding the structure, function and network of community-based urban land management. *Cities and the Environment, 1*, 1–32.

Tansley, A. G. (1935). The use and abuse of vegetational concepts and terms. *Ecology, 16*, 284–307.

Thompson, G. E., & Steiner, F. R. (Eds.). (1997). *Ecological design and planning*. New York: Wiley.

Toepfer, K. (2005). *From the desk of Klaus Toepfer, United Nations Under-Secretary-General and Executive Director*, UNEP. Our Planet.

Troy, A. R., Grove, J. M., O'Neil-Dunne, J. P. M., Pickett, S. T. A., & Cadenasso, M. L. (2007). Predicting opportunities for greening and patterns of vegetation on private urban lands. *Environmental Management, 40*, 394–412.

Turner, B. L., II, Kasperson, R. E., Matosn, P. A., McCarthy, J. J., Corell, R. W., Christensen, L., Eckley, N., Kasperson, J. X., Luers, A., Martello, M. L., Polsky, C., Pulsipher, A., & Schiller, A. (2003). A framework for vulnerability analysis in sustainability science. *Proceedings of the National Academy of Sciences USA, 100*, 8074–8079.

van der Leeuw, S. E. (1998). *The ARCHAEOMEDES Project B understanding the natural and anthropogenic causes of land degradation and desertification in the Mediterranean Basin*. Luxemburg: Office of Publications of the European Union.

Vandruff, L. W., Bolen, E. G., & San Julian, G. J. (1994). Management of urban wildlife. In T. A. Bookhout (Ed.), *Research and management techniques for wildlife and habitats* (pp. 507–530). Bethesda: The Wildlife Society.

Vincent, G., & Bergeron, Y. (1985). Weed synecology and dynamics in urban environment. *Urban Ecology, 9*, 161–175.

Wiens, J. A. (1995). Landscape mosaics and ecological theory. In L. Hansson, L. Fahrig, & G. Merriam (Eds.), *Mosaic landscapes and ecological processes* (pp. 1–26). New York: Chapman & Hall.

Wu, J., & Loucks, O. L. (1995). From balance of nature to hierarchical patch dynamics: A paradigm shift in ecology. *The Quarterly Review of Biology, 70*, 439–466.

Zhou, W., Troy, A. R., & Grove, J. M. (2009a). Modeling residential lawn fertilization practices: Integrating high resolution remote sensing with socioeconomic data. *Environmental Management, 41*, 742–752.

Zhou, W., Grove, J. M., Troy, A., & Jenkins, J. C. (2009b). Can money buy green? Demographic and socioeconomic predictors of lawncare expenditures and lawn greenness in urban residential areas. *Society and Natural Resources, 22*, 744–760.

Zimmerman, J. K., Scatena, F. N., Schneider, L. C., Gragson, T., Boone, C., & Grove, J. M. (2009). *Challenges for the implementation of the decadal plan for long-term ecological research: Land and water use change: Report of a workshop* (LTER Network Office Report).

Zipperer, W. C., Foresman, T. W., Sisinni, S. M., & Pouyat, R. V. (1997). Urban tree cover: An ecological perspective. *Urban Ecosystems, 1*, 229–247.

# Chapter 17
# Development of LTSER Platforms in LTER-Europe: Challenges and Experiences in Implementing Place-Based Long-Term Socio-ecological Research in Selected Regions

Michael Mirtl, Daniel E. Orenstein, Martin Wildenberg, Johannes Peterseil, and Mark Frenzel

**Abstract** This chapter introduces place-based, Long-Term Socio-Ecological Research (LTSER) Platforms conceptually and in practice. LTER-Europe has put strong emphasis on utilising the data legacy and infrastructure of traditional LTER Sites for building LTSER Platforms. With their unique emphasis on socio-ecological research, LTSER Platforms add a new and important dimension to the four pillars of LTER-Europe's science strategy (systems approach, process-oriented, long-term and site-based). In this chapter, we provide an overview of the regionalised or place-based LTSER concept, including experiences garnered from Platform models tested within LTER-Europe, and we discuss the current status of LTSER Platforms on the European continent. The experiences gathered in 6 years of practical work and development of regional socio-ecological profiles as conceptual frameworks in the Austrian Eisenwurzen LTSER Platform will be used to assess weaknesses and strengths of two implementation strategies (evolutionary vs. strategically managed) and to derive recommendations for the future. The chapter represents the close

---

M. Mirtl, Ph.D (✉) • J. Peterseil, Ph.D.
Department of Ecosystem Research and Monitoring,
Environment Agency Austria, Vienna, Austria
e-mail: michael.mirtl@umweltbundesamt.at; johannes.peterseil@umweltbundesamt.at

D.E. Orenstein
Faculty of Architecture and Town Planning, Technion – Israel
Institute of Technology, Haifa, Israel
e-mail: dorenste@tx.technion.ac.il

M. Wildenberg, Ph.D.
GLOBAL 2000, Vienna, Austria
e-mail: martin.wildenberg@global2000.at

M. Frenzel, Ph.D.
Department of Community Ecology, Helmholtz Centre
for Environmental Research (UFZ), Leipzig, Germany
e-mail: mark.frenzel@ufz.de

of the first substantive loop of LTSER research that began in 2003 from conceptualisation to implementation and, through the introspective analysis here, a reconsideration of the central concepts.

**Keywords** LTER-Europe • LTSER Platforms • Eisenwurzen LTSER Platform • Socio-ecological research • Socio-ecological profiling • Fuzzy cognitive mapping • Critical ecosystem services

## 17.1 LTSER as an Intrinsic Element of the LTER-Europe Design

### 17.1.1 The Development of LTSER in Europe

The emergence of Long-Term Socio-Ecological Research (LTSER) in Europe represented a profound shift in professional perceptions regarding how policy-relevant, proactive research should and could be conducted. The historical development in thinking that led to this paradigm shift and the conceptual background are elaborated in detail by Redman et al. (2004) and in the introduction of this book. The following outline sets the stage for understanding the synergies and linkages between long-term ecosystem research (LTER) and LTSER as a crucial part of the next generation of LTER research.

Up to the 1990s, the LTER programme focused mainly on studying ecological structure and function. Small-scale sites (1 ha to 10 $km^2$) were selected according to ecosystem-specific design criteria (e.g. hydrological catchments of small rivers and lakes), preferably in semi-natural or natural ecosystems. Based on site measurements, traditional LTER has aimed to document and analyse ecosystem structures and processes in order to detect environmental change and its impacts on ecosystems and their natural resources (Mirtl et al. 2009; Mirtl 2010). Later in the LTER programme, urban LTER Sites were added, such as the US LTER Sites in Phoenix and Baltimore (Hobbie et al. 2003; Grove et al., Chap. 16 in this volume). However, due to the small scale of Sites and biases in Site selection, LTER was constrained in explaining cause-effect relationships and larger scale phenomena such as biodiversity loss, often induced by human activities (Metzger et al. 2010; MEA 2005).

At the end of the twentieth century, national and continental networks for LTER, established in the 1980s and 1990s, were assessed regarding their societal relevance. Reviews scrutinised the efficiency of knowledge dissemination and adequacy of current designs in tackling urgent policy questions, including those related to the sustainable use of ecosystem services and the effects imposed on them by global environmental change (Hobbie et al. 2003). In a review of two decades of US-LTER (2011), the reviewers elaborated a list of 27 recommendations, *inter alia* those to establish interdisciplinary and cross-site projects and comparisons, to focus on synthesis science and, importantly, to include a "human dimension" in LTER.

In response to these reviews and on-going self-evaluations, teams in the USA and Europe started to promote LTSER to consider socioeconomic drivers of ecological change observed in traditional LTER, such as historical changes in the economy, public perceptions of their environment, and land use (Haberl et al. 2006; Mirtl and Krauze 2007; Singh et al. 2010). These efforts also built upon earlier publications advocating interdisciplinary research (IDR) among natural scientific disciplines (Pickett et al. 1999).

Given the timing of the developments outlined above, it became evident to the developing European regional group of the International Long-term Ecosystem Research (ILTER) network, which started in 2003, that they had to seize the window of opportunity in order to integrate socio-ecological research from the start of their activities (ILTER 2011). In Europe, as elsewhere, researchers were increasingly considering their landscapes as the ecological products of human activity – "cultural" landscapes that are contingent upon, and are the historical outcome of, the interplay between socioeconomic and biophysical forces (Wrbka et al. 2004). Thus, it had become widely accepted that current structures and states of the environment across the European continent could not be properly interpreted without taking social, environmental and land-use history into account (EEA 2010). Research demanded a new package of variables, including population density, land ownership settings, and patterns of use of ecosystem services at various scales by diverse and competing stakeholders and interactions with nature protection efforts. Accordingly, a range of applied interdisciplinary research approaches would be required, along with new questions regarding ecosystem valuation (Hein et al. 2006). Thus, the lessons from the aforementioned reviews of LTER found fertile and receptive ground within the European LTER community.

A second key factor facilitated the establishment of LTER-Europe and its LTSER component: The Sixth Research Framework Programme (FP6) of the European Commission, launched in 2004, promoted a new type of project, called "Networks of Excellence" (NoE), which aimed to overcome disciplinary fragmentation and foster interdisciplinary integration in the European Research Area. The NoE A Long-term Biodiversity and Ecosystem Research and Awareness Network (ALTER-Net), focused on biodiversity in the ecosystem context as a topical trigger, and provided a unique framework for (i) integrating the strengths of the existing, but fragmented, LTER infrastructure at the site level, (ii) developing a framework for identifying interdisciplinary research ideas, planning proposals and delivering syntheses on complex socio-ecological problems (Furman et al. 2009) and, (iii) working at the science-policy interface (Anon 2009).

Under the auspices of ALTER-Net, the European regional group of the global LTER network (ILTER website), LTER-Europe, was set up with a strong focus on LTSER. The next step was the establishment of "*LTSER Platforms*" in hot-spot areas of ecological research, which moved LTER-Europe on from conceptualisation to implementation.

## 17.1.2 Conceptual Common Denominators of LTSER and Traditional LTER

Socio-ecological research utilises inter- and transdisciplinary approaches and adopts a holistic conception of human-nature interactions in scrutinising complex cause-effect relationships and feed-back cycles. It does not necessarily imply a specific spatial scale or administrative level, nor must it necessarily extend over long periods of time. Framework models of socio-ecological research such as Press-Pulse Dynamics (PPD, Collins et al. 2011), the Ecosystem Service Initiative (Shibata and Bourgeron 2011) or the Driver-Pressure-State-Impact-Response scheme (DPSIR; EEA 1999) are – on the contrary – generic concepts that aim to maximise the applicability of the model(s) at varying dimensions in space and time. LTSER in Europe, in contrast, focused on the characteristics of the specific research setting in terms of time and space. Having emanated in the context of evolving the next generation of LTER, LTSER strongly mirrors the conceptual pillars of LTER, including (Mirtl et al. 2009; US-LTER 2011):

- **Systems approach:** LTER contributes to a better understanding of the complexity of natural ecosystems and coupled socio-ecological systems.
- **Focus on process:** LTER's research aims at identifying, quantifying and studying the interactions of ecosystem processes affected by internal and external drivers.
- **Temporally long-term:** LTER dedicates itself to the provisioning, documenting and continuous collection and use of long-term data on ecosystems with a time horizon of decades to centuries.
- **In situ:** LTER generates data at different spatial scales across ecosystem compartments of individual Sites and across European environmental zones and socio-ecological regions.

By definition, socio-ecological research deals with systems and processes beyond the functioning of natural ecosystems (i.e. coupled social-ecological systems), as well interactions with other systems and external factors (Grove et al., Chap. 16 in this volume).

As with the traditional LTER approach, the time dimension is a crucial component of the LTSER framework. Humans have been shaping the land and being shaped by the land throughout history. This interaction is complex, and includes feedbacks and legacies that would be overlooked without proper temporal depth of research. The interaction is dynamic at shorter time scales as well, which emphasises the need for temporally continuous research and data collection over time. Consequently, the paradigms of ecosystem services and sustainability are intrinsically linked with the time dimension (Nelson 2011; Lozano 2008) across human generations and therefore cannot be properly interpreted without consideration of the long term (WCED 1987; Costanza and Daly 1992).

As with time, LTSER research requires large spatial scales to capture drivers and pressures and their long-term impacts, which could not be comprehensively investigated on the spatial scale of hundreds of hectares (even in LTER-Europe's network of over 400 of Sites of that size covering Europe's environmental zones). Aside from large spatial scales, LTSER requires a different focus regarding the location of research sites. In order to support fundamental research on ecosystem processes while attempting to minimise the effects of anthropogenic drivers and management, the selection of locations for traditional LTER Sites was biased in favour of natural or semi-natural ecosystems (Metzger and Mirtl 2008; Metzger et al. 2010). But these anthropogenic drivers, sometimes perceived as 'disturbances' that should be excluded or at least minimised in LTER, are of special interest in LTSER. Thus, the characteristics of the LTER facilities as well as the disciplines involved in research do not suffice to investigate socio-ecological systems (Redman et al. 2004). LTSER research activities need to address spatial units on a sub-regional to regional scale that share a common land-use history and similar environmental conditions. Typically, such regions are in the range of $100-10,000$ km$^2$ and more.

Nonetheless, due to the similarities between LTER and LTSER programmes and the particular evolutionary development of LTSER, it was natural to implement place-based LTSER in the context of LTER, thereby adding a new dimension to the unique combination of the core characteristics above. The interdisciplinary expertise represented by the ALTER-Net consortium catalysed the development of the integrated networks of LTER Sites and LTSER Platforms under the umbrella of LTER-Europe.

An additional complimentarity between LTSER and LTER is that the former is context-driven, problem-focused and interdisciplinary (Mirtl et al. 2009). It involves multidisciplinary teams brought together for limited periods of time to work on specific, real-world problems collaboratively with stakeholders of concrete regions. Gibbons et al. (1994) labelled this type of work "**mode 2**" knowledge production as opposed to traditional "**mode 1**" research, which is academic, investigator-initiated and discipline-based knowledge production. By contrast, mode 2 is problem-focused, stakeholder-integrating and interdisciplinary. LTER-Europe, by initiating the LTSER programme alongside and complementing the continuing traditional LTER programme, provides an integrated framework for both types of knowledge production, maximising the use of existing infrastructure and data legacies.

## 17.2 From Conceptualisation to Regional Application: Place-Based LTSER Platforms

This section focuses on the creation of LTSER Platforms in which principles of socio-ecological research were put to practice in specific geographic regions. "Socio" in this context refers to disciplinary approaches from the economic, social, and cultural sciences as well as the humanities. As the major advance here is the application of socio-ecological research in a specific location, hereafter we distinguish between i) socio-ecological research as a conceptual framework as described in the introduction and part I of this book (concepts and methods) and ii) place-based

LTSER in regions at the scale of European landscapes representing units in terms of environmental history, land use and economic interactions as well as cultural identity (in the range of hundreds to thousands of square kilometres). LTER-Europe contributes to both by implementing LTSER on a regional scale and iteratively feeding practical experiences back into conceptual work.

Regionalising socio-ecological research in LTSER Platforms signifies a paradigm shift regarding the methods and goals of research. This shift is not on the level of individual research projects, but refers to the cooperative and collective goal of developing a detailed and holistic understanding of how spatially explicit socio-ecological systems work by integrating many projects across disciplines and over long time periods. This, of course, includes the investigation of socioeconomic components of the system and their interaction with the environment. The knowledge that the research aims to generate pertains to (1) sustainable use of resources and (2) development of adaptive policies for study regions whose systems are changing due to anthropogenic local and global environmental change (e.g. climate change adaptation).

The quest for this knowledge leads to one of the fundamental components of LTSER: the two-directional flow of information between actors in the region (stakeholders) and researchers (scientists). The actors are any members of the regional population, or those who are not from the region but have a distinct interest in the region's ecosystem services. They include any individuals or groups who have a vested interest in the area under research – whether that is economic, political, or social. The role of such stakeholders in LTSER is threefold: Firstly, the subjectively perceived knowledge gaps regarding sustainable use of ecosystem services have to be collected across actor groups (which is a scientific challenge in itself, and distinguishes the two major approaches of LTSER implementation in Europe discussed further below). Secondly, stakeholders assist in defining the key research questions, such that these questions are not solely generated from the scientific point of view of individual disciplines, but in the framework of an agreed interdisciplinary and stakeholder-informed research agenda. Thirdly, in order to identify realistic options and limitations for dealing with global changes (e.g. climate change) at the regional/local level, the region's social and economic environment must be identified and analysed (threshold interactions across scales and sectors, see below). This final step responds to the apparently contradictory requests for regionalisation on the one hand and internationalisation on the other, both on the continental European scale and internationally. Developing ILTER global comparisons are attracting increasing interest as the LTSER approach is adopted and implemented by a growing range of networks (national LTERs and other LTER regional groups, Global Land Project).

The process of moving from conceptualising LTSER to the implementation of actual regionalised research platforms has proven to be profoundly challenging. In fact, each phase of implementation carries with it its own unique challenges, from identifying appropriate regions and defining their boundaries to developing the common language indispensable for proper interdisciplinary research (Furman et al. 2009). In many cases, even the underlying concepts of LTSER are revisited and modified by regional teams. Thus, the physical implementation of LTSER Platforms has been a major long-term effort and requires both a shared vision and a division of tasks on the European scale.

At the network level, the strategic research intention of the LTSER component in LTER-Europe was to establish an infrastructure to facilitate and strengthen socio-ecological research capacity in the European Research Area. The major socio-ecological systems of the European continent (see socio-ecological stratification below, Metzger et al. 2010) would be represented by at least one LTSER Platform each, where exemplary research could take place including the participation in assessments and forecasts of changes in structure, functions and dynamics of ecosystems and their services, and defining the socio-ecological implications of those changes. Regionalised LTSER also has as a goal to define and address key management issues according to local and regional settings. Aside from the research goals emphasised above, regionally implemented LTSER should support testing and further development of tools and mechanisms for the communication and dissemination of knowledge across different cultural contexts and social gradients.

It is important to note that several additional research bodies have advocated the establishment of such a socio-ecological, place-based research programme (Carpenter et al. 2009). The interdisciplinary Programme on Ecosystem Change and Society (PECS) of the International Council of Science (ICSU website) has recently advocated "Place-Based Long-Term Social-Ecological Research" as being key in investigating society-nature interactions. In co-operation with UNESCO, this concept shall be fostered according to a 10-year action plan (ICSU 2010, Programme on ecosystem change and society (PECS) – A 10-year research initiative of ICSU and UNESCO – Workplan 2010: draft technical paper, Steve Carpenter, chair of PECS, personal communication).

## 17.3 Functional Components of LTSER Platforms

Analysing the challenges outlined above, physical infrastructure, actors and stakeholders, research activities and co-ordination/management have been identified as key components in the design of LTSER Platforms (Fig. 17.1).

In a nutshell, LTSER Platforms are regional hot spots of data and expertise, where infrastructure and monitoring, multiple research projects and regional stakeholders interact synergistically in order to (i) increase knowledge of socio-ecological interactions relevant for a sustainable use of environmental resources and (ii) feed this knowledge into local and regional decision making and management in the pursuit of long-term sustainability. This implies a high level of co-ordination embedding individual projects in a research framework and supporting them with data and relevant contacts.

The required components of LTSER Platforms are defined according to broad research demands to represent functionally and structurally relevant scales and levels on the one hand and characteristics specific to the region on the other. Specifically, the definition of the components depends on individual regions' landscape, habitat

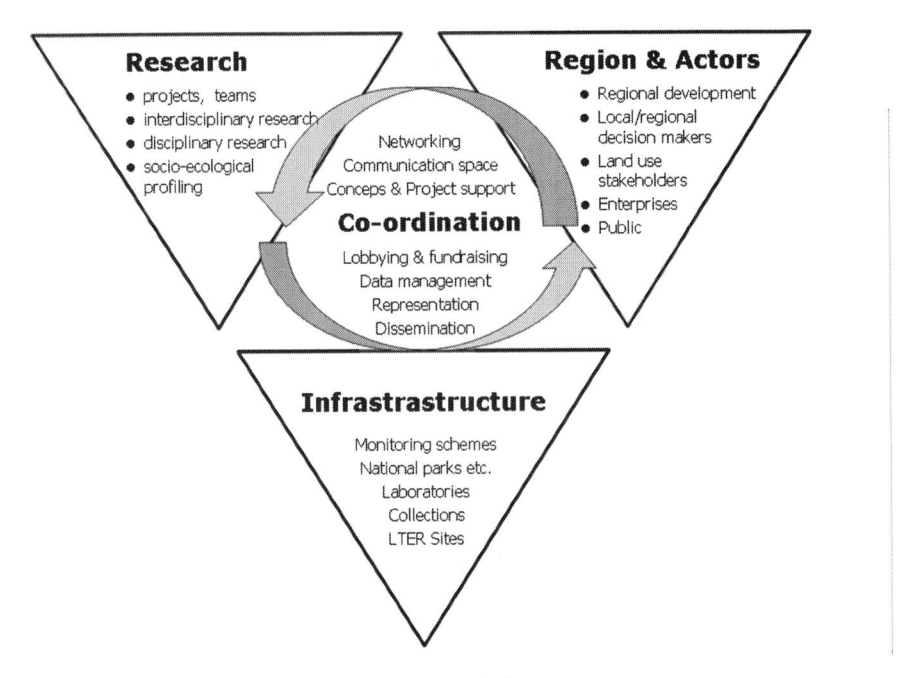

**Fig. 17.1** The functional components of LTSER Platforms

types and administrative structures as well as economic, social and natural gradients within the target region.

The designs of LTSER Platforms that have been established so far in principle combine elements of these four functional components with varying priorities reflected in several chapters on regionalised LTSER in part II and part III of this book (Lavorel et al., Peterseil et al., Tappeiner et al., Furman and Peltola in Europe and Grove et al. and Chertow et al. in the USA). These priorities and the relative importance of individual components also reflect existing settings of research (e.g. data availabilities) and the complexity of targeted issues.

## *17.3.1 Physical Infrastructure and Spatial Design*

Regarding physical infrastructure, LTSER Platforms represent clusters of facilities supporting LTER activities and providing data. In much previous socio-ecological research, studies designed to address interactions between society and natural resources suffered from a mismatch between the observed spatial units and the related spatial scale of management and political response (Dirnböck et al., Chap. 6 in this volume). LTSER Platforms seek to avoid these flaws by developing nested, scale- and level-explicit designs according to comprehensive socio-ecological profiles (example below).

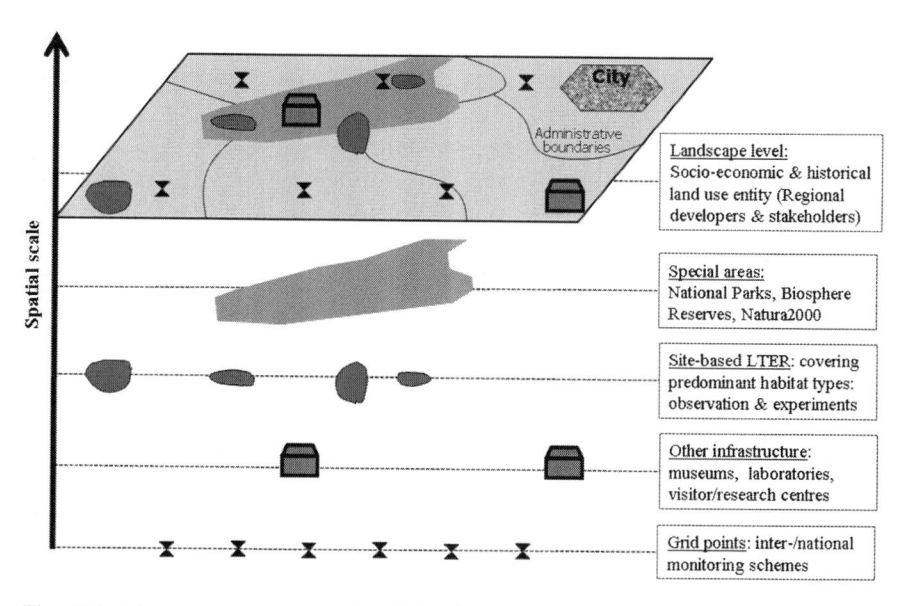

**Fig. 17.2** Infrastructural elements of LTSER Platforms across spatial scales within a LTSER Platform region

With respect to infrastructure, LTSER Platform design distinguishes between (i) grid points of regional, national or international monitoring schemes, (ii) local infrastructure, such as research centres, museums or laboratories (iii) site-level activities representing in-depth ecological research and monitoring in primary habitat types of the Platform region, containing specific sampling or experimental plots at finer spatial scales, (iv) intermediate-scale elements such as national parks, biosphere reserves or meso-catchments, and finally (v) landscapes (Fig. 17.2). The hierarchical design from the site- to the landscape-level and cascaded, harmonised sampling and parameter sets enable the systematic assessment of the representativeness of individual plots or sites. Elements belonging to higher scale activities, including national and international monitoring schemes, are functionally linked for further up- and downscaling and crosswise validation (e.g. biodiversity indicators).

The adequacy and appropriate composition of existing research infrastructures is assessed by means of land cover statistics, habitat and landscape type distributions, and environmental parameter gradients (e.g. predominant land use sectors like agriculture ought to be covered by applied research on the effects of current and alternative management practices).

In terms of socio-ecological interactions, administrative units such as municipalities, districts and provinces offer alternatives for delineating the boundaries of the LTSER Platform, or research units within them. The target is to provide correlating economic, demographic and environmental data with best possible resolution, better than the European Units for Territorial Statistics geocode standard NUTS-3 (Nomenclature d'Unités Territoriales Statistiques; 0.15–0.8 Mio inhabitants) and preferably LAU-2, representing the level of individual municipalities (NUTS 2011). In one promising example, the project IP SENSOR (Sixth Research Framework Programme, European Commission) has managed to collate and integrate national census data with national environmental monitoring data on the scale of the entire Eisenwurzen LTSER Platform. Based on that matching, project researchers have developed sustainability indicators for former mining areas (Putzhuber and Hasenauer 2010).

### 17.3.2 Actors and Stakeholders

Actor/stakeholder integration into research is one of the most important characteristics that distinguish LTSER research from LTER work. In order to identify relevant actors and stakeholders, the geographic extent of the LTSER must be identified, as discussed in the section above and in further practical detail below. Actor analyses identify the corresponding interest groups engaged in regional and local decision-making, management, administration, regional development, education, monitoring, primary research, enterprises, and stakeholders of predominant economic and land use sectors. In consultation with key actors/stakeholders, socio-ecological profiling (see below) is used to reveal key ecosystem services, environmental and economic sectors and social factors and trends driving changes in the system. Structured access to these key groups allows for the efficient identification of research demands. Special attention should be given to established social networks and multipliers (e.g. regional development associations) and their media, which can provide substantive support in recruiting stakeholders, as well as disseminating research findings to the public.

Through the integration of stakeholders, LTSER Platforms encourage a process of reconciling national and international top down research priorities and policies with bottom up, stakeholder-defined research needs of the particular region with regard to nature protection, economic development, and assessment and reporting of environmental conditions. Collaboration is essential at every stage of the process of identifying knowledge gaps, defining research needs, analysing results and translating results into policy recommendations. Considering that environmental policy making is a social process that should reflect political realities, social values and economic needs in order to maximise potential for success (Cohen 2006), the importance of integrating stakeholders into LTSER (ranging from local decision makers to regional developers to global conservation institutions) is self-evident.

### 17.3.3 The Research Component of LTSER Platforms

The research component of LTSER Platforms consists of research projects with best possible complementarity (Fig. 17.3), ranging from specific disciplinary projects to

**Fig. 17.3** Hierarchy of research projects in LTSER Platforms

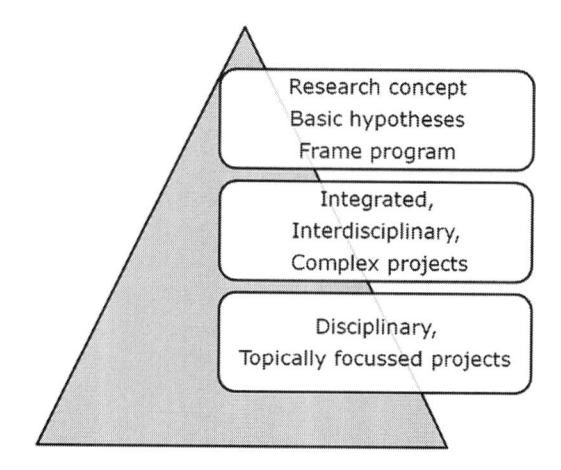

- Research concept
  Basic hypotheses
  Frame program
- Integrated, Interdisciplinary, Complex projects
- Disciplinary, Topically focussed projects

complex synthesis projects, both anchored in a research framework customised for the socio-ecological profile of a specific region (see section on socio-ecological profiling below).

LTSER's two major principles guiding its research programme are i) transdisciplinarity, i.e. the involvement of non-scientific stakeholders into the research process aiming to support regional decisions towards sustainability (Haberl et al. 2006, Haas et al., Chap. 22 in this volume) and, ii) interdisciplinarity, integrating natural sciences, social sciences and the humanities. Transdisciplinarity is particularly important in the definition phase of projects and in the translation of results into knowledge-based guidelines for administration and management, which might be supported by accessory implementation projects funded from sources other than the research itself (e.g. LEADER, LIFE+, Interreg in Europe). LTSER Platforms and their multidirectional space for communication are specifically constructed for developing interdisciplinary research (IDR) on complex socio-ecological questions. The research programmes of Platforms involve more than one disciplinary approach and their research teams closely scrutinise the particular roles of each discipline and, crucially, their interlinkages. ALTER-Net has developed a framework for identifying interdisciplinary research ideas, planning proposals and delivering syntheses (Furman et al. 2009). This framework can be used when developing research strategies in the LTSER Platforms.

Thematic areas of research in LTSER Platforms include (i) process-oriented ecosystem research (basic scientific research; investigation of functionally and structurally important ecosystem components; long-term impacts of drivers and combinations of drivers upon ecosystem functions and services), (ii) biodiversity and conservation research (documentation of the status, trend and functional relationships of species; safeguarding the long-term survival of species, their genetic diversity, and ecological integrity; functionality of habitats and ecosystems) and (iii) socio-ecological research (basic socio-ecological research: Society-nature interaction, socio-ecological transitions; land-use/land-cover change; social perceptions of environment and environmental change, changes in resource utilisation; environmental history and historical sustainability research; transdisciplinary and participative research; integrated socio-ecological modelling and scenarios) (Mirtl et al. 2010).

Hardly any project of one thematic area does not overlap with others when dealing with socio-ecological questions (e.g. the impact of game management on tree regeneration and forest species composition). In this sense, LTSER is an approach which challenges and changes the routines of academia. Although there have been many mainly programmatic discussions about these principles for decades, ("mode 2-", "mode 3-" and "post-normal science", e.g. Funtowitz and Ravetz 1992), major parts of the scientific community in general still remain sceptical regarding the potential for interdisciplinary research to contribute to our understanding of the world. Inter- and transdisciplinarity are sometimes seen as competing against disciplinary excellence (experiences from the ALTER-Net project mentioned above and managers of European LTSER Platforms, according to the LTSER workshop held in Helsinki, June 2011). LTSER Platforms have the potential to serve as experimental laboratories in which classic disciplinary research is combined with inter- and transdisciplinary research towards both scientific excellence and relevance to real-life challenges. Continuous accompanying research (Kämäräinen 1999) within LTSER Platforms is one of the key instruments to enable that combination and to support researchers from different disciplines and stakeholders from several societal groups in moving LTSER forward.

In contrast to conventional evaluation processes, accompanying research within LTSER should be seen as the common responsibility of all researchers and stakeholders involved and should focus on integration rather than on quantifying output. The responsibility for stimulating integration can be assigned to several individual researchers, who would be required for a certain time to travel and, by means of participant observation, to learn about the main points of research at different Sites. Although LTSER focuses on research relevant to its geographic area, coordination between LTSER Platform teams is crucial for maintaining a minimal level of comparability between Platforms. Project teams within and between LTSER Platforms should observe each other, looking for potentially conflicting basic assumptions and for paradigms underlying their research, and functioning as an internal "quality control" body. Further, teams should attempt to maintain a degree of commonality between the Platforms, which is crucial to the larger continental and global goals of comparability and assessing the impact of global processes in the local setting (Mirtl et al. 2009).

Conventional evaluation processes measure the scope of scientific output (e.g. published papers in academic journals). LTSER Platforms could and should be evaluated with conventional instruments of this kind. However, the appropriate evaluation and competitive chances of LTSER projects are constrained, as long as the ability of LTSER to generate realistic environmental/natural resource policy recommendations for stakeholders – based on both their input and research results – is not, in addition to that, acknowledged based on its societal relevance.

### 17.3.4 LTSER Platform Management, Co-ordination and Communication Space

It has been recognised by the LTER community that LTSER requires a Platform management and co-ordination team, secured for the long term and providing a wide range of services implied from the sections above. Amongst these services are communication space (meetings, website, bilateral contacts, local to global contacts), conceptual work (see following chapter "Socio-ecological profiling"), project development, networking across interest groups, disciplines and stakeholders both nationally and internationally, results dissemination, communication with the broader public, education, youth and researcher training, data integration and policy, data management, development and provisioning of IT tools, representation (nationally, internationally), lobbying and fundraising. An example for how these services have been implemented in detail is given by Peterseil et al. (Chap. 19 in this volume).

Successful LTSER depends strongly on internal factors, and first and foremost on the quality and content of scholarly exchange within the community. The conscious design of communication processes between different disciplines and between science and the public is crucial. Therefore, the "platform communication space," must be a multidimensional environment that allows for people from different technical and cultural backgrounds to understand one another. It uses a variety of media and communication formats to support the implementation of the transdisciplinary and participatory approaches necessary to adopt research agendas to regional and local needs and to achieve access to and involvement of the regional population, key stakeholders and decision-makers, all of whom can be seen as beneficiaries of the knowledge produced.

The same is true for science when it comes to the required data access and data flows. Without central facilitation, providing required data for complex LTSER projects alone would exhaust individual projects, even if these data were available for free. LTSER requires quick data exchange, ideally based on IT solutions, and the integration of dispersed data sources (ontologies, tools for semantic mediation). The LTSER Platform must therefore secure funding for numerous aspects of management, initiating, supporting and documenting research. Basic funding has to be ensured by the committing institution or by national funding programmes. Additional funding may be necessary for instrumentation, data and projects running on the Platform. Once the LTSER Platform is up and running, periodic funding will also be needed for synthesis projects.

## 17.4 Socio-ecological Profiling – Applying Tools and Various Conceptual Models to Socio-ecological Systems

The initiation of an LTSER Platform is aided by the adoption of conceptual models through which a socio-ecological profile can be developed. Such a profile distils the multiple social and ecological variables and their complex interactions operating within the Platform into the primary components important to study. These components are defined primarily through expert knowledge of the LTSER team and the local knowledge accrued through stakeholder mapping. The process of creating the profile can be considered as part of the scientific co-ordination activities of the Platform management. As a collective approach involving all actor groups, it is a typical outcome of the Platform's communication space. Participants perceive it as a common reference point to anchor their activities and projects.

Mapping the socio-ecological profile of a LTSER Platform region with the assistance of several conceptual models has increased the robustness of the profile in terms of acceptance and collective ownership by different disciplines and stakeholders. Unifying concepts increase the potential to parameterise additional socio-ecological models and helps establish a common research denominator across Platforms. Last but not least, the qualitative and semi-quantitative knowledge represented by several regionalised conceptual models form a sound basis for inter-Platform comparisons, nationally, continentally and on the global scale. This is exemplified below for the case of Eisenwurzen LTSER Platform in Austria (Peterseil et al., Chap. 19 in this volume), where researchers have shown how a regional socio-ecological profile can facilitate the identification of system properties (e.g. relevant ecosystem services) as well as similarities with other socio-ecological systems.

## 17.4.1 Overview

The entire research community and several stakeholders were involved in at least one of the following steps extending over 3 years:

**Step 1: Fuzzy cognitive mapping** was used to develop an integrated view of key elements and their interactions (direction, strength), based on mindmaps of individual actors and stakeholders. The results of the cognitive mapping reflected a collective perception of the region, which then served as a primary input for parameterising an Integrated Science for Society and Environment (ISSE) model (Collins et al. 2007).

**Step 2: Critical ecosystem services**: Identification of the critical ES, direction of change, primary drivers of change, public awareness of the ES, and institution(s) that manage the ES.

**Step 3:** The **ISSE model** (Integrated Science for Society and the Environment): This framework has been proven to provide an excellent basis for interdisciplinary teams working in a region (Collins et al. 2007, 2011; Grove et al., Chap. 16 in this volume). As the Eisenwurzen Platform showed, an LTSER Platform is unlikely to conduct just one project covering all socioeconomic and ecological systems. More likely there will be several potential projects on defined interfaces between the socioeconomic and the ecological system. The framework accentuates the shortcomings of disciplinary sciences. With the model available to all participants, the "bigger picture" of the system becomes clearer and linkages can be drawn by the scientists between their fields and their work.

**Step 4: Threshold interactions:** Identification of threshold interactions between environmental and socioeconomic dynamics at multiple scales, and forecasting the effects of these interactions on ecosystem services and ecological resilience (Kinzig et al. 2006; Holling 2001).

**Step 5: Use of the robust socio-ecological profile** in other conceptual models (e.g. DPSIR) and comparative assessments.

Steps 2–4 were done in the frame of the Ecosystem Services Initiative (ESI) within the ILTER Network. The Millennium Ecosystem Assessment utilised an approach to quantify ecosystem services in order to understand the value of ecosystems to humans (MEA 2005). In a similar vein, the ILTER Science Committee commenced the ESI to develop and apply threshold interaction models for selected biomes across the world. The initiative includes ISSE, Critical Ecosystem Services and Threshold Interactions as models and approaches for the understanding and rating of ecosystem services (Shibata and Bourgeron 2011). These models were applied in LTSER Platforms and LTER Sites across Europe (Kiskunság, Hungary; Donana, Spain; Eisenwurzen, Austria; Gascogne, France; Leipzig-Halle, Germany; Uckermark, Germany; Lake Päijänne, Finland; Central Poland, Poland). A synthesis within and among biomes of culture-specific, socioeconomic dynamics leading to increases or decreases in resilience of ecosystems is still ongoing.

We now expand on each of these steps as applied in the case of Eisenwurzen LTSER Platform in Austria.

## 17.4.2 Step 1: Fuzzy Cognitive Mapping: Collecting Socio-ecological Data from Stakeholders

Fuzzy cognitive mapping is a participatory modelling approach which allows the depiction of causal relations between important elements of coupled society-nature systems as they are perceived by stakeholders. A cognitive map like a Fuzzy Cognitive Map describes a system by showing the central factors and their causal relations, represented by weighted arrows, as a *directed graph*. Fuzzy cognitive maps are drawn by the stakeholders in an interview setting. Maps of different stakeholders can be merged to gain a broader system view. Combined maps can also be used to run scenario-analysis (Kosko 1986; Özesmi and Özesmi 2004). From autumn 2007 to spring 2009, six case studies were conducted in different LTSER Platforms across Europe using Fuzzy Cognitive Mapping. In order to analyse the fuzzy cognitive maps a freely available software was developed (www.fcmapper.net). In the context of the study the LTSER Platforms proved to be excellent working environments for this purpose due to established communication structures and good access to stakeholders (Wildenberg et al. 2010).

Figure 17.4 shows a simplified FCM derived from two interviews in the Eisenwurzen LTSER Platform. If the 'area under intensive farming' increases, the 'income of farmers' will increase and 'biodiversity' is expected to decrease. On the other hand 'biodiversity' is influenced positively by a 'diverse landscape structure'

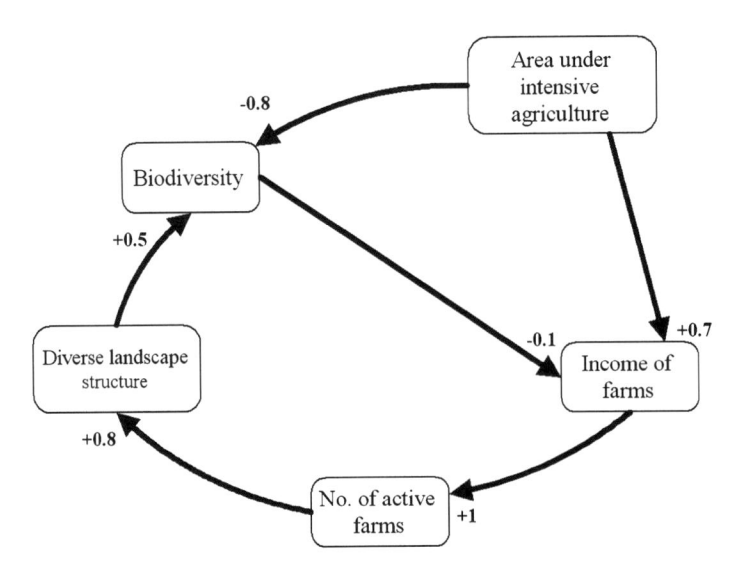

**Fig. 17.4** Schematic fuzzy cognitive map derived from two interviews in the Eisenwurzen LTSER Platform – Austria

which itself depends on the 'number of active farmers'. The relatively weak link between 'biodiversity' and the 'income of farmers', which drives the 'number of farmers', reflects a low level of subsidies for using extensive farming techniques.

Experiences from using FCM in the LTSER Platform context showed that it is a promising explorative method for LTSER, as it depicts complex socio-ecological systems in terms of the perceptions and mind models offered by people living in an area. They represent a vital component in every linked human-nature system. Another strength of FCM is its interactive and social learning component and its ability to handle all kinds of knowledge systems, making it suitable for "mode 2 research" (Gibbons et al. 1994) and stakeholder involvement for conservation planning or educational purposes. In the case of Eisenwurzen, FCM contributed to the development of decision trees for an agent based model (Gaube and Haberl, Chap. 3 in this volume) and the conceptual models presented below.

Figures 17.5, 17.6, 17.7 and 17.8 are schematic representations of diverse conceptual models. All were parameterised for the Eisenwurzen LTSER Platform region to assist in organising, clarifying and identifying the important elements and feedbacks within the socio-ecological system of the montainous post-mining area (described in detail by Peterseil et al., Chap. 19 in this volume).

We selected one critical process in the region, a case study, as trigger to demonstrate the stepwise elaboration and structured description of elements and their interactions across the models.

Due to: (i) the decline of the iron producing industry with its high energy demand, historically served by timber, and current low timber prices and (ii) land abandonment caused by depopulation, forests have been continuously reclaiming the central

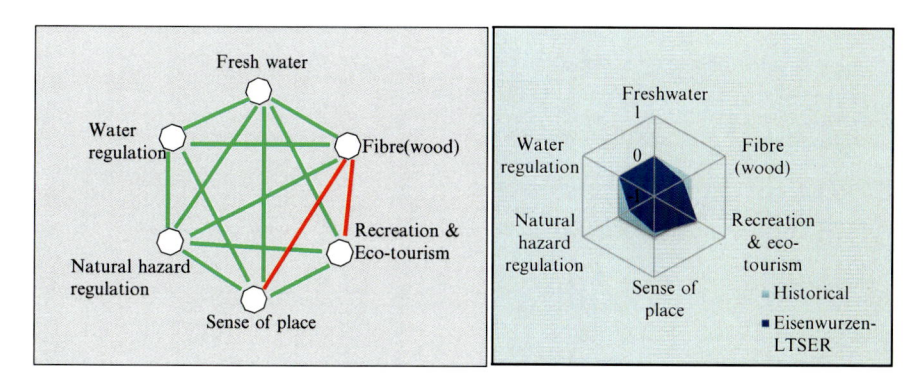

17 Development of LTSER Platforms in LTER-Europe: Challenges... 425

**Fig. 17.5** Simplified model of the Critical Ecosystem Services of the Eisenwurzen LTSER Platform and their interaction *left*: *green* = positive, *red* = negative) and scenario of their future importance (*right*: light *blue* = historical situation, dark *blue* = current situation) (Austrian contribution to the ILTER Ecosystem Service Initiative)

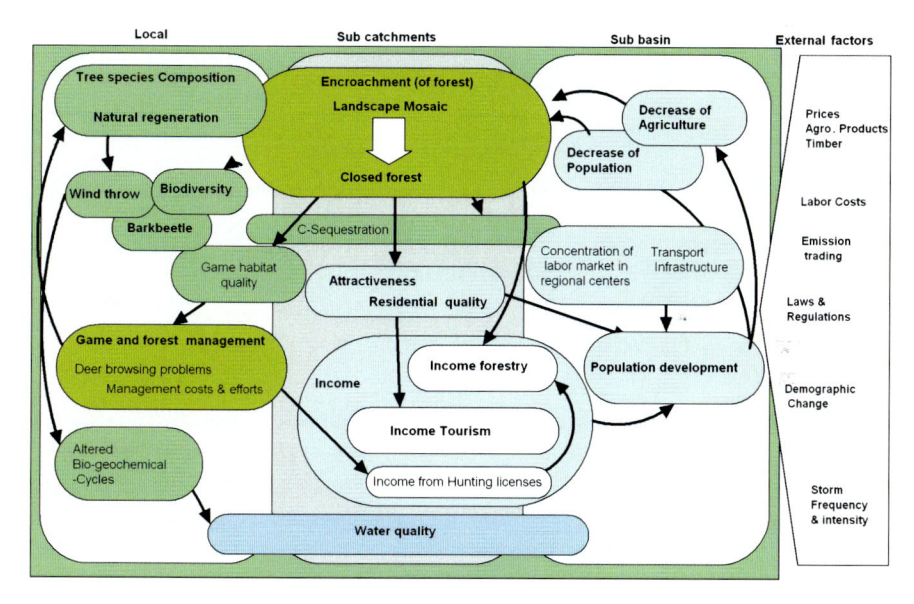

**Fig. 17.6** Interactions of key elements and factors in the socio-ecological system across sectors (environment in *greens* and *blue*, economy and society in *white* and *grey*) and scales in the LTSER Eisenwurzen (Austrian contribution to the ILTER Ecosystem Service Initiative)

parts of the Eisenwurzen region since the nineteenth century. Relying on the scenic cultural landscape, tourism has become an important alternative source of income. However, closed forests reduce the beauty of the area as it is subjectively perceived by tourists and also give local inhabitants the impression of being "overgrown by forest", a situation which is interpreted as signifying loss of importance and marginalisation.

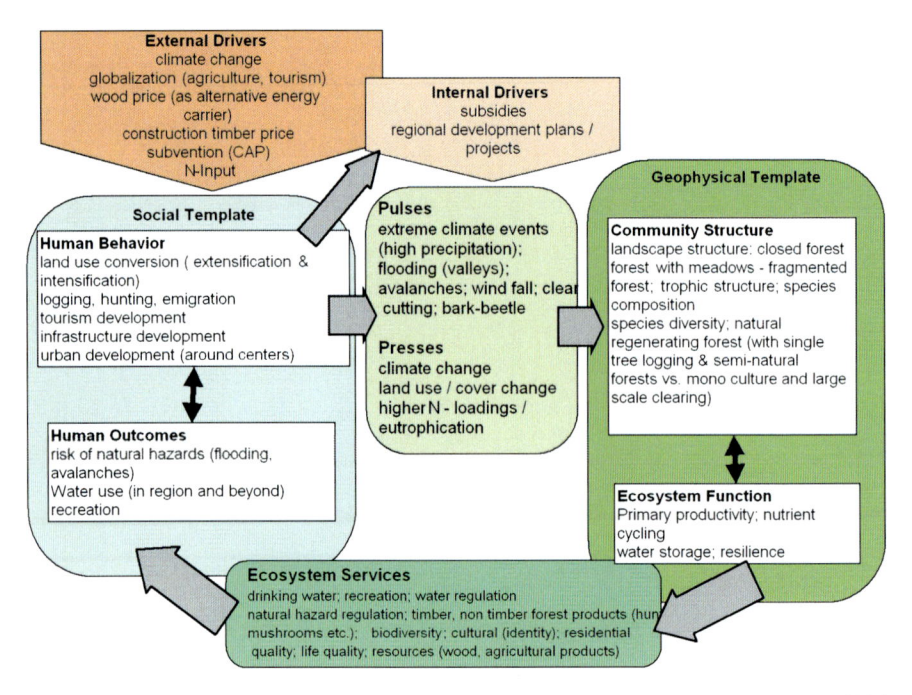

**Fig. 17.7** Socio-ecological profile of the LTSER Eisenwurzen Platform according to the ISSE/PPD framework (Collins et al. 2007, 2011): The conceptual elements, described by Grove et al. (Chap. 16 in this volume) are parameterised based on comprehensive analyses combining disciplinary scientific expertise and primary stakeholders perception (Fuzzy Cognitive Mapping)

## 17.4.3 Step 2: Critical Ecosystem Services

Ecosystem Services (ES) are widely regarded by LTSER teams as excellent common currency for cross-Platform comparisons (Dick et al. 2012). ES are appropriate in several ways to trigger interactions in socio-ecological systems. They represent objects of concern to stakeholders, and they are the conceptual link in models between the human and the natural sphere (Costanza et al. 1997). Key ecosystem services or "Critical Ecosystem Services" of the Eisenwurzen region were identified and scenarios for their relative importance in the future were developed, *inter alia*, by the use of information on relative importance, form of interaction and stakeholder expectations for the future from the Fuzzy Cognitive Mapping in combination with interdisciplinary expert knowledge (see above).

In terms of our case study, we can identify the negative relations between fibre production and recreation and local identity (feeling at home) (Fig. 17.5, left). Competing ecosystem services and related concerns of the local population with respect to afforestation are clearly reflected in the right part of Fig. 17.5, showing that sustainable income is, in the future, expected from eco-tourism rather than from timber production (Gaube and Haberl, Chap. 3 in this volume)

The concept of ecosystem services can be used to link social and ecological systems into an integrated, multi-scaled socio-ecological system. In preparation for

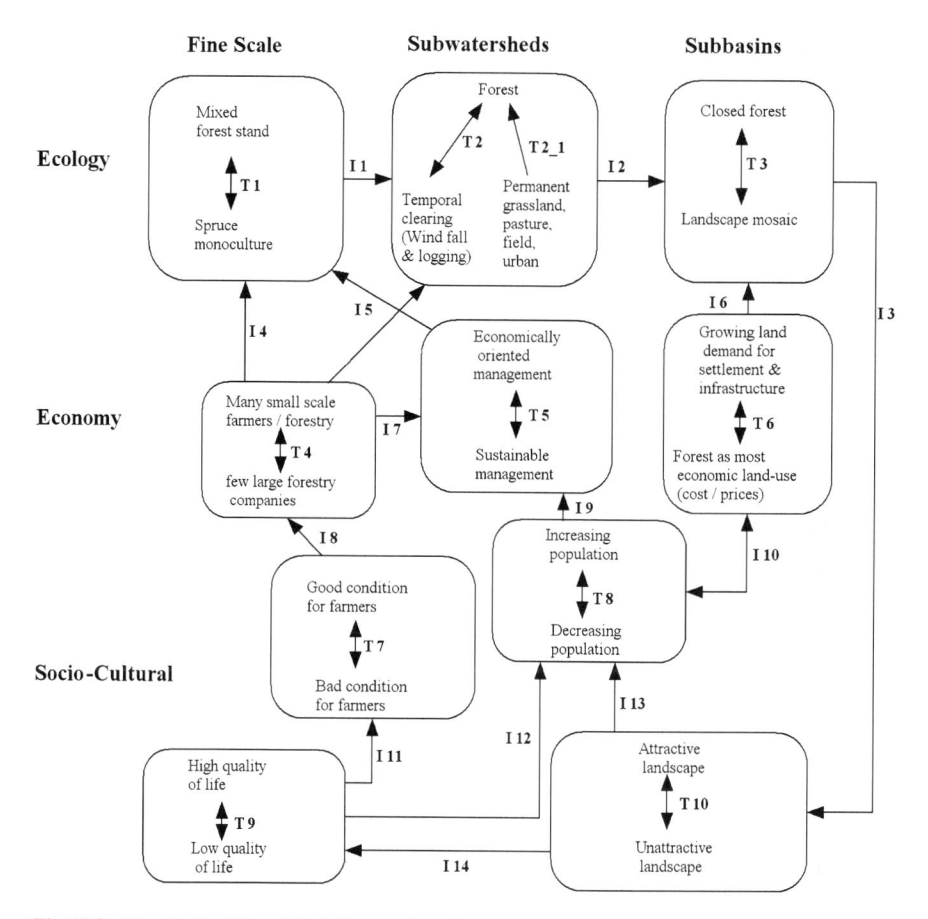

**Fig. 17.8** Thresholds (T) and their interactions (I) across sectors (environment, economy, society) and scales in the Eisenwurzen LTSER region (Austrian contribution to the ILTER Ecosystem Service Initiative according to Kinzig et al. (2006))

regionalising the ISSE model for the Eisenwurzen region, the interactions of key elements and factors in the socio-ecological system were mapped across sectors. The interplay described in our **case study** between forest encroachment, residential quality and income from tourism and forestry is again evident in the upper and central part of Fig. 17.6, but is now embedded in the broader context of human-nature interactions of the region. External drivers such as demographic change and transport infrastructure become visible (to the right).

## 17.4.4  Step 3: ISSE/PPD Feedback Loop Model

The ISSE (Integrative Science for Society and Environment) feedback loop model framework was developed in 2007 under the US-LTER strategic research initiative

"Integrative Science for Society and the Environment" (ISSE, Collins et al. 2007) and further developed into the PPD (Pulse and Pressure Dynamics) model (Collins et al. 2011). Graphs and a detailed theoretical overview are given by Grove et al. (Chap. 16 in this volume). The framework identifies two fundamental linkages between social and ecological systems. On the one hand, the social system, encapsulating political, economic and demographic trends among others, has a direct impact on ecological systems via presses (steady long-term changes, such as agricultural and urban expansion) and pulses (profound, non-routine changes, like wildfires and oil spills). On the other hand, modifications of ecological systems result in a change in the amount and types of ecosystem services provided to human societies. External factors, such as natural climate cycles, are also driving change in the ecological systems and therefore ecosystem services.

Regarding the case study, the elements presented needed to be assigned to the above categories. As seen in Fig. 17.7, land cover such as closed forests belong to the ecosystem structure (biophysical template to the left) providing the ecosystem services of timber production and recreation (bottom). The social template on the left contains the use of the services, including generation of income from tourism and creation of infrastructure, such as streets for commuting. These contribute to the disturbance regimes (pulses and presses) in the centre. Depopulation and changes in land use act as long-term presses that impact upon biophysical components such as land cover, which closes the loop. External drivers, such as market prices for timber and steel, seen above complete the ISSE modelling of the case study.

## 17.4.5 Step 4: Threshold Interactions

Even though ISSE is a feedback loop model, it still provides a static picture of the socio-ecological system. Socio-ecological profiling, however, aims at identifying potentially irreversible system alterations. Most accounts of thresholds between alternate regimes involve a single, dominant shift defined by one, often slowly changing variable in an ecosystem. Kinzig et al. (2006) develop a "general model" of threshold interactions in socio-ecological systems across spatial scales. Their generalized model of threshold interactions as parameterised for the Eisenwurzen region (Fig.17.8. 8) shows all possible combinations of domains and scales and the possible interactions between regime shifts for various domain-scale combinations.

Revisiting the case study, we identify the transition between closed forest and a landscape mosaic on the top right (box T 3). This interacts (arrow I 3 to the right) with the threshold between attractive and unattractive landscape (T 10, bottom right) impacting (I 14) upon quality of life (T 9) at the bottom. Population density (T 8, second row from bottom) drives timber use and land cover (T 5 and T 6) above. The generic picture in Fig. 17.5 has been detailed and structured to a level enabling systematic documentation and comparisons with other systems. Moreover, strengths of interactions and critical system conditions (thresholds) can be specified based on empirical regional knowledge (e.g. the critical level of forest coverage).

The elaboration of the Eisenwurzen LTSER socio-ecological profile represented a cornerstone in developing common ground for the LTSER community and regional stakeholders. Individual project leaders acknowledged the framework's value in anchoring their respective projects within the system context and affirming and reinforcing the social and policy relevance of the work. We stress that these models were either commandeered by or developed specifically to suit the research needs of LTSER. In this way, LTSER has served as a laboratory for the increasingly emphatic demand for societally relevant ecological research.

## 17.5 Implementation of Individual LTSER Platforms – Process and Experiences

Although the LTSER concept is still in its infancy, the European LTER network has accrued significant experience over the past decade in setting up LTSER Platforms. In this section, we present best-practice guidelines for establishing Platforms, based on the accumulating experience of Platform management teams across Europe.

Selecting a suitable region for the LTSER Platform is recommended as a first step, and such decisions are often made due to practical, rather than theoretical considerations (outlined in previous sections). Historically, the development of LTER-Europe was, at the request of the European Commission, to be based on existing infrastructure wherever possible. So it was logical that the first step in defining potential areas for LTSER capitalised on inventories of existing infrastructures at the national level such as LTER Sites, well-equipped sites of ecosystem monitoring schemes, protected areas, National Parks, Biosphere Reserves etc. carrying out traditional ecosystem research in habitats typical for the region (Mirtl and Krauze 2007; Mirtl et al. 2009, Mirtl 2010). LTSER Platforms are often, but not exclusively, established by building on existing LTER Sites, thus benefiting from the data legacy and associated facilities.

**Selection criteria** for appropriate LTSER Platform regions beyond the infrastructural component are:

- Well-documented land-use history, cultural and socioeconomic unity;
- Active, well-established institutions (research institutions, non-governmental agencies, private sector, and government agencies);
- Research covering ecosystem services of relevance for the region;
- Research on alternative management practices;
- Availability of reference areas (undisturbed natural habitat(s), or at least the most undisturbed possible, typical for the region);
- Coverage of socio-ecological gradients of the biogeographical regions;
- Interest among local stakeholders, government and policy makers for policy-oriented research;
- Eventual closure of network gaps on the European scale (see below).

Once potential localities for Platforms have been identified, further pre-selection ought to consider (i) the LTER-Europe criteria/descriptors for LTSER Platforms (comprising aspects of infrastructure, data and data availability, access to key actor groups and streamlined activities) (LTER-Europe website, key documents), (ii) the scientific interests and strengths of the national and local research communities and (iii) the importance of the environmental zone which the area represents (pressures, conflicts, ecosystem services). From the European perspective, national networks are expected to help improve the coverage of the network as far as possible and eventually all environmental zones (EnS) and socio-ecological zones (LTER Socio-Ecological Regions) should be represented by LTER Sites and LTSER Platforms. The coverage of European LTER facilities across 48 socio-ecological strata was tested by Metzger et al. (2010) and gaps identified. Each national decision on a new Platform enables the possible closure of such gaps in the network.

After the location of a new LTSER Platform is selected, the boundaries of the Platform region must be delineated. Because Platforms are to capture socio-ecological systems and their interactions, social (as well as ecological) boundaries must be considered. Therefore, Platforms may be delineated by political/administrative borders or by other existing borders (e.g. biospheres or national parks). Alternatively, the boundaries may be left only vaguely determined, and allow for individual research questions to determine boundaries.

In order to ensure the long-term administrative and economic stability of the Platform, a consortium of major regional research and policy institutions (e.g. universities, government agencies, major NGOs) often form the core group promoting and implementing LTSER Platforms. As their mission usually stretches over decades, they offer ideal settings for hosting LTSER Platform management. Ideally the Platform management is funded by the main beneficiaries of its services. Through promotional campaigns, workshops, and meetings with individuals and institutions, the LTSER concept and goals are advertised in relevant communities in order to invite interested parties and expand the LTSER Platform consortium. A Memorandum of Understanding (MoU) is written with the input of the growing management team and consortium, which should address the scientific and practical goals, governance structure and data policy of the Platform. Stakeholders may receive a feeling of empowerment and "buy-in" if they can contribute to the memorandum. This document will be useful, not only to clarify positions and aims, but also to lobby for the Platforms. The MoU will also guide the LTSER Platform management in setting up and providing specific services as specified in the document.

Spatial delineation will also drive data collection. As empirical socio-ecological research capitalises on data and information from different realms, these data need to refer to the same spatial units. In most cases the best available economic and census data are provided with a resolution at the level of municipalities. However, when moving from LTER Sites to LTSER Platforms, problems arise when ecological and social borders do not match. The Platforms, with boundaries also delineated by research questions and policy needs, provide a flexible framework to deal with this problem.

Defining the goals and scope of the LTSER Platform is a most crucial phase in the establishment of the Platform. While the entire process should be flexible and iterative, a careful set of research goals will assist the LTSER team in remaining focused on their objective, as well as in expressing themselves articulately to potential partners, funders and stakeholders. The LTSER team should match goals to the capacity of their team – academically, monetarily and taking uncertainties into account. A set of "meta-goals" will supplement and frame the local goals. These meta-goals shall serve as a common denominator for the comparison of data across LTSER Platforms. Meta-goals will be informed by the recommendations of the LTER-Europe Expert Panel Science Strategy and the international LTSER research agenda. This is crucial in the context of being part of a network, and for building the foundation for harmonisation of research activities and comparability of experiences and research results across regions. Two concepts important for LTSER goal-setting are that LTSER research programmes should adhere to the principles of sustainability science, and that LTSER research should be conducted using a common conceptual model (see previous sections).

There are two basic approaches in implementing LTSER Platforms:

- Strategically managed and all inclusive: as outlined above, especially in cases of high complexity in terms of Platform size, number of participating institutions, actor groups, etc., the inclusion of stakeholders from the beginning is important for developing a user-oriented research agenda. This approach requires substantial resources for co-ordination and central services.
- Project-based, evolutionary: An alternative approach is to start from the bottom in a project-oriented and iterative way. Here mainly research institutions develop a research strategy, plan research activities jointly and, if possible, build the monitoring infrastructure necessary for the planned research. This approach is particularly beneficial where innovations in research approaches are required. One risk of a top-down approach dominated by one group, e.g. traditional ecological research is that the framing of the research might not open space and build motivation for other disciplines to enter into LTSER research.

The LTSER Platforms established so far vary considerably in composition, size and targets. Whereas some follow an integrated regional approach considering the entire policy cycle from user-oriented knowledge generation to management and political measures, others are rather clusters of site-based research concentrated in a specific area. There is clear evidence of a trend towards integrated approaches. As pointed out earlier, only structured – and where necessary, formalised – access to key actor groups allows for the identification of research demands as regionally perceived and for the dissemination and implementation in practice of research findings.

## 17.6 LTSER Platforms Across Europe – Status of the LTER-Europe Network on the Continental Scale

Although implementing even a single LTSER Platform is a complex challenge, the European ambition was to build place-based socio-ecological research capacity in the European Research Area, where each of the major socio-ecological systems of the European continent (see below) would be represented by at least one LTSER

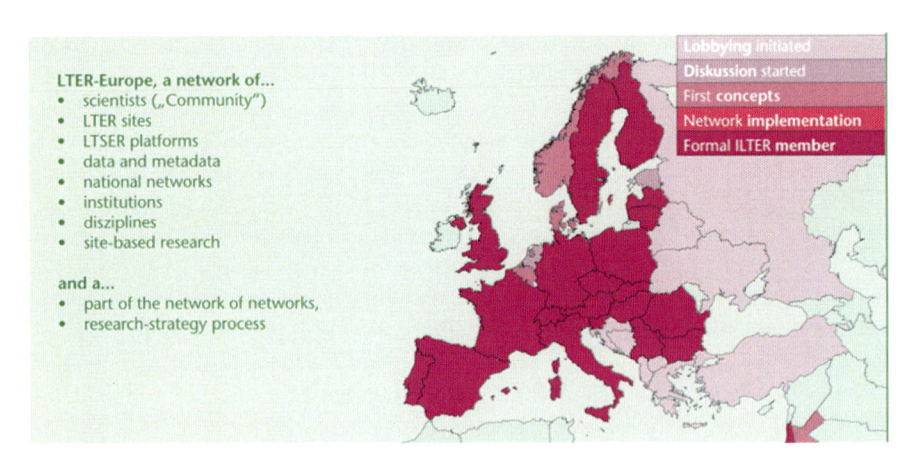

**Fig. 17.9** Geopolitical coverage of LTER-Europe (as of 2010; Mirtl et al. 2010)

Platform in order to exemplarily investigate socio-ecological interactions. LTER-Europe currently comprises formal national networks in 21 countries and emerging networks in about 5 countries (Fig. 17.9). The physical network consists of about 400 LTER Sites and 31 LTSER Platforms (as of 2010).

For the LTER Socio-Ecological Stratification (LTER-SER, Metzger et al. 2010), the European environmental zones (EnS) used in the Millennium Assessment (Metzger et al. 2005; Jongman et al. 2006) were combined with a newly developed socioeconomic stratification based on an economic density indicator. This enabled the LTER team to overcome both the limitations in data availability at the 1 km$^2$ resolution across Europe and in distortions caused by using administrative regions (NUTS 2011). The resulting 48 socio-ecological systems are reflected in the map of Europe depicted below in Fig. 17.10.

In recent years, the LTSER component of LTER-Europe has developed quickly; In 2008, only 23 LTSER Platforms (5 as emerging) were registered in the LTER-Europe Infobase (LTER-Europe website/information management). The 31 LTSER Platforms that are now operating are spread over 17 countries (Fig. 17.10) and cover all 48 socio-ecological regions (some Platforms are big enough to contain more than one socio-ecological region). A gap analysis in 2008 showed weak coverage in the Atlantic North because the few remaining countries without LTER were concentrated in this area (Belgium to Norway, see Fig. 17.9). Another gap in the Mediterranean South has started to close with the strong LTSER involvement of Israel. In addition, desert environments are now included through LTSER Platforms in Jordan (emerging network) and Israel.

According to the rules and governance of LTER-Europe, the national LTER networks are responsible for choosing the LTER Sites and LTSER Platforms in their respective countries. LTER-Europe provides a framework to assist in national network building and decision-making. Under the auspices of ALTER-Net, a set of criteria for LTER networks, LTER Sites and LTSER Platforms was developed in 2005 and formally adopted in 2008 (LTER-Europe website/key documents). Criteria

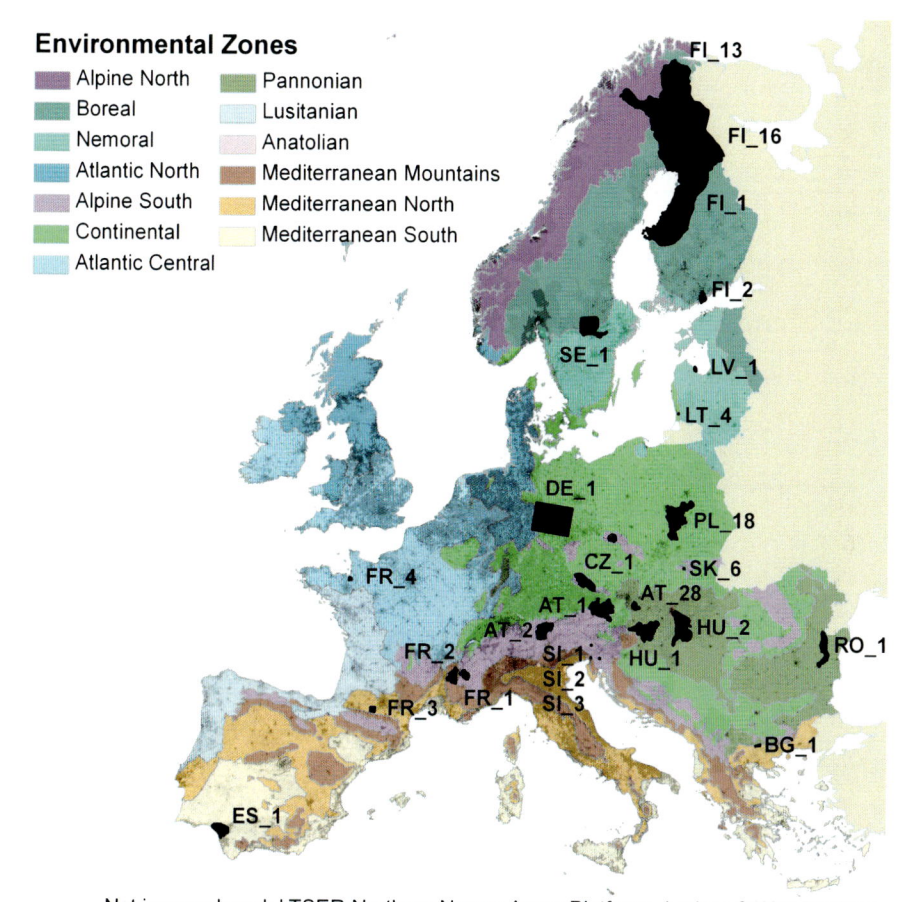

**Fig. 17.10** Location of 31 European LTSER Platforms in 2010 (including five preliminary Platforms). The map reflects the 48 socio-ecological systems of Europe (Metzger et al. 2010). Environmental zones are colour-coded. The brightness of each colour varies according to the economic density, varying between < 0.1 Mio €/km$^2$ (lightest) and > 0.1 Mio €/km$^2$ (darkest). The Platform labels are the unique LTER-Europe site codes. According to these site codes, details for each Platform can be found in Table 17.1

are continuously updated according to accumulated experiences on feasibility and identified weaknesses. In the case of LTSER Platforms and given the early stage of the application of the LTSER concept, these "criteria" have so far been applied as "descriptors", supporting comparative description of LTSER Platforms rather than as hard selection criteria.

LTER-Europe also provides cross-country analyses to promote decisions that optimise the division of tasks within the European Research Area. LTSER- and IDR-issues within LTER-Europe are governed by the Expert Panel on LTSER (www.lter-europe.net, Mirtl et al. 2009).

**Table 17.1** Overview of European LTSER Platforms, status as of 2010. The labels of platforms in Fig. 17.10 refer to the column "Site_Code" in this table

| LTER_Europe_Site_Code | LTSER Platform name | Country | Biogeographic region | Ecosystem type | Size km$^2$ |
|---|---|---|---|---|---|
| AT_001 | LTSER Platform Eisenwurzen (EW) | Austria | Alpine | Forest | 5,780 |
| AT_002 | LTSER Platform Tyrolean Alps (THA) | Austria | Alpine | Montane | 3,689 |
| AT_028 | LTSER Neusiedler See-Seewinkel | Austria | Pannonian | Fresh water | 634 |
| BG_002 | Belasitsa | Bulgaria | Sub-mediterranean | Forest | 111 |
| CZ_001 | LTSER Silva Gabreta (LTSER Silva Gabreta) | Czech Republic | Continental | Temperate forest | 3,337 |
| CZ_006 | LTSER Krkonose/Karkonosze (LTSER Krkonoše/Karkonosze) | Czech Republic | Continental | – | 871 |
| FI_001 | Bothnian Bay LTSER Platform | Finland | Boreal | Coastal | 58,439 |
| FI_002 | Helsinki Metropolitan Area | Finland | Boreal | Coastal | 879 |
| FI_008 | Northern LTSER Platform | Finland | Boreal | Forest | 118,656 |
| FI_013 | Kilpisjärvi LTSER | Finland | Alpine | – | 3,691 |
| FI_016 | Kuusamo LTSER | Finland | Boral | – | 5,790 |
| FR_001 | Alpes-Oisans | France | Alpine | Montane | 1,037 |
| FR_002 | Alpes-Vercors | France | Mediterranean | Montane | 1,890 |
| FR_003 | Côteaux de Gascogne | France | Atlantic | Agriculture | 441 |
| FR_004 | Pleine-Fougères | France | Atlantic | Agriculture | 132 |
| DE_001 | LTSER Leipzig-Hall | Germany | Continental | Fresh water | 22,781 |
| HU_001 | Balaton LTER | Hungary | Pannonian | NONE | 5,767 |
| HU_003 | KISKUN LTER | Hungary | Pannonian | Praire | 7,270 |
| IL_005 | LTSER Northern Negev | Israel | Mediterranean | Desert | – |
| IL_015 | Araval Platform (ARV) | Israel | Mediterranean | Desert | 981 |
| JO_001 | SAWA Platform | Jordan | Mediterranean | Desert | – |
| LV_001 | LTSER Engure | Latvia | Boreonemoral | – | 178 |
| LT_004 | Lithuanian Coastal Site (LT-04 Nagliai, Curonian Spit NP) | Lithuania | Boreal | Coastal | – |
| PL_018 | UNESCO/UNEP the Pilica River Demonstration Site | Poland | Continental | – | 9,256 |
| RO_001 | Danube Delta Biosphere Reserve | Romania | Steppic | Wetland | 3,120 |
| SK_006 | Tatra National Park | Slovakia | Alpine | Forest | – |
| SI_001 | Kras | Slovenia | Continental | Montane | – |
| SI_002 | Karst in the Ljubljanica River Basin | Slovenia | Continental | Montane | – |
| SI_003 | Alpine Karst | Slovenia | Alpine | Montane | – |
| ES_001 | Doñana/Huelva-Sevilla (ES-SNE) | Spain | Mediterranean | Wetland | 2,732 |
| SE_001 | Nora LTSER | Sweden | Boreal | – | 6,648 |

## 17.7 Lessons Learned and Outlook

By establishing LTSER Platforms in 17 European countries, both implementation approaches, "strategically managed all inclusive" and "project based evolutionary bottom up", could be tested. Here we summarise experiences as well as critical points and give recommendations. Figure 17.11 shows examples of conflicting priorities and purposes that LTSER Platforms are facing. Implementation of LTSER Platforms has to navigate between these poles.

LTSER Platforms represent a huge potential for both science and practice. The large number of Platform-specific projects and publications, also reported by several authors in this volume provide evidence of how this potential has been used scientifically in spite of the short operating time of LTSER Platforms to date. However, translating knowledge into practice presents a continuing, formidable challenge that is discussed further below.

So far, no comparable network has been set up to regionalise socio-ecological research and involve infrastructure, interdisciplinary research and regional actors and stakeholders in a collective process. However, setting up such a complex system is time- and resource-demanding. The complete "production cycle" of typical LTSER Platform products, from prioritising research questions to getting a research project accepted, producing the scientific findings, translating them into applicable measures, disseminating these recommendations in accompanying implementation projects and assessing the effects in terms of increased sustainability of ecosystem service use, might stretch over a decade. The more complex regions and questions

| POTENTIAL | PRODUCTS |
|---|---|
| Big/complex LTSER regions; strong socio-ecological gradients | Smaller LTSER regions, limited number of habitats |
| **Heterogeneity** | **Frictionless scientific work** |
| Many projects, data holders, institutions | Possibility to implement based on 1-few research project fundings |
| Wide range of relevant actors/users | |
| Necessity to disseminate | |
| | Simplicity, clear data ownership, possible reductionism |
| Increasing resource requirements for matrix functionalities/ central services | |

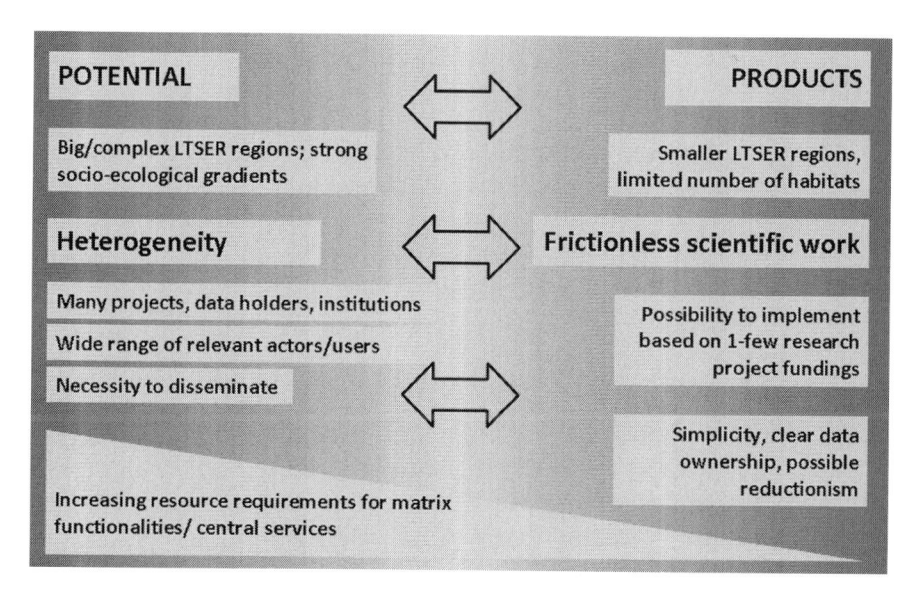

**Fig. 17.11** Conflicting priorities in LTSER Platform implementation. *Left side*: Cases requiring complex approaches in creating the framework for socio-ecological research. *Right side*: Less demand for matrix functionalities and supporting services due to simpler settings

are, the longer the latency period for tangible "products", if the successful establishment and operation of LTSER Platforms in itself is not accepted as a "product" in terms of increased scientific entropy.

Big and/or complex LTSER Platform regions featuring a wide variety of habitats, land-use forms, complementary stakeholders and use conflicts pose numerous interesting research questions. On the other hand, complexity hampers (i) quick progress in setting up a complete Platform communication space (comprising all relevant actors), (ii) agreement on the research framework and (iii) smooth division of tasks (e.g. competing research teams within the Platform). In smaller LTSER Platform regions, which cover less internal environmental, social and economic gradients, key problems might be more evident and could be tackled by one or a few institutions well integrated in the region and holding existing data without substantial additional efforts to establish a LTSER Platform.

The trade-off between frictionless scientific work and coping with heterogeneity could not be more evident than in LTSER Platforms:

- Up to a certain complexity of interdisciplinary research questions, the number of institutions and a few research projects, overhead costs and required central services can be kept to a minimum. Responsible and accessible funding instruments are clearer when questions are less complex and interdisciplinary. There are other advantages to smaller Platform teams. For example, established teams in one or a few institutions will most probably already have an interdisciplinary working culture established, reducing efforts needed to achieve a common language across disciplinarily specialised institutions. When requisite data are mainly kept within one institution, necessary information management will be broadly covered by the general institutional data infrastructure, including data use rights. If the research institutions involved are located in or closely connected to the concerned region, the required stakeholder interfaces might be few and may have already been developed by the institution, including communication spaces and mechanisms for information dissemination.
- With increasing heterogeneity, "small solutions" hit the wall due to increasing demand for services (actor analysis, stakeholder involvement, establishment of a transdisciplinary communication space, development of interdisciplinary research teams across institutional borders, data management and integration). There is a threshold size of LTSER Platforms for covering regional processes (e.g. commuting), heterogeneity of habitats, land use and related management practices. Large Platforms and a lack of substantial funding may lead to a lack of projects covering the entire region and the risk of scattered activities across scales, hampering possibilities of upscaling, downscaling and extrapolation. Research findings might not fit the scale of management measures and/or the level of local, sub-regional and regional decision-making.

The multiple experiences in setting up LTSER Platforms, with their pronounced heterogeneity of initial conditions across Platform regions and countries, suggest that no general formula regarding the "right" way to initiate a platform can be provided at this stage. Nevertheless, mid-term implications of chosen approaches and bottlenecks have become evident and merit precautionary advice and re-assessment of how LTSER Platforms are organised, managed and communicated:

- LTSER seems to be plausible and attractive to many stakeholders, but also has a tendency to create unrealistic expectations in terms of delivery time. Moreover, expectations of what LTSER can and can't deliver can vary wildly depending on the stakeholder. Regional development managers might after 2–3 years want to assess the relevance of project findings for sustainable regional development and the cost-efficiency of proposed alternative management practices. Provincial governments could after 2 years of work request a report on regional impact as a precondition for the continuation of funding, whereas a village mayor – inspired by a facilitated LTSER workshop with stakeholders – might expect customised delivery of results supporting the application for an additional bus stop in the village (compilation and analysis of related national statistics), and might ask in disappointment why he cannot find this on the Platform website.
- If performance criteria focus solely on the usual scientific output (publications, impact points per year) from the beginning, the project-based approach will support a traditional academic work routine. Only by considering the innovation and added values such as relevance for management, focus on stakeholder concerns, or open access to Platform data, will long-term, dependable support of the LTSER approach be assured. Because LTSER is not "traditional" science, Platform managers are encouraged to establish funding mechanisms and calls for tender specifically customised for the unique and innovative approach and goals of LTSER. Funding mechanisms must consider, for example, the intrinsic time lag between project implementation and the point at which society and – in the long term – science will benefit. Such a lag exists due to the unique combinations of expertise and data inherent in long-term, interdisciplinary socio-ecological research.
- Universities and other academic institutions have neither the resources nor the scope by themselves to provide Platform management and services. Neither do individual research projects foresee being able to pay for such services. Therefore, formalisation and institutionalisation are to some extent unavoidable in order to secure operation in the long term.
- The transdisciplinary component of LTSER requires a special skills portfolio for a wide range of non-scientific activities, which need to be carried out by specialists educated in communication, facilitation, and public relations. In LTSER, Platform scientists typically overstretch themselves with non-scientific work such as dissemination beyond scientific publishing, translation, production of stakeholder-specific material, participatory activities and lobbying beyond research proposals. Particularly idealistic and visionary people are therefore prone to self-exploitation, unrealistic planning assumptions and overload.
- Efforts in team building, integration between disciplines and institutions are hampered by competition between scientists applying for projects in the same funding mechanisms. Moreover, natural scientists or sociologists might perceive themselves as not receiving their due credit in interdisciplinary research, both in

terms of conceptual ownership and funding. This highlights the need for a truly open, inclusive and respectful working environment.

**In conclusion, we offer the following abbreviated advice to the LTSER novice:**

- Avoid unrealistic expectations with regard to both research and management goals and what topics can be successfully covered at which scale;
- Obtain a reliable picture of available funding for co-ordinating LTSER Platforms in the mid- to long-term to choose the most appropriate implementation model: What level of central services can be maintained over the long term?;
- Ensure that there is a critical mass of and balance between central services (e.g. data management) and the number of supported research projects ("products");
- Formalisation and institutionalisation are greatly assisted through the use of existing structures in the region (communication, dissemination);
- It is helpful to involve institutions with interdisciplinary teams that are already established (easier to achieve internally than across formal institutional borders due to common institutional language) – or very few flagship institutions with preferably one located in the region;
- Broaden the community and actively involve specialists in the required disciplines (e.g. allow a sociologist to develop the sociological component of LTSER). It is crucial not to assume that an ecologist, for example, can adequately apply the research and conceptual tools of an anthropologist. Respect all the participants in an interdisciplinary collaboration;
- Draw the line: you cannot please everyone (or meet their expectations) all the time;
- Co-operate internationally, taking advantage of experiences, tools and material developed in other LTSER Platforms;
- Regionalisation and transdisciplinarity do not work without toeholds in the region: identify key multipliers open to LTSER and involve scientists with personal connections to the region.

The identified weaknesses provide evidence that the research environment, performance indicators and scientific reward system still fail to provide the necessary framework for producing knowledge according to societal and political needs at the necessary pace. The interdisciplinary Programme on Ecosystem Change and Society (PECS) of the International Council of Science (ICSU website) addresses "Place-Based Long-Term Social-Ecological Research" as key to investigate society-nature interactions (ICSU 2010, Programme on ecosystem change and society (PECS) – A 10-year research initiative of ICSU and UNESCO – Workplan 2010: draft technical paper, Personal communication of PECS Chair, Steve Carpenter). So far, LTSER Platforms and their interdisciplinary teams are the only European test case for regionalised or place-based LTSER at the level of a continental network.

The LTSER concept was born in the midst of two profound upheavals. The first is the rapidly changing global environment as a result of unprecedented large human populations consuming an unprecedented amount of the earth's resources. The second, inspired in part by the first, is a major scientific paradigm shift away from traditional disciplinary approaches to environmental problem solving to an interdisciplinary, place-based science. Place-based LTSER Platforms confront global environmental challenges at the local and regional levels. They do so without compromising academic standards, but may be contributing to paradigm shifts in some areas. Not only does LTSER advance the state of knowledge, but it produces knowledge that matters to people and that is then translated into tangible environmental and natural resource policies for local and regional implementation. With less than a decade of practical experience, LTSER Platforms are emerging as living laboratories for socio-ecological research and a major contributor of policy/management relevant knowledge.

**Acknowledgments** We want to thank the LTSER communities in about 20 European countries for their enthusiasm in jointly developing and spreading a new scientific working culture and – at the same time – do appreciate their efforts in setting up LTSER Platforms in spite of all constraints related to the fact that LTSER is not yet rewarded by traditional scientific performance evaluations. The experiences of this group form the backbone of this chapter. Our thanks go also to our international – and especially US American – colleagues for an inspiring exchange of ideas and scientific cooperation.

# References

# Web Links

ALTER-Net Website, accessed 2011: http://www.alter-net.info/
ICSU Website, accessed 2011: www.icsu.org
ILTER Web-site: ILTER's aims and objectives, accessed 2011: http://www.ilternet.edu/about-ilter/mission
LTER-Europe Website, accessed 2011: http://www.lter-europe.net
LTER-Europe Website, key documents, accessed 2011: http://www.lter-europe.net/Organisation/key-documents
NUTS Website, accessed 2011: http://epp.eurostat.ec.europa.eu/portal/page/portal/nuts_nomenclature/introduction
US-LTER Website accessed 2011: http://www.lternet.edu/; LTER core areas accessed 2011: http://www.lternet.edu/coreareas/coreintro.htmlUS LTER review, accessed 2009: http://intranet.lternet.edu/archives/documents/reports/20_yr_review/

# Other

Anon. (2009). Integrated research to support biodiversity policies: The ALTER-Net approach. *ALTER-Net*. Retrieved from http://www.alter-net.info. Accessed June 2011.
Carpenter, S. R., Mooney, H. A., Agard, J., Capistrano, D., DeFries, R., Diaz, S., Dietz, T., Duriappah, A., Oteng-Yeboah, A., Pereira, H. M., Perrings, C., Reid, W. V., Sarukhan, J., Scholes, R. J., & Whyte, A. (2009). Science for managing ecosystem services: Beyond the

Millennium Ecosystem Assessment. *Proceedings of the National Academy of Sciences, 106*, 1305–1312.

Cohen, S. (2006). *Understanding environmental policy*. New York: Columbia University Press.

Collins, S. L., Swinton, S. M., Anderson, C. W., Benson, B. J., Brunt, J., Gragson, T. L., Grimm, N., Grove, J. M., Henshaw, D., Knapp, A. K., Kofinas, G., Magnuson, J. J., McDowell, W., Melack, J., Moore, J. C., Ogden, L., Porter, L., Reichman, J., Robertson, G. P., Smith, M. D., van de Castle, J., & Whitmer, A. C. (2007). *Integrated science for society and the environment: A strategic research initiative*. Albuquerque: LTER Network Office.

Collins, S. L., Carpenter, S. R., Swinton, S. M., Orenstein, D. E., Childers, D. L., Gragson, T. L., Grimm, N. B., Grove, J. M., Harlan, S. L., Kaye, J. P., Knapp, A. K., Kofinas, G. P., Magnuson, J. J., McDowell, W. H., Melack, J. M., Ogden, L. A., Robertson, G. R., Smith, M. D., & Whitmer, A. C. (2011). An integrated conceptual framework for long-term social-ecological research. *Frontiers in Ecology and the Environment, 9*, 351–357.

Costanza, R., & Daly, H. E. (1992). Natural capital and sustainable development. *Conservation Biology, 6*, 37–46.

Costanza, R., d'Arge, R., de Groot, R., Farber, S., Grasso, M., Hannon, B., Limburge, K., Neem, S., O'Neil, R., Parelo, J., Raskin, R., Sutton, P., & van den Belt, V. (1997). The value of the world's ecosystem services and natural capital. *Nature, 387*, 253–259.

Dick, J., Al-Assaf, A., Andrews, Ch., Díaz-Delgado, R., Groner, E., Halada, L., Izakovicova, Z., Kertész, M., Khoury, F., Krašić, D., Krauze, K., Matteucci, G., Melecis, V., Mirtl, M., Orenstein, D.E., Preda, E., Santos-Reis, M., Smith, R., Vadineanu, A., Veselić, S., & Vihervaara, P. (2012). *Ecosystem services: A rapid assessment method tested at 35 sites of the LTER-Europe network* (in preparation).

EEA. (1999). *Environmental indicators: Typology and overview* (Technical report No. 25). Copenhagen: EEA. http://www.eea.europa.eu/publications/TEC25. Accessed June 2011.

EEA. (2010). *Cultural landscapes and biodiversity heritage*. Copenhagen: EEA.

Funtowitz, S. O., & Ravetz, J. R. (1992). Three types of risk assessment and the emergence of post-normal science. In S. Krimsky & D. Golding (Eds.), *Social theories of risk* (pp. 251–274). Westport: Praeger.

Furman, E., Peltola, T., & Varjopuro, R. (Eds.). (2009). Interdisciplinary research framework for identifying research need. Case: Bioenergy-biodiversity interlinkages. In *The Finnish environment 17/2009, environmental protection*. Helsinki: Finnish Environment Institute.

Gibbons, M., Nowotny, H., Limoges, C., Schwartzman, S., Scott, P., & Trow, M. (1994). *The new production of knowledge: The dynamics of science and research in contemporary societies*. London: Sage.

Haberl, H., Winiwarter, V., Andersson, K., Ayres, R., Boone, C., Castillo, A., Cunfer, G., Fischer-Kowalski, M., Freudenburg, W.R., Furman, E., Kaufmann, R., Krausmann, F., Langthaler, E., Lotze-Campen, H., Mirtl, M., Redman, C. L., Reenberg, A., Wardell, A., Warr, B., & Zechmeister, H. (2006). From LTER to LTSER: Conceptualizing the socio-economic dimension of long-term socio-ecological research. *Ecology and Society, 11*, 13. [Online], Retrieved from http://www.ecologyandsociety.org/vol11/iss2/art13/-

Hein, L., van Koppen, K., de Groot, R. S., & van Iterland, E. C. (2006). Spatial scales, stakeholders and the valuation of ecosystem services. *Ecological Economics, 57*, 209–228.

Hobbie, J. E., Carpenter, S. R., Grimm, N. B., Gosz, J. R., & Seastedt, T. R. (2003). The U.S. long term ecological research program. *BioScience, 53*, 21–32.

Holling, C. S. (2001). Understanding the complexity of economic, ecological, and social systems. *Ecosystems, 4*, 390–405.

Jongman, R. H. G., Bunce, R. G. H., Metzger, M. J., Mücher, C. A., Howard, D. C., & Mateus, V. L. (2006). Objectives and applications of a statistical environmental stratification of Europe. *Landscape Ecology, 21*, 409–419.

Kämäräinen, P. (1999). The role of "accompanying research" within initiatives for VET development – Reflections on national and European developments. In J. Lasonen (Ed.), *EERA – Network 2 programme: Network on vocational education and training research (VETNET) and academy of human resource development (AHRD)* (pp. 185–198). Jyväskylä: University of Jyväskylä. Retrieved from http://www.leeds.ac.uk/educol/documents/000001147.htm. Accessed June 2011.

Kinzig, A. P., Ryan, P., Etienne, M., Allison, H., Elmqvist, T., & Walker B. H. (2006). Resilience and regime shifts: Assessing cascading effects. *Ecology and Society, 11*, 20. [Online] http://www.ecologyandsociety.org/vol11/iss1/art20/. Accessed June 2011.

Kosko, B. (1986). Fuzzy cognitive maps. *International Journal of Man–Machine Studies, 24*, 65–75.

Lozano, R. (2008). Envisioning sustainability three-dimensionally. *Journal of Cleaner Production, 16*, 1838–1846.

MEA. (2005). *Millennium ecosystem assessment: Ecosystems and human well being – Synthesis*. Washington, DC: Island Press.

Metzger, M. J., & Mirtl, M. (2008). *LTER Socio-Ecological Regions (LTER-SER) – Representativeness of European LTER facilities – Current state of affairs* (AlterNet Technical Paper. p. 5). http://www.alter-net.info. Accessed June 2011.

Metzger, M. J., Bunce, R. G. H., Jongman, R. H. G., Mücher, C. A., & Watkins, J. W. (2005). A climatic stratification of the environment of Europe. *Global Ecology and Biogeography, 14*, 549–563.

Metzger, M. J., Bunce, R. G. H., van Eupen, M., & Mirtl, M. (2010). An assessment of long term ecosystem research activities across European socio-ecological gradients. *Journal of Environmental Management, 91*, 1357–1365.

Mirtl, M. (2010). Introducing the next generation of ecosystem research in Europe: LTER-Europe's multi-functional and multi-scale approach. In F. Müller, C. Baessler, H. Schubert, & S. Klotz (Eds.), *Long-term ecological research: Between theory and application* (pp. 75–93). Berlin: Springer.

Mirtl, M., & Krauze, K. (2007). Developing a new strategy for environmental research, monitoring and management: The European Long-Term Ecological Research Network's (LTER-Europe) role and perspectives. In T. J. Chmielewski (Ed.), *Nature conservation management – From idea to practical results*. Lublin/Lodz/Hesinki/Aarhus: ALTERnet.

Mirtl, M., Boamrane ,M., Braat, L., Furman, E., Krauze, K., Frenzel, M., Gaube, V., Groner, E., Hester, A., Klotz, S., Los, W., Mautz, I., Peterseil, J., Richter, A., Schentz, H., Schleidt, K., Schmid, M., Sier, A., Stadler, J., Uhel, R., Wildenberg, M., & Zacharias, S. (2009). *LTER-Europe design and implementation report – Enabling "next generation ecological science": Report on the design and implementation phase of LTER-Europe under ALTER-Net & management plan 2009/2010*. Vienna: Umweltbundesamt (Environment Agency Austria). First circulation on-line, 220 p. ISBN 978-3-99004-031-7. Retrieved from http://www.lter-europe.net. Accessed June 2011.

Mirtl, M., Bahn, M., Battin, T., Borsdorf, A., Englisch, M., Gaube, V., Grabherr, G., Gratzer, G., Haberl, H., Kreiner, D., Richter, A., Schindler, S., Tappeiner, U., Winiwarter, V., & Zink, R. (2010). *Next generation LTER in Austria – On the status and orientation of process oriented ecosystem research, biodiversity and conservation research and socio-ecological research in Austria*. LTER-Austria Series (Vol. 1). Retrieved from http://www.lter-austria.at. Accessed June 2011.

Nelson, G. C. (2011). Untangling the environmentalist's paradox: Better data, better accounting and better technology will help. *BioScience, 61*, 9–10.

Özesmi, U., & Özesmi, S. L. (2004). Ecological models based on people's knowledge: a multi-step fuzzy cognitive mapping approach. *Ecological Modelling, 176*, 43–64.

Pickett, S. T. A., Burch, W. R., & Grove, M. (1999). Interdisciplinary research: Maintaining the constructive impulse in a culture of criticism. *Ecosystems, 2*, 302–307.

Putzhuber, F., & Hasenauer, H. (2010). Deriving sustainability measures using statistical data: A case study from the Eisenwurzen, Austria. *Ecological Indicators, 10*, 32–38.

Redman, C. L., Grove, J. M., & Kuby, L. H. (2004). Integrating social science into the long-term ecological research (LTER) network: Social dimensions of ecological change and ecological dimensions of social change. *Ecosystems, 7*, 161–171.

Shibata, H., & Bourgeron, P. (2011). Challenge of international long-term ecological research network (ILTER) for socio-ecological land sciences. *Global Land Project NEWS, 7,* 13–14. Retrieved from, http://www.globallandproject.org/Newsletters/GLP_NEWS_07.pdf

Singh, S. J., Haberl, H., Gaube, V., Grünbühel, C. M., Lisivieveci, P., Lutz, J., Matthews, R., Mirtl, M., Vadineanu, A., & Wildenberg, M. (2010). Conceptualising Long-term Socio-ecological Research (LTSER): Integrating socio-economic dimensions into long-term ecological research. In F. Müller, C. Baessler, H. Schubert, & S. Klotz (Eds.), *Long-term ecological research: Between theory and application* (pp. 377–398). Berlin: Springer.

WCED. (1987). *Our common future* (1st ed.). Oxford: Oxford University Press.

Wildenberg, M., Bachhofer, M., Adamescu, M., De Blust, G., Diaz-Delagdo, R., Isak, K., Skov, F., & Varjopuro, R. (2010). *Linking thoughts to flows – Fuzzy cognitive mapping as tool for integrated landscape modelling.* Proceedings of the 2010 international conference on integrative landscape modelling – Linking environmental, social and computer sciences. Symposcience, Cemagref, Cirad, Ifremer, Inra, Montpellier.

Wrbka, T., Erb, K.-H., Schulz, N. B., Peterseil, J., Hahn, C., & Haberl, H. (2004). Linking pattern and process in cultural landscapes: An empirical study based on spatially explicit indicators. *Land Use Policy, 21,* 289–306.

## Chapter 18
# Developing Socio-ecological Research in Finland: Challenges and Progress Towards a Thriving LTSER Network

**Eeva Furman and Taru Peltola**

**Abstract** At the time of planning the national LTER network (FinLTSER) in Finland, the approach of linking social and ecological issues in solving environmental problems was already well embedded in science and policy institutions and practices. A broad community of environmental, natural and social scientists had been carrying out problem-oriented research related to environmental issues for many years before the concept of LTSER platforms raised wide interest among Finnish research institutes. In this article, we analyse the research culture leading to this high level of interest and enthusiasm regarding socio-ecological research during the development phase of the FinLTSER network. By using interview and other materials from the process of establishment of the FinLTSER, this chapter analyses the initiation of the network, the very first steps taken by the platforms and the challenges faced during this period.

**Keywords** Socio-ecological • Interdisciplinary • FinLTSER Platform • Long term ecological research • Long term socio-ecological research

E. Furman, Ph.D. (✉)
Finnish Environment Institute, Helsinki, Finland
e-mail: eeva.furman@ymparisto.fi

T. Peltola, Ph.D.
Finnish Environment Institute, Joensuu, Finland
e-mail: taru.peltola@ymparisto.fi

## 18.1 Introduction

The protection of nature is a problem for which researchers are strongly encouraged to develop integrative approaches. Problem-oriented, interdisciplinary socio-ecological research which integrates various disciplinary perspectives is deemed to be the kind of research that is politically relevant and capable of generating valid knowledge. Conceptual papers have been published both in the US and in Europe on the need to move from Long-Term Ecological Research (LTER) to socio-ecological research (Redman et al. 2004; Haberl et al. 2006). There has also been pressure from the policy side for assessments linking societal and ecological issues (e.g. environmental impact assessment, EIA). In addition, integration of non-scientific knowledge has been promoted to advance the goal of "knowing nature", i.e. data collection on the distribution, health, and status of natural species and natural habitats to allow strategies, plans or programmes for the conservation and sustainable use of nature (Ellis and Waterton 2005).

The concept of LTER was introduced in the USA more than 25 years ago. Since then, the LTER programme has gradually developed into a global network that aims to respond to continuing long-term large-scale global change. In 1993, the International LTER (ILTER) network was formally founded (ILTER 1993). During its period of operation, the ILTER network has expanded to include 41 national networks. Three issues are of particular importance: comprehensiveness of the international network, comparability of the research and broadening of the research agenda from narrow ecology to synthetic socio-ecology.

In Europe, the Long-Term Biodiversity, Ecosystem and Awareness Research Network project, ALTER-Net (2005), which was the driving force of LTER-Europe, emphasised the socio-ecological approach. Long-Term Socio-Ecological Research (LTSER) platforms have been introduced as a means to engage in integrative knowledge production and to go beyond classical disciplinary research (Haberl et al. 2006). LTSER platforms are modular facilities consisting of sites which are located in a defined area. The 23 LTSER platforms established in Europe aim to enhance interdisciplinary collaboration, both between various natural sciences and between natural and social sciences (Mirtl 2010) through agreed research agendas, interdisciplinary and stakeholder forums and research facilities. In addition, collaborative research aiming at cooperation between scientists and nature managers, volunteer citizens and other actors is regarded as an asset (Mirtl et al. 2009).

Although interdisciplinarity is envisioned as a key feature of long-term socio-ecological research, it is not easily achieved. The problems and obstacles of interdisciplinary research have been widely discussed in the literature. In particular, it appears that collaboration between ecologists and social scientists has been rare until recent years, although ecology has been described as "an outward-looking science, open to models and methods from other natural sciences and concerned with human-environment interactions" (Phillipson et al. 2009). The reasons may be various practical, institutional, cognitive and cultural barriers (see McCallin 2006; Uiterkamp and Vlek 2007; Corley et al. 2006). Established institutions, conceptual frameworks and technologies of inquiry create epistemological and ontological commitments and structure interaction between researchers belonging to specific communities (Petts et al. 2007).

As these intellectual and practical structures direct scientific work within the communities, they may prevent possibilities for collaboration with others. For example, basic philosophical differences may prevent collaboration between natural and social scientists, and especially between the various social sciences (Phillipson et al. 2009). Moreover, the fact that acts, strategies and policies of knowing are culturally diverse even within a single discipline has been shown by science study scholar Karin Knorr Cetina. By studying disciplines such as high energy physics and molecular biology, she has demonstrated that there are methodological differences among the disciplines that are historically constituted and dependent on, for example, instrumental, linguistic, conceptual and organisational frameworks (Knorr Cetina 1999). Such diversity makes communicating at the intradisciplinary level difficult enough as it is, let alone at the interdisciplinary level.

The challenges of interdisciplinary integration relate in particular to the negotiation of the division of labour between sciences or between scientific and other knowledge practices. For example, Endter-Wada et al. (1998) reported that frustration may emerge from such imposed roles and predefined frameworks from natural sciences which do not enable meaningful entries and problem identification for social scientists. Ecologists, in turn, may feel frustrated if prevented from doing "good" ecology while having to adapt to the demands of policy-relevant research settings (Phillipson et al. 2009).

Against this background, the Finnish LTER network (FinLTSER) offers an interesting window to view the development of socio-ecological research on the national level. The concept of LTSER platforms (Haberl et al. 2006; Mirtl 2010) raised wide interest among Finnish research institutes and originally ten out of 14 applications to join the network were submitted specifically for LTSER status and not for LTER status, which was the other option. The final network includes four LTSER platforms and five LTER sites. In this article we analyse the research culture leading to high degree of interest and enthusiasm shown towards socio-ecological research during the development phase of the FinLTSER network. We also outline lessons learned from the early steps towards integrating social and natural sciences within the national LTSER Platform, Lepsämänjoki. Since the network and the platform were established only recently, they enable us to study the issues and concerns related to the initiation of scientific cooperation.

The analysis is based on 12 interviews, more than 20 documents including correspondence, agendas, minutes and policy documents, and on our personal experiences. Eeva Furman was involved in the process of creating the FinLTSER network. Her personal notes relating to the FinLTSER's initiation process, process documents including emails, written material and meeting memoranda, and interviews with the coordinators of five of the LTSER applicants form core materials of the analysis. Taru Peltola has conducted seven thematic interviews with the partners forming one of the LTSER platforms in Finland as part of a comparative ALTER-Net study on knowledge production in LTSERs (Mauz et al. 2012). The material also includes communications and various documents such as project proposals to which she was given access by the Lepsämänjoki consortium. This material is analysed here to identify the institutional and cultural factors that supported the formation of the FinLTSER network and its specific characteristics.

## 18.2 Driving Forces: What Has Boosted Interest Towards a Long-Term Socio-ecological Research Network?

### 18.2.1 Internal Forces: Traditions of Problem-Oriented Environmental Research in Finland

In Finland, the culture of knowledge production for sustainability started to evolve in the 1970s following the international environmental awakening and the formation of environmental administration (Haila 2001a). Following the international trend, science had an important role in making environmental issues visible in society since many environmental problems exist beyond everyday experience and observation (Haila 2001b). In particular, the loss of biodiversity has been made evident through extensive monitoring practice (Bowker 2000). Much of the progress in Finnish environmental research has involved basic ecological research. During the past 30 years, this field has become one of the strongest fields of science in Finland and has attained high standards internationally. This research has mainly been inspired by environmental problems and finding ways to manage them, including fields such as conservation biology, island biology and metapopulation biology.

On the whole, the connection between scientists and policy makers has been informal but close for a long time in Finland, unlike many EC member states. Researchers have frequently been invited to planning committees of various authorities and to parliamentary hearings. Since the 1990s, research funding programmes have been built with support from discussions at seminars and workshops attended by researchers from different disciplines, policy makers and other stakeholders (Furman et al. 2006). These forums have provided directions for the structure of such programmes but have also served as opportunity for participants to learn about the other parties and their approaches and knowledge on various issues. These forums have developed into interdisciplinary projects funded by the programmes in question and furthermore led to collaboration beyond the given theme and funding channel.

Environmental social science began to undergo significant expansion from the early 1990s in Finland (Lehtinen 2005). The early steps taken by environmental social scientists attracted strong criticism from the more established research fields, leading to a kind of competition over "academic markets" (Lehtinen and Rannikko 1994a). However, nowadays, environmental social sciences are regarded as one of the areas worth focusing on in terms of future research funding.

**Table 18.1** Interdisciplinary thematic research funding programmes of the Academy of Finland

| Funding programme | Funding period | |
|---|---|---|
| Finnish research programme on climate change | 1990–1995 | |
| Sustainable development research programme | 1991–1995 | |
| Finnish biodiversity research programme | 1997–2002 | |
| Studies on science and science policy | 1997–1999 | |
| Finnish global change research programme | 1999–2002 | |
| Sustainable use of natural resources | 2001–2004 | |
| Environment and law research programme | 2003–2008 | Funding for FinLTSER starts in 2005 |
| Sustainable energy | 2008–2011 | |
| Baltic sea research programme | 2010–2016 | |
| Research programme on climate change | 2010–2014 | |

The rise of environmental social science was supported by extensive funding programmes of the Academy of Finland, aimed at interdisciplinary, problem-oriented research. These programmes introduced a diversity of environmental issues (Table 18.1). The increased research funding was a primary reason for the growth of environmental social science: While environmental social science had been an interest of some early pioneering scholars, research funding instruments enabled doctoral training and thus gave birth to a new generation of environmental social scientists. Between 1991 and 2006, 87 doctoral degrees were taken in environmental social science at 13 Finnish universities (Kotakorpi 2007). The institutional basis of doctoral training was further strengthened when the YHTYMÄ graduate school in environmental social science was founded in 2002.

Because the early pioneers of environmental social science came from different scientific backgrounds, the thematic scope of environmental social science has had a broad range. Since environmental social science has been interdisciplinary in nature from the beginning and many of those involved in early research have collaborated across disciplinary boundaries, it is sometimes difficult to delineate between the various fields. Many themes, for example, environmental awakening and conflicts (Lehtinen and Rannikko 1994b), risk society (Kamppinen et al. 1995) and ecological modernisation (Massa and Rahkonen 1995) have been explored from various perspectives, ranging from sociology and geography to environmental economics and environmental policy and law. More recently, interdisciplinary explorations have been made into forest politics and environmental justice (e.g. Lehtinen et al. 2004; Lehtinen and Rannikko 2003), the legitimacy of natural resources governance (Rannikko and Määttä 2009) and ecosystem services (Hiedanpää et al. 2010). The Yearbook of Environmental Policy and Law (Määttä 2007) also offers a forum for environmental social scientific debate.

Broad interest in environmental issues has triggered the establishment of new curricula at Finnish universities and furthered the development of new research units. Courses in environmental sociology have become available at several universities (Helsinki, Turku, Oulu, Jyväskylä, Joensuu and Lapland), environmental law has been introduced in Tampere and Joensuu Universities in addition to Turku and Helsinki Universities and corporate environmental management is now studied at Helsinki Business School and Tampere University. Tampere University established a professorship in environmental policy in 1994, followed by Helsinki and Joensuu Universities.

The Finnish Society for Environmental Social Sciences (YHYS), founded in 1994, and the group of environmental sociologists who have convened during the annual meeting of the Westermarck Society – Finnish Sociological Society since 1990 demonstrate the high interest in environmental social science among researchers in Finland in both universities and research institutes. The YHYS organizes annual conferences, publishes a magazine and operates a mailing list for academic discussions and job announcements in the field. The annual conference of the society is a forum for international collaboration, with invited foreign guest speakers. A winter seminar is organized together with other scientific societies on a topical theme, and the spring colloquium, in turn, facilitates discussion between the environmental administration and researchers on topical issues.

Finnish environmental social scientists are linked with other Nordic scholars through the biannual NESS (Nordic Environmental Social Science) conferences and there is also a link to the Nordic Environmental History Society (NEHN). Many Finnish environmental sociologists have participated in world congresses organized by the International Sociological Association and the conferences of the European Sociological Association in which environmental sociology groups convene. EU funding has furthermore facilitated international networking in particular among governmental research organizations.

In 2002, the Finnish Environment Institute (SYKE) established a Research Programme for Environmental Policy and later in 2010 a Centre for Environmental Policy. These are among the largest environmental social science research units in Finland, with personnel of 20 and 50, respectively. In 2010, SYKE joined a Partnership for Research on Natural Resources and the Environment formed by six governmental research institutes. The social scientists in the partnership are in the process setting up the organization and a network of almost 100 researchers is planned by the end of 2011.

Networking and collaboration are among the strengths of Finnish environmental social science. Although research groups compete with each other for research funding, the scarce and scattered resources have successfully been focused on expanding the field of environmental social science. Links have also been built with natural scientists. In the early stages, some pioneering scholars from natural sciences moved into the social sciences. For instance, a professor of environmental policy, the long-term chair of the Finnish Society for Environmental Social Sciences and head of YHTYMÄ Graduate School, Yrjö Haila, moved from ecology to social science and began to develop theory on eco-social dynamics (Haila and Levins 1992; Haila and Dyke 2006). At the same time, some ecologists also adopted a strong interdisciplinary approach in their research, focusing on, e.g., urban ecology (Yli-Pelkonen and Niemelä 2006).

The interdisciplinary nature of environmental research is also reflected by the fact that not all related professorships and lecturing posts are based within social science departments. For example, the Laboratory of Environmental Protection, which has a distinctive social scientific focus, was founded in the Helsinki University of Technology as early as the 1990s. Some posts are also shared by departments, including the professorship in environmental policy at the University of Helsinki (shared by the faculties of Social Science, Biosciences and Agriculture and Forestry). The recently established professorship in natural resources policy in Joensuu University was jointly announced with the Finnish Environment Institute, a governmental research institute.

Although environmental social science has gained a foothold and those involved in early research have succeeded in broadening the scope of environmental social science, its institutional basis in many Finnish universities is not particularly strong. This is evident in the dynamics of research groups. New groups have emerged, but while researchers have moved between universities, in some cases the development has led to a negative trend. For example, in Turku University environmental social science barely existed for a while. Since 2006, the academic campus in Joensuu city has played a major role in developing collaboration around environmental social sciences and interdisciplinary research. The local university introduced an interdisciplinary curriculum for environmental science in the early 1990s. In 2006 it founded the Centre of Competence for Forest, Environment and Society, which strives to enhance the role of social and cultural research into the environment and forests. This network is led by a professor in environmental social science and it brings together various university departments and subjects as well as three governmental research organisations, the Finnish Environment Institute, the Forest Research Institute and the Game and Fisheries Research Institute. Although the university has now placed environmental social science on its strategic agenda, it is too early to judge whether this is really helping strengthen the institutional basis for research in environmental social science.

The environmental social scientific research carried out in Finland has been extensive in scope and has employed various theoretical and methodological approaches, drawing from the fields of geography, sociology, business, law and policy studies and collaborating with ecologists, forestry science, engineering and environmental sciences. However, there are only a few examples of long-term research orientation. In particular, social sciences appear to be oriented towards short-term research strategies as it is not easy to obtain funding for long-term approaches. Relevant project funding typically extends to a maximum of four years, and long-term research is often seen as being merely repetitive rather than innovative and productive. The leading, and exceptional work in this respect is the long-term research carried out in North Karelia during the past 30 years, focusing on the society-nature relationship in two remote villages (Knuutila et al. 2008).

In addition to concerns regarding project orientation, the relatively weak institutional status of environmental social science hinders the development of long-term research strategies. Where researchers are forced to move between universities, research groups are in a constant state of transformation and may also disappear. When the institutional status of research groups is not stable, the research team only have short-term contracts and team members change from year to year, creating a basis for long-term research interests is a challenge.

## 18.2.2 External Forces: Receiving Impulses and Achieving Action Through International Networks

In a small country such as Finland, external signals can play a major role in implementing new institutions. In 2002, a delegation from the LTER Network Office of the US visited Finland and the Baltic States in the hopes of increasing membership in ILTER. After the visit, a planning process began in Finland which led to an agreement that SYKE would coordinate a preliminary plan for the Finnish LTER network. Moreover, it was agreed that with the input from seven research institutions[1] a proposal would be submitted to the Board of Governors of the Academy of Finland on the process for developing the network and for increasing the involvement of Finnish scientists in the international LTER collaboration.

Discussions during 2003 revealed that there was a high level of awareness in Finland of the importance of long-term ecological and ecology-related socioeconomic research, and that there was a strong interest among the Finnish scientific community in long-term socio-ecological research platforms. Nationally, the centralisation of research infrastructures had already started to evolve, through evaluations of activities within the research station network in Finland.

The second external force was the European Research Framework Programme project, ALTER-Net, in which SYKE was a partner 2004–2009. At that time, it was the largest European project trying to combine biodiversity, ecosystem and long-term socioeconomic research to serve the needs of biodiversity management in the European context. One of the main aims of ALTER-Net was to ensure the development of the European LTER network. The role of SYKE both in ALTER-Net and in the establishment of the national LTER network ensured that Finland was among the nations that have placed a clear focus on socio-ecology.

## 18.3 Empowering Forces: The Comprehensive Planning Process for the Network

In spring 2005, the Academy of Finland took a decision to fund the development of the national LTER network. Thereafter SYKE, together with the other seven national institutions in environmental research, initiated a feasibility study regarding ILTER membership. The planning group consisted of 13 members representing different disciplines and sectors and the eight research bodies having made the original application to the Academy. Based on the accepted plan, the planning group's tasks were to ensure that the development process was open and transparent, that the network

---

[1] Finnish Environment Institute, Finnish Forest Research Institute, Finnish Meteorological Institute, University of Helsinki, University of Oulu, University of Joensuu, University of Turku.

would build on existing infrastructure strategies both nationally and internationally, and that it would take an interdisciplinary approach to ecological research. Regular contact between the members of the planning group was maintained throughout the entire planning period of 2005–2006 through five physical meetings and weekly e-mail communication.

Interviews with those participating revealed that the planning group's mandate for developing national infrastructure was not understood in the same way by the entire planning group. Some members viewed this first and foremost through the lens of their own organisation's interests, whereas others had a broader approach, and some based their approach predominantly on the present setting whereas others focused more on the future. However, the face-to-face brainstorming sessions of the planning group were seen as the key to finding a shared understanding of the process and its future benefits for the research community as a whole.

The planning group organised two stakeholder workshops, one for representatives of the research community and the other for potential interest groups and funding agencies to discuss the possibilities and options for setting up and maintaining the Finnish LTER network. These events were seen as the most rewarding activities resulting from the bottom-up approach applied by the planning group. The meetings served as a forum to disseminate knowledge of the LTER concept and the planning process, to share information of relevant research calls and to provide an opportunity for preliminary networking. Most importantly, they created an opportunity to develop new insights for all the participants concerning the feasibility of implementing the LTER/LTSER concept in Finland, and other issues that had to be taken into account in further planning. These included other ongoing processes such as the potential outcomes from the development of the network of biological stations, from the infrastructure strategy planning process and from the coordination of environmental research.

A website facilitated communication (FinLTSER 2005). In the interviews, the website was mentioned as providing transparency for the process; it gave impartial information about the ongoing planning process and stimulated discussion among the research community. It also served as a link to the international level, where many national networks were being developed at the same time and were looking to share experiences with one another.

After negotiations to appoint the responsible body, the Coordination Group for Environmental Research, led by the Ministry of the Environment, was appointed to decide on the mandate and structure of the Finnish LTER network. SYKE was asked to carry out the practical management of the process.

An open invitation was made for research consortia to apply for either an LTER site or an LTSER platform status within the Finnish LTSER network. The invitation, which was open for 3 months, was advertised through the LTER website and disseminated nationally via social and institutional networks and planning seminars. The criteria for successful applications were directly linked to the criteria of the ILTER. Taking on the status required a commitment of 6 years, and an LTER site should have a focus on ecological, small scale research while an LTSER platform required a transdisciplinary, large or multi-scale research focus. The Italian

LTER evaluation was used as a reference, as Italy was only a few months ahead of Finland in their process. No direct funding was guaranteed as a result of receiving the status.

Fourteen applications were received in total. The invitation encouraged the Finnish research community to build consortia bringing together existing facilities and stakeholder networks. Two of the accepted platform consortia based their applications on an existing network and shared a research agenda, whereas three based their applications on an earlier collaboration, which they then broadened while preparing the application. The process of building a consortium proposal for LTSER status was typically initiated by a senior person with responsibilities for strategic research management, who then invited others to participate. In one case, however, the building of a consortium was initiated by an individual researcher who learned about the invitation from the Internet and then informed some colleagues about it. This also led eventually to a successful application.

An international interdisciplinary evaluation panel was formed and the applications were provided to them for scrutiny. The outcomes of the evaluation panel were synthesized and forwarded to the Coordination Group for Environmental Research, which finally decided on the content and structure of the Finnish LTER network. The Finnish network, FinLTSER, was approved on 12 December 2006. It includes four LTSER platforms and five LTER sites, of which two are parts of an LTSER platform. The network covers the main geographical regions and habitats of Finland. FinLTSER was granted membership of ILTER in August 2007, in Beijing.

According to the feedback received from applicants, advertising via the internet, the international evaluation and the ongoing national research were issues which raised interest and confidence in the LTER network within the research community in Finland. In addition, some research institutions had been slower than others to commit themselves to the process, leading them later to raise questions concerning further possibilities to join the network.

The process of writing the proposals varied from one platform to another, which gave direction to the form of collaboration employed in the formalised platform. In one case, thematic groups were formed to enhance specific parts of the proposal but most of the writing was carried out either by one individual or by a small task force. The drafts were circulated to the large consortia community involving tens of researchers. Another proposal was initiated with a seminar at which the main lines of approach were agreed by all members, although the actual proposal was later written by the leader of the consortium alone. During the drafting process, linking the E (ecological) and the S (social) was the cornerstone. As most of the consortia were dominated by natural scientists, including the social (S) dimension required extra motivation. One consortium, however, was dominated by social scientists and in this case the situation was reversed: motivation for including the E (ecological) dimension in their research agenda required extra effort and careful attention to the criteria in the invitation.

There was a particular need to find research collaboration from social sciences and economics, given that most of the initiators were from the natural science community.

Considerable effort was also required to build collaboration with stakeholders such as municipalities and regional authorities during the development of the proposals.

## 18.4 The Risks of Interdisciplinary Collaboration: Indicators for Staying in or Stepping Out of the Lepsämänjoki LTSER

While in the above section we documented the increasing interest in the LTSER concept and integrated socio-ecological research, the insights from the Lepsämänjoki LTSER platform illuminate the debates, challenges and progress in interdisciplinarity in practice. The platform is based on previous collaboration between the partner organisations in projects focusing on agroecological research in the Lepsämänjoki River catchment area, situated in Southern Finland close to the Helsinki metropolitan area. The consortium involves ten partners. It is coordinated by the Department of Applied Biology at the University of Helsinki, which took the initiative in creating the LTSER consortium and started to recruit other partners, including other departments of the same university, the governmental research organisations SYKE and Agrifood Finland and more practically-oriented organisations such as regional level environmental administration and a local water protection association.

After receiving LTSER status, the consortium prepared funding applications to obtain resources for joint activities while the partners identified new research topics and developed a conceptual framework. The partners share a common goal to study sustainable agricultural practice and the impacts of agriculture on ecosystems, biodiversity and water quality, these being the major problem areas. One of the proposed projects focused on building a model which would enable the effectiveness of agri-environmental policies to be studied. One of the partners had previous expertise in modelling the effect of policies on farming practices. Combined with information about the real environmental impacts, the model was planned to turn the Lepsämänjoki river basin into an "agro-ecological research laboratory" in which policy changes could be tested.

The consortium partners were motivated by the possibilities that LTSER status could bring with it. These benefits included contacts and collaboration, new and interesting research topics, improved quality and reliability of knowledge, synergies between research and practical spheres, and contacts with end users of the relevant knowledge. In particular, data sharing was a strong motivation for collaboration. The consortium partners expected collaboration to create new possibilities for using existing data or to improve the quality of data. These ideas reflect the adoption of international discourse on what is considered as legitimate and relevant ecology. Still, analysis of the construction of the Lepsämänjoki LTSER platform reveals that there tensions were created in putting the new scientific ideals into practice. These tensions are highlighted by the different ways in which the consortium partners talked about the goals of data sharing as a means of advancing collaboration between organisations and disciplines.

The interviews revealed differences between the partners, and between their epistemic cultures, that is, the historically developed ideas and practices of doing science (Knorr Cetina 1999). One indicator of these differences is the relationship between the organisations involved in research and those not undertaking research. Both the non-research organisations, Uudenmaa Regional Environment Centre and its close collaboration partner Vantaanjoki Water Protection Association, carry out monitoring activities. They have constantly emphasised the difference between themselves and the university and the governmental research organisations by talking about "science discourse". For example, according to one of the interviewees, the scientists involved seemed to spend much time elaborating the reliability of knowledge and methodological questions while the practically-oriented actors appeared to have a far more straightforward relationship with data. The recognition of this difference indicates that although all the partners were involved in doing agro-ecological research, they sometimes had communication difficulties. However, in the end, the partners seemed to regard this difference not as an obstacle to collaboration but as a possibility to learn from each other.

Diverging institutional strategies and goals are another example of how differences in epistemic cultures affect data sharing and may require a negotiated approach to resolve practical problems of collaboration. One of the consortium partners of the Lepsämänjoki LTSER planned investments in research infrastructures in a neighbouring river catchment area. The platform has no permanent institutional structures, such as research stations or protected core areas. The research collaboration is loosely focused around the same geographical region, the river catchment area. The geographical focus was generally considered to be an asset by all partners, but in practice it seemed necessary to extend research beyond the physical boundaries of the Lepsämänjoki catchment area. In fact, it transpired that not all the researchers involved with the LTSER had visited the Lepsämänjoki area; they had been included on the basis of their valuable expertise gained through research conducted at similar regions and sites. Thus, restricting collaboration between the partners to those having undertaken research only at the particular geographical area was not advisable and would have led to the exclusion of some partners and their expertise. The solution was to allow the geographical boundaries of the Lepsämänjoki LTSER to become blurred, allowing partners to continue research activities in neighbouring regions and the inclusion of experts with important skills but no history of research activities within the region.

A third example of the significance of epistemological cultures in research collaboration relates to the integration of the social aspects of ecological research within the consortium. Long-term datasets are an important means of collaboration in biological research. However, the interviewees pointed out that the region of the Lepsämänjoki river basin and its population might be too small for significant long-term datasets for socioeconomic variables to be developed. Furthermore, it might also be difficult to form relevant, data-driven socioeconomic research questions. The region is, however, regarded as suitable for other kinds of socio-ecological research: for example, micro-level analysis or action research on the transformation of farming practices were mentioned by the partners as possible approaches. These approaches do not build on extensive datasets. The central idea of quantitative, long-term data as a means of coordinating collaboration thus proved problematic for the development of socio-ecological research in the context of the Lepsämänjoki LTSER. The difficulty of developing socio-ecological research is also related to the history of the platform, since there had been no previous contact with research partners having expertise in qualitative social scientific research.

On the basis of the experience gained from this particular LTSER platform it can be concluded that the concept of LTSER supports collaboration. One of the interviewees summarised the added value by stating that the planning of research activities had already improved communication between the different branches of science and had thus been very fruitful. Despite the clear progress made, it appears that collaboration has been selective. Data-driven, quantitative approaches seem to fit more easily together, thus directing research towards particular research strategies, such as agro-ecological modelling, in which the partners have prior expertise.

## 18.5 Looking into the Future: Challenging Potentials

For FinLTSER it was important that a connection to the international community had been maintained from the very start. International conferences hosted by LTER-Europe, LTER-USA and ILTER were attended by SYKE staff as well as by many others representing sites or platforms. FinLTSER has taken part in the development of LTER Europe in the context of data management and socio-ecological research. There is also close research collaboration between US-LTER and FinLTSER.

Nationally, the network has maintained its bottom-up operating approach. The FinLTSER steering group, led by SYKE, has been constituted and operationalised. An extranet and electronic mailing lists serve as the major means of communication, not only within the steering group but also towards the entire community and for certain specific activities such as data management. The platforms have organised annual internal workshops, during which shared practices and networking routines can be developed. The interviews revealed that during the first years, a degree of cohesion has developed between the social and the ecological dimensions in all platforms. Interestingly, those two LTSER applications which had failed have also taken new steps towards interdisciplinary collaboration with partners within their consortia in the form of joint project planning, and are interested in the possibility of joining FinLTSER at a later stage.

FinLTSER was given priority in 2008 in funding for research infrastructure in Finland, but the funding has yet to materialise. Achieving the institutionalisation of such long-term networks and research programmes usually requires input from state institutions (see Kwa 2005) rather than ordinary research funding. The lack of and uncertainty regarding funding now and in the future has taken its toll on the progress of LTSERs. No larger scale planning has taken place and the projects are more ad hoc than long-term in nature. After acceptance by ILTER in summer 2007, FinLTSER concentrated on gaining resources for joint research activities as well as actively contributing to the development of research infrastructure strategies in Europe and in Finland. The Academy of Finland is funding a 4-year research project led by

Dr Helena Karasti, in which she uses FinLTSER as the case study for her research on the development of knowledge infrastructures (see Karasti and Kuitunen 2011). This can be seen as the greatest contribution so far to the development of FinLTSER. The researchers have used the status of FinLTSER in support of their research proposals, and two platforms have been contacted from abroad in the search for further collaboration. The FinLTSER-based Vaccia project, which analyses the socio-environmental resilience of ecosystem services under conditions of climate change, received funding from the Life+programme of the EC and all participants in the network are taking part in this project.

A crucial question remains as to what the Finnish research infrastructure will achieve in concrete terms and when. Will further project proposals be developed and how successful will they be? What support will there be for the national management of FinLTSER from the Finnish authorities after the profound restructuring that is currently taking place? How will the LTSER be linked with the broader picture of research around natural resource management? What connection will there be to international structures?

Although it appears that the Finnish LTSER network grew at least partially from an established tradition and culture of interdisciplinary environmental research, bringing together the strong and theoretically ambitious work in environmental social science and long-term ecological research remains a particular challenge. Much (qualitative) social science is excluded from LTSERs, i.e. much of the potential arising from the tradition of environmental social science is not utilised. In the future it would be important to build links that take better advantage of this basis. By incorporating approaches from theoretically-oriented social science, LTSERs could grow stronger. Opening up possibilities for social scientists to carry out concrete long-term research in an interdisciplinary context would, in turn, strengthen the theoretical and empirical basis of environmental social science. There are many common starting points.

LTSERs offer possibilities to develop case-based strategies such as:

- problem-based approaches,
- multi-method research, as well as
- triangulation (conceptual, methodological).

This requires that social and natural scientists are seen as equal partners and collaborators in the platforms, and are allowed to take initiative for developing research ideas. Problem-oriented approaches could be helpful in developing further work that is genuinely interdisciplinary in character.

## 18.6 Conclusions

Moving from LTER to LTSER means a radical change in the way research is developed and research infrastructure is built. In the case of the latter, the spatial scale is many times larger or multi-scale, the optimal area is no longer necessarily the site most remote from human impact and the research requires inclusion of social scientists into the development and implementation of research strategy.

Why was the LTSER concept seen as a good strategic choice for Finland? When reflecting against the theoretical background on the difficulties of moving towards long-term socio-ecological research, certain characteristics can be identified as driving forces to catalyse the gradual move towards adopting the ideals of LTSER in the Finnish context.

As we showed, problem-oriented interdisciplinary environmental research, including environmental social science and informal and formal connections with knowledge producers and users, developed comparatively early in Finland. The LTSER approach was introduced to Finnish scientific institutions and the scientific community in an effective manner. Our analyses revealed that the message concerning the value and future of the LTSER concept was communicated to the research and stakeholder community in a way that created enthusiasm among the participants and that made its future potential apparent.

Many challenges still arise from the LTSER concept. Although problem-oriented environmental research is highly topical and its value is stressed by many funding agencies, there is also epistemological and methodological friction in practice when moving from research with a restricted time span to long-term research in social and interdisciplinary research on the one hand, and from long-term ecological research to socio-ecological research on the other.

The benefits of a strong LTSER approach are gradually starting to show and appear to validate the strategic decisions to undertake a bottom-up process and to adopt a strong interdisciplinary approach to the development of the FinLTSER network made by the planning group in 2004. A top-down process in a situation where no funding was available for the sites and platforms could have either left the process stuck at the planning stage or ended with a network taking action on paper only. Interdisciplinarity, on the other hand, has enabled Finland to function as a model for the existing and forthcoming European sister-networks through the LTER-Europe Expert Panel on LTSER (LTER-Europe 2005).

The main task of the individual LTSER platforms and of the network of platforms is to enhance the production of long-term knowledge to support the sustainability of socio-ecological systems, from the local to the global level. This requires further methodological development of long-term interdisciplinary research and close collaboration within universities and research institutes in Finland as well as internationally. This could be a potential focus of interest for the researchers in the platforms.

## References

ALTER-Net. (2005). *A long term biodiversity, ecosystem and awareness research network*. www.alter-net.info. Accessed July 15, 2011.

Bowker, G. (2000). Biodiversity datadiversity. *Social Studies of Science, 30*(5), 643–683.

Corley, E. A., Boardman, P. C., & Bozeman, B. (2006). Design and the management of multi-institutional research collaborations: Theoretical implications from two case studies. *Research Policy, 35*, 975–993.

Ellis, R., & Waterton, C. (2005). Caught between the cartographic and the ethnographic imagination: The whereabouts of amateurs, professionals, and nature in knowing biodiversity. *Environment and Planning D, 23*, 673–693.

Endter-Wada, J., Blahna, D., Krannich, R., & Brunson, M. (1998). A framework for understanding social science contributions to ecosystem management. *Ecological Applications, 8*, 891–904.
FinLTSER. (2005). Finnish long-term socio-ecological research network. *FinLTSER*. www.environment.fi/syke/lter. Accessed July 15, 2011.
Furman, E., Kivimaa, P., Kuuppo, P., Nykänen, M., Väänänen, P., Mela, H., & Korpinen, P. (2006). *Experiences in the management of research funding programmes for environmental protection*. Finnish Environment Institute (SYKE) 43. Retrieved January 28, 2011, from http://www.skep-era.net/site/files/WP3_best_practice_guidelines_final.pdf
Haberl, H., Winiwarter, V., Andersson, K., Ayres, R. U., Boone, C. G., Castillio, A., Cunfer, G., Fischer-Kowalski, M., Freudenburg, W. R., Furman, E., Kaufmann, R., Krausmann, F., Langthaler, E., Lotze-Campen, H., Mirtl, M., Redman, C. A., Reenberg, A., Wardell, A. D., Warr, B., & Zechmeister H. (2006). From LTER to LTSER: Conceptualizing the socio-economic dimension of long-term socio-ecological research. *Ecology and Society, 11*(2), 13. [Online] Retrieved from http://www.ecologyandsociety.org/vol11/iss2/art13/
Haila, Y. (2001a). Ympäristöherätys. In Y. Haila & P. Jokinen (Eds.), *Ympäristöpolitiikka: Mikä ympäristö, kenen politiikka* (pp. 21–46). Tampere: Vastapaino.
Haila, Y. (2001b). Tieteellisen tiedon merkitys. In Y. Haila & P. Jokinen (Eds.), *Ympäristöpolitiikka: Mikä ympäristö, kenen politiikka*. Vastapaino: Tampere.
Haila, Y., & Dyke, C. (Eds.). (2006). *How nature speaks: The dynamics of the human ecological condition*. Durham: Duke University Press.
Haila, Y., & Levins, R. (1992). *Humanity and nature: Ecology, science and society*. London: Pluto Press.
Hiedanpää, J., Suvantola, L., & Naskali, A. (Eds.). (2010). *Hyödyllinen luonto: Ekosysteemipalvelut, hyvinvointimme perustana*. Tampere: Vastapaino. (Nature as a benefit. Ecosystem services as the basis for our well-being).
ILTER. (1993). *International long term ecological research*. www.ilternet.edu/. Accessed July 15, 2011
Kamppinen, M., Raivola, P., Jokinen, P., & Karlsson, H. (1995). *Riskit yhteiskunnassa: Maallikot ja asiantuntijat päätösten tekijöinä*. Helsinki: Gaudeamus. (Risks in Society. Laymen and experts as decision makers).
Karasti, H., & Kuitunen, P. (2011). Socio-technical considerations: Initiating information management within the Finnish LTSER Network. In K. Krauze, M. Mirtl, & M. Frenzel (Eds.), *LTER Europe – The next generation of ecosystem research: A guide through European Long-Term Ecological Research Networks, Sites and Processes*. Cambridge: Cambridge University Press.
Knorr Cetina, K. (1999). *Epistemic cultures: How the sciences make knowledge*. Cambridge, MA: Harvard University Press.
Knuutila, S., Rannikko, P., Oksa, J., Hämynen, T., Itkonen, H., Kilpeläinen, H., et al. (2008). *Kylän paikka: Uusia tulkintoja Sivakasta ja Rasimäestä*. Helsinki: SKS.
Kotakorpi, E. (2007). *Yhteiskunnallisen ympäristötutkimuksen alalta väitelleiden työelämään sijoittuminen (1991–2006)* (The career development of PhD's in environmental social science (1991–2006). Retrieved January 27, 2011 from YHTYMÄ graduate school Web site: www.uta.fi/laitokset/yhdt/artikkelit/yhtyma_Ellin%20selvitys.pdf
Kwa, C. (2005). Local ecologies and global science: Discourses and strategies of the international geosphere-biosphere programme. *Social Studies of Science, 35*(6), 923–950.
Lehtinen, A. (2005). Esipuhe (Prologue). In A. Lehtinen (Ed.), *Maantiede, tila, luontopolitiikka – Johdatus yhteiskunnalliseen ympäristötutkimukseen*. Joensuu: Joensuu University Press.
Lehtinen, A., & Rannikko, P. (1994a). Esipuhe (Prologue). In: A. Lehtinen & P. Rannikko (Eds.), *Pasilasta Vuotokselle: Ympäristökamppailujen uusi aalto*. Helsinki: Gaudeamus (From Pasila to Vuotos. The new wave of environmental conflicts).
Lehtinen, A., & Rannikko, P. (Eds.) (1994b). *Pasilasta Vuotokselle: Ympäristökamppailujen uusi aalto*. Helsinki: Gaudeamus (From Pasila to Vuotos. The new wave of environmental conflicts).
Lehtinen, A., & Rannikko, P. (Eds.). (2003). *Oikeudenmukaisuus ja ympäristö*. Helsinki: Gaudeamus (Justice and environment).

Lehtinen, A., Donner-Amnell, J., & Sæther, B. (Eds.). (2004). *Politics of forests: Northern forest-industrial regimes in the age of globalisation*. Aldershot: Ashgate.

LTER-Europe. (2005). *European Long-term ecological research network*. www.lter-europe.net. Accessed 15 July 2011

Määttä, T. (2007). Lukijalle (To the reader). In T. Määttä (Ed.), *Ympäristöpolitiikan ja oikeuden vuosikirja 2007 (p. 5)*. Joensuu: Joensuu university press.

Massa, I., & Rahkonen, O. (Eds.). (1995). *Riskiyhteiskunnan talous: Suomen talouden ekologinen modernisaatio*. Helsinki: Gaudeamus. (The economy of risk society. The ecological modernization of the Finnish Economy).

Mauz, I., Peltola, T., Granjou, C., Buijs, A., & Van Bommel, S. (2012). How scientific visions matter. Insights from three long-term socio-ecological research (LTSER) platforms under construction in Europe. *Environmental Science and Policy 19–20*, 90–99.

McCallin, A. (2006). Interdisciplinary researching: Exploring the opportunities and risks of working together. *Nursing and Health Sciences, 8*, 88–94.

Mirtl, M. (2010). Introducing the next generation of ecosystem research in Europe: LTER-Europe's multifunctional and multi-scale approach. In F. Müller, C. Baessler, H. Schubert, & S. Klotz (Eds.), *Long-term ecological research: Between theory and application* (pp. 75–93). Dordrecht: Springer.

Mirtl, M., Boamrane,M., Braat, L., Furman, E., Krauze, K., Frenzel, M., Gaube, V., Groner, E., Hester, A., Klotz, S., Los, W., Mautz, I., Peterseil, J., Richter, A., Schentz, H., Schleidt, K., Schmid, M., Sier, A., Stadler, J., Uhel, R., Wildenberg, M., & Zacharias, S. (2009). *LTER-Europe design and implementation report – Enabling "next generation ecological science": Report on the design and implementation phase of LTER-Europe under ALTER-Net & management plan 2009/2010*. Vienna: Umweltbundesamt, Environment Agency Austria.

Petts, J., Owens, S., & Bulkeley, H. (2007). Crossing boundaries: Interdisciplinarity in the context of urban environments. *Geoforum, 39*, 593–601.

Phillipson, J., Lowe, P., & Bullock, J. M. (2009). Navigating the social sciences: Interdisciplinarity and ecology. *Journal of Applied Ecology, 46*, 261–264.

Rannikko, P., & Määttä, T. (Eds.). (2009). *Luonnonvarojen hallinnan legitimiteetti*. Tampere: Vastapaino. (The legitimacy of natural resources governance).

Redman, C. L., Grove, J. M., & Kuby, L. H. (2004). Integrating social science into the long-term ecological research (LTER) network: Social dimensions of ecological change and ecological dimensions of social change. *Ecosystems, 7*, 161–171.

Uiterkamp, A., & Vlek, C. (2007). Practice and outcomes of multidisciplinary research for environmental sustainability. *Journal of Social Issues, 63*, 175–197.

Yli-Pelkonen, V., & Niemelä, J. (2006). Use of ecological information in urban planning: Experiences from the Helsinki metropolitan area, Finland. *Urban Ecosystems, 9*, 211–226.

# Chapter 19
# The Eisenwurzen LTSER Platform (Austria) – Implementation and Services

Johannes Peterseil, Angelika Neuner, Andrea Stocker-Kiss, Veronika Gaube, and Michael Mirtl

**Abstract** The Austrian Eisenwurzen LTSER Platform has been implementing the concepts of long-term socio-ecological research since 2004 to foster sustainable regional development by facilitating scientific research according to regional and local needs. In order to achieve this, scientific expertise from various disciplines for a given region is concentrated and bundled to allow for a better integration of primarily disciplinary approaches into an inter- and transdisciplinary research framework. A better understanding of feedbacks and information flows between the region and science is the main task of the platform. The Eisenwurzen is a region with a long history but fuzzy boundaries, defined rather by the cultural identity of the local population than by natural characteristics. The platform is located in the borderland of the federal provinces of Upper Austria, Lower Austria and Styria, with a total area of approximately 5,776 km$^2$.

This chapter describes the characteristics and challenges of the region, the implementation process, and the structure and services of the Austrian Eisenwurzen LTSER Platform.

---

J. Peterseil, Ph.D. (✉) • A. Neuner, M.Sc. • A. Stocker-Kiss, M.Sc. • M. Mirtl, Ph.D.
Department of Ecosystem Research and Monitoring, Environment Agency Austria, Vienna 1090, Spittelauer Lände 5, Austria
e-mail: johannes.peterseil@umweltbundesamt.at; michael.mirtl@umweltbundesamt.at; angelika.neuner@umweltbundesamt.at; andrea.stocker-kiss@umweltbundesamt.at

V. Gaube, Ph.D.
Institute of Social Ecology Vienna (SEC), Alpen-Adria Universitaet Klagenfurt, Wien, Graz, Schottenfeldgasse 29/5, Vienna 1070, Austria
e-mail: veronika.gaube@aau.at

**Keywords** LTSER Platform • Eisenwurzen • Participatory approaches • Communication space • Data sharing • Data policy • Stakeholder integration • LTER Austria

## 19.1 Introduction

Up to the 1990s, long-term ecological research (LTER) focused mainly on processes and patterns of semi-natural and natural ecosystems. Small-scale sites (1–10 km$^2$) are selected according to specific ecosystem criteria (e.g. hydrological catchments of small rivers and lakes) and preferably in semi-natural or natural areas. Based on site measurements, LTER aims to document and analyse ecological patterns and processes in order to detect environmental change and its impacts on ecosystems and it provides relevant information on ecosystem change. However, LTER has its limits when it comes to explaining the drivers of ecosystem and biodiversity loss on a large scale, often induced by human activities. Consequently, the integration of aspects of society, economy, public opinion and land-use history led to the inclusion of the social dimension in such research, which was then termed long-term socio-ecological research (LTSER) (Redman et al. 2004; Haberl et al. 2006; Singh et al. 2010). Evidently, most drivers and pressures and their impacts cannot be comprehensively investigated on the spatial scale of hundreds of hectares of LTER sites. LTSER applies its research on a regional scale, sharing a common land-use history and similar environmental conditions. Beyond the requirement that LTSER should be interdisciplinary, LTSER is also context-driven and problem-focused which implies the involvement of stakeholders from the region in terms of transdisciplinary research approaches (for further details, see Mirtl et al., Chap. 17 in this volume). The European regional group of the global LTER network, LTER-Europe, was set up with a strong focus on the LTSER component and the implementation of LTSER in hot-spot areas, with so-called "LTSER Platforms" in the range of 100–10,000 km$^2$, ideally containing a number of LTER Sites carrying out traditional ecosystem research in habitats typical for the region (Mirtl and Krauze 2007; Mirtl et al. 2009; Mirtl 2010). Although the management of an effective LTSER platform is a complex challenge, nevertheless, in order to build up a socio-ecological research capacity across Europe, the LTER-Europe network wants each of the major socio-ecological systems (Metzger et al. 2010) of the continent be represented by at least one LTSER platform.

In this paper, we introduce the region of Eisenwurzen in Austria as one of the first LTSER platform implemented in accordance with the definitions and criteria of LTER-Europe. The long land-use history of the Eisenwurzen region together with a high density of research facilities at different scales were the main criteria for selecting the region as an LTSER platform. In this chapter, we first outline the history of the Eisenwurzen region and the challenges it has faced in terms of socioeconomic and ecological changes in past centuries, before presenting the implementation of the LTSER platform. Finally, we discuss the management structure, services and future perspectives of the Eisenwurzen LTSER Platform.

## 19.2 The Eisenwurzen Region in Austria

The Eisenwurzen is a region with a long history but fuzzy boundaries (Mejzlik 1935), defined by the cultural identity of the local population rather than by natural characteristics (Heintel and Weixlbaumer 1998, 1999; Roth 1997, 1998). It is situated at the border between three federal provinces, with rather evenly sized parts situated in Upper Austria, Lower Austria and Styria.

The Eisenwurzen represents a biogeographic gradient from the northern Alpine foothills in the north to the Northern and Central Alps in the south (see Fig. 19.2). It shows a pronounced relief, with altitudes ranging from 167 to 2,512 m ASL and having an average altitude of 840 m ASL. The climate is continental and exhibits a geology of limestone and flysch material (see Table 19.1).

The hilly landscape is dominated by forests and pastoral agriculture. The settlements are mainly concentrated in the valleys. Table 19.2 provides an overview on the existing land-cover types.

The main land-cover type is forest, which is also a major part of the regional economy. The potential natural forest vegetation especially in the montane altitudinal range (600–1,450 m ASL) are spruce-fir-beech forests. In addition, on drier sites scots pine (*Pinus sylvestris* L.) can also be found (Kilian et al. 1994). The natural forest types mainly remain on steeper slopes. Most of the forests are managed forests and are dominated by spruce (*Picea abies* L.). Spruce in the managed forests is either planted or resulting from natural regeneration. In the subalpine altitudinal range (1,450–1,900 m ASL) natural spruce-dominated forest with a higher amount of larch (*Larix decidua* L.) can be found. The timber line is formed by mountain pine (*Pinus mugo* L.). Along the smaller rivers especially gray alder (*Alnus incana* L.) and willow (*Salix* spp.) species can be found (Kilian et al. 1994).

The grassland in the valleys was historically dominated by wet meadows and bogs from which only some remain today. On drier parts of the landscapes, extensively managed and low-productivity dry grasslands, e.g. with *Bromus* species, can be found. Today, most of the grassland in the valleys can be seen as improved grassland due to melioration. In the mountainous parts of the region, species-rich subalpine pastures and alpine habitats such as screes, rocks and snow beds can be found.

### *19.2.1 History*

In the twelfth century, iron ore mining became established and reached its peak in the sixteenth century, when the region contributed 15% of the total European iron production (Sandgruber 1997a, b). For centuries, the entire region was characterised by complex interactions between mining, metallurgy and agriculture in the supplying hinterland. One consequence was the need for a large amount of timber for metallurgy (char burning) as well as for the construction of river rafts for iron transport (Heintel and Weixlbaumer 1998). Provision of food supply for the large number of industrial workers and energy supply for industries were major challenges

**Table 19.1** Biophysical characteristics of the Eisenwurzen region

| Characteristic | Min-Max | Mean | StdDev |
|---|---|---|---|
| Elevation | 167–2,512 m. a.s.l. | 840 m a.s.l. | 367.87 |
| Mean annual temperature | −0.8 to 9.14 °C | 6.1 °C | 1.44 |
| Mean temperature January | −8.3 to −1.4 °C | −3.3 °C | 0.82 |
| Mean temperature August | 6.8–18.3 °C | 14.7 °C | 1.69 |
| Annual precipitation | 730–2,202 mm | 1,382 mm | 240.10 |
| Precipitation summer | 443–1,306 mm | 834 mm | 136.72 |
| Precipitation winter | 274–939 mm | 571 mm | 117.57 |

**Table 19.2** Land-cover characteristics of the Eisenwurzen region

| Land cover | % |
|---|---|
| Forest | 63.8 |
| Managed grassland | 13.3 |
| Cropland | 6.3 |
| Complex agriculture | 3.7 |
| Constructed | 2.4 |
| Sparsely vegetated | 3.4 |
| Heathland | 5.5 |
| Natural grassland | 1.2 |
| Other (wetland, water bodies, shrubland) | 0.4 |

Source: CORINE Land Cover (2000)

for agriculture and forestry and a primary driver of environmental changes. At the turn of the seventeenth and eighteenth centuries, food was mainly imported into the area. High energy demands resulted in widespread deforestation. As rivers were used for transport and hydropower use, production sites were mainly located in the valleys and the forests surrounding these areas were used for energy and construction, with infrastructure in terms of roads for industrial activities and trade provided by the state. Around 1930, a high demand for roads due to the increase of automobiles and public transport resulted in the development of the road infrastructure. Relocation of existing roads with a larger curve radius once again altered the landscape (Kreuzer 1998).

In the 1850s, a Europe-wide economic crisis, technical improvements within the English metallurgy industry, and difficulties in restructuring the production from decentralised small-scale business to mass production all contributed to the decline of the industry in the Eisenwurzen region (Kropf 1997). This resulted in unemployment, the spread of poverty and depopulation of large parts of the Eisenwurzen region (Sandgruber 1998). The need for timber also declined and woodlands began to recover, in particular at higher altitudes. Afforestation and shrub encroachment were the first steps of landscape change.

Today, tourism, agriculture and forestry are the region's main economic bases. Tourism in the alpine region is highly dependent on the region's accessibility. After 1948, the number of tourists increased due to the construction of roads, rail connections,

**Table 19.3** Inhabitants and population density in the Eisenwurzen region

|  | 2001 | 2005 | 2008 |
|---|---|---|---|
| Total population | | | |
| Part Upper Austria | 171,346 | 172,948 | 173,651 |
| Part Lower Austria | 70,119 | 67,531 | 67,870 |
| Part Styria | 68,286 | 66,537 | 64,760 |
| Eisenwurzen | 309,751 | 307,016 | 306,281 |
| Austria | 8,078,225 | 8,206,524 | 8,342,746 |
| Population density | | | |
| Part Upper Austria | 72.62 | 73.34 | 73.64 |
| Part Lower Austria | 43.41 | 41.81 | 42.02 |
| Part Styria | 37.86 | 36.90 | 35.91 |
| Eisenwurzen | 53.61 | 53.15 | 53.02 |
| Austria | 96.31 | 97.84 | 99.46 |

Compiled from Statistik Austria (2010)

hiking paths and ski lifts (Dutzler 1998), thus placing new pressures on the biodiversity in the Eisenwurzen region. Within the agricultural sector, two opposing trends can be observed: Central areas of the region with a rough terrain are experiencing rapid afforestation and land abandonment, whereas land-use intensity is increasing in flat areas (foothills, foreland), partly driven by increased crop demand due to the promotion of biofuels.

In the late 1980s, efforts were made to establish a network of nature protection sites in the region. Today the region shows the highest proportion of nature protection areas in Austria, covering 6% of the total area (without taking into consideration nature protection sites of lower ranks, e.g. nature park areas ("Naturparke")). The Eisenwurzen region has two nature reserves (the Kalkalpen National Park, the Gesäuse National Park), a wilderness area (Dürnstein), and several nature park and Natura 2000 areas.

## 19.2.2 Socioeconomic and Ecological Challenges

The Eisenwurzen region has about 307,000 inhabitants. The population density of 53 cap/km$^2$ (Statistik Austria 2010, see Table 19.3) is lower than the Austrian national average of 99 cap/km$^2$. The inner part ("Innereisenwurzen", Styria and southern Upper and Lower Austria) differs significantly from the northern flat parts of the region. For example the city of Steyr (in the northern part) shows a population density of 1,481 cap/km$^2$, whereas Hieflau in Styria has only 18 cap/km$^2$. Similar patterns were identified for commuting, infrastructure, migration and demographic structure. The municipalities in the northern part have a moderately positive to positive migration rate (e.g. Dietach +9.9% 1991–2001, +9.1% 2002–2009), whereas the inner parts of the Eisenwurzen region show the highest negative migration rates in Austria (e.g. Vordernberg −20.9% 1991–2001, −10.1% 2002–2009). A similar trend is shown in the birth rate, with positive developments in the northern part of the region (e.g. Dietach +5.2% 1991–2001, +3.1% 2002–2009) and negative trends in the inner parts (e.g. Vordernberg −3.2% 1991–2001, −8.6% 2002–2009). The demographic structure is characterised by a rising ageing population in the inner parts of the Eisenwurzen (e.g. Eisenerz: 43.4% of the inhabitants were aged above 60 years in 2009), whereas figures for the northern parts are within the Austrian average. In addition, in the Styrian part of the Eisenwurzen region, e.g. in the Vordernberg municipality, more than 76% (Statistik Austria 2011) of the working population commute to work, whereas in the northern parts, e.g. in the district Steyr in Upper Austria, the proportion is only 30% (Statistik Austria 2011).

The dependence of the Eisenwurzen region on forestry, agriculture and tourism as the main sources of income makes the economic situation of the people living there vulnerable to the negative effects of climate change. Not only human society has to deal with the consequences of climate change but the ecosystem has to face new challenges too. In combination with other pressures on the ecosystem, the composition of habitats and ecosystems and therefore the character of the landscape and the value of ecosystem services are affected. Climate change increases the risk of disastrous weather phenomena, as mountain areas are vulnerable to periods of persistent heavy rainfall or snowfall, leading to a higher incidence of flooding of settlements and agricultural land in the valleys and endangerment by mudflows and avalanches. Studies looking at subalpine grasslands from the area show different responses of vegetation to invasion and shrub encroachment (Dullinger et al. 2004; Dirnböck and Dullinger 2004). Today, even the national parks within the region of Eisenwurzen are facing challenges due to land abandonment (Hasitschka 2007). Furthermore, species adapted for living in high mountain areas are at risk of losing their ecological niches and becoming extinct, with negative effects on the richness of native species (Grabherr et al. 1994; Pauli et al. 2003). Nitrogen and sulphur emissions increased dramatically during the second half of the twentieth century and caused excess deposition of N and in the late twentieth century a decreasing deposition of S in natural and semi-natural ecosystems. Excess N deposition causes soil eutrophication and the decrease of S deposition results in a significant, but soil-specific, recovery from acidification. Detected trends of soil properties were not unambiguously reflected in changes of forest floor vegetation (Dirnböck et al. 2007).

## 19.3 Eisenwurzen LTSER Platform

Massive environmental changes have been taking place in the region of Eisenwurzen over centuries. The long history of iron ore mining and processing shaped the current landscape between the twelfth and the nineteenth century (Sandgruber 1997b). Moreover, global change is affecting the state of the environment in the region today. These changes have a negative effect upon natural resources and ecosystem services (ecosystem protection, livelihoods, etc.). The necessary governance measures related to the sustainable use of ecosystem services and the effects imposed on them by global change require adequate research to provide knowledge on socio-ecological interactions for decades to come. A long-term socio-ecological research approach enables researchers to properly investigate how ecosystems react to natural and human alterations (drivers). The Eisenwurzen LTSER Platform was developed to integrate existing knowledge and results with the requirements of the region and provide services both for research and for local and regional users.

The aim of the Eisenwurzen LTSER Platform is to research the complex interconnectedness between the natural landscape of the region of Eisenwurzen, its historical and current land use, and the controlling variables of this use. Taking supra-regional constraints such as global change and socioeconomic framework conditions into account, the intention is to develop the scientific basis for sustainable management (LTER Management Austria 2009).

### 19.3.1 Process and Development of the Eisenwurzen LTSER Platform

In 2004, the Eisenwurzen LTSER Platform ("Forschungsplattform Eisenwurzen") was established. The most important issue of the implementation process was to integrate existing elements, such as monitoring and research infrastructure as well as relevant stakeholders and institutions. The goal was to establish a long-term involvement of research and stakeholders in order to set up a well-considered collaboration. The process of implementing the LTSER platform went through several phases. An overview on the process is given in Fig. 19.1.

*Concept development*: The first steps towards implementation of the Eisenwurzen LTSER Platform were taken back to back with the concept of LTSER itself in the mid-2000s. This parallel development shaped the LTSER design substantially according to the requirements of the region. Experts implementing the Eisenwurzen LTSER Platform were at the same time substantially involved in developing the actual LTSER concept, which was first presented under the title "Multifunctional Research Platforms (MFRP)" at a joint US-European LTER workshop in 2003 in Motz, France, and at the global LTER conference in Manaus, Brazil, (Mirtl 2004) in 2004. Outcomes of these first concept development events were put together on a broader conceptual basis (Haberl et al. 2006). Under the auspices of the Sixth Research Frame Programme (FP6) Network of Excellence ALTER-Net, the concept was then further specified in accordance with experiences gathered in establishing concrete LTSER platforms across Europe. From the very beginning, the Eisenwurzen LTSER Platform has been one of the flagships in that iterative process (Mirtl et al., Chap. 17 in this volume).

*Stakeholder involvement*: As LTSER is a problem-focused and context-driven approach, the involvement of local stakeholders as partners was indispensable. Therefore, in the beginning of the LTSER process starting in 2003, small-scale bilateral workshops with regional stakeholders (local mayors, regional development

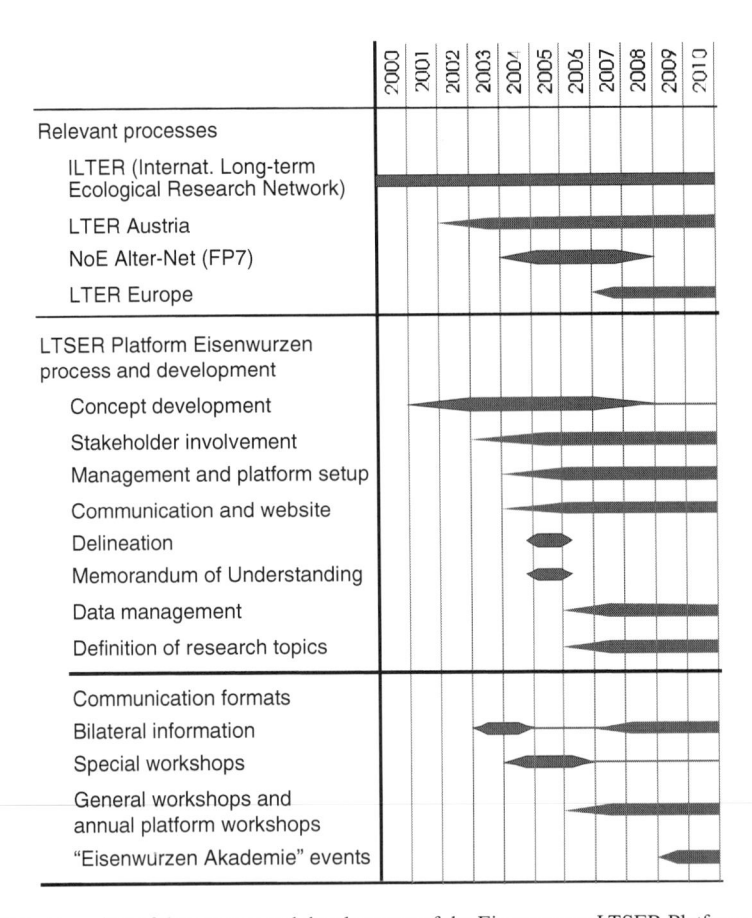

**Fig. 19.1** Overview of the process and development of the Eisenwurzen LTSER Platform

agencies, national parks, etc.) as well as the provincial governments of Lower Austria, Styria and Upper Austria were held in order to introduce the concept and discuss involvement options. The concept of the LTSER platform was attractive and plausible for both groups and led to a funding contract for the set-up and management of the Eisenwurzen LTSER Platform, first with the Federal Ministry of Science, in 2008 joined by the provincial governments with a 3-year contract.

In 2004, a special workshop in the context of the Network of Excellence ALTER-Net (FP6) took place in Gumpenstein (Styria). This was the starting point for joint cooperation between researchers with a primarily ecological background undertaking their individual research in the Eisenwurzen region, representatives of the Institute of Social Ecology, (SEC) Vienna (Alpen Adria University Klagenfurt) and the Umweltbundesamt (Environment Agency Austria), who discussed options for the implementation of the LTSER concept in Eisenwurzen that emerged simultaneously (for details, see Mirtl et al., Chap. 17 in this volume). This led to a general commitment to the LTSER concept within the research community dealing with the region of Eisenwurzen. In the first years of the platform's existence, natural scientists remained in the majority, which led the platform management, acknowledging the necessity of interdisciplinary research, to make efforts to involve other disciplines, such as landscape science, sustainability science, economics, history and sociology.

The general workshops in 2008 and 2009 revealed the success of these endeavours in achieving parity in the number of social and natural scientists involved in the platform. The focus of these workshops lay with synthesizing individual projects or project ideas focusing on the region and the knowledge gained as a result. The 2009 workshop led to three new inter- and transdisciplinary project outlines, providing evidence of the progress made towards developing a common language based on harmonised notions among disciplines and across professional boundaries (i.e. involving other actors than researchers).

By 2010, the platform could draw upon a broad pool of expertise and valuable cooperation experiences, providing the basis for quick and efficient project developments. In response to questions arising from regional stakeholders, established teams and institutional partnerships are currently involved in the development of various project proposals. National and international funding calls are of course a prerequisite for these activities. However, the platform is not an exclusive community but one that welcomes the involvement of new stakeholder and expert groups.

*Management and platform setup:* The platform management as the integrative element of the LTSER platform was formed and established at the Umweltbundesamt GmbH, Austria (Environment Agency Austria) in the Department for Ecosystem Research and Monitoring in 2004, with funding from the Ministry of Science and Research. At the first annual workshop of the Eisenwurzen LTSER Platform in 2006, the focus lay with prioritisation of tasks for the management according to research, regional development and policy makers' requirements.

*Delineation:* The region of Eisenwurzen presents important prerequisites for the implementation of a LTSER platform: (1) it has the densest network of protected areas in Austria; (2) it already contained a large number of LTER-Sites (in rather different terrains and habitats) and (3) it has always been of special interest to researchers from various backgrounds due to its history.

In addition to the existence of research and monitoring infrastructure, the delineation of the research area is based on socioeconomic and scientific criteria as well as on criteria of cultural identification. In the first instance, a literature review of scientific and grey literature was undertaken, searching for descriptions of the "Eisenwurzen" region. Secondly, references to the region made by local associations and initiatives (culture, slow food, etc.) carrying the name "Eisenwurzen" or "Eisenstraße" were collected. Thirdly, projects and initiatives mainly related to regional development (e.g. LEADER[1] regions) were taken into consideration.

---

[1] See http://europa.eu/legislation_summaries/regional_policy/provisions_and_instruments/g24208_en.ht$^m$

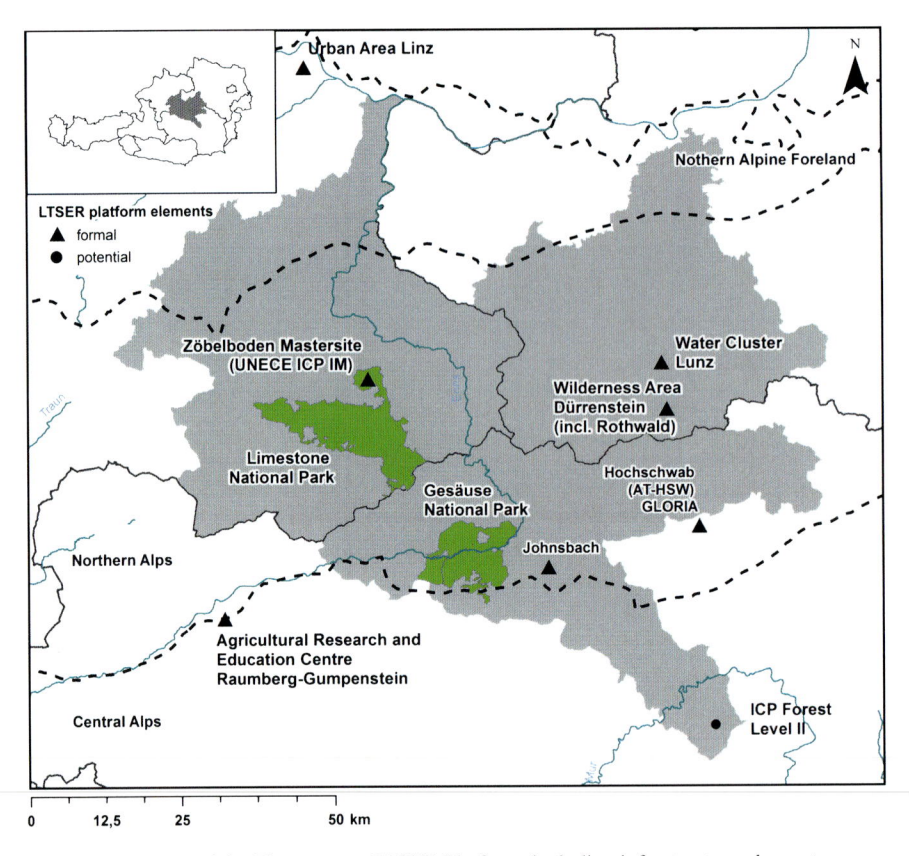

**Fig. 19.2** The area of the Eisenwurzen LTSER Platform, including infrastructure elements

In order to provide easily accessible and spatially relevant social and economic data, the delineation followed municipal boundaries. Finally, a total of 100 municipalities, extending over 5,776 km² in the border areas of the provinces of Lower Austria, Styria and Upper Austria were defined as constituting the Eisenwurzen LTSER platform (see Fig. 19.2).

*Memorandum of Understanding (MoU)*: In 2005, a Memorandum of Understanding was prepared, which by 2010 had been signed by over 35 parties. Signing the MoU meant a general agreement to focus on research closely connected to the needs of the population in the Eisenwurzen region, to link up isolated research projects, to joint knowledge generation and a general willingness to support data sharing among the participating institutions.[2]

---

[2] For full version, see http://www.umweltbundesamt.at/en/mfrp_eisenwurzen/

19 The Eisenwurzen LTSER Platform (Austria) – Implementation and Services 471

```
INFRASTRUCTURE
- LTER Sites
- Collections
- National parks
- LTEM

RESEARCH
- projects, teams, institutions
- basic ecosystem research
- nat. conservation research
- socio-ecological research

STAKEHOLDER
- regional development
- municipalities
- federal government
- etc.
```

```
          INFRASTRUCTURE / RESEARCH
                LTSER Platform
                 Management
              STAKEHOLDER
```

```
LTSER PLATFORM
MANAGEMENT

provided services
- Communication space
    Contacts and events
    Research topics
    Projects and results
- Networking
- Metadata and data
    management
```

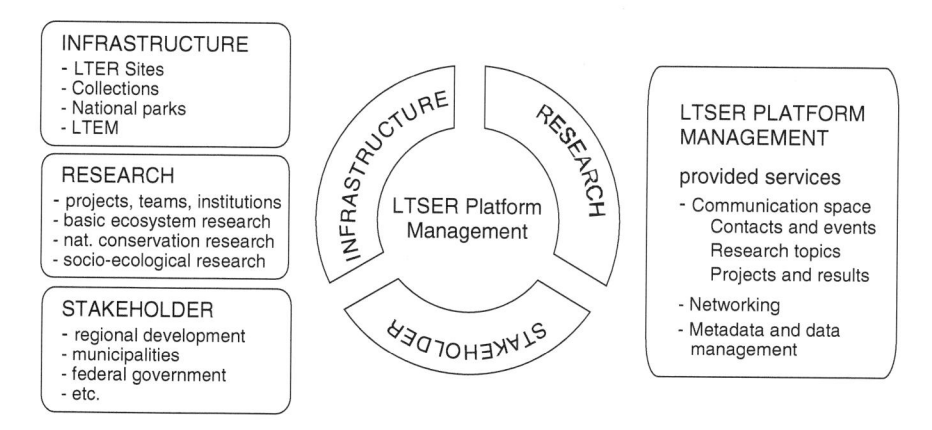

**Fig. 19.3** Interlinkage between the elements of the Eisenwurzen LTSER Platform

## 19.3.2 Elements of the Eisenwurzen LTSER Platform

To foster the aim of regional integrated research, a LTSER platform consists of functional elements that interact with each other (Mirtl et al. 2009, Mirtl et al., Chap. 17 in this volume). The *infrastructure* encompasses the existing monitoring and research facilities that provide one part of the data sets. The *research* encompasses the ongoing research projects in the LTSER platform. The *stakeholder* element encompasses the local and regional stakeholders, which on the one hand provide topics for research but on the other hand are key to the success of the research as they provide regional knowledge as regional experts and at the same time function as users of the results. These three elements may exist without a LTSER platform but will be lacking in knowledge exchange among each other. The *platform management* acts as a service point and communication space to interlink these existing elements and to enable a new type of research (see Fig. 19.3 Interlinkage between the elements of the Eisenwurzen LTSER Platform). It provides the basic services communication, networking and data management for information exchange.

This interlinkage and the services of the central platform management for the different elements of the LTSER platform are described in the following sections.

### 19.3.2.1 Infrastructure

The infrastructure addresses different spatial, temporal and thematic scales. This ranges from small-scale process-oriented research and monitoring of regionally or nationally relevant ecosystem types to a large-scale assessment on catchment, landscape or regional scale addressing parts or the whole of the LTSER platform. Sites such as national parks, biosphere reserves or smaller catchments act as intermediaries between the analyses of ecosystem processes and the large-scale assessments.

Comparison and integration of results from different scales and harmonisation of parameters and measurement methods across sites can be enhanced in terms of representation and validity of single sites.

Nevertheless, these research infrastructures typically encompass mostly monodisciplinary long-term research sites as well as research organisations and stakeholders from the region. Sites may be managed by a regional institution within the region of Eisenwurzen, such as the Kalkalpen National Park or Gesäuse National Park authorities (in Molln and Admont, respectively). It may also be part of a national monitoring network like the forest inventory sites of the ICP Forest Level II network (Federal Research and Training Centre for Forests, Natural Hazards and Landscape, Vienna), which is managed by an organisation outside the region.

Concerning funding schemes, the sites differ clearly from one other ranging from sites that are funded on a project basis with short-term contracts to sites which have a mid- to long-term funding commitment. The mid- to long-term version is the most common funding arrangement in the Eisenwurzen region. Those sites have a permanent management, such as the Lunz Water Cluster/WasserCluster Lunz (University of Natural Resources and Life Sciences, Vienna) or the Zöbelboden ICP Integrated Monitoring site (Umweltbundesamt GmbH, Vienna).

Figure 19.2 gives an overview of the existing infrastructure in the Eisenwurzen LTSER Platform. Data support is a major theme. With the eCatalogue and the LTER InfoBase, progress towards ensuring access to metadata is established. In the long run, direct access to data should be feasible. Metadata about the existing infrastructures is collected with the LTER InfoBase.[3]

### 19.3.2.2 Research

The participating research institutions are public as well as private organisations with headquarters mainly outside of the region in the cities of Vienna, Graz, and St. Pölten. The research disciplines extend from natural sciences (ecology, biodiversity research, forestry, veterinary science, limnology, etc.), landscape sciences (geography, landscape and rural development, etc.) and cultural sciences (environmental history, agricultural history) to interdisciplinary sciences (sustainable science, socio-ecological science, climate science). Additionally, technical sciences such as remote sensing and GIS offer methods used within various projects.

Questions including how do/will human actions and economy affect an ecosystem? Which constraints will arise from the modified ecosystem? How do changes in the ecosystem impact upon humans' well-being in the region? are within the remit of LTSER. Potential developments of regional and supra-regional parameters (global change, socioeconomic conditions, etc.) are analysed and visualised with the help of models and scenario-building. The Eisenwurzen LTSER Platform encompasses the following research areas:

---

[3] See https://secure.umweltbundesamt.at/eMORIS/

*Ecosystem research* – basic ecosystem research deals with the complex matter of mechanisms and functional chains within an ecosystem such as the effect of nitrogen deposition on the biodiversity and soil processes in terrestrial environments.

*Applied science on biodiversity and nature protection* – applied science is based on the knowledge and facts gained from different fields of natural science. The focus lies with problems occurring in the areas of nature protection and biodiversity.

*Socio-ecological research* – the main focus of this particular field of research comprises the biophysical interactions between society and the natural environment, and involves examining material and energy flows, land use and time use.

Within the Eisenwurzen LTSER Platform, a number of projects have been carried out across these research fields. Table 19.4 gives an exemplary overview of research projects in the Eisenwurzen region.

### 19.3.2.3 Stakeholder

Within the Eisenwurzen LTSER Platform, different stakeholder groups can be identified. Major players for the LTSER platform management are the *provincial governments*, both in terms of funding and formalisation. In addition, they are key parties responsible for the regional and local development. They are also stakeholders providing expert knowledge during the research process in terms of defining questions and problems to be investigated. As research findings – ideally – ought to support decision making at various scales, provinces are among the most important users of the results.

Secondly, *local decision makers and the public*, as municipalities or regional development associations. These are relevant for both the definition of research questions and for the transformation of results from research projects and analysis. Furthermore, as experts on the region they provide highly valuable information and knowledge that is particularly needed for transdisciplinary research work.

These two *local and regional stakeholder* groups are the main contact points for the platform management and interact, although generally not directly, with the following stakeholder groups.

Thirdly, the *managing institutions* which run research and monitoring sites in the region often work on various thematic and spatial scales and in different ecosystem types. The location of the managing institutions can be either within the region and normally with a high administrative commitment to the region (e.g. national park authority of the Kalkalpen National Park, Molln) or outside, with a lower administrative commitment to the region (e.g. Umweltbundesamt GmbH, Vienna).

Fourthly, the *research institutions* that have a special focus in the region can be identified (e.g. Institute of Social Ecology, University of Klagenfurt, Vienna). These research institutions often co-operate with the platform management on a project basis as they use data of existing long-term monitoring and research sites on the one hand and established contacts with regional and local stakeholders.

**Table 19.4** Exemplary list of research projects within the Eisenwurzen LTSER Platform allocated to the three research areas

| | Project | Description | Scale |
|---|---|---|---|
| Ecosystem research | Integrated monitoring | Long-term air pollution effects have been monitored since 1992 at the ICP Integrated Monitoring Zöbelboden site (90 ha catchment area) in the Northern Kalkalpen region (see www.umweltbundesamt.at/im/). Results of the monitoring programme on nitrogen deposition and leaching (Jost et al. 2011), effects of nitrogen and sulfur deposition on forest biodiversity (Dirnböck and Mirtl 2009; Diwold et al. 2010) and heavy metal (Kobler et al. 2010) are shown | Site |
| | ExtremAqua | ExtremAqua (Influences of Extreme Weather Conditions on Aquatic Ecosystems) is a long-term research programme of the WasserCluster Lunz. Central to the programme is the investigation of possible effects of weather events on aquatic ecosystems and their constituent organisms. The overall aim of ExtremAqua is to provide continuous data sets with high temporal and spatial resolution to form the basis for research questions. Both components are necessary to understand fluctuations in aquatic ecosystems and their coupling with the terrestrial milieu. Given that the hydrological cycle is a major topic which will doubtlessly affect various aspects of natural environments and public life, ExtremAqua focuses on basic research in ecosystem ecology to provide advice to the public sector | Site |
| Applied science on biodiversity and nature protection | IP Sensor (FP6) | The project analysed the region's strengths and weaknesses with respect to sustainability (Hasenauer et al. 2007; Putzhuber and Hasenauer 2010). The aim was to develop instruments to assess the effects of multifunctional land use upon environmental social and economic aspects in Europe which can also be used in a decision support system (SIAT). The Eisenwurzen region was one of the test areas in the SENSOR project | Regional |
| | Geoland (GMES) | For the protection of the Alps, comprehensive up-to-date, repeatable and comparable data about land cover and land use are required by the different agencies and organisations, e.g. national park management, for planning and monitoring purposes. High resolution satellite images (IKONOS) were used to assess the potentials for remote sensing-based habitat monitoring (http://www.gmes-geoland.info) | National Park |
| | Biodiversity on avalanche tracks (INTERGEG IIB) | Analysis of the nature conservation value of active avalanche tracks in providing information for the discussion concerning tension protection of natural processes on the one hand and protection against natural hazards on the other hand (Bohner et al. 2009). For this purpose, 16 plant stands on three different avalanche tracks in the Gesäuse National Park were examined | Regional |

| | | | |
|---|---|---|---|
| | Land abandonment and vegetation change (ALTER-Net, FP6) | The project aimed to compare the change in biodiversity between highly cultivated and abandoned alpine meadows and to give options for mitigation. Besides a vegetation survey, the abundance of zoological indicator groups as well as interviews with keepers and former alpine farmers were included. The study was carried out in the Gesäuse National Park | National Park |
| | Mostviertel Integrative Land-Use Model (proVISION) | Cultural landscapes are the result of a complex interplay of natural and socioeconomic factors. Industrial agriculture and the opening of the markets resulted in sustained changes over recent decades. The project uses an integrative model system to assess the effects of different support and price scenarios upon farm incomes, landscape and environment | Regional |
| | RIVAS – Regional Integrated Vulnerability Assessment for Austria | Based on a review of international model projects, an advanced operational and transferable conceptual, methodological and procedural framework for regional vulnerability assessment will be prepared. The framework will include the design of an improved science-based stakeholder dialogue, sets of vulnerability indicators for selected key sectors (forestry, agriculture, water, governance structures) and assessment methodologies. The participatory vulnerability assessment framework will be applied, demonstrated and its performance evaluated in the Mostviertel region, resulting in the improved re-design of the framework and implementation guidelines | Regional |
| Socio-ecological research | Public attitudes to nature and nature conservation (ALTER-Net, FP6) | The pilot study aims to identify public attitudes to nature and nature conservation in eight European regions. In Austria, the region of Eisenwurzen was surveyed. In particular, the recognition of changes in nature, the loss of endangered species and habitats, and nature conservation activities was surveyed and analysed | Regional |
| | Fuzzy Cognitive Mapping (ALTER-Net, FP6) | Fuzzy cognitive mapping as a semi-qualitative method deals with complex systems and questions. It can be used in a participatory process and allows the calculation of trends. The method was evaluated and tested in the region of Eisenwurzen | Regional |
| | Eisenwurzen LTSER – Reichraming case study (pro-VISION) | A transdisciplinary project (Gaube et al. 2009a, b; Singh et al. 2010) uses the databases made available through the LTSER platform to construct an integrated model of an Eisenwurzen municipality, Reichraming. This model can simulate socio-ecological trajectories resulting from decisions of actors, land-use change and socioeconomic and ecological material and energy flows dependent on changes in external social, political and economic conditions (e.g. CAP subsidies) | Municipality |

Further details on the projects can be found at http://www.umweltbundesamt.at/en_projekte_ew/

As the last stakeholder group, the *networks* can be identified that have a special focus on the research in the region (e.g. LTER Austria). These act as an interface between science and environmental policy.

#### 19.3.2.4 Platform Management

To cope with the diversity and heterogeneity of a LTSER region in terms of infrastructure, research, and stakeholders, the platform management provides a range of different services that allow for a new level of interaction and knowledge generation. The platform management acts as both a service provider and an interface between research institutions, regional institutions, infrastructure and funding organisations (see Fig. 19.3 Interlinkage between the elements of the Eisenwurzen LTSER Platform).

Formally, the Eisenwurzen LTSER Platform is a section of the registered association LTER Austria, which defines the legal framework. It can therefore act as a legal body. The management is undertaken by the Umweltbundesamt GmbH (Environment Agency Austria) and funded by the Ministry of Science and Research and the provincial governments of Upper Austria, Lower Austria and Styria. The funding was guaranteed by a 3-year contract starting in 2008.

Mirtl et al. (2009) list the categories of the core services for a LTSER platform which can be grouped into three main topics: (a) conceptual work and formalisation, (b) communication and networking (networking across interest groups, disciplines & stakeholders; communication space; representation (nationally, internationally); communication with the public; education, youth and researcher training; lobbying and fundraising), and (c) data management and integration (data integration and policy, IT tools).

As mentioned above, in the 2006 workshop, tasks for the platform management of the Eisenwurzen LTSER Platform were specified and developed in the succeeding years. These services are specified here for the case of Eisenwurzen in the following section. An overview of the way in which a typical research project can be supported by the platform management is given: from the establishment of the research consortium until the dissemination of the results.

Firstly, the *conceptual work and formalisation* of the Eisenwurzen LTSER Platform is an important task of the platform management. Based on the concepts of LTSER, this provides the concrete conceptual and legal framework for interaction.

This includes first of all the establishment of the label "Eisenwurzen LTSER Platform" within the research community and the local and regional stakeholders, together with the development of a "sales strategy" for this label. This requires ongoing definition and operationalisation of the services provided by the platform management to form an operational body. A steering committee was put in place to define the work plan. The steering committee consists of selected stakeholders from the provincial governments as funders, of regional development institutions as users, and the Umweltbundesamt GmbH (Environment Agency Austria) as the management.

Phasing out the usefulness of the LTSER concept in terms of the emerging problems in LTSER implementation is an important task. At the beginning of the implementation process of the Eisenwurzen LTSER Platform, a preliminary analysis of relevant stakeholders and main research topics was carried out. The goal was to identify the link between research and the potential users of the results produced by the Eisenwurzen LTSER Platform. This practical experience feeds back into the further development of the LTSER concept through international links and project involvement. Aspects of this work were discussed during the LTSER workshop organised by LTER Europe in 2009 in Sumava National Park (Czech Republic), leading to a LTSER best practice guide.

Secondly, the creation of a *communication space* was crucial due to the large number of people and institutions involved in the LTSER platform, some of them coming from very different backgrounds (regional context, scientists). A common language is not even spoken among scientists from similar disciplines using the same terms. Every discipline has its own epistemology, working culture, network and rules. In order to communicate effectively, different stakeholders have to get used to each other and learn to express themselves in a mutually understandable way, using common terminology and concepts, which first need to be elaborated. The greatest challenge involved rather simple and apparently self-explanatory terms, which – lacking proper discussion and clarification – sometimes hampered group progress for several months. Participants are invited to talk about their projects and ideas in workshops and small-scale meetings, which leads to lively discussions and learning processes among the participants with different backgrounds. This is also an aspect of the communication space provided by the platform. Especially where gaps in knowledge have been identified at a regional level, analyses of the research results and their translation into guidelines is an essential part of the communication process. Creating a proper framework and facilitating events at which regional stakeholders and researchers can exchange ideas and views on certain topics, but – very importantly – also expectations and hidden agendas is a challenging task for a platform manager.

A website for the Eisenwurzen LTSER Platform was set up to provide information and communication space.[4] Communication with the region has an additional feature since 2009: not only are questions from the region taken into account when defining research projects, but conversely the results of research projects are also discussed in the region at regular public events under the heading "Eisenwurzen-Academy".

Effective support provided by the platform for a project throughout the research process can include:

- Preparation phase of the studies

    - Provision of an overview of research questions and topics defined by the practitioners as being of relevance to the region
    - Contact to regional partners, as well as to international project partners
    - Organisation of meetings between stakeholders and researchers to enable common interests to be shared
    - Lobbying for research projects with funding bodies; informing researchers about possible funding opportunities
    - Checking availability of data

---

[4] See http://www.umweltbundesamt.at/umweltsituation/lter_allgemein/mfrp_eisenwurzen/

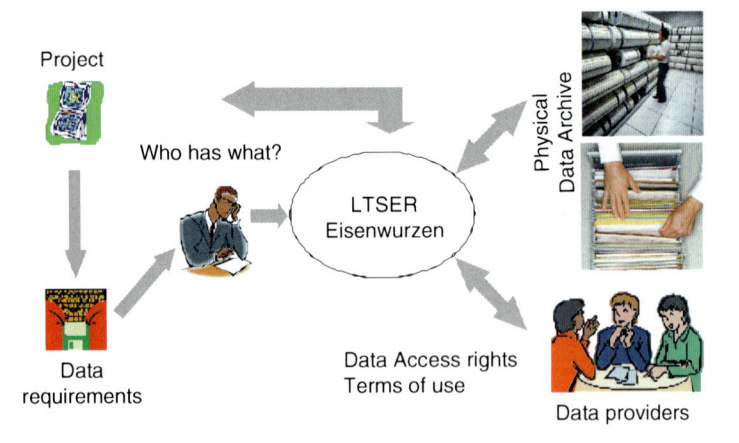

**Fig. 19.4** Data management concept for Eisenwurzen LTSER Platform

- Survey and analysis
  - Provision of basic data on the Eisenwurzen region
  - Provision of access to requisite data (through the data platform and in accordance with data sharing contracts)
- Reporting
  - Propagation of results within the community and beyond
  - Public relations support in terms of reporting back the results to the region in an event organised by platform management and regional partners, including the translation of reports and results for regional purposes

Thirdly, the *central facilitation of data flow* and access provided by the platform management represents an excellent means of support for complex LTSER projects. Data management includes providing an overview of available data sets as well as metadata on finished projects, literature and stakeholders, etc. The overview is made available through an online catalogue[5] accessible to all active stakeholders of the LTSER platform. Data management within a LTSER platform also involves setting up data sharing contracts with relevant institutions in order to secure clear criteria for data access and (re-)use (IPR) (see Fig. 19.4). A template of a data sharing contract based on common rules and according to the commitment in the Memorandum of Understanding (MoU) is bilaterally adjusted for each data holding partner, in accordance with institutional data policies and technical criteria.

The added value of the platform management is the facilitation of information flows (contacts, projects, etc.) between the different user groups. It provides support inter-alia in finding partners for project consortia as well as in integrating and utilising existing initiatives, data and knowledge for further research within and for the region.

---

[5] See https://secure.umweltbundesamt.at/eMORIS/

## 19.4 Lessons Learned

After 2 years of implementation of the Eisenwurzen LTSER Platform, an analysis of Strengths, Weaknesses, Options and Threats (SWOT) was carried out. It focused on the main topics of *generation of knowledge, communication*, and *data management*.

The *generation of knowledge* aims at providing information for regional and local decision makers at different levels. Difficulties arising from the difference between scientifically motivated questions and the questions of stakeholders and from the issue of time needed to generate the knowledge sought are crucial points to be solved.

The integration of local and regional stakeholders raising relevant (research) questions emerged as an important strength of the LTSER platform concept. This includes the implementation of procedures to identify specific questions from the stakeholders, the recruiting of researchers to address such questions and the development of a research concept. The implementing phase allowed trust to be built up between the research community and stakeholders over the years through platform management and personal contacts. This is also supported by the fact that many researchers have a personal background in the Eisenwurzen region and thus have a strong interest in the development of the region. Working for years in the region often results in a long-lasting relationship between researchers, stakeholders and the region itself. Furthermore, the "language" barriers (between disciplines and research/region) are becoming less dominant over the years through working together and generating a common understanding for problems and work routines.

The delineation of the region took account of borders in the European funding scheme for regional development, LEADER. The resulting spatial coherence makes it possible to use the results from research projects to formulate applied regional development funded by this programme. Compared to other LTSER platforms already implemented in Europe, the area of the Eisenwurzen LTSER Platform comprises "unspoiled nature" and industrialised settlements in similar environments, thus allowing the comparison of trends and developments.

Given the great demand for quick responses to the urgent questions arising from local and regional stakeholders, one of the weaknesses in the area of knowledge generation is the time lag that occurs between formulating a research question and obtaining the results. This reflects the contrast between long-term monitoring and research, which addresses trends over long time periods, versus regional development with its need for immediate answers. The development of a consistent strategy for "translating" research results into immediately usable information or even concrete advice for stakeholders is currently on the agenda. Within most of the research projects, however, no funding for this communication work is foreseen and researchers responsible for the outcomes of the projects generally do not discuss the results and options for implementation with the relevant stakeholders.

The spatial extent of the Eisenwurzen LTSER Platform is both a strength and a weakness. Its very large area of over 5,700 $km^2$ makes it impossible for most of the individual projects to cover the whole region. This raises the issue of up-scaling results, since many of the small projects only cover small areas or special topics.

Concepts for the up-scaling of results need to be developed to avoid vague and inaccurate generalised information for the region.

Finally, the lack of funding schemes for regional research in LTSER platforms is a significant weakness in the implementation of the Eisenwurzen LTSER Platform.

*Communication* is one of the major services provided by the management of the Eisenwurzen LTSER Platform. To keep everyone up-to-date, formats such as annual workshops, the "Eisenwurzen-Akademie", events to present project results, a newsletter and a website were employed. These activities support the formulation of transdisciplinary projects by providing the communication space to build contacts and define research questions together with stakeholders or other research teams. The research community represents various institutes, many different research topics, research methods and approaches. The Eisenwurzen LTSER Platform brings together the different disciplines and facilitates discussions. The resulting quality of the research is a great advantage and strength of the LTSER concept.

The fragmentation of the actor groups in the Eisenwurzen LTSER compared to other LTSER platforms is one of the largest challenges and weaknesses. As researchers are working in different institutions, large communication gaps have to be bridged in comparison to e.g., the two LTSER Platforms in southwest France (Alpes Oisans and Alpes Vercours), where researchers mostly derive from one (interdisciplinary) institute, CEMAGREF (Research Institute for sustainable land and water management, France), where a continuous exchange of thoughts takes place easily.

Furthermore, the large spatial extent of the Eisenwurzen LTSER Platform, spread across the administrative boundaries of three provincial states, is a major challenge, as different policies and emphasis occur in every part or the region, and different contact points must be negotiated with. Furthermore, on the practical level too, people are not used to working together as their work is usually limited to their federal areas. Despite the close personal relations between many of the researchers working on the Eisenwurzen region, many research institutions are situated outside the region, which makes communication difficult even with the help of modern technologies, since local stakeholders in particular generally prefer to communicate in person or directly by telephone.

Providing *data management* for the Eisenwurzen LTSER Platform is a huge task in terms of efforts and costs. The strength of the platform is the large pool of available research and monitoring data from the large number of monitoring facilities and historic and ongoing research. Progress has been made in setting up services of the platform management as a data platform (including data sharing contracts and metadata on sites and research projects). This also includes the integration of the Eisenwurzen sites in the LTER-Europe metadata catalogue as a central documentation point for the sites and infrastructure.

The fragmentation in terms of the storage of monitoring data across many institutions is one of the challenges and weaknesses of the platform. It is difficult to collect metadata (research projects, literature, etc.) and to obtain an overview of the existing data. A weakness also lies in the structure of existing data sets. Demographic and economic data mostly refer to administrative units on different scales, whereas data sets on the natural conditions mostly refer to analytical units as plots or smaller regions. This raises the issue of cross-scale integration of the data, which is an important scientific question to be solved for the Eisenwurzen LTSER Platform.

The outcome of the SWOT analysis forms a valuable basis for defining the actions required to improve the Eisenwurzen LTSER Platform. Key requirements are: enhancement of data accessibility, provision of data management tools, lobbying for funding of regional research projects, further development of the LTSER concept, and securing the long-term funding of the platform management and the existing research and monitoring infrastructures in the region.

## 19.5 Conclusion

In 2004, the Eisenwurzen LTSER Platform was launched and formalised through a Memorandum of Understanding signed by 35 parties. In comparison to other LTSER platforms, which are more or less dominated by one large research institute or national park, etc., the Eisenwurzen LTSER Platform consists of several research institutions both in and outside the region, national parks and other protected areas, regional development organisations and municipalities. This special setting implies a more comprehensive and integrated approach but also brings with it many challenges. Bringing the different players together and keeping them motivated has taken and continues to demand a lot of time and personal effort by those involved.

The Eisenwurzen LTSER Platform provides a framework for transdisciplinary socio-ecological research in the region. One of the main tasks of the platform management is to facilitate communication between researchers and local communities which are mainly represented by regional development organisations. Therefore, providing an adequate communication space and relevant information for the different actor groups is crucial for the platform's success. Information on national and international research funding programmes, conferences and other events, new region-relevant topics, calls for cooperation, research projects and their results etc. have to be collected, screened and disseminated via different communication channels. In principle, this information management is time- and resource-intensive and largely dependent on the willingness of the different players to provide information.

This is especially true for information on existing data. Data on the region are essential for further research as well as for decision makers aiming to create sustainable regional development. But the majority of existing data is neither stored centrally nor easily accessible. Therefore, a common data strategy and data management have proven to be essential. The general commitment to make one's own data available for research projects within the platform was signed by the partners as part of the Memorandum of Understanding. Effective accessibility of data sources can be crucial for making informed decisions about where to undertake a research project and as a consequence, for the flow of research funds into the region.

The time factor should never be neglected when establishing a LTSER platform – in several respects. On the one hand, it takes a great deal of time and effort to inform all key people and institutions of a new research approach. Although those involved are usually open-minded and prepared to devote some time to the new "LTSER idea", expectations are unfortunately often too high and unrealistic, given the lengths of the cycle that begins with identifying a research question, and moves through developing a project, getting the funding, implementing the project, analyses, publication, dissemination of results, translation of results in management and/or political measures and – finally – to the assessment of effects. This process may take 5–10 years or more, but the longer the particular players have to wait for the anticipated benefits, the higher is the risk that their interest or trust will be lost and the greater the effort needed to keep them motivated. In addition, complex questions on society-nature interactions cannot be answered by individual interdisciplinary projects but involve the support of disciplinary research and monitoring to provide results and data and clusters of concerted projects in the long term, ideally specific research framework programmes and funding mechanisms, especially at the national level.

During the start-up phase over the last 5 years, financial support from the ministry concerned as well as from the three federal provinces involved in the platform management and a few appropriate national and European research programmes proved essential for the successful establishment of the Eisenwurzen LTSER Platform. Without an independent and neutral platform management, this challenge would have been unlikely to have gained such broad acceptance within a relatively short time. The platform management provides services to the stakeholders of the LTSER platform rather than being active in research projects. Therefore its work cannot be directly measured in scientific terms. But good research projects are the focal aim of all these efforts. The more projects develop from the transdisciplinary platform context based on real knowledge needs, the more research results enabling sustainable regional development can be delivered and the more the key participants remain satisfied and motivated to engage actively in the LTSER process. One of the major concerns for the platform management is the existence of funding opportunities for project ideas and those projects already delineated. If these are not available, the platform is in danger of losing its associates. Nevertheless, a Europe-wide comparison shows that the Eisenwurzen LTSER Platform is among the most highly developed and structured LTSER platforms.

## References

Bohner, A., Habeler, H., Starlinger, F., & Suanjak, M. (2009). Artenreiche montane Rasengesellschaften auf Lawinenbahnen des Nationalparks Gesäuse (Österreich). *Tuexenia, 29*, 97–120.

Dirnböck, T., & Dullinger, S. (2004). Habitat distribution models, spatial autocorrelation, functional traits and dispersal capacity of alpine plant species. *Journal of Vegetation Science, 15*(1), 77–84.

Dirnböck, T., & Mirtl, M. (2009). Integrated monitoring of the effects of airborne nitrogen and sulfur in the Austrian Limestone Alps: Is species diversity a reliable indicator? *Mountain Research and Development, 29*, 153–160.

Dirnböck, T., Mirtl, M., Dullinger, S., Grabner, M., Hochrathner, P., Hülber, K., Karrer, G., Kleinbauer, I., Mayer, W., Peterseil, J., Pfefferkorn-Dellali, V., Reimoser, F., Reimoser, S., Türk, R., Willner, W., & Zechmeister, H. (2007). *Effects of nitrogen and sulphur deposition on forests and forest biodiversity. Austrian Integrated Monitoring Zöbelboden*. Vienna: Umweltbundesamt.

Diwold, K., Dullinger, S., & Dirnböck, T. (2010). Effect of nitrogen availability on forest understorey cover and its consequences for tree regeneration in the Austrian limestone Alps. *Plant Ecology, 209*, 11–22.

Dullinger, S., Dirnböck, T., & Grabherr, G. (2004). Modelling climate change-driven treeline shifts: relative effects of temperature increase, dispersal and invasibility. *Journal of Ecology, 92*(2), 241–252.

Dutzler, A. (1998). Alpinismus und Fremdenverkehr. In J. Stieber (Ed.), *Land der Hämmer – Heimat Eisenwurzen* (pp. 130–135). Salzburg: Residenz-Verlag.

Gaube, V., Kaiser, C., Wildenberg, M., Adensam, H., Fleissner, P., Kobler, J., Lutz, J., Schaumberger, A., Schaumberger, J., Smetschka, B., Wolf, A., Richter, A., & Haberl, H. (2009a). Combining agent-based and stock-flow modelling approaches in a participative analysis of the integrated land system in Reichraming, Austria. *Landscape Ecology, 24*(9), 1149–1165.

Gaube, V., Reisinger, H., Adensam, H., Aigner, B., Colard, A., Haberl, H., Lutz, J., Maier, R., Punz, W., & Smetschka, B. (2009b). Agentenbasierte Modellierung von Szenarien für Landwirtschaft und Landnutzung im Jahr 2020, Traisental, Niederösterreich. *Verhandlungen der Zoologisch-Botanischen Gesellschaft in Österreich, 146*, 79–101.

Grabherr, G., Gottfried, M., & Pauli, H. (1994). Climate effects on mountain plants. *Nature, 369*, 448.

Haberl, H., Winiwarter, V., Andersson, K., Ayres, R. U., Boone, C., Castillo, A., Cunfer, G., Fischer-Kowalski, M., Freudenburg, W. R., Furman, E., Kaufmann, R., Krausmann, F., Langthaler, E., Lotze-Campen, H., Mirtl, M., Redman, C. L., Reenberg, A., Wardell, A., Warr, B., & Zechmeister, H. (2006). From LTER to LTSER: Conceptualizing the socioeconomic dimension of long-term socioecological research. *Ecology and Society, 11*(2), 13. Retrieved from http://www.ecologyandsociety.org/vol11/iss2/art13/

Hasenauer, H., Putzhuber, F., Mirtl, M., & Wenzel, W. (2007). Multifunctional land use: The Eisenwurzen region of the Austrian Alps. In Ü. Mander, H. Wiggering, & K. Helming (Eds.), *Multifunctional land use: Meeting future demands for landscape goods and services* (pp. 341–354). Heidelberg: Springer.

Hasitschka, J. (2007). *Die Geschichte der Almen und Halten im Gesäusetal*. Bericht im Auftrag der Nationalpark Gesäuse GmbH, p. 27.

Heintel, M., & Weixelbaumer, N. (1998). Region Eisenwurzen: Ein geographisch-kulturräumlicher Begriff. In J. Stieber (Ed.), *Land der Hämmer – Heimat Eisenwurzen* (pp. 16–23). Salzburg: Residenz-Verlag.

Heintel, M., & Weixelbaumer, N. (1999). *Oberösterreichische Eisenwurzen/Eisenstraße III: Langzeitstudie (1995–1999) zur räumlichen Abgrenzung, Akzeptanz und regionale Identität der Region Eisenwurzen bzw. der Eisnestraßenidee: Endergebnisse*. Vienna: Institut für Geographie und Regionalentwicklung der Universität Wien.

Jost, G., Dirnböck, T., Grabner, M.-T., & Mirtl, M. (2011). Nitrogen leaching of two forest ecosystems in a Karst watershed. *Water Air and Soil Pollution, 218*, 633–649.

Kilian, W., Müller, F., & Starlinger, F. (1994). *Die forstlichen Wuchsgebiete Österreichs. Eine Naturraumgliederung nach waldökologischen Gesichtspunkten. FBVA Berichte 82*. Vienna: Forstliche Bundesversuchsanstalt.

Kobler, J., Fitz, J. F., Dirnböck, T., & Mirtl, M. (2010). Soil type affects migration pattern of airborne Pb and Cd under a spruce-beech forest of the UN-ECE Integrated Monitoring site Zöbelboden, Austria. *Environmental Pollution, 158*, 849–854.

Kreuzer, B. (1998). Eine Region wird mobil. In J. Stieber (Ed.), *Land der Hämmer – Heimat Eisenwurzen Weyer* (pp. 114–154). Steyr: Ennsthaler Verlag

Kropf, R. (1997). Die Krise der Kleineisenindustrie in der oberösterreichischen Eisenwurzen im 19. Jahrhundert. In *Heimat Eisenwurzen. Beiträge zum Eisenstraßensymposium Weyer* (pp. 114–154). Steyr: Ennstaler Verlag.

LTER Management Austria. (2009). *Memorandum of understanding. Multifunctional research platform Eisenwurzen.* http://www.umweltbundesamt.at/fileadmin/site/umweltthemen/oekosystem/MFRP_Eisenwurzen/MFRP_EW_MoU.pdf. Accessed January 25, 2011.

Mejzlik, H. (1935). *Die nördlichen Eisenwurzen in Österreich*. Berlin, Vienna: Carl Heymanns Verlag, Österreichischer Wirtschaftsverlag.

Metzger, M. J., Bunce, R. G. H., van Eupen, M., & Mirtl, M. (2010). An assessment of long term ecological research activities across European socio-ecological gradients. *Journal of Environmental Management, 91*(6), 1357–1365.

Mirtl, M. (2004). *LTER in Austria: Implementing multifunctional research platforms (MFRPs) for LTER*. ILTERN coordinating committee meeting 7–9 July, Manaus, Brazil. Retrieved from http://www.ilternet.edu/about/ilter-annual-meetings

Mirtl, M. (2010). Introducing the next generation of ecosystem research in Europe: LTER-Europe's multi-functional and multi-scale approach. In F. Müller, C. Baessler, H. Schubert, & S. Klotz (Eds.), *Long-term ecological research, between theory and application* (pp. 75–94). Dordrecht, Heidelberg, London, New York: Springer.

Mirtl, M., & Krauze, K. (2007). Developing a new strategy for environmental research, monitoring and management: The European Long-Term Ecological Research Network's (LTER-Europe) role and perspectives. In T. J. Chmielewski (Ed.), *Nature conservation management – From idea to practical results* (pp. 36–52). Lublin-Lodz-Hesinki-Aarhus: ALTER-Net.

Mirtl, M., Boamrane, M., Braat, L., Furman, E., Krauze, K., Frenzel, M., Gaube, V., Groner, E., Hester, A., Klotz, S., Los, W., Mautz, I., Peterseil, J., Richter, A., Schentz, H., Schleidt, K., Schmid, M., Sier, A., Stadler, J., Uhel, R., Wildenberg, M., & Zacharias, S. (2009). *LTER-Europe design and implementation report – Enabling "next generation ecological science": Report on the design and implementation phase of LTER-Europe under ALTER-Net & management plan 2009/2010*. Vienna: Umweltbundesamt (Environment Agency Austria).

Pauli, H., Gottfried, M., Dirnböck, T., Dullinger, S., & Grabherr, G. (2003). Assessing the long-term dynamics of endemic plants at summit habitats. In L. Nagy, G. Grabherr, C. Körner, & D. B. A. Thompson (Eds.), *Alpine biodiversity in Europe – A Europe-wide assessment of biological richness and change* (pp. 195–207). Berlin: Springer.

Putzhuber, F., & Hasenauer, H. (2010). Deriving sustainability measures using statistical data: A case study from the Eisenwurzen, Austria. *Ecological Indicators, 10*, 32–38.

Redman, C. L., Grove, J. M., & Kuby, L. H. (2004). Integrating social science into long-term ecological research (LTER) network: social dimensions or ecological change and ecological dimensions of social change. *Ecosystems, 7*, 161–171.

Roth, P. W. (1997). Die Steirische Eisenstraße: Von der Industrie- zur Museumslandschaft. In: *Heimat Eisenwurzen. Beiträge zum Eisenstraßensymposium Weyer* (pp. 169–173). Steyr: Ennstaler Verlag.

Roth, P. W. (1998). Die Eisenwurzen: Eine Region in drei Ländern. In J. Stieber (Ed.), *Land der Hämmer – Heimat Eisenwurzen* (pp. 36–39). Salzburg: Residenz-Verlag.

Sandgruber, R. (1997a). Eine Einleitung. Die Wurzel des Berges. In *Heimat Eisenwurzen. Beiträge zum Eisenstraßensymposium Weyer* (pp. 9–24). Steyr: Ennsthaler Verlag.

Sandgruber, R. (1997b). *Die Eisenwurzen. Landschaft – Kultur – Industrie*. Vienna: Pichler-Verlag

Sandgruber, R. (1998). Netzwerk Eisenwurzen. In J. Stieber (Ed.), *Land der Hämmer – Heimat Eisenwurzen* (pp. 94–107). Salzburg: Residenz-Verlag.

Singh, S. J., Haberl, H., Gaube, V., Grünbühel, C. M., Lisievici, P., Lutz, J., Matthews, R., Mirtl, M., Vadineanu, A., & Wildenberg, M. (2010). Conceptualising long-term socio-ecological research (LTSER): Integrating the social dimension. In F. Müller, C. Baessler, H. Schubert, & S. Klotz (Eds.), *Long-term ecological research, between theory and application* (pp. 377–398). Dordrecht, Heidelberg, London, New York: Springer.

Statistik Austria. (2010). *Statistisches Jahrbuch Österreichs 2011* (612 pp.). Vienna: Statistik Austria. ISBN: 978-3-902703-62-0.

Statistik Austria. (2011). *Ein Blick auf die Gemeinde*. http://www.statistik.at/blickgem/index.jsp. Accessed January 24, 2011.

URL: http://www.plattform-eisenwurzen.at

# Chapter 20
# Fostering Research into Coupled Long-Term Dynamics of Climate, Land Use, Ecosystems and Ecosystem Services in the Central French Alps

Sandra Lavorel, Thomas Spiegelberger, Isabelle Mauz, Sylvain Bigot, Céline Granjou, Laurent Dobremez, Baptiste Nettier, Wilfried Thuiller, Jean-Jacques Brun, and Philippe Cozic

**Abstract** The Central French Alps long-term socio-ecological research Platform (Central French Alps LTSER) focuses on the coupled dynamics of alpine ecosystems, their uses and climate. The creation of the Platform has provided a unique opportunity to initiate and strengthen collaborative transdisciplinary research involving a range of natural and social scientists (ecologists, agronomists, climatologists, sociologists) and key regional stakeholders from the agriculture, tourism and nature conservation sectors. The main research questions were built on existing long-term research projects at two sites. They include climate change effects on biodiversity and ecosystem functioning, and coupled dynamics of grassland management, biodiversity and ecosystem functioning through ecosystem services, using not only observations of

---

S. Lavorel (✉) • W. Thuiller, Ph.D.
Laboratoire d'Ecologie Alpine (LECA, UMR 5553), Centre for National
Scientific Research (CNRS), Grenoble, France
e-mail: sandra.lavorel@ujf-grenoble.fr; wilfried.thuiller@ujf-grenoble.fr

T. Spiegelberger, Ph.D. • J.-J. Brun, Ph.D. • P. Cozic
Research Unit Mountain Ecosystems, National Research Institute for Environmental
and Agricultural Sciences and Technologies (Irstea / Cemagref), Grenoble, France
e-mail: thomas.spiegelberger@irstea.fr; jean-jacques.brun@irstea.fr;
philippe.cozic@hotmail.com

I. Mauz, Ph.D. • C. Granjou, Ph.D. • L. Dobremez • B. Nettier
Research Unit Development of Mountain Territories, National Research Institute
for Environmental and Agricultural Sciences and Technologies
(Irstea / Cemagref), Grenoble, France
e-mail: isabelle.mauz@irstea.fr; celine.granjou@irstea.fr;
laurent.dobremez@irstea.fr; baptiste.nettier@irstea.fr

S. Bigot
Laboratoire Transferts, Hydrologie et Environnement (LTHE, UMR 5564),
Joseph Fourier University, Grenoble, France
e-mail: sylvain.bigot@ujf-grenoble.fr

natural and human systems, but also manipulative experiments of climate, management and plant and soil diversity to feed models. The LTSER Platform has fostered three important types of advances: (1) Long-term data consolidation and sharing. (2) Invigorating interdisciplinary projects (e.g. coupled transformations of economic functioning of farming systems and mountain summer pastures dynamics; mutations of alpine tourism in the face of climate change). (3) New transdisciplinary projects, including climate change adaptation of mountain territories, integrated carbon cycle modelling in response to historical land-use change and climate; a sociological study of the process of construction of the LTSER Platform.

**Keywords** Mountain system • Land use dynamics • Climate change • Biodiversity • Ecosystem services

## 20.1 Introduction

In the last few decades, social and ecological scientists have worked and developed their theories in more or less complete isolation from each other. While the former have often thought "as if nature did not matter" (Murphy 1995), the latter were particularly interested in the biological and physical dimension of nature. Moreover, human influence was and is still today often interpreted by ecologists as a "perturbation" (White and Jentsch 2001) and is more rarely seen as part of the complex interactions in nature (Liu et al. 2007), while many sociologists regard nature as a purely social construct. Today's environmental challenges, in particular the effects of global change and the predicted considerable human impact via land-use changes on biodiversity (Sala et al. 2000) require in particular that sociologists, economists, climatologists and ecologists develop common approaches in order to collaborate on one of the most pressing problems of the twenty-first century. Recognition has grown in all fields over recent years that neither human nor biophysical systems can be studied autonomously. This recognition is the impetus for studying the dynamics of complex and interdependent social and ecological systems as it is undertaken in LTSER research (Haberl et al. 2006).

In France, several research institutes and universities decided to participate in the creation of an LTSER Platform in the Alps in order to address the complex question of interactions between humans and the environment, in the context of mountain systems. Here, we describe and reflect on the process of creation and implementation of a newly established long-term socio-ecological research Platform in the French Alps seeking to promote the integration of interdisciplinarity in research that initially focused on ecology. The project emerged simultaneously with a Europe-wide initiative to create and implement LTER sites and LTSER Platforms (Mirtl 2010), but the construction process took place in parallel, rather than being driven by the European initiative.

We first describe the process of creating the LTSER Platform, including how the genesis of the LTSER Platform itself became an object of research for sociologists, and the key collaborative tools that were built to facilitate data exchange. We then illustrate with recent research projects how the creation of a long-term socio-ecological research Platform provides an important framework to formalise transdisciplinary collaboration.

## 20.2 Constructing the Central French Alps LTSER Platform

### 20.2.1 Identification of Potential Partners

When the idea of creating a LTSER Platform first emerged, both lead partners, CEMAGREF (Institute of agricultural and environmental engineering research; a government-funded research organisation under the administrative authority of France's Ministries of Agriculture and Research) and CNRS (a government-funded research organization under the administrative authority of France's Ministry of Research), were already involved within their respective organisations in establishing their own priority research area in the French Alps: CEMAGREF within the framework of Alter-Net, while CNRS was integrated in the French scheme of the "Zones Ateliers" (literal translation: "workshop areas"). In each of these organizations, one laboratory was particularly involved: the Alpine Ecology Laboratory (LECA) at CNRS, and the Mountain Ecosystems Research Unit (RU EM) at CEMAGREF. As they were already collaborating as part of ongoing projects and regional infrastructure development, the idea of merging the two processes into a larger one was rapidly adopted. However, as both laboratories within each institute were largely dominated by ecologists, significant effort was made to enlist researchers from other disciplines in the project (Table 20.1). In a second step, non-scientific actors such as local representatives from forestry and agriculture, and important stakeholders in the field of nature conservation including national and regional parks were contacted (Table 20.1).

### 20.2.2 Choice of the Main Research Topic and the Study Region

#### 20.2.2.1 The Study Region

The Central French Alps have a number of key assets for the study of the dynamics of land systems in the context of global change. It is well known that mountain regions in general, and the Alps in particular, have developed under strong natural constraints. The rugged topography as well as the north–south direction of the ranges results in steep environmental gradients, as well as in a high spatial heterogeneity in climatic, topographic and soil conditions. Such constraints promote the existence of a large diversity of natural habitats, making the Alps a biodiversity hot

**Table 20.1** Partners and their disciplines involved in the Central French Alps LTSER Platform Aps

| Scientific partners | Disciplines |
|---|---|
| CNRS Alpine Ecology Laboratory* | Ecology |
| Cemagref Research Unit (RU) Mountain Ecosystems* | Ecology, agronomy, forestry |
| CNRS Transfers, Hydrology and Environment Laboratory | Climatology, hydrology |
| CNRS Alpine Research Station Joseph Fourier | Ecology |
| Cemagref RU Development of Mountain Areas | Agronomy, sociology, economics |
| Cemagref RU Public policies, politic actions and land management | Geography |
| Local and socioeconomic partners | |
| Ecrins National Park | Nature conservation, education |
| Vercors Regional Nature Park | Nature conservation, education |
| Nature Reserve Hauts Plateaux du Vercors | Nature conservation |
| National Alpine Botanical Conservatory | Nature conservation, botany |
| Several professional bodies in the field of agriculture | Agriculture |
| Office for Agriculture of Hautes-Alpes | Agriculture |
| National Forest Bureau | Forestry |

Lead partners are indicated with an asterisk

---

**Box 20.1** Structure and Governance of the Central French Alps LTSER Platform

Zone Atelier Alpes, the Central French Alps LTSER Platform, was established jointly by CNRS and CEMAGREF in 2009, following a 2-year construction process based on exchanges between both these two institutions and the leading laboratories, and a participative construction process among academic partners, and between these partners and key regional stakeholders (see Table 20.1 for the list of academic and non-academic partners). French Zones Ateliers are an administrative entity of CNRS, which are certified for a period of 4 years, with an annual review and a full re-examination, leading to continuation, restructure or even discontinuation, every 4 years. As such they receive an annual budget (ca. 50 k€/year in the case of Zone Atelier Alpes, based on pooled contributions from CNRS and CEMAGREF), allocated on a year-to-year basis depending on a requested budget and availability of funds, and which they then manage in a subsidiary manner.

In the case of the Central French Alps LTSER Platform, the scientific strategy and the budget are managed by a steering committee made of the two coordinators (currently, one from CNRS and one from CEMAGREF) and the three scientists in charge of each of the scientific axes (see main text – Sect. 20.2.2.2, "Scientific Questions"). This group communicates on a regular basis by email for the management of projects, information dissemination and other issues such as data management, and meets on a regular basis (about

(continued)

> **Box 20.1** (continued)
>
> every trimester) to monitor the implementation of scientific strategy, organise internal meetings and meetings with outside scientific or societal parties, and coordinate the budget and its allocation across projects and other actions. The management structure was designed to be light and flexible so as to adapt to the day-to-day dynamics of research. In addition, a formal meeting with representatives of each of the Platform partners (Table 20.1), including in particular non-academic partners, is organised yearly. Annual science meetings, when possible associated with the field trips, are organised to enable sharing of project results. In addition, dissemination meetings for regional stakeholders from the different sectors relevant to the project (agriculture, forestry, nature conservation, tourism) have taken place on a 2-yearly basis.
>
> The Platform and the projects that are run therein involves a total of 11 full-time equivalent (FTE) scientists, 6 FTE technical staff and 17 FTE contract staff (mostly Ph.D.s and postdoctoral students but also some short-term technical staff). It draws upon an annual operating budget for research projects (salaries of permanent staff not included) which amounts on average to 1.5 million EUR per year, and comprises funds obtained from the local (e.g. Conseil Général de l'Isère, Région Rhône-Alpes), national and European levels.

spot (Körner 1999). This is particularly true for the Central French Alps, given their geographic position at the boundary between the external Alps with rather more Atlantic conditions, continental influences in the internal Alps, and those areas with Mediterranean influences.

At the same time, these natural constraints have been reflected in human development, and especially in agricultural development, in, for example, obstacles to mechanisation and limited competitivity in trade due to high labour requirements and transportation costs. Over history, climatic and other natural constraints and their spatial heterogeneity have become integrated into cultural diversity and activities such as agriculture, forestry, tourism, and hydroelectricity (Tappeiner and Bayfield 2004). Therefore, we hypothesise that ecological and human co-adaptation has taken place. Among other features of the coupled human-environment system, important aspects that have shaped today's landscapes include: (i) climate variability over a large variety of scales (centuries to interannual); (ii) the development of land-use systems that are tightly linked with the provision of a variety of ecosystem services (Girel et al. 2010); and (iii) a strong multi-functionality and regional diversity in land-use systems, but also competition among alternative uses. Such complex interactions among climate, ecosystems and land-use systems raise the question of whether the Central French Alps land system is expected to be resilient or vulnerable to global environmental change, and which adaptation strategies may be developed to respond to environmental as well as to socio-political changes.

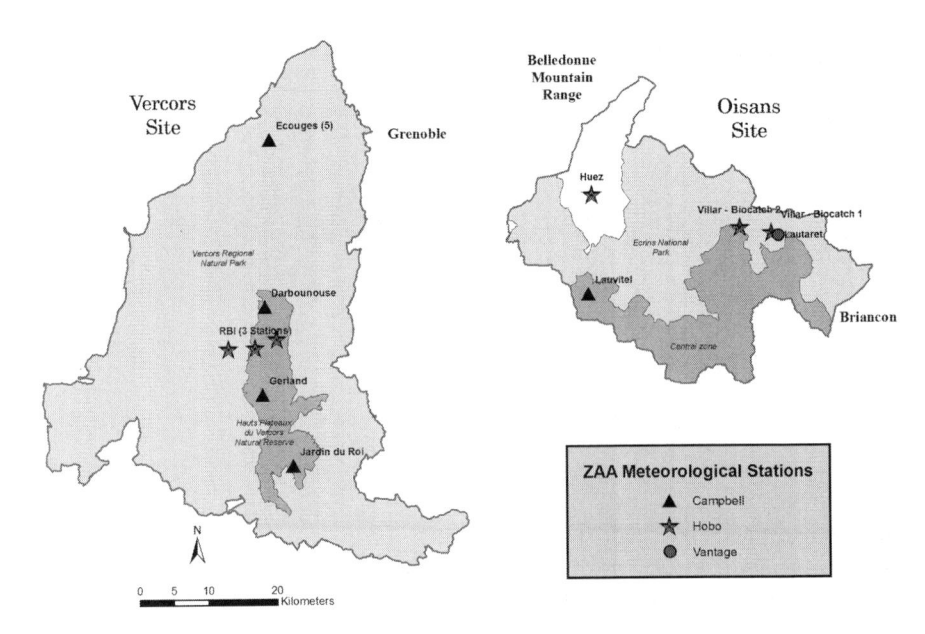

**Fig. 20.1** Location map for the Central French Alps LTSER Platform and meteorological stations. The Platform includes areas with strict protection status (*dark grey*): the Vercors High Plateaux Natural Reserve and some of the core area of the Ecrins National Park, as well as inhabited areas managed by agriculture and forestry (*light grey*): the Vercors Natural Regional Park and part of the peripheral area of the Ecrins National Park. Meteorological monitoring stations set up by the LTSER Platform are located using different symbols depending on the equipment in place

The specific study areas for the Central French Alps LTSER Platform were selected among sites already studied by at least one of the institutional partners, and thus where significant data sets were available and much knowledge had already been acquired regarding ecological (for the main part), climatic, economic, and social processes. We cross-checked the most promising regions with the recommendations given by Ohl et al. (2007). Finally, two study regions with three main monitoring sites were selected, and form the Central French Alps LTSER Platform (Fig. 20.1). In economic terms as well, the Platform contains a wide diversity. Intensification and extensification processes are ongoing within the agricultural sector, and, with the ski resort "Alpe d'Huez" and the Ecrins National Park, the tourist sector is highly developed. Moreover, site-specific policies addressing biodiversity issues are implemented and governmental investments in biodiversity conservation are present as a national park (Ecrins) and a regional park (Vercors) which are included in the Platform. Finally, the Platform encompasses two NUTS 2 levels (Rhône-Alpes and Provence-Alpes-Côte d'Azur) and three NUTS 3 levels (departments of Isère, Drôme and Hautes-Alpes) which may help us to understand how European or national decisions may influence ecological and socioeconomic factors at a regional level.

#### 20.2.2.2 Scientific Questions

It seems important to emphasise that the choice of the core research theme and of the study areas was a dynamic process, in which several other partners were contacted and the research topic and the study area redrawn several times. However, we only describe the process and its outcome, and deliberately leave out (in spite of the substantial efforts entailed) any description of the processes required to obtain funding and resources.

Right from the beginning, the intention of all partners was to create not an observatory of current environmental and social development, but an area where most of the research is and will be carried out. We therefore deliberately favoured a bottom-up approach, where the core research topic was developed based on the interests of the majority of the participants and on the characteristics of the study area, and not, as proposed by others, imposed following a more academic, top-down, approach (Redman et al. 2004). All partners agreed that research themes may not be permanent, but will evolve as more data and knowledge are accumulated concerning the study system. The core theme finally selected is "research on coupled dynamics of alpine ecosystems, their uses and the climate". It has already consolidated over the 3 years since the Platform was established, with transdisciplinary projects converging towards the analysis of processes of adaptation of alpine territories to global change and societal transformations (Fig. 20.2). Three main research axes were then jointly selected to implement this core theme, and consider (i) the assessment of climate changes and its impacts on biodiversity and uses, (ii) the analysis of the coupled dynamics of grassland management, biodiversity, and ecosystem services, and (iii) the development of new interdisciplinary studies in the study area. This joint approach is implemented with a simple governance structure where each research axis is coordinated by a climatologist, an ecologist, and a sociologist respectively.

### 20.2.3 Data Pooling and Standardisation as a Key Element of Success

#### 20.2.3.1 Climate Monitoring and Climate Data

The quantification of spatial and temporal climate variation is essential to the characterisation of the studied environments, both in terms of quantifying climate trends and of providing essential variables for the analyses of biodiversity, biophysical processes and resources for human use. Since 2006, the Central French Alps LTSER Platform has developed a network of stations to monitor climate at the three intensive research sites and to characterise key gradients of variation (Fig. 20.1).

Such data will make it possible to produce fine-scale decadal analyses of climate trends, as was undertaken at the regional scale for the Vercors central plateau (Fig. 20.3). There the analysis of long-term data series from Météo-France stations

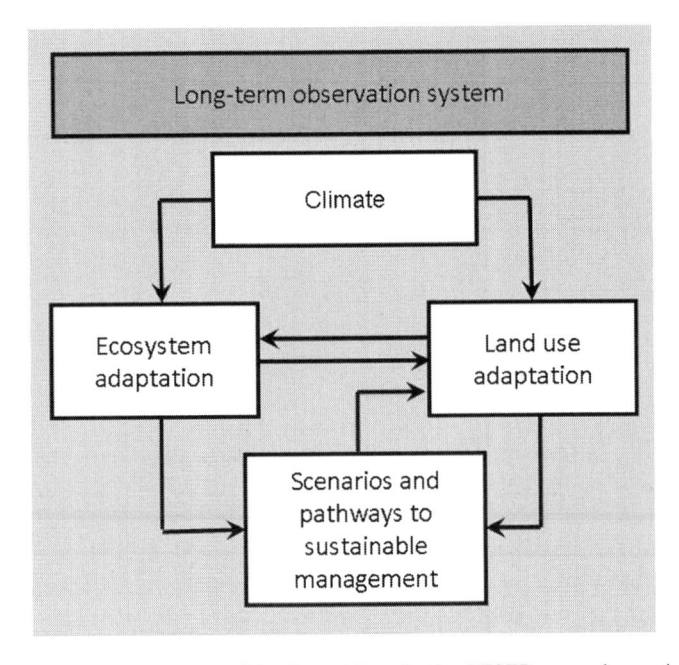

**Fig. 20.2** Conceptual presentation of the Central French Alps LTSER research questions

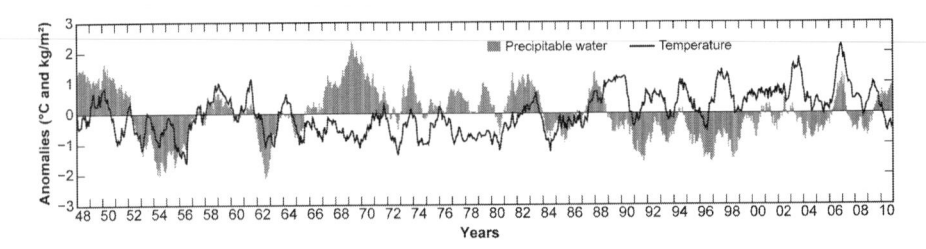

**Fig. 20.3** Recent climatic trends over the Vercors High Plateaux. Interannual variability of air temperature (at 850 hPa level; in °C) and water precipitation (in kg/m$^2$) calculated for the Vercors site on the 1948–2010 period (anomalies were calculated from NCAR-NCEP reanalysis; time-series anomalies are smoothed using a moving average over 12 months)

within the study region demonstrated a clear warming trend of +1 °C over the last 30 years, while there was no clear trend for total annual precipitation in spite of a shift to more Mediterranean regimes (Bigot and Rome 2010). The new network will complement such analyses with stations at high altitude, making it possible to incorporate the effects of complex terrain into regional-scale climate analyses, as well as providing detailed information to be linked with other social and ecological variables monitored at the study sites.

### 20.2.3.2 Plant Species Distributions and Plant Functional Traits

The understanding of the spatial distribution of plant species and their functional attributes in relation to environmental drivers requires a comprehensive distribution and plant functional trait database. The Alpine National Botanical Conservatory (CBNA) and the Alpine Ecology Laboratory (LECA) spent a considerable amount of time crosschecking and validating existing vegetation surveys from the original CBNA database, in order to keep only recent (>1980) and sufficiently precise (resolution >200 m) survey plots. LECA maintains this subset of the CBNA database with a comprehensive taxonomy (and associated synonymy), knowledge of the local habitat of each survey plot and percentage of bare-soil. In the mean time, LECA has coordinated the development of the ANDROCASE plant functional trait database that aims to collect existing plant functional traits from various sources (open database, literature, own measurements) for the regional flora of the entire French Alps (3,400 taxa). CBNA but also the Ecrins National Park have participated in developing and contributing to the database. ANDROSACE now contains about one million entries and some traits (e.g. maximum vegetation height, reproduction type, life form) or ecological indicators (Ellenberg values) have been collected for all known species in the region. The area covered by the Central French Alps LTSER Platform encompasses 1,457 plant species, present in a total of 4,717 georeferenced botanical surveys. For these species, trait values are known exhaustively for common traits such as plant height, phenological types, as well as their ecological preferences (Ellenberg or Landoldt indices). Traits requiring more intensive data collection are for instance SLA which is known for 687 species, LNC for 207 species, or seed mass for 999 species, using both data collected from Platform sites and sourced from external data bases.

### 20.2.3.3 Coupling Botanical and Agronomic Data on Alpine Grasslands to Investigate Land-Use and Climate Change Responses

Observed vegetation changes between two sampling periods may reflect responses to changes in land-use, but also to climate. The Research Unit Mountain Ecosystems (EMGR) started in the 1970s a systematic inventory of summer grazing pastures in the Vercors in order to characterise the vegetation and its trajectories. Relevés were repeated more or less regularly and have recently been compiled into the database FlorEM containing more than 5,000 geolocalized line transects and plot data with records of the abundance of each species. In addition, at each of the two sites of the Central French Alps LTSER 12 paired permanent plots, one grazed and one where grazing has been excluded for more than 20 years (Haut-Plateaux du Vercors) or 30 years (Alpe d'Huez), have been monitored over the years. These plots may serve as a baseline under conditions where the vegetation follows its natural successional trajectory only influenced by current climatic conditions and past grazing patterns.

However, as little expertise for assessing agro-economic aspects was available at RU EMGR, changes in land use were not always recorded. Recently, agronomists from the RU Development of Mountain Areas have repeated field investigations already conducted several years ago in order to assess changes in the management and use of summer pastures and of lowland pastures used by the same herds. These data are also stored in a spatially explicit database.

Further research will link these two currently independent databases in order to provide a ready overview of changes in the vegetation and, in areas where socioeconomic data is available, information about changes in land use of the summer pastures under study. In a further step, these two databases should be linked to the already existing climatic database in order to distinguish land use from climate signals in the observed vegetation dynamics.

#### 20.2.3.4 Building a Meta-Data Base

In order to facilitate the comparison of data sets on different components of the human-environmental system, one of the first common actions in the LTSER Platform was the creation of a meta-database (MDB). The description of existing data within the meta-database provides a first approach to supporting and promoting collective use of the data collected by different partners. In particular, the quality of old data we wished to analyse in a new context (global change) was quite heterogeneous, especially because methods have evolved and original data sets were collected in contexts that have changed. In the future, data collection will be planned and coordinated as much as possible by standardising measurements, collectively establishing protocols, and sharing data management tools. The MDB also aims to rationalise the researchers' activities through avoiding the duplication of work or the search of data from external sources, without knowing it has already been conducted by somebody else. For example, current efforts focus on inventorying and geo-localising botanical and agronomic data within a common GIS Platform, and on consolidating a database of remote sensing and aerial photographic images.

### *20.2.4 Integrated Approaches from Field Observations to Experiments and Modelling*

One of the strengths of the French Central Alps LTSER Platform lies in the integration of the full range of approaches from field observations and monitoring to experiments and modelling. The Platform includes monitoring of climate (see above), land use, vegetation and a range of ecosystem processes such as biomass production, soil water transfers and fertility over a set of permanent grasslands plots located across the three sites. This data baseline is complemented by spot surveys, e.g. vegetation and soil maps, over entire sites, which we aim to repeat on a 30-year basis. More intensive experimental field manipulations address the effects of specific drivers such as grassland management (through the use of long-term grazing exclosures – e.g. two experiments running for over 30 years), mesotopography (comparisons of ridges and snowbeds), or climate change (climate warming and drought experiments) on vegetation composition and biogeochemistry (C and N cycles), as well as on agronomic characteristics such as digestibility or phenology of grass flowering. These manipulations are complemented by garden experiments, which aim to identify detailed ecophysiological, demographic or biochemical mechanisms and are made possible by the Lautaret Alpine Botanical Garden located at 2,100 m, which allows experiments under realistic mountain conditions. Finally long-term data analyses and process studies feed into modelling, including macroecological modelling, ecophysiological modelling and spatial modelling of ecosystem services (Baptist and Choler 2008; Albert et al. 2010; Lavorel et al. 2011).

## 20.3 The LTSER Platform Itself Becomes a Research Object

### 20.3.1 Objectives and Research Questions

The establishment of the LTSER Platform has been accompanied by a reflexive evaluation of the process whereby the LTSER Platform itself is seen as a research object. The LTSER Platform constitutes a very good case in which to study scientific cooperation in practice, from a science and technology studies (STS) perspective, looking at "science in the making" rather than in "ready-made science" (Latour 1987; Knorr-Cetina 1999). Indeed, several types of scientific cooperation are present: cooperation between research groups tackling closely related research questions in ecology but belonging to organisations focusing on 'fundamental' (CNRS) or 'applied' (CEMAGREF) research; cooperation between various disciplines, within natural sciences but also between natural and social sciences; cooperation between scientists and nature managers; cooperation among scientists themselves, with different research projects and individual aims. Beyond the claim that cooperation is good for science, we aim at understanding how, to what extent, and under which conditions it takes place, and at understanding its effects. We addressed specifically the question of whether the LTSER Platform brings something new, in the context of an overall objective of our current work of analysing how the climate and biodiversity 'crises' transform ecological research.

### 20.3.2 Methods

We used a survey relying mainly on semi-structured interviews with the members of the LTSER Platform, during which the following topics were addressed: history and assigned objectives of the structure, biographical information on the interviewee, creation of the Platform and personal involvement, ongoing activities with the Platform, expectations and difficulties encountered. A targeted survey focused on the relationships between the CBNA and CNRS scientists around the exchange of botanical data and the construction of the ANDROSACE data base (see above). We also drew on various other materials such as participating as observers during meetings, analysis of communication and of documents in order to analyse the perspectives of different partners and individual participants, the topics of debate, and meeting and interviewing participants' about their expectations. For more details on the interviews, see Granjou and Mauz (2011) and Mauz and Granjou (2011). Here we briefly summarise outcomes in terms of reported perceptions on how the LTSER Platform started, why it started and what it changes in the type of work and working relationships among scientific partners, and with partner nature managers.

### 20.3.3 Key Lessons

How it started: The interviews revealed that a meeting was organized by the Ecrins national park as early as 2004, hence long before the project of creating a LTSER Platform was considered. This meeting gathered researchers from CNRS and CEMAGREF and staff of the park's scientific service, hence anticipating the association of different groups of actors (researchers, nature managers and some local stakeholders) in the Platform (Table 20.1). Citing a lack of cooperation between the research teams in ecology working on its grounds, the park was eager to promote exchanges and offered itself as a possible common research field. However, one of the labs had just been restructured and felt a need to consolidate its own research project before embarking on a broad cooperation with the other labs. It therefore took some time before building the Platform as a medium-term, cooperatively constructed process. Similarly, the Vercors reserve managers expressed the same need for scientific work on its territory to be promoted and rationalised, which illustrates the importance of the nature managers' initiative in this process.

Why it started: A majority of ecologists involved in the LTSER Platform seek to understand ecosystems' responses to changes in climate and land use. They rely very much on long-term data and access to databases, unanimously described as difficult and costly, which is thus undoubtedly one of the main aims of the Platform. For that matter, the construction of a meta-database has been one of the first activities carried out by the Platform (see Sect. 20.2.3). However, the partners' relationship to data is rather ambivalent: their need for data encourages them to focus on a few common research sites, preferably where long-term[1] data are available, and to

---

[1] It is important to note that "long-term" does not mean the same for all participants. Ecologists working on plant adaptation to climate change and land-use changes may need data extending back to the nineteenth century, whereas a single decade may be considered as "long-term" by researchers studying phenomena developing over short spans of time.

pool and share their data with their colleagues. On the other hand, they are sometimes reluctant to hand over data obtained through a long period of painstaking work, which, as Bowker states, is certainly not "raw" (Bowker 2000, p. 670). The meta-database illustrates this ambiguity: it records what kind of data is available without giving access to the data itself, and the conditions under which these can be exchanged, shared and used are negotiated between individuals. Nevertheless the construction of the relationship, e.g. between CBNA and CNRS, illustrates the process of mutual building of trust. It is hoped that this example will pave the way for facilitated data exchange and access between members of the LTSER.

<u>What it changes</u>: For natural scientists, the LTSER Platform seems to affirm existing relationships and data exchanges more than it creates new ones. The collaborations were already active between at least some of the natural scientists, and between them and the nature managers. For example, young researchers from the two main labs have worked together since their PhD studies, and an agreement was signed by the LECA and CBNA in 2007, by which the lab has obtained access to the conservatory's exceptional floristic database, under agreed conditions. According to many participants, collaborations were hence already well under way and the Platform construction does not appear to have been a revolutionary event. However, official labels, such as a "Zone Atelier"[2] and a LTSER Platform, are said by some respondents to be important in terms of visibility and recognition. This sign of recognition is particularly important for nature managers, who thereby obtain official acknowledgement of their direct involvement in research projects.

The situation is different with social scientists, since there was very little previous collaboration: the Platform could theoretically play an important role for them. However, the recruitment of social scientists into projects has turned out to be problematic, despite the efforts of the Platform leaders and despite the existence of the aforementioned projects. Moreover, interdisciplinarity between social and natural sciences seems scarcely desired by many. Significantly, an ecologist describing this relationship used the term "forced marriage": it is seen as imposed from the top rather than pushed from the ground up and researchers underline that this kind of interdisciplinarity is difficult to achieve, demands very firm disciplinary grounds and does not lead to easily publishable results. Some interviewees recollected unpleasant experiences, where they had submitted research projects with a clear interdisciplinary orientation that had been rejected (contrary to more monodisciplinary projects), although the calls had strongly encouraged interdisciplinary contributions. Interdisciplinarity was therefore regarded as a rather risky adventure upon which only well-established scholars could afford to embark.

---

[2] Zone Atelier sites make up the French LTER network, coordinated by CNRS's Environment and Sustainable Development Institute. Currently there are ten sites, whose membership of the network and management are the responsibility of an international Scientific Committee. The Zone Atelier network promotes long-term interdisciplinary research at sites representative of the main French river basins and of different type of natural and managed environments. See: http://www.za.univ-nantes.fr

## 20.4 Emerging Transdisciplinary Projects of the Central French Alps LTSER Platform

In spite of these initial teething pains, the process of collectively constructing inter- and trandisciplinarity through the LTSER Platform has been fruitful. New projects offer an added-value to the sets of already ongoing disciplinary projects, which they are expected to reinforce in the long term. The construction of the Platform has made it possible to retrieve "old" collaborative projects sometimes designed for other study areas, which will now use the LTSER Platform as a study area. Other projects have been created *ex nihilo* thanks to the existence of the Platform. In the following, we briefly describe three projects which have already been implemented.

### 20.4.1 Integrated Analyses of Ecosystem Services

Ecosystem services have recently developed as a concept that makes it possible to interface the ecological analysis of natural or managed systems and analyses of their uses and associated human dynamics (Turner et al. 2007). Recent reviews and individual studies have highlighted the complexity of ecosystem services analyses in that they require the availability and coupling of a large variety of types of data including land-use/land-management data, abiotic data e.g. on topography and soils, biodiversity data and data on human uses and preferences (de Groot et al. 2010). A LTSER Platform thus appears to be an ideal basis for detailed analyses of ecosystem service provision and its relationships to demand by stakeholders.

We quantified ecosystem services provided to farmers and other locals, as well as to tourists, by the south-facing grasslands of Villar d'Arène. The site's land-use history since the eighteenth century has been documented and is referenced in a Geographic Information System (Girel et al. 2010). Locally relevant ecosystem services were identified on the basis of interviews with local farmers on their needs from and uses of grasslands (Quétier et al. 2010). Grassland functions identified by stakeholders were translated by researchers into measurable indicators (Quétier et al. 2007). Vegetation, plant functional trait and environmental data, including soil parameters, have been collected since 2003 for 57 permanent plots stratified by land-use type, landscape sector, and altitude. We measured biomass production, biomass Crude Protein Content (CPC) and digestibility, litter accumulation, flowering phenology (date of flowering onset and duration of flowering), and then calculated indices of plant species and functional diversity. This data allowed us to build statistical models for the supply by the grasslands of the range of services of interest to the stakeholders including agronomic value, pollination, soil carbon storage and cultural value from aesthetics and conservation of plant diversity (see

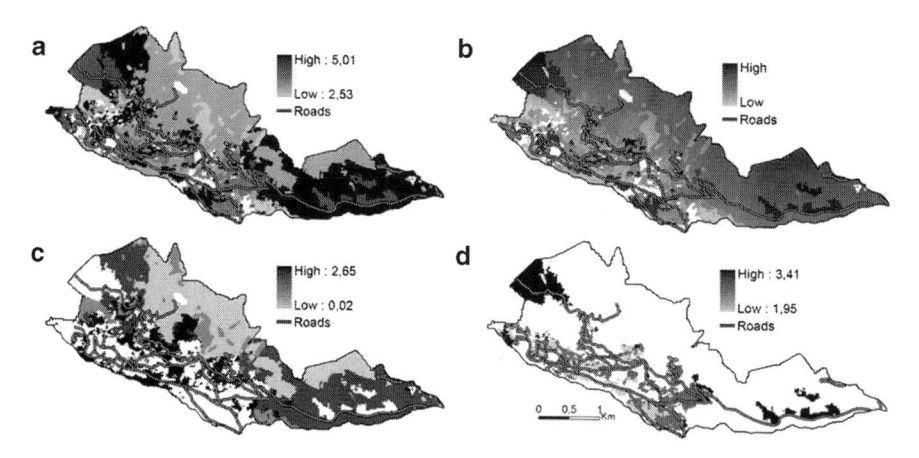

**Fig. 20.4** Potential ecosystem service supply and actual provision of agronomic services to farmers of Villar d'Arène (Hautes Alpes). (**a**) Green biomass (tons/ha) (**b**) Potential agronomic value (unitless) calculated as a combination of different functions (green biomass, digestibility and phenology), and actual benefits: (**c**) the number of days of livestock units/ha and (**d**) hay production (tons/ha). Roads and tracks are added on maps as they are important elements of analysis (Modified from Lamarque et al. 2011)

Lavorel et al. (2011) for details of the field measurements and models) (Fig. 20.4). These analyses also highlighted tradeoffs and synergies among these different ecosystem services. Traditional land uses such as organic fertilisation and mowing, or high altitude summer grazing, were linked with ecosystem services hot spots because functional characteristics supporting fodder production and quality are compatible with species and functional diversity. By building on state-of-the-art ecological knowledge, our analyses made it possible to relate patterns of association among ecosystem services to the dominant plant traits underlying different ecosystem properties. For example, the functional decoupling between height and leaf traits provided alternative pathways for providing high agronomic value (through high biomass production and/or high quality and delayed flowering), as well as determining hot and cold spots of ecosystem services. Finally, we were able to compare this estimated supply of ecosystem services to the actual uses made by the farmers for grazing and hay making – i.e. to actual ecosystem service provision, showing that, depending on grassland location in the landscape with respect to the village and to altitude, and ultimately to actual use as part of the farming systems, different dimensions of modelled agronomic value such as biomass production, quality or timing of flowering determined use value to the farmer (Lamarque et al. 2011).

Similar analyses coupling historical land-use analysis, vegetation and plant trait data and estimates of the components of carbon cycling through ground-based measurements and remote sensing are in progress for the Vercors Plateau as part of a new funded project.

## 20.4.2 Adaptation of Mountain Livestock Farming Systems to Socioeconomic and Climate Change

The LTSER Platform Central French Alps began with the asset of a large amount of pre-existing data from ecological monitoring, in particular for summer pastures which host flocks managed by transhumance every summer. Several studies on management practices have afforded us a good knowledge of the functioning and of the organisation of mountain summer pasture management, but little is known about the role and value of these summer pastures as part of farm system management. However, the socioeconomic environment has undergone and is still undergoing significant changes (Europe's new Common Agricultural Policy, volatility of prices for agricultural raw materials, land access difficulties, urbanisation, etc.), as well as farm organization (increasing labour costs, less manpower due to competition or complementarity with other activities within the household). In this context, one can hypothesise that farming systems will change radically, which in turn could have important consequences for the management of mountain summer pastures with resulting substantial changes in their biodiversity and the ecosystem services they provide (see below). At the same time, climate change and the associated dynamics of natural habitats may also induce changes in farm management.

This study involves several areas of expertise represented by different specialists within the LTSER Platform. Agroeconomists study the role of the mountain summer pastures as part of farm systems, and on socioeconomic changes within recent years and their impacts on entire farm systems, whether located within the mountains or in the foothills. The potential for adaptation to drought of the Vercors and Villar d'Arène farms has been explored through semi-directed interviews, showing a diversity of possible responses depending on, among other factors, type of livestock production (sheep, suckler or dairy cattle), dependency on hay stocks, diversity of available grassland types, and potential to access new areas for grazing or hay making (Nettier et al. 2011).

Long-term ecological observations on plants and animals by ecologists and natural area managers make it possible to record changes in the fauna and flora and to relate these to past climate data monitored by climatologists. Experimental climate manipulations have also been set up in order to analyse the effects of extreme drought expected from climate scenarios on plant diversity, components of agronomic value and soil functioning. For instance, first results indicate a strong resistance of the studied subalpine grasslands on the Vercors Plateaux and at Lautaret to an extreme drought event, but the consequences of longer and/or even more intense events are unknown yet.

Scenarios of future farm system organisation and resulting changes in grassland management are developed together with agronomists, sociologists and stakeholders. Their impacts on social, economic and ecological components of the human-environment system will be assessed through participative approaches and modelling, (see Gaube and Haberl, Chap. 3 in this volume) in particular using ecosystem service models as presented above in the case of Villar d'Arène.

### 20.4.3 Mutations of Alpine Tourism in the Face of Climate Change

Tourism is a major economic and social component in the territory of the Central French Alps LTSER Platform. Tourism activities have strong direct and indirect impacts on ecosystems, on which they also depend, especially in the case of winter sports. A new project therefore focuses on the coupled dynamics of tourism and ecosystems in the context of climate change. Alpine tourism is facing a severe structural crisis and its future is full of uncertainties. Climate change is one of them, and is often an accelerator of other crises (Elsasser and Bürki 2002). For many observers, the current tourism system established during the second half of the twentieth century, which is mainly concentrated on skiing, is largely exhausted and condemned to undergoing drastic changes. In this context, research questions regard two complementary levels and at the same time highlight current representations of alpine tourism and possible new actions for this sector: (i) What are the themes and indicators by which climate change is recognised by different stakeholders (visitors, actors from the tourism industry, members of the wildlife or administrative sectors)? (ii) What are the current observable impacts of climate change on recreation and professional tourism uses that are involved in outdoor activities? and (iii) To what extent and in what ways do actors from the tourism industry adapt their practices to ongoing and foreseeable changes due to rising concerns about climate change and biodiversity loss? Addressing such questions requires the strengthening of social sciences within the LTSER Platform and promotion of interdisciplinarity.

A survey questionnaire among individual participants of the LTSER Platform in order to identify ongoing and planned interdisciplinary projects on the interactions between ecosystems and tourism in the context of climate change highlighted the rarity of research on this question in spite of a broad recognition of its importance for the future of the study area. This survey also made it possible to build a better understanding within the Platform of ongoing projects on related topics and of their approaches and methods, and to identify research questions potentially conducive to collaboration between scientists from different disciplines. Based on this understanding, we called for internal proposals and two new projects are now due to start.

The first project focuses on the consideration of natural ecosystems around ski resorts, with a study of wetlands in Val Thorens (Savoie). Novel wetland management methods are being experimented with in order to preserve these ecosystems and, simultaneously, to use their potential to support tourism, both through their cultural value and as critically linked to snowmaking through competition for water supply. A survey will be carried out to analyse different approaches, their implementation and their ecological, economic and social impacts, as well as the role of different involved parties, including that of scientists specialising in wetland ecology. This research will involve one ecologist, one economist and one sociologist.

The second project analyses current and future linkages between tourism and natural ecosystems as seen by specialists of each of these sectors, using the Vercors Regional Natural Park as a case study. Semi-open interviews will be conducted with members of the tourism industry, ecologists and environmental advocates. This work will involve a geographer and a sociologist specialised in tourism and with a background in ecology.

## 20.5 Conclusion

Although the construction of a LTSER Platform requires much patience, only a short time after the creation of the Central French Alps LTSER Platform, individual participants and partner institutions conclude that it has provided an important framework to formalise already ongoing collaborations between different disciplines. Moreover, already within the first years of the Platform's existence, a process of de-compartmentalisation has started and had led to the initiation of several common projects in which social sciences are truly incorporated into the design of research agendas. In particular climate change has appeared as a common issue fostering transdisciplinary collaboration. Nevertheless, as in many similar long-term ecological Sites (Ohl et al. 2007), ecologists are still the most numerous group in our LTSER Platform. Incorporating a social sciences dimension into ecological research will ensure consistency of knowledge and the collection of data reflecting the complexity and changing nature of the coupled natural-human environment system, in order to enhance our ability to project global change impacts. The focus on ecosystem services and nature's values to land users is particularly promising in terms of enhancing integration. Research results will open new perspectives for management as well as for territorial development and policies.

**Acknowledgments** This research was conducted on the long-term research site Zone Atelier Alpes, a member of the ILTER-Europe network. ZAA publication no. 5. This paper was initially prepared within the framework of "A Long-Term Biodiversity, Ecosystem and Awareness Research Network" (ALTER Net), a partnership of 24 organisations from 17 European countries, funded by the EU's 6th Framework Programme. It contributes to MEDDTL GICC2 project SECALP. We thank Karl Grigulis, Pénélope Lamarque and Nicole Sardat for the artwork.

## References

Albert, C. H., Thuiller, W., Yoccoz, N. G., Soudant, A., Boucher, F., Saccone, P., & Lavorel, S. (2010). Intraspecific functional variability: extent, structure and sources of variation within a French alpine catchment. *Journal of Ecology, 98*, 604–613.

Baptist, F., & Choler, P. (2008). A simulation of the importance of growing season length and plant functional diversity on the seasonal Gross Primary production of temperate alpine canopies. *Annals of Botany, 101*, 549–559.

Bigot, S., & Rome, S. (2010). Contraintes climatiques dans les Préalpes françaises: évolution récente et conséquences potentielles futures. *Echogéo, 14*, Retrieved from http://echogeo.revues.org/12160

Bowker, G. C. (2000). Biodiversity datadiversity. *Social Studies of Science, 30*, 643–683.
de Groot, R. S., Alkemade, R., Braat, L., Hein, L., & Willemen, L. (2010). Challenges in integrating the concept of ecosystem services and values in landscape planning, management and decision making. *Ecological Complexity, 7*, 260–272.
Elsasser, H., & Bürki, R. (2002). Climate change as a threat to tourism in the Alps. *Climate Research, 20*, 253–257.
Girel, J., Quetier, F., Bignon, A., & Aubert, S. (2010). *Histoire de l'Agriculture en Oisans*. Grenoble: Station Alpine Joseph Fourier.
Granjou, C., & Mauz, I. (2011). L'équipement du travail de production de données en écologie. L'exemple de la constitution de la Zone Atelier Alpes. *Revue d'anthropologie des connaissances, 5*, 287–301.
Haberl, H., Winiwarter, V., Andersson, K., Ayres, R. U., Boone, C., Castillo, A., Cunfer, G., Fischer-Kowalski, M., Freudenburg, W. R., Furman, E., Kaufmann, R., Krausmann, F., Langthaler, E., Lotze-Campen, H., Mirtl, M., Redman, C. L., Reenberg A., Wardell, A., Warr, B., & Zechmeister, H. (2006). From LTER to LTSER: Conceptualizing the socioeconomic dimension of long-term socioecological research. *Ecology and Society, 1*, 13. Retrieved from http://www.ecologyandsociety.org/vol11/iss2/art13/
Knorr-Cetina, K. (1999). *Epistemic cultures: How the sciences make knowledge*. Harvard: Harvard University Press.
Körner, C. (1999). *Alpine plant life: Plant ecology of high mountain ecosystems*. Heidelberg: Springer.
Lamarque, P., Quétier, F., & Lavorel, S. (2011). The diversity of the ecosystem services concept: implications for quantifying the value of biodiversity to society. *Compte-Rendus de l'Académie des Sciences, Biologie, 334*, 441–449.
Latour, B. (1987). *Science in action: How to follow scientists and engineers through society*. Milton Keynes: Open University Press.
Lavorel, S., Grigulis, K., Lamarque, P., Colace, M.-P., Garden, D., Girel, J., Douzet, R., & Pellet, G. (2011). Using plant functional traits to understand the landscape-scale distribution of multiple ecosystem services. *Journal of Ecology, 99*, 135–147.
Liu, J. G., Dietz, T., Carpenter, S. R., Alberti, M., Folke, C., Moran, E., Pell, A. N., Deadman, P., Kratz, T., Lubchenco, J., Ostrom, E., Ouyang, Z., Provencher, W., Redman, C. L., Schneider, S. H., & Taylor, W. W. (2007). Complexity of coupled human and natural systems. *Science, 317*, 1513–1516.
Mauz, I., & Granjou, C. (2011). Rendre visibles les "travailleurs invisibles"? Vers de nouveaux collectifs de travail en écologie. *Terrains et travaux, 18*, 121–139.
Mirtl, M. (2010). Introducing the next generation of ecosystem research in Europe: LTER-Europe's multi-functional and multi-scale approach. In F. Müller, C. Baessler, H. Schubert, & S. Klotz (Eds.), *Long-term ecological research: Between theory and application*. Part 3 (pp. 75–93). Dordrecht: Springer.
Murphy, R. (1995). Sociology as if nature did not matter: An ecological critique. *The British Journal of Sociology, 46*, 688–707.
Nettier, B., Dobremez, L., Coussy, J.-L., & Romagny, T. (2011). Attitudes des éleveurs et sensibilité des systèmes d'élevage face aux sécheresses dans les Alpes françaises. *Revue de Géographie Alpine 98-4*. [Online] http://rga.revues.org/index1294.html
Ohl, C., Krauze, K., & Grünbühel, C. (2007). Towards an understanding of long-term ecosystem dynamics by merging socio-economic and environmental research: Criteria for long-term socio-ecological research sites selection. *Ecological Economics, 63*, 383–391.
Quétier, F., Lavorel, S., Thuiller, W., & Davies, I. D. (2007). Plant trait-based assessment of ecosystem service sensitivity to land-use change in mountain grasslands. *Ecological Applications, 17*, 2377–2386.
Quétier, F., Rivoal, F., Marty, P., De Chazal, J., & Lavorel, S. (2010). Social representations of an alpine grassland landscape and socio-political discourses on rural development. *Regional Environmental Change, 10*, 119–130.

Redman, C. L., Grove, J. M., & Kuby, L. H. (2004). Integrating social science into the Long-Term Ecological Research (LTER) Network: Social dimensions of ecological change and ecological dimensions of social change. *Ecosystems, 7*, 161–171.

Sala, O. E., Chapin, F. S., III, Armesto, J. J., Berlow, E., Bloomfeld, J., Dirzo, R., Huber-Sannwald, E., Huenneke, L. F., Jackson, R. B., Kinzig, A. P., Leemans, R., Lodge, D. M., Mooney, H. A., Oesterheld, M., Poff, L., Sykes, M. T., Walker, B. H., Walker, M., & Wall, D. H. (2000). Global biodiversity scenarios for the year 2100. *Science, 287*, 1770–1774.

Tappeiner, U., & Bayfield, N. (2004). Management of mountainous areas. In W. Verheye (Ed.), *Encyclopedia of life support systems (EOLSS)* (pp. 1–17). Oxford: UNESCO, Eolss Publishers [online].

Turner, B. L., Lambin, E. F., & Reenberg, A. (2007). The emergence of land change science for global environmental change and sustainability. *Proceedings of the National Academy of Sciences, 104*, 20666–20671.

White, P. S., & Jentsch, A. (2001). The search for generality in studies of disturbance and ecosystem dynamics. In K. Esser, U. Lüttge, Beyschlag, W., et al. (Eds.), *Progress in botany 62* (pp. 399–449). Berlin/Heidelberg: Springer.

# Chapter 21
# Long-Term Socio-ecological Research in Mountain Regions: Perspectives from the Tyrolean Alps

**Ulrike Tappeiner, Axel Borsdorf, and Michael Bahn**

**Abstract** Mountain habitats have been classified as particularly sensitive to changes in land use and climate, which are occurring at increasingly high rates. The Tyrolean Alps host a strong tradition of research on a range of ecological processes in mountain environments, and how they are affected by changing environmental conditions. Research topics, partly studied over several decades, include responses of organisms and of biogeochemical processes to extreme life conditions and to global changes in both terrestrial and aquatic ecosystems. Research sites in the Tyrolean Alps span a vast range in altitude (1,000–3,450 m) and climate. For two valleys/valley sections, socio-economic changes have been documented and past, current and possible future landscape changes have been assessed, evaluating also effects on ecosystem services. The recent research history at the Tyrolean Alps LTSER Platform has shown that a monitoring of the biogeochemistry of target ecosystems combined with an experimental unravelling of global change effects on processes, and the consideration of socioeconomic developments together constitute a fruitful way forward, increasing the value of LTSER sites also for international projects and networks.

**Keywords** Land-use • Climate change • Long term socio-ecological research • Mountain ecosystems • Semi-natural and natural ecosystems

---

U. Tappeiner, Ph.D. (✉) • M. Bahn
Institute of Ecology, University of Innsbruck, Innsbruck, Austria
e-mail: ulrike.tappeiner@uibk.ac.at; michael.bahn@uibk.ac.at

A. Borsdorf, Ph.D.
Department of Geography, University of Innsbruck, Innsbruck, Austria
e-mail: axel.borsdorf@uibk.ac.at

## 21.1 Introduction

Mountains cover almost a fifth of the global land mass. Twenty percent of the world's population live in mountain regions and half of humanity depends on resources (esp. water) from mountain regions. Mountain habitats have an intricate structure and feature extreme living conditions that are the cause of high biodiversity. For these reasons, however, mountain ecosystems recover only slowly if disturbed. Thus, mountain habitats must be classified as particularly sensitive to change (e.g. EEA 2004; Becker et al. 2007). Today there is an increasing awareness that highlands and lowlands are strongly connected: mountains not only influence the lowlands by down-slope physical processes but there are increasing human relationships linking the two (Ives et al. 1997). Hence changes in mountain communities can exert enormous impacts not only on the adjacent but also on far distant lowlands.

In recent decades, the dynamics of global change have increased dramatically. $CO_2$ concentrations in the atmosphere and their rates of increase are higher than ever before within the last two million years, a global rise in temperature is no longer in dispute and there is an increase in climate extremes and related droughts and intense precipitation events (IPCC 2007). In addition, a rapid change in economic and social systems has led to changes in land use on a large scale. As a result, ecosystems and landscapes and their socially relevant services have irrevocably changed and continue to change, all the way down to the regional level (Millennium Ecosystem Assessment 2005). The impacts of such global changes differ by region and season, but mountain regions are among the areas most affected (Schröter et al. 2005; Beniston 2006). In the Alps, temperature change during the last 100 years was 1.4 °C, i.e. twice as high as the global average (0.7 °C, cf. Beniston 2006; Auer et al. 2007). Due to socioeconomic and socio-cultural changes in the last 20 years, 40% of farms in the Alps have been abandoned (Streifeneder et al. 2007). Today, on average 20%, and in some regions up to 70%, of formerly agricultural land in the Alps has been taken out of production (Tappeiner et al. 2006; Zimmermann et al. 2010).

The Central Alps in Tyrol are exceptionally well suited for ecological long-term research that can yield important insights into climate, economic, demographic and social change as well as globalisation effects. The area offers unique ecological features in terms of geology, relief energy, morphodynamics, climate region, climate gradients, hydrology, cryosphere, limnology, soils, vegetation gradients and glacier foreland as well as a case in point for the severe impacts of direct socioeconomic activities, such as winter and summer tourism, hydropower generation, agriculture and changes in land use, transport, settlement, etc. The insights gained are an indispensible prerequisite for initiating and implementing adaptation strategies and concrete measures, not just for the study area but also for similarly structured mountain ecosystems elsewhere.

The objective of this chapter is to present the Tyrolean Alps LTSER Platform with respect to its regional specifications, the research undertaken so far, and – to a certain degree – the challenges for future research. Long-term ecological research has a long tradition in this region, whereas socio-ecological and socioeconomic studies have been much less undertaken. In the following paragraphs we demonstrate that the Tyrolean Alps are a unique setting for LTSER research. On this basis, the research in these regions will be discussed, taking into account the long-term research tradition as well as perspectives for the future.

## 21.2 The Tyrolean Alps – A Unique Setting for LTSER Research

Although Austrians are highly engaged in the international LTER and LTSER communities, it must be stated that – caused by a lack of funding by the Austrian government for the programme – research on long-term ecological and socio-ecological topics has been more or less concentrated upon some individual sites and is dependent on occasional funding by other programmes. This is why Austria has so far only managed to install one official LTSER platform (Eisenwurzen), whereas research in the Tyrolean Alps has not yet been formally linked to either the LTER or the LTSER programme (Mirtl et al. 2010). However, in 2010 the Austrian government decided to provide funds for the formal implementation of the LTSER programme. The national LTSER consortium decided to install a second platform (after Eisenwurzen), called Tyrolean Alps. Given the relatively small funding basis, the challenges are to implement the platform with an appropriate management, to form interdisciplinary research teams, to define the concrete research area and to install transdisciplinary processes between scientists, stakeholders and the local population.

There is already a good basis for addressing these challenges, as ecological – and to a certain degree even socio-ecological research has been carried out in the Tyrolean Alps over many decades (see Sect. 21.3.1). Databases exist on climate, glacier balances, permafrost, hydrology, biodiversity, greenhouse gas fluxes, historical land-use changes, tourism, demography, agro-economy and a wealth of comprehensive but singular studies in the region. Furthermore, the excellent contacts to stakeholders in the Tyrolean Alps and the existing network of researchers, formally cooperating in the research focus "Alpine Space – Man and Environment" of the University of Innsbruck form excellent preconditions for the implementation of the new LTSER platform.

### 21.2.1 Geography and Geology

The Tyrolean Alps LTSER Platform extends southwards to the border of the federal province of Tyrol and its northern and western boundaries are formed by the valley of the river Inn and the municipal borders around Mt. Patscherkofel (Fig. 21.1). The

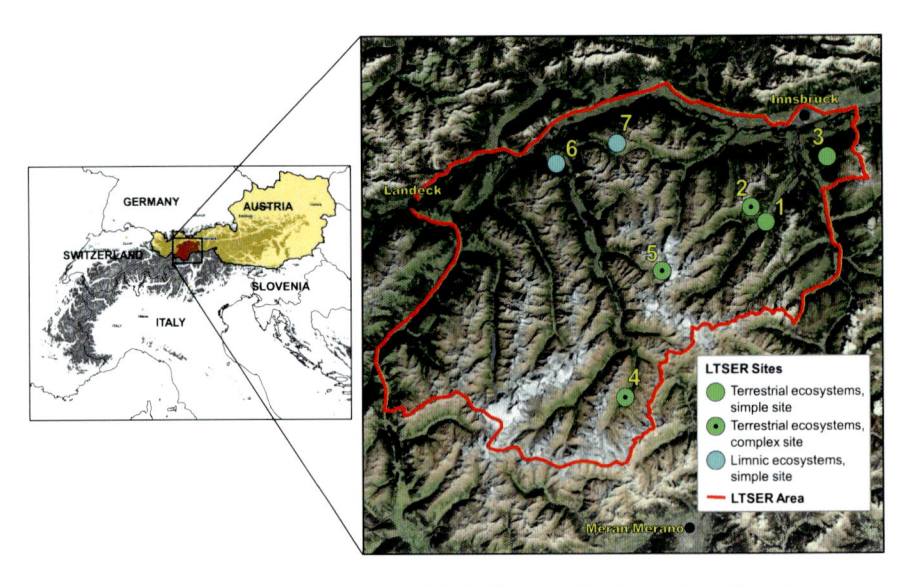

**Fig. 21.1** LTER Sites in the Tyrolean Alps LTSER Platform. *Numbers* refer to Table 21.1

total area is 3.7 million hectares, with a 3,200 m altitudinal span from 550 m in the Inn valley at Innsbruck to 3,750 m at the Wildspitze summit. In this area, dominated by high mountains and their sensitive ecosystems, eight LTER sites (some of them composed of several ecosystems) are embedded, including two lakes, six grasslands at different altitudes, a treeline site, a glacier foreland, and several glaciers (Fig. 21.1, Table 21.1). These sites span a vast range in altitude (1,000–3,450 m) and climate, mean annual temperature and precipitation of the terrestrial sites covering a range of 5.5 °C and 900 mm, respectively, and for the similarly structured grassland ecosystems of 3.6 °C and 570 mm, respectively.

The Tyrolean Alps are essentially part of the Eastern Central Alps with their crystalline basement rocks, which do include remnants of sedimentary rock nappes (so-called "Kalkkögel", near Innsbruck, of dolomite and limestone). Their great petrovariance includes anatexic granites, metamorphous rock such as gneiss, schist, phyllitic schist, quartz phyllite, marble, amphibolite, eclogite, as well as sedimentary rocks such as limestone and dolomite. In tectonic terms, the Ötztal Mass dominates the area, onto which remnants of the Err-Bernina nappe have slid. Near Mt. Patscherkofel, the Innsbruck quartz phyllite stratum has become exposed.

In the course of three relief generations, the great relief energy has created striking and highly diverse surface features. The highest level shows the fairly even, extensive forms of the Tertiary, generated in a wet-dry tropical climate. Below the summits, flattened areas can still be detected, many of which became cirques during the Pleistocene. Such areas are most prominent in the less metamorphous rocks south of the Inn valley, which experienced drastic reshaping during the ice ages. The crests were sharpened, glacial horns and cirques formed as a result of intensive physical weathering and of the excoriating force of the glaciers. These glacial

**Table 21.1** Location, site characteristics and research focus of current research sites at the Tyrolean Alps LTSER Platform

| Site | Location | Land cover type/management | Altitude (m) | MAT (°) | MAP (mm) | Current research focus |
|---|---|---|---|---|---|---|
| 1. Stubai Neustift | 47° 7′ N, 11° 18′ E | Grassland: meadow; 3–5 cuts, fertilised | 1,000 | 6.3 | 850 | Carbon cycle, productivity, greenhouse gas emissions, volatile organic compounds, microclimate, energy exchange |
| 2. Stubai Kaser-stattalm | 47° 7′ N, 11° 18′ E | Grasslands: lightly fertilised meadow (1 cut), pasture, abandoned | 1,820–1,970 | 3.0 | 1,097 | Carbon cycle, productivity, greenhouse gas emissions, water balance, nutrient dynamics, plant-soil interactions (carbon, nutrients), land use and climate change effects, potential risks (surface runoff, snow gliding) |
| 3. Patscherkofel | 47° 13′ N, 11° 28′ E | Treeline | 1,950 | 2.4 | 950 | Treeline ecophysiology, dendrology, climate change effects, stress physiology |
| 4. Obergurgl | 46° 50′ N, 11° 02′ E | Grazed versus ungrazed ecosystems, glacier foreland and glaciers | 2,240–3,028 | −1.3 to 2.8 | 830–1,460 | Land use and climate change effects on species richness, colonisation dynamics, glacial dynamics, rock glacier dynamics, meteorology |
| 5. Schrankogel | 47° 3′ N, 11° 6′ E | Alpine grassland, snowbeds | 2,900–3,450 | 1.0 | n.a. | Vegetation composition, climate warming |
| 6. Piburger See | 47°11′ N, 10°50′ E | Lake | 913 | n.a. | n.a. | Hydrology, water chemistry, sediment, phytoplankton, zooplankton, zoobenthos, fish ecology, microbial food webs |
| 7. Gossenköllesee | 47°14′N, 11°01′E | Lake | 2,413 | n.a. | n.a. | Atmospheric deposition, snow and water chemistry, hydrology, phyto- and zooplankton, zoobenthos, fish ecology, ecotoxicology, microbial food webs, paleolimnology, aquatic photobiology |

Not available where indicated by n.a.
*MAT* mean annual temperature, *MAP* mean annual precipitation

processes also gave the valleys their typical deep U-shape, often with distinct steps in the longitudinal profile. In the Inn valley, the ice layer was up to 1,900 m thick. The main glaciers in the valleys of Inn and Adige were linked by an almost complete network of ice flows, with only the highest summits jutting out as nunataks. Towards the end of the Pleistocene, when the glaciers retreated, ice lakes formed at the edge of the ice mass, while cirques and other hollow forms filled with water.

During the Holocene, large rockslides occurred on the slopes of the mountains once the pressure of the ice had gone. To this day, the Ötz valley is a prime example of such mass dynamics. In morphodynamic terms, the appearance of the landscape is characterised today by fluvial erosion at the valley floor, especially at the mouths of the hanging valleys, as well as by sediment deposition in the Inn valley and the emergence of terraces and alluvial fans. At higher levels, denudation still dominates. The snow line today ranges from 2,200 to 2,700 m, depending on exposition. Thus summits of the Ötz valley and Stubai Alps that jut out beyond this are still covered by glaciers. Below today's ice areas, we often find permafrost, which has created impressive rock glaciers.

## 21.2.2 *Climate, Soils and Vegetation*

At high altitudes, up to 1,600 mm annual precipitation feeds the glaciers. At lower altitudes, precipitation is only half of this amount. The south foehn in North Tyrol brings warm, sunny but also blustery days, especially in transition seasons. The high climate variability in the Tyrolean Alps (Fig. 21.2) is also reflected by the range of LTER sites (Table 21.1).

Cambisols and, at higher altitudes, also Leptosols are characteristic for the pedological milieu of the area. On steep slopes in particular, there is always the threat of landslides and soil erosion. In some places, such processes have resulted in striking earth pyramids, e.g. at the Ritten and in the Wipp valley.

The altitudinal span of 3,200 m within the Tyrolean Alps not only affects relief, climate and soils but expresses itself in a variety of vegetation forms. Even the untrained eye will notice their change over the altitudinal range. Parts of the Vinschgau are still in the colline belt. Grapes and sweet chestnuts are typically cultivated at this level. The lower levels of the large valleys belong to the montane belt. In virgin condition, this would be deciduous forest (oak, beech, maple). However, this forest has largely been cleared, the valleys are dominated by arable farming and the steep slopes by spruce plantations (Fig. 21.2b).

At around 1,000 m, the subalpine belt starts with relatively homogenous spruce forests, interspersed with lighter larch stands. These are not natural in origin but rather the result of a former use as woodland pasture, maintained today for its scenic appeal. At the boundary to the alpine belt, loose stands of arolla pine dominate. The mugo pine forests typical for the Limestone Alps are scarce in the crystalline milieu. The forest has an important protective function for the lower areas, but landslides and avalanche chutes have rent this cover in many places.

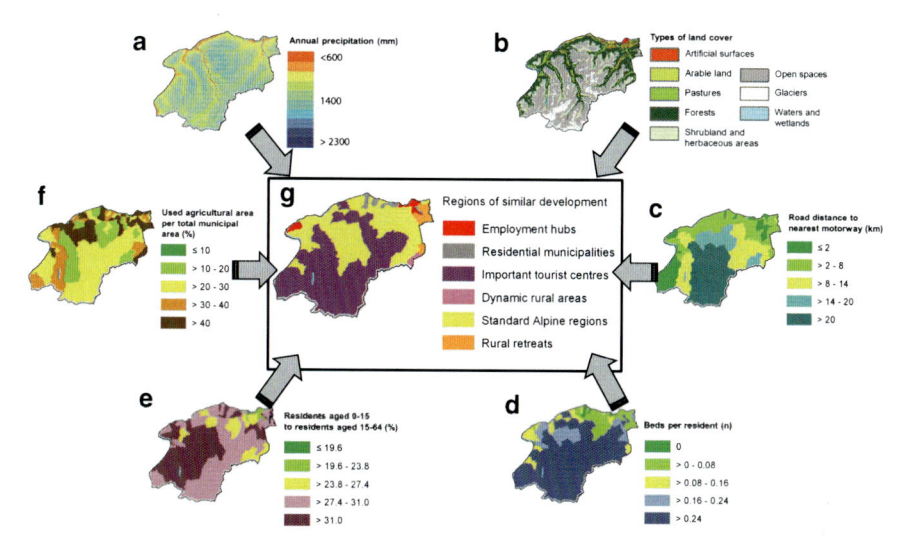

**Fig. 21.2** Climate, land cover and indicators of development in the Tyrolean Alps LTSER Platform (For details see Tappeiner et al. 2008b)

Intensive morphodynamics are also found in the alpine belt. Above the treeline at around 1,700–2,000 m, there are extensive grasslands (Fig. 21.2), interspersed with mugo pine and raised bogs. Some meadows disappear under rock fall and become more typical of the subnival belt. In the highest, nival, belt, glaciation permits hardly any vegetation. Ice-free patches show at most rudimentary vegetation in the form of lichen and mosses.

## 21.2.3 Land Use and Economic Sectors

In their traditional land use, humans adapted well to the various altitudinal belts (Fig. 21.2f). The High Alps are situated at the upper limit of the arable area, the highest permanent settlements are the Rofenhöfe farms at ca. 2,200 m. Agricultural land use features into distinct zones along altitudinal gradients. In North Tyrol, the valley floors today are mainly used to grow fodder but – showing early effects of climate change – special cultures such as fruit and recently even wine growing are expanding, as is intensive market gardening in the Inn valley.

In North Tyrol, the base farmsteads of the alpine *Staffelwirtschaft* (multi-level pasture system) are situated in the valleys, and in South Tyrol at middle elevations. In May, livestock is driven on to higher pastures and in June further on to the high mountain pastures. In September, with stops at intermittent stages, it returns to the base. In former times, cheese was produced on many high mountain pastures. Today, as a result of EU agricultural policy, these are increasingly stocked with heifers or yearlings.

Agriculture thus presents trends towards both intensification and extensification. In North Tyrol, the number of full-time farms is falling while the farmed acreage is growing. Many farmers are looking for an additional off-farm income, mainly in tourism.

Tourism is the dominant economic factor today. It started in the nineteenth century, mainly in the spa of Merano, but not until after the First World War did it reach a sufficient level to overcome the region's former extreme poverty, which had forced people to rent out their children as labour into southern Germany. Initially, tourism took the form of summer holidays and mountaineering (Alpine Association huts). Today it is dominated by winter sports, with the season extended considerably by means of artificial snowmaking and skiing on glaciers. Recently, extreme and fashionable sports have been added, such as paragliding, rock climbing, mountain biking and rafting.

Industrialisation started late in the Tyrolean Alps and is today exposed to severe global competition. Despite the very good accessibility of the valleys (Fig. 21.2c), passes and winter sports resorts, the industrial sites suffer from high transport costs. Altogether, traffic has become a major burden for humans and the environment. The route across the Brenner Pass is by far the most heavily frequented connection between the Northern and the Southern Alps.

The highly developed transport infrastructure and the resulting increased mobility have, especially in North Tyrol, resulted in a concentration of retail and service units at the mouths of the valleys. Within the Tyrolean Alps area, however, there has not been significant abandonment of higher elevations or marginalisation of tributary valleys through depopulation and overageing to date. On the contrary, the Tyrolean Alps score highly in demographic terms, with almost 30% of people aged under 15 years (Fig. 21.2e). This is both due to economically very active rural areas and a very strong tourism economy in this area (Fig. 21.2d). In its wake, there has been an immigration of extra-alpine people, some as workers, some as amenity migrants, who have found a new home in the High Alps and need to be integrated. However the area shows a distinct trend towards looser/dispersed settlement. Suburbanisation and post-suburbanisation trends are occurring, with central functions increasingly moving into formerly rural areas and triggering new transport relations (Borsdorf 2004). They also siphon off purchasing power from South Tyrol. North Tyrolean shopping malls and department stores fill with Italian customers not only on Italian bank holidays but all year round. Italians tend to have more mobile homes than Austrians and are particularly keen on city breaks. This puts severe pressure on parking spaces and North Tyrolean towns within the study area, particularly Innsbruck, are required to find solutions.

All these elements are interrelated and changes in one element have effects on the others. This is why Wöhlke (1969) understood the system as a function, constituted by different parameters (Fig. 21.3). The challenge for long-term socio-ecological research (LTSER) lies in identifying precisely both the interactions between the socioeconomic systems and ecosystems and the impact this has on the natural environment and on society, due to direct local activities but also exogenous global

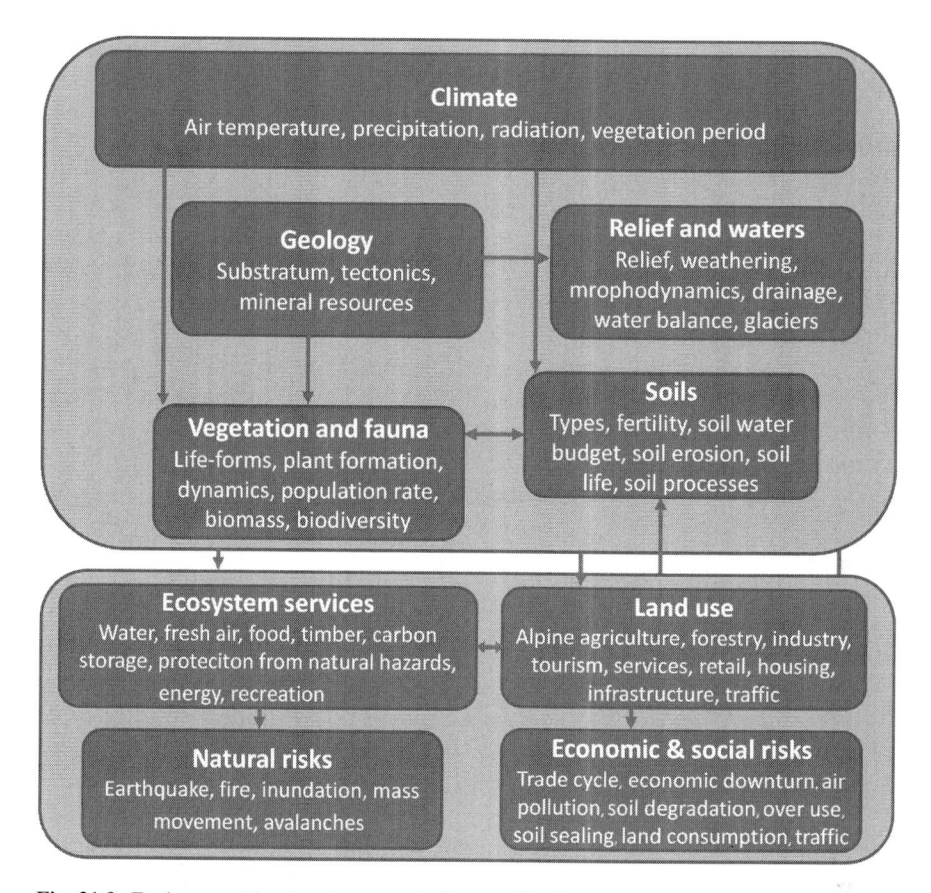

**Fig. 21.3** Environmental and socioeconomic factors affecting ecosystems in the Tyrolean Alps

change. At the same time, the challenge also involves using these findings to create adaptation strategies for ecologically sound, sustainable regional development.

## 21.3 Research in the Tyrolean Alps LTSER Platform

### 21.3.1 *Past and Ongoing Long-Term Ecological Research*

A characteristic feature of high mountains is their vertical zonation into altitudinal belts with distinct climates and life zones. This sequence of belts and their differences and interactions along elevation gradients contribute to the high environmental and biological diversity of mountain areas. Furthermore, topographic effects such as the influence of exposure on microclimate, the slope-specific intake of radiation and its influence on the energy balance or the effects of relief on snow distribution create intricate small-scale patterns of living conditions in mountains. All this makes mountain regions like the Alps an ideal open-air laboratory. It is therefore hardly surprising that generations of biologists and climatologists from the University of Innsbruck, located as it is in the heart of the Alps, have been fascinated by the natural test areas in their vicinity. In 1971, Artur Pisek wrote, "Hardly any botanical institute in the world at that time had such favourable conditions gratia loci to study how plants and plant societies cope with the circumstance of their location". The Tyrolean Alps LTSER Platform thus has its roots in more than a century of research into mountain ecology at the University of Innsbruck. As early as the late nineteenth century, Anton Kerner taught at the University of Innsbruck, as a classic scholar of plant geography and alpine botany (Gärtner 2004) as well as a pioneer of chemical ecology (Hartmann 2008). He studied the impact of geological, climatic and biotic factors on plant distribution and survival. In addition, he carried out early experimental analyses with transplants from low altitude to high altitude in Tyrol. Kerner cultivated over 300 perennial and annual taxa from homogenous seed origin in parallel in his trial gardens in Vienna (180 m a.s.l), Innsbruck (569 m a.s.l), in the Gschnitz valley (1,215 m a.s.l) and on Mt. Blaser (2,195 m a.s.l). He noticed that with increasing altitude, the plants become stockier, with fewer flowers in deeper colours. Contrary to his hypothesis, however, he had to concede after a 6-year trial that such adaptation characteristics to the conditions of the alpine zone were not hereditary. With this finding, Anton Kerner was one of the first to document non-hereditary changes in organisms that are caused by the environment, making him a trailblazer for the concept of genotype and phenotype that later entered common usage (Ehrendorfer 2004).

In the early 1930s, Arthur Pisek and Engelbert Cartellieri began to study various functional plant groups and plant societies on the mountains around Innsbruck and were the first to systematically combine field studies with controlled environmental studies (Pisek 1971; Körner 2003). In the mid-twentieth century, they became the founders of a modern comparative and experimental ecology of alpine plants.

In the 1960s, and building on this long tradition at Innsbruck in experimental plant ecology, Walter Larcher introduced the ecosystem approach. He initiated and directed a broad mountain ecology research programme at the treeline on Mt. Patscherkofel near Innsbruck and in the nival zone in the Tyrolean Central Alps in the valleys between Stubai valley and Ötz valley (Larcher 1977a, b; Moser et al. 1977). This constituted a major first step towards today's Tyrolean Alps LTSER Platform. These investigations were part of the Ecosystem Analysis Studies of the Tundra Biome Programme of the International Biological Programme (IBP), the first global attempt to coordinate large-scale ecological and environmental studies, focusing on the productivity of biological resources, human adaptability to environmental change and environmental change itself. The Innsbruck studies aimed to capture as fully as possible the biogeochemical cycles of the dwarf shrub heath of the alpine zone and of vascular plants in the nival zone in their dependence on environmental circumstances. Findings from this first comprehensive ecosystem analysis in mountain areas are still being used in major textbooks today (e.g. Körner 2003; Larcher 2003). From the mid-1960s, extensive alpine research remained focused on

Mt. Patscherkofel. Logistic and scientific support has been provided by the Mountain Research Station of the Botanical Institute at the University of Innsbruck as well as by the Alpine Timberline Research Station Patscherkofel (Klimahaus) of the Research and Training Centre for Forests, Natural Hazards and Landscape (BFW). Long-term ecological research at the Patscherkofel Station includes climate monitoring at the treeline and a study of the impact of climate change on growth, stability and biogeochemical cycles within the treeline ecotone (Wieser and Tausz 2007; Oberhuber et al. 2008), studies on eco- and stress-physiology of alpine plant species, their resistance to climatic extremes like frost, heat and drought and their mechanisms of avoidance, tolerance and recovery (e.g. Taschler et al. 2004; Neuner and Pramsohler 2006), as well as reproduction-biology studies of high mountain plants (e.g. Escaravage and Wagner 2004).

The Ötz valley, with several sites, also developed into a key research area within the Tyrolean Alps LTSER Platform. Ecological studies started with the Nebelkogel IBP project. In addition, the University of Innsbruck runs two long-term climate stations, one in Vent, started in 1935, and the station in Obergurgl, in operation since 1951. The Ötztal Alps are also a key research area for studying glaciers, an important aspect of the current climate debate. Two sites are used for mass balance measurements, investigations on the Hintereisferner glacier have been going on for more than 100 years, whereas the Kesselwandferner glacier has been studied since 1952/1953 (Fischer and Markl 2009; Fischer 2010). The time series of mass balance at Hintereisferner is one of the longest worldwide and thus crucial for developing glacier-climate models. These allow the interpretation of glaciers as indicators of climate change as well as an estimation of the contribution of glacier melt to present and future rises in sea level. Moreover, because of the excellent data available, the Hintereisferner often serves as a test glacier for new remote sensing or ground-based methods. At the Kesselwandferner, in addition to mass balance measurements, a stake network has been maintained since 1962, which allows the local investigation of velocity vectors in three dimensions. With a time series of more than 40 years, this is unique in Austria.

An important LTER site is Mt. Schrankogel (3,497 m a.s.l.), which is integrated in GLORIA, the worldwide long-term observation network in alpine environments (http://www.gloria.ac.at; Grabherr et al. 2010). Around 1,000 permanent plots, established in 1994, are distributed at altitudes between 2,900 m and 3,450 m a.s.l. near the summit area, spanning the alpine-nival ecotone from the upper margin of closed alpine grassland to the nival zone (Pauli et al. 2007). These permanent plots were set up in response to evidence of upward shifts of alpine plants on high peaks in the Alps. Other research themes on Mt. Schrankogel were the influence of domestic and wild ungulates (Huelber et al. 2005), nitrogen gradients, permafrost patterns, flowering phenology and photoperiodism of alpine and nival vascular plants (Keller and Körner 2003; Huelber et al. 2006). One of the most important additional datasets at the Schrankogel Site are temperature time series, measured at around 40 positions distributed over the mountain's southern slope system since 1997.

Further to the terrestrial sites, the Tyrolean Alps also include two limnological LTER sites. Lake Piburger See in the Ötz valley at 913 m a.s.l. is protected since

1929 and part of a Nature Reserve since 1983. During the 1960s, Piburger See suffered from eutrophication due to increasing recreational activities and application of fertiliser on nearby fields. In 1970, lake restoration started with exporting anoxic and nutrient-rich hypolimnetic waters by means of a deep-water siphoning tube (Olszewski tube) and reducing external nutrient loading (fertilisers on nearby fields, domestic sewage from a public beach). The restoration of Piburger See has been accompanied by a monitoring programme covering hydrology, water chemistry, sediment, phytoplankton, zooplankton, zoobenthos, fish ecology and microbial food webs.

For more than 30 years, the University of Innsbruck has also been studying Lake Gossenköllesee, situated above the treeline in the Stubai Alps at 2,413 m. The station was established by Roland Pechlaner, who moved it there from the Finstertal Lakes, where the first limnological station monitoring high elevation lakes had been installed by Steinböck in 1956. The lake is part of a UNESCO biosphere reserve established in 1976. The lake and its surroundings were the central study objects in a series of EU projects focused on atmospheric deposition and global warming issues. The intricate relationship between climate and water acidification was also detected in Tyrolean high altitude lakes (Psenner and Schmidt 1992; Sommaruga-Wögrath et al. 1997) as well as the mechanisms permitting plankton to cope with extreme UV radiation levels in lakes above the timberline (Sommaruga and Augustin 2006; Rose et al. 2009). Gossenköllesee is not only the highest lake with a fully equipped clean research laboratory in the Alps, it is also unique in harbouring a population of pure Danubian brown trout stocked in the sixteenth century (Kamenik et al. 2000).

Another landmark in the development of the Tyrolean Alps LTSER Platform came about in 1992, when Alexander Cernusca established an altitudinal transect of LTER grassland sites in the Stubai valley, near the village of Neustift, as part of the EU-FP3 project Integralp. They include a meadow on the valley floor at 970 m a.s.l. as well as three grasslands of differing land use (meadow, pasture, abandoned area) at the Kaserstattalm (1,820–1,970 m a.s.l.). A range of EU research projects (Integralp, Cernusca et al. 1992; Ecomont, Cernusca et al. 1999; Carbomont, Cernusca et al. 2008; and the ongoing projects Vital, Carbo-Extreme and GHG Europe), as well as numerous further international and national projects have contributed studies on ecosystem processes. These have addressed issues of productivity, C sequestration and greenhouse gas fluxes with a focus on $CO_2$, and more recently also methane, $N_2O$ and VOC, but also nitrogen cycling, the water balance and potential risks such as erosion and snow gliding (e.g. Tasser et al. 2003; Wohlfahrt et al. 2005, 2008; Bahn et al. 2008; Wieser et al. 2008; Tenhunen et al. 2009; Bamberger et al. 2010; Leitinger et al. 2010; Schmitt et al. 2010). Documentation of management history and vegetation dynamics of the whole Stubai valley is available all the way back to 1865, together with detailed information on the current socioeconomic situation and future land-use scenarios. Since 2001, the net exchanges of $CO_2$ and water vapour between the meadow on the valley floor and the atmosphere have been monitored at a high temporal resolution and contribute to the international Fluxnet database (e.g. Wohlfahrt et al. 2008; Groenendijk et al.

2011). On some grasslands in the Stubai valley, experiments have been carried out to underpin mechanisms determining the functioning of mountain grassland and to assess the impact of global changes (climate, land use) on ecological processes (e.g. Bahn et al. 2006, 2009). In consideration of the LTSER concept, the human dimension has been largely included in the Stubai valley research programme (see Sect. 21.4).

## 21.3.2 Global Change Effects on Semi-natural and Natural Ecosystems

Land-use and climate changes have been affecting the natural environment in the Tyrolean Alps in multiple ways. Land use has shaped landscape patterns over centuries, creating one of the most conspicuous human-shaped features of the Alps: species-rich semi-natural grasslands in the subalpine vegetation belt that have been used for hay-making and as pastures. Over the past decades, for socioeconomic reasons, increasing numbers of grasslands have been abandoned, often resulting in a regrowth of forests, which reduces biodiversity locally, both on the levels of species and landscape (Tasser and Tappeiner 2002). Such a reforestation of former open larch-meadows (where lightly interspersed larch trees were used as timber) has been well documented including in the area surrounding the LTER site in the Stubai valley (Tappeiner et al. 2008a). On the other hand, intensification of land management has been likewise shown to reduce species richness in subalpine grassland (Niedrist et al. 2009).

At higher altitudes, in the alpine and subnival zone, climate change has been shown to affect biodiversity by favouring an upward migration of plants typical for comparatively lower vegetation belts (Grabherr et al. 1994), as e.g. documented for the Mt. Schrankogel LTER Site (Pauli et al. 2007) and at Mt. Glungezer, a neighbouring peak of the Mt. Patscherkofel LTER Site (Bahn and Körner 2003). While such an upward migration of species may potentially increase species richness, it may also lead to the loss of species adapted specifically to subnival and nival environments by competitive exclusion and possibly due to niche shifts caused by global warming (Gottfried et al. 2002). As a particularly drastic example, Grabherr et al. (1995) pointed out that an upward shift of the treeline on Mt. Patscherkofel to the very peak of the mountain would eliminate 13 % of the (mostly grassland) species on this particular peak.

Changes in species composition, whether related to land-use or climate changes, may strongly affect biogeochemical cycles and related feedbacks to the environment. For example, abandonment of grasslands has been shown to affect the nutrient cycle, slowing down nitrogen mineralisation by changing the quality of plant litter and microbial community composition (Zeller et al. 2000, 2001). In combination with a cessation of fertilisation after abandonment, this leads to a reduction in nutrient availability, which also slows down carbon dynamics and reduces the net ecosystem exchange of $CO_2$ between grasslands and the atmosphere, as has been demonstrated for the complex grassland LTER site at Stubai valley (Schmitt et al. 2010). This is partly explained by physiological changes of plant species, which determine both the $CO_2$ fixation capacity and $CO_2$ losses to the atmosphere (Wohlfahrt et al. 2003; Bahn et al. 2006). Change in plant functional composition may also affect the impacts of climate extremes on the water cycle and the regional climate system. During heatwaves, grasslands initially increase evaporation and thus regional cooling more than do forests, however with increasing duration of hot and dry periods, they deplete soil moisture more strongly, which leads to stronger heating (Teuling et al. 2010).

Last but not least, global changes may also affect potential risks in mountain ecosystems. Surface runoff, the contribution of which to mountain torrent runoff and erosion is frequently underestimated, may be increased on pastures especially during extreme rainfall events. Managed meadows and pastures have been shown to be significantly less erodible than abandoned grasslands (Tasser et al. 2003). However, it is not the land-use activities themselves that lead to changes in erosion risks, but rather the direct or indirect effects on vegetation and soil properties. These include relative cover of grasses, herbs and dwarf shrubs as well as the total root length and the rooting density in main fracture depth (Tasser et al. 2003), abandonment initially leading to an increase in grass cover and lateron to an increase in the cover of dwarf shrubs and related effects on surface roughness (decreasing initially and increasing lateron). Snow gliding is a key component leading to natural hazards, i.e. avalanches and erosions, and due to ongoing global changes, has become a topic of major concern (Leitinger et al. 2008). Through changes in species composition, and related effects on surface roughness, abandonment of mountain grassland has been shown to affect the probability of snow gliding (Leitinger et al. 2008).

Effects of climate and land-use changes on ecosystems eventually feed back to human society. Ecosystem services, such as productivity, carbon sequestration, reduction of potential risks, the value of biodiversity for recreation (also feeding back to the attractiveness of municipalities and regions for tourism) may all be important players in the socio-ecological status and development of mountain areas. Current threats and mitigation opportunities of climate and land-use changes on natural ecosystems are manifold.

## 21.4 Relevance of an LTSER Perspective in the Tyrolean Alps

So far, the LTSER-related research in the Tyrolean Alps has been underdeveloped in comparison to natural science research. Two notable examples can be mentioned in this direction. One concerns the sub-region of Obergurgl. When UNESCO implemented the International Biological Program (IBP) around the world, much of today's research in ecology started, many new methods were developed and introduced in the field in order to monitor and analyse ecosystem processes and patterns. There was, however, one major problem within the IBP. Based almost entirely on ecological research, it could not contribute to finding solutions for problems of socio-ecological systems and their sustainable development. Hence, the follow-up programme of UNESCO, Man and the Biosphere (MAB), tried to include an understanding of socioeconomic activities that use and change ecosystems. This was a very modern approach for the 1970s and is reflected today in the LTSER concept of Long-Term Socio-Ecological Research. In the Austrian contribution to the MAB-6 project, led by Walter Moser, Obergurgl was chosen as a case in point. Scientists and local stakeholders worked together to generate data for a computer model. This was intended to serve to model the ecological and economic options for the future of the region as an aid to decision making about regional development (Price 1995). Even then, a team from disciplines as diverse as anthropology, botany, ethnology, geography, meteorology, microbiology, sociology and the soil sciences tried to capture and analyse the complex interplay of ecological and social dynamics (Patzelt 1987). MAB-6 in Obergurgl ran out of funding in 1979, but to this day it provides an excellent starting point for the Obergurgl Alpine Research Centre. This centre is the logistic and organisational hub of numerous studies on glacier and glacier foreland ecology, climate and geology of the high alpine sites in the inner Ötz valley area. For example, at the glacier foreland it was found that colonisation in bare-ground plots is limited by a lack of safe sites and is also dispersal-limited, and that during the primary succession on glacier foreland, species behave demographically like late-successional or climax species in secondary successions, mainly relying on survival of adult individuals (Erschbamer et al. 2008; Marcante et al. 2009).

A second example of an LTSER perspective in the Tyrolean Alps concerns research being undertaken in the Stubai valley where experiments have been carried out to underpin mechanisms determining the functioning in mountain grassland and to assess the impact of global changes (climate, land use) on ecological processes (e.g. Bahn et al. 2006, 2009). In view of an LTSER perspective extending beyond the pure natural sciences approach pursued at the LTER sites, in the Stubai valley the human dimension has been largely included in the research process. The intention has been to analyse the interaction of society with nature at various spatial and temporal scales and their cumulative effects, and to assess ecosystem and landscape services of the whole Stubai valley (e.g. Bayfield et al. 2008; Tappeiner et al. 2008a). This means integrating non-scientists in the projects and retranslating scientific findings into practice and education by means of transdisciplinary and participatory approaches. Major challenges to science lie in overcoming disciplinary boundaries and in combining experimental studies with modelling approaches and scenario techniques across several scales. Responding to the requirements of LTSER, in the Stubai valley we are trying to analyse the long-term interaction of ecological and social systems (including politics, economy and society), to identify current and future problem areas and to develop sustainable solutions.

## 21.5 Conclusions: Future Perspectives of the Tyrolean Alps LTSER Platform

Given the need for an integrated science-based monitoring and experimental evaluation of effects of global changes in climate and land use on mountain ecosystems, it would be highly desirable to strengthen the overall structure and the activities of the Tyrolean Alps LTSER Platform, based on the considerable research efforts made over the last two and more decades (cf. previous section), and to integrate it as a full-fledged platform in line with the criteria and structure laid down in LTSER Europe. Some of the sites contained in the Tyrolean Alps Platform fulfil the key criteria for LTSER site selection, as defined by Ohl et al. (2007), very well, including economic diversity, conservation relevant policy and participation by stakeholders, and internationally accessible data availability on demographic trends and different land-use and land-cover types and their past and future dynamics. Most importantly, some sites also fulfil the vulnerability criteria, which are of particular relevance in mountain environments (Körner 2003), including not only biodiversity aspects (Ohl et al. 2007), but expanding to encompass a range of biogeochemical processes concerning the carbon, water and nutrient cycles and related ecosystem services, whose relationships have been increasingly recognised (e.g. Chapin et al. 2000; Díaz et al. 2007). This has been particularly well documented for the sites at the Stubai Valley, which have been integrated into a larger-scale framework by stakeholder-based scenario development (Bayfield et al. 2008) and its implementation in assessments of the past and future landscape level carbon and water balance (Tappeiner et al. 2008a; Tenhunen et al. 2009). The recent research history at the Tyrolean Alps LTSER Platform has shown that a combination of sound monitoring of biogeochemistry, the experimental unravelling of underlying processes and the consideration of socioeconomic developments is a fruitful way forward, also increasing the value of LTSER sites for international projects and networks (e.g. Fluxnet, COST Action SIBAE, EU-FP7 and ERA-Net projects). Given this international context and perspective, all three priority research themes identified by the recently published White Paper "Next generation LTER" in Austria (Mirtl et al. 2010) should and will likely be key elements of future process-oriented ecosystem research activities in the Tyrolean Alps, including (1) the regulation of primary production, removal and accumulation of dead organic material in terrestrial and aquatic ecosystems, taking particular account of the problem posed by greenhouse gases, (2) recycling and transformation of carbon and other nutrients in natural and disturbed ecosystems, and (3) the impact of spatial-temporal patterns and the intensity of disturbances (including weather extremes) upon the stability of ecosystems. In view of this considerable task, however, the major challenge remains to take LTSER-relevant research from a project-by-project funding basis to a reliable long-term funding basis, which is inevitably required for developing perspectives that go beyond the scope of what is achievable by a patchwork of typical short-term research projects. Only based on a long-term perspective of secured funding will it be possible to develop and pursue a long-term perspective in assessing long-term effects of global changes on the highly sensitive Alpine ecosystem.

# References

Auer, I., Böhm, R., Jurkovic, A., Lipa, W., Orlik, A., Potzmann, R., Schoner, W., Ungersbock, M., Matulla, C., Briffa, K., Jones, P., Efthymiadis, D., Brunetti, M., Nanni, T., Maugeri, M., Mercalli, L., Mestre, O., Moisselin, J. M., Begert, M., Muller-Westermeier, G., Kveton, V., Bochnicek, O., Stastny, P., Lapin, M., Szalai, S., Szentimrey, T., Cegnar, T., Dolinar, M., Gajic-Capkaj, M., Zaninovic, K., Majstorovic, Z., & Nieplova, E. (2007). HISTALP – Historical instrumental climatological surface time series of the greater Alpine region 1760–2003. *Journal of Climatology, 27*, 17–46.

Bahn, M., & Körner, C. H. (2003). Recent increases in summit flora caused by warming in the Alps. In L. Nagy, G. Grabherr, C. H. Körner, & D. B. A. Thompson (Eds.), *Alpine biodiversity in Europe* (Ecological Studies 167, pp. 437–442). Berlin: Springer.

Bahn, M., Knapp, M., Garajova, Z., Pfahringer, N., & Cernusca, A. (2006). Root respiration in temperate mountain grasslands differing in land use. *Global Change Biology, 12*, 995–1006.

Bahn, M., Rodeghiero, M., Anderson-Dunn, M., Dore, S., Gimeno, S., Drösler, M., Williams, M., Ammann, C., Berninger, F., Flechard, C., Jones, S., Balzarolo, M., Kumar, S., Newesely, C., Priwitzer, T., Raschi, A., Siegwolf, R., Susiluoto, S., Tenhunen, J., Wohlfahrt, G., & Cernusca, A. (2008). Soil respiration in European grasslands in relation to climate and assimilate supply. *Ecosystems, 11*, 1352–1367.

Bahn, M., Schmitt, M., Siegwolf, R., Richter, A., & Brüggemann, N. (2009). Does photosynthesis affect grassland soil respired $CO_2$ and its carbon isotope composition on a diurnal timescale? *New Phytologist, 182*, 452–460.

Bamberger, I., Hörtnagl, L., Schnitzhofer, R., Graus, M., Ruuskanen, T., Müller, M., Dunkel, J., Wohlfahrt, G., & Hansel, A. (2010). BVOC fluxes above mountain grassland. *Biogeosciences, 7*, 1413–1424.

Bayfield, N., Barancok, P., Furger, M., Sebastia, M. T., Domínguez, G., Lapka, M., Cudlinova, E., Vescovo, L., Gianelle, D., Cernusca, A., Tappeiner, U., & Drösler, M. (2008). Stakeholder Perceptions of the Impacts of Rural Funding Scenarios on Mountain Landscapes across Europe. *Ecosystems, 11*, 1368–1382.

Becker, A., Körner, C. H., Björnsen Gurung, A., Brun, J., Guisan, A., Haeberli, W., & Tappeiner, U. (2007). Altitudinal Gradient Studies and Highland-lowland Linkages in Mountain Biosphere Reserves. *Mountain Research and Development, 27*, 58–65.

Beniston, M. (2006). Mountain weather and climate: A general overview and a focus on climatic change in the Alps. *Hydrobiologia, 562*, 3–16.

Borsdorf, A. (2004). Innsbruck. From city to cyta? Outskirt development as an indicator of spatial, economic and social development. In G. Dubois-Taine (Ed.), *From Helsinki to Nicosia. Eleven case studies & synthesis. European Cities. Insights on Outskirts* (pp. 75–96). Brussels: COST Office C10.

Cernusca, A., Tappeiner, U., Agostini, A., Bahn, M., Bezzi, A., Egger, R., Kofler, R., Newesely, C., Orlandi, D., Prock, S., Schatz, H., & Schatz, I. (1992). Ecosystem research on mixed grassland/woodland ecosystems. First results of the EC-STEP-project INTEGRALP on Mt. Bondone. *Studi trentini di scienze naturali, Acta Biologica, 67*, 99–133.

Cernusca, A., Tappeiner, U., & Bayfield, N. (Eds.). (1999). *Land-use changes in European mountain ecosystems. ECOMONT – Concept and results*. Berlin: Blackwell Science.

Cernusca, A., Bahn, M., Berninger, F., Tappeiner, U., & Wohlfahrt, G. (2008). Effects of land use changes on sources, sinks and fluxes of carbon in European mountain grasslands. *Ecosystems, 11*, 1335–1337.

Chapin, F. S., III, Zavaleta, E. S., Eviner, V. T., Naylor, R. L., Vitousek, P. M., Reynolds, H. L., et al. (2000). Consequences of changing biodiversity. *Nature, 405*, 234–242.

Díaz, S., Lavorel, S., de Bello, F., Quétier, F., Grigulis, K., & Robson, M. T. (2007). Incorporating plant functional diversity effects in ecosystem service assessments. *Proceedings of the National Academy of Sciences, 104*, 20684–20689.

EEA. (2004). *Impacts of Europe's changing climate – An indicator-based assessment* (EEA report No. 2). Copenhagen: EEA.

Ehrendorfer, F. (2004). Anton Kerner von Marilaun als Pionier der botanischen Evolutionsforscher. In M. Petz-Grabenbauer & M. Kiehn (Eds.), *Anton Kerner von Marilaun (1831–1898)* (pp. 65–76). Vienna: Verlag der Österr. Akademie der Wiss.

Erschbamer, B., Ruth, N. S., & Winkler, E. (2008). Colonization processes on a central Alpine glacier foreland. *Journal of Vegetation Science, 19*, 855–862.

Escaravage, N., & Wagner, J. (2004). Pollination effectiveness and pollen dispersal in a Rhododendron ferrugineum (Ericaceae) population. *Plant Biology, 6*, 606–615.

Fischer, A. (2010). Glaciers and climate change: Interpretation of 50 years of direct mass balance of Hintereisferner. *Global and Planetary Change, 71*, 13–26.

Fischer, A., & Markl, G. (2009). Mass balance measurements on Hintereisferner, Kesselwandferner and Jamtalferner 2003 to 2006: database and results. *Zeitschrift für Gletscherkunde und Glazialgeologie, 42*, 47–83.

Gärtner, G. (2004). Anton Kerner und die Botanik an der Universität Innsbruck in den Jahren 1860–1878. In M. Petz-Grabenbauer & M. Kiehn (Eds.), *Anton Kerner von Marilaun (1831–1898)* (pp. 27–36). Vienna: Verlag der Österr. Akademie der Wiss.

Gottfried, M., Pauli, H., Reiter, K., et al. (2002). Potential effects of climate change on alpine and nival plants in the Alps. In C. Körner & E. M. Spehn (Eds.), *Mountain biodiversity – A global assessment* (pp. 213–223). London/New York: Parthenon Publishing.

Grabherr, G., Gottfried, M., & Pauli, H. (1994). Climate effects on mountain plants. *Nature, 369*, 448–448.

Grabherr, G., Gottfried, M., Gruber, A., et al. (1995). Patterns and current changes in alpine plant diversity. In F. S. Chapin & C. Körner (Eds.), *Arctic and Alpine biodiversity: Patterns, causes and ecosystem consequences* (Ecological Studies 113, pp. 167–181). Berlin: Springer.

Grabherr, G., Gottfried, M., & Pauli, H. (2010). Climate change impacts in Alpine environments. *Geography Compass, 4*, 1133–1153.

Groenendijk, M., Dolman, A. J., van der Molen, M. K., Arneth, A., Delpierre, N., Gash, J. H. C., Leuning, R., Lindroth, A., Richardson, A. D., Verbeek, H., & Wohlfahrt, G. (2011). Assessing parameter variability in a photosynthesis model within and between plant functional types using global Fluxnet eddy covariance data. *Agricultural and Forest Meteorology, 151*, 22–38.

Hartmann, T. (2008). The lost origin of chemical ecology in the late 19th century. *Proceedings of the National Academy of Sciences, 105*, 4541–4546.

Huelber, K., Ertl, S., Gottfried, M., Reiter, K., & Grabherr, G. (2005). Gourmets or gourmands? Diet selection by large ungulates in high-alpine plant communities and possible impacts on plant propagation. *Basic and Applied Ecology, 6*, 1–10.

Huelber, K., Gottfried, M., Pauli, H., Reiter, K., Winkler, M., & Grabherr, G. (2006). Phenological responses of snowbed species to snow removal dates in the Central Alps: Implications for climate warming. *Arctic, Antarctic, and Alpine Research, 38*, 99–103.

IPCC. (2007). Climate change 2007. In M. L. Parry, O. F. Canziani, J. P. Palutikof, P. J. van der Linden, & C. E. Hanson (Eds.), *Fourth assessment report of the intergovernmental panel on climate change*. Cambridge: Cambridge University Press.

Ives, J. D., Messerli, B., & Spiess, E. (1997). Mountains of the world: A global priority: Chapter 1. In B. Messerli & J. D. Ives (Eds.), *Mountains of the world: A global priority* (pp. 1–16). London, New York: Parthenon.

Kamenik, C., Koinig, K. A., Schmidt, R., Appleby, P. G., Dearing, J. A., Lami, A., Thompson, R., & Psenner, R. (2000). Eight hundred years of environmental changes in a high Alpine lake (Gossenköllesee, Tyrol) inferred from sediment records. *Journal of Limnology, 59*(Suppl. 1), 43–52.

Keller, F., & Körner, C. (2003). The role of photoperiodism in alpine plant development. *Arctic, Antarctic, and Alpine Research, 35*, 361–368.

Körner, C. (2003). *Alpine plant life – Functional plant ecology of high mountain ecosystems*. Heidelberg: Springer.

Larcher, W. (1977a). Ergebnisse des IBP-Projektes "Zwergstrauchheide Patscherkofel". *Sitzungsbericht der Österreichischen Akademie der Wissenschaften (Wien) Math Naturwiss Kl I, 186*, 301–371.

Larcher, W. (1977b). Produktivität und Überlebensstrategien von Pflanzen und Pflanzenbeständen im Hochgebirge. *Sitzungsbericht der Österreichischen Akademie der Wissenschaften (Wien) Math Naturwiss Kl I, 186*, 373–386.

Larcher, W. (2003). *Physiological plant ecology*. Berlin/New York: Springer.

Leitinger, G., Höller, P., Tasser, E., Walde, J., & Tappeiner, U. (2008). Development and validation of a spatial snow-glide model. *Ecological Modelling, 211*, 363–374.

Leitinger, G., Tasser, E., Newesely, C., Obojes, N., & Tappeiner, U. (2010). Seasonal dynamics of surface runoff in mountain grassland ecosystems differing in land use. *Journal of Hydrology, 385*, 95–104.

Marcante, S., Winkler, E., & Erschbamer, B. (2009). Population dynamics along a primary succession gradient: do alpine species fit into demographic succession theory? *Annals of Botany, 103*, 1129–1143.

MEA. (2005). *Millennium ecosystem assessment*. Washington, DC: Island Press.

Mirtl, M., Bahn, M., Battin, T., Borsdorf, A., Englisch, M., Gaube, V., Grabherr, G., Gratzer, G., Haberl, H., Kreiner, D., Richter, A., Schindler, S., Tappeiner, U., Winiwarter, V., & Zink, R. (2010) *"Next Generation LTER" in Austria – On the status and orientation of process oriented ecosystem research, biodiversity and conservation research and socio-ecological research in Austria* (Vol. 1). LTER-Austria Series. ISBN 978-3-901347-94-8. Retrieved from http://www.lter-austria.at

Moser, W., Brzoska, W., Zachhuber, K., & Larcher, W. (1977). Ergebnisse des IBP-Projektes "Hoher Nebelkogel 3184 m". *Sitzungsbericht der Österreichischen Akademie der Wissenschaften (Wien) Math Naturwiss Kl I, 186*, 378–419.

Neuner, G., & Pramsohler, M. (2006). Freezing and high temperature thresholds of photosystem 2 compared to ice nucleation, frost and heat damage in evergreen subalpine plants. *Physiologia Plantarum, 126*, 196–204.

Niedrist, G., Tasser, E., Lüth, C., Dalla, V. J., & Tappeiner, U. (2009). Plant diversity declines with recent land use changes in European Alps. *Plant Ecology, 202*, 195–210.

Oberhuber, W., Kofler, W., Pfeifer, K., Seeber, A., Gruber, A., & Wieser, G. (2008). Long-term changes in tree-ring climate relationships at Mt. Patscherkofel (Tyrol, Austria) since the mid-1980s. *Trees, 22*, 31–40.

Ohl, C., Krauze, K., & Grünbühel, C. (2007). Towards an understanding of long-term ecosystem dynamics by merging socio-economic and environmental research: Criteria for long-term socio-ecological research sites selection. *Ecological Economics, 63*, 383–391.

Patzelt, G. (Ed.). (1987). *MaB-Projekt Obergurgl* (Veröffentlichung des Österreichischen MaB-Programm 10). Innsbruck: Universitätsverlag Wagner.

Pauli, H., Gottfried, M., Reiter, K., Klettner, C., & Grabherr, G. (2007). Signals of range expansions and contractions of vascular plants in the high Alps: Observations (1994–2004) at the GLORIA master site Schrankogel, Tyrol, Austria. *Global Change Biology, 13*, 147–156.

Pisek, A. (1971). Zur Geschichte der experimentellen Ökologie (besonders des in Innsbruck hierzu geleisteten Beitrages). *Berichte der Deutschen Botanischen Gesellschaft, 84*, 365–379.

Price, M.F. (1995). *Mountain research in Europe: An overview of MAB research from the Pyrenees to Siberia* (MAB Book Series No. 14). Paris: UNESCO, and Carnforth: Parthenon.

Psenner, R., & Schmidt, R. (1992). Climate-driven pH control of remote alpine lakes and effects of acid deposition. *Nature, 56*, 781–783.

Rose, K. C., Williamson, C. E., Saros, J. E., Sommaruga, R., & Fischer, J. M. (2009). Differences in UV transparency and thermal structure between alpine and subalpine lakes: Implications for organisms. *Photochemical and Photobiological Sciences, 8*, 1244–1256.

Schmitt, M., Bahn, M., Wohlfahrt, G., Tappeiner, U., & Cernusca, A. (2010). Land use affects the net ecosystem $CO_2$ exchange and its components in mountain grasslands. *Biogeosciences, 7*, 2297–2309.

Schröter, D., Cramer, W., Leemans, R., Prentice, I. C., Araújo, M. B., Arnell, N. W., Bondeau, A., Bugmann, H., Carter, T. R., Gracia, C. A., de la Vega-Leinert, A. C., Erhard, M., Ewert, F., Glendining, M., House, J. I., Kankaanpää, S., Klein, R. J. T., Lavorel, S., Lindner, M., Metzger,

M. J., Meyer, J., Mitchell, T. D., Reginster, I., Rounsevell, M., Sabaté, S., Sitch, S., Smith, B., Smith, J., Smith, P., Sykes, M. T., Thonicke, K., Thuiller, W., Tuck, G., Zaehle, S., & Zierl, B. (2005). Ecosystem service supply and vulnerability to global change in Europe. *Science, 310,* 1333–1337.

Sommaruga, R., & Augustin, G. (2006). Seasonality in UV transparency of an alpine lake is associated to changes in phytoplankton biomass. *Aquatic Sciences, 68,* 129–141.

Sommaruga-Wögrath, S., Koinig, K. A., Schmidt, R., Tessadri, R., Sommaruga, R., & Psenner, R. (1997). Temperature effects on the acidity of remote alpine lakes. *Nature, 387,* 64–67.

Streifeneder, T., Tappeiner, U., Ruffini, F. V., Tappeiner, G., & Hoffmann, C. (2007). Selected aspects of agro-structural change within the Alps: A comparison of harmonised agro-structural indicators on a municipal level. *Journal of Alpine Research, 3,* 27–40.

Tappeiner, U., Tasser, E., Leitinger, G., & Tappeiner, G. (2006). Landnutzung in den Alpen: historische Entwicklung und zukünftige Szenarien. In R. Psenner & R. Lackner (Eds.), *Die Alpen im Jahr 2020.* Alpine Space – Man & Environment (Vol. 1, pp. 23–39). Innsbruck: Innsbruck University Press.

Tappeiner, U., Tasser, E., Leitinger, G., Cernusca, A., & Tappeiner, G. (2008a). Effects of historical and likely future scenarios of land use on above- and below-ground vegetation carbon stocks of an Alpine valley. *Ecosystems, 11,* 1383–1400.

Tappeiner, U., Borsdorf, A., & Tasser, E. (Eds.). (2008b). *Mapping the Alps.* Heidelberg: Spektrum Verlag.

Taschler, D., Beikircher, B., & Neuner, G. (2004). Frost resistance and ice nucleation in leaves of five woody timberline species measured in situ during shoot expansion. *Tree Physiology, 24,* 331–337.

Tasser, E., & Tappeiner, U. (2002). The impact of land-use changes in time and space on vegetation distribution in mountain areas. *Applied Vegetation Science, 5,* 173–184.

Tasser, E., Mader, M., & Tappeiner, U. (2003). Effects of land use in alpine grasslands on the probability of landslides. *Basic and Applied Ecology, 4,* 271–280.

Tenhunen, J., Geyer, R., Adiku, S., Reichstein, M., Tappeiner, U., Bahn, M., Cernusca, A., Dinh, N. Q., Kolcun, O., Lohila, A., Otieno, D., Schmidt, M., Schmitt, M., Wang, Q., Wartinger, M., & Wohlfahrt, G. (2009). Influences of changing land use and $CO_2$ concentration on ecosystem and landscape level carbon and water balances in mountainous terrain of the Stubai valley, Austria. *Global and Planetary Change, 67,* 29–43.

Teuling, A. J., Seneviratne, S. I., Stöckli, R., Reichstein, M., Moors, E., Ciais, P., Luyssaert, S., van den Hurk, B., Ammann, C., Bernhofer, C., Dellwik, E., Gianelle, D., Gielen, B., Grünwald, T., Klumpp, K., Montagnani, L., Moureaux, C., Sottocornola, M., & Wohlfahrt, G. (2010). Contrasting response of European forest and grassland energy exchange to heat waves. *Nature Geoscience, 3,* 722–727.

Wieser, G., & Tausz, M. (Eds.). (2007). *Trees at their upper limit. treelife limitation at the Alpine timberline* (Series: Plant Ecophysiology Vol. 5). Berlin: Springer.

Wieser, G., Hammerle, A., & Wohlfahrt, G. (2008). The water balance of grassland ecosystems in the Austrian Alps. *Arctic, Antarctic, and Alpine Research, 40,* 439–445.

Wohlfahrt, G., Bahn, M., Newesely, C., Sapinsky, S., Tappeiner, U., & Cernusca, A. (2003). Canopy structure versus physiology effects on net photosynthesis of mountain grasslands differing in land use. *Ecological Modelling, 170,* 407–426.

Wohlfahrt, G., Anfang, C., Bahn, M., Haslwanter, A., Newesely, C., Schmitt, M., Drösler, M., Pfadenhauer, J., & Cernusca, A. (2005). Quantifying nighttime ecosystem respiration of a meadow using eddy covariance, chambers and modelling. *Agricultural and Forest Meteorology, 128,* 141–162.

Wohlfahrt, G., Hammerle, A., Haslwanter, A., Bahn, M., Tappeiner, U., & Cernusca, A. (2008). Seasonal and inter-annual variability of the net ecosystem $CO_2$ exchange of a temperate mountain grassland: Effects of weather and management. *Journal of Geophysical Research, 113,* D08110. doi:10.1029/2007JD009286.

Wöhlke, W. (1969). Kulturlandschaft als Funktion von Veränderlichen: Überlegungen zur dynamischen Betrachtung in der Kulturgeographie. *Geographische Rundschau, 21*, 298–308.

Zeller, V., Bahn, M., Aichner, M., & Tappeiner, U. (2000). Impact of land-use changes on nitrogen mineralization in subalpine grasslands in the Southern Alps. *Biology and Fertility of Soils, 31*, 441–448.

Zeller, V., Bardgett, R. D., & Tappeiner, U. (2001). Site and management effects on soil microbial properties of subalpine meadows: A study of land abandonment along a north–south gradient in the European Alps. *Soil Biology and Biochemistry, 33*, 639–649.

Zimmermann, P., Tasser, E., Leitinger, G., & Tappeiner, U. (2010). Effects of land-use and land-cover pattern on landscape-scale biodiversity in the European Alps. *Agriculture, Ecosystem and Environment, 139*, 13–22.

# Chapter 22
# Integrated Monitoring and Sustainability Assessment in the Tyrolean Alps: Experiences in Transdisciplinarity

Willi Haas, Simron Jit Singh, Brigitta Erschbamer, Karl Reiter, and Ariane Walz

**Abstract** The chapter is an experience in transdisciplinarity illustrated by the case of the Upper Ötztal, part of the Tyrolean LTSER Platform in the Austrian Alps. In this effort, the search was for an effective framework for integrated monitoring that would not be limited to observing and monitoring the state of nature alone, but one that would assess and guide overall (regional) sustainability with a focus on the interaction between the natural and the social realms. To this end, the chapter proposes an integrated monitoring and sustainability assessment scheme that has been developed in the context of biosphere reserves by our team and might potentially be useful for many LTSER Sites. Applying this scheme to the Upper Ötztal, this chapter offers various scientific insights into the social, interaction and natural sphere of the study area. The transdisciplinary component is captured in the scenario workshop where these insights were discussed with local stakeholders to better understand their views, interests and developmental perspectives. Despite challenges that underlie transdisciplinary processes, the chapter highlights the relevance of engaging local communities as part of a self-organising and self-maintaining socio-ecological system when it comes to addressing questions of regional sustainability.

---

W. Haas (✉) • S.J. Singh, Ph.D.
Institute of Social Ecology Vienna (SEC), Alpen-Adria Universitaet Klagenfurt,
Wien, Graz, Schottenfeldgasse 29/5, Vienna 1070, Austria
e-mail: willi.haas@aau.at; simron.singh@aau.at

B. Erschbamer, Ph.D.
Institute of Botany, Innsbruck, Austria
e-mail: brigitta.erschbamer@uibk.ac.at

K. Reiter, Ph.D.
Department of Conservation Biology, Vegetation and Landscape Ecology,
University of Vienna, Vienna, Austria
e-mail: karl.reiter@univie.ac.at

A. Walz, Ph.D.
Potsdam Institute for Climate Impact Research (PIK), Potsdam, Germany
e-mail: ariane.walz@pik-potsdam.de

**Keywords** Sustainability assessment • Integrated monitoring • Biosphere reserve • Participatory scenario development • Transdisciplinary research • Ötztal • Alps

## 22.1 Introduction

In recent years there has been a growing acceptance within sustainability science (Kates et al. 2001) of the relevance of engaging social actors in addressing complex societal problems. It is argued that the engagement of a wide range of stakeholders[1] helps to achieve research results that are more meaningful in terms of their adequacy to inform policy and socially responsible behaviour. The emergence of what later came to be called 'transdisciplinary research' is an outcome of such deliberations. Transdisciplinary research goes beyond transgressing scientific disciplines within academia to also engage stakeholders in problem definition and solution seeking. This is based on the premise that knowledge exists and is produced in societal fields other than science (Klein 2004; Pereira and Funtowicz 2006; Hirsch Hadorn et al. 2008).

Transdisciplinary approaches have often proved to be beneficial in the context of regional development. In order to foster sustainable development at local or sub-national scales (e.g. in LTSER Platforms), perspectives from science, management and other stakeholders need to be well integrated to allow for viable outcomes in terms of research, policy and monitoring (Stoll-Kleemann and Welp 2008). In this contribution, we present an example of an application of transdisciplinarity and its challenges to a region in the Austrian Alps, the Gurgler Kamm Biosphere Reserve (Upper Ötztal), which is part of the Tyrolean Alps LTSER Platform (see Tappeiner et al., Chap. 21 in this volume). To a large extent, both biosphere reserves and LTSER Platforms share common goals: to promote sustainability research and education and advance a showcase model for sustainable development.

We begin with a brief overview of the various 'transdisciplinary' schools of thought and their main features within the broader scheme of science-society interaction. We then present our transdisciplinary framework for integrated monitoring and stakeholder involvement at the local level, with possible engagement of stakeholders from higher levels. The following part of the chapter focuses on the Austrian Gurgler Kamm Biosphere Reserve case study where we present outcomes from the integrated monitoring and sustainability assessment by stakeholders.

---

[1] In this chapter, the term "stakeholders" and "non-science actors" is used interchangeably. While in a broad sense, scientists are also stakeholders, here this notion refers only to social actors or interest groups outside the scientific community.

## 22.2 The Emergence and Main Facets of Transdisciplinary Science

Several notable attempts have been made to reconstruct the history of transdisciplinary research (e.g. Balsiger 2003; Bogner et al. 2010; Farrell et al. 2011, Hirsch-Hadorn et al. 2008; Klein 1990; Mobjörk 2009). However, as a variety of terms and labels have been used in the past to refer to science-society interactions across research traditions, the outcome of such reviews lacks coherence. Hence the brief review provided here is far from being complete. We focus only on some selected "variants of transdisciplinary research", the scholars of which explicitly discuss their own approach under this label. Research traditions that are inherently organised around the science-society interface but do not explicitly label themselves as 'transdisciplinary' are excluded from this discussion. These are, for example, anthropology (Speed 2006; Balter 2010), sociology (Hale 2008; Burawoy 2004), development studies (Cooke and Kothari 2001; Irvine et al. 2004) and public health (Wilcox and Kueffer 2008).

The emergence of transdisciplinary research fields as it is now called had its own predecessors (see Table 22.1). An early approach was "action research" (Lewin 1951), characterised by a belief in socially constructed and subjective multiple realities. Ranging from unemployment, religious, racial and educational to sustainability concerns later on, action researchers engaged with social realities in a cyclical process of action and reflection. In the 1970s, a group of scholars focussed on seeking solutions to societal problems, and in doing so, emphasised the integration of a wide range of disciplines. In Europe, the main advocates of "trandisciplinarity" (as it was then termed) were Erich Jantsch (1972), Jean Piaget (1973) and Jürgen Mittelstraß (1992). In order to rescue science from the threat of becoming irrelevant, these scholars felt the need to engage research in planning and innovation for society at large "with a focus on the sphere outside the research community" (Mittelstraß 1992). In other words, here the research questions are of societal relevance, but the assumption is that science can substantially contribute to these questions and to solution seeking. Mainly deliberative, their efforts laid the foundations for later transdisciplinary research that was applied to a greater extent in nature. In the US, such approaches were and still are referred to as "integrative studies" (Klein 1996).

In more recent decades, a new variant of transdisciplinarity is "post-normal science" (Funtowicz and Ravetz 1993) that explicitly emphasises the inclusion of voices from civil society. Sustainability problems are complex societal problems and manifest themselves in forms such as the loss of biodiversity and the depletion of natural resources, climate change, bovine spongiform encephalopathy (BSE), and soil degradation. This often implies that when using science for policy-making, long term consequences may persist and scientists and policy-makers are confronting issues where, "facts are uncertain, values in dispute, stakes high and decisions urgent" (Funtowicz and Ravetz 1994). In such cases, scientists cannot provide any useful input without interacting with the rest of society and the rest of the society cannot perform any sound decision making without interacting with scientists.

Table 22.1 Variants of "transdisciplinary" ("td") research and main features

| Variants within "td" research | Main advocates | Main features of "td" research | | | |
|---|---|---|---|---|---|
| | | Integration of disciplines | Focus on societal problems | Engaging non-science actors | Support in decision making |
| Action research | Jahoda & Lazarsfeld (1933), Lewin (1951), Greenwood & Levin (1998), Reason (2001), Burns (2007) | ++ | ++ | − | − |
| Early transdisciplinary research tradition | Jantsch (1972), Piaget (1973), Kockelmans (1979), Mittelstraß (1992), Nicolescu (1992) | ++ | ++ | − | − |
| Integrative studies | Miller (1982), Newell (1983), Klein (1996) | ++ | ++ | − | − |
| Post-normal science | Funtowitz & Ravetz (1993), Munda (2004) | + | ++ | ++ | + |
| Participatory technology assessment | Sclove (1995), Durant (1999), Joss & Bellucci (2002) | ++ | ++ | ++ | ++ |
| Mode 2 research | Gibbons, Limoges & Nowotny (1994) | ++ | ++ | ++ | − |
| Current transdisciplinary research tradition | Balsiger et al. (1996), Scholz (2000), Klein (2001), Pohl & Hirsch Hadorn (2007), McDonald, Bammer & Deare (2009), Jahn (2010) | ++ | ++ | ++ | + |

++ Strong feature with wide acceptance amongst scholars of this variant
+ Week feature of this variant with scholars putting more or less attention to it
− Scholars of this variant place little or no emphasis on this feature

Critiques and doubts concerning new scientific and technological developments and the possibility that these might lead to sometimes unintended side effects (such as higher inequalities in terms of access and benefit distribution, or environmental risks) have meant that technology assessments have gained more and more importance (Joss and Bellucci 2002). This process, first triggered by political requests in the US senate in the 1970s, evolved further over the years with Europe playing a leading role thereafter. A wide range of schools of thought have contributed significantly to its development, including system analysis, policy science, democratic theory, sociology of scientific knowledge and communication theory. While in the US the approach taken in assessments was rather analytical, European technology assessments made efforts to ensure the fruitful inclusion of values and interests by organising participatory procedures that included those who were potentially being affected by such developments. The aim of Participatory Technology Assessments (PTA) is to provide advice to policy-makers and to encourage wider public debate about socio-technological developments. With the participation of those affected, PTA move beyond disciplines working together to create a transdisciplinary approach with a strong political dimension.

In 1994, Gibbons and colleagues anticipated the emergence of a new mode of knowledge production that they called 'Mode 2' as opposed to the traditional disciplinary production of scientific discovery (which they call 'Mode 1') (Gibbons et al. 1994). While the production of Mode 1 knowledge is characterised by "the hegemony of theoretical or, at any rate, experimental science; by an internally driven taxonomy of disciplines; and by the autonomy of scientists and their host institutions, the universities" (Nowotny et al. 2003, p. 179), Mode 2 is "knowledge which emerges from a particular context of application with its own distinct theoretical structures, research methods and modes of practice, but which may not be locatable on the disciplinary map." (Gibbons et al. 1994, p. 168) Their use of the term 'transdisciplinarity' reflected what they considered to be an increasing body of research dealing with problems emerging in the context of application, closely attuned to social needs and aspirations and not circumscribed in any existing disciplinary field.

In current European scientific debate on transdisciplinarity, a growing stock of publications is provides a continuous contribution to a formidable body of knowledge. Several networks in German speaking countries, with increasing international participation, play a crucial role in the formation of this cross-cutting field (especially the Swiss td-net with its annual conferences and a data base of publications comprising about 2,000 titles http://www.transdisciplinarity.ch/e/Bibliography/Publications.php). Following up on these efforts, the foundation of an international platform for discussion and promotion of interdisciplinary and transdisciplinary research, teaching, and policy was launched in 2011. Founding institutions are mainly from the US (Association of Integrative Studies) and the Swiss td-net. Two prominent scholars of this initiative, Christian Pohl and Gertrude Hirsch Hadorn (2007), suggest that transdisciplinary research should deal with socially relevant problems in such a way that it can: "(a) grasp the complexity of problems, (b) take into account the diversity of life-world and scientific perceptions of problems, (c) link abstract and case-specific knowledge, and (d) constitute knowledge and practices that promote what is perceived to be the common good. Participatory research and collaboration between disciplines are the means of meeting requirements (a)–(d) in the research process."(p. 30)

In most of the transdisciplinary variants, engaging stakeholders is seen to be essential. To this end, several methods to include society in participatory processes – from problem framing to seeking solutions – are in use. Within the field of participatory technology assessment, the involvement of non-science actors has taken the most elaborated and reflected forms, including Citizens' Panels, Scenario Workshops, Round Tables and Consensus Conferences, 21st Century Town Meetings, Charrettes, Citizens' Juries, Technology Festivals, World Cafés, Deliberative Polling, Expert Panels, Focus Groups, Planning Cells, and PAME (Participatory Assessment, Monitoring and Evaluation) (for an excellent overview of these approaches, see Elliott et al. 2005). In addition to these participation tools, Societal Multi-Criteria Evaluation (SMCE) in post-normal science has developed formalised procedures for the entire research process aiming at a participatory strategic integrated assessment of a problem. The emphasis is explicitly on the quality of the "decision process" rather than on the "final choice" (Munda 2004, 2008).

Among the various methods used to engage stakeholders, scenario workshops have proven to be both practical and useful (e.g. Loukopoulos and Scholz 2004; Biggs et al. 2007; Walz et al. 2007). Scenarios are best employed if big changes are anticipated and if expected challenges appear to be rather complex. Such a situation requires thorough consideration of the many driving forces and alternative pathways. Scenario workshops are interactive and are perfectly suited to the development of shared story lines and visions of alternative futures. They promote cross-group communication on the pros and cons of various scenarios, and if the goal be such, to finally arrive at a consensus on a future scenario or perspective. This method of engaging stakeholders is best suited to developing an informed understanding of future local or regional sustainability.

While an inclusive science with an emphasis on integration across disciplines holds enormous potential for addressing sustainability challenges, the limitations of these approaches must be seriously considered (Wallner and Wiesmann 2009). Seeking consensus among a variety of perspectives and interests can be frustrating. For example, in the case of nature protection, opinions vary among local inhabitants, businesses, tourists, and political actors on whether the protected area is a constraint or a resource for local development (Siddiq Khan and Bhagwat 2010). Finding answers to such questions is not easy. Ideally, in such an instance, a successful transdisciplinary process must make mutual benefits visible to all stakeholders.

## 22.3 Integrated Monitoring and Assessment for LTSER Platforms: A Transdisciplinary Approach

Integration of disciplines and engagement of non-science actors can be a highly challenging task if not aided by a sound conceptual framework. In our study of the Gurgler Kamm region, the search was for an effective framework that would not be limited to addressing environmental concerns alone by observation and monitoring the state of nature, but that would also provide guidance for a transdisciplinary process with a focus on overall (regional) sustainability. The need, therefore, was for an integrated monitoring (by integrating disciplinary knowledge) and sustainability assessment (by engaging stakeholders) scheme. Without appropriate mechanisms for integrated monitoring and sustainability assessment by stakeholders, it is impossible to tell whether a region is becoming more or less sustainable over a given period of time, and to understand how this translates into future management decisions. This is a far more complex task than monitoring the various features of the state of the environment alone and requires a focus on the interaction between social and natural systems. In this section, we present an integrated monitoring and assessment scheme that has been developed in the context of biosphere reserves by our team (Fischer-Kowalski et al. 2008) and might potentially be useful for many LTSER Sites.

In the context of evaluating sustainability, it is useful to concentrate monitoring efforts on those elements of the social sphere that have a direct causal effect on the ecosystem. In other words, to focus on those socioeconomic activities that strongly relate to and alter the natural environment. Much of the interaction between the social and the natural system can be conceptualised as an exchange process of investment (intervention) and benefits that cause environmental pressures. Human interventions in designated areas such as an LTSER Platform can be varied: building tourist infrastructure, fishing, farming, simply walking for pleasure, or undertaking a conservation measure. These interventions from a social point of view are performed in pursuit of certain benefits: a harvest, a beautiful view, an income from the sale of souvenirs or stable ecosystems that attract tourists. These benefits may be achieved because ecosystems, by their very functioning, allow for certain 'services' that can be utilised by humans (such as soil fertility, beautiful landscapes or slopes for skiing). Looking at the very same intervention from an ecological perspective, it always represents an impact on the ecosystem, be it stabilising or destabilising the system.

In Fig. 22.1, we propose to monitor exactly these use-related interactions between the social and the natural systems since they link societal benefits of services with impacts on ecosystems. Examples for such indicators include, for instance, construction activities or visitor frequency at remote places. However, to undertake a sustainability evaluation we also need a set of indicators related to the state of the environment that are affected by human interventions. Thus, the distribution of certain plant species is of interest if these are strongly influenced by certain anthropogenic pressures or by conservation measures. The other set of indicators required are those that relate to activities in the social sphere. These indicators help to understand those inner societal dynamics that matter most for increasing or decreasing pressures (such as population, economic activities, consumption patterns, or even conservation policies). These indicators would be subject to scientific analysis as non-routine scientific efforts to answer specific research questions. Monitoring the natural, social and interactive sphere requires integration of disciplines, most effective if structured around a common conceptual and theoretical framework.

In the absence of universally agreed benchmarks, assessing the ecological sustainability of various economic activities is difficult, if not impossible; hence it is more rewarding to focus on dynamics. As unsustainable trends concerning the use

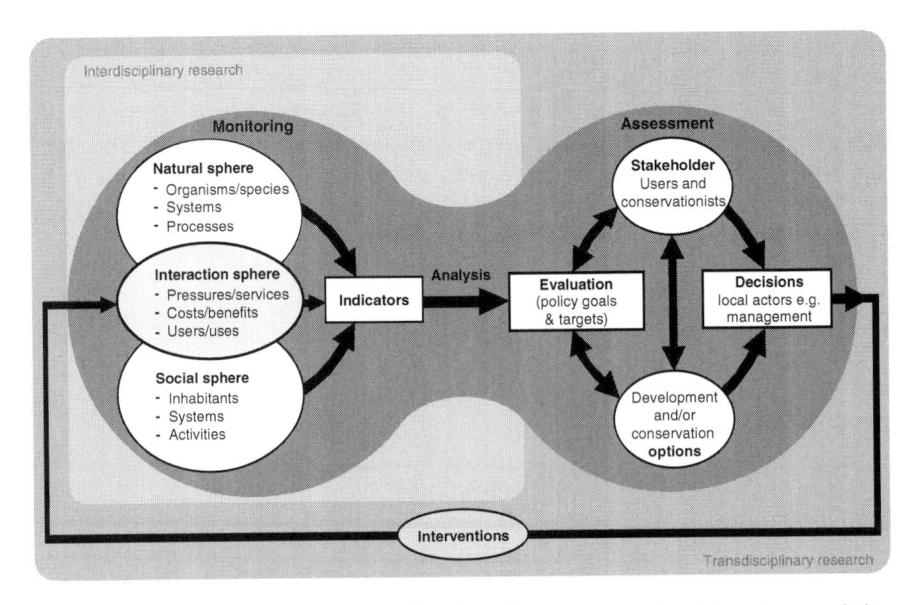

**Fig. 22.1** Integrated monitoring of natural and social spheres as a foundation for negotiating development options (Adapted from Fischer-Kowalski et al. 2008)

of nature continues, an important quest is in understanding the dynamics that lead to such aggravations and how to steer them for more sustainable outcomes. Such insights may be gained by observing a set of relevant system characteristics over longer periods of time. However, comprehensively monitoring all system variables is neither possible nor even necessary. It is far more meaningful to identify those development trends that seem to be crucial for future sustainability. Thus, indicators that capture dynamics towards or away from sustainability are of primary interest.

The scientific analysis of monitoring results will have to be assessed by the relevant stakeholders including management in the light of policy goals and targets. In the best case, this assessment will lead to a 'no-need-to-do-anything' option. However, there might be a need for decisions to, e.g., reduce pressures on the environment. An alternative route to trigger decisions might be the intention of some stakeholders to take new steps in regional economic development. In this case, development scenarios can be discussed in the light of a scientific analysis based on monitoring results in consideration of possible future impacts. Effective and accepted solutions in both emerging problems and newly intended economic developments greatly depend on a decision-making process that includes a wide range of perspectives and knowledge, e.g. from science, use and management.

Sustainable development often involves balancing conflicting goals: reduction of environmental pressures depends on restricting uses, while development depends on allowing and even supporting further uses of nature. The internal sustainability of local systems depends on resolving such conflicts, or at least keeping them at bay. Sustainability from the perspective of social systems and actors can be secured if they feel their costs/investments are balanced by benefits. If this is not the case, then actors will try to make a change, and this may well be at the expense of reducing environmental pressures. For example, if farmers feel that in a sustainability model region they have to work harder for less income than elsewhere without any beneficial return, or if hotel proprietors feel their investment in tourism is wasted, or if tourists feel the trip was not worth the effort, then the maintenance of such a model region will be threatened. Equally, if uses exert pressures upon the ecosystems that substantially reduce ecosystem services, then ecological sustainability is failing.

In dealing with the right-hand part of Fig. 22.1, transdisciplinary knowledge comes to play an important role. The expertise required in guiding social processes and the methodological skills to go along with engaging non-science actors effectively are of great relevance to sustainability assessment (Vilsmayer 2010).

## 22.4 Description of the Study Area and Problem Definition

In this section we present experiences and outcomes of applying transdisciplinary science to the Upper Ötztal, in particular the Gurgler Kamm biosphere reserve and the two neighbouring villages of Gurgl and Vent with about 500 inhabitants between them, both belonging to the municipality of Sölden. Detailed results will only be presented for the larger village of Gurgl with 327 inhabitants. Located in the federal province of Tyrol, Gurgler Kamm is a high-alpine landscape in the Upper Ötztal. The biosphere reserve came into existence as a result of the Man and Biosphere (MaB) research activities in Obergurgl (Patzelt 1987). In 1977, Walter Moser, then director of the Alpine Research Centre Obergurgl (an extra-mural station of the Innsbruck University) made an application to UNESCO in Paris for recognition of the area as a biosphere reserve to allow some protection at a time when there was none (Lange 2005, p. 48). The biosphere reserve covers an area of 1,500 ha$^2$ at an altitude of between 1,900 and >3,400 m. Since 1981, about 90% of the biosphere reserve was notified as a "Ruhegebiet" (tranquillity zone) of the Ötztal Alps and in 1995 the "Ruhegebiet" was declared a Natura 2000 area. Until the 1950s, the majority of local inhabitants were farmers. With more and more reliable access routes into the Upper Ötztal, a dynamic development of the tourism sector began. A well-known ski resort was gradually established to the north of the biosphere reserve. From the 1950s until recent years, farming lost more and more in importance and today plays only a minor role.

With the Seville strategy of 2005, it became crucial for all biosphere reserves to follow a specific zonation pattern to include a core, transition and development zone with varying degrees of use. The core zone was required to be specially protected and monitored following the superimposition of other stronger legislations such as

---

[2]New plans suggest that the new biosphere reserve should have a size of more than 20,000 ha or 200 km$^2$.

that of Natura 2000 or under the Ramsar convention. At the time the Seville strategy came into force, the Gurgler Kamm had no zonation and was at risk of losing its designation as a biosphere reserve. In response, the Austrian Man and Biosphere committee (MaB) asked the responsible stakeholders to develop a zonation plan in order to re-constitute the biosphere park in line with the Seville criteria. Along with this, the MaB committee also commissioned a new research project to investigate the status quo of the biosphere reserve and to compare results with those from previous research projects in the 1970s.

Thus, the goal of this research project was not only scientific, but had high societal relevance insofar as the investigation would have to take into account current socioeconomic trends and future development options compatible with regional sustainability. The explicit goal of the MaB committee was to engage the interested public in a discussion about the changes that had taken place in the region over the last 30 years with an exploration of possible future scenarios. Such a challenge to address the overall sustainability of a region required not only the integration of disciplines, but also meant engaging stakeholders in generating context-specific insights and seeking solutions for future decisions. As such, the process required a transdisciplinary approach.

## 22.5 Integration of Disciplines: Outcomes from Monitoring and Scientific Analysis

The first challenge was to form an appropriate team of scholars from relevant disciplines that could effectively contribute to this process. Four different research institutions became engaged in the project, covering topics such as biology and alpine research, conservation biology and vegetation/landscape ecology, snow and avalanche research, as well as social ecology.[3] The core exercise was to come up with a joint research concept and framework. The integrated monitoring and assessment framework (see Fig. 22.1) provided reasonable guidance for developing research questions and for communicating between the research teams. For integrated monitoring, we agreed on observing a set of variables for each of the three spheres in the left-hand part of Fig. 22.1 with a focus on only those that provide insights on local sustainability (see Table 22.2). Furthermore, a special analysis of the snow cover was performed including a projection of snow security under climate change scenarios as an important input to the transdisciplinary activity of scenario development. Each of the spheres is discussed in more detail below.

---

[3] These were: University of Innsbruck (Faculty of Biology and Alpine Research Centre Obergurgl), University of Vienna (Department for Conservation Biology, Vegetation Ecology and Landscape Ecology), Swiss Federal Institute of Snow and Avalanche Research, and the Alpen-Adria Universität (Institute of Social Ecology).

Table 22.2 Critical indicators investigated in the biosphere reserve Gurgler Kamm

| Sphere | Observed variables |
|---|---|
| Social sphere | Population dynamics |
| | Employment and income options |
| | Economic activities and tourism development |
| | Buildings and beds |
| Interaction sphere | Construction and surface sealing |
| | Expansion of skiing infrastructure |
| | Visitor frequency on hiking trails |
| | Skiing and related vegetation changes |
| Natural sphere | Biodiversity/vegetation assessment |
| | Diversity of landscapes |
| | Snow cover (projections) |

## 22.5.1 Dynamics Within the Social Sphere

Within the social sphere, we identified four variables crucial to understanding the internal dynamics. The number of inhabitants was seen as an indicator that is of importance for two quite different reasons: firstly, it refers to pressures caused by a growing population and related impacts caused by consumption patterns and lifestyle in general; secondly, it enables us to explore future perspectives for young people in a region dominated by tourism, after they have received higher education outside the valley. In other words, we asked whether there would be income options and expected quality of life upon their return.

A modelling project in the 1970s (Franz and Holling 1974) projected a decline in the local population due to lack of income perspectives. They argued that income greatly depends on tourism and that tourism is faced with two crucial primary limitations. These included the limitation upon safe land for buildings due to the hazard of avalanches. Moreover it was argued that a further expansion of buildings would lead to environmental degradation and would consequently reduce the natural attraction for the landscape among the summer tourists. However, inhabitants found creative ways to better utilise the existing built-up area (see Fig. 22.3). At the same time, the attraction of the landscape for winter tourists was not affected so much since most of the new developments revolved around winter skiing tourism where land degradation is not visible, being covered by snow. Moreover, the skiing infrastructure was positively welcomed by tourists, who benefited from it. Consequently, there was an increase in both the number of *buildings* and guest *beds* (see Fig. 22.2a). With increased tourist infrastructure, a doubling of overnight stays in the winter season since the late 1970s was made possible (Fig. 22.2b). In the period since the 1950s, farming as the dominant *economic activity* has been pushed back by tourism. The growing winter tourism in particular emerged as the main driver for development and attracted the younger generation due to equal or even better *income options* than in nearby urban areas (see inhabitants in Fig. 22.2a). Consequently, this supported a moderate *population growth* rate of about 0.8% per annum.

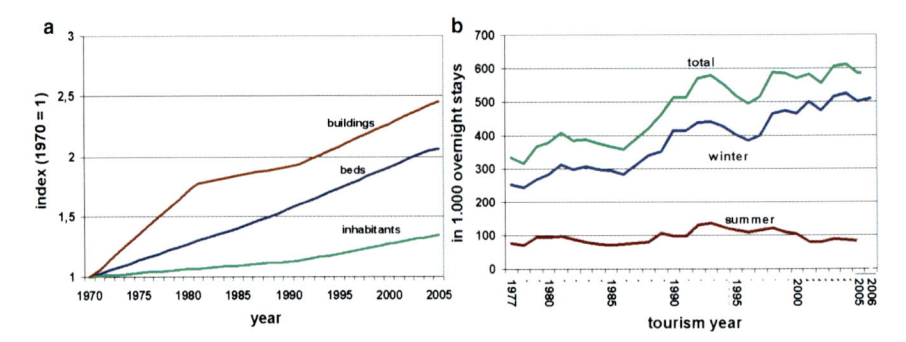

**Fig. 22.2** (**a**) Development of number of inhabitants, buildings and beds in relation to the index year 1970 (equals 1) up to 2005 (Source: Haas & Weber 2008a, p. 36). (**b**) Number of overnight stays for summer and winter tourism and the total of both of these in 1,000 stays from 1977 to 2006

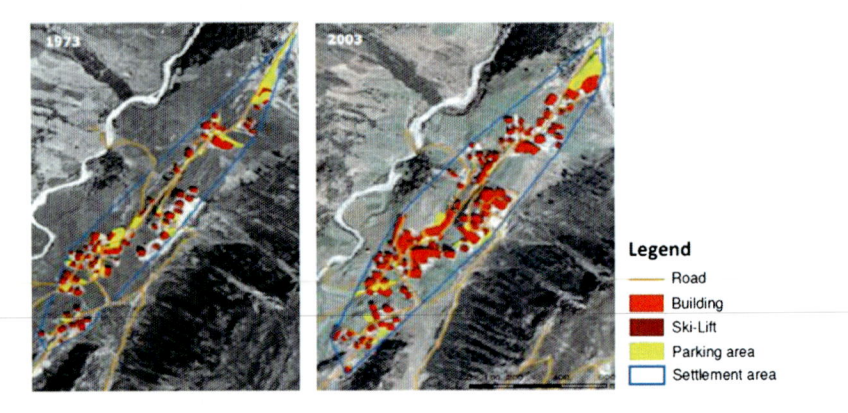

**Fig. 22.3** Settlement area in 1973 and 2003. The *red spots* in the aerial photos are buildings and the *blue line* is the boundary of the settlement area (Source: Walz and Ryffel 2008, p. 19)

## 22.5.2 Dynamics Within the Society-Nature Interaction Sphere

To capture the complex interaction between social and ecological systems, gaining insights into land use dynamics is increasingly recognised in the notion of 'integrated land-system science' (GLP 2005; Turner et al. 2007; Gaube and Haberl, Chap. 3 in this volume). Changes in land use due to natural and socioeconomic factors contribute to environmental problems such as loss of ecosystem services, biodiversity loss and greenhouse-gas emissions (Millennium Ecosystem Assessment 2005). The using and shaping of the land is a good representation of the relations between changes in socio-economic organisation, land use and land cover. Since both farming and tourism depend on land in the Gurgler Kamm case study, its availability, usability and productivity will determine future developments in the region. Therefore, to understand society-nature interactions over time we focus on the changes in the *settlement area* and changes in the surrounding mountainous landscape between the early 1970s and the 2000s.

**Fig. 22.4** (**a**) Comparing a photomontage from 1974 projecting the year 2000 (*left*), with (**b**) an actual photo taken in 2007 (*right*) (Sources: (**a**) Reproduction of laminated photograph, inventory of Alpine Research Centre Obergurgl, Kaufmann 2007; (**b**) Kaufmann 2007)

Figure 22.3 shows the expansion of the built-up area in the same period due to an increasing demand for more tourist infrastructure. An analysis of aerial images shows the changes in the built-up area in more detail. The ground area for buildings has increased by 151% and the settlement area by 38% from 1973 to 2003. This clearly reflects the tension between the need for more buildings in a region that has limited availability of safe land for building purposes.

The new constructions utilised some of the surrounding grasslands considered safe for this purpose. Interestingly, in 1974 a group of geographers, in consultation with some local stakeholders, had attempted to construct a photomontage projecting the future expansion of the village area and its built stocks for the year 2000 (Fig. 22.4, *left*). In retrospect, it can be stated that the photomontage for 2000 is almost identical with the situation as it actually was in 2007.

The mountains are the main tourist attraction in the area. Tourism activities are mainly skiing in winter and hiking in summer. As compared to summer activities, winter skiing is associated with major environmental impacts for which continuous monitoring of investments and environmental pressures is necessary. These impacts mainly originate from the construction and expansion of ski infrastructure, such as lifts, pipes for artificial snow making and slope preparation.

The vegetation assessments on the *ski-slopes* showed that the vegetation outside and within ski-slopes is significantly different (Erschbamer and Mayer 2011a). Dwarf shrubs and lichens are hardly present on the ski-slopes whereas the abundance of mosses was found to be increased (Mayer and Erschbamer 2009). This difference was mainly caused by levelling of the slopes 25–30 years ago as well as by artificial seed mixture application and fertilisation. However, also actual slope preparation and management (snow-making in winter, fertilising and sowing every summer) are responsible for the differences between ski piste and adjacent slope vegetation. Mayer and Erschbamer (2009) recommended the reassessment of sowing measures above 2,300 m since seed mixtures used until the present were not autochthonous. Other relevant skiing-related indicators useful for monitoring are the length of ski-slopes, snow-making performance (snowing area and volume of used water, annual balance of kilometres prepared with machinery, etc.).

*Hiking* is the most common activity during the summer season and is generally thought to exert far less pressure on ecosystems. To see whether this holds true, we surveyed the hiking activities in summer. Researchers interviewed hikers over 11 days during the 2008 hiking season along the four most popular starting points. Interviews included questions on the intended route and activities, their motivation for doing so and for doing so here. People predominantly went for one-day hikes with "being in nature" as the main attraction and motivation. There were no incidents of particularly damaging tourist activities like quad-biking, down-hill racing, etc. As such, summer tourism activities were assessed as low-impact activities with visitor frequencies rising to over 75 per day only in the close proximity of the village (Ackermann and Walz 2012).

When discussing the sustainability of future options, it became quite evident that the relation between winter and summer tourism is influential. While at present the capacity is utilised quite efficiently in winter, it is under-utilised in summer. Consequently, further growth of tourism in winter would require an extension of infrastructure, while tourism growth in summer could be well served by existing facilities. Ultimately this means that from the perspective of both nature protection and economy, a shift to more summer tourism would be preferable.

### 22.5.3 Changes of the Natural Sphere

The Upper Ötztal Valley has been a research site for the natural sciences since 1863 (Erschbamer and Mayer 2012b); consequently there exists a rich body of literature on the natural attributes of the region. Studies range from biological and agricultural surveys to expert opinions in nature protection affairs. One of the first steps in the present research was to systematically document, archive and analyse existing research (Schwienbacher et al. 2007). Almost 60 studies were analysed according to their relevance to scenario development, biosphere reserve, regional development and their reproducibility.

On the basis of this body of research, two vegetation studies were performed. One focussed on summarising the botanical studies from the 1990s to the present in the inner Ötz Valley to estimate the actual diversity within 13 subalpine/alpine plant communities (Erschbamer and Mayer 2012a). To obtain an overview for all these types, *diversity indices* were calculated. The second study tried to assess the changes over time in two alpine plant communities (snow bed, alpine grassland), which were investigated during the MaB project at Mt Hohe Mut (2,650 m a.s.l.) in the 1970s (Erschbamer and Mayer 2012b). Vegetation plot sampling performed by Duelli (1977) was repeated in 2006. Unfortunately, Duelli (1977) did not permanently mark the plots, however, it was possible to reconstruct the sampling locality properly. Within the alpine grassland the quantitative dissimilarity amounted to 0.58 (Bray Curtis Distance) showing a high loss of species since the 1970s (Fig. 22.5, *left bar*).

**Diversity changes**

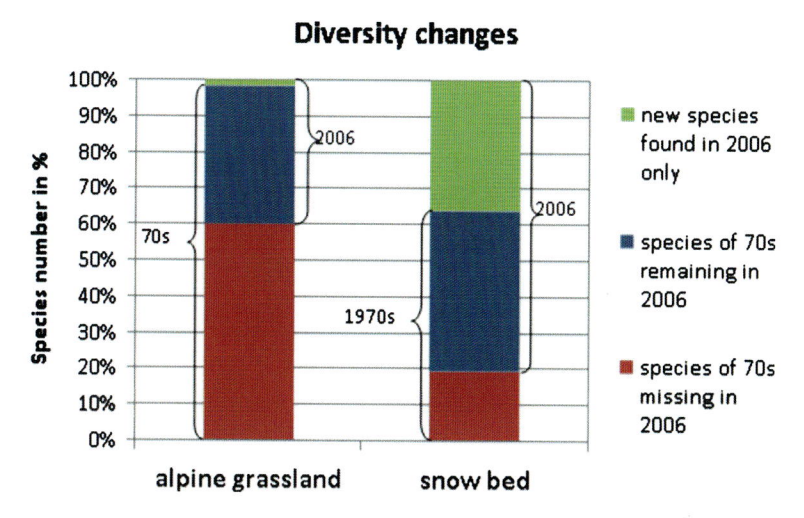

**Fig. 22.5** Comparison of the occurrence of species of two alpine plant communities (snow bed, alpine grassland with Carex curvula) in 1970s and 2006

In the snow bed, dissimilarity was 0.46 and species numbers increased until 2006. In both the communities about 40–50% of the species found in the 1970s were still present in 2006 (Fig. 22.5).

Consequently we were confronted with some open questions: one was to better understand the role of possible influential factors such as land use and climate change, as these questions are crucial for discussing management options. Another was to assess crucial research gaps and to develop a research plan that can provide information needed for discussing local sustainability in the long run.

Researchers also generated maps of *land cover changes* (Reiter et al. 2008). Orthophotos were analysed with the help of a digital elevation model and categorised into 11 land cover classes (like Swiss stone pine or dwarf-shrub heath land). To better match appearance in the orthophotos with effective land cover class, field studies of numerous randomly chosen study sites were performed. With this method, it was possible to map the entire study area of approximately 275 km$^2$. Furthermore, it allowed for a first comparison with aerial photos from earlier times. In one specific comparison below (Fig. 22.6), we can see that there was an increase in fodder meadow and a decrease in Swiss stone pine. Furthermore, river training measures were applied, the farmstead was expanded and one shed was removed.

Given that climate change, in particular rising temperatures, could become a threat for winter tourism in Austria, a forecast of probable *changes in snow cover* was essential in the study for the future planning of the region. A study was undertaken to model the most probable number of days that feature snow cover of more than 30 cm (Walz and Ryffel 2008). Such days are defined as alpine winter sporting days. The general rule of thumb is that a ski resort needs at least 100 alpine winter sport days to be attractive for tourists and economically viable, as first postulated by Abegg (1996). The study suggested that even in warm winters in the lowest parts of

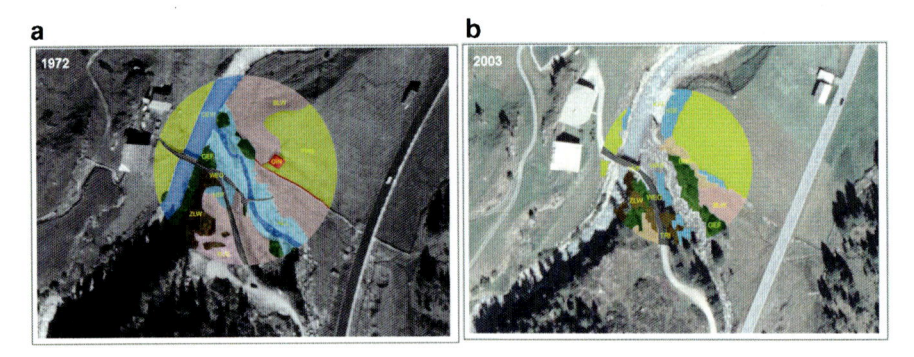

**Fig. 22.6** Land cover change visible in aerial photos for 1972 and 2003 (Source: Reiter 2007)

**Table 22.3** Alpine winter sporting days for village and mountains of Obergurgl for a recent decade and the decade starting in 2041

|  |  | 1996–2005 average measured | Average 2041–2050 | |
|---|---|---|---|---|
|  |  |  | Cold winter+2°C | Warm winter+4°C |
| Obergurgl village (1950 m altitude) | Alpine winter sporting days | 143 | 127 | 116 |
|  | Potential days for snowmaking | 79 | 53 | 33 |
| Obergurgl mountains (2,650 m altitude) | Alpine winter sporting days | 166 | 154 | 143 |

the ski resort, snow cover will be well above this limit in the decade starting in 2041 (115 days). In addition, there are 33 days with temperatures that allow for snowmaking with the use of currently available technologies (Table 22.3). Therefore, Obergurgl is in the fortunate situation of being able to plan its winter tourism with no imminent threat from climate change in the future. It is even more likely that Obergurgl will benefit from snow scarcity occurring in neighbouring ski resorts.

## 22.6 Engaging Society: The "Transdisciplinary" Process

The transdisciplinary process required engaging non-science actors in a dialogue related to the sustainability assessment of their region and the future it entails. In other words, the process required the communication of the outcomes from monitoring of social, natural and interactive spheres to the stakeholders and working with scenarios (Fig. 22.1, *right side*). In preparation for an interactive process such as this, it is essential to first undertake a stakeholder analysis. A second step was then to actively engage local stakeholders in discussions on the past, present and future of their village. People are for the most part interested in the history of their surroundings, be it the economic or cultural development or any changes or continuities in nature. For this reason, the research team was confident that feeding back the outcomes from monitoring and scientific analysis on the past decades would play a crucial role in helping to build scientific credibility with stakeholders and to establish a shared communication base to discuss the future sustainability of the region. The research team chose the scenario workshop as the most appropriate method.

### 22.6.1 *Stakeholder Analysis*

A sound stakeholder analysis goes beyond simply mapping the various groups, but also requires an understanding of their interests, preferences and the resources/competencies which could either facilitate or hinder the process. This is an important dimension of transdisciplinary science and a prerequisite for any participatory process to decide which groups are relevant in providing contextual knowledge, have high interests, are likely to be affected, are important for seeking solutions and taking decisions towards the end. In this section we focus on the main orientation of stakeholders with respect to regional sustainability. The following grouping has proven to be useful:

- Stakeholder description according to their relations to nature
  - Mainly protection-oriented: These are persons or groups who are mainly concerned with the protection of nature
  - Mainly use-oriented: These are persons or groups who are mainly concerned with the use of nature, such as forest or tourism enterprises
  - Science-oriented: These are persons or groups who have mainly scientific interests, be it concerning nature, society or their interrelations
- Stakeholder description according to their predominant scale of action
  - Local level: These are persons or groups who predominantly decide and act at the local level, in our case this refers to the villages of Gurgl and Vent.
  - Municipal level: These are persons or groups who predominantly decide and act at the municipal level with their actions influencing the local level, in our case this refers to the municipality of Sölden.
  - Provincial level: in analogy to the municipal level, in our case the federal state of Tyrol.
  - National level: in analogy to the municipal level, in our case Austria.

After performing interviews and experiencing stakeholders in various situations, the following stakeholder map was generated (see Fig. 22.7). The stakeholder map and discussions around their interests proved to be very useful for the research team in preparing for the scenario workshop. For instance the clustering of actors into stakeholder groups guided the selection process to achieve a fair representation of the diverse interests.

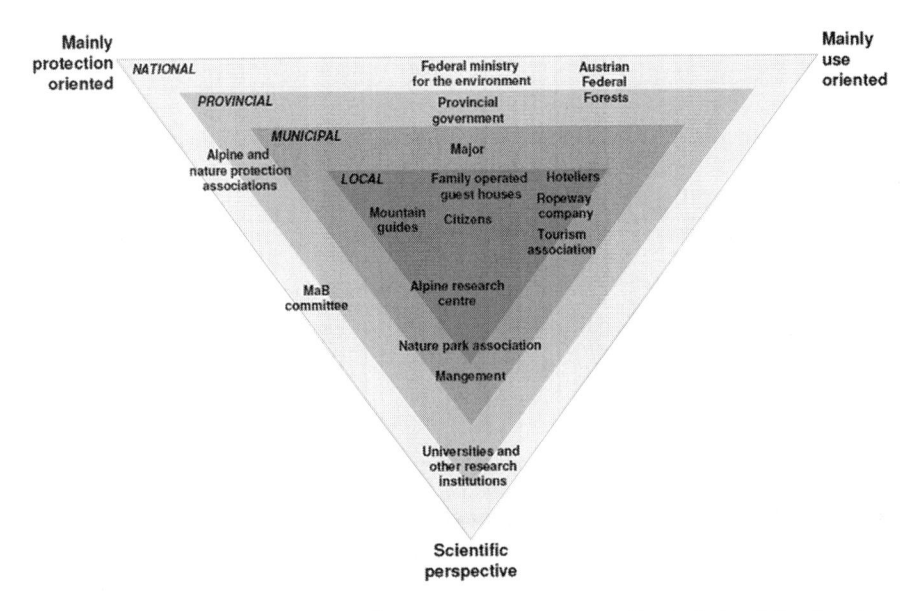

**Fig. 22.7** Map of stakeholders, grouping persons or groups according to their relation to nature and their predominant scale of action (Source: graph based on Haas & Weber 2007, p. 21)

Despite the divergent interests, orientation and competencies of the various actors it seemed possible to engage them in a dialogue on local sustainability. In some way or other, all stakeholders would benefit from the sustainable use of local resources. For example, the tourist industry indeed had concerns regarding a possible reduction in tourism in the region if the landscape became degraded or lost its aesthetic appeal. Institutions oriented towards nature protection were deeply aware that conservation efforts would be less effective without the local support of communities that draw their sustenance from the surrounding environment. In effect, local people could play a very important role in monitoring and management of local ecosystems if they stand to benefit from the healthy state of nature. The federal administration has to balance both objectives of protection and of use, and therefore sustainable use of local resources is inherent in their organisation's mandate. Last but not least, the nostalgia of local people inspiring the revival or at least maintenance of their local environment as an important part of their history and identity also plays a crucial role in the discussion of regional sustainability.

Despite the common interest in local sustainable development within this compilation of stakeholders, not all of them are actively involved in the decision of whether and how to reconstitute the Gurgler Kamm biosphere reserve. Here, we need to also differentiate between the stakeholders that are in a position to take decisions and action regarding the reconstitution of the biosphere reserve, and other that are "only" affected, but that are without such authority. However, in line with the idea of biosphere reserves which explicitly promotes the sustainable use of nature, decision-makers have a strong interest in a high level of acceptance of and support for the (reconstituted) biosphere reserve. Key decision-makers in the case of the Gurgler Kamm include the MaB Committee at the national level (building the bridge to UNESCO), the provincial administration of Tyrol (particularly the environmental department, but with a strong orientation towards policy) and the authorities of the municipality (with the mayor playing a central role). It is remarkable that the local stakeholders, who constitute a primary focus according to the initially proposed research by the MaB, formally play a minor role in decision-making. Nonetheless, they are the most relevant group when it comes to the actual management of the land and the natural resources within the area.

As initially planned, a first step in the research project was to get the local actors involved and identify how they see and value possible futures of their region. In a second step it was planned to feed these finding into the overall decision-making process including different hierarchical levels.

## 22.7 Local Level Scenario Development

Based on the assumption that people are interested in the history and future development of their own area, the scientific team decided to run a joint scenario workshop between scientists and local stakeholders called "Gurgl and Vent – yesterday, today, tomorrow". The development of scenarios for possible futures can provide multiple functions. It can help each of the different stakeholder groups to better understand the various and different perspectives of others on drivers, main issues and contingencies for development as well as anticipated future problems (Elliott et al. 2005). Thus it allows for the creation of a common communication basis for discussing complex sustainability problems and finally for the ground to be prepared for decision-making.

The workshop participants comprised of the biosphere park management, researchers, and representatives of several small to medium size enterprises and associations (guest houses and hotels, mountain guides, grocery shop, the local tourism association, the church and the alpine research station). To allow for better trust building higher level stakeholders from outside the region were not approached for this first event. The first part of the workshop was to describe the developments over the last 30 years by using mainly images and maps. This triggered the contribution of personal views. Since snow cover is crucial in determining the prosperity of winter tourism, the projection on snow cover for the years 2041–2050 was presented. It was received with some satisfaction, given that it suggested there will be no threat for winter tourism within the region.

Participants received a short introduction to the scenario approach (Elliott et al. 2005). The example of the IPCC scenarios (IPCC Working Group III 2000, p. 4) was used as an illustration. Stakeholders and scientists worked on the scenarios in small mixed groups, following three steps: First, to define the variables they consider as highly relevant for future development of the region. The two groups listed a total of 13 influencing factors crucial for the future. These were then grouped into external

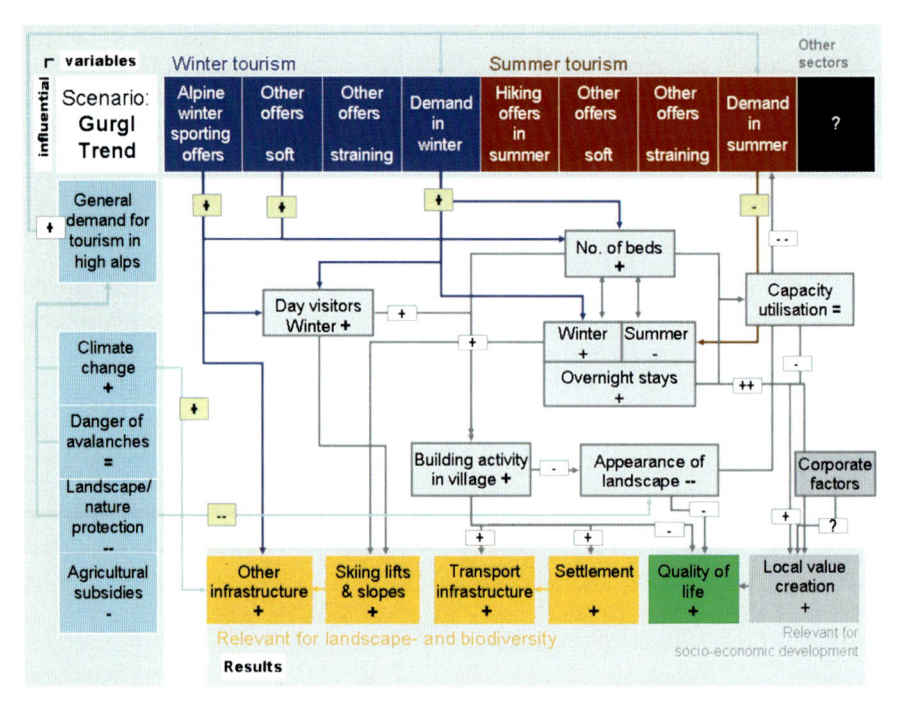

**Fig. 22.8** Causal model linking influential external and internal variables – 13 in all – with system parameters and ultimately with the outcome dimension of sustainability

framework conditions (over which they have little or no control) and those internal to the system over which they have more power to decide upon (Fig. 22.8). These ranged from the various forms of tourist attractions to be created in the region and demands by tourists to the danger of avalanches. As a second step, participants developed short storylines for four different scenarios: a trend, a desired, a horror and a surprise scenario.

In the "trend" scenario, participants envisaged a moderate increase in winter tourism and stagnation in summer. They argued that the required additional annual capacities for the winter season entail more construction activities both in the village and on the ski slopes in summer and thereby substantially diminish the village's attractiveness for summer tourism, which would limit summer tourism. In contrast, the "desired" scenario envisioned freezing the number of overnight stays in winter at the current level and increasing summer tourism. Since the village has sufficient excess capacities in summer, building activity could be restricted to maintenance work on already existing infrastructure, hence increasing the attractiveness of the village to allow for an increase in summer tourism. The "horror" scenario took the trend scenario to the extreme, with an emphasis on the development of winter tourism. Due to the risk of avalanches, there are no additional building areas available. Bed capacities could only be increased by moving staff out of the village to a less attractive and cheaper town at a lower altitude. In this scenario, construction activities would increase significantly to adapt staff quarters for the new purpose and to

**Table 22.4** Sustainability indicators for describing and assessing the scenarios

| | Sustainability indicators | Impact | Relevant for |
|---|---|---|---|
| Economic | Local value creation | Local value creation determines income options for employees and entrepreneurs and is an influential factor persuading young people to stay in the village | Socioeconomic development |
| Social | Quality of life | Quality of life is the other influential factor for young people to stay in the village. It comprises attractiveness of landscape and vitality of village outside tourism seasons | Socioeconomic development |
| Environ. pressure | Settlement | Extension of settlement areas | Biodiversity and land cover |
| | Transport infrastructure | Extension of infrastructure (landscape fragmentation) | |
| | Ski-lifts, slopes and facilities | Extension and intensification of use | |

develop new ski slopes and ski-lifts to keep waiting times down. Consequently this would mean a near collapse of summer tourism. The "surprise" scenario was connected with the rediscovery of the "*Sommerfrische*", meaning that due to climate change, tourists would increasingly be longing for refreshing holidays instead of hot summer holidays further to the South. Gurgl could offer a good mix of wellness, health services and walking in restorative mountain air for all four seasons. Special tourist groups such as allergy sufferers could be addressed. This scenario would require little in the way of construction activities in summer and would consequently lead to a stagnation of winter tourism.

In the final phase of the workshop, the groups discussed the scenarios according to the list of influential variables and agreed for each variable on the direction of change needed in order to achieve the situation described in each scenario for the year 2020/2021. After the workshop, the scenarios were assessed using site-specific sustainability indicators developed by the research team Haas and Weber (2008b) (see Table 22.4).

For the assessment itself, a causal model was developed to link influential variables (external and internal ones) with certain system parameters to ultimately assess how sustainability indicators will be altered over time (Fig. 22.8). Finally, the summary of the scenarios, the causal model and the sustainability assessment were discussed in interviews with selected stakeholders.

In broad terms, the assessment shows that the "desired" scenario is the most sustainable one, offering increased local value generation and a much better quality of life with no intensification of land use required (Table 22.5). Local stakeholders and scientists shared this view. The "horror" scenario developed by the local stakeholders was the worst in terms of sustainability. The "surprise" scenario favoured by some local stakeholders achieved a mixed score. While the assessment shows an intensification of pressures on the environment due to energy intensive wellness

**Table 22.5** Initial sustainability assessment of the four scenarios for Gurgl 2020

| Scenario Gurgl 2020 | Trend | Desired | Horror | Surprise |
|---|---|---|---|---|
| Description of scenarios | Intensification of winter tourism; Steady state summer tourism | A balance between summer and winter tourism | Winter tourism only; No summer tourism | Health- and wellness- offers; especially for allergy sufferers (high altitude) |
| Economic: Local value creation | ↘ | ↗ | ↗ | ↗ |
| Social: Quality of life | ↗↘ | ↗↗ | ↘↘ | ↗↘ |
| Ecological: Land use intensification | ↘ | → | ↘↘ | ↘ |

activities and greater per capita indoor space, it would possibly increase the overall local value creation. In terms of quality of life the huge and possibly risky investments required could lead to a split among the village population. While small family-operated guest houses would not be able to afford the investments needed and might therefore be forced to close their business and consequently move out of the village, bigger hotels might benefit strongly from this development. So while the latter would gain continuous improvements in their quality of life, the former would not be able to remain in their desired living environment. In further discussions of the scenarios and interviews, the different interests of the two groups emerged more clearly. Those running large hotels with many employees favoured winter tourism, while smaller enterprises wished for a balance between winter and summer tourism. The former preferred to have a heavy workload in one season and to spend the rest of the year renovating and relaxing. In contrast, small family-run enterprises where the whole family is involved were interested in a more even distribution of the workload throughout the seasons. These are differences that will require further attention if the idea of a model for sustainability in the region is to be pursued in future.

## 22.8 Conclusions

Transdisciplinary approaches can generate valuable knowledge not only in scientific terms, but also in terms of providing decision-making support for a sustainability transition. At the same time, sharing experiences in transdisciplinarity and comparisons with other projects engaged in similar endeavours enhances learning and improved outcomes. In general there is no recipe for transdisciplinary processes, but there are several lessons and insights that can be concluded from this and other research experiences (Box 22.1).

The chances of success appear to increase if transdisciplinary research is conceived of as a joint learning process with as yet unclear outcomes. All participants have their specific incentives for dealing with an issue that is not yet properly defined. And if participants find that the process will not serve their interests (not even in the long run), then they are always free to drop out. An honest communication base is the best precondition. Any lack of clarity about the intention to participate usually signals the beginning of an asymmetrical and possibly unpleasant experience of collaboration. Clear roles and responsibilities would appear to be another precondition. One of these concerns an agreement on who will steer and facilitate the process. There are several models available, such as the "mixed steering group" or a "third party facilitator". Whichever one is chosen, clarity and, at the least, no rejection by the participating groups is essential. Another role to be clarified is that of science. The main question concerns whether scientists have a degree of self-interest concerning the outcome of the process or whether their interest is focussed on formulating a feasible and promising research question and the sound application of methods with open outcomes. The openness to outcomes seems to be another crucial prerequisite for success.

Another rather important ingredient for the success of the process seems to be that participants have a minimum degree of willingness to reflect on the collaborative process and to re-adjust their actions based on insights gained from such reflections. A useful starting point is for scientists or site managers to organise their own critical reflection in discussions with peers from other protected landscapes. Last but not least, transdisciplinary research needs careful planning. This planning does not only start with the project, but sometimes includes careful thought that is required even at the point of initialising the research.

---

**Box 22.1** Challenges of Transdisciplinary Research

Transdisciplinary research poses several challenges to both researchers and stakeholders.

*No Unity of Science*

If interdisciplinary research is taken seriously, this implies different scientific viewpoints on one and the same problem. These different viewpoints, e.g. between natural and social scientists, enrich science. However, lay persons might see scientists as the provider of true answers and different or incompatible opinions might confuse them.

*Conflicting Interests*

Transdisciplinary research implies participatory processes. Since stakeholders most probably have conflicting interests already, the joint process of defining the problem to be investigated might become a stumbling block. Different resources, goals and values at stake and their social representation pose serious challenges to researchers.

(continued)

**Box 22.1** (continued)

### *External and Internal Power Play*

Usually problems such as unsustainable developments have evolved over time and there are good reasons why societal counter-movements have not taken place. In many cases, a likely reason is that the most influential stakeholders might underestimate the adverse effects of such trends. In some cases it might even be that the most influential groups benefit from the unequal distribution of environmental resources and are reluctant to change. Transdisciplinary research needs to understand how to motivate stakeholders with various backgrounds and how to establish a communication base where scientific arguments can be considered openly. However, attention needs to be paid to the fact that the transdisciplinary process should not become the ball in a game between the stakeholders involved, in such a way that this further perpetuates the problem.

### *Need for Collaborative Integration*

Throughout the transdisciplinary process, integration is a precondition for success. This starts with the joint problem definition and ends with a joint formulation of results that might be used differently in scientific communities and in practice. However, if scientific results are entirely detached from joint findings and if stakeholders' efforts are limited to communicative action only, this lack of integration is a strong indication that the process has failed. Non-collaborative behaviour is a powerful means to undermine such a process.

### *Uncertainties*

Transdisciplinary research processes are highly dependent on collaboration and findings relevant to the joint problem definition. This entails a high level of uncertainties concerning the outcome. While stakeholders might easier be motivated to engage in a process when this seems to promise a pleasant outcome, uncertainty might be a stumbling block for the engagement of both stakeholders and scientists. Half-hearted collaborations may lead to diffuse, unarticulated, disputed, over-interpreted or over-generalised outcomes.

### *Limited Resources and Unlimited Expectations*

Scientists and hired professionals face both limited resources and high production pressure, e.g. peer reviewed publications. At the same time, stakeholders may have diverse and even unlimited expectations with a wish to benefit from continued services from science.

*Based on authors' experience, observations and literature including Wallner and Wiesmann (2009, p. 49) and Wiesmann et al. (2008, p. 440).*

Clearly, transdisciplinary science is always confronted with some sort of power play. Stepping into any social system, one is soon confronted by established power structures and hierarchies that are by no means a matter of chance. Those who benefit most from existing arrangements do have a great interest in reproducing these power relations and might therefore resist change. Economic disparities, unequal access to land, knowledge and political processes greatly influence the way these power structures are organised. While some power play is quite obvious, some is played out behind closed doors. Furthermore, if not cautious or well-informed, research efforts too might face the danger of being selectively used by some stakeholders to promote their own interest. Due to these complex power dynamics, in the Ötztal case, it was not possible to engage local, provincial and relevant national stakeholders in a dialogue for local sustainability. Still, for the research team, this was quite a useful learning exercise. We discovered, for example, that groups we had believed to be very homogenous were actually quite heterogeneous. Nevertheless, these very different perspectives allowed for resonance and increased interaction and enabled progress on issues which at first seemed to be stumbling blocks. This is true not only with stakeholders, but also while working in an interdisciplinary team. Natural scientists had been primarily concerned with the degradation of natural richness, viewing humans as a disturbance to nature, whereas social scientists had focussed on the study and survival of human societies. Only an interdisciplinary approach that conceptualises local communities as part of a self-organising and self-maintaining socio-ecological system allows for integrative views based on both perspectives. In the Ötztal case, this view was crucial to the achievement of successful integration.

**Acknowledgments** The authors would like to thank colleagues at the Institute of Social Ecology whose efforts and insights provided a sound conceptual basis for this chapter. They are Marina Fischer-Kowalski and Karlheinz Erb. Special thanks goes to the Austrian Man and Biosphere Committee, which funded the research to foster both sustainability at local level and better integration between natural and social sciences.

# References

Abegg, B. (1996). *Klimaänderung und Tourismus – Klimafolgenforschung am Beispiel des Wintertourismus in den Schweizer Alpen*. Schlussbericht NFP31. Zürich: vdf Hochschulverlag AG.

Ackermann, M., & Walz, A. (2012). Auswertung der Befragung von Wanderern und anderen Freizeitsportlern im oberen Ötztal. In: W. Haas, et al. (Eds.), *Endbericht (Kapitel 2) MaB Biosphärenpark Ötztal (Phase 3)* (pp. 7–25). Wien: IFF Soziale Ökologie.

Balsiger, P. W. (2003). Supradisciplinary research practices: History, objectives and rationale. *Futures, 36*, 407–421.

Balsiger, P. W., Defila, R. & Di Giulio, A. (Eds.). (1996). *Ökologie und Interdisziplinarität – eine Beziehung mit Zukunft? Wissenschaftsforschung zur Verbesserung der fachübergreifenden Zusammenarbeit*. Basel: Birkhäuser.

Balter, M. (2010). Anthropologist brings worlds together. *Science, 329*, 743–745.

Biggs, R., Raudsepp-Hearn, C., Atkinson-Palombo, C., Bohensky, E., Boyd, E., Cundill, G., Fox, H., Ingram, S., Kok, K., Spehar, S., Tengo, M., Timmer, D., & Zurek, M. (2007). Linking futures across scales: A dialog on multiscale scenarios. *Ecology and Society, 12*, [online] URL: http://www.ecologyandsociety.org/vol12/iss1/art17/

Bogner, A., Kastenhofer, K., & Torgersen, H. (Eds.). (2010). *Inter- und Transdisziplinarität im Wandel? Neue Perspektiven auf problemorientierte Forschung und Politikberatung*. Baden-Baden: Nomos.

Burawoy, M. (2004). Public sociologies: Contradictions, dilemmas, and possibilites. *Social Forces, 82*, 1603–1618.

Burns, D. (2007). *Systemic action research. A strategy for whole system change*. Bristol: Policy Press.

Cooke, B., & Kothari, U. (Eds.). (2001). Participation: The new tyranny. London: Zed Books Ltd.

Duelli, M. (1977). *Die Vegetation des Gaisbergtales. Ein Versuch, das Datenmaterial mit Hilfe der EDV-Anlage zu bearbeiten*. Univ. Innsbruck.

Durant, J. (1999). Participatory technology assessment and the democratic model of the public understanding of science. *Science and Public Policy, 26*(5), 313–319.

Elliott, J., Heesterbeek, S., Lukensmeyer, C. J., & Slocum, N. (2005). *Leitfaden partizipativer Verfahren. Ein Handbuch für die Praxis*. Brüssel-Wien: kbs-ITA.

Erschbamer, B., & Mayer, R. (2012a). Schipisten im Raum Obergurgl footprints. In: Haas, W. et al. (Eds.), *Endbericht (Kapitel 5) MaB Biosphärenpark Ötztal (Phase 3)* (pp. 131–181). Wien: IFF Soziale Ökologie.

Erschbamer, B., & Mayer, R. (2012b). Diversitätserhebungen im hinteren Ötztal – Raum Obergurgl, Vent/Rofental. In: W. Haas, et al. (Eds.), *Endbericht (Kapitel 4) MaB Biosphärenpark Ötztal (Phase 3)* (pp. 131–181). Wien: IFF Soziale Ökologie.

Farrell, K., van den Hove, S., & Luzzati, T. (Eds.). (2011). *Beyond reductionism. A passion for interdisciplinarity*. London: Routledge.

Franz, H., & Holling, C. S. (Eds.). (1974). *Alpine areas workshop. Proceedings*. 13–17 May 1974, Laxenburg, Austria.

Fischer-Kowalski, M., Erb, K.-H., & Singh, S. J. (2008). Extending BRIM to BRIA: Social monitoring and integrated sustainability assessment. In: T. Chmielewski (Eds.), *Nature conservation management: From idea to practical results* (pp. 208–219). Lublin: PWZN "Print 6".

Funtowicz, S. O., & Ravetz, J. R. (1993). Science for the post-normal age. *Futures, 25*, 739–755.

Funtowicz, S. O., & Ravetz, J. R. (1994). The worth of a songbird: Ecological economics as a post-normal science. *Ecological Economics, 10*, 197–207.

Gibbons, M., Limoges, C., & Nowotny, H. (Eds.). (1994). *The new production of knowledge. The dynamics of science and research in contemporary societies*. London: Sage.

GLP. (Eds.). (2005). *Global land project. Science plan and implementation strategy*. Stockholm: IGBP Secretariat.

Greenwood, D. J., & Levin, M. (1998). *Introduction to action research*. Thousand Oaks: Sage Publications.

Haas, W., & Weber, M. (2008a). Sozio-ökonomische Entwicklung Obergurgl/Vent. In: W. Haas, et al. (Eds.), *Endbericht (Kapitel 4) MaB Bisophärenpark Ötztal (Phase 2)* (pp. 35–39). Wien: IFF Soziale Ökologie.

Haas, W., & Weber, M. (2008b). Szenarien für Gurgl und Vent 2020/21. In: W. Haas, et al. (Eds.), *Endbericht (Kapitel 8) MaB Bisophärenpark Ötztal (Phase 2)* (pp. 66–97). Wien: IFF Soziale Ökologie.

Haas, W., & Weber, M. (2007). Integriertes Monitoring im Ötztal. In: W. Haas, et al. (Eds.), *Endbericht MaB Bisophärenpark Ötztal (Phase 2)* (pp. 20–27). Wien: IFF Soziale Ökologie.

Hale, C. R. (Ed.). (2008). *Engaging contradictions. Theory, politics, and methods of activist scholarship. Global, area, and international archive*. Berkeley: University of California Press.

Hirsch Hadorn, G., Hoffmann-Riem, H., Biber-Klemm, S., Grossenbacher-Mansuy, W., Joye, D., Pohl, C., Wiesmann, U., & Zemp, E. (Eds.). (2008). *Handbook of transdisciplinary research*. Stuttgart/Berlin/New York: Springer.

IPCC working group III. (2000). *IPCC special report: Emission scenarios – Summary for policy makers*. Geneva: Intergovernmental Panel on Climate Change.

Irvine, R., Chambers, R., & Eyben, R. (Eds.). (2004). *Learning from poor people's experience: Immersions*. Brighton: Institute of Development Studies/University of Sussex.

Jahn, T. (2010). Transdisciplinarity in the practice of research. Translation from "*Transdisziplinäre Forschung. Integrative Forschungsprozesse verstehen und bewerten*". Matthias Bergmann, Engelbert Schramm (Hg.). Frankfurt am Main/New York: Campus Verlag.

Jahoda-Lazarsfeld, M. (1933). The influence of unemployment on children and young people in Austria. In: The Save the Children International Union (Eds.), *Children, young people and unemployment (vol. 2, Part II)* (p. 115–137). Geneva: Eigenverlag.

Jantsch, E. (1972). Towards Interdisciplinarity and Transdisciplinarity in Education and Innovation. In: Centre for Educational Research and Innovation (CERI) (Eds.), *Problems of teaching and research in universities* (pp. 97–121). Paris: OECD.

Joss, S., & Bellucci, S. (Eds.). (2002). *Participatory technology assessment: European perspectives*. London: Centre for Study of Democracy/University of Westminster.

Kates, R. W., Clark, W. C., Corell, R., Hall, J. M., Jaeger, C. C., Lowe, I., McCarthy, J. J., Schellnhuber, H. J., Bolin, B., Dickson, N. M., Faucheux, S., Gallopin, G. C., Grübler, A., Huntley, B., Jäger, J., Jodha, N. S., Kasperson, R. E., Mabogunje, A., Matson, P. A., Mooney, P., Mooney, H. A., Moore III, B., O'Riordan, T., & Svedin, U. (2001). Environment and development: Sustainability science. *Science, 292,* 641–642.

Klein, J. T. (Eds.). (1990). *Interdisciplinarity: History, theory, and practice*. Detroit: Wayne State UP.

Klein, J. T. (Ed.). (1996). *Crossing boundaries: Knowledge, disciplinarities & interdisciplinarities*. Virginia: University Press.

Klein, J. T. (2001). *Transdisciplinarity: Joint problem solving among science, technology, and society: An effective way for managing complexity*. Basel: Birkhäuser.

Klein, J. T. (2004). Prospects for transdisciplinarity. *Futures, 36,* 515.

Kockelmans, J. (1979). Why interdisciplinarity? *Interdisciplinarity in higher education*. University Park: Pennsylvania State University Press.

Lange, S. (Ed.) (2005). *Inspired by diversity – UNESCO's biosphere reserves as model regions for sustainable interaction between man and nature*. Vienna: Austrian Academy of Science Press.

Lewin, K. (Ed.) (1951). *Field theory in social science: Selected theoretical papers*. New York: Harper & Row.

Loukopoulos, P., & Scholz, R. W. (2004). Sustainable future urban mobility: using 'area development negotiations' for scenario assessment and participatory strategic planning. *Environment and Planning, 36,* 2203–2226.

Mayer, R., & Erschbamer, B. (2009). Die Vegetation von Schipisten im Vergleich zur angrenzenden Vegetation im inneren Ötztal (Zentralalpen, Nordtirol). *Verhandlungen der Zoologisch-Botanischen- Gesellschaft in Österreich, 146,* 139–157.

McDonald, D., Bammer, G., & Deane, P. (2009). *Research integration using dialogue methods*. Canberra: ANU Press.

Millennium Ecosystem Assessment. (Eds.) (2005). *Ecosystems and human well-being – Our human planet. Summary for decision makers*. Washington, D.C.: Island Press.

Miller, R. C. (Ed.) (1982). *Issues in integrative studies. An occasional publication of the association for integrative studies*. Oxford: Association for Integrative Studies.

Mittelstraß, J. (1992). Auf dem Wege zur Transdisziplinarität. *GAIA, 5,* 250.

Mobjörk, M. (2009). *Crossing boundaries: The framing of transdisciplinarity*. Örebro: Örebro University and Mälardalen University/Centre for Housing and Urban Research Series.

Munda, G. (2004). Social multi-criteria evaluation: Methodological foundations and operational consequences. *European Journal of Operational Research, 158,* 662–677.

Munda, G. (Eds.). (2008). *Social multi-criteria evaluation for a sustainable economy*. Berlin/Heidelberg: Springer-Verlag.

Newell, W. H. (1983). The case for interdisciplinary studies: Response to professor Benson's five arguments. *Issues in Integrative Studies, 2,* 1–19.

Nicolescu, B. (1992). *Manifesto of transidisciplinarity*. Albany: State University of New York.

Nowotny, H., Scott, P., & Gibbons, M. (2003). Mode 2 revisited: The new production of knowledge. *Minerva The International Review of Ancient Art & Archaeology, 41*, 179–194.
Patzelt, G. (Eds.). (1987). *MaB-Projekt Obergugl. Veröffentlichung des. Österreichischen MaB-Programm*. Innsbruck: Univ. Verlag Wagner.
Pereira, A. G., & Funtowicz, S. (2006). Knowledge representation and mediation for transdisciplinary frameworks: Tools to inform debates, dialogues & deliberations. *International Journal of Transdisciplinary Research, 1*, 34–50.
Piaget, J. (Eds.). (1973). *Der Strukturalismus*. Olten: Walter-Verlag.
Pohl, C., & Hirsch Hadorn, G. (2007). Systems, targets and transformation knowledge. In: G. Hirsch Hadorn, H. Hoffmann-Riem, S. Biber-Klemm, W. Grossenbacher-Mansuy, D. Joye, C. Pohl, U. Wiesmann, & E. Zemp (Eds.), *Principles for designing transdisciplinary research.* (pp. 36–40). Munich: Oekom-Verlag.
Reason, P. (2001). *Handbook of action research. Participative inquiry and practice*. London: Sage-Publications.
Reiter, K. (2007). Landschaftsökologie im oberen Ötztal. In: W. Haas, et al. (Eds.), *Endbericht, MaB Biosphärenpark Ötztal (Phase 1)* (pp. 28–36). Wien: IFF, Institut für Soziale Ökologie.
Reiter, K., Wrbka, T., & Prinz, M. (2008). Die Landschaften des inneren Ötztals, In: W. Haas, et al. (Eds.), *Endbericht (Kapitel 9), MaB Biosphärenpark Ötztal (Phase 2)* (pp. 98–133). Wien: IFF Soziale Ökologie.
Scholz, R. W., Mieg, H. A., & Ostwald, J. E. (2000). Transdisciplinarity in groundwater management – Towards mutual learning of science and society. *Water Air and Soil Pollution, 123*, 477–487.
Schwienbacher, E., Glaser, F., Kaufmann, R., & Erschbamer, B. (2007). Dokumentation naturwissenschaftlicher Studien. In: W. Haas, et al. (Eds.), *Endbericht (Kapitel 7) MaB Biosphärenpark Ötztal (Phase 1)* (pp. 37–45). Wien: IFF Soziale Ökologie.
Sclove, R. (1995). *Democracy and technology*. New York: Guilford Press.
Siddiq Khan, M., & Bhagwat, S. A. (2010). Protected areas: A resource or constraint for local people? A study at Chitral Gol National Prak, North-West Frontier Province, Pakistan. *Mountain Research and Development (MRD), 30*, 14–24.
Speed, S. (2006). At the crossroads of human rights and anthropology: Toward a critically engaged activist research. *American Anthropologist, 108*, 66–76.
Stoll-Kleemann, S., & Welp, M. (2008). Participatory and integrated management of biosphere reserves. Lessons from case studies and a global survey. *GAIA, 17*, 161–168.
Turner, B. L. I., Lambin, E. F., & Reenberg, A. (2007). The emergence of land change science for global environmental change and sustainability. *Proceedings of the National Academy of Sciences of the United States of America, 104*, 20666–20671.
Vilsmayer, U. (2010). Transdisciplinarity and protected areas: A matter of research horizon. Eco.mont, *mont – Journal on Protected Mountain Areas Research, 2*, 37–44.
Wallner, A., & Wiesmann, U. (2009). Critical issues in managing protected areas by multi-stakeholder participation – Analysis of a process in the Swiss Alps. *eco.mont – Journal on Protected Mountain Areas Research, 1*, 45–50.
Walz, A., Lardelli, C., Behrendt, H., Grêt-Regamey, A., Lundstrom, C., Kytzia, S., & Bebi, P. (2007). Participatory scenario analysis for integrated regional modelling. *Landscape and Urban Planning, 81*, 114–131.
Walz, A., & Ryffel, A. (2008). Landschaftseingriffe durch die Siedlungsentwicklung im Oberen Ötztal 1973 – 2003. In: W. Haas, et al. (Eds.) *Endbericht (Kapitel 3) MaB Biosphärenpark Ötztal (Phase 2)* (pp. 17–34). Wien: IFF Soziale Ökologie.
Wiesmann, U., Biber-Klemm, S., Grossenbacher-Mansuy, W., Hirsch Hadorn, G., Hoffmann-Riem, H., Joyce, D., Pohl, C., & Zemp, E. (Eds.). (2008). Enhancing transdisciplinary research: A synthesis in fifteen propositions. In: Anonymous. (Eds.), *Handbook of transdisciplinary research* (pp. 433–442). Stuttgart/Berlin/New York: Springer Verlag.
Wilcox, B., & Kueffer, C. (2008). Transdisciplinarity in ecohealth: Status and future prospects. *Ecohealth, 5*, 1–3.

# Chapter 23
# Conclusions

**Marian Chertow, Simron Jit Singh, Helmut Haberl, Michael Mirtl, and Martin Schmid**

A human ecologist, a systems ecologist, a biogeochemist, an environmental historian, and an industrial ecologist have come together to present this book, offering at least some evidence of the great multi-disciplinary interest in the still maturing approach of Long-Term Socio-Ecological Research. Indeed, we see that LTSER did not spring fully grown from a single source, but is very much developing as an inter- and transdisciplinary field of inquiry that might evolve into a meta-discipline. While these chapters stand by themselves, when they are reviewed together, themes emerge and fields come into alignment, even those encompassing different ontological roots. One clear, but heretofore implicit message suggested by this compilation is the need for some standardisation of LTSER models and methods. As the field advances, there is a need for sufficient standardisation to achieve comparability – without diminishing the creative sparks that celebrate the differences of time and place. This conclusion seeks to identify messages from the 21 contributions over three sections that contribute to the advancement of the LTSER idea and concludes with a short research agenda.

M. Chertow, Ph.D. (✉)
Center for Industrial Ecology, Yale School of Forestry and Environmental Studies,
Yale University, New Haven, CT, USA
e-mail: marian.chertow@yale.edu

S.J. Singh, Ph.D. • H. Haberl, Ph.D. • M. Schmid, Ph.D.
Institute of Social Ecology Vienna (SEC), Alpen-Adria
Universitaet Klagenfurt, Wien, Graz, Vienna, Austria
e-mail: simron.singh@aau.at; helmut.haberl@aau.at; martin.schmid@aau.at

M. Mirtl, Ph.D.
Department of Ecosystem Research and Monitoring,
Environment Agency Austria, Vienna, Austria
e-mail: michael.mirtl@umweltbundesamt.at

Section 23.1 introduces LTSER concepts, methods, and linkages, some adapted for LTSER from other fields and some developing from within. A common element in these chapters is the importance of measuring a system's metabolism as a central approach of LTSER – tracking the flows of materials and energy through a specified spatial system at a particular time and, in some instances, collecting sufficient data to look at a system over historical time periods (Fischer-Kowalski et al., Chap. 4). As we are reminded in Chap. 7 by biohistorian Stephen Boyden, the "human species is now using about 12,000 times as much energy and emitting 12,000 times as much $CO_2$ as was the case when our ancestors started farming some 10,000 years ago." Tracking the impacts of our activities, not only the activities themselves, provides a critical starting point for assessing how we might do things differently in future.

Several authors have sought to move the conversation regarding these long-term studies from a view of *socioeconomic* metabolism that focuses on cataloguing the material and energy flows of economic activity to a broader view of *socio-ecological* metabolism that also incorporates the ecological impacts of human-environment interactions. A brass tacks question from ecology asks: how do we know how much nature humans have altered? While we have the ambition to fully analyse human and natural systems, we fall short. A method based in ecology prescribed by Haberl and colleagues (Chap. 2) that moves LTSER further into the natural realm is the measurement of "human appropriation of net primary production." HANPP entails an evaluation of the extent to which human activities alter the amount of biomass available to ecosystems as a means of quantifying how much humans modify natural systems.

In another metric of human impact on nature, Krausmann (Chap. 11) calculates the area needed to grow forests for fuel wood with the area needed for the equivalent coal energy in a method he – following German environmental historian Rolf Peter Sieferle – calls the "subterranean forest" or "virtual forest area." In the case of Vienna, he finds that if today's energy requirement had to be met with fuel wood, there simply would not be enough area in the city nor, indeed, in the entire nineteenth-century Austrian forest, highlighting the need for ecological as well as economic measures. Many authors encourage modelling approaches that can embrace both the more common socioeconomic as well as the more integrative socio-ecological components through systems dynamics modelling including agent-based modelling such as in Gaube and Haberl (Chap. 3).

In advancing LTSER it is important to ask: having examined this new approach, what does it have in common with other, pre-existing areas of study, particularly in the social sciences? How can LTSER accrete the valuable lessons from these disciplines and sub-disciplines including environmental history (Winiwarter et al., Chap. 5), geography (Zimmerer, Chap 8), and anthropology (Gragson, Chap. 9)? These chapters show how social sciences and the humanities can deepen and contextualise LTSER through the use of both quantitative and qualitative data as well as historical sources and narratives. In one excellent example, we see how other scholars have paved the way for a more realistic starting point for LTSER, avoiding paths already disproven.

In this case, Zimmerer describes how disciplines grounded in field work and analysis such as geography help to debunk the myth that nature is 'pristine' by demonstrating how nature shapes and is shaped by human interactions.

With respect to LTSER applications across ecosystems, time and space (Sect. 23.1.1), a LTSER Platform provides a means of conducting socio-ecological research at a specified starting point. Some of the studies in the book reflect transdisciplinary research of a single place while others are cross-cutting reviews in time and space. The common call for multi-scalar research by Dirnböck et al. (Chap. 6) in the first section of the book is addressed in the section on LTSER applications, but through accumulating the lessons of individual chapters rather than presenting and analysing them in a comprehensive way.

There are many variations when applying LTSER to single sites and in particular, this section looks at (1) the transformation of a specific activity – agriculture – within a single region over time (Gingrich et al., Chap. 13), (2) changes in the urban metabolism of a single European city but over 200 years (Krausmann, Chap. 11), and (3) a city within an expansive region in the U.S. (Grimm et al., Chap. 10). At broader spatial scales comparisons are possible through (1) a chapter looking across studies of four different island systems (Chertow et al., Chap. 14); (2) a chapter looking at a single activity, agriculture again, but reviewing it across hemispheres (Cunfer and Krausmann, Chap. 12), and (3) a global view of transitions described in socio-metabolic terms (Krausmann and Fischer Kowalski, Chap. 15).

Looking across this broad variety of projects, the importance of a socio-ecological perspective for examining the drivers and impacts of technological change emerges, providing insights of a different character beyond those typically offered by historians and economists. This alternative perspective places a great deal of attention on changing patterns of energy use and transportation as critical elements of societal transformations. The tracking of material and energy flows provides evidence that technology development and efficiency gains have not reduced metabolism, rather these advances have added to industrial metabolism even in the recent transformation to information and communication technology (Krausmann and Fischer Kowalski, Chap. 15).

With respect to urban LTSER, Krausmann's study of Vienna takes more of an environmental history approach grounded in government statistics over time, while Grimm et al. take a more interdisciplinary approach in the Central Arizona-Phoenix LTER – combining, for example, ecology, public health, and demography (Chap. 10). Despite the differing approaches, both studies reveal the 'hidden costs' of urbanisation in terms of resource mobilisation. The broad question for Grimm et al. is: "how do the services provided by evolving urban ecosystems affect human outcomes and behaviour and how does human action (response) alter patterns of ecosystem structure and function and, ultimately, urban sustainability in a dynamic environment?" Ecosystem services, then, are recognised as the key focal point of interaction between people and the environment and, in the case of the Central Arizona-Phoenix project, how ecosystem services play out in built and highly modified landscapes.

Apart from the output of long-term socio-ecological research, there are other basic questions to be addressed by LTSER. There are now 31 LTSER platforms in the EU across 21 countries and encompassing 48 socio-ecological regions. In addition to 26 LTER sites in the US, in the last 2 years 21 new exploratory urban sites were selected through the Urban Long-term Research Area programme (ULTRA-Ex), with much explicit encouragement to combine social and ecological study. So it is important to ask: How are such programmes formed and funded? What paths of development have they followed? What methods and frameworks have been deployed and which aspects have been successful thus far and could be taken up by others? Sect. 23.1.2 on LTSER formations assembles contributions that outline some of these efforts and experiences that could provide inspiration and useful guidance for ongoing and future LTSER processes.

Although there are always numerous strands and pieces that come together in any new endeavour, many authors point to the influence of the US long-term ecological research (LTER) programme on the eventual emergence of the long-term socio-ecological research concept and LTSER Platforms. With the exception of the two urban US LTER sites in Phoenix and Baltimore (Chaps. 10 and 16), however, most of the other LTER sites were selected to be remote from human impact rather than inclusive of it. These sites have emphasised themes and protocols more in line with science than with social science. Indeed, Furman and Peltola state in Chapter 18 that transitioning from LTER to LTSER requires a "radical change in the way research is developed and research infrastructure is built."

LTSER, therefore, brings many challenges in its wake as discussed by the authors involved with European projects here (Chaps. 17, 18, 19, 20, 21 and 22) including: the differences in epistemological approaches of social and natural scientists; the need for large and/or multi-scale sites; the importance of stakeholders as well as researchers to LTSER; data management issues; shortages of long-term funding and the competition for funding among scientists; and coordination between practitioners and academics, with the latter group much more focused on data reliability and validity than the former.

Curiously, some of the projects discussed in this section evolved from pre-existing collaborations and these seemed to go more smoothly than, for example, the first Austrian LTSER in Eisenwurzen (Peterseil et al., Chap. 19), which had to be developed from scratch. In addition, the Finnish project drew from its active history of interdisciplinary sustainability research especially in the area of environmental social science. The US cases, too, involved longer term participation by social scientists prior to project formation as well as multi-year funding commitments which also contributed to the positive experiences recorded there.

Haas et al. (Chap. 22) uses the insights of transdisciplinary studies in a case that includes engaging civil society from the outset to reveal a potential solution to a tourism problem. The subject of that study, the alpine "Gurgler Kamm" Biosphere Reserve, has been transitioning from agriculture to tourism for 35 years. The study reveals that while the summer tourism capacity is currently under-utilised, the winter capacity is facing potential resource and infrastructure constraints if winter tourism grows. Rebalancing winter and summer tourism, though complex, could reduce strains on resources and infrastructure.

A final lesson brings together themes discussed across the chapters and sections, including: (1) learning from history, (2) the value of tracing metabolic flows, and (3) the importance of a socio-ecological framework. Returning to Chap. 12 and the story of the seeming richness of the Great Plains of the US, we see that overlooking an ecological understanding of agricultural history could lead to the misconception that farmers could endlessly achieve crop surpluses. Rather, farmers in the Great Plains circa 1880 were taking advantage of a vast quantity of "stockpiled soil nutrients" that would be washed away by the Depression years and later by substituting "oil for soil", through the use of chemical fertilisers.

Indeed, many scholars are reinterpreting what had previously been seen as a march of history dominated by economics by applying the understanding of the enormous role ecology plays. Drawing from a newly published work, Charles Mann's (2011) "1493: Uncovering the New World Columbus Created", what Peruvian export, for example, was more influential in altering European destiny following the Columbian Exchange: silver or potatoes? While the silver served to back up the currency, the introduction of potatoes allowed most of Europe to feed itself, thereby increasing political stability and providing "the fuel for the rise of Europe."

## 23.1 Looking Ahead

There is enough evidence, as this volume has shown, to claim that efforts in LTSER science spanning over a decade are already in the process of convergence, bringing together what has been, thus far, fragmented scientific and practical knowledge of the long-term dynamics of society-nature interactions. A cumulative understanding of these processes across space and time holds great promise in addressing sustainability concerns on a firmer footing, not only for scientific analysis, but also in its appeal to policy. Already, LTSER appears to be gaining in substance and attention. As noted in Mirtl et al. (Chap. 17):

> With less than a decade of practical experience, LTSER Platforms are emerging as living laboratories for socio-ecological research and a major contributor of policy/management relevant knowledge.

Nonetheless, there are still challenges ahead. On a scientific level, they relate to the further refinement and standardisation of LTSER concepts, methods and analytical tools and the need to test these in a wide range of settings. At the same time, there is an urgency to consolidating and broadening the existing LTSER community, not only in Europe and the US, but also to integrate scholars from other continents of the world, in particular, the emerging economies of Asia and Latin America. How these regions will organise their economic structure and development policies will determine future patterns of global resource use and sustainability. Stronger cooperation among such regions will be advantageous to LTSER especially since socio-ecological dynamics are not determined by activities and policies on any one scale, but are largely influenced by and in return exert influence upon other scales through current global outreach and international division of labour. Expanding the network of LTSER Platforms (or the like) beyond Europe and the US will enrich our understanding of society-nature dynamics across space, scale and time and how these Platforms interrelate with one another. Finally, the need to enhance the visibility of LTSER research in social and policy fields is critical for establishing an effective interface between science and practice.

Tappeiner et al. from the Tyrolean Alps LTSER Platform (Chap. 21) summarise well both the purpose and complexity of LTSER as follows:

> The challenge for long-term socio-ecological research (LTSER) lies in identifying precisely both the interactions between the socioeconomic systems and ecosystems and the impact this has on the natural environment and on society, due to direct local activities but also exogenous global change. At the same time, the challenge also involves using these findings to create adaptation strategies for ecologically sound, sustainable regional development.

Grappling with these multi-faceted challenges informs the opportunity to think ahead to topics needed for the next generation of LTSER at two levels, within and across LTSER platforms as discussed below.

### 23.1.1 Within LTSER Platforms

1. Further understanding of the process of constructing inter- and transdisciplinary research, including whether the structure is more top-down or bottom-up as discussed in Furman and Peltola (Chap. 18).
2. Proactively addressing what was described as the "forced marriage" in Lavorel et al. (Chap. 20) between ecologists, who have traditionally dominated LTER research, and social scientists.
3. For urban LTSER, embracing the shift from a mode of Ecology *in* Cities to an Ecology *of* Cities as discussed by Grove et al. (Chap. 16), recognising that cities are not simply "urban and non-urban areas", but distinct ecosystems.

### 23.1.2 Across LTSER Platforms

1. Coordinating across LTSER Platforms focusing on lessons learned and the beginning of meta-analyses and increased cross-scalar research.
2. Convening a group to consider what aspects of LTSER methodology can be standardised and protocols developed across projects to allow inter-site comparisons.
3. Systematically examining how non-material changes in society, such as changes in the system of meaning, or shared expectations, legal regulation, currency or cultural heritage affect physical and social outcomes at LTSER Platforms.

23 Conclusions

4. Developing models to generalize from results derived within LTSER Platforms to larger (regional or even global) scales, as well as scenario development as a means of informing policies not only at local (as in Gaube and Haberl, Chap. 3) but also at broader scales.

Within a short period of time, LTSER has come a long way and there are reasons to believe that this emerging field of research has contributed richly to the field of global environmental change. These efforts, however, need to be honed and expanded to be useful in a wide range of settings and societal dynamics to enhance its applicability and efficacy. We hope that this volume in its effort to crystallise some of the relevant streams in LTSER is a leap forward and will serve to provide further impetus in this direction.

# About the Contributors

**Weslynne Ashton** (Ph.D. in Environmental Studies) is an assistant professor of environmental management and sustainability at the Illinois Institute of Technology in Chicago, USA. Her research and teaching focus on corporate sustainability, industrial ecology, social entrepreneurship and sustainable industrial development. She previously worked as an associate research scientist and lecturer at Yale University, with visiting scholar appointments at the National University of Singapore and The Energy and Resources Institute (TERI) University in India. She has also worked extensively in the Caribbean, as well as in Hawaii and India. Contact: washton@iit.edu

**Michael Bahn** is an associate professor at the Institute of Ecology of the University of Innsbruck, Austria. His research focus is on linking plant, soil and ecosystem processes in a global change context. He has co-edited a book on soil carbon dynamics (Cambridge) and is an editor of the journal Biogeosciences. He has been principle investigator/work package leader of several national and European projects, and has authored more than 50 peer-reviewed publications in international journals and books. He contributed to a recently published white paper on research perspectives for LTER in Austria. Contact: michael.bahn@uibk.ac.at

**Peter Bezák** (Ph.D. in Regional Geography) works at the Institute of Landscape Ecology, Slovak Academy of Sciences, Slovakia. His research interests include socio-ecological interactions in the landscape, landscape and biodiversity changes and driving forces behind these, landscape management and policy, sustainable development, participatory approaches in landscape and nature protection research. He participated in several research projects funded under EU Framework Programmes (EVALUWET, BIOSCENE, BIOPRESS, ALTER-NET) and he has monitored LIFE programme projects in Central and Eastern Europe dedicated to the support of the Natura 2000 sites. He is the main coordinator of the Landscape Europe network. Contact: peter.bezak@savba.sk

**Sylvain Bigot** is Professor of Physical Geography at the Joseph Fourier University (Grenoble, France) and Researcher in the LTHE (Laboratory of study of Transfers in Hydrology and Environment). Specialised in climatology and in Remote Sensing, his experience is dedicated to climate variability diagnostics and land cover/land use monitoring in the French Prealps (Long-Term Ecological Research site) and in tropical Africa. Director of the international review Climatologie (publication of the International Association of Climatology) since 2006, his expertise is on global and regional climate teleconnections and environmental impacts. Contact: sylvain.bigot@ujf-grenoble.fr

**Christopher Boone** (Ph.D. in Geography) is a professor with joint appointments in the School of Sustainability and the School of Human Evolution and Social Change at Arizona State University, USA. For more than a decade he has participated in the Long-Term Ecological Research programme in the United States. His work examines social and ecological drivers and consequences of urbanisation, with a special focus on environmental justice. Contact: cgboone@asu.edu

**Axel Borsdorf** (Doctorate and Habilitation in Geography) is a full professor at the Geography Department of the University of Innsbruck and the director of the Institute for Mountain Research: Man and Environment of the Austrian Academy of Sciences at Innsbruck, Austria. He is also a full member of the Austrian Academy of Sciences and the Vice-President of the Austrian Latin-America Institute. Axel Borsdorf edits several scientific journals and has led EU-funded projects on urbanisation, alpine development and mountain research. Contact: axel.borsdorf@uibk.ac.at

**Stephen Boyden** obtained a degree in Veterinary Science in London in 1947. From 1949 to 1965 he carried out research in immunology in Cambridge, New York, Paris, Copenhagen and Canberra. From 1965 onwards he pioneered human ecology and biohistory at the Australian National University. In the 1970s he directed the Hong Kong Human Ecology Programme. Since retirement in 1991 he has been actively involved in the work of the *Nature and Society Forum* – a community-based organisation concerned with the future wellbeing of humankind and the natural environment. He has published a series of books on human biohistory. Contact: sboyden@netspeed.com.au

**Jean-Jacques Brun** (Ph.D.) is a research professor at Cemagref Grenoble Centre, France. Leader of the team "Ecosystem assessment and Conservation", he works as a soil ecologist in a landscape ecological perspective. He focuses on soil biodiversity in an above ground/below ground approach and studies the role of humus forms, organic matter and soil fauna as indicators of global change impact on mountain ecosystems. He is the current President of the International Committee on Alpine Research (ISCAR). He is project partner in the European INTERREG IV B project ECONNECT "Improving ecological connectivity in the Alps" and was involved in the Sixth European Framework Programme GLOCHAMORE (Global Change in Mountain Regions) which aims to develop an integrative working plan for environmental and social monitoring in mountain regions that will facilitate the implementation of global change research strategies in selected UNESCO MAB Biosphere Reserves. Contact: jean-jacques.brun@cemagref.fr

**Mary L. Cadenasso** (Ph.D. in Ecology) is an associate professor and ecologist in the Department of Plant Sciences at the University of California, Davis, USA, with research interests in landscape, ecosystem, and plant ecology specifically focused on the role of spatial heterogeneity in system dynamics. She is involved in research on urban land cover and the link to ecosystem and hydrologic processes, urban agriculture and environmental justice, the influence of vegetation structure on nitrogen dynamics in California oak savannas, and the structure and function of riparian zones in Kruger National Park, South Africa. She is co-principal investigator of the Baltimore Ecosystem Study Long-Term Ecological Research Project (BES), co-principal investigator on the Fresno-Clovis Urban Long-Term Research Areas Exploratory Project and lead principal investigator on two projects in Sacramento investigating urban land cover and ecosystem services. Contact: mlcadenasso@ucdavis.edu

**Stephen R. Carpenter** (Ph.D. in Botany and Oceanography & Limnology) directs the Center for Limnology at the University of Wisconsin-Madison, USA. His research interests include integrated social-ecological analysis of watersheds, large-scale and long-term field experiments, and early warnings of regime shifts. Currently, he is Chair of the Scientific Committee for the Program on Ecosystem Change and Society, a new interdisciplinary initiative from the International Council of Science. Carpenter served as co-chair of the Scenarios Working Group of the Millennium Ecosystem Assessment, and as lead principal investigator of the North Temperate Lakes Long-Term Ecological Research Site. Contact: srcarpen@wisc.edu

**Marian Chertow** (Ph.D. in Environmental Studies) is a professor of industrial environmental management at the Yale University School of Forestry and Environmental Studies, USA. Her research and teaching focus is on industrial ecology, business/environment issues, waste management, and environmental technology innovation. Her current research compares two urbanised socio-ecological systems on Hawai'i Island. She is also appointed at the Yale School of Management and the National University of Singapore. She serves on the National Advisory Council for Environmental Policy and Technology (NACEPT) that advises US EPA and is incoming President of the International Society of Industrial Ecology. Contact: marian.chertow@yale.edu

**Daniel L. Childers** (Ph.D. in Marine Science) is a professor in the School of Sustainability at Arizona State University, USA. He is also director of the CAP LTER and co-director of the Urban Sustainability Research Coordination Network. His research focuses on systems ecology (primarily in wetland, aquatic, and urban ecosystems) and sustainability science. He has studied many different freshwater and estuarine systems around the world, recently expanding to urban wetlands and "ciénega" systems of arid southwestern streams. Dan came to ASU after 15 years at Florida International University, where he was director of the Florida Coastal Everglades LTER Program. Contact: dan.childers@asu.edu

**Philippe Cozic** is a research director and agronomist, specialised in agro-ecology of mountain pastures and in ecological engineering. His research is characterised by two main topics: applied and participative research implicating local mountain stakeholders, and interdisciplinary studies that he has conducted (ecology, economics, agronomy, forestry, geography) for a sustainable development of mountain territories. For over 20 years he has managed several teams and research units of Cemagref (UR Ecosystèmes Montagnards, UR Agricultures et Milieux Montagnards). He has been co-founder and leader of the LTSER Central French Alps with Sandra Lavorel. Contact: philippe.cozic@hotmail.fr

**Wolfgang Cramer** (Ph.D. in Vegetation Science, Professor of Global Ecology) has been working since 2011 at CNRS in France, and is Research Director of the newly created Mediterranean Institute for Biodiversity and Ecology (IMBE) in Aix-en-Provence and Marseille, France. For 18 years, he helped establish the Potsdam Institute for Climate Impact Research in Potsdam, Germany, leading work on impacts on natural systems as well as on Earth system analysis. His work concerns modelling terrestrial ecosystem dynamics and the assessment of services provided by ecosystems under global change. He has had leading roles in several projects of the International Council of Science (currently as member of the Scientific Committees of DIVERSITAS and PECS), has contributed to the IPCC (since its Second Assessment Report, currently as convening lead author for the WG2 Chapter on Detection and Attribution) and the Millennium Ecosystem Assessment, and now helps establish the Intergovernmental Panel on Biodiversity and Ecosystem Services (IPBES). Contact: wolfgang.cramer@imbe.fr

**Geoff Cunfer** (Ph.D. in History) is an environmental historian of the North American Great Plains in the Department of History and School of Environment and Sustainability at the University of Saskatchewan, Canada. He directs the Historical GIS Laboratory where he researches agricultural land use, dust storms and wind erosion, material and energy flows in agricultural landscapes, and historical geography. He holds a Ph.D. in U.S. History from the University of Texas and is the author of *On the Great Plains: Agriculture and Environment* (2005) and *As a Farm Woman Thinks: Life and Land on the Texas High Plains, 1890–1960* (2010). Contact: geoff.cunfer@usask.ca

**Thomas Dirnböck** (Ph.D. in Botany) works at the Environment Agency Austria, mainly within Long-Term Ecosystem Research and Monitoring. Major research activities focus on ecosystem response to air pollution and climate change, and on modelling biodiversity changes as impacted by air pollutants, climate and land use impacts. He is national focal point for the UN/ECE task forces on Integrated Monitoring of Air Pollution Effects on Ecosystems (ICP IM) and on Reactive Nitrogen (TFRN). He was and is involved in a number of EU research and infrastructure projects (EU Network of Excellence ALTER-Net, Life+ EnvEurope, EXPEER, LTER Europe) and was expert adviser in the EEA initiative, Streamlining European Biodiversity Indicators (SEBI2010). Contact: thomas.dirnboeck@umweltbundesamt.at

About the Contributors

**Laurent Dobremez** is an agronomist researcher at Cemagref Development of Mountain Territories Research Unit, France. The main research question concerns how to reconcile environmental issues and agricultural objectives in farming systems. His field of activity is mountain agriculture and pastoralism, implementing a whole-farm approach, spatial organisation and agricultural practices analysis. Contact: laurent.dobremez@cemagref.fr

**Stefan Dullinger** (Ph.D. in Ecology) is a professor of vegetation science at the Department of Conservation Biology, Vegetation and Landscape Ecology at the University of Vienna, Austria. He is a member of the Vienna Institute for Nature Conservation and Analyses. His research concentrates on analysing spatial patterns of plant species distribution and diversity, their origin and their possible dynamics under global change. He is currently involved in several projects focusing on possible impacts of climate warming on biodiversity and on the spatio-temporal spread of invasive organisms. Contact: stefan.dullinger@univie.ac.at

**Karl-Heinz Erb** (Ph.D. in Ecology, Habilitation in Social Ecology) works at the Institute of Social Ecology, Vienna, Austria. His research focuses on interactions between humans and global environmental systems, global land system science, analysing changes in land use and land cover, and its consequences for ecosystem structures and functioning, e.g. carbon stocks and flows. He is a member of the Scientific Steering Committee of the Global Land Project, member of the Young Curia at the Austrian Academy of Sciences, and member of the Commission on Ecosystem Management (CEM) at The World Conservation Union (IUCN). In 2010, he was awarded an ERC Starting Independent Researcher Grant by the European Research Council, for the project "Land Use Intensity from a Socio-Ecological Perspective". Contact: karlheinz.erb@aau.at

**Brigitta Erschbamer** (Ph.D. and Habilitation in Botany) is professor at the Institute of Botany, Innsbruck, Austria. She is also scientific director of the Alpine Research Centre Obergurgl. Her research concentrates on alpine ecology and population biology. Her main projects have dealt with factors and processes governing colonisation in glacier forelands. A major focus of her research concerns diversity changes due to land use and climate changes. She is a partner in the worldwide project GLORIA (Global Observation Research Initiative in Alpine) Environments. Contact: brigitta.erschbamer@uibk.ac.at

**Marina Fischer-Kowalski** (Ph.D. in Sociology) is professor at the Alpen-Adria University, Austria, and founding director of the Vienna-based Institute of Social Ecology. Her research focus involves socio-economic metabolism, sustainability transitions, and the interdependence of core features of social systems with their nature relations, in a comparative perspective across time and space. She has been, *inter alia*, Chair of the Scientific Advisory Board of the Potsdam Institute of Climate Impact Research (PIK); currently, she is Vice President of the European Society for Ecological Economics and expert member of UNEP's International Resource Panel. Contact: marina.fischer-kowalski@aau.at

**Mark Frenzel** (Ph.D. in Ecology) is an animal ecologist at the Department of Community Ecology of the Helmholtz Centre for Environmental Research (UFZ), Germany. The main areas of his work are insect-plant interactions and permanent monitoring activities dealing with breeding birds and pollinating insects in the TERENO (Terrestrial Environmental Observatories) site Harz/Central German Lowland Observatory run by the UFZ. Furthermore he is engaged in cross-European networking in the field of LTER (Long-Term Ecosystem Research), with a special focus on selection of recommended parameters for LTER sites based on the Ecological Integrity concept. This is reflected in his leadership of related tasks in several recent EU projects (EnvEurope, EXPEER) and the Expert Panel Standardization and Technology of LTER-Europe. Contact: mark.frenzel@ufz.de

**Ezekiel Fugate** (M.Sc. in Environmental Engineering) teaches environmental science and physics at the Renaissance School and agro-ecology at CHEC in Charlottesville, USA. He is the founder of Full Hands Farm, an educational farm designed around industrial ecology principles and offering hands-on learning opportunities for children of all ages. In graduate school at Yale he studied water systems, dynamic modelling, and industrial ecology. Contact: ezekiel.fugate@gmail.com

**Eeva Furman** (Ph.D. in Marine Biology) works at the Finnish Environment Institute, Helsinki, Finland as the director for the Centre for Environmental Policy. Her research interests include conceptualising and operationalising of ecosystem services as well as collaboration between cultures, research disciplines and stakeholders in knowledge production for sustainability-driven decision-making. She chairs the Council of ALTER-Net (The European long term biodiversity, ecosystem and awareness research network) and the expert panel for socio-ecological research in LTER-Europe (long-term ecological research network). Contact: eeva.furman@ymparisto.fi

**Veronika Gaube** (Ph.D. in Social Ecology) works at the Institute of Social Ecology, Vienna, Austria. Her research focuses on sustainable rural and urban development and the impacts of multiparty decision-making on land use, material, substance and energy flows at the regional level. Methodologically, she is experienced in interlinkages of spatially explicit (GIS) models, dynamic system models and agent based models for socio-ecological systems. Other research interests include the integration of socioeconomic and ecological parameters in land-use models, material, energy and substance flow assessments in (agro) ecosystems. Recently she contributed to several projects involving participative approaches ("participative modelling"). Contact: veronika.gaube@aau.at

**Simone Gingrich** (Ph.D. in Social Ecology) works at the Institute of Social Ecology, Vienna, Austria. Her research focus is on the biophysical dimensions of industrialisation. She has participated in several research projects quantifying material, energy and carbon flows of different national economies. In addition, she is interested in inter- and transdisciplinary communication. She is member of the Centre for Environmental History in Vienna. Contact: simone.gingrich@aau.at

About the Contributors

**Markus Gradwohl** (M.Sc. in Ecology) is currently working on his Ph.D. in Environmental History at the Centre for the Study of Agriculture, Food and Environment, University of Otago, New Zealand. His main research interests are long-term changes in agricultural systems, development of material and energy flows over time and transitional processes. Contact: grama999@student.otago.ac.nz

**Ted L. Gragson** (Ph.D. in Ecological Anthropology) is professor and head of the Department of Anthropology at the University of Georgia, Athens, USA. His research centres on the behavioural and historical aspects of human-environment interaction and in particular disturbance processes over large temporal and spatial scales. He has conducted extensive field research in Lowland South America, the southern Appalachian Mountains and the north-facing Pyrenees. He is the lead principal investigator of the Coweeta LTER in the southern Appalachian Mountains, a socio-ecological project examining the impact of exurbanisation and climate change on diverse ecosystem services. Contact: tgragson@uga.edu

**Céline Granjou** (Ph.D. in Sociology) is tenured researcher at Cemagref-Irstea (Institute for Sciences and technologies in Agriculture and Environment), France. Her Ph.D. focuses on the normalisation of scientific expertise for action on public sanitation and risk assessment in France. Her current research interests include the present transformations of environmental institutions and professions; the construction of a new research community on biodiversity in France since the end of the 1980s, focusing on changes in practices and cultures in scientific ecology; on the current creation of IPBES, the International Platform on Biodiversity and Ecosystem Services. She currently coordinates the project PAN-Bioptique, "the new institutions of biodiversity: inventorying, digitizing, expertising nature" (https://panbioptique.cemagref.fr), funded by the French National Agency for Research. Contact: celine.granjou@cemagref.fr

**Nancy B. Grimm** (Ph.D. in Zoology) is a professor in Arizona State University's School of Life Sciences, USA. She is currently on assignment to the US National Science Foundation as an interdisciplinary programme liaison, programme director in Ecosystem Science, and senior scientist with the US National Climate Assessment. An ecosystem scientist, she was the founding principal investigator of the Central Arizona–Phoenix (CAP) LTER program, an interdisciplinary study of the Phoenix urban socio-ecosystem, from 1997 to 2010. Her research focuses on ecology and biogeochemistry of desert, urban, and stream-riparian ecosystems, and integrating social and ecological thinking toward understanding urban socio-ecosystems. Contact: nbgrimm@asu.edu

**J. Morgan Grove** (Ph.D. in Social Ecology) works for the Northern Research Station of the US Department of Agriculture Forest Service, USA. He is a research scientist and team leader of the Baltimore Field Station and a co-principal investigator of the Baltimore Ecosystem Study Long-Term Ecological Research Project (BES) and the District of Columbia – Baltimore City Urban Long-Term Research Areas Exploratory Projects (D.C.-B.C. ULTRA-Ex). Grove is the lead principal investigator for the social science research in BES and D.C.-B.C. ULTRA-Ex projects. He has worked in Baltimore since 1989, focusing in particular on the social and ecological dynamics of urban watersheds and residential land management. Contact: mgrove@fs.fed.us

**Willi Haas** (graduate engineer in mechanical engineering, science of management and ergonomics) works at the Institute of Social Ecology, Vienna, Austria. His research focuses on society-nature interactions at micro (local), meso (organisations) and macro (countries) level as well as scale interactions within the framework of sustainability. He pays special attention to both how to organise interdisciplinary research that deals with complex societal sustainability problems and how to make scientifically researched insights effective in societal practice (transdisciplinary research). Contact: willi.haas@aau.at

**Helmut Haberl** (Ph.D. in Ecology, Habilitation in Human Ecology) is the director of the Institute of Social Ecology, Vienna, Austria. His research interests include integrated land-change science, energy flow analysis, sustainability indicators, integrated socio-ecological modelling, climate-change mitigation in land use and the study of long-term changes in society-nature interaction. He was lead author of two chapters in the Global Energy Assessment and currently serves as lead author of the land-use chapter in the IPCC's Fifth Assessment Report. He has served on the Scientific Committee of the European Environment Agency as well as the Scientific Steering Committee of the Global Land Project. Contact: helmut.haberl@aau.at

**Gertrud Haidvogl** (Ph.D. in History) is senior scientist at the Institute of Hydrobiology and Aquatic Ecosystem Management, University of Natural Resources and Life Sciences, Vienna, Austria. Gertrud Haidvogl works especially on the environmental history of Austrian rivers with a focus on the development of riverine landscapes, fish and fisheries as well as ecological conditions. Contact: gertrud.haidvogl@boku.ac.at

**Sharon L. Harlan** (Ph.D. in Sociology) is on the faculty of the School of Human Evolution and Social Change and is a senior sustainability scientist in the Global Institute of Sustainability, Arizona State University, USA. Her research is about neighbourhood effects on environmental health disparities in cities. She directs a multi-year research and education project on urban vulnerability to climate change as a dynamic feature of coupled natural and human systems. She is a member of the Research Applications Laboratory Advisory Panel, National Center for Atmospheric Research and the American Sociological Association Task Force on Global Climate Change. Contact: sharon.harlan@asu.edu

**Severin Hohensinner** (Ph.D. in Landscape Ecology/Planning) has been a research assistant at the University of Natural Resources and Life Sciences Vienna, Austria, since 2001, with a focus on the reconstruction of historical river/floodplain hydromorphology and morphodynamic processes. His specific scientific interest is on the historical development of the Danube in the context of applied river restoration projects. The results of his studies contribute to the identification of historical living conditions of the biocoenoses in riverine ecosystems. Contact: severin.hohensinner@boku.ac.at

**Fridolin Krausmann** (Ph.D. in Human Ecology, Habilitation in Social Ecology) works at the Institute of Social Ecology, Vienna, Austria. In his research he focuses on socio-ecological transition processes. He has studied changes in socioeconomic use of energy, materials and land during the last centuries in local rural and urban systems, national economies and at the global scale. His work has also contributed to the development of methods of socio-ecological research (e.g. material flow analysis, human appropriation of net primary production) and their adaptation for application in environmental history. Contact: fridolin.krausmann@aau.at

**Sandra Lavorel** is Director of Research at the Center for National Scientific Research (CNRS) Grenoble, France, and is an ecologist with training in agronomy and ecological science. Her research focuses on global change impacts on ecosystems and ecosystem services. Her current projects seek to advance two directions: firstly, understanding how ecosystem services contribute to the coupling of the human-environment system, both as expressions of responses of ecosystems to human-related forcings and as components of human decision-making in response to environmental change: secondly, projecting the impacts of adaptation of land use and management to climate change on ecosystems and their services. These interdisciplinary projects are conducted on long-term research sites. Contact: sandra.lavorel@ujf-grenoble.fr

**Hermann Lotze-Campen** (Ph.D. in Agricultural Economics) studied Agricultural Sciences and Agricultural Economics in Kiel (Germany), Reading (UK) and Minnesota (USA). In a previous position at Astrium/InfoTerra, a European space company, he developed applications of satellite remote sensing information for agricultural statistics and large-scale modelling, precision farming and forestry. At the Potsdam Institute for Climate Impact Research he leads a research group on global land use modelling. Major topics are the interactions between climate change and food production, land and water use, bioenergy production and technological change in agriculture. Contact: lotze-campen@pik-potsdam.de

**Isabelle Mauz** (Ph.D. in Sociology) works at the National Research Institute for Environmental and Agricultural Sciences and Technologies, in Grenoble, France. Her research is grounded in empirical surveys and draws mainly on environmental sociology and science and technology studies. It aims to grasp and analyse the evolution of the work of nature scientists and nature managers in the biodiversity era. She is particularly interested in interactions and exchanges between these two professional groups. Contact: isabelle.mauz@cemagref.fr

**Michael Mirtl** (M.Sc. in Environmental engineering, Ph.D. in Ecology) heads the Department for Ecosystem Research and Monitoring at the Environment Agency Austria. As focal point for the UNECE Integrated Monitoring of Air Pollution Effects on Ecosystems, his work focused at the monitoring and assessment of critical deposition loads, including the development of object-relational information systems and ontologies. He has undertaken conceptual work on the integration of ecological and socioeconomic research in LTSER since 2003 and is co-initiator of LTER-Austria (Chair since 2008) and the Eisenwurzen LTSER Platform, and goal lead for LTER in ALTER-Net. He is also the first chairman (since 2007) of LTER-Europe, with 400 sites in 21 countries. Contact: michael.mirtl@umweltbundesamt.at

**Baptiste Nettier** is an agronomist. He studies farming systems transformations in response both to a changing environment (such as the increase of climatic and economic uncertainties, new social demands, etc.) and to internal issues (resolving work problems or developing new projects). He mainly focuses on mountain agriculture and pastoralism, and he studies farm functioning viewed from the forage perspective. He is interested in the conditions for integration of environmental issues on farm management and investigates ways to reconcile biodiversity preservation with the production function of the farms. Contact: baptiste.nettier@cemagref.fr

**Angelika Neuner** (M.Sc. in Environmental Science, majoring in Geography) is at the University of Graz, Austria. Since 2005, she has worked in the field of environmental education and education for sustainable development both in Great Britain and Austria. From 2007 to 2010 she carried out project management for the Eisenwurzen LTSER Platform and research assistance in the Network of Excellence ALTER-Net, LTER-Europe and LTER-Austria at the Environment Agency Austria. In 2011, she became coordinating editor within the Joint Programming Initiative *Climate Knowledge for Europe* at the University of Natural Resources and Life Sciences, Vienna. Since 2007, she has been a board member of Gartenpolylog association, initiating and implementing communal urban gardening projects in Vienna. Contact: angelika.neuner@umweltbundesamt.*at*

**Daniel E. Orenstein** is Senior Lecturer in the Faculty of Architecture and Town Planning at the Technion – Israel Institute of Technology, Israel. His research and teaching focuses on social assessment of ecosystem services, the dynamics and implications of urban sprawl, and environmental and land-use policy. He is also a researcher with the Israel Long-Term Ecological Research (LTER) Network, assisting in the establishment of Israel's LTSER Platforms in the Dead Sea/Arava Valley, and in the Northern Negev Desert and a member of the European LTER expert committee on socio-ecological research. Contact: dorenste@tx.technion.ac.il

**Taru Peltola** (Ph.D. in Environmental Policy) works at the Finnish Environment Institute, Joensuu, Finland. Her research interests include knowledge practices in natural resource policy, human-animal cohabitation and transition of natural resource use. Her current ethnographic fieldwork focuses on the integration of ecological knowledge into forestry in Finland. She is on the management board of ALTER-Net (The European Long-Term Biodiversity, Ecosystem and Awareness Research Network). Contact: taru.peltola@ymparisto.fi

**Johannes Peterseil** (Ph.D. in Ecology/Landscape Ecology). Since 2003 he has worked as senior expert at the Environment Agency Austria. The main fields of activity are long-term ecosystem monitoring (UNECE ICP Integrated Monitoring), the development of monitoring systems on the national as well as the European scale (Framework Programme 7 EBONE), and data management and knowledge transfer using semantics. Currently he leads the Expert Panel on Information Management of LTER-Europe. Contact: johannes.peterseil@umweltbundesamt.at

**Steward T. A. Pickett** (Ph.D. Botany) is a Distinguished Senior Scientist and plant ecologist at the Cary Institute of Ecosystem Studies, in Millbrook, New York, USA. He directs the Baltimore Ecosystem Study Long-Term Ecological Research programme. His research focuses on the ecological structure of urban areas and the temporal dynamics of vegetation. In addition to his work in Baltimore, research has taken him to the primary forests of western Pennsylvania, the post-agricultural oldfields of New Jersey, and the riparian woodlands and savannas of Kruger National Park, South Africa. He has edited or written books on ecological heterogeneity, humans as components of ecosystems, conservation, the linkage of ecology and urban design, and the philosophy of ecology. Contact: picketts@caryinstitute.org

**Charles L. Redman** (Ph.D. in Anthropology) is the Virginia Ullman Professor of Natural History and the Environment and works at the School of Sustainability, Arizona State University, USA. His research focuses on the integration of social and ecological perspectives, the dynamics underlying rapid urbanisation, the long-term aspects of human impacts on the environment and the application of resilience theory. He has conducted archaeological research in the Near East, North Africa and the American Southwest as well as co-directing contemporary interdisciplinary projects in Central Arizona. He was co-director of the CAP LTER from 1997 to 2010. Contact: Charles.redman@asu.edu

**Stephan Redpath** is Professor and chair of Conservation Science at the Aberdeen Centre for Environmental Sustainability (ACES), University of Aberdeen, UK. His core research interests lie in ecology and conservation. He focused on long-term and large-scale field systems, using experiments to tease out the impact of population processes and land use on individual behaviour, populations and communities. Much of this work has taken place within the uplands of Great Britain. Within ACES he is linking natural, social and physical sciences together with policy makers and stakeholders to tackle key research questions in environmental sustainability. Contact: s.redpath@abdn.ac.uk

**Karl Reiter** (Ph.D. in Botany) works at the Department of Conservation Biology, Vegetation and Landscape Ecology, Vienna, Austria. His research interests are focused on vegetation distribution modelling, sampling design based on GIS tools, definition of landscape-types in alpine environments, classification of landscapes to landscape-types by the use of digital spatial decision systems and planning and zonation of biosphere reserves. He is member of the core team of the international GLORIA programme. Contact: karl.reiter@univie.ac.at

**Martin Schmid** (Ph.D. in Environmental History) is head of the Centre for Environmental History at the Institute of Social Ecology, Vienna, Austria. As an environmental historian, he takes a genuinely interdisciplinary approach that crosses the great divide between natural sciences and the humanities. He has been involved in the development of theories and concepts for such an interdisciplinary environmental history with conceptual frameworks like "socio-natural sites". He has published on the long-term history of agro-ecosystems, on changing cultural attitudes towards nature in early modern times and on the environmental history of Austria after World War II. Currently he is working on an environmental history of the Danube from 1500 onwards. Contact: martin.schmid@aau.at

**Simron Jit Singh** (Ph.D. in Human Ecology) is assistant professor at the Institute of Social Ecology, Vienna, Austria. His research focuses on the theoretical, conceptual and empirical aspects of society-nature interactions across time and space within the framework of sustainability and development discourses. He has conducted extensive field research in the Indian Himalayas and the Nicobar Islands in the Bay of Bengal. He is on the management board of Europe's Network of Excellence – ALTER-Net (A Long Term Biodiversity, Ecosystem and Awareness Research Network) and deputy leader of LTER-Europe's expert panel on Long-Term Socio-Ecological Research (LTSER). Contact: simron.singh@aau.at

**Barbara Smetschka** (M.A. in Social Anthropology, post-graduate degree as "Science communicator") works at the Institute of Social Ecology, Vienna, Austria. Her research interests include gender and sustainability studies, time-use studies, integrated socio-ecological modelling, participatory research and inter- and transdisciplinary research. She has experience with participatory modelling in transdisciplinary research projects, where time-use serves as an integrating concept. She is teaching competencies for inter- and transdisciplinary sustainability research. Contact: barbara.smetschka@aau.at

**Thomas Spiegelberger** (Ph.D. in Biology) is interested in conserving grassland biodiversity in mountain regions. He has participated in several projects on the effect of land-use changes on vegetation and plant-soil interactions. He has developed sound knowledge in construction and exploiting long-term observational data sets and works on the impact of climate change on the dynamics of mountain grassland vegetation. Particularly interested in long-term observations and vegetation dynamics including invasive species, he has work on both aboveground (vegetation composition, impact of fertiliser and C-addition) and belowground diversity (soil microbial communities, mycorrhizae). At present he is co-chair of the Northern French Alps LTSER Platform. Contact: thomas.spiegelberger@epfl.ch

**Andrea Stocker-Kiss**, (M.Sc. in ecology with special focus on vegetation and landscape ecology) was involved in several national and international projects at the University of Vienna until 2007. Since 2006, she has worked at the Department of Ecosystem Research and Monitoring of the Environment Agency Austria. Her main task is the management of the Eisenwurzen LTSER Platform. Current activities also involve assistance in LTER-Austria, LTER-Europe and projects at the Zöbelboden LTER Site. Contact: Andrea.Stocker-Kiss@umweltbundesamt.at

**Ulrike Tappeiner** (Ph.D. in Biology/Informatics, Habilitation in Ecology) is Full Professor at the Institute of Ecology of the University of Innsbruck and head of the Institute of Alpine Environment at the European Academy Bozen/Bolzano, Italy. She carries out ecological research in mountain environments on various temporal and spatial scales ranging from ecosystem to landscape level with a special focus on global change, ecosystem services, and sustainable development. She has been principal investigator and work package leader in several EU and nationally funded projects and has authored more than 70 peer-reviewed publications in international journals and books. Contact: ulrike.tappeiner@uibk.ac.at

**Wilfried Thuiller** (Ph.D. in Biology) is a senior research scientist at the Laboratoire d'Ecologie Alpine (LECA, CNRS), France, and has major interests in global change biology, community ecology and species co-existence and functional ecology. Part of his research focuses on the developments of eco-evolutionary models of biodiversity to provide more reliable estimates of the effects of global environmental changes on biodiversity. Contact: wilfried.thuiller@ujf-grenoble.fr

**Justin Travis** (Ph.D. in Theoretical Spatial Ecology) works at the Institute of Biological and Environmental Sciences, Aberdeen, UK. His group uses models to study the population and evolutionary dynamics of spatially structured populations. Current work focuses on: (1) developing models of invasions that incorporate increased biological and environmental detail such as genetics and habitat variability; (2) understanding the dynamics of populations living at biogeographic range margins; and (3) using models to inform the development of management strategies for spatially structured populations, especially range-expanding and range-shifting species. Contact: justin.travis@abdn.ac.uk

**Billie L. Turner II** (Ph.D. in Geography) works in the School of Geographical Sciences and Urban Planning and in the School of Sustainability at Arizona State University, USA. His research specialty is land-change science, applied to themes from the ancient Maya, to tropical agriculture, to contemporary deforestation. He has conducted extensive field research in Latin America, especially in the Yucatán Peninsula, Africa, and Bangladesh. He currently sits on the scientific committees of the Global Land Project and DIVERSITAS. Contact: billie.l.turner@asu.edu

**Ariane Walz** (Ph.D. in Natural Sciences) works at the Potsdam Institute for Climate Impact Research (PIK), Germany. Her research focuses on human-environmental systems with an emphasis on investigations of land-use pressures, participatory approaches to scenario analysis, and changes in ecosystem services. During her time at the WSL Institute for Snow and Avalanche Research SLF, she has conducted much of her research in the Swiss and Austrian Alps, and recent research increasingly addresses European and global scale problems. Contact: ariane.walz@pik-potsdam.de

**Ali Whitmer** (Ph.D. in Botany) works for Georgetown University, USA. Her scholarly interests are in population and urban ecology, science literacy and environmental justice. She is the lead principal investigator of the District of Columbia – Baltimore

City Urban Long Term Research Areas Exploratory Projects (D.C.-B.C. ULTRA-Ex) and a co-principal investigator on a sustainability-focused Research Coordination Network (RCN) and the Environmental Literacy Math-Science Partnership (MSP) project. Whitmer is the lead for education research and programmes in the Santa Barbara Coastal Long Term Ecological Research (SBC LTER) programme. Contact: whitmer@georgetown.edu

**Sander M. J. Wijdeven** (M.Sc. in Forest Ecology) is researcher at ALTERRA, Wageningen. His main field of interest is in forest dynamics and disturbance in relation to management and diversity. Some recent activities include field and modelling studies on spontaneous forest developments vs. contrasting management regimes, and projects on forest dynamics, dead wood and natural regeneration. Contact: sander.wijdeven@wur.nl

**Martin Wildenberg** (Ph.D. in Social Ecology) leads the team on sustainable development at GLOBAL 2000/Friends of the Earth Austria, in Vienna, Austria. His current work focuses on the development and use of sustainability indicators in food supply chains with the aim to increase the sustainability of the mass-market. He previously worked at the Institute of Social Ecology on participatory and integrated modelling. He participated in the ALTER-Net project and co-developed FCMappers, the first freely available software to analyse fuzzy cognitive maps (www.fcmappers.net). He is interested in the dynamics of social-ecological systems and transdisciplinary approaches. Contact: martin.wildenberg@global2000.at

**Verena Winiwarter** (Ph.D. in History, Habilitation in Human Ecology) is Professor for Environmental History at the Institute of Social Ecology, and Head of the Centre for Environmental History, Vienna, Austria. Her research interests include environmental history of agrarian societies (with particular interest in knowledge about soils in these societies), environmental history of river systems, theory and methodology of environmental history, epistemology of interdisciplinary research. Contact: verena.winiwarter@uni-klu.ac.at

**Donald Worster** currently holds the Hall Distinguished Professorship Chair in American History at the University of Kansas, USA. Formerly the president of the American Society for Environmental History, he is president-elect of the Western History Association and an elected member of the Society of American Historians and the American Academy of Arts and Sciences. His principal areas of research and teaching include North American and world environmental history and the history of the American West. Among his books are: A Passion for Nature, Rivers of Empire, Dust Bowl, and Nature's Economy. Contact: dworster@ku.edu

**Karl S. Zimmerer** (Ph.D. in Social-Ecological Geography) is in the Department of Geography and global sustainability programmes at Pennsylvania State University, USA. His research and teaching focus is on: agrobiodiversity and global change (Environment and Resources, 2010); the conservation-agriculture interface (Latin American Research, 2011); and landscape-based social-ecological models (Knowing

Nature, Transforming Ecologies, 2011). He has published four books – most recently, Globalization and New Geographies of Conservation (2006) – and more than 70 scientific and scholarly articles and chapters, and he has held faculty appointments in geography, anthropology, and environmental science at the University of North Carolina at Chapel Hill, University of Wisconsin—Madison, and Yale University. Contact: ksz2@psu.edu

# Index

**A**
Agent-based modelling, 13, 15, 56, 58–60, 66, 68, 71, 72, 131, 132, 556
Agrarian-industrial transition, 19, 99, 340
Agrarian sociometabolic regime, 35, 340, 342, 345
Agricultural history, 18, 472, 559
Agrobiodiversity, 179
Agro-ecological system, 273, 283, 299
Agro-ecology, 274, 488
A Long-Term Biodiversity and Ecosystem Research and Awareness network (ALTER-Net), 10, 11, 411, 413, 419, 420, 432, 444, 445, 450, 467, 468, 475, 487
Alpine National Botanical Conservatory (CBNA), 493, 496, 497
Alpine Research Centre Obergurgl, 535, 536, 539
Alpine Space–Man and Environment, University of Innsbruck, 507
Alpine Timberline Research Station Patscherkofel (Klimahaus), 515
Alpine tourism, 501
Alpine tourism, impact of climate change upon, 486, 501–502
ALTER-Net. *See* A Long-Term Biodiversity and Ecosystem Research and Awareness network (ALTER-Net)
Altitudinal belts, alpine landscape, 300, 513
Andean watersheds of the upper Amazon, 168, 176–179
Andes, 165–181
ANDROSACE (plant functional trait database), 493, 496
Anthropogenic climate change, 150, 151, 414

Anthropology
  contribution to LTSER research, 6, 17, 155, 170, 191–210, 519
  as problem-oriented discipline, 17, 196–199
Applied interdisciplinary research, 411
Area productivity, 263, 279, 281, 283, 285–287, 291, 309, 349
Arrangements, 16, 18, 105–108, 111–119, 141, 142, 153, 154, 159, 297–311, 472, 551
Austria-Hungary, 108, 270, 271, 273
Austria, sociometabolic transition of, 39

**B**
Baltimore Ecosystem Study (BES), 19, 220, 238, 369–402
Behavioural decisions, and ecosystem change, 194
BES. *See* Baltimore Ecosystem Study (BES)
BES data temple, 395
Biodiversity
  conservation, 125, 126, 128, 131, 133, 173, 490
  hotspot, 117, 176, 180
  human impact upon, 237
  loss, 14, 16, 30, 33, 55, 124, 152, 462, 501, 538
  management, 126–128, 131, 450, 491
Biogeochemical cycles, 30, 33, 38, 224, 233, 234, 344, 360, 514, 515, 517, 520
Biogeophysical processes, 132
Biohistory, 16, 139–160
Biohistory-conceptual framework, 142, 143
Biometabolism, 149–150
Biophysical flows, 31, 32, 34–36, 45

Biophysical processes, 140, 166, 491
Biophysical structures of society, 33, 56, 57
Biorealism, 143
Biosensitive society, 16, 152–154, 160
Bolivia, 176–178, 180, 181

## C

Cadastral records, 261, 279, 308
Cadastre, 276
CAP LTER. *See* Central Arizona–Phoenix Long-Term Ecological Research programme (CAP LTER)
Carbon accounting, 33, 40, 41
Carbon balance, 40, 41, 43, 219, 234
Carbon capture and storage (CCS), 399
Carbon cycle, 33, 394, 486, 495, 509, 520
Carpenter's table
  experimentation, comparative analyses, modelling, 19, 393–395
  long-term monitoring, 393–395
Cary Institute of Ecosystem Studies, 370–372
Causal loop model, 44
CBD. *See* Convention of Biodiversity (CBD)
CBNA. *See* Alpine National Botanical Conservatory (CBNA)
CCS. *See* Carbon capture and storage (CCS)
Census, 126, 207, 222, 224, 251, 271–273, 277, 279, 319, 326, 387, 389, 390, 418, 430
Central Arizona–Phoenix Long-Term Ecological Research programme (CAP LTER), 17, 217–239, 372, 557
Central Arizona-Phoenix SES, 224
Central French Alps LTSER, 20, 487–495, 498–502
Centre for Environmental History, Vienna, 105
Centre of Competence for Forest, Environment and Society (Finland), 449
Chains of explanation, 166
Changes in land use and climate, 21, 61, 203, 493
Chlorofluorocarbons, ban on, 359
Cities
  in arid lands (or arid-land cities), 228
  as dominant global human habitat, 373
  ecological efficiencies, 373, 374, 385
  environmental impact of, 17, 374
  relation to resource-providing hinterland, 78, 248
  as socio-ecological systems, 78, 218, 238, 248
  vulnerability to climate change, 223, 374
Climate change and spatial/temporal variation, 12

Climate change deniers, 147
$CO_2$ enrichment, 345
Co-evolutionary process between natural and human systems, 322, 324
Cognitive anthropology, 199, 206
Collective action, 159, 192, 198
Colonisation of natural processes, 57
Colonisation of natural systems, 83
Colonisation of nature, 340, 342
Comparative research, 17, 18, 192, 193, 207, 208
Complexity in social-ecological systems, 377
Conceptual framework for urban SES, 220–223, 238
Controlled solar energy system, 341, 343
Convention of Biodiversity (CBD), 30, 124, 126
Coordination Group for Environmental Research (Finland), 451, 452
Coupled dynamics of alpine ecosystems and society, 491
Coupled human-environment interactions, 16, 165–167, 175, 177, 179
Coupled human-environment system, 18, 273, 316, 489
Coupled social-ecological systems, 166, 412
Coupled socio-environmental resource systems, 171
Coupling of socio-ecological activity, tighter/looser, 316, 321–325
Coweeta LTER project, Southern Appalachia
  ecosystem diversity, 203
  land use, land cover change, 203
  Water quantity, water quality, 203
Critical ecosystem services, 422, 423, 425–427
Cross-cultural comparison
  anthropology as, 200–201
Cross-disciplinary research
  contribution to LTSER, 191
Cross-scale interaction, 125, 129
Cross-scale research, 17, 192, 207, 208
Cultural adaptation, 146
Cultural anthropology, 82, 194
Cultural-historical ecology, 172
Cultural maladaptation, 16, 145, 146, 151–152
Cultural models, 199
Cultural reform, 146–147, 154

## D

Danube-Black Sea Canal, 117
Danube Delta, 434
Danube Environmental History Initiative (DEHI), 104, 105

Index

Danube River Basin (DRB), 103–120
DEC. *See* Domestic energy consumption (DEC)
Decision trees, 62, 63, 69, 424
De-growth concept, 362
DEHI. *See* Danube Environmental History Initiative (DEHI)
Deliberative processes, 197
Department of Ecosystem Research and Monitoring, Umweltbundesamt, 469
Differentiated temporal scaling, 168, 175
Disamenities, 219, 223, 235, 384
Disciplinary research, contribution to LTSER, 190, 191
District of Columbia/Baltimore City Urban Long-Term Ecological Research Area exploratory Project (DC-BC ULTRA-Ex Project), 370
Diversity indices, 540
Domestic energy consumption (DEC), 249, 253, 254, 256, 264
DRB. *See* Danube River Basin (DRB)
Drivers of long-term change, 220
Dürnstein wilderness area, 465
Dynamics of socio-ecological change, 316, 330–333

**E**

*Early farming phase*, 144
*Early urban phase*, 144
Ecological footprint analysis, 374
Ecologically scaled landscape indices (ELSIs), 130
Ecological modelling, 130, 131
Ecological phases, 143–145, 148, 150, 152
Ecology *in* cities, 19, 376–377, 385, 560
Ecology *of* cities, 19, 248, 376–377, 383, 385, 387, 560
Ecosystem and landscape services, 519
Ecosystem functioning, 37, 222, 318, 419
Ecosystem health needs, 156, 158
Ecosystem metabolism of cities, 219
Ecosystem processes, 9, 208, 209, 218, 222, 224–225, 227, 273, 279, 289, 412, 413, 471, 494, 516, 518–519
Ecosystem research, 429, 462, 469, 473, 474, 520
Ecosystem services
  sustainable use of, 8, 104, 410, 414, 466–467, 544
Ecosystem structure and function, 13, 218, 220, 222, 223, 238, 382, 410, 557
Ecosystem transformation
  risks of, 223, 238, 353, 447, 466, 516, 518

Ecrins National Park, 488, 490, 493, 496
EFA. *See* Energy flow analysis (EFA)
eHANPP. *See* Embodied HANPP (eHANPP)
EIA. *See* Environmental impact assessment (EIA)
Eisenwurzen
  biophysical characteristics, 464
  land-cover characteristics, 463, 464
Eisenwurzen LTSER Platform, 13, 19–20, 59, 298, 299, 418, 422–426, 461–482
ELSIs. *See* Ecologically scaled landscape indices (ELSIs)
Embodied HANPP (eHANPP), 39, 40
Energy flow analysis (EFA), 14, 17, 32, 33, 35, 298, 301
Energy history, 33
Energy procurement, 112
Energy return on investment (EROI), 35, 39, 43, 342
Energy sink
  agriculture as an, 355
Energy transition, 253–255, 257, 260, 261, 264, 345
Energy use by humankind, 341
Environmental administration, 125, 446, 448, 453
Environmental anthropology, 3–6, 13, 17, 168, 170, 172, 176, 192, 195–196, 198, 200, 201, 209, 556
Environmental geography, 1–3, 5, 6, 16, 17, 31, 166, 168–176, 179, 181, 209, 418, 447, 449, 452, 556
Environmental governance, 17, 36, 167, 168, 171–173, 177, 179, 181, 200, 447, 466–467, 491
Environmental history, 5, 6, 16, 18, 31, 103–120, 170, 174, 176, 258, 273, 274, 276, 298, 341, 348, 393, 395–396, 414, 420, 448, 472, 555–557
Environmental impact assessment (EIA), 360, 444
Environmental justice, 173, 228, 235, 384, 394, 447
Environmental landscape history and ideas, 17, 167, 168, 174, 179, 181
Environmental policy, 9, 16, 125, 176, 418, 447–449, 476
Environmental scientific concepts (in models, management and policy), 17, 167, 168, 175, 179, 181
Environmental social science, 170, 176, 446–449, 456, 457, 558
ERA. *See* European Research Area (ERA)

EROI. *See* Energy return on investment (EROI)
Esther Boserup, 342
Ethnoscience, 172, 201
EU research projects Integralp
  Carbo-Extreme, 516
  Carbomont, 516
  Ecomont, 516
  GHG Europe, 516
  Vital, 516
European long-term ecosystem research network (LTER-Europe ), xi, 7, 10, 11, 13, 19, 409–439, 444, 455, 457, 462, 468, 477, 480, 566, 568, 572–574
European Research Area (ERA), 411, 415, 431, 433
European Sturgeon, 114
European Water Framework Directive (WFD), 104, 120
Evolutionary health principle, 147, 148, 158
Exergy harvest, 106, 107, 111, 114, 116, 119
External ecosystems, urban dependence on, 219
Exurbanisation, and climate change, 207, 208

## F

Farm-scale evaluation (FSE), 128
Farm size, 62, 67, 68, 279, 281–285, 301, 302, 304, 306
Federal Research and Training Centre for Forests, Natural Hazards and Landscape, 472
Fertiliser use, 43, 64, 68, 114, 307, 349, 390, 499, 539, 559
Final energy, 249, 253, 258
Final energy use, 33
FinLTSER Network (Finland), 20, 445, 446, 452, 455–457
Flood protection, 104, 114, 118, 119
FlorEM botanical database, 493
Foot-binding, 145
*Fordism*, 350
Forestry, 30, 36, 37, 41, 57, 60, 61, 64, 172, 173, 175, 250, 371, 427, 449, 463–464, 466, 472, 475, 487–490
Fossa carolina, 109
Fossil fuel, 16–18, 35, 40, 41, 43, 60, 61, 119, 120, 144, 151, 249, 262, 263, 265, 270, 290, 292, 293, 303, 308, 310, 323, 327–331, 359, 360
Fossil fuel-powered industrialisation of agriculture, 263
Four-field geography, 166, 179

FSE. *See* Farm-scale evaluation (FSE)
Fuzzy cognitive mapping, 422–426, 475

## G

Gabcikovo dam, Slovakia, 116
Gender perspective in sustainability research, 70
Geo-engineering, 361
Gesäuse National Park, 465, 472, 474
GHG emission. *See* Greenhouse gases (GHG) emission
Global Observation Research Initiative in Alpine Environments (GLORIA), 515
GLORIA. *See* Global Observation Research Initiative in Alpine Environments (GLORIA)
Grain yield, 279, 281–283, 285–287, 304, 349
Grassland management, 36, 464, 491, 494, 500
Great Plains
  agricultural development of, 283, 287, 292
Great Regulation of the Viennese Danube, 118
Greenhouse gases (GHG) emission, 15, 36, 43, 55, 60, 68, 360, 538
Green revolution, 176, 233, 354–355
Gurgler Kamm Biosphere Reserve, Austria (AT), 21, 528, 535, 537, 544, 558

## H

Habitat distribution modeling, 130
HANPP. *See* Human appropriation of net primary production (HANPP)
Hawai'i Island, 18, 316, 317, 319–328, 330, 331
HE-NS geography. *See* Human-environment and nature-society (HE-NS) geography
HERO project. *See* Human-Environment Regional Observatory (HERO) project
Hierarchy of decision-making, 198
High consumption phase, 144, 150, 152
Hintereisferner and Kesselwandferner glacier, 515
Human adaptability, 6, 146, 514
Human and natural systems integration of, 502
Human appropriation of net primary production (HANPP), 14, 29–46, 57, 132, 556

Index  583

Human behaviour, 46, 141, 149, 151, 192, 205, 206, 381, 382
Human culture, 16, 140, 142, 143, 145–146, 155, 156
  a force in nature, 145–146
Human ecosystem framework, 378–381, 383
Human-ecosystem interactions/feedbacks, 224–238
Human-environment and nature-society (HE-NS) geography, 18, 164–169, 171–174, 177, 179
Human-Environment Regional Observatory (HERO) project, 164, 168
Human-environment relations, 2, 3, 176, 206
Human health needs, 156, 158
Human-nature interaction, 18, 315–333, 412, 427
Human outcomes and actions, 218, 220, 223, 238, 557
Human population, 79, 83, 140, 142–144, 147, 149–153, 159, 218, 220, 223, 227, 237, 288, 317, 321, 343, 357, 372, 384, 439
Hunter-gatherer phase, 144, 149, 158
Hydromorphological change, 104
Hydropower plants, 110, 112, 255

**I**
IBP. *See* International Biological Programme (IBP)
IBW. *See* Indian Bend Wash (IBW)
ICPDR. *See* International Commission for the Protection of the Danube River (ICPDR)
ICP Forest Level II network, forest inventory sites, 472
ILTER-Net. *See* International Long-Term Ecosystem Research network (ILTER-Net)
Implementation models and assessment, 438
Income, 60–64, 66–68, 70, 85, 89, 90, 152, 228, 238, 239, 319, 350, 352, 356, 358, 359, 423–428, 466, 475, 512, 533, 535, 537, 547
Indian Bend Wash (IBW), 231, 232
Industrial ecology, 2, 6, 31, 69, 249, 316, 318
Industrial metabolism, 19, 219, 264, 265, 318, 320, 358–360, 557
Industrial regime, 340, 342, 359
Industrial revolution, 144, 332, 340–345, 360
Information and communication technology revolution in, 354
Institute of Social Ecology (SEC), 10, 468, 473, 536

Institutions
  as social controls, 197
Integrated land-system science, 55, 538
Integrated monitoring, 21, 472, 474, 527–551
Integrated research, 125, 227, 396, 471
Integrated river basin management, 104
Integrated Science for Society and the Environment (ISSE) model, 19, 422, 427–428
Integrated socio-ecological modelling, 14, 56–59, 66, 69, 70, 132, 381, 420
Integration of disciplines, 530, 532, 533, 536–542
Integrative research, 191, 310
Integrative research tools, 19
Integrative studies, 155, 529–531
Interdisciplinarity, 20, 166, 173–174, 191, 192, 419, 444, 453, 457, 486, 497, 501
Interdisciplinary research, 2, 11, 16, 17, 20, 21, 46, 129, 190–193, 298, 377, 411, 414, 419, 420, 435–437, 444, 449, 457, 469, 496, 507, 549
Interdisciplinary scientific collaboration, 6, 55, 191, 438, 444, 453–455
Interlinked social-ecological scales, 167
International Biological Programme (IBP), 6, 514, 515, 518, 519
International Commission for the Protection of the Danube River (ICPDR), 104
International Long-Term Ecosystem Research network (ILTER-Net), 7, 8, 411, 423, 444, 450
Iron Gate (I and II) hydropower plants, Serbia/Romania, 118
Iron ore mining, 463, 466
Islands
  ecosystems, 325
  isolation from, connectivity to global economy, 316, 325–327
  as 'model systems', 317
  natural resource use, 316, 327–330
  as socio-ecological systems, 316, 318, 325, 330, 333
  tourism, 326
  water, energy, waste assimilation capacity, 327
ISSE model. *See* Integrated Science for Society and the Environment (ISSE) model

**J**
Joseph Fourier Alpine Research Station (SAJF), Lautaret, 488

# K

Kalkalpen National Park, 60, 64, 465, 472, 511
Kalkalpen (Limestone) region, 60, 301
Kansas, 18, 270–274, 276, 277, 279, 283, 285, 287, 290–292
Kyoto Protocol, 360

# L

Laboratory of Environmental Protection, Helsinki University of Technology, 448–449
Labour productivity, 18, 97–98, 262, 263, 265, 279, 281–283, 286, 291, 292, 349, 354
Land availability, 279, 281–284, 304
Land-based energy systems, 261
Land change science (LCS), 16–17, 53, 59, 69, 165, 166, 169, 175, 177, 179, 570, 575
Landscape
 metrics, 130
 multifunctionality, 123–124
 scale research, 167
 change, 15, 37, 40, 44, 45, 55, 56, 59, 126, 131, 167, 237, 299, 390, 475, 486, 496, 507, 518, 574
 change models, 55
 dynamics, 576
 history, 274, 277, 411, 413, 429, 462, 498
 intensity, 39, 61, 98, 360, 465, 567
 legacies, 202, 225, 235
Land-use and land-cover change (LUCC), 16–17, 165, 166, 169, 177, 179, 227, 419–420
LCA. *See* Life cycle analysis (LCA)
LCS. *See* Land change science (LCS)
Lepsämänjoki LTSER (Finland), 445–446, 453–455
Life cycle analysis (LCA), 32
Linkage between decision-making, monitoring & assessment, and science, 396–398
Livestock, 32, 33, 43, 45, 61, 65, 68, 79, 83, 87, 167, 262, 271, 273, 274, 277–284, 287–293, 298, 300–304, 306, 307, 309, 341–344, 351, 354, 355, 499, 500, 511
 density, 281–283, 289–291, 304, 307
 management, 279, 281, 288–290
Local level scenario development, 545–548
Long-term data consolidation and sharing, 20
Long-term research, 125, 129, 449, 456, 457, 472, 474, 507, 571
Long-term socio-ecological research (LTSER)
 actors/stakeholders, 418, 435
 data management/sharing, 421, 436, 438, 455, 471, 476, 478, 480, 481, 488, 494, 558
 management, communication and setup, 421
 physical infrastructure and spatial design, 416–418
 platform, 57–58, 133, 409–439
 region delineation, selection criteria, 429
 research component, 418–420
LTER. *See* Long-term ecological research (LTER)
LTSER. *See* Long-term socio-ecological research (LTSER)
LUCC. *See* Land-use and land-cover change (LUCC)
Lunz Water Cluster, 472

# M

Machland dam, Austria, 109, 117–118, 120
Marketable crop production, 279, 281, 282, 286, 287
Market environmentalism, 171
Mass Lifted (ML), transport indicator, 80, 82, 90–91
Mass Moved (MM), transport indicator, 80, 82, 90–91
Material and Energy Flow Accounting (MEFA), 13, 30, 249, 273, 274, 320, 321
Material and Energy Flow Analysis (MEFA), 31–36
Material Flow Analysis (MFA), 33, 34, 132, 320, 332
Material mass balance, 219, 238
MEA. *See* Millennium Ecosystem Assessment (MEA)
Meliors, 148
Melior-stressor concept, 148
Metabolic profile, 79, 80, 248, 359
Metabolic reconfiguration, and development, 324
Methane, $N_2O$ and VOC, 516
MFA. *See* Material Flow Analysis (MFA)
MFRPs. *See* Multifunctional Research Platforms (MFRPs)
Migration, 18, 78, 131, 176–179, 205, 270–274, 276, 310, 465, 517
Millennium Ecosystem Assessment (MEA), 1, 13, 37, 38, 55, 125, 133, 218, 220, 358, 410, 423, 506, 538, 565, 566

Index 585

Mode 2, 413, 420, 424, 530, 531
Modelling approaches, 58, 59, 69, 78, 130–133, 169, 423, 519, 556
Modelling as a participatory process, 64–66
Motown Cluster, 350
Mountain livestock farming systems, 500
Mountain Research Station, Botanical Institute University of Innsbruck, 514–515
Multidisciplinarity, 155
Multidisciplinary integrated models, 131–133
Multifunctional landuse system, 262, 474
Multifunctional Research Platforms (MFRPs), 10, 467
Multi-level material flows, 320
Multi-scale approaches, 133, 166
Multi-site social-environmental networks, 179

**N**
Natura 2000, 465, 535–536, 563
NEHN. *See* Nordic Environmental History Society (NEHN)
Neolithic revolution, 340
NESS. *See* Nordic Environmental Social Science (NESS)
Net ecosystem exchange of $CO_2$, 517–518
Network of Excellence
New World farm system, 274–277
New York Botanical Garden (NYBG), 370–371
Nitrogen
  cycling, 394, 495, 516
  drainage, 263
  enrichment, 30
  flow, 15, 33, 38, 68, 69, 132, 281, 293
  return, 263, 281–283, 289, 290
Nordic Environmental History Society (NEHN), 448
Nordic Environmental Social Science (NESS), 448
Nutrient
  flows, 36, 98, 239, 279
  management, 279, 281, 288–290
NYBG. *See* New York Botanical Garden (NYBG)

**O**
O'ahu, 18, 316, 319, 320, 322, 323, 326–330
Obergurgl Alpine Research Centre, 519, 567

Old World farm system, 274–277, 285, 288, 293
Open-air laboratory, 514
Ötz valley research area, 510, 514–516, 519, 540

**P**
Participatory Assessment, Monitoring and Evaluation (PAME), 532
Participatory research, 64, 531–532, 574
Participatory Technology Assessments (PTA), 530, 531
Partnership for Research on Natural Resources and the Environment (Finland), 448
Pasteur's quadrant, 375, 376, 395, 396
Patch dynamics, 383–384
Per capita energy use, 256–258, 357
Pesticide use, 171
Phylogenetic maladjustment, 147, 149
Physical health needs, 148
Place-based LTSER, 17, 19, 193, 413–415, 431, 438, 439
Place-based research, 17, 192, 193, 207
Policy-relevant knowledge generation, 431
Political ecology, 4, 17, 165, 166, 169–171, 173, 175–177, 179
Pollution control, 321
Population
  density, 9, 39, 81, 82, 87, 88, 93, 159, 252, 273, 274, 279, 281–285, 288, 304–306, 316, 342, 349, 411, 428, 465
  and energy use, 149, 255–257, 265, 347
  metabolism, 149
  Viability Analysis (PVA), 130
Post-normal science, 420, 529, 530, 532
Potential risks in mountain ecosystems, 518
Power play, 341, 550, 551
Practices, 16, 18, 105–108, 112–119, 170, 191, 225, 237, 276, 283, 285, 292, 298, 299, 304–306, 309, 310, 349, 374, 417, 429, 437, 445, 453–455, 500, 501, 531
Pre-industrial agriculture, 263, 274, 299, 475
Pre-industrial societies, 77–99, 249, 341
Press-Pulse Dynamics framework, 13, 14, 17, 380–383, 394, 412, 426
Price, 12, 15, 36, 55, 60, 62, 65, 66, 68, 92, 169, 195, 200, 206, 223, 250, 253, 257–259, 265, 323, 328, 351, 355, 356, 424, 428, 475, 500
Primary energy
  sources, 323, 331, 341, 352
  supply, 33, 255, 351

Problem-focused research, 462
Problem-oriented environmental research, 20, 190, 446–449, 457
Productivity, vii, 18, 37, 38, 60, 88, 90, 93, 96–99, 195, 233, 234, 262, 263, 265, 279, 281–283, 285–287, 291, 292, 304, 306, 308–310, 343, 349, 354, 388, 389, 463, 509, 514, 518, 538
Programme on Ecosystem Change and Society (PECS), 415, 438, 566
Pro-growth development, and poverty, 197
Psychosocial health needs, 148
PTA. *See* Participatory Technology Assessments (PTA)
Puerto Rico
 'Operation Bootstrap' development programme, 324

### R

Rapid urbanization as press event, 224–227, 257
Regional integrated research, 431, 471
Reichraming, municipality of, 56, 60
Renewable energy
 flows, 341
Research funding, for single-discipline/interdisciplinary research, 191, 446, 447
Residential landscapes, 223, 225, 226, 237
Resource availability, accessibility and transport, 260
Resource density, 80–83
Resource-providing hinterland, 78, 248
Riverine landscape, interventions, 108–116
River transport, 257

### S

Scalable data platforms, 377, 387–390, 394, 395, 397
Scale
 spatial, 55, 79, 84, 85, 95, 127, 128, 131, 132, 164, 166, 203, 310, 412, 413, 416, 417, 428, 456, 462, 473, 557
 temporal, 11, 125, 132, 387, 471, 520
Scale mismatch, 16, 126–128
Scenario workshops, 532, 543, 545
SEC. *See* Institute of Social Ecology (SEC)
SERD. *See* Simulation of ecological compatibility of regional development (SERD)
Seville Strategy for Biosphere Reserves, 535, 536
SFA. *See* Substance flow analysis (SFA)
Silent Spring (Rachel Carson), 146, 358
Simulation of ecological compatibility of regional development (SERD), 59–62
Singapore
 Trans-shipment, 326
SMCE. *See* Societal multi-criteria evaluation (SMCE)
Social-ecological adaptive capacity and vulnerability, 16, 17, 165, 167–168, 177
Social metabolism, 17, 18, 79, 249, 277, 310, 311
Societal arrangements, 142, 154, 159
Societal metabolism, 340, 343, 347, 348, 351, 352, 355, 357
Societal multi-criteria evaluation (SMCE), 532
Society-nature interaction, 1, 5, 6, 9, 12, 13, 18, 21, 31, 45, 46, 53, 54, 57, 58, 298, 340, 415, 419, 438, 482, 538–540, 559
Socio-ecological
 indicators, 37, 57, 61, 130, 132, 223, 279, 292
 metabolism, 13, 14, 30, 36–40, 45, 58, 60, 298, 299, 556
 profiling, 19, 416–417, 419, 421–429
 research, vi, ix, 53–70, 103–120, 123–134
 system, 9–13, 16, 17, 21, 36, 37, 40, 53–70, 78, 98, 129, 132, 166, 168, 170, 173, 192, 193, 202, 207, 208, 217–239, 248, 265, 298, 310, 311, 316, 318, 321, 322, 325, 330, 333, 340, 372, 373, 376, 377, 381, 383, 385–387, 393, 396, 397, 402, 412–415, 421–433, 457, 462, 486, 519, 538, 551
 transitions, 31, 44, 175, 310, 321, 385, 419
Socioeconomic drivers of ecological change, 36, 411
Socioeconomic metabolism, 6, 11, 14, 15, 29–46, 56, 57, 132, 318, 556
Sociometabolic regime, 33, 36, 340, 342, 345, 351, 356, 362
Sociometabolic transition, 35, 36, 39
Sociometabolism, 17, 18, 79, 249, 277, 310, 318, 341–345, 347, 348, 351, 352, 355, 357–362
Socio-natural sites, 16, 18, 105, 112, 119, 299, 304, 309–311
 carbon, 40, 42, 43, 288, 291, 498
 fertility, 156, 264, 281, 284, 287–292, 533
 nutrients, 85, 273, 284, 287–289, 293, 559
 organic carbon (SOC), 42, 234, 235, 288, 291
 surveys, 234

Index                                                                                                                          587

Sol Tax, 195, 196, 200
Space and time scales, 220–222
Spatial and temporal coupling, 7, 11, 12, 16,
    39, 44, 55, 125, 127, 128, 130, 132,
    165, 167, 168, 179, 202, 203, 224, 377,
    387, 471, 491, 519, 520
Spatial imprint of urban consumption, 18, 248,
    258–264
Spatial-network approach, 177
Spatial scales, 44, 55, 79, 84, 85, 95, 127, 128,
    131, 132, 164, 166, 203, 310, 412, 413,
    416, 417, 428, 456, 462, 473, 557
Species-area relationship, 130
Stakeholder
    analysis of, 542–545
    informed research, 414, 551
    involvement, inclusion, 15, 46, 424, 436,
        467–469, 498, 528, 534
    map, 422–426, 543, 544
    workshop, 437, 446, 451, 467, 532, 543,
        545, 547
Stock-flow framework, 29
Stormwater management, 231
Structuration (socioenvironmental
    structuration), 179
Stubai valley research area, 516, 517, 519, 520
Subsidies, 15, 35, 55, 60, 62, 65, 66, 68,
    424, 475
Substance flow analysis (SFA), 33
Subterranean forest, 261, 348, 556
Sustainability
    assessment, 21, 69, 527–551
    science, 2, 16, 35, 45, 54, 56–59, 64, 69,
        165–168, 174, 177, 179, 220, 273, 431,
        469, 528
    triangle, 61
SustainabilityA-Test, EU project, 132
Sustainable development, 4, 8, 10, 114,
    116, 124, 248, 362, 447, 519, 528,
    534, 544
SYKE, Finnish Environment Institute, 448
    Centre for Environment Policy, 448
1950s Syndrome, 258, 351, 352
System boundary, 252, 320, 378
System-dynamic modelling, 31, 40–45, 69


**T**

Technoaddiction, 188–189
Technological change
    socio-ecological significance of, 340, 557
Technology cluster, 349, 350
Technometabolism, 148–150
Terrestrial and aquatic ecosystems, 520

Thick description, 194
Threshold interactions, 423, 428–429
Time use, 62, 64, 299, 473
Total primary energy supply (TPES), 33,
    351, 353
Toxic substances, emissions, 30
TPES. *See* Total primary energy supply
    (TPES)
Trade-offs, 61, 200, 207, 219, 223, 227, 231,
    238, 436, 499
Transdisciplinarity
    challenges, 528, 549–550
    research, 5, 528–532, 548–551, 557
Transport
    in agrarian societies, 83–92, 94, 98
    energy, 78, 79, 98, 99
    in hunting and gathering societies, 80–83
    infrastructure, 15, 79, 83, 84, 92, 99
    networks, 260, 265
    technology, 18, 79, 80, 87, 99, 265
Trophic dynamics
    urban, 226, 237
TTPES. *See* Total primary energy supply
    (TPES)
Tyrolean Alps LTSER Platform, 20, 21,
    506–520, 528, 532–536, 560


**U**

Umweltbundesamt, Environment Agency
    Austria, 468, 469, 473, 476
UNESCO
    biosphere reserve Gossenköllesee
    Man and the Biosphere (MAB)
        programme, 6, 519, 535, 545
Urban
    biogeochemical processes, 233–235
    ecology, 374, 398, 402, 448
    footprint, 219, 260, 261
    footprint size, 260, 261
    growth, 78, 99, 226, 255, 257, 262–265,
        273, 348
    heat island (UHI), 224, 227, 228, 231
    institutional drivers of urban
        growth, 226
    landscape, 226, 227, 394
    LTSER, 369–403, 557, 560
    metabolism, 150, 170, 219, 248, 251,
        370, 557
    resource use, 18, 248
    socio-ecological system (SES),
        217–239, 265
    sustainability strategies, 17, 98, 217–239,
        377, 397, 557

Urbanisation
    physical constraints upon, 78
    and spatial concentration, 78
    and sustainability, 238–239
    and transport requirements, 78, 80, 93–98
Urban-Rural Gradient Ecology (URGE) project, 371
USDA Forest Service, 371, 372

**V**

Vaccia project, analysing socio-environmental resilience of ecosystem services under conditions of climate change (Finland), 456
Vegetation
    diversity (urban), 222, 224, 238
    phenology (urban), 228, 495
Vegetation assessment, 537, 539
Vercors high plateaux natural reserve, 490, 492
Vercors Natural Regional Park, 490
Virtual forest area, 250, 261, 262, 346, 348, 556

**W**

Wallsee-Mitterkirchen hydropower plant, 117
Waste management, 12, 321, 329
Water
    dynamics, 224, 227–233
    quality, protection, 116, 200, 203
    use in desert city, 224, 228–233

Watersheds, 140, 163–179, 204, 205, 222, 231, 233, 323, 372, 377, 384, 387, 390–392, 394, 398–400
WFD. *See* European Water Framework Directive (WFD)
World Summit on Sustainable Development (WSSD), 124
WSSD. *See* World Summit on Sustainable Development (WSSD)

**Y**

Yale Center for Industrial Ecology, 316, 318
Yale University School of Forestry and Environmental Studies, 371
Ybbs-Persenbeug hydropower plant, 116, 117
Yearbook of European Environmental Law, 447
YHTYMÄ graduate school in environmental social science, 447, 448
YHYS, Finnish Society for Environmental Social Sciences, 448
Yield, 38, 39, 43, 78, 83, 87, 90, 98, 231, 233, 250, 261–263, 276, 277, 284, 286, 287, 289, 290, 292, 293, 304, 306, 307, 342, 344, 349, 354, 390, 506

**Z**

Zöbelboden ICP Integrated Monitoring site, 472, 474
Zone Atelier Alpes, 488, 497

Printed by Publishers' Graphics LLC
BT20130405.15.01.281